Analytical Geometry

Vittal's

Analytical Geometry

2D and 3D

P. R. Vittal
Visiting Professor
Department of Statistics
University of Madras
Chennai

Pearson

Original edition, Analytical Geometry - 2D and 3D, 2nd Edition Vittal, published by Pearson Education Ltd.
Copyright © 2013.

ISBN 9788131773604
First Impression, 2013
Second Impression, 2023
Third Impression, 2024

Published by Pearson India Education Services Pvt. Ltd, CIN: U72200TN2005PTC057128.

Head Office: 1st Floor, Berger Tower, Plot No. C-001A/2, Sector 16B, Noida - 201 301, Uttar Pradesh, India
Registered Office: Featherlite, 'The Address' 5th Floor, Survey No 203/10B, 200 Ft MMRD Road, Zamin Pallavaram, Chennai – 600 044.
Website: in.pearson.com; Email: companysecretary.india@pearson.com

Printed in India at: Manipal Technologies Limited, Manipal

To my grandchildren
Aarav and Advay

Brief Contents

	About the Author	xix
	Preface	xxi
1	Coordinate Geometry	1-1
2	The Straight Line	2-1
3	Pair of Straight Lines	3-1
4	Circle	4-1
5	System of Circles	5-1
6	Parabola	6-1
7	Ellipse	7-1
8	Hyperbola	8-1
9	Polar Coordinates	9-1
10	Tracing of Curves	10-1
11	Three Dimension	11-1
12	Plane	12-1
13	Straight Line	13-1
14	Sphere	14-1
15	Cone	15-1
16	Cylinder	16-1
	Index	I-1

Contents

About the Author xix

Preface xxi

1 Coordinate Geometry 1-1

1.1 Introduction 1-1

1.2 Section Formula 1-3

Illustrative Examples 1-7

Exercises 1-27

2 The Straight Line 2-1

2.1 Introduction 2-1

2.2 Slope of a Straight Line 2-3

2.3 Slope-intercept Form of a Straight Line 2-3

2.4 Intercept Form 2-4

2.5 Slope-point Form 2-5

2.6 Two Points Form 2-5

2.7 Normal Form 2-5

2.8 Parametric Form and Distance Form 2-6

2.9 Perpendicular Distance on a Straight Line 2-7

2.10 Intersection of Two Straight Lines 2-8

2.11 Concurrent Straight Lines 2-10

2.12 Angle between Two Straight Lines 2-11

2.13 Equations of Bisectors of the Angle between Two Lines 2-12

Illustrative Examples 2-12

Exercises 2-45

3 Pair of Straight Lines 3-1

3.1 Introduction 3-1

3.2 Homogeneous Equation of Second Degree in x and y 3-2

3.3 Angle between the Lines Represented by $ax^2 + 2hxy + by^2 = 0$ 3-2

3.4 Equation for the Bisector of the Angles between the Lines Given by $ax^2 + 2hxy + by^2 = 0$ 3-3

3.5 Condition for General Equation of a Second Degree Equation to Represent a Pair of Straight Lines 3-5

Illustrative Examples 3-9

Exercises 3-35

4 Circle 4-1

4.1 Introduction 4-1

4.2 Equation of a Circle whose Centre is (h, k) and Radius r 4-1

4.3 Centre and Radius of a Circle Represented by the Equation $x^2 + y^2 + 2gx + 2fy + c = 0$ 4-1

4.4 Length of Tangent from Point $P(x_1, y_1)$ to the Circle $x^2 + y^2 + 2gx + 2fy + c = 0$ 4-2

4.5 Equation of Tangent at (x_1, y_1) to the Circle $x^2 + y^2 + 2gx + 2fy + c = 0$ 4-3

4.6 Equation of Circle with the Line Joining Points $A(x_1, y_1)$ and $B(x_2, y_2)$ as the ends of Diameter 4-3

4.7 Condition for the Straight Line $y = mx + c$ to be a Tangent to the Circle $x^2 + y^2 = a^2$ 4-4

4.8 Equation of the Chord of Contact of Tangents from (x_1, y_1) to the Circle $x^2 + y^2 + 2gx + 2fy + c = 0$ 4-5

4.9 Two Tangents can Always be Drawn from a Given Point to a Circle and the Locus of the Point of Intersection of Perpendicular Tangents is a Circle 4-6

4.10 Pole and Polar 4-7
4.11 Conjugate Lines 4-8
4.12 Equation of a Chord of Circle $x^2 + y^2 + 2gx + 2fy + c = 0$ in Terms of its Middle Point 4-9
4.13 Combined Equation of a Pair of Tangents from (x_1, y_1) to the Circle $x^2 + y^2 + 2gx + 2fy + c = 0$ 4-10
4.14 Parametric Form of a Circle 4-11
Illustrative Examples 4-11
Exercises 4-46

5 System of Circles 5-1

5.1 Radical Axis of Two Circles 5-1
5.2 Orthogonal Circles 5-2
5.3 Coaxal System 5-3
5.4 Limiting Points 5-4
5.5 Examples (Radical Axis) 5-6
5.6 Examples (Limiting Points) 5-18
Exercises 5-26

6 Parabola 6-1

6.1 Introduction 6-1
6.2 General Equation of a Conic 6-1
6.3 Equation of a Parabola 6-2
6.4 Length of Latus Rectum 6-3
6.5 Different Forms of Parabola 6-3
Illustrative Examples Based on Focus Directrix Property 6-5
6.6 Condition for Tangency 6-8
6.7 Number of Tangents 6-9
6.8 Perpendicular Tangents 6-9
6.9 Equation of Tangent 6-10
6.10 Equation of Normal 6-11

6.11 Equation of Chord of Contact 6-12
6.12 Polar of a Point 6-13
6.13 Conjugate Lines 6-13
6.14 Pair of Tangents 6-14
6.15 Chord Interms of Mid-point 6-15
6.16 Parametric Representation 6-16
6.17 Chord Joining Two Points 6-16
6.18 Equations of Tangent and Normal 6-17
6.19 Point of Intersection of Tangents 6-17
6.20 Point of Intersection of Normals 6-18
6.21 Number of Normals from a Point 6-18
6.22 Intersection of a Parabola and a Circle 6-19
Illustrative Examples Based on Tangents and Normals 6-20
Illustrative Examples Based on Parameters 6-41
Exercises 6-62

7 Ellipse **7-1**
7.1 Standard Equation 7-1
7.2 Standard Equation of an Ellipse 7-1
7.3 Focal Distance 7-4
7.4 Position of a Point 7-4
7.5 Auxiliary Circle 7-5
Illustrative Examples Based on Focus-directrix Property 7-6
7.6 Condition for Tangency 7-11
7.7 Director Circle of an Ellipse 7-12
7.8 Equation of the Tangent 7-13
7.9 Equation of Tangent and Normal 7-14
7.10 Equation to the Chord of Contact 7-16
7.11 Equation of the Polar 7-17
7.12 Condition for Conjugate Lines 7-18
Illustrative Examples Based on Tangents, Normals, Pole-polar and Chord 7-19

7.13 Eccentric Angle 7-45
7.14 Equation of the Chord Joining the Points 7-45
7.15 Equation of Tangent at 'θ' on the Ellipse $\frac{x^2}{a^2}+\frac{y^2}{b^2}=1$ 7-47
7.16 Conormal Points 7-48
7.17 Concyclic Points 7-49
7.18 Equation of a Chord in Terms of its Middle Point 7-51
7.19 Combined Equation of Pair of Tangents 7-52
7.20 Conjugate Diameters 7-54
7.21 Equi-conjugate Diameters 7-84
Illustrative Examples Based on Conjugate Diameters 7-85
Exercises 7-100

8 Hyperbola **8-1**
8.1 Definition 8-1
8.2 Standard Equation 8-1
8.3 Important Property of Hyperbola 8-3
8.4 Equation of Hyperbola in Parametric Form 8-3
8.5 Rectangular Hyperbola 8-3
8.6 Conjugate Hyperbola 8-4
8.7 Asymptotes 8-28
8.8 Conjugate Diameters 8-35
8.9 Rectangular Hyperbola 8-41
Exercises 8-61

9 Polar Coordinates **9-1**
9.1 Introduction 9-1
9.2 Definition of Polar Coordinates 9-1
9.3 Relation between Cartesian Coordinates and Polar Coordinates 9-3
9.4 Polar Equation of a Straight Line 9-4
9.5 Polar Equation of a Straight Line in Normal Form 9-4
9.6 Circle 9-9

9.7 Polar Equation of a Conic 9-17
Exercises 9-51

10 Tracing of Curves 10-1

10.1 General Equation of the Second Degree and Tracing of a Conic 10-1
10.2 Shift of Origin without Changing the Direction of Axes 10-1
10.3 Rotation of Axes without Changing the Origin 10-2
10.4 Removal of XY-term 10-3
10.5 Invariants 10-3
10.6 Conditions for the General Equation of the Second Degree to Represent a Conic 10-4
10.7 Centre of the Conic Given by the General Equation of the Second Degree 10-6
10.8 Equation of the Conic Referred to the Centre as Origin 10-6
10.9 Length and Position of the Axes of the Central Conic whose Equation is $ax^2 + 2hxy + by^2 = 1$ 10-8
10.10 Axis and Vertex of the Parabola whose Equation is $ax^2 + 2hxy + by^2 + 2gx + 2fy + c = 0$ 10-10
Exercises 10-16

11 Three Dimension 11-1

11.1 Rectangular Coordinate Axes 11-1
11.2 Formula for Distance between Two Points 11-4
11.3 Centroid of Triangle 11-7
11.4 Centroid of Tetrahedron 11-8
11.5 Direction Cosines 11-8
Illustrative Examples 11-13
Exercises 11-25

12 Plane 12-1

12.1 Introduction 12-1
12.2 General Equation of a Plane 12-1

12.3 General Equation of a Plane Passing Through a Given Point 12-2
12.4 Equation of a Plane in Intercept Form 12-2
12.5 Equation of a Plane in Normal Form 12-3
12.6 Angle between Two Planes 12-5
12.7 Perpendicular Distance from a Point on a Plane 12-5
12.8 Plane Passing Through Three Given Points 12-7
12.9 To Find the Ratio in which the Plane Joining the Points (x_1, y_1, z_1) and (x_2, y_2, z_2) is Divided by the Plane $ax + by + cz + d = 0$. 12-8
12.10 Plane Passing Through the Intersection of Two Given Planes 12-9
12.11 Equation of the Planes which Bisect the Angle between Two Given Planes 12-10
12.12 Condition for the Homogenous Equation of the Second Degree to Represent a Pair of Planes 12-11
Illustrative Examples 12-12
Exercises 12-33

13 Straight Line 13-1

13.1 Introduction 13-1
13.2 Equation of a Straight Line in Symmetrical Form 13-1
13.3 Equations of a Straight Line Passing Through the Two Given Points 13-3
13.4 Equations of a Straight Line Determined by a Pair of Planes in Symmetrical Form 13-4
13.5 Angle between a Plane and a Line 13-5
13.6 Condition for a Line to be Parallel to a Plane 13-6
13.7 Conditions for a Line to Lie on a Plane 13-6
13.8 To Find the Length of the Perpendicular from a Given Point on a Line 13-7
13.9 Coplanar Lines 13-7
13.10 Skew Lines 13-9

13.11 Equations of Two Non-intersecting Lines 13-13
13.12 Intersection of Three Planes 13-14
13.13 Conditions for Three Given Planes to Form a Triangular Prism 13-16
Illustrative Examples 13-17
Illustrative Examples (Coplanar Lines and Shortest Distance) 13-34
Exercises 13-47

14 Sphere 14-1

14.1 Definition of Sphere 14-1
14.2 The equation of a sphere with centre at (a, b, c) and radius r 14-1
14.3 Equation of the Sphere on the Line Joining the Points (x_1, y_1, z_1) and (x_2, y_2, z_2) as Diameter 14-2
14.4 Length of the Tangent from $P(x_1, y_1, z_1)$ to the Sphere $x^2 + y^2 + z^2 + 2ux + 2vy + 2wz + d = 0$ 14-3
14.5 Equation of the Tangent Plane at (x_1, y_1, z_1) to the Sphere $x^2 + y^2 + z^2 + 2ux + 2vy + 2wz + d = 0$ 14-3
14.6 Section of a Sphere by a Plane 14-4
14.7 Equation of a Circle 14-5
14.8 Intersection of Two Spheres 14-5
14.9 Equation of a Sphere Passing Through a Given Circle 14-6
14.10 Condition for Orthogonality of Two Spheres 14-6
14.11 Radical Plane 14-7
14.12 Coaxal System 14-9
Illustrative Examples 14-11
Exercises 14-50

15 Cone 15-1

15.1 Definition of Cone 15-1
15.2 Equation of a Cone with a Given Vertex and a Given Guiding Curve 15-1

15.3 Equation of a Cone with its Vertex at the Origin 15-4
15.4 Condition for the General Equation of the Second Degree to Represent a Cone 15-7
15.5 Right Circular Cone 15-13
15.6 Tangent Plane 15-23
15.7 Reciprocal Cone 15-26
Exercises 15-34

16 Cylinder 16-1

16.1 Definition 16-1
16.2 Equation of a Cylinder with a Given Generator and a Given Guiding Curve 16-1
16.3 Enveloping Cylinder 16-2
16.4 Right Circular Cylinder 16-3
Illustrative Examples 16-4
Exercises 16-18

Index I-1

About the Author

P. R. Vittal was a postgraduate professor of Mathematics at Ramakrishna Mission Vivekananda College, Chennai, from where he retired as Principal in 1996. He was a visiting professor at Western Carolina University, USA, and has visited a number of universities in the USA and Canada in connection with his research work. He is, at present, a visiting professor at the Department of Statistics, University of Madras; Institute of Chartered Accountants of India, Chennai; The Institute of Technology and Management, Chennai; and National Management School, Chennai, besides being a research guide in Management Science at BITS, Ranchi.

Professor Vittal has published 30 research papers in journals of national and international repute and guided a number of students to their M.Phil. and Ph.D. degrees. A fellow of Tamil Nadu Academy of Sciences, his research topics are probability, stochastic processes, operations research, differential equations and supply chain management. He has authored about 30 books in mathematics, statistics and operations research.

Preface

A successful course in analytical geometry must provide a foundation for future work in mathematics. Our teaching responsibilities are to instil certain technical competence in our students in this discipline of mathematics. A good textbook, as with a good teacher, should accomplish these aims. In this book, you will find a crisp, mathematically precise presentation that will allow you to easily understand and grasp the contents.

This book contains both two-dimensional and three-dimensional analytical geometry. In some of the fundamental results, vector treatment is also given and therefrom the scalar form of the results has been deduced.

The first 10 chapters deal with two-dimensional analytical geometry. In Chapter 1, all basic results are introduced. The concept of locus is well explained. Using this idea, in Chapter 2, different forms for the equation of a straight line are obtained; all the characteristics of a straight line are also discussed. Chapter 3 deals with the equation of a pair of straight lines and its properties. In Chapters 4 and 5, circle and system of circles, including coaxial system and limiting points of a coaxial system, are analysed.

Chapters 6, 7 and 8 deal with the conic sections—parabola, ellipse and hyperbola. Apart from their properties such as focus and directrix, their parametric equations are also explained. Special properties such as conormal points of all conics are described in details. Conjugate diameters in ellipse and hyperbola and asymptotes of a hyperbola and rectangular hyperbola are also analysed with a number of examples. A general treatment of conics and tracing of conics is also provided.

In Chapter 9, we describe polar coordinates, which are used to measure distances for some special purposes. Chapter 10 examines the conditions for the general equation of the second degree to represent the different types of conics.

In Chapters 11 to 16, we study the three-dimensional analytical geometry. The basic concepts, such as directional cosines, are introduced in Chapter 11. In Chapter 12, all forms of plane are analysed with the help of examples. Chapter 13 introduces a straight line as an intersection of a pair of planes. Different forms of a straight line are studied; especially, coplanar lines and the shortest distance between two skew lines. Chapter 14 deals with spheres and system of spheres. In Chapters 15 and 16, two special types of conicoids—cone and cylinder—are discussed.

A number of illustrative examples and exercises for practice are given in all these 16 chapters, to help the students understand the concepts in a better manner.

I hope that this book will be very useful for undergraduate students and engineering students who need to study analytical geometry as part of their curriculum.

Acknowledgements

I express my sincere thanks to Pearson Education, India, especially to K. Srinivas, Sojan, Charles, and Ramesh for their constant encouragement and for successfully bringing out this book.

P. R. Vittal

CHAPTER

1 Coordinate Geometry

1.1 INTRODUCTION

Let XOX' and YOY' be two fixed perpendicular lines in the plane of the paper. The line OX is called the axis of X and OY the axis of Y. OX and OY together are called the coordinate axes. The point O is called the origin of the coordinate axes. Let P be a point in this plane. Draw PM perpendicular to XOX'. The distance OM is called the x-coordinate or abscissa and the distance MP is called the y-coordinate or ordinate of the point P.

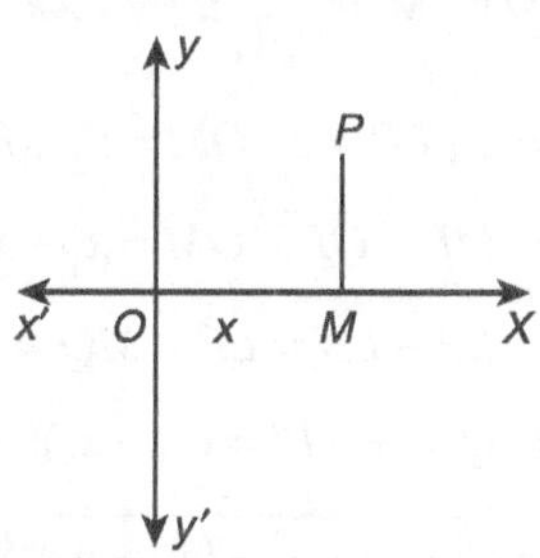

If $OM = x$ and $MP = y$ then (x, y) are called the coordinates of the point P. The coordinates of the origin O are (0, 0). The lines XOX' and YOY' divide the plane into four quadrants. They are XOY, YOX', $X'OY'$ and $Y'OX$. The lengths measured in the directions OX and OY are considered positive and the lengths measured in the directions OX' and OY' are considered negative. The nature of the coordinates in the different quadrants is as follows:

Quadrant	*x-coordinate*	*y-coordinate*
First	+	+
Second	–	+
Third	–	–
Fourth	+	–

The method of representing a point by means of coordinates was first introduced by Rena Descartes and hence this branch of mathematics is called the rectangular Cartesian coordinate system.

Using this coordinate system, one can easily find the distance between two points in a plane, the coordinates of the point that divides a line segment in a given ratio, the centroid of a triangle, the area of a triangle and the locus of a point that moves according to a given geometrical law.

1.1.1 Distance between Two Given Points

Let P and Q be two points with coordinates (x_1, y_1) and (x_2, y_2).

Draw PL and QM perpendiculars to the x-axis, and draw QN perpendicular to PL. Then,

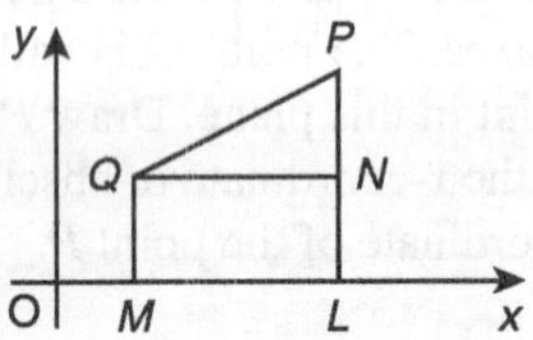

$$OL = x_1,\ LP = y_1,\ OM = x_2,\ MQ = y_2$$

$$QN = ML = OL - OM = x_1 - x_2$$

$$NP = LP - LN = LP - MQ = y_1 - y_2$$

In ΔPQN, $PQ^2 = QN^2 + NP^2 = (x_1 - x_2)^2 + (y_1 - y_2)^2$

$$\therefore\ PQ = \sqrt{(x_1 - x_2)^2 + (y_1 - y_2)^2}$$

Note 1.1.1: The distance of P from the origin O is $OP = \sqrt{x_1^2 + y_1^2}$

Example 1.1.1

If P is the point (4, 7) and Q is (2, 3), then

$$PQ = \sqrt{(4-2)^2 + (7-3)^2} = \sqrt{4+16} = \sqrt{20} = 2\sqrt{5} \text{ units}$$

Example 1.1.2

The distance between the points $P(2, -5)$ and $Q(-4, 7)$ is

$$PQ = \sqrt{(2+4)^2 + (-5-7)^2} = \sqrt{36+144} = \sqrt{180} = 6\sqrt{5} \text{ units}$$

1.2 SECTION FORMULA

1.2.1 Coordinates of the Point that Divides the Line Joining Two Given Points in a Given Ratio

Let the two given points be $P(x_1, y_1)$ and $Q(x_2, y_2)$.

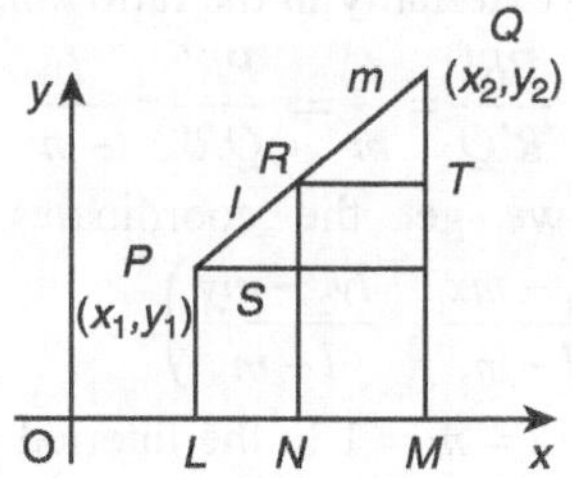

Let the point R divide PQ internally in the ratio $l:m$. Draw PL, QM and RN perpendiculars to the x-axis. Draw PS perpendicular to RN and RT perpendicular to MQ. Let the coordinates of R be (x, y). R divides PQ internally in the ratio $l:m$. Then,

$$OL = x_1,\ LP = y_1,\ OM = x_2,\ MQ = y_2,\ ON = x,\ NR = y$$

$$PS = LN = ON - OL = x - x_1;\ RT = NM = OM - ON = x_2 - x$$

$$SR = NR - NS = NR - LP = y - y_1;\ TQ = MQ - MT = y_2 - y_1$$

Triangles PSR and RTQ are similar.

$$\therefore \quad \frac{PS}{RT} = \frac{SR}{TQ} = \frac{PR}{RQ} = \frac{l}{m}$$

$$\frac{PS}{RT} = \frac{l}{m} \Rightarrow \frac{x - x_1}{x_2 - x} = \frac{l}{m} \Rightarrow m(x - x_1) = l(x_2 - x)$$

$$\therefore x(l + m) = lx_2 + mx_1 \qquad \text{or } x = \frac{lx_2 + mx_1}{l + m}$$

Also

$$\frac{SR}{TQ} = \frac{l}{m} \Rightarrow \frac{y - y_1}{y_2 - y} = \frac{l}{m} \Rightarrow y = \frac{ly_2 + my_1}{l + m}$$

Hence, the coordinates of R are $\left(\frac{lx_2 + mx_1}{l + m}, \frac{ly_2 + my_1}{l + m}\right)$.

1.2.2 External Point of Division

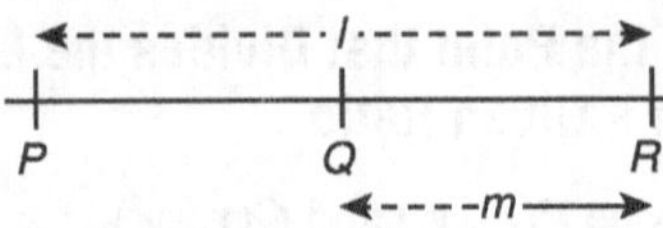

If the point R' divides PQ externally in the ratio $l{:}m$, then

$$\frac{PR'}{R'Q} = \frac{l}{m} \Rightarrow \frac{PR'}{QR'} = \frac{l}{-m}$$

Choosing m negative, we get the coordinates of R'. Therefore, the coordinates of R' are $\left(\dfrac{lx_2 - mx_1}{l-m}, \dfrac{ly_2 - my}{l-m}\right)$.

Note 1.2.2.1: If we take $l = m = 1$ in the internal point of division, we get the coordinates of the midpoint. Therefore, the coordinates of the midpoint of PQ are $\left(\dfrac{x_1 + x_2}{2}, \dfrac{y_1 + y_2}{2}\right)$.

1.2.3 Centroid of a Triangle Given its Vertices

Let ABC be a triangle with vertices $A(x_1, y_1)$, $B(x_2, y_2)$ and $C(x_3, y_3)$.

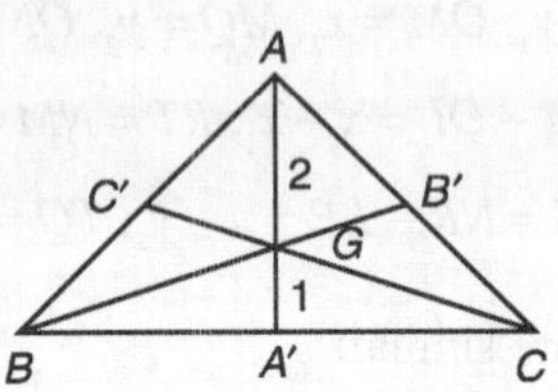

Let AA', BB' and CC' be the medians of the triangle. Then A', B', C' are the midpoints of the sides BC, CA and AB, respectively. The coordinates of A' are $\left(\dfrac{x_2 + x_3}{2}, \dfrac{y_2 + y_3}{2}\right)$. We know that the medians of a triangle are concurrent at the point G called the centroid and G divides each median in the ratio 2:1. Considering the median AA', the coordinates of G are

$$\left(\frac{1\cdot x_1 + 2\cdot\left(\dfrac{x_2 + x_3}{2}\right)}{1+2}, \frac{1\cdot y_1 + 2\cdot\left(\dfrac{y_2 + y_3}{2}\right)}{1+2}\right)$$

$$\text{(i.e.)} \quad \left(\frac{x_1 + x_2 + x_3}{3}, \frac{y_1 + y_2 + y_3}{3}\right)$$

1.2.4 Area of Triangle *ABC* with Vertices $A(x_1, y_1)$, $B(x_2, y_2)$ and $C(x_3, y_3)$

Let the vertices of triangle *ABC* be $A(x_1, y_1)$, $B(x_2, y_2)$ and $C(x_3, y_3)$.

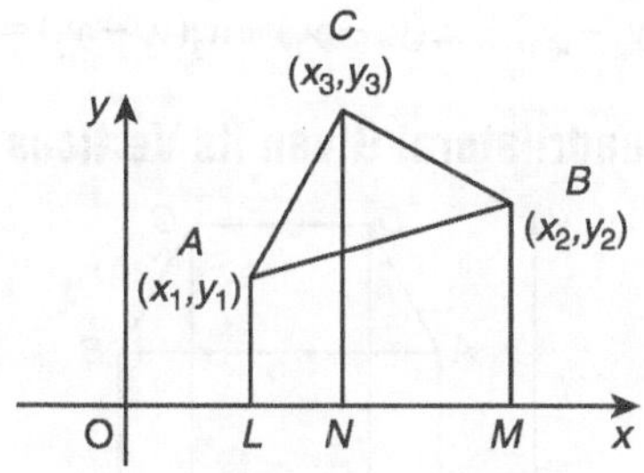

Draw *AL*, *BM* and *CN* perpendiculars to *OX*. Then, area Δ of triangle *ABC* is calculated as

Δ = Area of trapezium *ALNC* + Area of trapezium *CNMB* − Area of trapezium *ALMB*

$$= \frac{1}{2}(LA+NC)\cdot LN+\frac{1}{2}(NC+MB)\cdot NM-\frac{1}{2}(LA+MB)\cdot LM$$

$$=\frac{1}{2}(y_1+y_3)(x_3-x_1)+\frac{1}{2}(y_3+y_2)(x_2-x_3)-\frac{1}{2}(y_1+y_2)(x_2-x_1)$$

$$=\frac{1}{2}[x_1(y_1+y_2-y_1-y_3)+x_2(y_3+y_2-y_1-y_2)+x_3(y_1+y_3-y_3-y_2)]$$

$$=\frac{1}{2}[x_1(y_2-y_3)+x_2(y_3-y_1)+x_3(y_1-y_2)]$$

$$=\frac{1}{2}\begin{vmatrix} x_1 & y_1 & 1 \\ x_2 & y_2 & 1 \\ x_3 & y_3 & 1 \end{vmatrix}$$

Note 1.2.4.1: The area is positive or negative depending upon the order in which we take the points. Since scalar area is always taken to be a positive quantity, we take

$$\Delta=\frac{1}{2}\left|x_1(y_2-y_3)+x_2(y_3-y_1)+x_3(y_1-y_2)\right|$$

Note 1.2.4.2: If the vertices of the triangle are (0, 0), (x_1, y_1) and (x_2, y_2), then $\Delta=\frac{1}{2}|x_1y_2-x_2y_1|$.

Note 1.2.4.3: If the area of the triangle is zero, i.e. $\Delta = 0$, then we note that the points are collinear. Hence, the condition for the points (x_1, y_1), (x_2, y_2) and (x_3, y_3) to be collinear is

$$x_1(y_2 - y_3) + x_2(y_3 - y_1) + x_3(y_1 - y_2) = 0$$

1.2.5 Area of the Quadrilateral Given its Vertices

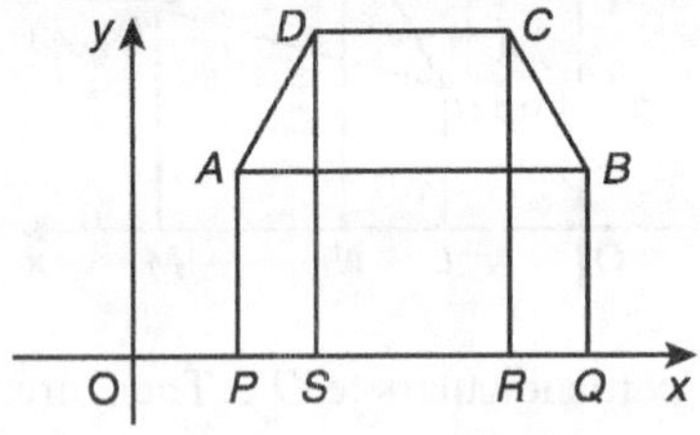

Let $ABCD$ be the quadrilateral with vertices $A(x_1, y_1)$, $B(x_2, y_2)$, $C(x_3, y_3)$ and $D(x_4, y_4)$. Draw AP, BQ, CR and DS perpendiculars to the x-axis. Then,

Area of quadrilateral $ABCD$ = Area of trapezium $APSD$ + Area of trapezium $DSRC$ + Area of trapezium $CRQB$ – Area of trapezium $APQB$

$$= \frac{1}{2}(PA + SD)PS + \frac{1}{2}(SD + RC)SR + \frac{1}{2}(RC + QB)RQ - \frac{1}{2}(PA + QB)PQ$$

$$= \frac{1}{2}\begin{bmatrix}(x_4 - x_1)(y_1 + y_4) + (x_3 - x_4)(y_3 + y_4) + (x_2 - x_3)(y_3 + y_2) - \\ (x_2 - x_1)(y_1 + y_2)\end{bmatrix}$$

$$= \frac{1}{2}[(x_1y_2 - x_2y_1) + (x_2y_3 - x_3y_2) + (x_3y_4 - x_4y_3) + (x_4y_1 - x_1y_4)]$$

Note 1.2.5.1: This result can be extended to a polygon of n sides with vertices (x_1, y_1), (x_2, y_2)...... (x_n, y_n) as

$$\text{Area} = \frac{1}{2}[(x_1y_2 - x_2y_1) + (x_2y_3 - x_3y_2) + \ldots + (x_ny_1 - x_1y_n)]$$

Locus

When a point moves so as to satisfy some geometrical condition or conditions, the path traced out by the point is called the locus of the point.

For example, if a point moves keeping a constant distance from a fixed point, the locus of the moving point is called circle and the fixed distance is called the radius of the circle. Moreover, if a point moves such that its distance from two fixed points are equal, then the locus of the point is the perpendicular bisector of the line joining the two fixed points. If A and B are two fixed points and point P moves such that $\angle APB = \frac{\pi}{2}$ then the locus of P is a circle with AB as the diameter. It is possible to represent the locus of a point by means of an equation.

Suppose a point $P(x, y)$ moves such that its distance from two fixed points $A(2, 3)$ and $B(5, -3)$ are equal. Then the geometrical law is $PA = PB \Rightarrow PA^2 = PB^2$

$$\Rightarrow (x-2)^2 + (y-3)^2 = (x-5)^2 + (y+3)^2$$
$$\Rightarrow x^2 - 4x + 4 + y^2 + 9 - 6y = x^2 + 25 - 10x + y^2 + 9 + 6y$$
$$\Rightarrow 6x - 12y - 21 = 0 \quad \Rightarrow 2x - 4y - 7 = 0$$

Here, the locus of P is a straight line.

ILLUSTRATIVE EXAMPLES

Example 1.1

Find the distance between the points (4, 7) and (–2, 5).

Solution

Let P and Q be the points (4, 7) and (–2, 5), respectively.

$$PQ^2 = (x_1 - x_2)^2 + (y_1 - y_2)^2 = (4+2)^2 + (7-5)^2 = 36 + 4 = 40$$
$$PQ = \sqrt{40} \text{ units}$$

Example 1.2

Prove that the points (4, 3), (7, –1) and (9, 3) are the vertices of an isosceles triangle.

Solution

Let $A(4, 3)$, $B(7, -1)$, $C(9, 3)$ be the three given points.

Then $AB^2 = (4 - 7)^2 + (3 + 1)^2 = 9 + 16 = 25$

$$BC^2 = (7 - 9)^2 + (-1 - 3)^2 = 4 + 16 = 20$$

$$AC^2 = (4 - 9)^2 + (3 - 3)^2 = 25 + 0 = 25$$

$$\therefore AB = 5, BC = \sqrt{20}, AC = 5$$

Since the sum of two sides is greater than the third, the points form a triangle. Moreover, $AB = AC = 5$. Therefore, the triangle is an isosceles triangle.

Example 1.3

Show that the points (6, 6), (2, 3) and (4, 7) are the vertices of a right angled triangle.

Solution

Let A, B, C be the points (6, 6), (2, 3) and (4, 7), respectively.

$$AB^2 = (6-2)^2 + (6-3)^2 = 16 + 9 = 25$$

$$BC^2 = (2-4)^2 + (3-7)^2 = 4 + 16 = 20$$

$$AC^2 = (6-4)^2 + (6-7)^2 = 4 + 1 = 5$$

$$AC^2 + BC^2 = 20 + 5 = 25 = AB^2$$

Hence, the points are the vertices of a right angled triangle.

Example 1.4

Show that the points (7, 9), (3, –7) and (–3, 3) are the vertices of a right angled isosceles triangle.

Solution

Let A, B, C be the points (7, 9), (3, –7), (–3, 3), respectively.

$$AB^2 = (7-3)^2 + (9+7)^2 = 16 + 256 = 272$$

$$BC^2 = (3+3)^2 + (-7-3)^2 = 36 + 100 = 136$$

$$AC^2 = (7+3)^2 + (9-3)^2 = 100 + 36 = 136$$

$$BC^2 + AC^2 = 136 + 136 = 272 = AB^2$$

Hence, the points are vertices of a right angled triangle. Also, $BC = AC$. Therefore, it is a right angled isosceles triangle.

Example 1.5

Show that the points (4, –4), (–4, 4) and $(4\sqrt{3}, 4\sqrt{3})$ are the vertices of an equilateral triangle.

Solution

Let A, B, C be the points (4, –4), (–4, 4) and $(4\sqrt{3}, 4\sqrt{3})$, respectively.

$$AB^2 = (4 + 4)^2 + (-4 - 4)^2 = 64 + 64 = 128$$

$$BC^2 = (-4-4\sqrt{3})^2 + (4-4\sqrt{3})^2 = 16+48+32\sqrt{3}+16+48-32\sqrt{3} = 128$$

$$AC^2 = (4-4\sqrt{3})^2 + (-4-4\sqrt{3})^2 = 16+48-32\sqrt{3}+16+48+32\sqrt{3} = 128$$

$$AB = BC = AC = \sqrt{128}$$

Hence, the points A, B and C are the vertices of an equilateral triangle.

Example 1.6

Show that the set of points (–2, –1), (1, 0), (4, 3) and (1, 2) are the vertices of a parallelogram.

Solution

Let A, B, C, D be the points (–2, –1), (1, 0), (4, 3) and (1, 2), respectively. A quadrilateral is a parallelogram if the opposite sides are equal.

$$AB^2 = (-2 - 1)^2 + (-1 - 0)^2 = 9 + 1 = 10$$

$$BC^2 = (1 - 4)^2 + (0 - 3)^2 = 9 + 9 = 18$$

$$CD^2 = (4 - 1)^2 + (3 - 2)^2 = 9 + 1 = 10$$

$$AD^2 = (-2 - 1)^2 + (-1 - 2)^2 = 9 + 9 = 18$$

$$AB = CD = \sqrt{10} \quad BC = AD = \sqrt{18}$$

Since the opposite sides of the quadrilateral $ABCD$ are equal, the four points form a parallelogram.

Example 1.7

Show that the points (2, –2), (8, 4), (5, 7) and (–1, 1) are the vertices of a rectangle taken in order.

Solution

A quadrilateral in which the opposite sides are equal and the diagonals are equal is a rectangle. Let $A(2, -2)$, $B(8, 4)$, $C(5, 7)$ and $D(-1, 1)$ be the four given points.

$AB^2 = (2-8)^2 + (-2-4)^2 = 36 + 36 = 72 \quad AC^2 = (2-5)^2 + (-2, -7)^2 = 9 + 81 = 90$

$BC^2 = (8-5)^2 + (4-7)^2 = 9 + 9 = 18 \quad BD^2 = (8+1)^2 + (4-1)^2 = 81 + 9 = 90$

$CD^2 = (5+1)^2 + (7-1)^2 = 36 + 36 = 72 \quad AD^2 = (2+1)^2 + (-2-1)^2 = 9 + 9 = 18$

$$AB = CD = \sqrt{72} \qquad BC = AD = \sqrt{18} \qquad AC = BD = \sqrt{90}$$

Thus, the opposite sides are equal and the diagonals are also equal. Hence, the four points form a rectangle.

Example 1.8

Prove that the points (3, 2), (5, 4), (3, 6) and (1, 4) taken in order form a square.

Solution

A quadrilateral in which all sides are equal and diagonals are equal is a square. Let A, B, C, D be the points (3, 2), (5, 4), (3, 6), (1, 4), respectively.

$AB^2 = (3-5)^2 + (2-4)^2 = 4 + 4 = 8 \quad CD^2 = (3-1)^2 + (6-4)^2 = 4 + 4 = 8$

$BC^2 = (5-3)^2 + (4-6)^2 = 4 + 4 = 8; \quad AD^2 = (3-1)^2 + (2-4)^2 = 4 + 4 = 8$

$AC^2 = (3-3)^2 + (2-6)^2 = 0 + 4^2 = 16 \quad BD^2 = (5-1)^2 + (4-4)^2 = 16 + 0 = 16$

$$AB = BC = CD = AD = \sqrt{8} \qquad AC = BD = \sqrt{16} = 4$$

Thus, all sides are equal and also the diagonals are equal. Hence, the four points form a square.

Example 1.9

Find the coordinates of the circumcentre of a triangle whose vertices are $A(3, -2)$, $B(4, 3)$ and $C(-6, 5)$. Also, find the circumradius.

Solution

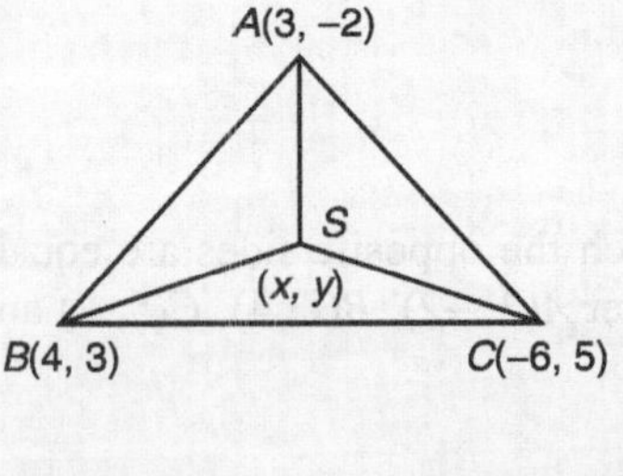

Let $A(3, -2)$, $B(4, 3)$, and $C(-6, 5)$ be the given points. Let $S(x, y)$ be the circumcentre of ΔABC. Then $SA = SB = SC$ = circumradius. Now

$$SA^2 = (x-3)^2 + (y+2)^2 = x^2 + y^2 - 6x + 4y + 13$$

$$SB^2 = (x-4)^2 + (y-3)^2 = x^2 + y^2 - 8x - 6y + 25$$

$$SC^2 = (x+6)^2 + (y-5)^2 = x^2 + y^2 + 12x - 10y + 61$$

$$SA^2 = SB^2 \Rightarrow 2x + 10y = 12 \qquad (1.1)$$

$$SA^2 = SC^2 \Rightarrow 18x - 14y = -48 \qquad (1.2)$$

$$\begin{aligned} (1.1)\times 9 &\Rightarrow \cancel{18x} + 90y = 108 \\ (1.2)\times 1 &\Rightarrow \cancel{18x} - 14y = -48 \\ \hline & \qquad 104y = 156 \\ & \qquad y = \frac{156}{104} = \frac{78}{52} = \frac{3}{2} \end{aligned}$$

$$(1.1) \Rightarrow 2x + 10\times\frac{3}{2} = 12 \Rightarrow 2x = 12 - 15 = -3$$

$$\Rightarrow x = -\frac{3}{2}$$

Hence, the circumcentre is $\left(-\frac{3}{2}, \frac{3}{2}\right)$.

Now $SA^2 = \left(3+\frac{3}{2}\right)^2 + \left(-2-\frac{3}{2}\right)^2 = \frac{81}{4} + \frac{49}{4} = \frac{130}{4}$

Therefore, circumradius $= \sqrt{\frac{130}{4}} = \frac{\sqrt{130}}{2}$ units.

Example 1.10

Show that the points (3, 7), (6, 5) and (15, −1) lie on a straight line.

Solution

Let $A(3, 7)$, $B(6, 5)$ and $C(15, -1)$ be the three points. Then

$$AB^2 = (3-6)^2 + (7-5)^2 = 9 + 4 = 13$$

$$BC^2 = (6-15)^2 + (5+1)^2 = 81 + 36 = 117$$

$$AC^2 = (3-15)^2 + (7+1)^2 = 144 + 64 = 208$$

$$\therefore AB = \sqrt{13}$$

$$BC = \sqrt{117} = \sqrt{9 \times 13} = 3\sqrt{13}$$

$$AC = \sqrt{208} = \sqrt{16 \times 13} = 4\sqrt{13}$$

$$AB + BC = AC = 4\sqrt{13}$$

Hence, the three given points lie on a straight line.

Example 1.11

Show that (4, 3) is the centre of the circle that passes through the points (9, 3), (7, –1) and (1, –1). Find its radius.

Solution

Let $A(9, 3)$, $B(7, -1)$, $C(1, -1)$ and $P(4, 3)$ be the given points.

Then $PA^2 = (9-4)^2 + (3-3)^2 = 25 + 0 = 25$

$$PB^2 = (7-4)^2 + (-1-3)^2 = 9 + 16 = 25$$

$$PC^2 = (1-4)^2 + (-1-3)^2 = 9 + 16 = 25$$

$$\therefore PA = PB = PC = \sqrt{25} = 5$$

Hence, P is the centre of the circle passing through the points A, B, C; its radius is 5.

Example 1.12

If O is the origin and the coordinates of A and B are (x_1, y_1) and (x_2, y_2), respectively, prove that $OA \cdot OB \cos\theta = x_1x_2 + y_1y_2$ where $\underline{|AOB} = \theta$.

Solution

By cosine formula

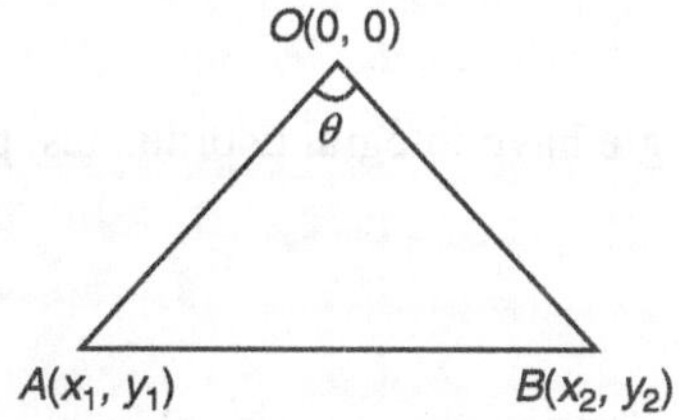

$$2OA \cdot OB \cos\theta = OA^2 + OB^2 - AB^2$$

$$= x_1^2 + y_1^2 + x_2^2 + y_2^2 - [(x_1 - x_2)^2 + (y_1 - y_2)^2]$$

$$= 2(x_1x_2 + y_1y_2)$$

Hence, $OA \cdot OB \cos\theta = x_1x_2 + y_1y_2$

Example 1.13

If $\tan\alpha$, $\tan\beta$ and $\tan\gamma$ be the roots of the equation $x^3 - 3ax^2 + 3bx - c = 0$ and the vertices of the triangle ABC are $(\tan\alpha, \cot\alpha)$, $(\tan\beta, \cot\beta)$ and $(\tan\gamma, \cot\gamma)$ show that the centroid of the triangle is (a, b).

Solution

Given $\tan\alpha$, $\tan\beta$ and $\tan\gamma$ are the roots of the equation

$$x^3 - 3ax^2 + 3bx - c = 0 \tag{1.3}$$

Then $$\tan\alpha + \tan\beta + \tan\gamma = 3a \tag{1.4}$$

$$\tan\alpha\tan\beta + \tan\beta\tan\gamma + \tan\gamma\tan\alpha = 3b \tag{1.5}$$

$$\tan\alpha\tan\beta\tan\gamma = 1 \tag{1.6}$$

Then dividing (1.5) by (1.6), $\cot\alpha + \cot\beta + \cot\gamma = 3b$ (1.7)

The centroid of ΔABC is $\left(\dfrac{\tan\alpha + \tan\beta + \tan\gamma}{3}, \dfrac{\cot\alpha + \cot\beta + \cot\gamma}{3}\right)$

(i.e.) (a, b) from (1.4) and (1.7).

Example 1.14

If the vertices of a triangle have integral coordinates, prove that it cannot be an equilateral triangle.

Solution

The area of the triangle with vertices (x_1, y_1), (x_2, y_2) and (x_3, y_3) is

$$\Delta = \frac{1}{2}[x_1(y_2 - y_3) + x_2(y_3 - y_1) + x_3(y_1 - y_2)] \tag{1.8}$$

Also, the area of ΔABC is

$$\Delta = \frac{\sqrt{3}}{4}a^2 \tag{1.9}$$

where a is the side of the equilateral triangle. If the vertices of the triangle have integral coordinates, then Δ is a rational number. However, from (1.9) we infer that the area is $\sqrt{3}$ times a rational number. Hence, if the vertices of a triangle have integral coordinates, it cannot be equilateral.

Example 1.15

If t_1, t_2 and t_3 are distinct, then show that the points $(at_1^2, 2at_1)$, $(at_2^2, 2at_2)$ and $(at_3^2, 2at_3)$, $a \neq 0$ cannot be collinear.

Solution

$$\text{The area of the triangle } \Delta = \frac{1}{2}\begin{vmatrix} x_1 & y_1 & 1 \\ x_2 & y_2 & 1 \\ x_3 & y_3 & 1 \end{vmatrix} = \frac{1}{2}\begin{vmatrix} at_1^2 & 2at_1 & 1 \\ at_2^2 & 2at_2 & 1 \\ at_3^2 & 2at_3 & 1 \end{vmatrix}$$

$$= a^2\begin{vmatrix} t_1^2 & t_1 & 1 \\ t_2^2 & t_2 & 1 \\ t_3^2 & t_3 & 1 \end{vmatrix} = a^2(t_1 - t_2)(t_2 - t_3)(t_3 - t_1) \neq 0$$

since t_1, t_2 and t_3 are distinct. Hence, the three given points cannot be collinear.

Example 1.16

The vertices of a triangle ABC are (2, 3), (4, 7), (–5, 2). Find the length of the altitude through A.

Solution

The area of ΔABC is given by

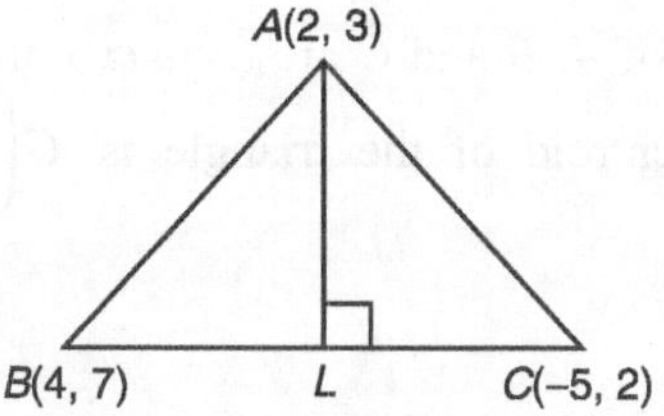

$$\Delta = \frac{1}{2}[x_1(y_2 - y_3) + x_2(y_3 - y_1) + x_3(y_1 - y_2)]$$

$$\Delta = \frac{1}{2}[2(7-2) + 4(2-3) - 5(3-7)] = \frac{1}{2}[10 - 4 + 20] = 13 \text{ square units}$$

$$BC = \sqrt{(4+5)^2 + (7-2)^2} = \sqrt{81+25} = \sqrt{106}$$

We know that $\Delta = \frac{1}{2} BC \cdot AL \Rightarrow AL = \frac{2\Delta}{BC} = \frac{2\times 13}{\sqrt{106}} = \frac{26\sqrt{106}}{106}$

$$\Rightarrow AL = \frac{13}{53}\sqrt{106} \text{ units}$$

Example 1.17

The vertices of a triangle ABC are $A(x_1, x_1 \tan\alpha)$, $B(x_2, x_2 \tan\beta)$ and $C(x_3, x_3 \tan\gamma)$. If $H(\bar{x}, \bar{y})$ is the orthocentre and $S(0,0)$ is the circumcentre, then show that $\frac{\bar{y}}{\bar{x}} = \frac{\sin\alpha + \sin\beta + \sin\gamma}{\cos\alpha + \cos\beta + \cos\gamma}$.

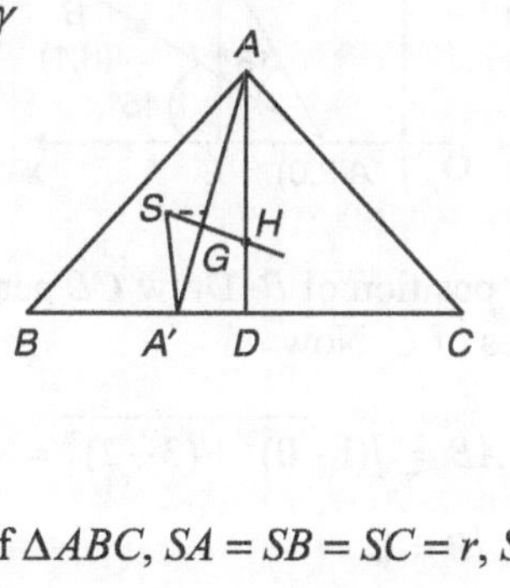

Solution

If r is the circumradius of ΔABC, $SA = SB = SC = r$, $SA^2 = r^2$

(i.e.) $\quad x_1^2 + x_1^2 \tan^2\alpha = r^2$

(i.e.) $x_1^2(1 + \tan^2\alpha) = r^2 \Rightarrow x_1 = r\cos\alpha$

Similarly, $x_2 = r\cos\beta$ and $x_3 = r\cos\gamma$

Then, the coordinates of A, B and C are ($r\cos\alpha$, $r\sin\alpha$), ($r\cos\beta$, $r\sin\beta$) and ($r\cos\gamma$, $r\sin\gamma$). The centroid of the triangle is $G\left(\dfrac{\cos\alpha+\cos\beta+\cos\gamma}{3}, \dfrac{\sin\alpha+\sin\beta+\sin\gamma}{3}\right)$.

The orthocentre is $H(\bar{x},\bar{y})$ and the circumcentre is $S(0,0)$. Geometrically we know that H, G and S are collinear. Therefore, the slope of SG and GH are equal.

$$\therefore \frac{\sin\alpha+\sin\beta+\sin\gamma}{\cos\alpha+\cos\beta+\cos\gamma} = \frac{\bar{y}}{\bar{x}}$$

Example 1.18

A line joining the two points $A(2, 0)$ and $B(3, 1)$ is rotated about A in the anticlockwise direction through an angle of 15°. If B goes to C in the new position, find the coordinate of C.

Solution

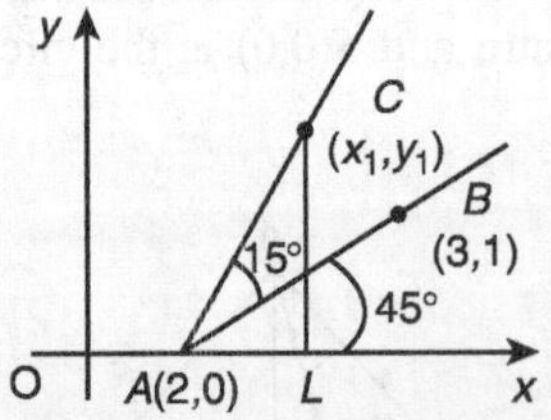

Given that C is the new position of B. Draw CL perpendicular to OX and let (x_1, y_1) be the coordinates of C. Now

$$AB = \sqrt{(1-0)^2 + (3-2)^2} = \sqrt{2}$$

Slope of $AB = 1$

AB makes 45° with x-axis and $\underline{|CAC} = 60°$. Then

$$OL = OA + AL$$
$$= 2 + AC\cos 60^\circ$$
$$= 2 + \sqrt{2} \times \frac{1}{2} = 2 + \frac{\sqrt{2}}{2} = \frac{4+\sqrt{2}}{2}$$
$$LC = AC\sin 60^\circ = \frac{\sqrt{2}\times\sqrt{3}}{2} = \frac{\sqrt{6}}{2}$$

Hence, the coordinates of C are $\left(\frac{4+\sqrt{2}}{2}, \frac{\sqrt{6}}{2}\right)$.

Example 1.19

The coordinates of A, B and C are (6, 3), (–3, 5) and (4, –2), respectively, and P is any point (x, y). Show that the ratio of the area of ΔPBC and ΔABC is $\left|\frac{x+y-2}{7}\right|$.

Solution

The points A, B, C and P are (6, 3), (–3, 5), (4, –2) and (x, y), respectively.

$$\text{Area of } \Delta\, ABC = \frac{1}{2}[6(5+2) - 3(-2-3) + 4(3-5)]$$
$$= \frac{1}{2}[42+15-8] = \frac{49}{2} \text{ square units}$$
$$\text{Area of } \Delta\, PBC = \frac{1}{2}[x(5+2) - 3(-2-y) + 4(y-5)]$$
$$= \frac{1}{2}[7x+6+3y+4y-20] = \frac{1}{2}\left[\frac{7x+7y-14}{1}\right]$$
$$\frac{\text{Area of } \Delta\, PBC}{\text{Area of } \Delta\, ABC} = \frac{1}{2}\left|\frac{7x+7y-14}{\frac{49}{2}}\right| = \left|\frac{x+y-2}{7}\right|$$

Example 1.20

Find the coordinates of the point that divides the line joining the points (2, 3) and (–4, 7) (i) internally (ii) externally in the ratio 3:2.

Solution

Let R and R' respectively divide PQ internally and externally in the ratio 3:2.

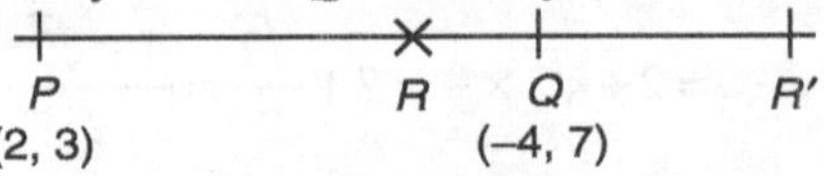

(i) The coordinates of R are $\left(\dfrac{lx_2+mx_1}{l+m}, \dfrac{ly_2+my_1}{l+m}\right)$

(i.e.) $\left(\dfrac{3\times(-4)+2\times 2}{3+2}, \dfrac{3\times 7+2\times 3}{3+2}\right)$

(i.e.) $\left(\dfrac{-8}{5}, \dfrac{27}{5}\right)$

(ii) The coordinates of R' are $\left(\dfrac{3\times(-4)-2\times 2}{3-2}, \dfrac{3\times 7-2\times 3}{3-2}\right)$

(i.e.) $(-16, 15)$

Example 1.21

Find the ratio in which the line joining the points (4, 7) and (–3, 2) is divided by the y-axis.

Solution

Let the y-axis meet the line joining the joints $P(4, 7)$ and $Q(-3, 2)$ at R. Let the coordinates of R be $(0, y)$. Let R divide PQ in the ratio k:1.

The coordinates of R is given by $\dfrac{k(-3)+1\times 4}{k+1} = 0$. $\therefore 3k = 4 \Rightarrow k = \dfrac{4}{3}$

Hence, the ratio in which R divides PQ is 4:3.

Example 1.22

Show that the points (–2, –1), (1, 0), (4, 3) and (1, 2) form the vertices of a parallelogram.

Solution

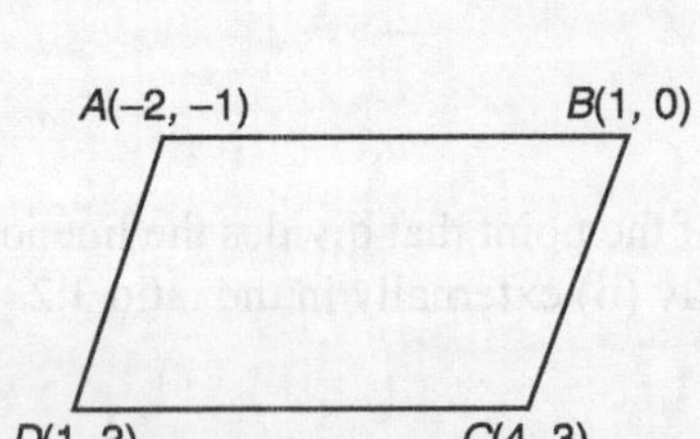

A quadrilateral in which the diagonals bisect each other is a parallelogram.

The midpoint of AC is $\left(\frac{x_1+x_2}{2},\frac{y_1+y_2}{2}\right)$ (i.e.) $\left(\frac{-2+4}{2},\frac{-1+3}{2}\right)$ (i.e.) (1,1).

The midpoint of BD is $\left(\frac{1+1}{2},\frac{2+0}{2}\right)$ (i.e.) (1,1).

Since the diagonals bisect each other, $ABCD$ is a parallelogram.

Example 1.23

Find (x, y) if (3, 2), (6, 3), (x, y) and (6, 5) are the vertices of a parallelogram taken in order.

Solution

Let the four points be A, B, C and D, respectively. Since $ABCD$ is a parallelogram, the midpoint of AC is the same as the midpoint of BD.

The midpoint of AC is $\left(\frac{3+x}{2},\frac{2+y}{2}\right)$. The midpoint of BD is $\left(\frac{6+6}{2},\frac{3+5}{2}\right)$ (i.e.) (6,4).

$$\therefore \frac{3+x}{2}=6 \quad \text{and} \quad \frac{2+4}{2}=4 \Rightarrow x=9, y=6.$$

Hence (x, y) is (9, 6).

Example 1.24

The midpoints of the sides of a triangle are (6, –1), (–1, –2) and (1, 4). Find the coordinates of the vertices.

Solution

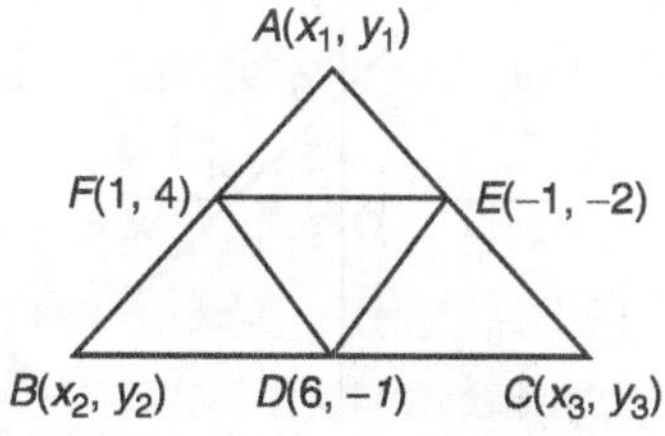

Let D, E and F be the midpoints of the sides BC, CA and AB, respectively. Then, (6, –1), (–1, –2), (1, 4) are the points D, E and F, respectively.

Let $A(x_1, y_1)$, $B(x_2, y_2)$ and $C(x_3, y_3)$ be the vertices of the triangle. Then *BDEF* is a parallelogram.

The midpoint of *DF* is $\left(\frac{6+1}{2}, \frac{-1+4}{2}\right)$ (i.e.) $\left(\frac{7}{2}, \frac{3}{2}\right)$.

The midpoint of *BE* is $\left(\frac{x_2-1}{2}, \frac{y_2-2}{2}\right)$.

$$\therefore \frac{x_2-1}{2} = \frac{7}{2} \text{ and } \frac{y_2-2}{2} = \frac{3}{2} \quad \therefore x_2 = 8,\ y_2 = 5 \quad \therefore B\ (8,5).$$

Since *F* is the midpoint of *AB*, $\left(\frac{8+x_1}{2}, \frac{5+y_1}{2}\right) = (1,4)$

$$\therefore \frac{8+x_1}{2} = 1,\ \frac{5+y_1}{2} = 4 \Rightarrow x_1 = -6, y_1 = 3 \ \therefore A(-6,3)$$

Since *E* is the midpoint of *AC*, $\frac{-6+x_3}{2} = -1,\ \frac{3+y_3}{2} = -2$

$\therefore x_3 = 4, y_3 = -7.$ $\therefore C$ is the point $(4, -7)$.

Hence, the vertices of the triangle are (−6, 3), (8, 5) and (4, −7).

Example 1.25

Show that the axes of coordinates trisect the straight line joining the points (2, −2) and (−1, 4).

Solution

Let the line joining the points (2, −2) and (−1, 4) meet *x*-axis and *y*-axis at *A* and *B*, respectively. Let the coordinates of *A* and *B* be $(x, 0)$ and $(0, y)$, respectively. Let *A* divide the line in the ratio k:1. Then the *x*-coordinate of *A* is given by $\frac{k(-1)+1\times 2}{k+1} = 0$.

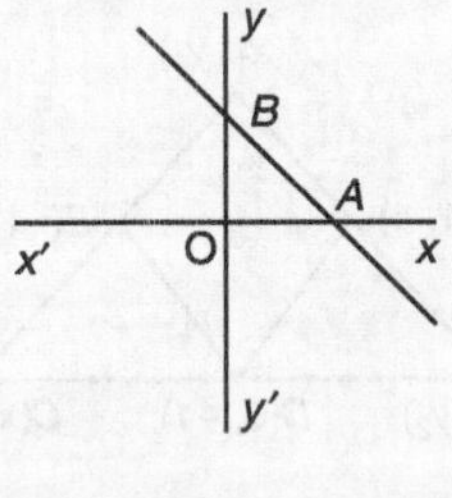

$$\therefore -k+2 = 0 \Rightarrow k = 2.$$

Hence, *A* divides the line in the ratio 2:1.

Let B divide the line in the ratio 1:1. Then, $\frac{\lambda(4)+1(2)}{\lambda+1}=0$.

Hence, $4\lambda=2 \Rightarrow \lambda=\frac{1}{2}$

B divides the line in the ratio 1:2. Hence, A and B trisect the line joining the points (2, –4) and (–1, 4).

Example 1.26

The vertices of a triangle are $A(3, 5)$, $B(-7, 9)$ and $C(1, -3)$. Find the length of the three medians of the triangle.

Solution

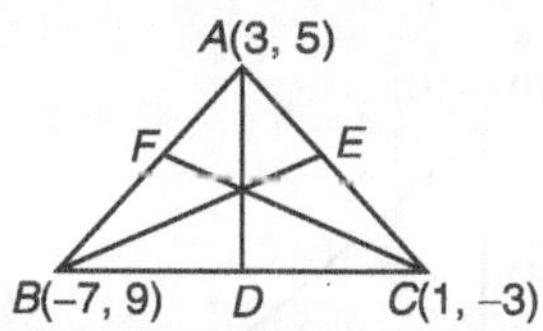

Let D, E and F be the midpoints of the sides of BC, CA and AB, respectively. The coordinates of D are $\left(\frac{-7+1}{2}, \frac{9-3}{2}\right)$ (i.e.) (–3, 3). The coordinates of E are $\left(\frac{3+1}{2}, \frac{5-3}{2}\right)$(i.e.) (2, 1). The coordinates of F are $\left(\frac{3-7}{2}, \frac{5+9}{2}\right)$ (i.e.) (–2, 7).

$$AD^2=(3+3)^2+(5-3)^2=36+4=40$$
$$BE^2=(-7-2)^2+(9-1)^2=81+64=145$$
$$CF^2=(1+2)^2+(-3-7)^2=9+100=109$$

Hence, the lengths of the medians are $AD=2\sqrt{10}$, $BE=\sqrt{145}$ and $CF=\sqrt{109}$ units.

Example 1.27

Two of the vertices of a triangle are (4, 7) and (–1, 2) and the centroid is at the origin. Find the third vertex.

Solution

Let the third vertex of the triangle be (x, y). Then

$$\frac{4-1+x}{3}=0 \text{ and } \frac{7+2+y}{3}=0 \quad \therefore x=-3,\ y=-9.$$

Hence, the third vertex is (–3, –9).

Example 1.28

Show that the midpoint of the hypotenuse of the right angled triangle whose vertices are (8, –10), (7, –3) and (0, –4) is equidistant from the vertices.

Solution

Let the three given points be $A(8, -10)$, $B(7, -3)$ and $C(0, -4)$.

$$AB^2 = (8, 7)^2 + (-10 + 3)^2 = 1 + 49 = 50$$

$$BC^2 = (7 - 0)^2 + (-3 + 4)^2 = 49 + 1 = 50$$

$$AC^2 = (8 - 0)^2 + (-10 + 4)^2 = 64 + 36 = 100$$

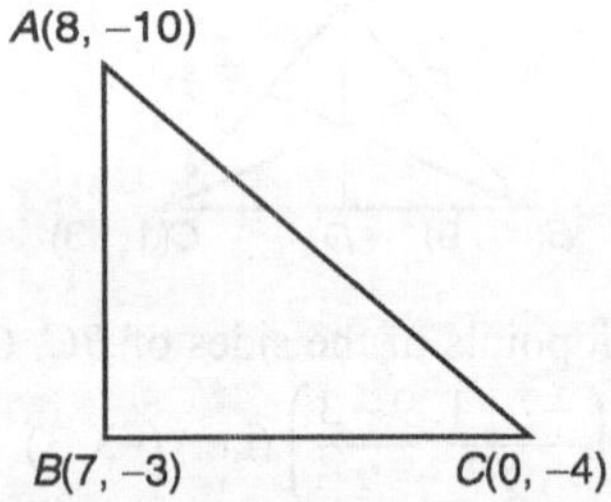

$$AB^2 + BC^2 = AC^2$$

Hence, ABC is a right angled triangle with AC as hypotenuse.

The midpoint of AC is $P\left(\frac{8+0}{2}, \frac{-10-4}{2}\right)$ (i.e.) (4, –7).

$$PA^2 = (8 - 4)^2 + (-10 + 7)^2 = 16 + 9 = 25$$

$$PB^2 = (7 - 4)^2 + (-3 + 7)^2 = 9 + 16 = 25$$

$$PC^2 = (0 - 4)^2 + (-4 + 7)^2 = 16 + 9 = 25$$

$$PA = PB = PC$$

Hence, the midpoint of the hypotenuse is equidistant from the vertices.

Example 1.29

Find the ratio in which the line joining the points (1, –1) and (4, 5) is divided by the point (2, 1).

Solution

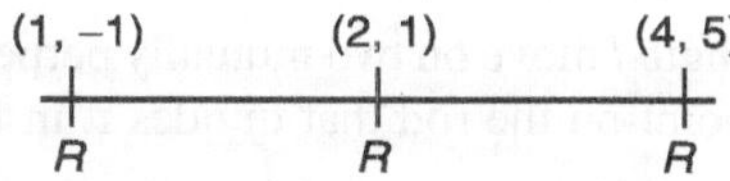

Let the point $R(2, 1)$ divide the line joining the points $P(1, -1)$ and $Q(4, 5)$ in the ratio k:1. Then

$$\frac{(k\times 4)+(1\times 1)}{k+1} = 2 \Rightarrow 4k+1 = 2k+2$$

$$\Rightarrow 2k = 1 \Rightarrow k = \frac{1}{2}$$

Therefore, $R(2, 1)$ divides PQ in the ratio 1:2.

Example 1.30

Find the locus of the point that is equidistant from two given points (2, 3) and (–4, 1).

Solution

Let $P(x, y)$ be a point such that $PA = PB$ where A and B are the points (2, 3) and (–4, 1), respectively.

$$PA^2 = PB^2 \Rightarrow (x-2)^2 + (y-3)^2 = (x+4)^2 + (y-1)^2$$

$$\Rightarrow x^2 + 4 - 4x + y^2 - 6y + 9 = x^2 + 16 + 8x + y^2 - 2y + 1$$

$$\Rightarrow 12x + 4y = 4 \text{ or } 3x + y = 1$$

Example 1.31

Find the locus of the point that moves from the point (4, 3) keeping a constant distance of 5 units from it.

Solution

Let $C(4, 3)$ be the given point and $P(x, y)$ be any point such that $CP = 5$. Then

$$CP^2 = 25 \Rightarrow (x-4)^2 + (y-3)^2 = 25$$

$\Rightarrow x^2 + y^2 - 8x - 6y = 0$. This is the required locus.

Example 1.32

The ends of a rod of length l move on two mutually perpendicular lines. Show that the locus of the point on the rod that divides it in the ratio 1:2 is $9x^2 + 36y^2 = l^2$.

Solution

Let AB be a rod of length l whose ends A and B are on the coordinate axes. Let the coordinates of A and B be $A(a, 0)$ and $B(0, b)$. Let the point $P(x_1, y_1)$ divide AB in the ratio 1:2.

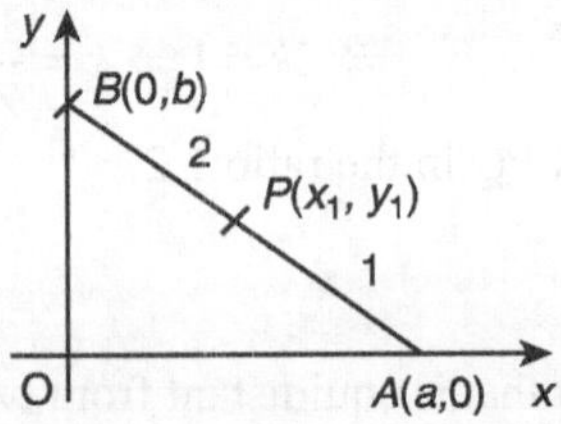

Then the coordinates of P are $\left(\frac{2a}{3}, \frac{b}{3}\right)$ $\therefore x_1 = \frac{2a}{3}, y_1 = \frac{b}{3}$

(i.e) $a = \frac{3x_1}{2}$ and $b = 3y_1$

$$OA^2 + OB^2 = AB^2 \Rightarrow a^2 + b^2 = l^2 \Rightarrow \left(\frac{3x_1}{2}\right)^2 + (3y_1)^2 = l^2$$

or $\quad 9x_1^2 + 36y_1^2 = l^2$

Hence, the locus of (x_1, y_1) is $9x^2 + 36y^2 = l^2$.

Example 1.33

A point moves such that the sum of its distances from two fixed points $(al, 0)$ and $(-al, 0)$ is always $2a$. Prove that the equation of the locus is $\frac{x^2}{a^2} + \frac{y^2}{a^2(1-l^2)} = 1.$

Solution

Let the two fixed points be $A(al, 0)$ and $B(0, -al)$. Let $P(x_1, y_1)$ be a moving point such that $PA + PB = 2a$.

Given that $$\sqrt{(x_1 - al)^2 + y_1^2} + \sqrt{(x_1 + al)^2 + y_1^2} = 2a. \qquad (1.10)$$

Then

$$\left[(x_1-al)^2+y_1^2\right]-\left[(x_1+al)^2+y_1^2\right]=2a\left[\sqrt{(x_1-al)^2+y_1^2}-\sqrt{(x_1+al)^2+y_1^2}\right]$$

$$-4ax_1l=2a\left[\sqrt{(x_1-al)^2+y_1^2}-\sqrt{(x_1+al)^2+y_1^2}\right]$$

or $\sqrt{(x_1-al)^2+y_1^2}-\sqrt{(x_1+al)^2+y_1^2}=-2x_1l$ (1.11)

Adding (1.10) and (1.11),

$$2\sqrt{(x_1-al)^2+y_1^2}=2(a-lx_1)$$

Squaring on both sides, we get

$$(x_1-al)^2+y_1^2=(a-lx_1)^2$$

$$\text{(i.e.)}\quad x_1^2(1-l^2)+y_1^2=a^2(1-l^2)$$

Dividing by $a^2(1-l^2)$, we get

$$\frac{x_1^2}{a^2}+\frac{y_1^2}{a^2(1-l^2)}=1$$

Therefore, the locus of (x_1, y_1) is $\frac{x^2}{a^2}+\frac{y^2}{a^2(1-l^2)}=1$.

Example 1.34

A right angled triangle having the right angle at C with $CA = a$ and $CB = b$ moves such that the angular points A and B slide along the x-axis and y-axis, respectively. Find the locus of C.

Solution

Let the points A and B be on the x-axis and y-axis, respectively. Let A and B have coordinates $(\alpha, 0)$ and $(0, \beta)$. Let C be the point with coordinates (x_1, y_1).

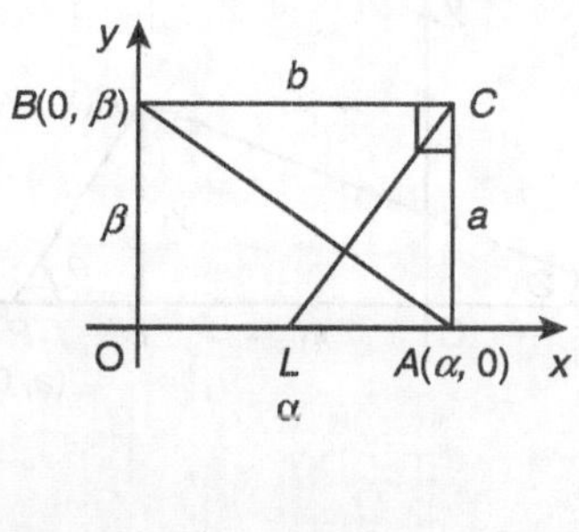

Then $$(x_1-\alpha)^2+y_1^2=a^2 \tag{1.12}$$

$$x_1^2+(y_1-\beta)^2=b^2 \tag{1.13}$$

Then $AB^2 = a^2 + b^2$. Also $AB^2 = \alpha^2 + \beta^2$.

$$\therefore\ a^2+b^2=\alpha^2+\beta^2 \tag{1.14}$$

$$\alpha = x_1-\sqrt{a^2-y_1^2}$$

$$\beta = y_1-\sqrt{b^2-x_1^2}$$

Hence, $\alpha^2+\beta^2=a^2+b^2$

$$\Rightarrow \left(x_1-\sqrt{a^2-y_1^2}\right)^2+\left(y_1-\sqrt{b^2-x_1^2}\right)^2=a^2+b^2$$

$$\Rightarrow 2x_1\sqrt{a^2-y_1^2}-2y_1\sqrt{b^2-x_1^2}=0$$

$$\text{or}\quad x_1^2(a^2-y_1^2)=y_1^2(b^2-x_1^2)\Rightarrow a^2x_1^2-b^2y_1^2=0$$

Hence, the locus of $c(x_1, y_1)$ is $a^2x^2 - b^2y^2 = 0$.

Example 1.35

Two points P and Q are given. R is a variable point on one side of the line PQ such that $\underline{|RPQ} - \underline{|RQP}$ is a positive constant 2α. Find the locus of the point P.

Solution

Let PQ be the x-axis and the perpendicular through the midpoint of PQ be the y-axis. Let P and Q be the points $(a, 0)$ and $(-a, 0)$, respectively. Let R be the point (x_1, y_1). Let $\underline{|RPQ} = \theta$ and $\underline{|RQP} = \phi$. Then $\underline{|RPQ} - \underline{|RQP} = 2\alpha$.

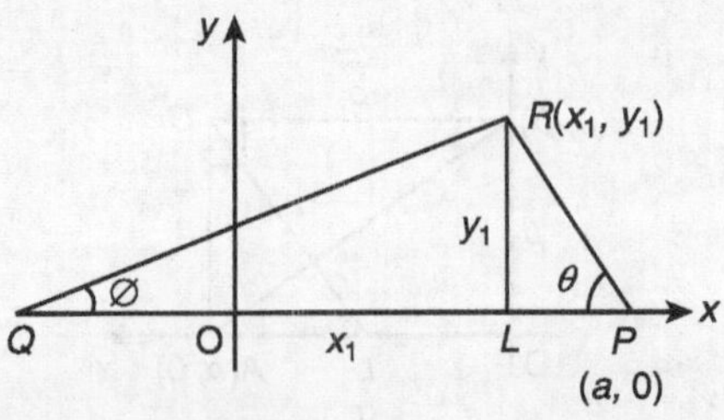

(i.e.) $\theta - \phi = 2\alpha$. Then $\tan(\theta - \phi) = \tan 2\alpha$

$$\frac{\tan\theta - \tan\phi}{1+\tan\theta\tan\phi} = \tan 2\alpha$$

$$\Rightarrow \frac{\frac{y_1}{a-x_1} - \frac{y_1}{a+x_1}}{1+\frac{y_1}{a-x_1}.\frac{y_1}{a+x_1}} = \tan 2\alpha$$

$$\Rightarrow \frac{y_1(a+x_1-a+x_1)}{a^2-x_1^2+y_1^2} = \tan 2\alpha$$

$$\Rightarrow 2x_1y_1 = (a^2-x_1^2+y_1^2)\tan 2\alpha \text{ or } x_1^2 - y_1^2 + 2x_1y_1\cot 2\alpha = a^2$$

Hence, the locus of (x_1, y_1) is $x^2 - y^2 - 2xy\cot 2\alpha = a^2$.

Exercises

1. Show that the area of the triangle with vertices (a, b), (x_1, y_1) and (x_2, y_2) where a, x_1 and x_2 are in geometric progression with common ratio r and b, y_1 and y_2 are in geometric progression with common ratio s is $\frac{1}{2}|ab(r-1)(s-1)(b-r)|$.

2. If $P(1, 0)$, $Q(-1, 0)$ and $R(2, 0)$ are three given points, then show that the locus of the point S satisfying the relation $SQ^2 + SR^2 - 2SP^2$ is a straight line parallel to the y-axis.

3. Show that the points $(p+1, 1)$, $(2p+1, 3)$ and $(2p+2, 2)$ are collinear if $p = 2$ or $-1/2$.

4. Show that the midpoint of the vertices of a quadrilateral coincides with the midpoint of the line joining the midpoint of the diagonals.

5. Show that if t_1 and t_2 are distinct and nonzero, then $(t_1^2, t_1), (t_2^2, t_2)$ and $(0,0)$ are collinear.

6. If the points $\left(\frac{a^3}{a-1}, \frac{a^2-3}{a-1}\right), \left(\frac{b^3}{b-1}, \frac{b^3-3}{b-1}\right)$ and $\left(\frac{c^3}{c-1}, \frac{c^3-3}{c-1}\right)$ are collinear for three distinct values a, b and c, then show that $abc - (bc + ca + ab) + 3(a + b + c) = 0$.

7. Perpendicular straight lines are drawn through the fixed point $C(a, a)$ to meet the axes of x and y at A and B. An equilateral triangle is described with AB as the base of the triangle. Prove that the equation of the locus of C is the curve $y^2 = 3(x^2 + a^2)$.

8. The ends A and B of a straight line segment of constant length c slides upon the fixed rectangular axes OX and OY, respectively. If the rectangle $OAPB$ is completed, then show that the locus of the foot of the perpendicular drawn from P to AB is $x^{2/3} + y^{2/3} = c^{2/3}$.

9. The point A divides the line joining $P(1, -5)$ and $Q(3, 5)$ in the ratio k:1. Find the two values of k for which the area of the triangle ABC is equal to 2 units in magnitude when the coordinates of B and C are (1, 5) and (7, –2), respectively.

10. The line segment joining $A(3, 0)$ and $B(0, 2)$ is rotated about a point A in the anticlockwise direction through an angle of 45° and thus B moves to C. If point D be the reflection of C in the y-axis, find the coordinates of D.

Ans.: $\left(-3, 2\sqrt{2}\right)$

11. If (a, b), (h, k) and (p, q) be the coordinates of the circumcentre, the centroid and the orthocentre of a triangle, prove that $3h = p + 2\alpha$.

12. Prove that in a right angled triangle, the midpoint of the hypotenuse is equidistant from its vertices.

13. If G is the centroid of a triangle ABC, then prove that $3(GA^2 + GB^2 + GC^2) = AB^2 + BC^2 + CA^2$.

14. Show that the line joining the midpoint of any two sides of a triangle is half of the third side.

15. Prove that the line joining the midpoints of the opposite sides of a quadrilateral and the line joining the midpoints of the diagonals are concurrent.

16. If Δ_1 and Δ_2 denote the area of the triangles whose vertices are (a, b), (b, c), (c, a) and $(bc - a^2, ca - b^2)$, $(ca - b^2, ab - c^2)$ and $(ab - c^2, bc - a^2)$, respectively, then show that $\Delta_2 = (a + b + c)^2\Delta_1$.

17. Prove that if two medians of a triangle are equal, the triangle is isosceles.

18. If a, b and c be the pth, q^{th} and r^{th} terms of a HP, then prove that the points having coordinates (ab, r), (bc, p) and (ca, q) are collinear.

19. Prove that a point can be found that is at the same distance from each of the four points $\left(at_1, \frac{a}{t_1}\right)\left(at_2, \frac{a}{t_2}\right), \left(at_3, \frac{a}{t_3}\right)$ and $\left(\frac{a}{t_1t_2t_3}, at_1t_2t_3\right)$.

20. If (x_1, y_1) (x_2, y_2) (x_3, y_3) and (x_4, y_4) be the vertices of a parallelogram and $x_1x_3 + y_1y_3 = x_2x_1 + y_2y_1$ then prove that the parallelogram is a rectangle.

21. In any ΔABC, prove that $AB^2 + AC^2 = 2(AD^2 + DC^2)$ where D is the midpoint of BC.

22. If G is the centroid of a triangle ABC and O be any other point, then prove that

    (i) $AB^2 + BC^2 + CA^2 = 3(GA^2 + GB^2 + GC^2)$

    (ii) $OA^2 + OB^2 + OC^2 = GA^2 + GB^2 + GC^2 + 3GO^2$

23. Find the incentre of the triangle whose vertices are (20, 7), (−36, 7) and (0, −8).

Ans.: $\left(2\sqrt{2}, 2-2\sqrt{2}\right)$

24. If A, B and C are the points (−1, 5), (3, 1) and (5, 7), respectively, and D, E and F are the midpoints of BC, CA and AB, respectively, prove that area of ΔABC is four times that of ΔDEF.

25. If D, E and F divide the sides BC, CA and AB of ΔABC in the same ratio, prove that the centroid of ΔABC and ΔDEF coincide.

26. A and B are the fixed points $(a, 0)$ and $(-a, 0)$. Find the locus of the point P that moves in a plane such that

    (i) $PA^2 + PB^2 = 2k^2$

    (ii) $PA^2 - PB^2 = 2PC^2$ where C is the point $(c, 0)$

Ans.: (i) $2ax + k^2 = 0$
(ii) $2cx = c^2 - a^2$

27. If (x_i, y_i), $i = 1, 2, 3$ are the vertices of the ΔABC and a, b and c are the lengths of the sides BC, CA and AB, respectively, show that the incentre of the triangle ABC is $\left(\dfrac{ax_1+bx_2+cx_3}{a+b+c}, \dfrac{ay_1+by_2+cy_3}{a+b+c}\right)$.

28. Show that the points $(-a, -b)$, $(0, 0)$, (a, b) and (a^2, b^2) are either collinear, the vertices of a parallelogram or the vertices of a rectangle.

29. The coordinates of three points O, A and B are (0, 0), (0, 4) and (6, 0), respectively. A point P moves so that the area of ΔPOA is always twice the area of ΔPOB. Find the equation of the locus of P.

Ans.: $x^2 - 9y^2 = 0$

30. The four points $A(x_1, 0)$, $B(x_2, 0)$, $C(x_3, 0)$ and $D(x_4, 0)$ are such that x_1, x_2 are the roots of the equation $ax^2 + 2hx + b = 0$ and x_3, x_4 are the roots of the equation $a^1x^2 + 2h^1x + b^1 = 0$. Show that the sum of the ratios in which C and D divide AB is zero, provided $ab^1 + a^1b = 2hh^1$.

31. If $\begin{vmatrix} x_1 & y_1 & 1 \\ x_2 & y_2 & 1 \\ x_3 & y_3 & 1 \end{vmatrix} = \begin{vmatrix} a_1 & b_1 & 1 \\ a_2 & b_2 & 1 \\ a_3 & b_3 & 1 \end{vmatrix}$, then show that the triangle with vertices (x_i, y_i), $i = 1, 2, 3$ and (a_i, b_i), $i = 1, 2, 3$ are congruent.

32. The point (4, 1) undergoes the following three transformations successively:

(i) Reflection about the line $y = x$

(ii) Transformation through a distance of 2 units along the positive direction of x-axis

(iii) Rotation through an angle of $\frac{\pi}{4}$ about the origin in the anticlockwise direction.

Find the final position of the point.

33. Show that the points $P(2, -4)$, $Q(4, -2)$ and $R(1, 1)$ lie on a straight line. Find (i) the ratio $PQ:QR$ and (ii) the coordinates of the harmonic conjugation of Q with respect to P and R.

34. If a point moves such that the area of the triangle formed by that point and the points (2, 3) and (−3, 4) is 8.5 square units, show that the locus of the point is $x + 5y - 34 = 0$.

35. Show that the area of the triangle with vertices $(p + 5, p - 4)$, $(p - 2, p + 3)$ and (p, p) is independent of p.

CHAPTER

2 The Straight Line

2.1 INTRODUCTION

In the previous chapter, we defined that the locus of a point is the path traced out by a moving point according to some geometrical law. We know that the locus of a point which moves in such a way that its distance from a fixed point is always constant.

2.1.1 Determination of the General Equation of a Straight Line

Suppose the point $P(x, y)$ moves such that $P(x, y)$, $A(4, -1)$, and $B(2, 3)$ form a straight line. Then, $\Delta = \begin{vmatrix} x & y & 1 \\ 4 & -1 & 1 \\ 2 & 3 & 1 \end{vmatrix} = 0$

$$\Rightarrow x(-4) - y(2) + 14 = 0$$

(i.e.) $4x + 2y - 14 = 0$ or $2x + y - 7 = 0$, which is a first degree equation in x and y that represents a straight line.

The general equation of a straight line is $ax + by + c = 0$. Suppose $ax + by + c = 0$ is the locus of a point $P(x, y)$. If this locus is a straight line and if $P(x_1, y_1)$ and $Q(x_2, y_2)$ be any two points on the locus then the point R which divides PQ with ratio $\lambda:1$ is also a point on the line. Since $P(x_1, y_1)$ and $Q(x_2, y_2)$ lie on the locus $ax + by + c = 0$,

$$ax_1 + by_1 + c = 0 \tag{2.1}$$

$$ax_2 + by_2 + c = 0 \tag{2.2}$$

On multiplying equation (2.2) by λ and adding with equation (2.1), we get

$$\lambda(ax_2 + by_2 + c) + (ax_1 + by_1 + c) = 0.$$

$$\text{(i.e.)} \quad a(\lambda x_2 + x_1) + b(\lambda y_2 + y_1) + c(\lambda + 1) = 0.$$

On dividing by $\lambda + 1$, we get

$$a\left(\frac{\lambda x_2 + x_1}{\lambda + 1}\right) + b\left(\frac{\lambda y_2 + y_1}{\lambda + 1}\right) + c = 0. \tag{2.3}$$

Equation (2.3) shows that the point $\left(\frac{\lambda x_2 + x_1}{\lambda + 1}, \frac{\lambda y_2 + y_1}{\lambda + 1}\right)$ lies on the locus $ax + by + c = 0$. This shows that the point which divides PQ in the ratio $\lambda:1$ also lies on the locus which is the definition for a straight line.

$\therefore$ $ax + by + c = 0$ always represents a straight line.

Note 2.1.1.1: The above equation can be written in the form $\frac{a}{c}x + \frac{b}{c}y + 1 = 0$ which is of the form $Ax + By + 1 = 0$. Hence, there are two independent constants in equation of a straight line.

Now, we look into various special forms of the equation of a straight line.

2.1.2 Equation of a Straight Line Parallel to *y*-axis and at a Distance of *h* units from *x*-axis

Let PQ be the straight line parallel to y-axis and at a constant distance h units from y-axis. Then every point on the line PQ has the x-coordinate h.

Hence the equation of the line PQ is $x = h$.

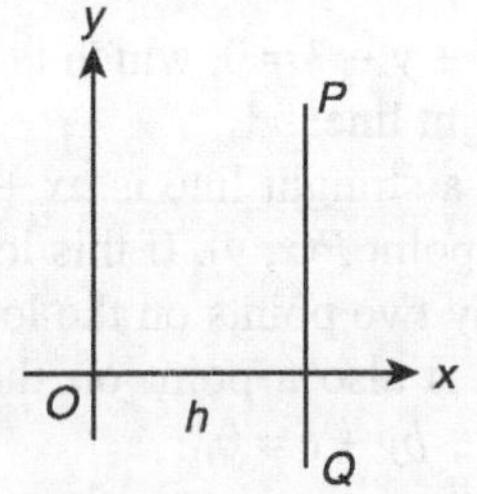

Note 2.1.2.1:

1. Similarly, the equation of the line parallel to x-axis and at a distance k from it is $y = k$.
2. The equation of x-axis is $y = 0$.
3. The equation of y-axis is $x = 0$.

2.2 SLOPE OF A STRAIGHT LINE

If a straight line makes an angle θ with the positive direction of x-axis then $\tan\theta$ is called the slope of the straight line and is denoted by m.

$$\therefore m = \tan\theta.$$

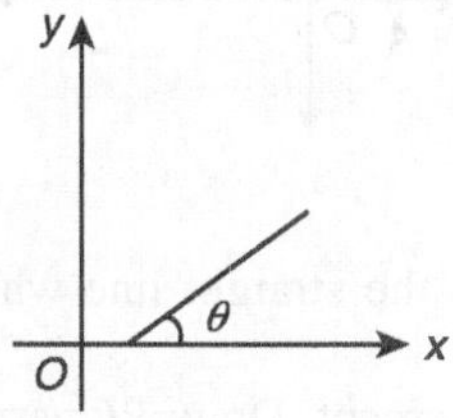

We can now determine the slope of a straight line in terms of coordinates of two points on the line. Let $P(x_1, y_1)$ and $Q(x_2, y_2)$ be the two given points on a line. Draw PL and QM perpendiculars to x-axis. Let PQ make an angle θ with OX.

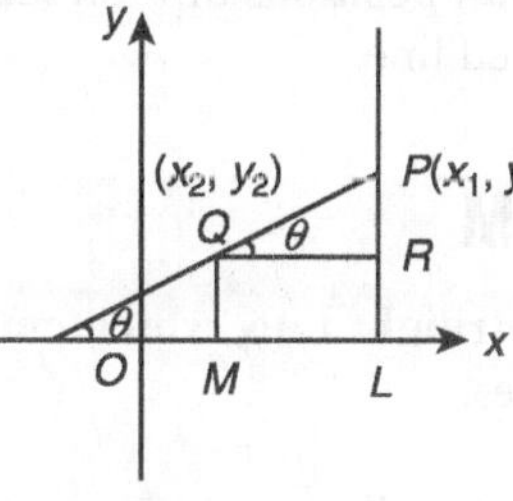

Draw QR perpendicular to LP. Then $\angle RQP = \theta$.

$$\therefore \quad QR = ML = OL - OM = x_1 - x_2.$$

$$RP = LR - LP = LP - QM = y_1 - y_2.$$

In ΔPQR, $\tan\theta = \dfrac{RP}{QR}$ (i.e.) $m = \dfrac{y_1 - y_2}{x_1 - x_2}$

2.3 SLOPE-INTERCEPT FORM OF A STRAIGHT LINE

Find the equation of the straight line, which makes an angle θ with OX and cuts off an intercept c on the y-axis.

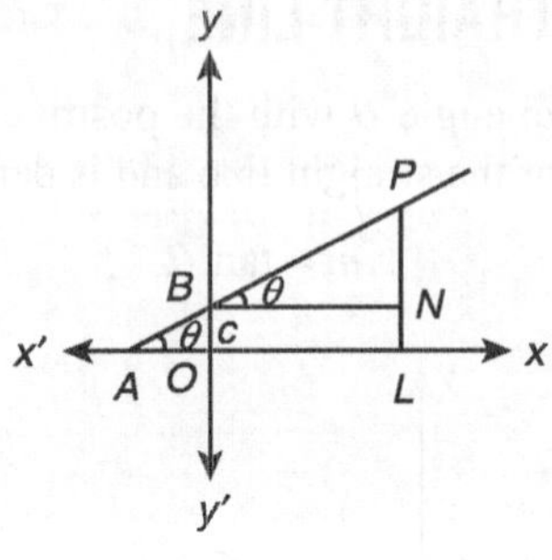

Let $P(x, y)$ be any point on the straight line which makes an angle θ with x-axis.

$\underline{|OAP} = \theta$, $OB = c = y$-intercept. Draw PL perpendicular to x-axis and BN perpendicular to LP. Then, $\underline{|NBP} = \theta$. $BN = OL = x$.

$$\therefore \quad NP = LP - LN = LP - OB = y - c.$$

In $\Delta NBP, \tan\theta = \dfrac{NP}{BN}$ (i.e.) $m = \dfrac{y-c}{x} \Rightarrow y = mx + c.$

This equation is true for all positions of P on the straight line. Hence, this is the equation of the required line.

2.4 INTERCEPT FORM

Find the equation of the straight line, which cuts off intercepts *a* and *b*, respectively on *x* and *y* axes.

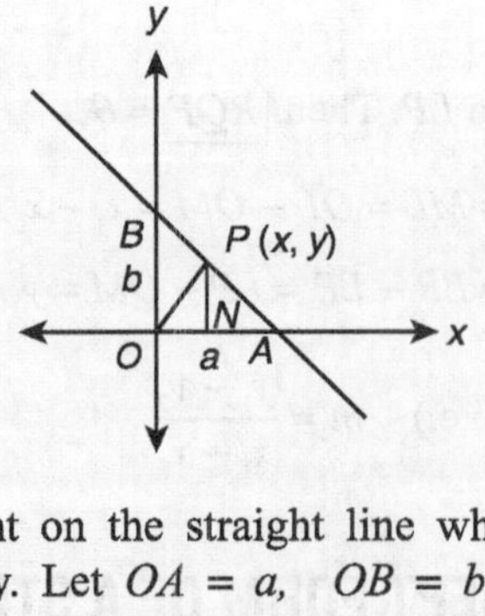

Let $P(x, y)$ be any point on the straight line which meets x and y axes at A and B, respectively. Let $OA = a$, $OB = b$, $ON = x$, and $NP = y$; $NA = OA - ON = a - x$. Triangles PNA and BOA are similar. Therefore, $\dfrac{PN}{OB} = \dfrac{NA}{OA}$ (i.e.) $\dfrac{y}{b} = \dfrac{a-x}{a} \Rightarrow \dfrac{y}{b} = 1 - \dfrac{x}{a}$ or $\dfrac{x}{a} + \dfrac{y}{b} = 1$. This result is true for all positions of P on the straight line and hence this is the equation of the required line.

2.5 SLOPE-POINT FORM

Find the equation of the straight line with slope m and passing through the given point (x_1, y_1).

The equation of the straight line with a given slope m is

$$y = mx + c \tag{2.4}$$

Here, c is unknown. This straight line passes through the point (x_1, y_1). The point has to satisfy the equation $y = mx + c$.

$\therefore y_1 = mx_1 + c$. Substituting the value of c in equation (2.4), we get the equation of the line as

$$y = mx + y_1 - mx_1 \Rightarrow y - y_1 = m(x - x_1).$$

2.6 TWO POINTS FORM

Find the equation of the straight line passing through two given points (x_1, y_1) and (x_2, y_2).

$$y - y_1 = m(x - x_1) \tag{2.5}$$

where, m is unknown. The slope of the straight line passing through the points

$$(x_1, y_1) \text{ and } (x_2, y_2) \text{ is } m = \frac{y_1 - y_2}{x_1 - x_2} \tag{2.6}$$

By substituting equation (2.6) in equation (2.5), we get the required straight line $y - y_1 = \dfrac{y_1 - y_2}{x_1 - x_2}(x - x_1)$.

$$\text{(i.e.)} \quad \frac{y - y_1}{x - x_1} = \frac{y_1 - y_2}{x_1 - x_2}.$$

2.7 NORMAL FORM

Find the equation of a straight line in terms of the perpendicular p from the origin to the line and the angle that the perpendicular line makes with axis.

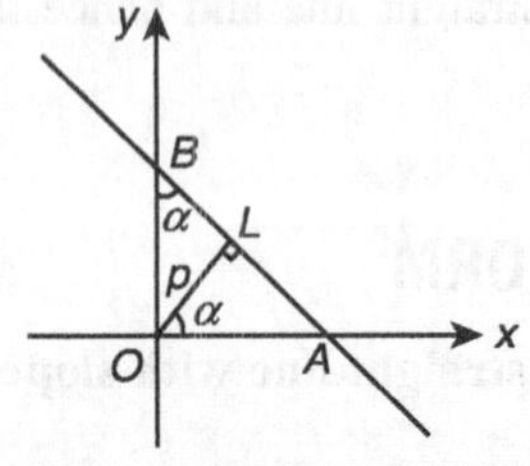

Draw $OL \perp AB$. Let $OL = p$.

Let $\angle AOL = \alpha$

$$\therefore \angle OBA = \alpha$$

$$\frac{OA}{OL} = \sec\alpha$$

$$OA = OL\sec\alpha = p\sec\alpha$$

$$\frac{OB}{OL} = \operatorname{cosec}\alpha$$

$$\therefore OB = OL\operatorname{cosec}\alpha = p\operatorname{cosec}\alpha.$$

Therefore, the equation of the straight line AB is $\dfrac{x}{p\sec\alpha} + \dfrac{y}{p\operatorname{cosec}\alpha} = 1$.

(i.e.) $x\cos\alpha + y\sin\alpha = p$

2.8 PARAMETRIC FORM AND DISTANCE FORM

Let a straight line make an angle θ with x-axis and $A(x_1, y_1)$ be a point on the line. Draw AL, PM perpendicular to x-axis and AQ perpendicular to PM. Then, $\angle PAQ = \theta$.

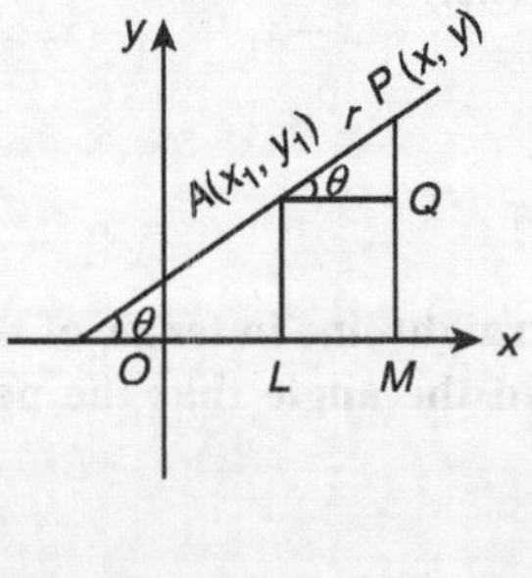

$$AQ = LM = OM - OL = x - x_1.$$

$$QP = MP - MQ = MP - LA = y - y_1. \text{ Let } AP = r.$$

In ΔPAQ, $x - x_1 = r\cos\theta$; $y - y_1 = r\sin\theta$.

$$\therefore \frac{x - x_1}{\cos\theta} = \frac{y - y_1}{\sin\theta} = r.$$

These are the parametric equations of the given line.

Note 2.8.1: Any point on the line is $x = x_1 + r\cos\theta$, $y = y_1 + r\sin\theta$.

Note 2.8.2: r is the distance of any point on the line from the given point $A(x_1, y_1)$.

2.9 PERPENDICULAR DISTANCE ON A STRAIGHT LINE

Find the perpendicular distance from a given point to the line $ax + by + c = 0$.

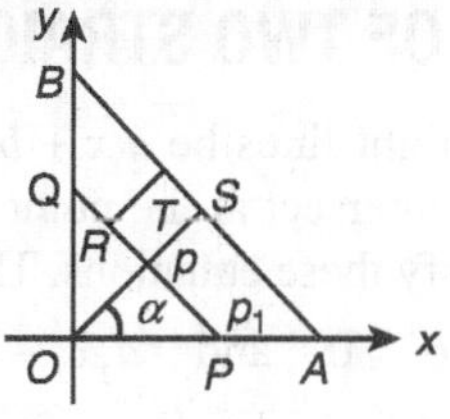

Let $R(x_1, y_1)$ be a given point and $ax + by + c = 0$ be the given line. Through R draw the line PQ parallel to AB. Draw OS perpendicular to AB meeting PQ at T. Let $OS = p$ and $PT = p_1$. Let $\angle AOS = \alpha$. Then the equation of AB is

$$x\cos\alpha + y\sin\alpha = p \tag{2.7}$$

which is the same as

$$ax + by = -c \tag{2.8}$$

Equations (2.7) and (2.8) represent the same line and, therefore, identifying

we get $\dfrac{\cos\alpha}{a} = \dfrac{\sin\alpha}{b} = \dfrac{-p}{c} = \pm\dfrac{\sqrt{\cos^2\alpha + \sin^2\alpha}}{\sqrt{a^2 + b^2}}$

$$\therefore \cos\alpha = \pm\frac{a}{\sqrt{a^2 + b^2}}; \quad \sin\alpha = \pm\frac{b}{\sqrt{a^2 + b^2}}.$$

$$p = \mp \frac{c}{\sqrt{a^2 + b^2}}.$$

The equation of the line PQ is $x\cos\alpha + y\sin\alpha = p$. Since the point $R(x_1, y_1)$ lies on the line $x_1\cos\alpha + y_1\sin\alpha - p_1 = 0$.

$$\therefore p_1 = x_1\cos\alpha + y_1\sin\alpha.$$

Then, the length of the perpendicular line from R to AB

$$= p - p_1 = p - x_1\cos\alpha - y_1\sin\alpha = \mp\frac{c}{\sqrt{a^2+b^2}} \mp \frac{ax_1}{\sqrt{a^2+b^2}} \mp \frac{by_1}{\sqrt{a^2+b^2}}$$

$$= \mp\frac{ax_1 + by_1 + c}{\sqrt{a^2+b^2}} = \left|\frac{ax_1 + by_1 + c}{\sqrt{a^2+b^2}}\right|$$

Note 2.9.1: The perpendicular distance from the origin on the line $ax + by + c = 0$ is $\frac{|c|}{\sqrt{a^2+b^2}}$.

2.10 INTERSECTION OF TWO STRAIGHT LINES

Let the two intersecting straight lines be $a_1x + b_1y + c_1 = 0$ and $a_2x + b_2y + c_2 = 0$. Let the straight lines intersect at the point (x_1, y_1). Then (x_1, y_1) lies on both the lines and hence satisfy these equations. Then

$$a_1x_1 + b_1y_1 + c_1 = 0 \quad \text{and} \quad a_2x_2 + b_2y_2 + c_2 = 0.$$

Solving the equations, we get

$$\frac{x_1}{b_1c_2 - b_2c_1} = \frac{y}{c_1a_2 - c_2a_1} = \frac{1}{a_1b_2 - a_2b_1}$$

Therefore, the point of intersection is $\left(\frac{b_1c_2 - b_2c_1}{a_1b_2 - a_2b_1}, \frac{c_1a_2 - c_2a_1}{a_1b_2 - a_2b_1}\right)$.

Find the ratio at which the line $ax + by + c = 0$ divides the line joining the points (x_1, y_1) and (x_2, y_2).

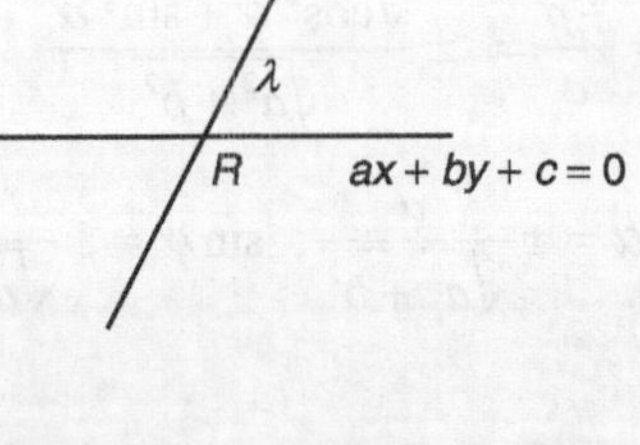

Let the line $ax + by = c = 0$ divide the line joining the points $P(x_1, y_1)$ and $Q(x_2, y_2)$ in the ratio λ:1. Then, the coordinates of the point of division R are $\left(\dfrac{x_1+\lambda x_2}{1+\lambda}, \dfrac{y_1+\lambda y_2}{1+\lambda}\right)$.

This point lies on the line $ax + by + c = 0$.

$$\therefore a\left(\frac{x_1+\lambda x_2}{1+\lambda}\right)+b\left(\frac{y_1+\lambda y_2}{1+\lambda}\right)+c=0.$$

$$\text{(i.e.)}\quad a\,(x_1+\lambda x_2)+b\,(y_1+\lambda y_2)+c\,(1+\lambda)=0.$$

$$ax_1+by_1+c+\lambda\,(ax_2+by_2+c_2)=0.$$

$$\Rightarrow \lambda = -\frac{ax_1+by_1+c}{ax_2+by_2+c_2}.$$

Note 2.10.1:

1. If λ is positive then the points (x_1, y_1) and (x_2, y_2) lie on the opposite sides of the line $ax + by + c = 0$.
2. If λ is negative then the points (x_1, y_1) and (x_2, y_2) lie on the same side of the line $ax + by + c = 0$.
3. In other words, if the expressions $ax_1 + by_1 + c$ and $ax_2 + by_2 + c_2$ are of opposite signs then the point (x_1, y_1) and (x_2, y_2) lie on the opposite sides of the line $ax + by + c = 0$. If they are of the same sign then the points (x_1, y_1) and (x_2, y_2) lie on the same side of the line $ax + by + c = 0$.

Find the equation of a straight line passing through intersection of the lines $a_1x + b_1y + c = 0$ and $a_2x + b_2y + c = 0$.

Consider the equation $a_1x + b_1y + c_1 + l\,(a_2x + b_2y + c_2) = 0$. (2.9)
This is a linear equation in x and y and hence this equation represents a straight line. Let (x_1, y_1) be the point of intersection of the lines $a_1x + b_1y + c_1 = 0$ and $a_2x + b_2y + c_2 = 0$. Then (x_1, y_1) has to satisfy the two equations:

$$a_1x_1+b_1y_1+c_1=0 \tag{2.10}$$

$$a_2x_2+b_2y_2+c_2=0 \tag{2.11}$$

On multiplying equation (2.11) by λ and adding with equation (2.10) we get,

$$(a_1x_1+b_1y_1+c_1)+\lambda\,(a_2x_2+b_2y_2+c_2)=0.$$

This equation shows that the point $x = x_1$ and $y = y_1$ satisfies equation (2.9). Hence the point (x_1, y_1) lies on the straight line given by the equation (2.9), which is a line passing through the intersection of the lines $a_1x + b_1y + c_1 = 0$ and $a_2x + b_2y + c_2 = 0$.

2.11 CONCURRENT STRAIGHT LINES

Consider three straight lines given by equations:

$$a_1x + b_1y + c_1 = 0 \tag{2.12}$$

$$a_2x + b_2y + c_2 = 0 \tag{2.13}$$

$$a_3x + b_3y + c_3 = 0 \tag{2.14}$$

The point of intersection of lines given by equations (2.12) and (2.13) is $\left(\frac{b_1c_2 - b_2c_1}{a_1b_2 - a_2b_1}, \frac{c_1c_2 - c_2a_1}{a_1b_2 - a_2b_1}\right)$.

If the three given lines are concurrent, the above point should lie on the straight line given by equation (2.14).

$$\text{Then, } a_3\left(\frac{b_1c_2 - b_2c_1}{a_1b_2 - a_2b_1}\right) + \left(\frac{c_1a_2 - c_2a_1}{a_1b_2 - a_2b_1}\right)b_3 + c_3 = 0.$$

$$\text{(i.e.)} \quad a_3(b_1c_2 - b_2c_1) + b_3(c_1a_2 - c_2a_1) + c_3(a_1b_2 - a_2b_1) = 0.$$

This is the required condition for the three given lines to be concurrent. The above condition can be expressed in determinant form $\begin{vmatrix} a_1 & b_1 & c_1 \\ a_2 & b_2 & c_2 \\ a_3 & b_3 & c_3 \end{vmatrix} = 0.$

If l, m, and n are constants such that $l(a_1x + b_1y + c_1) + m(a_2x + b_2y + c_2) + n(a_3x + b_3y + c_3)$ vanishes identically then prove that the lines $a_1x + b_1y + c_1 = 0$, $a_2x + b_2y + c_2 = 0$, and $a_3x + b_3y + c_3 = 0$ are concurrent.

Let the lines $a_1x + b_1y + c_1 = 0$ and $a_2x + b_2y + c_2 = 0$ meet at the point (x_1, y_1).

$$\text{Then} \quad a_1x_1 + b_1y_1 + c_1 = 0 \tag{2.15}$$

$$\text{and} \quad a_2x_2 + b_2y_2 + c_2 = 0 \tag{2.16}$$

For all values of x and y given that,

$$l(a_1x + b_1y + c_1) + m(a_2x + b_2y + c_2) + n(a_3x + b_3y + c_3) = 0 \tag{2.17}$$

Then it will be true for $x = x_1$ and $y = y_1$.

$\therefore\ l(a_1x_1 + b_1y_1 + c_1) + m\,(a_2x_1 + b_2y_1 + c_2) + n\,(a_3x_1 + b_3y_1 + c_3) = 0.$ Using equations (2.15) and (2.16), we get $a_3x_1 + b_3y_1 + c_3 = 0$. That is, the point (x_1, y_1) lies on the line $a_3x + b_3y + c_3 = 0$.

Therefore, the lines $a_1x + b_1y + c_1 = 0$, $a_2x + b_2y + c_2 = 0$, $a_3x + b_3y + c_3 = 0$ are concurrent at (x_1, y_1).

2.12 ANGLE BETWEEN TWO STRAIGHT LINES

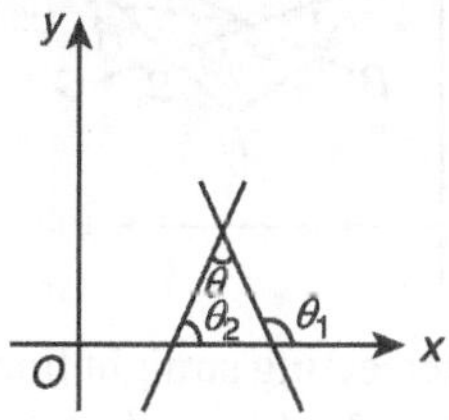

Let θ be the angle between two straight lines, whose slopes are m_1 and m_2. Let the two lines with slopes m_1 and m_2 make angles θ_1 and θ_2 with x-axis. Then, $m_1 = \tan\,\theta_1$, $m_2 = \tan\,\theta_2$. Also, $\theta = \theta_1 - \theta_2$

$$\tan\theta = \tan(\theta_1 - \theta_2) = \frac{\tan\theta_1 - \tan\theta_2}{1 + \tan\theta_1\tan\theta_2} = \frac{(m_1 - m_2)}{1 + m_1m_2}.$$

If the RHS is positive, then θ is the acute angle between the lines. If RHS is negative, then θ is the obtuse angle between the lines.

$$\therefore \tan\theta = \pm\frac{(m_1 - m_2)}{1 + m_1m_2} \Rightarrow \theta = \tan^{-1}\left(\pm\frac{(m_1 - m_2)}{1 + m_1m_2}\right)$$

Note 2.12.1: If the lines are parallel then $\theta = 0$ and $\tan\,\theta = \tan 0 = 0$.

$$\therefore \frac{m_1 - m_2}{1 + m_1m_2} = 0 \Rightarrow m_1 = m_2.$$

Note 2.12.2: If the lines are perpendicular then, $\theta = \frac{\pi}{2}$.

$$\tan\frac{\pi}{2} = \pm\frac{m_1 - m_2}{1 + m_1m_2} \Rightarrow 1 + m_1m_2 = 0 \Rightarrow m_1m_2 = -1.$$

Therefore,

(1) If two lines are parallel then their slopes are equal.
(2) If the two lines are perpendicular then the product of their slopes is −1.

2.13 EQUATIONS OF BISECTORS OF THE ANGLE BETWEEN TWO LINES

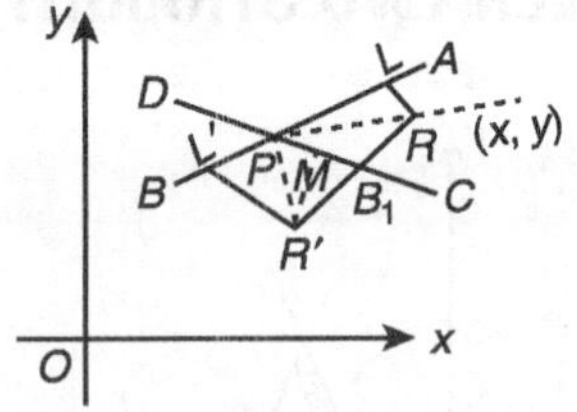

Let AB and CD be the two intersecting straight lines intersecting at P. Let these lines be represented by the equations $a_1x + b_1y + c_1 = 0$ and $a_2x + b_2y + c_2 = 0$. Let PR and PR' be the bisectors of angles $\underline{|APC}$ and $\underline{|CPB}$, respectively. Then the perpendicular distances from R (or R') AB, and CD are equal.

$$\therefore \frac{a_1x+b_1y+c}{\sqrt{a_1^2+b_1^2}} = \pm\frac{a_2x+b_2y+c_2}{\sqrt{a_2^2+b_2^2}}$$

If c_1 and c_2 are positive, then the equations of the bisector containing the origin is given by

$$\frac{a_1x+b_1y+c_1}{\sqrt{a_1^2+b_1^2}} = \frac{a_2x+b_2y+c_2}{\sqrt{a_2^2+b_2^2}} \tag{2.18}$$

The equation of the bisector not containing the origin is

$$\frac{a_1x+b_1y+c_1}{\sqrt{a_1^2+b_1^2}} = -\frac{a_2x+b_2y+c_2}{\sqrt{a_2^2+b_2^2}}. \tag{2.19}$$

If c_1 and c_2 are not positive then the equations of lines should be written in such a way that c_1 and c_2 are positive.

Note 2.13.1: We can easily observe that the two bisectors are at right angles.

ILLUSTRATIVE EXAMPLES

Example 2.1

Find the equation of the straight line which is at a distance of 10 units from x-axis.

Solution

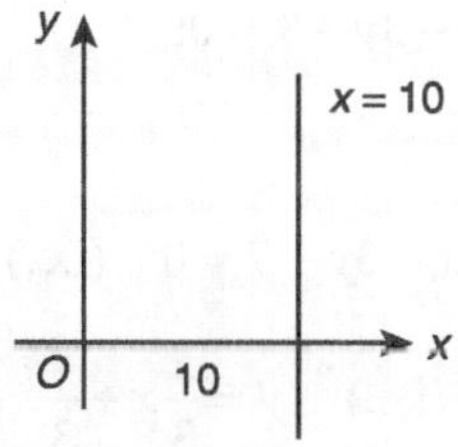

The equation of the required straight line is $x = 10$ or $x - 10 = 0$.

Example 2.2

Find the equation of the straight line which is at a distance of –15 units from *y*-axis.

Solution

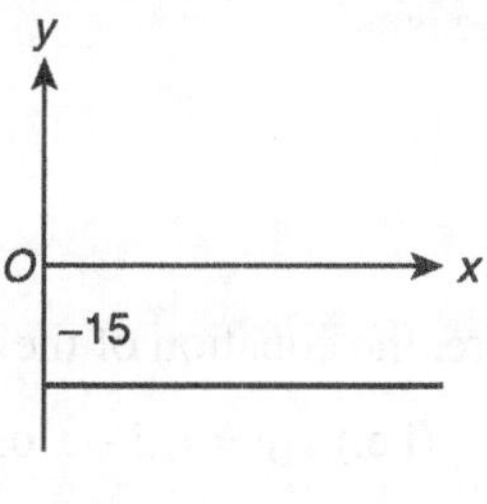

The equation of the required line is $y = -15$ or $y + 15 = 0$.

Example 2.3

Find the slope of the line joining the points (2, 3) and (4, –5).

Solution

The slope of the line joining the two given points (x_1, y_1) and (x_2, y_2) is $m = \dfrac{y_2 - y_1}{x_2 - x_1}$.

Therefore, the slope of the line joining the two given points is $m = \dfrac{-5-3}{4-2} = \dfrac{-8}{2} = -4$.

Example 2.4

Find the slope of the line $2x - 3y + 7 = 0$.

Solution

The equation of the line is $2x - 3y + 7 = 0$ (i.e.) $3y = 2x + 7$.

$$\text{(i.e.)} \quad y = \frac{2}{3}x + \frac{7}{3}.$$

Therefore, slope of the line $= \frac{2}{3}$.

Example 2.5

Find the equation of the straight line making an angle 135° with the positive direction of x-axis and cutting of an intercept 5 on the y-axis.

Solution

The slope of the straight line is

$$\begin{aligned} m &= \tan\theta = \tan 135^\circ \\ &= \tan(180 - 45) \\ &= -\tan 45^\circ \\ &= -1. \end{aligned}$$

y intercept $= c = 5$. Therefore, the equation of the straight line is

$$y = mx + c \quad \text{(i.e.)} \quad y = +5 - x \text{ or } x + y = 5.$$

Example 2.6

Find the equation of the straight line cutting off the intercepts 2 and −5 on the axes.

Solution

The equation of the straight line is $\frac{x}{a} + \frac{y}{b} = 1$. Here, $a = 2$ and $b = -5$.

Therefore, the equation of the straight line is $\frac{x}{2} - \frac{y}{5} = 1$ or $5x - 2y = 10$.

Example 2.7

Find the equation of the straight line passing through the points (7, −3) and cutting off equal intercepts on the axes.

Solution

Let the equation of the straight line be $\frac{x}{a}+\frac{y}{b}=1.$

$$\text{(i.e.)} \quad x+y=a.$$

This straight line passes through the point (7, –3).

Therefore, $7-3=a$ (i.e.) $a=4$.

$\therefore$ The equation of the straight line is $x+y=4$.

Example 2.8

Find the equation of the straight line, the portion of which between the axes is bisected at the point (2, –5).

Solution

Let the equation of the straight line be $\frac{x}{a}+\frac{y}{b}=1.$

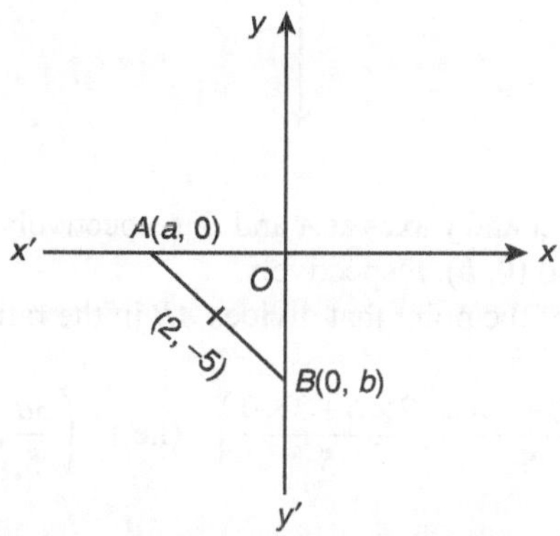

Let the line meet the x and y axes at A and B, respectively. Then the coordinates of A and B are $(a, 0)$ and $(0, b)$. The midpoint of AB is $\left(\frac{a}{2},\frac{b}{2}\right)$. However, the midpoint is given as (2, –5).

Therefore, $\frac{a}{2}=2$ and $\frac{b}{2}=-5$.

$$\therefore a=4 \quad \text{and} \quad b=-10.$$

Hence, the equation of the straight line is $\frac{x}{4}-\frac{y}{10}=1.$

$$\text{(i.e.)} \quad 5x-2y=20.$$

Example 2.9

Find the equation of the straight line of the portion of which between the axes is divided by the point (4, 3) in the ratio 2:3.

Solution

Let the equation of the straight line be $\frac{x}{a}+\frac{y}{b}=1$.

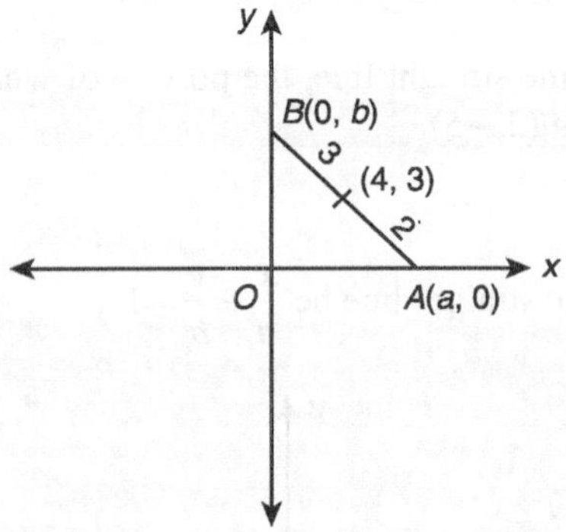

Let this line meet the x and y axes at A and B, respectively. The coordinates of A and B are $(a, 0)$ and $(0, b)$, respectively.

The coordinates of the point that divides AB in the ratio 2:3 are

$$\left(\frac{2\times 0+3\times a}{5},\frac{2\times b+3\times 0}{5}\right) \text{ (i.e.) } \left(\frac{3a}{5},\frac{2b}{5}\right)$$

This point is given as (4, 3).

Therefore, $\frac{3a}{5}=4$ and $\frac{2b}{5}=3$.

$$\therefore\ a=\frac{20}{3} \text{ and } b=\frac{15}{2}.$$

$\therefore$ The equation of the straight line is $\frac{3x}{20}+\frac{2y}{15}=1$. (i.e.) $9x+8y=60$.

Example 2.10

Find the equations to the straight lines each of which passes through the point (3, 2) and intersect the x and y axes at A and B such that $OA-OB=2$.

Solution

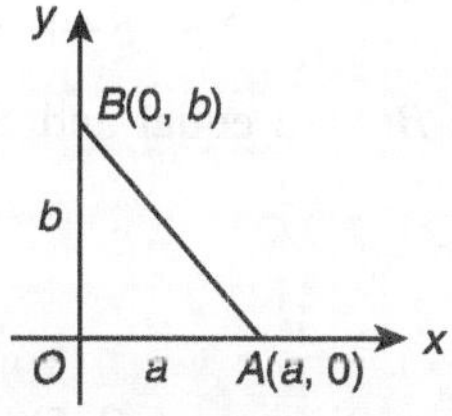

Let the equation of the straight line be $\frac{x}{a}+\frac{y}{b}=1$. This straight line passes through the point (3, 2).

$$\therefore \frac{3}{a}+\frac{2}{b}=1$$

$$\Rightarrow 3b+2a=ab \tag{2.20}$$

Also, given that $OA - OB = 2$

$$\text{(i.e.)} \quad a-b=2. \tag{2.21}$$

Therefore, $b = a - 2$. Substituting this in equation (2.20) we get $3(a - 2) + 2a = a(a - 2)$.

$$3a-6+2a=a^2-2a$$

$$\Rightarrow a^2-7a+6=0$$

$$\Rightarrow (a-1)(a-6)=0.$$

$$\therefore a=1 \text{ and } a=6.$$

$$b=-1 \text{ and } b=4.$$

$\therefore$ The two straight lines are $\frac{x}{1}+\frac{y}{-1}=1$ and $\frac{x}{6}+\frac{y}{4}=1$.

$$\text{(i.e.)} \quad x-y=1 \quad \text{and} \quad 2x+3y=12.$$

Example 2.11

Show that the points $A(1, 1)$, $B(5, -9)$, and $C(-1, 6)$ are collinear.

Solution

The slope of AB is $m_1=\frac{y_1-y_2}{x_1-x_2}=\frac{1+9}{1-5}=\frac{10}{-4}=-\frac{5}{2}$.

$$\therefore \text{ The slope of } BC \text{ is } m_2 = \frac{-9-6}{5+1} = \frac{-15}{6} = -\frac{5}{2}.$$

Since the slopes of AB and BC are equal and B is the common point, the points are collinear.

Example 2.12

Prove that the triangle whose vertices are (–2, 5), (3, –4), and (7, 10) is a right angled isosceles triangle. Find the equation of the hypotenuse.

Solution

Let the points be A (–2, 5), B (3, –4), and C (7, 10).

$AB^2 = (-2-3)^2 + (5+4)^2 = 25 + 81 = 106,$

$BC^2 = (3-7)^2 + (-4-10)^2 = 16 + 196 = 212.$

$AC^2 = (-2-7)^2 + (5-10)^2 = 81 + 25 = 106$. Therefore, $AB^2 + AC^2 = BC^2$ and $AB = AC$. Hence, the ΔABC is a right angled isosceles triangle.

The equation of the hypotenuse BC is

$$\frac{y-y_1}{x-x_1} = \frac{y_1-y_2}{x_1-x_2} \text{ (i.e.) } \frac{y+4}{x-3} = \frac{-4-10}{3-7} = \frac{-14}{-4} = \frac{7}{2}$$

$$\Rightarrow 7x - 21 = 2y + 8 \Rightarrow 7x - 2y - 29 = 0.$$

Example 2.13

Find the equation of the straight line which cuts off intercepts on the axes equal in magnitude but opposite in sign and passing through the point (4, 7).

Solution

Let the equation of the straight line cutting off intercepts equal in magnitude but opposite in sign be $\frac{x}{a} - \frac{y}{b} = 1$ (i.e.) $x - y = a$.

This passes through the point (4, 7).

Therefore, $4 - 7 = a$ (i.e.) $a = -3$.

Hence, the equation of the straight line is $x - y + 3 = 0$.

Example 2.14

Find the ratio in which the line $3x - 2y + 5 = 0$ divides the line joining the points (6, –7) and (–2, 3).

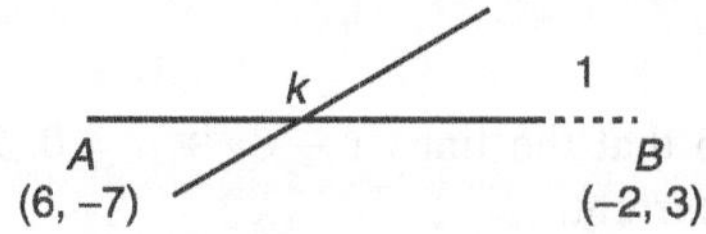

Solution

Let the line $3x - 2y + 5 = 0$ divide the line joining the points $A(6, -7)$ and $B(-2, 3)$ in the ratio k:1. Then the coordinates of the point of division are $\left(\frac{-2k+6}{k+1}, \frac{3k-7}{k+1}\right)$. Since this point lies on the straight line $3x - 2y + 5 = 0$, we get

$$3\left(\frac{-2k+6}{k+1}\right)-2\left(\frac{3k-7}{k+1}\right)+5=0.$$

$$\text{(i.e.)} \quad -6k+18-6k+14+5k+5=0.$$

$$\text{(i.e.)} \quad -7k+37=0 \Rightarrow k=\frac{37}{7}.$$

∴ The required ratio is 37:7.

Example 2.15

Prove that the lines $3x - 4y + 5 = 0$, $7x - 8y + 5 = 0$, and $4x + 5y = 45$ are concurrent.

Solution

Given

$$3x-4y=-5 \quad (2.22)$$
$$7x-8y=-5 \quad (2.23)$$
$$4x+5y=45 \quad (2.24)$$

Solving equations (2.22) and (2.23), we get the point of intersection of the two lines.

$$(2.22)\times 2 \quad 6x-8y=-10$$
$$(2.23)\times 1 \quad 7x-8y=-5$$
$$x=5$$

∴ From equation (2.22), $15 - 4y = -5$.

∴ $y = 5$.

Hence, the point of intersection of the lines is (5, 5). Substituting $x = 5$ and $y = 5$, in equation (2.24), we get $20 + 25 = 45$ which is true.

∴ The third line also passes through the points (5, 5). Hence it is proved that the three lines are concurrent.

Example 2.16

Find the value of a so that the lines $x - 6y + a = 0$, $2x + 3y + 4 = 0$, and $x + 4y + 1 = 0$ are concurrent.

Solution

Given
$$x - 6y + a = 0 \quad (2.25)$$
$$2x + 3y + 4 = 0 \quad (2.26)$$
$$x + 4y + 1 = 0 \quad (2.27)$$

Solving the equations (2.26) and (2.27) we get,

$$(2.26)\times 1: 2x + 3y = -4$$
$$(2.27)\times 2: 2x + 8y = -2$$

On subtracting, we get $5y = 2$

$$\therefore y = \frac{2}{5}$$

$$\therefore \text{From equation (2.27) } x = \frac{-8}{5} - 1 = \frac{-8-5}{5}$$

$$\therefore x = \frac{-13}{5}$$

Hence, the point of intersection of the lines is $\left(\frac{-13}{5}, \frac{2}{5}\right)$. Since the lines are concurrent this point should lie on $x - 6y + a = 0$.

$$\text{(i.e.)} \quad \frac{-13}{5} - \frac{12}{5} + a = 0.$$

$$\therefore a = 5.$$

Example 2.17

Prove that for all values of λ the straight line $x(2 + 3\lambda) + y(3 - \lambda) - 5 - 2\lambda = 0$ passes through a fixed point. Find the coordinates of the fixed point.

Solution

$x(2 + 3\lambda) + y(3 - \lambda) - 5 - 2\lambda = 0$. This equation can be written in the form

$$2x + 3y - 5 + \lambda(3x - y - 2) = 0 \quad (2.28)$$

This equation represents a straight line passing through the intersection of lines

$$2x + 3y - 5 = 0 \tag{2.29}$$

$$3x - y - 2 = 0 \tag{2.30}$$

for all values of λ.

(2.29) $\quad 2x + 3y = 5$

(2.28) × 3 $\quad 9x - 3y = 6$

On adding, we get $11x = 11 \Rightarrow x = 1$ and hence from equation (2.29) we get $y = 1$.

Therefore, the point of intersection of straight lines (2.29) and (2.30) is (1, 1). The straight line (2.28) passes through the point (1, 1) for all values of λ. Hence (2.28) passes through the fixed point (1, 1).

Example 2.18

Find the equation of the straight line passing through the intersection of the lines $3x - y = 5$ and $2x + 3y = 7$ and making an angle of 45° with the positive direction of x-axis.

Solution

Solving the equations,

$$3x - y = 5 \tag{2.31}$$

$$2x + 3y = 7 \tag{2.32}$$

We get,

(2.31) × 3 $\quad 9x - 3y = 15$

(2.32) $\quad 2x + 3y = 7$

On adding, we get $11x = 22$.

$$\therefore x = 2.$$

From equation (2.31), $6 - y = 5$.

$$\therefore y = 1.$$

Hence (2, 1) is the point of intersection of the lines (2.31) and (2.32).

The slope of the required line is $m = \tan\theta$, $m = \tan 45° = 1$. Therefore, the equation of the required line is $y - y_1 = m(x - x_1)$ (i.e.) $y - 1 = 1(x - 2) \Rightarrow x - y = 1$.

Example 2.19

Find the equation of the straight line passing through the intersection of the lines $7x + 3y = 7$ and $2x + y = 2$ and cutting off equal intercepts on the axes.

Solution

The point of intersection of the lines is obtained by solving the following two equations:

$$7x + 3y = 7 \tag{2.33}$$

$$2x + y = 2 \tag{2.34}$$

(2.33) $\quad 7x + 3y = 7$

(2.34) × 3 $\quad 6x + 3y = 6$

On subtracting, we get $x = 1$ and hence $y = 0$. Therefore, the point of intersection is (1, 0). The equation of the straight line cutting off equal intercepts is $\frac{x}{a} + \frac{y}{a} = 1$ (i.e.) $x + y = a$.

This straight line passes through (1, 0). Therefore, $1 + 0 = a$ (i.e.) $a = 1$. Hence, the equation of the required straight line is $x + y = 1$.

Example 2.20

Find the equation of the straight line concurrent with the lines $2x + 3y = 3$ and $x + 2y = 2$ and also concurrent with the lines $3x - y = 1$ and $x + 5y = 11$.

Solution

The point of intersection of the lines $2x + 3y = 3$ and $x + 2y = 2$ is obtained by solving the following two equations:

$$2x + 3y = 3 \tag{2.35}$$

$$x + 2y = 2 \tag{2.36}$$

(2.35) $\quad 2x + 3y = 3$

(2.36) × 2 $\quad 2x + 4y = 4$

On subtracting, we get $y = 1$ and hence $x = 0$.

Therefore, the point of intersection is (0, 1).

$$3x - y = 1 \tag{2.37}$$

$$x + 5y = 11 \tag{2.38}$$

$(2.37) \times 5$ $\quad 15x - 5y = 5$

$(2.38) \times 1$ $\quad x + 5y = 11$

On adding, we get $16x = 16$ which gives $x = 1$ and hence $y = 2$. The point of intersection of the second pair of lines is (1, 2). The equation of the line joining the two points (0, 1) and (1, 2) is

$$\frac{y-1}{x-0} = \frac{1-2}{0-1} = 1 \Rightarrow y - 1 = x \quad \text{(i.e.)} \quad x - y + 1 = 0.$$

Example 2.21

Find the angle between the lines $y = \sqrt{3}x + 4$ and $y = \frac{1}{\sqrt{3}}x + 2$.

Solution

The slope of the line $y = \sqrt{3}x + 4$ is $\sqrt{3}$. Therefore, $m_1 = \tan\theta_1 = \sqrt{3}$ (i.e.) $\theta_1 = 60°$. The slope the line $y = \frac{1}{\sqrt{3}}x + 2$ is $\frac{1}{\sqrt{3}}$. Therefore, $m_2 = \tan\theta_2 = \frac{1}{\sqrt{3}}$ (i.e.) $\theta_2 = 30°$. The angle between the lines is $\theta_1 - \theta_2 = 30°$.

Example 2.22

Find the equation of the perpendicular bisector of the line joining the points (–2, 6) and (4, –6).

Solution

The slope of the line joining the points (–2, 6) and (4, –6) is $m = \frac{6+6}{-2-4} = \frac{-12}{6} = -2$. Therefore, the slope of the perpendicular line is $\frac{1}{2}$.

The midpoint of the line joining the points (–2, 6) and (4, –6) is $\left(\frac{-2+4}{2}, \frac{6-6}{2}\right)$ (i.e.) (1, 0).

Therefore, the equation of the perpendicular bisector is $y - y_1 = m(x - x_1)$ (i.e.) $y - 0 = \frac{1}{2}(x - 1) \Rightarrow 2y = x - 1$ or $x - 2y - 1 = 0$.

Example 2.23

$A(4, 1)$, $B(7, 4)$, and $C(5, -2)$ are the vertices of a triangle. Find the equation of the perpendicular line from A to BC.

Solution

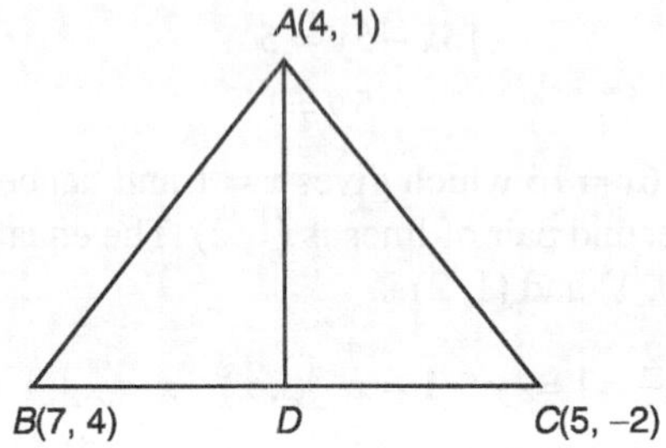

The slope of the line BC is

$$m = \frac{y_1 - y_2}{x_1 - x_2} = \frac{4+2}{7-5}$$

$$= \frac{6}{2} = 3.$$

Therefore, the slope of the perpendicular AD to BC is $-\frac{1}{3}$. Hence, the equation of the perpendicular from $A(4, 1)$ on BC is $y - y_1 = m(x - x_1)$

(i.e.) $$y - 1 = \frac{-1}{3}(x-4) \Rightarrow 3y - 3 = -x + 4$$

$$\therefore x + 3y = 7.$$

Example 2.24

The foot of the perpendicular from the point (1, 2) on a line is (3,–4). Find the equation of the line.

Solution

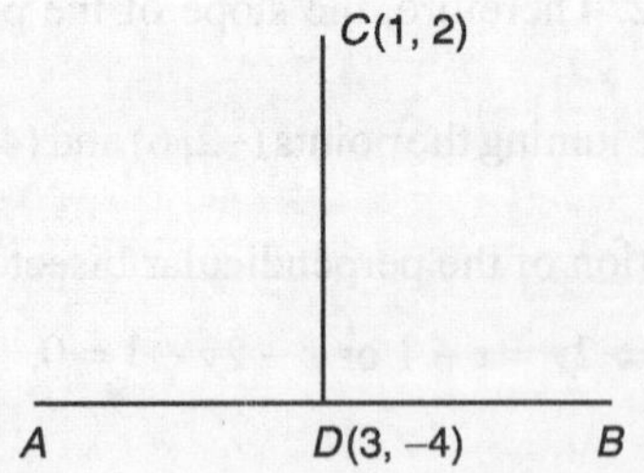

Let AB be the line and $D(3, -4)$ be the foot of the perpendicular from $C(1, 2)$

The slope of the line CD is $\frac{2+4}{1-3} = -3.$

Therefore, the slope of the line AB is $\frac{1}{3}$.

The equation of the line AB is $y - y_1 = m(x - x_1)$

$$\text{(i.e.)} \quad y+4=\frac{1}{3}(x-3) \Rightarrow 3y+12=x-3 \quad \text{(i.e.)} \quad x-3y=15.$$

Example 2.25

Find the equation of the right bisector of the line joining the points (2, 3) and (4, 5).

Solution

The right bisector is the perpendicular bisector of the line joining the points (2, 3) and (4, 5). The midpoint of the line is $\left(\frac{2+4}{2}, \frac{3+5}{2}\right)$ (i.e.) (3,4).

Therefore, the slope of the given line is $m = \frac{3-5}{2-4} = 1.$

$\therefore$ The slope of the right bisector is -1. The equation of the right bisector is $y - y_1 = m(x - x_1) \Rightarrow y - 4 = -1\ (x - 3)$ or $y - 4 = -x + 3$ or $x + y = 7$.

Example 2.26

Find the point on the line $3y - 4x + 11 = 0$ which is equidistant from the points (3, 2) and (−2, 3).

Solution

Let $P(x_1, y_1)$ be the point on the line $3y - 4x + 11 = 0$ which is equidistant from the points $A(3, 2)$ and $B(-2, 3)$.

$$PA^2 = (x_1-3)^2 + (y_1-2)^2 = x_1^2 + 9 - 6x_1 + y_1^2 + 4 - 4y_1 = x_1^2 + y_1^2 - 6x_1 - 4y_1 + 13$$

$$PB^2 = (x_1+2)^2 + (y_1-3)^2 = x_1^2 + 4 + 4x_1 + y_1^2 + 9 - 6y_1 = x_1^2 + y_1^2 + 4x_1 - 6y_1 + 13$$

Since, $PA^2 = PB^2, x_1^2 + y_1^2 - 6x_1 - 4y_1 + 13 = x_1^2 + y_1^2 + 4x_1 - 6y_1 + 13.$

$$\text{(i.e.)} \quad 10x_1 - 2y_1 = 0 \text{ or } 5x_1 - y_1 = 0 \tag{2.39}$$

Since the (x_1, y_1) lies on the line,

$$3y - 4x + 11 = 0,\ 3y_1 - 4x_1 + 11 = 0 \tag{2.40}$$

Substituting $y_1 = 5x_1$ in (2.40), we get $15x_1 - 4x_1 + 11 = 0$.

$\therefore x_1 = -1$ and hence $y_1 = -5$.

Therefore, the required point is $(-1, -5)$.

Example 2.27

Find the equation of the line passing through the point (2, 3) and parallel to $3x - 4y + 5 = 0$.

Solution

The slope of the line $3x - 4y + 5 = 0$ is $\frac{3}{4}$.

Therefore, the slope of the parallel line is also $\frac{3}{4}$.

Hence the equation of the parallel line through (2, 3) is $y - y_1 = m(x - x_1)$

(i.e.) $y - 3 = \frac{3}{4}(x - 2) \Rightarrow 4y - 12 = 3x - 6$. (i.e.) $3x - 4y + 6 = 0$.

Example 2.28

Find the equation of the line passing through the point (4, –5) and is perpendicular to the line $7x + 2y = 15$.

Solution

The slope of the line $7x + 2y = 15$ is $\frac{-7}{2}$.

Therefore, the slope of the perpendicular line is $\frac{+2}{7}$. The equation of the perpendicular line through (4, – 5) is $y - y_1 = m(x - x_1)$

$$\text{(i.e.)}\quad y + 5 = \frac{2}{7}(x - 4) \Rightarrow 7y + 35 = 2x - 8$$

$$\Rightarrow 2x - 7y = 43.$$

Example 2.29

Find the equation of the line through the intersection of $2x + y = 8$ and $3x + 7 = 2y$ and parallel to $4x + y = 11$.

Solution

The point of intersection of the lines $2x + y = 8$ and $3x + 7 = 2y$ is obtained by solving the following two equations:

$$2x + y = 8 \tag{2.41}$$

$$3x - 2y = -7 \qquad (2.42)$$

$$(2.41) \times 2 \qquad 4x + 2y = 16$$

$$(2.42) \qquad \underline{3x - 2y = -7}$$

On adding, we get $7x = 9 \Rightarrow x = \frac{9}{7}$.

$$\therefore y = \frac{-18}{7} + 8 = \frac{38}{7}.$$

Therefore, the point of intersection is $\left(\frac{9}{7}, \frac{38}{7}\right)$.

The slope of the line $4x + y = 11$ is -4. The slope of the parallel line is also -4. The equation of the parallel line is

$$y - y_1 = m(x - x_1) \Rightarrow y - \frac{38}{7} = -4\left(x - \frac{9}{7}\right) \Rightarrow 7y - 38 = -4(7x - 9)$$

$$\text{(i.e.)} \quad 28x + 7y = 74.$$

Example 2.30

Find the image of the origin on the line $3x - 2y = 13$.

Solution

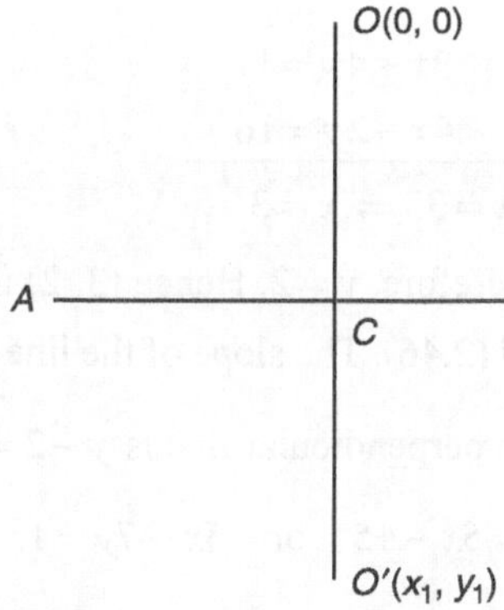

Let $O'(x_1, y_1)$ be the image of O on the line AB. Then C is the midpoint of OO'. The slope of the line OO' is $\frac{-2}{3}$.

The equation of the line OO' is $y - 0 = \frac{-2}{3}(x - 0) \Rightarrow 2x + 3y = 0$.

Solving the equations $\quad 3x - 2y = 13 \quad (2.43)$

$$2x + 3y = 0 \quad (2.44)$$

To get the coordinates of C:

$(2.43) \times 3 \quad 9x - 6y = 39$

$(2.44) \times 2 \quad 4x + 6y = 0$. On adding $13x = 39$. $\therefore x = 3$ and hence $y = -2$.

Therefore, C is $(3, -2)$. C being the midpoint of OO'.

$$\frac{0 + x_1}{2} = 3 \text{ and } \frac{0 + y_1}{2} = -2. \therefore x_1 = 6 \text{ and } y_1 = -4.$$

Therefore, the image is $(6, -4)$.

Example 2.31

Find the equation of the straight line passing through the intersection of the lines $3x + 4y = 17$ and $4x - 2y = 8$ and perpendicular to $7x + 5y = 12$.

Solution

$$3x + 4y = 17 \quad (2.45)$$

$$4x - 2y = 8 \quad (2.46)$$

$(2.45) \times 1 + (2.46) \times 2$ gives

$$3x + 4y = 17$$

$$4x - 2y = 16$$

$$11x = 33 \Rightarrow x = 3.$$

From (2.45), $9 + 4y = 17$. Therefore, $y = 2$. Hence $(3, 2)$ is the point of intersection of the lines (2.45) and (2.46). The slope of the line $7x + 5y = 12$ is $\frac{-7}{5}$.

Therefore, the slope of the perpendicular line is $y - 2 = \frac{5}{7}(x - 3)$.

$$7y - 14 = 5x - 15 \quad \text{or} \quad 5x - 7y = 1.$$

Example 2.32

Find the orthocentre of the triangle whose vertices are $(5, -2)$, $(-1, 2)$, and $(1, 4)$.

Solution

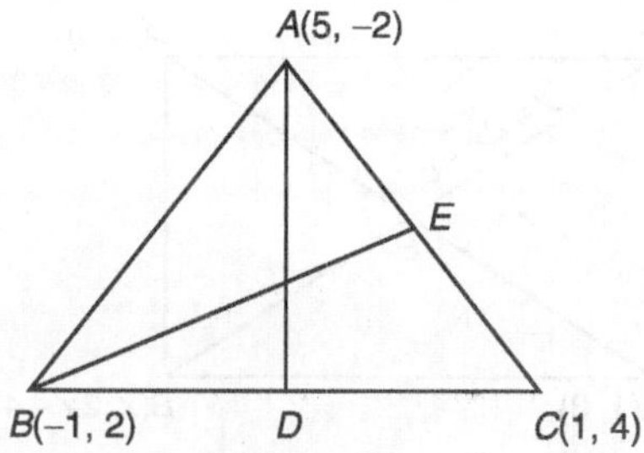

Slope of $BC = \dfrac{2-4}{-1-1} = 1$. Therefore, slope of the perpendicular AD is -1.

The equation of the line AD is $y + 2 = -1(x - 5)$.

$$y + 2 = -x + 5 \text{ or } x + y = 3 \tag{2.47}$$

Slope of AC is $\dfrac{4+2}{1-5} = \dfrac{-3}{2}$. Therefore, slope of BE is $\dfrac{2}{3}$.

The equation of BE is

$$y-2-\frac{2}{3}(x+1) \quad \text{or} \quad 3y - 6 = 2x + 2 \text{ or } 2x - 3y + 8 = 0 \tag{2.48}$$

Solving the equations (2.47) and (2.48), we get the coordinates of the orthocentre:

(2.47) × 3 $\qquad 3x + 3y = 9$

(2.48) × 1 $\qquad 2x - 3y = -8$

On adding $5x = 1$ or $x = \dfrac{1}{5}$.

From (2.47), $y = 3 - \dfrac{1}{5} = \dfrac{14}{5}$.

$\therefore$ The orthocentre is $\left(\dfrac{1}{5}, \dfrac{14}{5}\right)$.

Example 2.33

The points (1, 3) and (5, 1) are two opposite vertices of a rectangle. The other two vertices lie on the line $y = 2x + c$. Find c and the remaining two vertices.

Solution

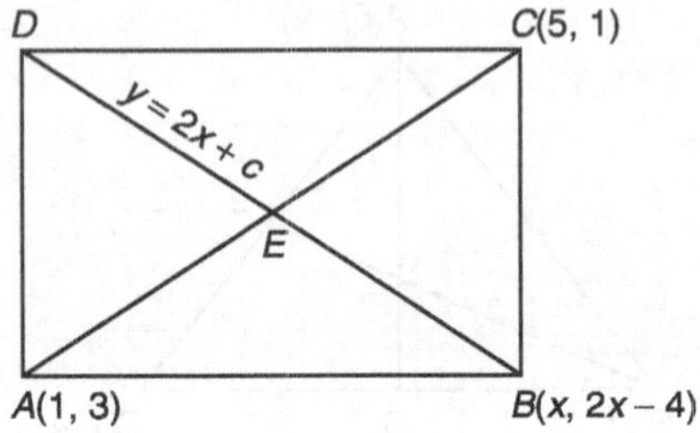

Let $ABCD$ be the rectangle with A and C as the points with coordinates (1, 3) and (5, 1), respectively. In a rectangle the diagonals bisect each other. The midpoint AC is $\left(\frac{1+5}{2}, \frac{3+1}{2}\right)$ (i.e.) (3, 2).

As this point lies on BD whose equation is $y = 2x + c$. We get $2 = 6 + c$ or $c = -4$. Therefore, the equation of the line BD is $y = 2x - 4$.

Therefore, the coordinates of any point on this line is $(x, 2x - 4)$. If this is the point B then $AB^2 + BC^2 = AC^2$.

$$(1-x)^2 + (3-2x+4)^2 + (x-5)^2 + (2x-4-1)^2 = (1-5)^2 + (3-1)^2$$

$$\text{(i.e.)} \quad \begin{aligned} &(1-x)^2 + (7-2x)^2 + (x-5)^2 + (2x-5)^2 = 16+4 \\ &\Rightarrow 10x^2 - 60x + 30 = 0 \Rightarrow x = 2, 4. \end{aligned}$$

As $y = 2x - 4$, the corresponding values of $y = 0, 4$.

Therefore, the coordinates of B and D are (2, 0) and (4, 4).

Example 2.34

If a, b, and c are distinct numbers different from 1 then show that the points $A\left(\frac{a^3}{a-1}, \frac{a^2-3}{a-1}\right)$, $B\left(\frac{b^3}{b-1}, \frac{b^2-3}{b-1}\right)$, and $C\left(\frac{c^3}{c-1}, \frac{c^2-3}{c-1}\right)$ are collinear if $ab + bc + ca - abc = 3(a + b + c)$.

Solution

Let A, B, and C lie on the straight line $px + qy + r = 0$.

Then the equation of the line satisfies the condition $p\left(\frac{t^3}{t-1}\right) + q\left(\frac{t^2-3}{t-1}\right) + r = 0$ where $t = a, b$, and c (i.e.) $pt^3 + qt^2 + rt - 3q - r = 0$. Here, a, b, and c are the roots of this equation.

$$a+b+c=\frac{-q}{p},\ ab+bc+ca=\frac{r}{p},\ abc=\frac{3q+r}{p}$$

$$\therefore abc-3\frac{q}{p}-\frac{r}{p}=0.$$

$$\text{or}\quad abc+3(a+b+c)-(ab+bc+ca)=0.$$

$$\therefore\quad ab+bc+ca-abc=3(a+b+c).$$

Example 2.35

A vertex of an equilateral triangle is at (2, 3) and the equation of the opposite side is $x+y=2$. Find the equations of the other sides.

Solution

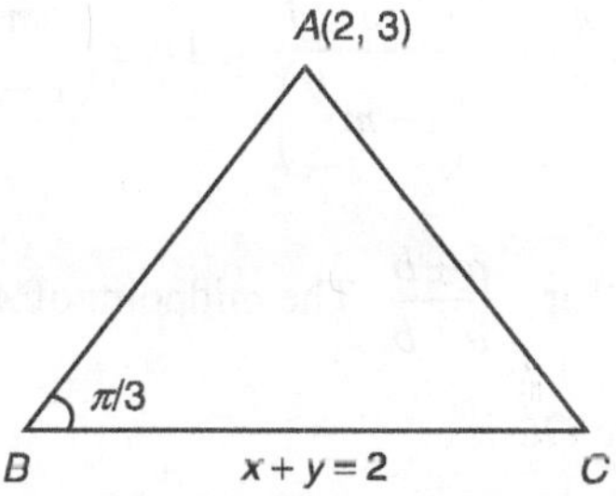

The slope of BC is -1. Let m be the slope of AB or AC. Then $\tan\frac{\pi}{3}=\pm\frac{(m+1)}{1-m}$

$$\text{(i.e.)}\quad \sqrt{3}(1-m)=\pm(m+1)$$

$$\therefore m=\frac{\sqrt{3}-1}{\sqrt{3}+1}\quad\text{or}\quad\frac{\sqrt{3}+1}{\sqrt{3}-1}=2-\sqrt{3}\quad\text{or}\quad 2+\sqrt{3}.$$

Therefore, the equation of other two sides are $y-3=(2+\sqrt{3})(x-2)$ and $y-3=(2+\sqrt{3})(x-2)$.

$$\text{(i.e.)}\quad (2-\sqrt{3})x-y+2\sqrt{3}-1=0\quad\text{and}\quad(2+\sqrt{3})x-y-(2\sqrt{3}+1)=0.$$

$$(2+\sqrt{3})x-y+3-2(2+\sqrt{3}=0)\quad(2-\sqrt{3})x-2(2-\sqrt{3})-y+3=0$$

$$(2+\sqrt{3})x-y-2\sqrt{3}-1=0)\qquad(2-\sqrt{3}x-y-1+2\sqrt{3}=0)$$

Example 2.36

One diagonal of a square is the portion of the line $\frac{x}{a}+\frac{y}{b}=1$ intercepted between the axes. Find the equation of the other diagonal.

Solution

The slope of AB is $-\dfrac{b}{a}$. Let m be slope of AC.

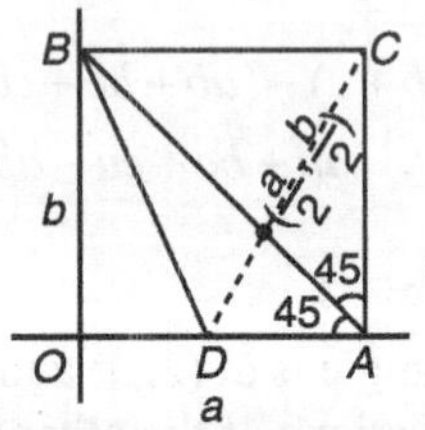

$$\therefore \tan\frac{\pi}{4} = \pm\frac{\left(m+\dfrac{b}{a}\right)}{\left(1-m\dfrac{b}{a}\right)} \Rightarrow 1 = \pm\left(\frac{am+b}{a-mb}\right)$$

$\therefore m = \dfrac{a-b}{a+b}$ or $\dfrac{a+b}{a-b}$. The midpoint of AB is $\left(\dfrac{a}{2}, \dfrac{b}{2}\right)$.

The equation of the side OC is

$$y - \frac{b}{2} = \frac{a-b}{a+b}x \tag{2.49}$$

The equation of the side BD is

$$y - \frac{b}{2} = -\left(\frac{a+b}{a-b}\right)x \tag{2.50}$$

The equation of the other diagonal is

$$y - \frac{b}{2} = \frac{a}{b}\left(x - \frac{a}{2}\right) \tag{2.51}$$

Example 2.37

If the vertices of ΔABC are (x_i, y_i) $i = 1, 2, 3$. Show that the equation of the median through A is given by $\begin{vmatrix} x & y & 1 \\ x_1 & y_1 & 1 \\ x_2 & y_2 & 1 \end{vmatrix} + \begin{vmatrix} x & y & 1 \\ x_1 & y_1 & 1 \\ x_3 & y_3 & 1 \end{vmatrix} = 0.$

Solution

The coordinates of the midpoint of BC are $D\left(\frac{x_2+x_3}{2}, \frac{y_2+y_3}{2}\right)$.

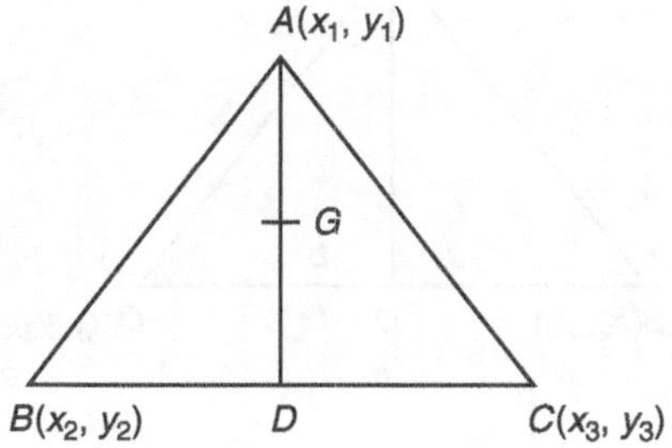

The equation of the median AD is given by $\begin{vmatrix} x & y & 1 \\ x_1 & y_1 & 1 \\ \frac{x_2+x_3}{2} & \frac{y_2+y_3}{2} & 1 \end{vmatrix} = 0$. Since the area of a line is zero.

$$\text{(i.e.)} \quad \begin{vmatrix} x & y & 1 \\ x_1 & y_1 & 1 \\ x_2+x_3 & y_2+y_3 & 1 \end{vmatrix} = 0$$

$$\therefore \quad \begin{vmatrix} x & y & 1 \\ x_1 & y_1 & 1 \\ x_2 & y_2 & 1 \end{vmatrix} + \begin{vmatrix} x & y & 1 \\ x_1 & y_1 & 1 \\ x_3 & y_3 & 1 \end{vmatrix} = 0.$$

Example 2.38

If (x, y) is an arbitrary point on the internal bisector of vertical angle A of ΔABC, where (x_i, y_i), $i = 1, 2, 3$ are the vertices of A, B, and C, respectively, and a, b, and c are the length of the sides BC, CA, and AB, respectively, prove that $b\begin{vmatrix} x & y & 1 \\ x_1 & y_1 & 1 \\ x_2 & y_2 & 1 \end{vmatrix} + c\begin{vmatrix} x & y & 1 \\ x_1 & y_1 & 1 \\ x_3 & y_3 & 1 \end{vmatrix} = 0$.

Solution

In ΔABC, AD is the internal bisector of $\underline{|A}$. We know that

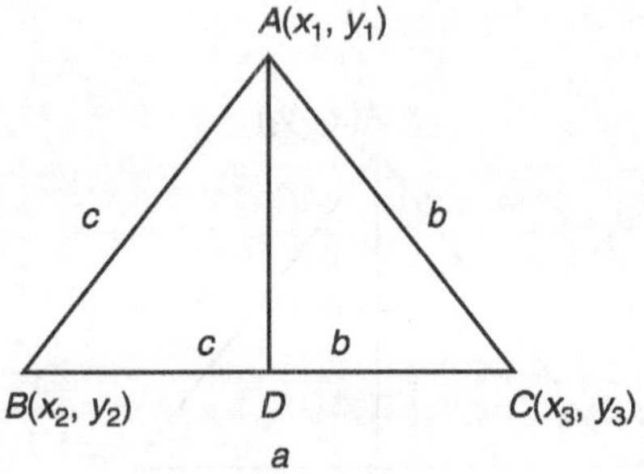

$$\frac{BD}{DC} = \frac{AB}{AC}, \therefore \frac{BD}{DC} = \frac{c}{b}.$$

The coordinates of D are $\left(\dfrac{bx_2 + cx_3}{b+c}, \dfrac{by_2 + cy_3}{b+c}\right)$

The equation of AD is given by $\begin{vmatrix} x & y & 1 \\ x_1 & y_1 & 1 \\ \dfrac{bx_2 + cx_3}{b+c} & \dfrac{by_2 + cy_3}{b+c} & 1 \end{vmatrix} = 0$

$$\text{(i.e.)} \quad \begin{vmatrix} x & y & 1 \\ x_1 & y_1 & 1 \\ bx_2 + cx_3 & by_2 + cy_3 & b+c \end{vmatrix} = 0.$$

$$\text{(i.e.)} \quad \begin{vmatrix} x & y & 1 \\ x_1 & y_1 & 1 \\ bx_2 & by_2 & b \end{vmatrix} + \begin{vmatrix} x & y & 1 \\ x_1 & y_1 & 1 \\ cx_3 & cy_3 & c \end{vmatrix} = 0.$$

$$\therefore \quad b\begin{vmatrix} x & y & 1 \\ x_1 & y_1 & 1 \\ x_2 & y_2 & 1 \end{vmatrix} + c\begin{vmatrix} x & y & 1 \\ x_1 & y_1 & 1 \\ x_3 & y_3 & 1 \end{vmatrix} = 0.$$

Example 2.39

Find the orthocentre of the triangle whose vertices are $(a, 0)$, $(0, b)$, and $(0, 0)$.

Solution

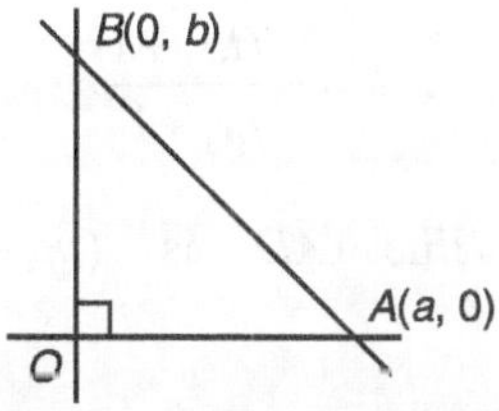

The orthocentre is the point of concurrence of altitudes. Since OA and OB are perpendicular to each other, OA and OB are the altitudes through A and B of ΔABC. Therefore, 0 is the orthocentre. Hence, the coordinates of the orthocentre is (0, 0).

Example 2.40

Prove that the orthocentre of the triangle formed by the three lines $y = xt_1 + \frac{a}{t_1}, y = xt_2 + \frac{a}{t_2},$ and $y = xt_3 + \frac{a}{t_3}$ lies on the line $x + a = 0$.

Solution

The equation of the line passing through the intersection of the lines

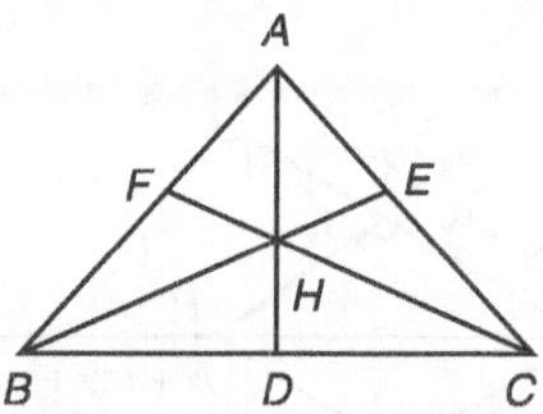

$$y = xt_1 + \frac{a}{t_1}$$

$$y = xt_2 + \frac{a}{t_2} \text{ and } y = xt_3 + \frac{a}{t_3} \text{ is}$$

$$\left(y - xt_2 - \frac{a}{t_2}\right) + k\left(y - xt_3 - \frac{a}{t_3}\right) = 0.$$

The slope of the line AD is $\frac{t_2 + kt_3}{1+k}$. The slope of the line BC is t_1. Since AD is perpendicular to BC, $\frac{t_2 + kt_3}{k+1} \times t_1 = -1$

$$\Rightarrow t_1t_2 + kt_1t_3 = -k - 1$$

$$\therefore k = \frac{-(t_1t_2 + 1)}{t_1t_3 + 1}.$$

The equation of the line AD is $(t_1t_3 + 1)\left(y - xt_2 - \frac{a}{t_2}\right) - (t_1t_2 + 1)\left(y - xt_3 - \frac{a}{t_3}\right) = 0.$

$$\text{(i.e.)}\quad t_1 t_2 t_3 y + t_2t_3x - a(1 + t_1t_3 + t_1t_2) = 0. \qquad (2.52)$$

Similarly the equation of the line BE is

$$t_1t_2t_3y + t_3t_1x - a(1 + t_1t_2 + t_2t_3) = 0 \qquad (2.53)$$

Subtracting equations (2.52) from (2.53), $t_3(t_1 - t_2)x + at_3(t_1 - t_2) = 0$. Since $t_1 \neq t_2$, $x + a = 0$ the orthocentre lies on the line $x + a = 0$.

Example 2.41

Show that the reflection of the line $px + qy + r = 0$, on the line $lx + my + n = 0$ is $(px + qy + r)(l^2 + m^2) - 2(lp + mq)(lx + my + n) = 0$.

Solution

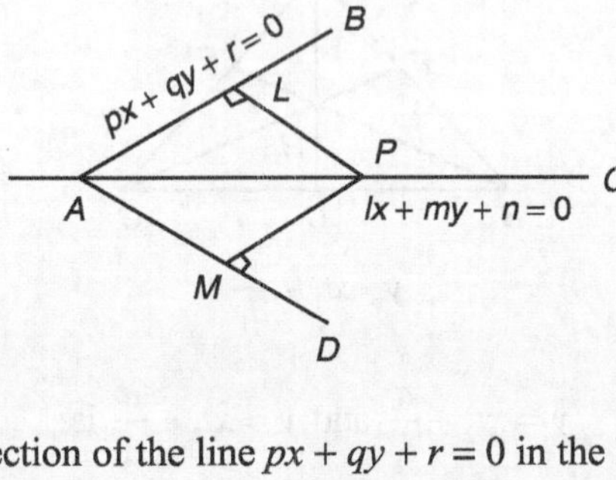

Let AD be the reflection of the line $px + qy + r = 0$ in the line $lx + my + n = 0$. Then the equation of line AD is $px + qy + r + k(lx + my + n) = 0$. Then the perpendicular from any point on AC to AB and AD are equal.

$$\therefore \frac{px + qy + r}{\sqrt{p^2 + q^2}} = \pm \frac{(px + qy + r) + k(lx + my + n)}{\sqrt{(p + kl)^2 + (q + km)^2}}$$

Since the point P lies on AC, $lx + my + n = 0$, $lx_1 + my_1 + n = 0$

$$\therefore p^2+q^2=(p+kl)^2+(q+km)^2$$
$$p^2+q^2=p^2+q^2+k^2(l^2+m^2)+2k(pl+qm)$$
$$\therefore k(l^2+m^2)+2(pl+qm)=0 \Rightarrow k=\frac{-2(pl+qm)}{l^2+m^2}$$

Hence the equation of the line AD is $(l^2+m^2)(px+qy+r)-2(pl+qm)(lx+my+n)=0$.

Example 2.42

The diagonals of the parallelogram are given by the sides $u = p$, $u=q$, $v=r$, $v=s$ where $u=ax+by+c$ and $v=a_1x+b_1y+c_1$. Show that the equation of the diagonal which passes through the points of intersection of $u=p$, $v=r$ and $u=q$ and $r=s$ is given by $\begin{vmatrix} u & v & 1 \\ p & r & 1 \\ q & s & 1 \end{vmatrix}=0$.

Solution

Consider

$$\begin{vmatrix} u & v & 1 \\ p & r & 1 \\ q & s & 1 \end{vmatrix}=0 \tag{2.54}$$

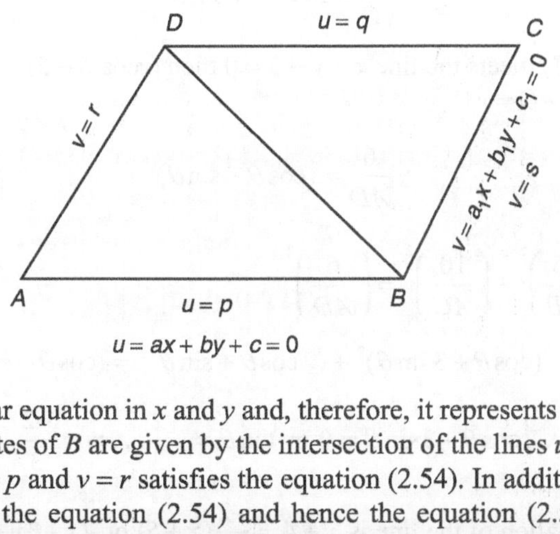

This is a linear equation in x and y and, therefore, it represents a straight line. The coordinates of B are given by the intersection of the lines $u=p$ and $v=s$. However, $u=p$ and $v=r$ satisfies the equation (2.54). In addition, $u=q$, and $v = s$ satisfy the equation (2.54) and hence the equation (2.54) is the line passing through B and D and represents the equation of the diagonal BD.

Example 2.43

A line through the point $A(-5, -4)$ meets the lines $x + 3y + 2 = 0$, $2x + y + 4 = 0$, and $x - y - 5 = 0$ at the points B, C, and D, respectively. If $\left(\frac{15}{AB}\right)^2 + \left(\frac{10}{AC}\right)^2 = \left(\frac{6}{AD}\right)^2$ find the equation.

Solution

The equation of the line passing through the point $(-5, -4)$ is

$$\frac{x+5}{\cos\theta} = \frac{x+4}{\sin\theta} = r \tag{2.55}$$

Any point on the line is $(r\cos\theta - 5, r\sin\theta - 4)$. The point meets the line $x + 3y + 2 = 0$ at B then $AB = r \cdot (r\cos\theta - 5) + 3(r\sin\theta - 4) + 2 = 0$.

$$r(\cos\theta + 3\sin\theta) = 15$$

$$\therefore r = \frac{15}{\cos\theta + 3\sin\theta} \quad \text{or} \quad \frac{15}{AB} = \cos\theta + 3\sin\theta \tag{2.56}$$

If the line (2.55) meets the line $2x + y + 4 = 0$ at C then $2(r\cos\theta - 5) + (r\sin\theta - 4) + 4 = 0$. Now, $r = AC$ and

$$\frac{10}{AC} = 2\cos\theta + \sin\theta \tag{2.57}$$

If the line (2.55) meets the line $x - y - 5 = 0$ then $(r\cos\theta - 5) - (r\sin\theta - 4) - 5 = 0$ and here $AD = r$.

$$\therefore \frac{6}{AD} = (\cos\theta - \sin\theta) \tag{2.58}$$

Given that $\left(\frac{15}{AB}\right)^2 + \left(\frac{10}{AC}\right)^2 = \left(\frac{6}{AD}\right)^2$

$$\text{(i.e.)} \quad (\cos\theta + 3\sin\theta)^2 + (2\cos\theta + \sin\theta)^2 = (\cos\theta - \sin\theta)^2$$

$$\therefore 2\cos\theta + 3\sin\theta = 0 \text{ or } \tan\theta = -\frac{2}{3}. \therefore m = \frac{-2}{3}.$$

Hence, the equation of the line is $y + 4 = \frac{-2}{3}(x+5)$ or $2x + 3y + 22 = 0$.

Example 2.44

A variable straight line is drawn through O to cut two fixed lines L_1 and L_2 at A_1 and A_2. A point A is taken on the variable line such that $\frac{l+q}{OA} = \frac{p}{OA_1} + \frac{q}{OA_2}$. Show that the locus of P is a straight line passing through the point of intersection of L_1 and L_2.

Solution

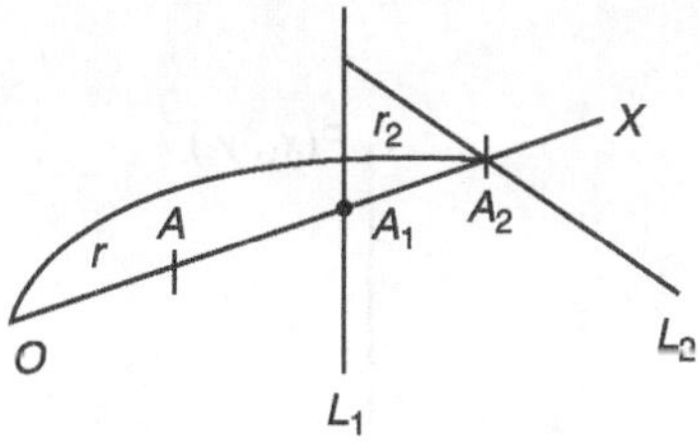

Let the equation of the line OX be $\frac{x}{\cos\theta} = \frac{y}{\sin\theta} = r$ where $OA = r$. Any point on this line is ($r\cos\theta$, $r\sin\theta$). Let $OA_1 = r_1$ and $OA_2 = r_2$. A_1 is ($r_1\cos\theta$, $r_1\sin\theta$) and A_2 is ($r_2\cos\theta$, $r_2\sin\theta$). Let the two fixed straight lines be L_1: $l_1x + m_1y - 1 = 0$ and L_2: $l_2x + m_2y - 1 = 0$. Since the points A_1 and A_2 lie on the two lines, respectively,

$$l_1(r_1\cos\theta) + m_1(r_1\sin\theta) = 1.$$

$$l_2(r_2\cos\theta) + m_2(r_2\sin\theta) = 1$$

$$\therefore \frac{1}{r_1} = l_1\cos\theta + m_1\sin\theta$$

$$\frac{1}{r_2} = l_2\cos\theta + m_2\sin\theta$$

$$\frac{p}{OA_1} + \frac{q}{OA_2} = \frac{p+q}{OA}$$

$$\text{(i.e.)}\quad p(l_1\cos\theta + m_1\sin\theta) + q(l_2\cos\theta + m_2\sin\theta) = \frac{p+q}{r}$$

$$p(l_1r\cos\theta + m_1r\sin\theta) + q(l_2r\cos\theta + m_2r\sin\theta) = p+q$$

$$\text{(i.e.)}\quad p(l_1x + m_1y) + q(l_2x + m_2y) - (p+q) = 0$$

(i.e.) $p(l_1x+m_1y-1)+q(l_2x+m_2y-1)=0$ which is a straight line passing through the point of intersection of the two fixed straight lines $L_1 = 0$ and $L_2 = 0$.

Example 2.45

If the image of the point (x_1, y_1) with respect to the line $my + lx + n = 0$ is the point (x_2, y_2) show that $\dfrac{x_2-x_1}{l}=\dfrac{y_2-y_1}{m}=\dfrac{-2(lx_1+my_1+n)}{l^2+m^2}$.

Solution

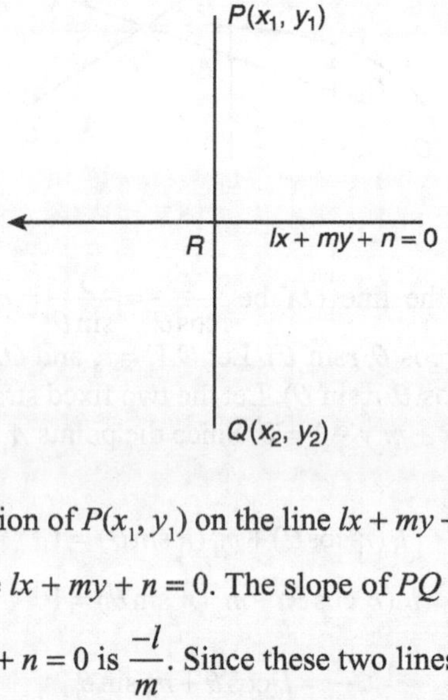

$Q(x_2, y_2)$ is the reflection of $P(x_1, y_1)$ on the line $lx + my + n = 0$. The midpoint of PQ lies on the line $lx + my + n = 0$. The slope of PQ is $\dfrac{y_1-y_2}{x_1-x_2}$. The slope of the line is $lx + my + n = 0$ is $\dfrac{-l}{m}$. Since these two lines are perpendicular,

$$\left(\frac{y_1-y_2}{x_1-x_2}\right)\left(\frac{-l}{m}\right)=-1 \text{ or } l(y_1-y_2)=m(x_1-x_2)$$

$$\text{or} \quad \frac{x_1-x_2}{l}=\frac{y_1-y_2}{m}=\frac{\sqrt{(x_1-x_2)^2+(y_1-y_2)^2}}{\sqrt{l^2+m^2}}=\frac{PQ}{\sqrt{l^2+m^2}}$$

$$=\frac{2PR}{\sqrt{l^2+m^2}}=\frac{2}{\sqrt{l^2+m^2}}\cdot\frac{lx_1+my_1+n}{\sqrt{l^2+m^2}}=\frac{2(lx_1+my_1+n)}{l^2+m^2}$$

$$\therefore \frac{x_2-x_1}{l}=\frac{y_2-y_1}{m}=\frac{-2(lx_1+my_1+n)}{\sqrt{l^2+m^2}}.$$

Example 2.46

Prove that the area of the triangle whose roots are $L_r = a_r x + b_r y + c_r$ $(r = 1, 2, 3)$ is $\dfrac{\Delta^2}{2C_1C_2C_3}$ where C_i is the cofactor of c_i $(i = 1, 2, 3)$ in Δ given by

$$\Delta = \begin{vmatrix} a_1 & b_1 & c_1 \\ a_2 & b_2 & c_2 \\ a_3 & b_3 & c_3 \end{vmatrix}.$$

Solution

Let A_r, B_r, and C_r be the cofactors of a_r, b_r, and c_r in D. The point of intersection of the lines $a_1x + b_1y + c_1 = 0$, $a_2x + b_2y + c_2 = 0$ is $\left(\dfrac{b_1c_2 - b_2c_1}{a_1b_2 - a_2b_1}, \dfrac{c_1a_2 - c_2a_1}{a_1b_2 - a_2b_1}\right)$

$$\text{(i.e.)} \quad \left(\frac{A_3}{C_3}, \frac{B_3}{C_3}\right)$$

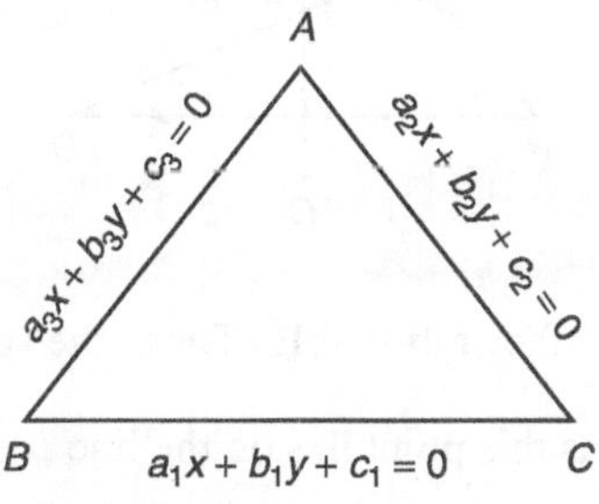

$\therefore$ The vertices of the triangle are $\left(\dfrac{A_1}{C_1}, \dfrac{B_1}{C_1}\right)$, $\left(\dfrac{A_2}{C_2}, \dfrac{B_2}{C_2}\right)$, and $\left(\dfrac{A_3}{C_3}, \dfrac{B_3}{C_3}\right)$. Then the area of the triangle is given by,

$$\Delta_1 = \frac{1}{2}\begin{vmatrix} \dfrac{A_1}{C_1} & \dfrac{B_1}{C_1} & 1 \\ \dfrac{A_2}{C_2} & \dfrac{B_2}{C_2} & 1 \\ \dfrac{A_3}{C_3} & \dfrac{B_3}{C_3} & 1 \end{vmatrix}$$

where D is the determinant formed by the cofactors.

$$= \frac{D}{2C_1C_2C_3} = \frac{\Delta^2}{2C_1C_2C_3} \quad \text{(Since we have the property } D = \Delta^2\text{).}$$

Example 2.47

A straight line L intersects the sides BC, CA, and AB of a triangle ABC in D, E, and F, respectively. Show that $\frac{BD}{DC}\cdot\frac{CE}{EA}\cdot\frac{AF}{FB} = -1$.

Solution

Let DEF be the straight line meeting BC, CA, and AB at D, E, and F, respectively. Let the equations of the line DEF be $lx + my + n = 0$.

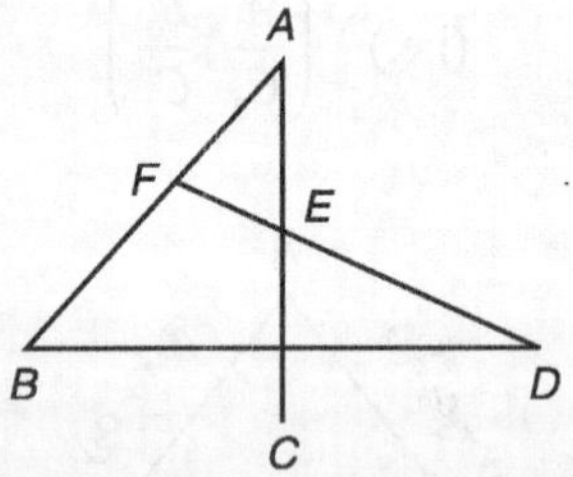

Let D divide BC in the ratio $\lambda{:}1$. Then the coordinates of D are $\left(\frac{\lambda x_3 + x_2}{\lambda+1}, \frac{\lambda y_3 + y_1}{\lambda+1}\right)$. As this point lies on the line $lx + my + n = 0$.

$$l\left(\frac{\lambda x_3 + x_2}{\lambda+1}\right) + m\left(\frac{\lambda y_3 + y_2}{\lambda+1}\right) + n = 0$$

$$\Rightarrow l(\lambda x_3 + x_2) + m(\lambda y_3 + y_2) + n(\lambda+1) = 0$$

$$\text{(i.e.)} \quad \lambda(lx_3 + my_3 + n) + lx_2 + my_2 + n = 0$$

$$\therefore \lambda = -\frac{(lx_2 + my_2 + n)}{lx_3 + my_3 + n}$$

$$\therefore \frac{BD}{DC} = -\frac{(lx_2 + my_2 + n)}{lx_3 + my_3 + n}$$

Similarly $\frac{CE}{EA} = -\frac{(lx_3 + my_3 + n)}{lx_1 + my_1 + n}; \frac{AF}{FB} = -1\frac{(lx_1 + my_1 + n)}{lx_2 + my_2 + n}$

Multiplying these three we get $\frac{BD}{DC} \times \frac{CE}{EA} \times \frac{AF}{FB} = -1.$

Example 2.48

A straight line is such that the algebraic sum of perpendiculars drawn upon it from any number of fixed points is zero. Show that the straight line passes through a fixed point.

Solution

Let $(x_1, y_1), (x_2, y_2), \ldots, (x_n, y_n)$, be n fixed points and $ax + by + c = 0$ be a given line. The algebraic sum of the perpendiculars from (x_i, y_i), $i = 1, 2, \ldots, n$ to this line is zero.

$$\therefore \sum_{i=1}^{n} \frac{ax_i + by_i + c}{\sqrt{a^2 + b^2}} = 0$$

$$\frac{a\Sigma x_i}{n} + \frac{b\Sigma y_i}{n} + \frac{nc}{n} = 0. \text{ Where } n = \sqrt{a^2 + b^2}$$

This equation shows that the point $\left(\frac{x_1 + x_2 + \cdots + x_n}{n}, \frac{y_1 + y_2 + \cdots + y_n}{n}\right)$ lies on the line $ax + by + c = 0$. Therefore, the line passes through a fixed point.

Example 2.49

Determine all the values of α for which the point (α, α^2) lies inside the triangle formed by the lines $2x + 3y - 1 = 0$, $x + 2y - 3 = 0$, and $5x - 6y - 1 = 0$.

Solution

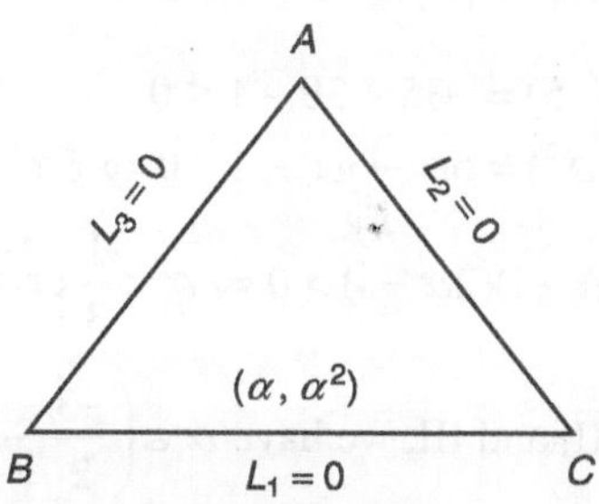

$$L_1: 2x + 3y - 1 = 0$$
$$L_2: x + 2y - 3 = 0$$
$$L_3: 5x - 6y - 1 = 0.$$

$2x+3y-1=0$	$5x-10y-15=0$	$4x+6y-2=0$
$\underline{2x+4y-6=0}$	$\underline{5x-6y-1=0}$	$\underline{5x-6y-1=0}$
$y=5$	$16y=14$	$9x=3$
$x=-7$	$y=\frac{7}{8}$	$x=\frac{1}{3}$
$\therefore C(-7, 5)$	$x=\frac{5}{4}$	$y=\frac{1}{4}$
	$\therefore A=\left(\frac{5}{4},\frac{7}{8}\right)$	Hence B is $\left(\frac{1}{3},\frac{1}{8}\right)$

$$L_1\,(A): 2\left(\frac{5}{4}\right)+3\left(\frac{7}{8}\right)-1>0.$$

$\therefore L_1\,(\alpha, \alpha^2) = 2a + 3\alpha^2 - 1 > 0$ if points A and (a, α^2) lies on the same side of the line. $3\alpha^2 + 2a - 1 > 0 \Rightarrow (3\alpha - 1)(\alpha + 1) > 0.$

$$\alpha < -1 \text{ or } \alpha > \frac{1}{3} \qquad \text{(I)}$$

$$L_2\,(B) = \frac{1}{3}+\frac{2}{3}-3<0;$$

$$\therefore L_2(a, \alpha^2) = \alpha + 2\alpha^2 - 3 < 0$$

$$\text{(i.e.)} \quad = 2\alpha^2 + \alpha - 3 < 0$$

$$= (2\alpha + 3)(\alpha - 1) < 0 \quad \therefore \frac{-3}{2} < \alpha < 1 \qquad \text{(II)}$$

$$L_3(-7, 5) = -35 - 30 - 1 < 0$$

$$\therefore L_3(\alpha, \alpha^2) = 5\alpha - 6\alpha^2 - 1 < 0 \Rightarrow 6\alpha^2 - 5\alpha + 1 > 0$$

$$\Rightarrow (3\alpha - 1)(2\alpha - 1) > 0 \Rightarrow \alpha < \frac{1}{3} \text{ or } \alpha > \frac{1}{2} \qquad \text{(III)}$$

From the conditions I, II, and III, we have $\alpha \in \left(\frac{-3}{2},-1\right)\cup\left(\frac{1}{2},1\right)$

Example 2.50

Find the direction in which a straight line must be drawn through the point (1, 4) so that its point of intersection with the line $x + y + 5 = 0$ may be at a distance $5\sqrt{2}$ units.

Solution

Let the equation of the line through the point (1, 4) be $\frac{x-1}{\cos\theta} = \frac{y-4}{\sin\theta} = r$.

Any point on this line is $(r\cos\theta + 1, r\sin\theta + 4)$. If this point lies on the line $x + y + 5 = 0$ then $r\cos\theta + 1 + r\sin\theta + 4 + 5 = 0$.

$$r(\cos\theta + \sin\theta) = -10. \text{ If } r = 5\sqrt{2} \text{ then } \cos\theta + \sin\theta = -\frac{10}{5\sqrt{2}} = -\sqrt{2}$$

$$\frac{1}{\sqrt{2}}\cos\theta + \frac{1}{\sqrt{2}}\sin\theta = -1$$

$$\cos\frac{\pi}{4}\cos\theta + \sin\frac{\pi}{4}\sin\theta = -1$$

$$\cos\left(\theta - \frac{\pi}{4}\right) = \cos\pi$$

$$\therefore \theta = \frac{\pi}{4} + \pi \text{ or } \theta = \frac{5\pi}{4}.$$

$\therefore$ The required straight line makes an angle of $\frac{5\pi}{4}$ with the positive directions of x-axis and passes through the point (1, 4).

Exercises

1. Find the area of triangle formed by the axes, the straight line L passing through the points (1, 1) and (2, 0) and the line perpendicular to the L and passing through $\left(\frac{1}{2}, 0\right)$.

   Ans.: $\frac{25}{16}$ sq. units

2. The line $3x + 2y = 24$ meets y-axis at A and x-axis at B. The perpendicular bisector of AB meets the line through (0, –1) parallel to x-axis at C. Find the area ΔABC.

   Ans.: 91 sq. units

3. If (x, y) be an arbitrary point on the altitude through A of ΔABC with vertices (x_i, y_i), $i = 1, 2, 3$ then the equation of the altitude through A is

$$b\sec B\begin{vmatrix} x & y & 1 \\ x_1 & y_1 & 1 \\ x_2 & y_2 & 1 \end{vmatrix} + c\sec C\begin{vmatrix} x & y & 1 \\ x_1 & y_1 & 1 \\ x_2 & y_2 & 1 \end{vmatrix} = 0.$$

4. A ray of light is sent along the line $x - 2y - 3 = 0$. Upon reaching the line $3x - 2y - 5 = 0$ the ray is reflected from it. Find the equation of the line containing the reflected ray.

Ans.: $29x - 2y - 31 = 0$

5. The extremities of the diagonals of a square are (1, 1) and (–2, –1). Obtain the equation of the other diagonal.

Ans.: $6x + 4y + 3 = 0$

6. The straight line $3x + 4y = 5$ and $4x - 3y = 15$ intersect at the point A. On this line, the points B and C are chosen so that $AO = AC$. Find the possible equations of the line BC passing through the point (1, 2).

Ans.: $x - 7y + 13 = 0$ and $7x + y - 9 = 0$

7. The consecutive sides of a parallelogram are $4x + 5y = 0$ and $7x + 2y = 0$. If the equation of one diagonal is $11x + 7y = 9$, find the equation of the other diagonal.

Ans.: $x - y = 0$

8. Show that the lines $ax \pm by \pm c = 0$ enclose a rhombus of area $\dfrac{2c^2}{ab}$.

9. If the vertices of a ΔOBC are $O(0, 0)$, $B(-3, -1)$, and $C(-1, -3)$, find the equation of the line parallel to BC and intersecting sides OB and OC whose perpendicular distance from (0, 0) is $\dfrac{1}{2}$.

Ans.: $2x + 2y + \sqrt{2} = 0$

10. Find the locus of the foot of the perpendicular from the origin upon the line joining the points $(a\cos\theta, b\sin\theta)$ and $(-a\sin\theta, b\cos\theta)$ where a is a variable.

Ans.: $a^2x^2 + b^2y^2 = 2(x^2 + y^2)^2$

11. Show that the locus given by $x + y = 0$, $(a - b)x + (a + b)y = 2ab$ and $(a + b)x + (a - b)y = 2ab$ form an isosceles triangle whose vertical angle is $2\tan^{-1}\left(\dfrac{a}{b}\right)$. Determine the centroid of a triangle.

Ans.: $\left(\dfrac{b}{3}, \dfrac{b}{3}\right)$

12. The sides of a quadrilateral have the equations, $x + 2y = 3$, $x = 1$, $x - 3y = 4$, and $5x + y + 12 = 0$. Show that the diagonals of the quadrilateral are at right angles.

13. Given n straight lines and a fixed point O. Through O a straight line is drawn meeting these lines in the point $A_1, A_2, \ldots, A_n$ and a point A such that $\frac{n}{OA} = \frac{1}{OA_1} + \frac{1}{OA_2} + \cdots + \frac{1}{OA_n}$. Prove that the locus of the point A is a straight line.

14. Find the equation of the line joining the point (3, 5) to the point of intersection of the lines $4x + y - 1 = 0$ and $7x - 3y - 35 = 0$ and prove that the line is equidistant from the origin and the points A, B, C, and D.

15. Find the equation of the line passing through the point (2, 3) and making intercepts of length 2 units and between the lines.

Ans.: $3x + 4y - 8 = 0$ and $x - 2 = 0$

16. If $x\cos\alpha + y\sin\alpha = p$ where $p = \frac{\sin^2\alpha}{\cos\alpha}$ be a straight line, prove that the perpendiculars p_1, p_2, and p_3 on the line from the point $(m^2, 2m)$, $(mm', m + m')$, and $(m'^2, 2m')$, respectively, are in G.P.

17. Prove that the points (a, b), (c, d), and $(a - c, b - d)$ are collinear if $(ad = bc)$. Also, show that the straight line passing through these points passes through the origin.

18. One diagonal of a square is along the line $8x - 15y = 0$ and one of its vertices is (1, 2). Find the equations of the sides of the square through this vertex.

Ans.: $2x + y = 4$ and $x - 2y + 3 = 0$

19. Find the orthocentre of a triangle formed by lines whose equations are $x + y = 1$, $2x + 8y = 6$, and $4x - y + 4 = 0$.

Ans.: $\left(\frac{-3}{7}, \frac{22}{7}\right)$

20. The sides of a triangle are $u_r = x\cos\alpha_r + y\sin\alpha_r - p_r = 0$, $r = 1, 2, 3$. Show that its orthocentre is given by $u_1\cos(\alpha_2 - \alpha_3) = u_2\cos(\alpha_3 - \alpha_1) = u_3\cos(\alpha_1 - \alpha_2)$.

21. Find the equation of straight lines passing through the point (2, 3) and having an intercept of length 2 units between the straight lines $2x + 3y = 3$ and $2x + y = 5$.

Ans.: $x = 2$, $3x + 4y = 18$

22. Let a line L has intercepts a and b on the coordinate axes. When the axes are rotated through an angle, keeping the origin fixed, the same line L has intercepts p and q. Obtain the relation between a, b, p, and q.

Ans.: $\frac{1}{p^2} + \frac{1}{q^2} = \frac{1}{a^2} + \frac{1}{b^2}$

23. A line through the variable point $A(k + 1, 2k)$ meets the line $7x + y - 16 = 0$, $5x - y + 8 = 0$, $x - 5y + 8 = 0$ at B, C, and D, respectively. Prove that AC, AB, and AD are in G.P.

24. Find the equation of the straight lines passing through (–2, –7) and having an intercept of length 3 between the straight lines $4x + 3y = 12$ and $4x + 3y = 3$.

Ans.: $x + 2 = 0$, $7x - 24y + 182 = 0$

25. A line is such that its segment between the straight lines $5x - y - 4 = 0$ and $3x + 4y - 4 = 0$ is bisected at the point (1, 5). Obtain its equation.

Ans.: $83x - 35y + 92 = 0$.

26. Prove that the $(a - b)x + (b - c)y + (c - a) = 0$, $(a - c)x + (c - a)y + (a - b) = 0$, and $(c - a)x + (a - b)y + (b - c) = 0$ are concurrent.

27. Two vertices of a triangle are (5, –1) and (–2, 3). If the orthocentre of the triangle is at the origin, find the coordinates of the third vertex.

Ans.: (–4, –7)

28. A line intersects *x*-axis at *A*(7, 0) and *y*-axis at *B*(0, –5). A variable line *PQ* which is perpendicular to *AB* intersects *x*-axis at *P* and *y*-axis at *Q*. If *AQ* and *BP* intersect at *R*, then find the locus of *R*.

Ans.: $x^2 + y^2 - 7x + 5y = 0$

29. A rectangle *PQRS* has its side *PQ* parallel to the line $y = mx$ and vertices *P*, *Q*, and *S* on the lines $y = a$, $x = b$, and $x = -b$, respectively. Find the locus of the vertex *R*.

Ans.: $(m^2 - 1)x - my + b(m^2 + 1) + am = 0$

30. Determine the condition to be imposed on β so that (O, β) should be on or inside the triangle having sides $y + 3x + 2 = 0$, $3y - 2x - 5 = 0$, and $4y + x - 14 = 0$.

Ans.: $\frac{5}{3} \leq \beta \leq \frac{7}{2}$

31. Show that the straight lines $7x - 2y + 10 = 0$, $7x + 2y - 10 = 0$, and $y = 2$ form an isosceles triangle and find its area.

Ans.: 14 sq. units

32. The equations of the sides *BC*, *CA*, and *AB* of a triangle *ABC* are $K_r = a_r x + b_r y + c_r = 0$, $r = 1, 2, 3$. Prove that the equation of a line drawn through *A* parallel to *BC* is $K_3(a_2 b_1 - a_1 b_2) = K_2(a_3 b_1 - a_1 b_3)$.

33. The sides of a triangle *ABC* are determined by the equation $u_r = a_r x + b_r y + c_r = 0$, $r = 1, 2, 3$. Show that the coordinates of the orthocentre of the triangle *ABC* satisfy the equation $\lambda_1 u_1 = \lambda_2 u_2 + \lambda_3 u_3$ where $\lambda_1 = a_2 a_3 + b_2 b_3$, $\lambda_2 = a_3 a_1 + b_3 b_1$, and $\lambda_3 = a_1 a_2 + b_1 b_2$.

34. Prove that the two lines can be drawn through the point *P*(*P*, *Q*) so that their perpendicular distances from the point *Q*(2*a*, 2*a*) will be equal to *a* and find their equations.

Ans.: $y = a$, $4x - 3y + 3a = 0$.

35. Find the locus of a point which moves such that the square of its distance from the base of an isosceles triangle is equal to the rectangle under its distances from the other sides.

Ans.: $\left[y^2\left(\frac{1}{a^2}+\frac{1}{b^2}\right)\right]=\pm\left[\left(\frac{y}{b}-1\right)^2-\frac{\lambda^2}{a^2}\right]$

36. Prove that the lines given by $(b+c)x-bcy=a(b^2+bc+c^2)$, $(c+a)x-cay=b(c^2+ca+a^2)$, and $(a+b)x-aby=c(a^2+ab+b^2)$ are concurrent.

37. Show that the area of the triangle formed by the lines $y=m_1x+c_1$, $y=m_2x+c_2$, and $y=m_3x+c_3$ is $\frac{1}{2}\left[\frac{(c_2-c_3)^2}{m_2-m_3}+\frac{(c_3-c_1)^3}{m_3-m_1}+\frac{(c_1-c_2)^2}{m_1-m_2}\right]$.

38. Find the bisector of the acute angle between the lines $3x+4y=1$ which is the bisector containing the origin.

Ans.: $11x+3y-17=0$ (origin lies in the obtuse angle between the lines.)

39. If $a_1a_2+b_1b_2>0$ prove that the origin lies at the obtuse angle between the lines $a_1x+b_1y+c_1=0$ and $a_2x+b_2y+c_2=0$, where c_1 and c_2 both being of the same sign.

40. Find the equation to the diagonals of the parallelogram formed by the lines $ax+by+c=0$, $ax+by+d=0$, $a'x+b'y+c'=0$, $a'x+b'y-d'=0$. Show that the parallelogram will be a rhombus if $(a^2+b^2)(c'-d')^2=(a'^2+b'^2)(c-d)^2$.

41. A variable line is at a constant distance p from the origin and meets coordinate axes in A and B. Show that the locus of the centroid of the ΔOAB is $x^{-2}+y^{-2}=p^{-2}$.

42. A moving line is $lx+my+n=0$ where l, m, and n are connected by the relation $al+bm+cn=0$, and a, b, and c are constants. Show that the line passes through a fixed point.

43. Find the equation of bisector of acute angle between the lines $3x-4y+7=0$ and $12x+5y-2=0$.

44. Q is any point on the line $x-a=0$ and O is the origin. If A is the point $(a, 0)$ and QR, the bisector $\angle OQA$ meets x-axis on R. Show that the locus of the foot of the perpendicular from R to OQ is the $(x-2a)(x^2+y^2+a^2x)=0$.

45. The lines $ax+by+c=0$, $bx+cy+a=0$, and $cx+ay+b=0$ are concurrent where a, b, and c are the sides of the ΔABC in usual notation and prove that $\sin^3 A+\sin^3 B+\sin^3 C=3\sin A\sin B\sin C$.

46. A variable straight line OPQ passes through the fixed point O, meeting the two fixed lines in points P and Q. In the straight line OPQ, a point R is taken such that OP, OR, and OQ are in harmonic progression. Show that the locus of point Q is a straight line.

47. A ray of light is set along the line $x - 2y - 3 = 0$. On reaching the line $3x - 2y - 5 = 0$, the ray is reflected from it. Find the equation of the line containing the reflected ray $2qx - 2y - 31 = 0$.

Ans.: $2qx - 2y - 31 = 0$.

48. Let ΔABC be a triangle with $AB = AC$. If D is the midpoint of BC, and E is the foot of the perpendicular drawn from D to AC and F is the midpoint of BE. Prove that AF is perpendicular to BE.

Ans.: $14x + 23y - 40 = 0$.

49. The perpendicular bisectors of the sides AB and AC of a triangle ABC are $x - y + 5 = 0$ and $x + 2y = 0$, respectively. If the point A is $(1, -2)$, find the equation of the line $14x + 23y - 40 = 0$.

50. A triangle is formed by the lines $ax + by + c = 0$, $lx + my + n = 0$, and $px + qy + r = 0$. Show that the straight line $\frac{ax + by + c}{cp + bq} = \frac{lx + my + n}{lp + mq}$ passes through the orthocentre of the triangle.

51. Prove that the diagonals of the parallelogram formed by the lines $ax + by + c = 0$, $ax + by + c' = 0$, $a'x + b'y + c = 0$, and $a'x + b'y + c' = 0$ will be at right angles if $a^2 + b^2 = a'^2 + b'^2$.

52. One diagonal of a square is the portion of the line $\frac{x}{a} + \frac{y}{b} = 1$ intercepted between the axes. Show that the extremities of the other diagonal are $\left(\frac{a+b}{2}, \frac{a+b}{2}\right)$ and $\left(\frac{a-b}{2}, \frac{a-b}{2}\right)$.

53. Show that the origin lies inside a triangle whose vertices are given by the equations $7x - 5y - 11 = 0$, $8x + 3y + 31 = 0$, and $x + 3y - 19 = 0$.

54. A ray of light travelling along the line OA, O being the origin, is reflected by the line mirror $x - y + 1 = 0$, the point of incidence A is $(1, 2)$. The reflected ray is again reflected by the mirror $x - y = 1$, the point of incidence being B. If the reflected ray moves along BC, find the equation of BC.

Ans.: $2x - y - 6 = 0$

55. If the lines $p_1x + q_1y = 1$, $p_2x + q_2y = 1$, and $p_3x + q_3y = 1$ are concurrent, prove that the points (p_1, q_1), (p_2, q_2), and (p_3, q_3) are collinear.

56. If p, q, and r be the length of the perpendiculars from the vertices A, B, and C of a triangle on any straight line, prove that $a^2(p-q)(p-r) + b^2(q-r)(q-p) + c^2(r-p)(r-q) = 4\Delta^2$.

57. Prove that the area of the parallelogram formed by the straight line $a_1x + b_1y + c_1 = 0$, $a_1x + b_1y + d_1 = 0$, $a_2x + b_2y + c_2 = 0$, and $a_2x + b_2y + d_2 = 0$ is $\left|\frac{(d_1 - c_1)(d_2 - c_2)}{a_1b_2 - a_2b_1}\right|$.

58. A ray of light is sent along the line $2x - 3y = 5$. After refracting across the line $x + y = 1$, it enters the opposite sides after turning by 15° away from the line $x + y = 1$. Find the equation of the line along which the refracted ray travels

Ans.: $(15\sqrt{3} - 20)x - (30 - 10\sqrt{3})y + (11 - 18\sqrt{3}) = 0.$

59. Two sides of an isosceles triangle are given by the equations $7x - y + 3 = 0$ and $x + y - 7 = 0$ and its third side passes through the point $(1, -10)$. Determine the equation of the third side.

Ans.: $x - 3y - 31 = 0,\ 3x + y + 7 = 0.$

60. Find all those points on the line $x + y = 4$ which are at c unit distance from the line $4x + 3y = 10$.

61. Are the points (3, 4) and (2, –6) on the same or opposite sides of the line $3x - 4y = 8$?

Ans.: opposite sides

62. How many circles can be drawn each touching all the three lines $x + y = 1, y = x,$ and $7x - y - 6$? Find the centre and radius of one of the circles.

Ans.: Focus: (0, 7) Incentre $\left(\frac{T}{12}, 1\right)$, $r = \frac{7}{12\sqrt{2}}$.

63. Show that $P\left(1 + \frac{t}{\sqrt{2}}, 2 + \frac{t}{\sqrt{2}}\right)$ be any point on a line then the range of values of t for which the point p lies between the parallel lines $x + 2y = 1$ and $2x + 4y = 15$ is $\left(\frac{-4\sqrt{2}}{5}, \frac{5\sqrt{2}}{6}\right)$.

64. Show that a, b, and c are any three terms of AP then the line $ax + by + c = 0$ always passes through a fixed point.

65. Show that if a, b, and c are in G.P., then the line $ax + by + c = 0$ forms a triangle with the axes, whose area is a constant.

CHAPTER

3 Pair of Straight Lines

3.1 INTRODUCTION

We know that every linear equation in x and y represents a straight line. That is $Ax + By + C = 0$, where A, B and C are constants, represents a straight line. Consider two straight lines represented by the following equations:

$$l_1x + m_1y + n_1 = 0 \tag{3.1}$$

$$l_2x + m_2y + n_2 = 0 \tag{3.2}$$

Also consider the equation

$$(l_1x + m_1y + n_1)\ (l_2x + m_2y + n_2) = 0 \tag{3.3}$$

If (x_1, y_1) is a point on the straight line given by (3.1) then

$$l_1x_1 + m_1y + n_1 = 0$$

This shows that (x_1, y_1) is also a point on the locus of (3.3). Therefore, every point on the line given by (3.1) is also a point on the locus of (3.3). Similarly, every point on the line given by (3.2) is also a point on the locus of (3.3). Therefore, (3.3) satisfies all points on the straight lines given by (3.1) and (3.2). Hence, we say (3.3) represents the combined equation of the straight lines given by (3.1) and (3.2). It is possible to rewrite (3.3) as

$$l_1l_2x^2 + (l_1m_2 + l_2m_1)xy + m_1m_2y^2 + (l_1n_2 + l_2n_1)x + (m_1n_2 + m_2n_1)y + n_1n_2 = 0 \tag{3.4}$$

(i.e.) $ax^2 + 2hxy + by^2 + 2gx + 2fy + c = 0$ (3.4) where $l_1l_2 = a$, $m_1m_2 = b$, $n_1n_2 = c$

$$l_1m_2 + l_2m_1 = 2h,\ l_1n_2 + l_2n_1 = 2g,\ m_1n_2 + m_2n_1 = 2f$$

The pair of straight lines given by (3.1) and (3.2) is in general represented in the form (3.4). However, we cannot say that every equation of this form will represent a pair of straight lines. We will find the condition that is necessary and sufficient for the equation of the form (3.4) to represent a pair of straight lines.

Before that we will see that every second degree homogeneous equation in x and y will represent a pair of straight lines.

3.2 HOMOGENEOUS EQUATION OF SECOND DEGREE IN *x* AND *y*

Every homogeneous equation of second degree in x and y represents a pair of straight lines passing through the origin.

Consider the equation $ax^2 + 2hxy + by^2 = 0, a \neq 0$.

Dividing by x^2, we get $b\left(\frac{y}{x}\right)^2 + 2h\left(\frac{y}{x}\right) + a = 0$. This is a quadratic equation in $\frac{y}{x}$ and hence there are two values for $\frac{y}{x}$, say m_1 and m_2. Then $b\left(\frac{y}{x}\right)^2 + 2h\left(\frac{y}{x}\right) + a = b\left(\frac{y}{x} - m_1\right)\left(\frac{y}{x} - m_2\right)$.

$$\text{(i.e.)} \quad b(y - m_1x)(y - m_2x) = 0.$$

But $y - m_1x = 0$ and $y - m_2x = 0$ are straight lines passing through the origin.

Therefore, $ax^2 + 2hxy + by^2 = 0$ represents a pair of straight lines passing through the origin.

Note 3.2.1: $ax^2 + 2hxy + by^2 = b(y - m_1x)\ (y - m_2x)$

Equating the coefficients of x^2 and xy, we get

$$m_1 + m_2 = \frac{-2h}{b}$$

$$m_1 m_2 = \frac{a}{b}$$

3.3 ANGLE BETWEEN THE LINES REPRESENTED BY $ax^2 + 2hxy + by^2 = 0$

Let $y - m_1x = 0$ and $y - m_2x = 0$ be the two lines represented by $ax^2 + 2hxy + by^2 = 0$.

Then $$m_1 + m_2 = \frac{-2h}{b} \tag{3.5}$$

and $$m_1 m_2 = \frac{a}{b} \tag{3.6}$$

Let θ be the angle between the lines given by $ax^2 + 2hxy + by^2 = 0$. Then the angle between the lines is given by

$$\tan\theta = \left|\frac{m_1 - m_2}{1 + m_1 m_2}\right| = \frac{\pm\sqrt{(m_1 + m_2)^2 - 4m_1 m_2}}{m_1 + m_2}$$

$$= \frac{\pm 2\sqrt{\left(\frac{-2h}{b}\right)^2 - 4\frac{a}{b}}}{1 + \frac{a}{b}} = \frac{\pm 2\sqrt{h^2 - ab}}{a + b} \quad (3.7)$$

$$\therefore \theta = \tan^{-1}\left(\frac{\pm 2\sqrt{h^2 - ab}}{a + b}\right)$$

The positive sign gives the acute angle between the lines and the negative sign gives the obtuse angle between them.

Note 3.3.1: If the lines are parallel or coincident, then $\theta = 0$. Then $\tan\theta = 0$. Therefore, from (3.7), we get $h^2 = ab$.

Note 3.1.3: If the lines are perpendicular then $\theta = \frac{\pi}{2}$ and so we get from (3.7) $\tan\frac{\pi}{2} = \frac{\pm 2\sqrt{h^2 - ab}}{a + b}$. This means $a + b = 0$. Hence, the condition for the lines to be parallel or coincident is $h^2 = ab$ and the condition for the lines to be perpendicular is $a + b = 0$ (i.e.) Coefficient of x^2 + Coefficient of $y^2 = 0$.

3.4 EQUATION FOR THE BISECTOR OF THE ANGLES BETWEEN THE LINES GIVEN BY $ax^2 + 2hxy + by^2 = 0$

We will now derive the equation for the bisector of the angles between the lines given by $ax^2 + 2hxy + by^2 = 0$. The combined equation of the bisectors of the angles between the lines given by $ax^2 + 2hxy + by^2 = 0$ is $\frac{x^2 - y^2}{a - b} = \frac{xy}{h}$.

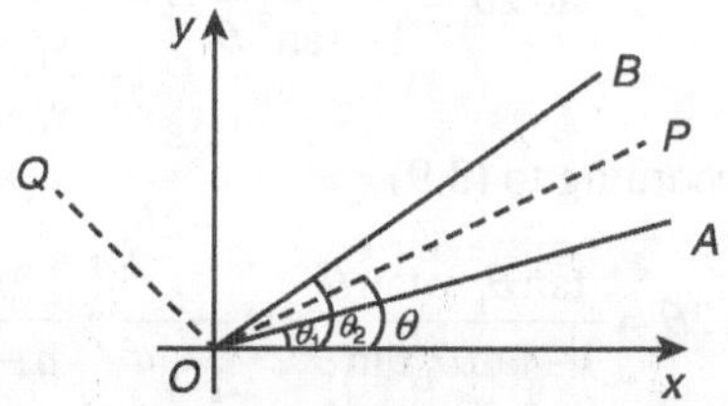

Let OA and OB be the two lines $y - m_1x = 0$ and $y - m_2x = 0$ represented by $ax^2 + 2hxy + by^2 = 0$. Let the lines OA and OB make angles θ_1 and θ_2 with the x-axis. Then, we know that

$$\tan\theta_1 + \tan\theta_2 = m_1 + m_2 = \frac{-2h}{b}$$

$$\tan\theta_1 \cdot \tan\theta_2 = m_1 m_2 = \frac{a}{b}$$

Let θ be the angle made by the internal bisector OP with OX.

Then $\theta + \frac{\pi}{2}$ is the angle made by the external bisector OQ with OX. The combined equation of the bisectors is

$$(y - x\tan\theta)\left(y - x\tan\left(\frac{\pi}{2} + \theta\right)\right) = 0$$

$$\Rightarrow \quad (y - x\tan\theta)(y + x\cot\theta) = 0$$

$$\Rightarrow \quad y^2 - x^2 + xy(\cot\theta - \tan\theta) = 0$$

$$\Rightarrow x^2 - y^2 = xy\left(\frac{1}{\tan\theta} - \tan\theta\right) = \frac{2xy(1 - \tan^2\theta)}{2\tan\theta} = 2xy\cot\theta \qquad (3.8)$$

But $\theta = \frac{\theta_1 + \theta_2}{2}$ or $2\theta = \theta_1 + \theta_2$

$$\tan 2\theta = \tan(\theta_1 + \theta_2) = \frac{\tan\theta_1 + \tan\theta_2}{1 - \tan\theta_1\tan\theta_2} = \frac{m_1 + m_2}{1 - m_1 m_2} = \frac{\frac{-2h}{b}}{1 - \frac{a}{b}} = \frac{2h}{a - b} \qquad (3.9)$$

From (3.8) and (3.9), we get $x^2 - y^2 = 2xy\left(\frac{a - b}{h}\right)$ or $\frac{x^2 - y^2}{a - b} = \frac{xy}{h}$.

Hence, the combined equation of the pair of bisectors is $\frac{x^2 - y^2}{a - b} = \frac{xy}{h}$.

Aliter:

Let (x_1, y_1) be a point on the bisector OP. Then

$$\tan\theta = \frac{y_1}{x_1} \qquad \therefore \tan 2\theta = \frac{2\tan\theta}{1 - \tan^2\theta} = \frac{\frac{2y_1}{x_1}}{1 - \frac{y_1^2}{x_1^2}} = \frac{x_1 y_1}{x_1^2 - y_1^2} \qquad (3.10)$$

Also, $2\theta = \theta_1 + \theta_2$. According to (3.9)

$$\tan 2\theta = \frac{\tan\theta_1 + \tan\theta_2}{1 - \tan\theta_1\tan\theta_2} = \frac{\frac{-2h}{b}}{1 - \frac{a}{b}} = \frac{2h}{a - b}$$

From (3.9) and (3.10),

$$\frac{x_1y_1}{x_1^2 - y_1^2} = \frac{2h}{a-b} \text{ or } \frac{x_1^2 - y_1^2}{a-b} = \frac{x_1y_1}{h}$$

The locus of (x_1, y_1) is $\dfrac{x^2 - y^2}{a-b} = \dfrac{xy}{h}$.

This is the combined equation of the bisectors.

3.5 CONDITION FOR GENERAL EQUATION OF A SECOND DEGREE EQUATION TO REPRESENT A PAIR OF STRAIGHT LINES

We will now derive the condition for the general equation of a second degree equation to represent a pair of straight lines. The condition for the general equation of the second degree $ax^2 + 2hxy + by^2 + 2gx + 2fy + c = 0$ to represent a pair of straight lines is $abc + 2fgh - af^2 - bg^2 - ch^2 = 0$.

Method 1:

Consider the general equation of the second degree

$$ax^2 + 2hxy + by^2 + 2gx + 2fy + c = 0 \tag{3.11}$$

Let $lx + my + n = 0$ and $l_1x + m_1y + n_1 = 0$ be the equations of two lines represented by (3.11). Then

$$ax^2 + 2hxy + by^2 + 2gx + 2fy + c = (lx + my + n)(l_1x + m_1y + n_1)$$

Comparing the coefficients, we get

$$\begin{aligned} ll_1 &= a & lm_1 + l_1m &= 2h \\ mm_1 &= b & ln_1 + l_1n &= 2g \\ nn_1 &= c & mn_1 + m_1n &= 2f \end{aligned} \tag{3.12}$$

We know that

$$\begin{vmatrix} l & l_1 & 0 \\ m & m_1 & 0 \\ n & n_1 & 0 \end{vmatrix} \times \begin{vmatrix} l_1 & l & 0 \\ m_1 & m & 0 \\ n_1 & n & 0 \end{vmatrix} = 0$$

By multiplying the two determinants, we get

$$\begin{vmatrix} 2ll_1 & lm_1 + l_1m & ln_1 + l_1n \\ l_1m + lm_1 & 2mm_1 & mn_1 + nm_1 \\ l_1n + n_1l & m_1n + nm_1 & 2nn_1 \end{vmatrix} = 0 \tag{3.13}$$

Substituting the values from (3.12) in (3.13), we get

$$\begin{vmatrix} 2a & 2h & 2g \\ 2h & 2b & 2f \\ 2g & 2f & 2c \end{vmatrix} = 0$$

(i.e.) $$\begin{vmatrix} a & h & g \\ h & b & f \\ g & f & c \end{vmatrix} = 0$$

Expanding the determinant, we get

$$a(bc - f^2) - h(hc - gf) + g(hf - bg) = 0$$

$$\Rightarrow abc - af^2 - ch^2 + ghf + ghf - bg^2 = 0$$

$$\Rightarrow abc + 2fgh - af^2 - bg^2 - ch^2 = 0$$

This is the required condition.

Method 2:

$$ax^2 + 2hxy + by^2 + 2gx + 2fy + c = 0$$

Writing this equation in the form $by^2 + 2hxy + 2fy + (ax^2 + 2gx + c) = 0$ and solving for y we get

$$y = \frac{(-2hx - 2f) \pm \sqrt{4(f + hx)^2 - 4b(ax^2 + 2gx + c)}}{2b}$$

$$y = \frac{-(hx + f) \pm \sqrt{(h^2 - ab)x^2 + 2(hf - bg)x + (f^2 - bc)}}{b} \qquad (3.14)$$

This equation will represent two straight lines if the quadratic expression under the radical sign is a perfect square. The condition for this is $4(hf - bg)^2 - 4(h^2 - ab)(f^2 - bc) = 0$

(i.e.) $$h^2f^2 + b^2g^2 - 2hfbg - (h^2f^2 - ab^2c - abf^2 - h^2bc) = 0$$

(i.e.) $$b(abc + 2fgh - af^2 - bg^2 - ch^2) = 0$$

Since $b \neq 0$,

$$abc + 2fgh - af^2 - bg^2 - ch^2 = 0$$

This is the required condition.

Method 3:
Let the equation $ax^2 + 2hxy + by^2 + 2gx + 2fy + c = 0$ represent a pair of straight lines and let (x_1, y_1) be their point of intersection. Shifting the origin to the point (x_1, y_1), we get

$$\begin{aligned} &a(X + x_1)^2 + 2h(X + x_1)(Y + y_1)^2 + b(Y + y_1) + \\ &2g(X + x_1) + 2f(Y + y_1) + c = 0 \end{aligned} \quad (3.15)$$

where the new axes OX and OY are parallel to (Ox, Oy).

$$\begin{aligned} &aX^2 + 2hXY + bY^2 + 2X(ax_1 + hy_1 + g) + 2Y(hx_1 + by_1 + f) + \\ &ax_1^2 + 2hx_1y_1 + 2gx_1 + 2fy_1 + c = 0 \end{aligned} \quad (3.16)$$

As (3.16) represents a pair of straight lines passing through the new origin, it has to be a homogeneous equation in X and Y. Hence,

$$ax_1 + hy_1 + g = 0 \quad (3.17)$$

$$hx_1 + by_1 + f = 0 \quad (3.18)$$

$$ax_1^2 + 2hx_1y_1 + 2gx_1 + 2fy_1 + c = 0 \quad (3.19)$$

$$\text{(i.e.)} \quad (ax_1 + hy_1 + g)x_1 + (hx_1 + hy_1 + f)y_1 + gx_1 + fy_1 + c = 0 \quad (3.20)$$

Substituting (3.17) and (3.18) in (3.20), we get

$$gx_1 + fy_1 + c = 0 \quad (3.21)$$

Eliminating x_1 and y_1 from (3.17), (3.18) and (3.21), we get

$$\begin{vmatrix} a & h & g \\ h & b & f \\ g & f & c \end{vmatrix} = 0$$

Expanding, we get $abc + 2fgh - af^2 - bg^2 - ch^2 = 0$.

Note 3.5.1: Solving (3.17) and (3.18), we get

$$ax_1 + hy_1 + g = 0$$

$$hx_1 + by_1 + f = 0$$

$$\Rightarrow \frac{a}{hf - bg} = \frac{b}{gh - af} = \frac{1}{ab - h^2}$$

Hence, the point of intersection of the lines represented by (3.11) is $\left(\frac{hf-bg}{ab-h^2}, \frac{gh-af}{ab-h^2}\right)$.

Note 3.5.2: If $lx+my+n=0$ and $l_1x+m_1y+n_1=0$ are the two straight lines represented by (3.11), then $lx+my=0$ and $l_1x+m_1y=0$ will represent two straight lines parallel to the lines represented by (3.11) and passing through the origin. Their combined equation is

$$(lx+my)(l_1x+m_1y)=0$$

$$\text{(i.e.)}\quad ll_1x^2+(lm_1+l_1m)xy+mn_1y^2=0$$

$$\text{(i.e.)}\quad ax^2+2hxy+by^2=0$$

Therefore, if $ax^2+2hxy+by^2+2gx+2fy+c=0$ represents a pair of straight lines, then the equation $ax^2+2hxy+by^2=0$ will represent a pair of lines parallel to the lines given by (3.11).

We know that every homogeneous equation of second degree in x and y represents a pair of straight lines passing through the origin. We now use this idea to get the combined equation of the pair of lines joining the origin to the point of intersection of the curve $ax^2+2hxy+by^2+2gx+2fy+c=0$ and the line $lx+my=1$.

The equation of the curve $\quad ax^2+2hxy+by^2+2gx+2fy+c=0 \quad$ (3.22)

and the line $\quad lx+my=1. \quad$ (3.23)

will meet at two points say P and Q. Let (x_1, y_1) be one of the points of intersection, say P. Then

$$ax_1^2+2hx_1y_1+by_1^2+2gx_1+2fy_1+c=0 \quad (3.24)$$

and $lx_1+my_1=1$

Let us homogenise (3.22) with the help of (3.23). Then, we write

$$ax^2+2hxy+by^2+(2gx+2fy)(lx+my)+c(lx+my)^2=0 \quad (3.25)$$

If we substitute $x=x_1$ and $y=y_1$ in (3.25), we get

$$ax_1^2+2hx_1y_1+by_1^2+(2gx_1+2fy_1)(lx_1+my_1)+c(lx_1+my_1)^2=0$$

because of (3.23) and (3.24).

Therefore $P(x_1, y_1)$ lies on the locus of (3.25). Similarly we can show that the point $Q(x_2, y_2)$ also lies on the locus of (3.25). However, a second degree homogeneous equation represents a pair of straight lines passing through origin. Hence, (3.25) is the combined equation of the pair of lines OP and OQ.

Hence, homogensing the second degree equation (3.22) with the help of (3.23), we get a pair of straight lines passing through the origin.

ILLUSTRATIVE EXAMPLES

Example 3.1

The gradient of one of the lines $ax^2 + 2hxy + by^2 = 0$ is twice that of the other. Show that $8h^2 = 9ab$.

Solution

The equation $ax^2 + 2hxy + by^2 = 0$ represents a pair of straight lines passing through the origin. Let the lines be $y - m_1x = 0$ and $y - m_2x = 0$. Then

$$ax^2 + 2hxy + by^2 = b(y - m_1x)(y = m_2x)$$

Equating the coefficients of xy and x^2 on both sides, we get

$$m_1 + m_2 = \frac{-2h}{b} \tag{3.26}$$

$$m_1m_2 = \frac{a}{b} \tag{3.27}$$

Here, it has been given that $m_2 = 2m_1$.

From (3.26) and (3.27), we get $3m_1 = \dfrac{-2h}{b}$ or $m_1 = \dfrac{-2h}{3b}$

Also $2m_1^2 = \dfrac{a}{b} \quad \Rightarrow 2\left(\dfrac{-2h}{3b}\right)^2 = \dfrac{a}{b} \Rightarrow 8h^2 = 9ab$

Example 3.2

Prove that one of the lines $ax^2 + 2hxy + by^2 = 0$ will bisect an angle between the coordinate axes if $(a + b)^2 = 4h^2$.

Solution

Let $y - m_1x = 0$ and $y - m_2x = 0$ be the two lines represented by $a^2 + 2hxy + by^2 = 0$.

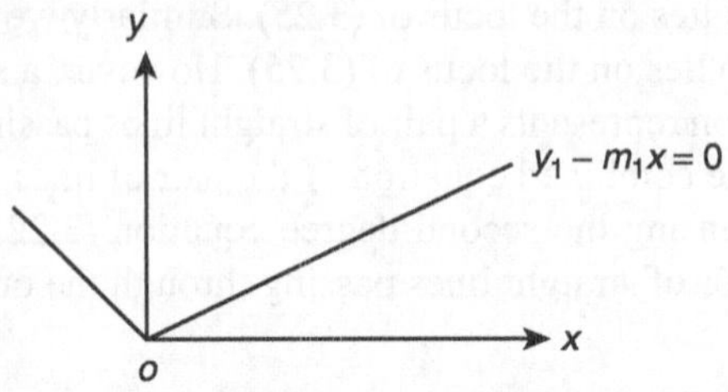

Then $m_1 + m_2 = \dfrac{-2h}{b}$, $m_1m_2 = \dfrac{a}{b}$

Since one of the lines bisects the angle between the axes, we take $m_1 = \pm 1$. Then

$$m_1m_2 = \frac{a}{b} \Rightarrow m_2 = \pm\frac{a}{b}$$

$$m_1 + m_2 = \frac{-2h}{b} \Rightarrow \pm 1 \pm \frac{a}{b} = \frac{2h}{b} \Rightarrow \pm(a+b) = 2h$$

$$\text{(i.e.)} \quad (a+b)^2 = 4h^2$$

Example 3.3

Find the centroid of the triangle formed by the lines given by the equations $12x^2 - 20xy + 7y^2 = 0$ and $2x - 3y + 4 = 0$.

Solution

$$12x^2 - 20xy + 7y^2 = 0 \Rightarrow 12x^2 - 6xy - 14xy - 7y^2 = 0$$

$$\Rightarrow 6x(2x - y) - 7y(2x - y) = 0$$

$$\Rightarrow (2x - y)(6x - 7y) = 0$$

Therefore, the sides of the triangle are represented by

$$2x - y = 0 \tag{3.28}$$

$$6x - 7y = 0 \tag{3.29}$$

$$2x - 3y + 4 = 0 \tag{3.30}$$

The point of intersection of the lines (3.28) and (3.29) is (0, 0). Let us solve (3.29) and (3.30).

$$\begin{aligned} 6x - 7y &= 0 \\ 2x - 3y &= -4 \end{aligned}$$

$$\begin{aligned} \cancel{6x} - 7y &= 0 \\ \text{(i.e.)} \quad \cancel{6x} - 9y &= -12 \\ \hline 2y = 12 &\Rightarrow y = 6 \\ \therefore x &= 7 \end{aligned}$$

Thus, the point of intersection of these two lines is (7, 6). Now, let us solve (3.28) and (3.30).

$$\begin{aligned} \cancel{2x} - y &= 0 \\ \cancel{2x} - 3y &= -4 \\ \hline 2y = 4 &\Rightarrow y = 2 \\ \therefore x &= 1 \end{aligned}$$

Thus, the point of intersection of these two lines is (1, 2). Then, the centroid of the triangle with vertices (0, 0), (7, 6) and (1, 2) is $\left(\frac{0+7+1}{3}, \frac{0+6+2}{3}\right)$ (i.e.) $\left(\frac{8}{3}, \frac{8}{3}\right)$.

Example 3.4

Find the product of perpendiculars drawn from the point (x_1, y_1) on the lines $ax^2 + 2hxy + by^2 = 0$.

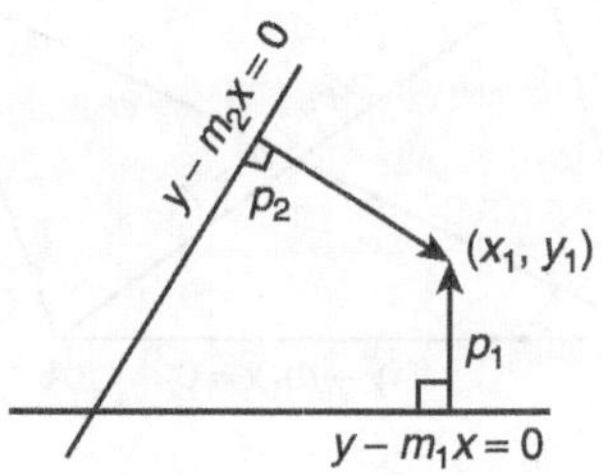

Solution

Let the lines be $y - m_1 x = 0$ and $y - m_2 x = 0$. Then

$$m_1 + m_2 = \frac{-2h}{b}; \; m_1 m_2 = \frac{a}{b}$$

Let p_1 and p_2 be the perpendicular distances from (x_1, y_1) on the two lines $y - m_1x = 0$ and $y - m_2x = 0$, respectively.

Then $$p_1 = \frac{|y_1 - m_1x_1|}{\sqrt{1+m_1^2}} \text{ and } p_2 = \frac{|y_1 - m_2x_1|}{\sqrt{1+m_2^2}}$$

Then $$p_1p_2 = \frac{|(y_1 - m_1x_1)(y_1 - m_2x_1)|}{\sqrt{1+m_1^2}\sqrt{1+m_2^2}} = \frac{|y_1^2 - (m_1+m_2)xy + m_1m_2x_1^2|}{\sqrt{1+(m_1^2+m_2^2)+(m_1m_2)^2}}$$

$$= \frac{\left|y_1^2 + \frac{2h}{b}x_1y_1 + \frac{a}{b}x_1^2\right|}{\sqrt{1+(m_1+m_2)^2 - 2m_1m_2 + (m_1m_2)^2}} = \frac{\left|y_1^2 + \frac{2h}{b}x_1y_1 + \frac{a}{b}x_1^2\right|}{\sqrt{1+\frac{4h^2}{b^2} - \frac{2a}{b} + \frac{a^2}{b^2}}}$$

$$= \frac{\sqrt{by_1^2 + 2hx_1y_1 + ax_1^2}}{\sqrt{b^2 + 4h^2 - 2ab + a^2}} = \frac{|ax_1^2 + 2hx_1y_1 + by_1^2|}{\sqrt{(a-b)^2 + 4h^2}}$$

Example 3.5

If the lines $ax^2 + 2hxy + by^2 = 0$ be the two sides of a parallelogram and the line $lx + my = 1$ be one of the diagonals, show that the equation of the other diagonal is $y(bl - hm)y = (am - h)lx$. Show that the parallelogram is a rhombus if $h(a^2 - b^2) = (a - h)lm$.

Solution

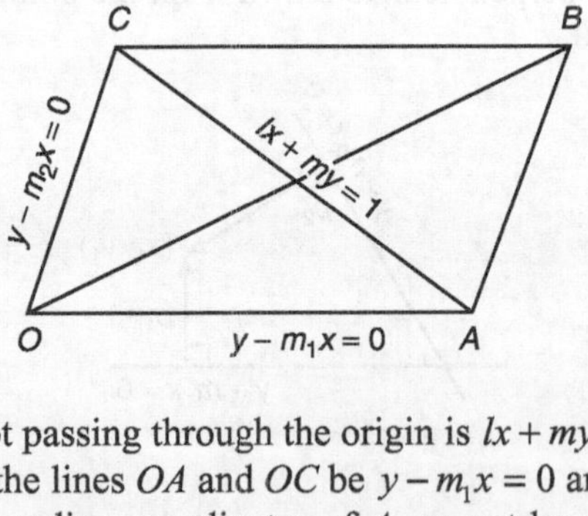

The diagonal AC not passing through the origin is $lx + my = 1$.

The equation of the lines OA and OC be $y - m_1x = 0$ and $y - m_2x = 0$.

Then the corresponding coordinates of A are got by solving $y - m_1x = 0$ and $lx + my = 1$.

$$lx + m(m_1x) = 1 \quad \text{or} \quad x = \frac{1}{l+mm_1} \quad y = \frac{m_1}{l+mm_1}$$

$$\therefore A \text{ is } \left(\frac{1}{l+mm_1}, \frac{m_1}{l+mm_1}\right)$$

Similarly, C is $\left(\frac{1}{l+mm_2}, \frac{m_2}{l+mm_2}\right)$. The midpoint of AC is

$$\left(\frac{\frac{1}{l+mm_1}+\frac{1}{l+mm_2}}{2}, \frac{\frac{m_1}{l+mm_1}+\frac{m_2}{l+mm_2}}{2}\right)$$

(i.e.) $$\left(\frac{2l+m(m_1+m_2)}{2(l+mm_1)(l+mm_2)}, \frac{2mm_1m_2+l(m_1+m_2)}{2(l+mm_1)(l+mm_2)}\right) = \text{say } (x_1, y_1)$$

Since diagonals bisect each other in a parallelogram, the equation of the diagonal OB is $\frac{y}{x} = \frac{y_1}{x_1}$

$$\frac{y}{x} = \frac{l(m_1+m_2)+2mm_1m_2}{2l+m(m_1+m_2)} = \frac{l\left(\frac{-2h}{b}\right)+2m\left(\frac{a}{b}\right)}{2l+m\left(\frac{-2h}{b}\right)}$$

$$= \frac{am-lh}{bl-mh} \quad \text{(i.e.)} \quad y(bl-hm) = x(am-lh)$$

If $OABC$ is a rhombus, then the diagonals are at right angles
Hence, the product of their slopes is -1.

(i.e.) $$\left(\frac{-l}{m}\right)\left(\frac{am-hl}{bl-hm}\right) = -1 \Rightarrow l(am-hl) = m(bl-hm)$$

or $$h(l^2-m^2) = (a-b)lm$$

Example 3.6

Prove that the area of the triangle formed by the lines $y = x + c$ and the straight lines $ax^2+2hxy+by^2=0$ is $c^2\left|\frac{\sqrt{h^2-ab}}{a+b+2h}\right|$.

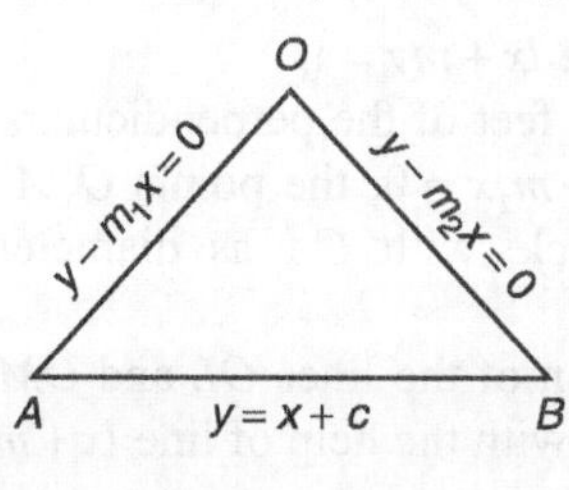

Solution

Let the two lines represented by $ax^2+2hxy+by^2=0$ be $y-m_1x=0$ and $y-m_2x=0$. Solving the equations $y-m_1x=0$ and $y=x+c$, we get the coordinates of A to be

$$A\left(\frac{c}{m_1-1}, \frac{m_1c}{m_1-1}\right). \text{ Similarly } B \text{ is } \left(\frac{c}{m_2-1}, \frac{m_2c}{m_2-1}\right).$$

$$\text{Area of } \Delta OAB = \frac{1}{2}|(x_1y_2-x_2y_1)|$$

$$= \frac{1}{2}\left|\frac{c}{m_1-1}\cdot\frac{m_2c}{m_2-1}-\frac{c}{m_2-1}\cdot\frac{m_1c}{m_1-1}\right|$$

$$= \frac{c^2}{2}\left|\frac{m_2-m_1}{(m_1-1)(m_2-1)}\right| = \frac{c^2}{2}\left|\frac{\sqrt{(m_2+m_1)^2-4m_1m_2}}{m_1m_2-(m_1+m_2)+1}\right|$$

$$= \frac{c^2}{2}\left|\frac{\sqrt{4h^2-4ab}}{a+2h+b}\right| = c^2\left|\frac{\sqrt{h^2-ab}}{a+b+2h}\right|$$

Example 3.7

L and M are the feet of the perpendiculars from $(c, 0)$ on the lines $ax^2+2hxy+by^2=0$. Show that the equation of the line LM is $(a-b)x+2hy+bc=0$.

Solution

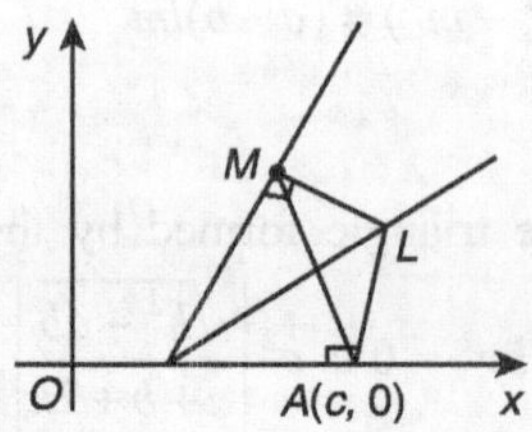

Let the equation of LM be $lx+my=1$.

Since L and M are the feet of the perpendiculars from $A(c, 0)$ on the two lines $y-m_1x=0$ and $y-m_2x=0$, the points O, A, L and M are concyclic. The equation of the circle with OA as diameter is $x(x-c)+y^2=0$ or $x^2+y^2-cx=0$.

The combined equation of the lines OL and OM is got by homogenising the equation of the circle with the help of line $lx+my=1$.

Hence, the combined equation of the lines OL and LM is

$$x^2 + y^2 - cx(lx + my) = 0$$

$$\text{(i.e.)} \quad (1 - cl)x^2 - cmxy + y^2 = 0$$

But the combined equation of the lines OL and OM is

$$ax^2 + 2hxy + by^2 = 0$$

Both these equations represent the same lines. Therefore identifying these equations, we get

$$\frac{1-cl}{a} = \frac{-cm}{2h} = \frac{1}{b} \Rightarrow 1 - cl = \frac{a}{b} \text{ or } l = \frac{b-a}{bc}$$

$$-cm = \frac{2h}{b} \text{ or } m = \frac{-2h}{bc}$$

Therefore, the line $lx + my - 1 = 0$ is $\frac{b-a}{bc}x - \frac{2h}{bc}y - 1 = 0.$

$$\text{(i.e.)} \quad (a-b)x + 2hy + bc = 0$$

Example 3.8

Show that for different values of p the centroid of the triangle formed by the straight lines $ax^2 + 2hxy + by^2 = 0$ are $x\cos\alpha + y\sin\alpha = p$ lies on the line $x(a\tan\alpha - h) + y(h\tan\alpha - b) = 0.$

Solution

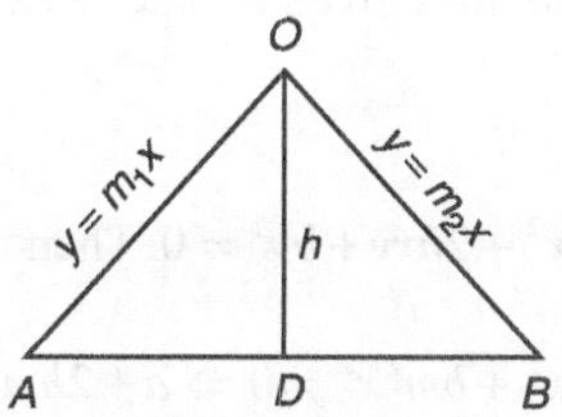

Let OA and OB be the lines represented by $ax^2 + 2hxy + by^2 = 0$ and their equations be $y - m_1x = 0$ and $y - m_2x = 0$.

The equation of the line AB is $x\cos\alpha + y\sin\alpha = p$. The coordinates of A are $\left(\frac{p}{\cos\alpha + m_1\sin\alpha}, \frac{m_1p}{\cos\alpha + m_1\sin\alpha}\right)$. The coordinates of B are $\left(\frac{p}{\cos\alpha + m_2\sin\alpha}, \frac{m_2p}{\cos\alpha + m_2\sin\alpha}\right)$.

The midpoint (x_1, y_1) of AB is

$$D\left(\frac{\dfrac{p}{\cos\alpha + m_1 \sin\alpha} + \dfrac{p}{\cos\alpha + m_2 \sin\alpha}}{2}, \frac{\dfrac{m_1 p}{\cos\alpha + m_1 \sin\alpha} + \dfrac{m_2 p}{\cos\alpha + m_2 \sin\alpha}}{2}\right).$$

The equation of OD is $\dfrac{y}{x} = \dfrac{y_1}{x_1}$.

$$\frac{y}{x} = \frac{m_1 p(\cos\alpha + m_2 \sin\alpha) + m_2 p(\cos\alpha + m_1 \sin\alpha)}{p(\cos\alpha + m_2 \sin\alpha) + p(\cos\alpha + m_1 \sin\alpha)}$$

$$= \frac{p\cos\alpha(m_1 + m_2) + 2m_1 m_2 p \sin\alpha}{2p\cos\alpha + p\sin\alpha(m_1 + m_2)} = \frac{p\cos\alpha\left(\dfrac{-2h}{b}\right) + 2p\sin\left(\dfrac{a}{b}\right)}{2p\cos\alpha + p\sin\alpha\left(\dfrac{-2h}{b}\right)}$$

$$\frac{y}{x} = \frac{-2h + 2a\tan\alpha}{2b + 2h\tan\alpha} = \frac{a\tan\alpha - h}{-(h\tan\alpha - b)}$$

$$\Rightarrow x(a\tan\alpha - h) + y(h\tan\alpha - b) = 0$$

Example 3.9

Find the condition that one of the lines given by $ax^2 + 2hxy + by^2 = 0$ may be perpendicular to one of the lines given by $a_1x^2 + 2h_1xy + by^2 = 0$.

Solution

Let $y = mx$ be a line of $ax^2 + 2hxy + by^2 = 0$. Then

$$ax^2 + 2hmx^2 + bm^2x^2 = 0 \Rightarrow a + 2hm + bm^2 = 0 \qquad (3.31)$$

Then $y = -\dfrac{1}{m}$ is a line of $a_1x^2 + 2hx_1y_1 + b_1y^2 = 0$.

Hence, $a_1x^2 - 2h\dfrac{mx_1^2}{m} + \dfrac{h_1}{m^2}x^2 = 0$ or $a_1m^2 - 2h_1m + b_1 = 0$ (3.32)

From (3.31) and (3.32), we get $\dfrac{m^2}{2hb_1 + 2h_1a} = \dfrac{m}{aa_1 - bb_1} = \dfrac{1}{-(2bh_1 + a_1h)}$.

Hence, the required condition is $(aa_1 - bb_1)^2 + 4(ha_1 + h_1b)(bh_1 + a_1h) = 0$.

Example 3.10

Two sides of a triangle lie along $y^2 - m^2x^2 = 0$ and its orthocentre is (c, d). Show that the equation of its third side is $(1-m^2)(cx+dy) = c^2 - m^2d^2$.

Solution

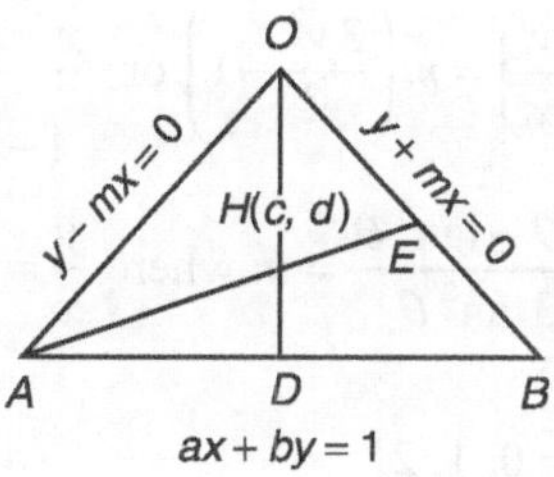

Let OA, OB and AB be the lines

$$y - mx = 0$$
$$y + mx = 0$$
$$ax + by = 1$$

Equation of OD is $bx - ay = 0$. This passes through $H(c, d)$. $\therefore bc = ad$. (1)

Equation of AH is $\quad x - my = c - md$, and $b = \dfrac{ad}{c}$ from (1) (3.33)

The coordinates of A are $\left(\dfrac{1}{a+bm}, \dfrac{m}{a+bm}\right)$

That point lies $x - my = c - md$

$$\therefore \frac{1-m^2}{a+bm} = c - md$$

$$a + bm = \frac{1-m^2}{c-md}$$

From (3.33), $\quad a = \dfrac{c(1-m^2)}{c^2 - m^2d^2}, b = \dfrac{d(1-m^2)}{c^2 - m^2d^2}$

Hence, the equation of the line AB $(ax + by = 1)$ becomes $(1 - m^2)(cx + dy) = c^2 - m^2d^2$.

Example 3.11

Show that the equation $m(x^3 - 3xy^2) + y^3 - 3x^2y = 0$ represents three straight lines equally inclined to one another.

Solution

$$y^3 - 3x^2y = m(3xy^2 - x^3)$$

Dividing by x^3, we get

$$\left(\frac{y}{x}\right)^3 - 3\left(\frac{y}{x}\right) = m\left(\frac{3y^2}{x^2} - 1\right) \text{ or } \frac{\frac{3y}{x} - \left(\frac{y}{x}\right)^3}{1 - \frac{3y^2}{x^2}} = m$$

or $$\frac{3\tan\theta - \tan^3\theta}{1 - 3\tan^2\theta} = m \text{ where } \frac{y}{x} = \tan\theta$$

$\Rightarrow \tan 3\theta = m = \tan 3\alpha$, say

$\therefore 3\theta = 3\alpha + n\pi$ where $n = 0, 1, 2$

$$\theta = \alpha + \frac{n\pi}{3}$$

$$\therefore \theta = \alpha, \alpha + \pi/3, \alpha + \frac{2\pi}{3}$$

These values of θ show that the lines are equally inclined to one another.

Example 3.12

Show that the straight lines $(A^2 - 3B^2)x^2 + 8ABx + (B^2 - 3A^2) = 0$ form with the line $Ax + By + C = 0$ an equilateral triangle of area $\frac{C^2}{(\sqrt{3}A^2 + B^2)}$.

Solution

The sides of the triangle are given by

$$(A + \sqrt{3}B)x + (B - \sqrt{3}A)y = 0 \quad (3.34)$$

$$(A - \sqrt{3}B)x + (B + \sqrt{3}A)y = 0 \quad (3.35)$$

$$Ax + by = c \quad (3.36)$$

The angle between the lines (3.34) and (3.36) is

$$\tan\theta = \frac{\pm\left[\left(\frac{A + \sqrt{3}B}{B - \sqrt{3}A}\right) - \frac{A}{B}\right]}{\left[1 + \frac{A + \sqrt{3}B}{B - \sqrt{3}A} \cdot \frac{A}{B}\right]} = \pm\sqrt{3}$$

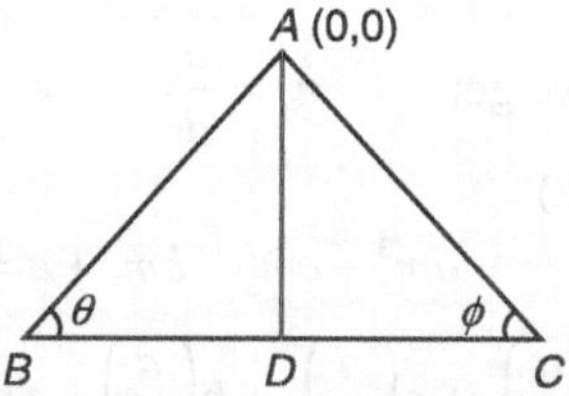

Similarly the angle between the lines (3.35) and (3.36) is $\tan\phi = \pm\sqrt{3}$

Since the three sides form a triangle, $\theta = \dfrac{\pi}{3}$ and $\phi = \dfrac{\pi}{3}$ is the only possibility. Hence, the triangle is equilateral.

$$\begin{aligned}\text{Area of triangle} &= \frac{1}{2}BC \cdot AD\\ &= \frac{1}{2}(2BD)\cdot AD\\ &= BD \cdot AD\\ &= AD \cdot AD \cdot \tan 30^\circ\\ &= \frac{AD^2}{\sqrt{3}}\\ &= \frac{C^2}{\sqrt{3}(A^2+B^2)}.\end{aligned}$$

Example 3.13

Show that two of the straight lines $ax^3 + bx^2y + cxy^2 + dy^3 = 0$ will be perpendicular to each other if $a^2 + d^2 + bd + ac = 0$.

Solution

$$ax^3 + bx^2y + cxy^2 + dy^3 = 0$$

This being a third degree homogeneous equation, it represents three straight lines passing through origin. Let the three lines be $y - m_1x = 0$, $y - m_2x = 0$ and $y - m_3x = 0$.

If m is the slope of any line then

$$dm^3 + cm^2 + bm + a = 0 \tag{3.37}$$

$$\therefore m_1m_2m_3 = \frac{-a}{d} \tag{3.38}$$

Since the two lines are perpendicular $m_1m_2 = -1$ (3.39)

From (3.38) and (3.39), we get $\quad m_3 = -\dfrac{a}{d}$

Since m_3 is a root of (3.37)

$$dm_3^3 + cm_3^2 + bm_3 + a = 0$$

$$d\left(\frac{a}{d}\right)^3 + c\left(\frac{a}{d}\right)^2 + b\left(\frac{a}{d}\right) + a = 0$$

$$\Rightarrow d^2 + bd + ca + a^2 = 0$$

Exercises

1. Show that the equation of pair of lines through the origin and perpendicular to the pair of lines $ax^2 + 2hxy + by^2 = 0$ is $bx^2 - 2hxy + ay^2 = 0$.
2. Through a point A on the x-axis, a straight line is drawn parallel to the y-axis so as to meet the pair of straight lines $ax^2 + 2hxy + by^2 = 0$ in B and C. If $AB = BC$, prove that $8h^2 = 9ab$.
3. From a point $A(1, 1)$, straight lines AL and AM are drawn at right angles to the pair of straight lines $3x^2 + 7xy - 2y^2 = 0$. Find the equation of the pair of lines AL and AM. Also find the area of the quadrilateral $ALOM$ where O is the origin of the coordinate.
4. Show that the area of the triangle formed by the lines $ax^2 + 2hxy + by^2 = 0$ and $lx + my = 1$ is $\dfrac{\sqrt{h^2 - ab}}{am^2 - 2hlm + bl^2}$.
5. Show that the orthocentre of the triangle formed by the lines $ax^2 + 2hxy + by^2 = 0$ and $lx + my = 1$ is given by $\dfrac{x}{l} = \dfrac{y}{m} = \dfrac{a+b}{am^2 - 2hlm + bl^2}$.
6. Show that the centroid (x_1, y_1) of the triangle formed by the lines $ax^2 + 2hxy + by^2 = 0$ and $lx + my = 1$ is $\dfrac{x}{hl - hm} = \dfrac{y}{am - hl} = \dfrac{2}{3(am^2 - 2hlm + bl^2)}$.
7. A triangle has the lines $ax^2 + 2hxy + by^2 = 0$ for two of its sides and the point (c, d) for its orthocentre. Prove that the equation of the third side is $(a + b)(cx + dy) = ad^2 - 2hbd + bc^2$.
8. If the slope of one of the lines given by $ax^2 + 2hxy + by^2 = 0$ is k times the other, prove that $4kh^2 = abc\ (1 + k)^2$.
9. If the distance of the point (x_1, y_1) from each of two straight lines through the origin is d, prove that the equation of the straight lines is $(x_1y - xy_1)^2 = d^2(x^2 + y^2)$.
10. A straight line of constant length $2l$ has its extremities one on each of the straight lines $ax^2 + 2hxy + by^2 = 0$. Show that the line of midpoint is $(ax + by)^2 (hx + by) + (ab - h^2)^2l^2a$.

11. Prove that the triangle formed by the lines $ax^2 + 2hxy + by^2 = 0$ and $lx + my = 1$ is right angled if $(a + b)/al^2 + 2hlm + bm^2) = 0$.
12. Show that if two of the lines $ax^3 + bx^2y + cxy^2 + dy^3 = 0$ make complementary angles with x-axis in anticlockwise direction, then $a(a - c) + d(b - d) = 0$.
13. If the slope of the lines given by $ax^2 + 2hxy + by^2 = 0$ is the square of the other, show that $ab(a + h) - 6ahb + 8h^3 = 0$.
14. Show that the line $ax + by + c = 0$ and the two lines given by $(ax + by)^2 = 3(bx - ay)^2$ form an equilateral triangle of area $\dfrac{c}{\sqrt{3}(a^2+b^2)}$.
15. If one of the line given by $ax^2 + 2hxy + by^2 = 0$ is common with one of the lines of $a_1x^2 + 2h_1xy + b_1y^2 = 0$. show that $(ab_1 - a_1b)^2 + 4(ah_1 - a_1h).(bh_1 - b_1h) = 0$.
16. A point moves so that its distance between the feet of the perpendiculars from it on the lines $ax^2 + 2hxy + by^2 = 0$ is a constant $2k$. Show that the locus of the point is $(x^2 + y^2)(h^2 - ab) = k^2[(a - b)^2 + 4h^2]$
17. Show that the distance from the origin to the orthocentre of the triangle formed by the lines $\dfrac{x}{p}+\dfrac{y}{q}=1$ and $ax^2 + 2hxy + by^2 = 0$ is $\dfrac{(a+b)pq\sqrt{p^2+q^2}}{ap^2-2hpq+bq^2}$.
18. A parallelogram is formed by the lines $ax^2 + 2hxy + by^2 = 0$ and the lines through (p, q) parallel to them. Show that the equation of the diagonal not passing through the origin is $(2x - p)(ap + hq) + (2y - q)(hp + bq) = 0$.
19. If the lines given by $lx + my = 1$ and $ax^2 + 2hxy + by^2 = 0$ form an isosceles triangle, show that $h(l^2 - m^2) = lm(a - b)$.

Example 3.14

Find λ so that the equation $x^2 + 5xy + 4y^2 + 3x + 2y + \lambda = 0$ represents a pair of lines. Find also their point of intersection and the angle between them.

Solution

Consider the second degree terms $x^2 + 5xy + 4y^2$.

$$x^2 + 5xy + 4y^2 = (x + y)(x + 4y)$$

Let the two straight lines be $x + y + l = 0$ and $x + 4y + m = 0$.

Then

$$x^2 + 5xy + 4y^2 + 3x + 2y + \lambda = (x + y + l)\,(x + 4y + m)$$

Equating the coefficients of x, y and constant terms, we get

$$m + l = 3 \tag{3.40}$$

$$m + 4l = 2 \tag{3.41}$$

$$lm = \lambda \tag{3.42}$$

Solving (3.40) and (3.41), we get $3l = -1 \Rightarrow l = \frac{-1}{3}$ $\quad \therefore m = 3 + \frac{1}{3} = \frac{10}{3}$

From (3.42), $\lambda = -\frac{10}{9}$

Then the two lines are $x + y - \frac{1}{3} = 0$, $x + 4y + \frac{10}{3} = 0$ (or) $3x + 3y - 1 = 0$ and $3x + 12y + 10 = 0$.

The angle between the lines is given by $\tan\theta = \left|\frac{m_1 - m_2}{1 + m_1 m_2}\right|$

$$\tan\theta = \left|\frac{1 - 1/4}{1 + 1.1/4}\right| = \frac{3}{5} \Rightarrow \theta = \tan^{-1}\left(\frac{3}{5}\right)$$

Example 3.15

Find the value of λ so that the equation $\lambda x^2 - 10xy + 12y^2 + 5x - 16y - 3 = 0$ represents a pair of straight lines. Find also their point of intersection.

Solution

$$\lambda x^2 - 10xy + 12y^2 + 5x - 16y - 3 = 0$$

Comparing with the equation $ax^2 + 2hxy + by^2 + 2gx + 2fy + c = 0$

we get $\quad a = \lambda,\ 2h = -10,\ b = 12,\ 2g = 5,\ 2f = -16,\ c = -3$

$$\Rightarrow a = \lambda,\ h = -5,\ b = 12,\ g = \frac{5}{2},\ f = -8,\ c = -3$$

The condition for the given equation to represent a pair of straight lines is $abc + 2fgh - af^2 - bg^2 - ch^2 = 0$

$$-36\lambda + 200 - 64\lambda - 75 + 75 = 0 \Rightarrow \lambda = 2$$

Then $2x^2 - 10xy + 12y^2 + 5x - 16y - 3 = (2x - 4y + l)(x - 3y + m)$

Equating the coefficients of x, y and constant terms,

$$2m + l = 5 \quad \text{or} \quad 4m + 2l = 10$$

$$4m + 3l = 16 \qquad\quad \underline{4m + 3l = 16}$$

$$lm = -3 \qquad\qquad\quad l = 6$$

$$\therefore m = \frac{-3}{6} = \frac{-1}{2}$$

Therefore, the two lines are $x - 2y + 3 = 0$ and $2x - 6y - 1 = 0$. Solving these two equations, we get the point of intersection as $\left(4, \frac{7}{2}\right)$.

Example 3.16

Find the value of λ so that the equation $x^2 - \lambda xy + 2y^2 + 3x - 5y + 2 = 0$ represents a pair of straight lines.

Solution

$$a=1,\ b=2,\ c=2,\ f=\frac{-5}{2},\ g=\frac{3}{2},\ h=\frac{-\lambda}{2}$$

$$abc+2fgh-af^2-bg^2-ch^2=0$$

$$\Rightarrow 4+2\left(-\frac{5}{2}\right)\times\frac{3}{2}\times\left(\frac{-\lambda}{2}\right)-1\times\frac{9}{4}-1\left(\frac{25}{4}\right)-2\left(\frac{3}{2}\right)-2\left(\frac{-\lambda^2}{4}\right)-2\times\frac{\lambda^2}{4}=0$$

$$\Rightarrow 2\lambda^2-15\lambda+27=0$$

$$\Rightarrow (2\lambda-9)(\lambda-3)=0 \qquad \Rightarrow \lambda=3,\frac{9}{2}$$

Example 3.17

Prove that the general equation of the second degree $ax^2+2hxy+by^2+2gx+2fy+c=0$ represents parallel straight lines if $h^2=ab$ and $bg^2=af^2$. Prove that the distance between the two straight lines is $2\sqrt{\frac{g^2-ac}{a(a+b)}}$.

Solution

Let the parallel lines be $lx+my+n=0$ and $lx+my+n_1=0$.

Then $ax^2+2hxy+by^2+2gx+2fy+c=(lx+my+n)\,(lx+my+n_1)$

Equating the like terms, we get

$$l^2=a \qquad ln_1+nl=2g$$

$$m^2=b \qquad mn_1+nm=2f$$

$$nn_1=c \qquad h^2=ab=l^2m^2$$

$$bg^2=m^2\left(\frac{l(n+n_1)}{2}\right)^2$$

$$\therefore bg^2=af^2$$

$$l^2\left[\frac{m(n+n_1)}{2}\right]^2=l^2f^2=af^2$$

Also, the distance between the lines $lx+my+n=0$ and $lx+my+n_1=0$ is

$$\left|\frac{n-n_1}{\sqrt{l^2+m^2}}\right|=\frac{\sqrt{(n+n_1)^2-4nn_1}}{\sqrt{l^2+m^2}}$$

$$=\frac{\sqrt{\left(\frac{2g}{l}\right)^2-4c}}{\sqrt{a+b}}=\frac{2\sqrt{g^2-ac}}{a(a+b)}$$

Example 3.18

If $ax^2 + 2hxy + by^2 + 2gx + 2fy + c = 0$ represents two straight lines equidistant from the origin, show that $f^4 - g^4 = c(bf^2 - ag^2)$.

Solution

Let the two lines represented by the given equation be $lx + my + n = 0$ and $l_1x + m_1y + n_1 = 0$. Then

$$ll_1 = a \qquad lm_1 + l_1m = 2h$$

$$mm_1 = b \qquad ln_1 + l_1n = 2g$$

$$nn_1 = c \qquad mn_1 + m_1n = 2f$$

Perpendicular distances from the origin to the two lines are equal. Therefore,

$$\left|\frac{n}{\sqrt{l^2 + m^2}}\right| = \left|\frac{n_1}{\sqrt{l^2 + m_1^2}}\right|$$

$$n^2(l_1^2 + m_1^2) = n_1^2(l^2 + m^2)$$

$$n^2l_1^2 - n_1^2l^2 = n_1^2m^2 - n^2m_1^2$$

$$(nl_1 + n_1l)(nl_1 - n_1l) = (n_1m + nm_1)(n_1m - nm_1)$$

Squaring

$$(nl_1 + n_1l)^2[(nl_1 + n_1l)^2 - 4n_1nll_1] = (n_1m + nm_1)^2\left[(n_1m + nm_1)^2 - 4mm_1nn_1\right]$$

$$\Rightarrow 4g^2(4g^2 - 4ac) = 4f^2[4f^2 - 4bc]$$

$$\Rightarrow f^4 - g^4 = c(bf^2 - ag^2)$$

Example 3.19

If the equation $ax^2 + 2hxy + by^2 + 2gx + by^2 + 2gx + 2fy + c = 0$ represents two straight lines, prove that the product of the lengths of the perpendiculars from the origin on the straight lines is $\dfrac{|c|}{\sqrt{(a-b)^2 + 4h^2}}$.

Solution

Let the two lines be $lx + my + n = 0$ and $l_1x + my + n = 0$. Therefore,

$$ll_1 = a \qquad lm_1 + l_1m = 2h$$

$$mm_1 = b \qquad ln_1 + l_1 n = 2g$$

$$nn_1 = c \qquad mn_1 + m_1 n = 2f$$

The product of the perpendiculars from the origin on these lines

$$p.p_L = \left|\frac{n}{\sqrt{l^2+m^2}}\right| \cdot \left|\frac{n_1}{\sqrt{l_1^2+m_1^2}}\right| = \left|\frac{nn_1}{\sqrt{l^2l_1^2+m^2m_1^2+l^2m_1^2+l_1^2m^2}}\right|$$

$$= \frac{|nn_1|}{\sqrt{(ll_1)^2+(mm_1)^2+(lm_1+l_1m)^2-2ll_1mm}},$$

$$= \frac{|c|}{\sqrt{a^2+b^2-2ab+4h^2}} = \frac{|c|}{\sqrt{(a-b)^2+4h^2}}$$

Example 3.20

If $ax^2 + 2hxy + by^2 + 2gx + by^2 + 2gx + 2fy + c = 0$ represents two straight lines, prove that the square of the distance of their point of intersection from the origin is $\frac{c(a+b)-f^2-g^2}{ab-h^2}$. Further, if the two given lines are perpendicular, then prove that the distance of their point of intersection from the origin is $\frac{f^4+g^4}{b^2+h^2}$.

Solution

Let the two straight lines be $lx + my + n = 0$ and $l_1x + m_1y + n_1 = 0$.

Their point of intersection is $\left(\frac{mn_1 - m_1n}{lm_1 - l_1m}, \frac{nl_1 - n_1l_1}{lm_1 - l_1m}\right)$.

Hence, the distance of this point from the origin is given by

$$d^2 = \frac{(mn_1 - m_1n)^2 + (nl_1 - n_1l)^2}{(lm_1 - l_1m)^2}$$

$$= \frac{[(mn_1 + m_1n)^2 - 4mm_1nn_1] + [(nl_1 + n_1l)^2 - 4nn_1ll_1]}{(lm_1 + l_1m)^2 - 4ll_1mm_1}$$

$$= \frac{(4f^2 - 4bc) + (4g^2 - 4ac)}{4h^2 - 4ab} = \frac{f^2 + g^2 - c(a+b)}{h^2 - ab}$$

$$= \frac{c(a+b) - f^2 - g^2}{ab - h^2}$$

If the lines are perpendicular then $(a + b) = 0$. Then

$$d^2 = \frac{0-(f^2+g^2)}{-ab-h^2} = \frac{f^2+g^2}{ab-h^2}$$

Example 3.21

Show that the lines given by $12x^2 + 7xy - 12y^2 = 0$ and $12x^2 + 7xy - 12y^2 - x + 7y - 1 = 0$ are along the sides of a square.

Solution

$$12x^2 + 7xy - 12y^2 = 0 \tag{3.42}$$

$$12x^2 + 7xy - 12y^2 - x + 7y - 1 = 0 \tag{3.43}$$

The second degree terms in (3.42) and (3.43) are the same. This implies that the two lines represented by (3.42) are parallel to the two lines represented by (3.43). Hence, these four lines from a parallelogram. Also, in each of the equations coefficient of x^2 + coefficient of $y^2 = 0$.

Hence, each equation forms a pair of perpendicular lines. Thus, the four lines form a rectangle. The two lines represented by (3.42) are $3x + 4y = 0$ and $4x - 3y = 0$. The two lines represented by (3.43) are $3x + 4y - 1 = 0$ and $4x - 3y + 1 = 0$.

The perpendicular distance between $2x + 4y = 0$ and $3x + 4y - 1 = 0$ is $\frac{1}{5}$.

The perpendicular distance between $4x - 3y = 0$ and $4x - 3y + 1 = 0$ is $\frac{1}{5}$.

Hence, the four lines form a square.

Exercises

1. Show that the equation $6x^2 + 17xy + 12y^2 + 22x + 31y + 20 = 0$ represents a pair of straight lines and find their equations.

   Ans.: $2x + 3y + 4 = 0$
   $3x + 4y + 5 = 0$

2. Prove that the equations $8x^2 + 8xy + 2y^2 + 26x + 13y + 15 = 0$ represents two parallel straight lines and find the distance between them.

   Ans.: $\frac{7\sqrt{5}}{2}$

3. Prove that the equation $3x^2 + 8xy - 7y^2 + 21x - 3y + 18 = 0$ represents two lines. Find their point of intersection and the angle between them.

   Ans.: $\left(\frac{-3}{2}, \frac{-5}{2}\right), \frac{\pi}{2}$

4. If $ax^2 + 2hxy + by^2 + 2gx + 2fy + c = 0$ and $ax^2 + 2hxy + by^2 - 2gx - 2fy + c = 0$ each represents a pair of lines, prove that the area of the parallelogram enclosed is $\frac{2c}{\sqrt{h^2 - ab}}$.

5. Show that the equation $3x^2 + 10xy + 8y^2 + 14x - 22y + 15 = 0$ represents two straight lines intersecting at an angle $\tan^{-1}\left(\frac{2}{11}\right)$.

6. The equation $ax^2 - 2xy - 2y^2 - 5x + 5y + c = 0$ represents two straight lines perpendicular to each other. Find a and c.

Ans.: $a = 2, c = -3$

7. Find the distance between the parallel lines given by $4x^2 + 12xy + 9y^2 - 6x - 9y + 1 = 0$.

Ans.: $\frac{3}{2}$

8. Show that the four lines $2x^2 + 3xy - 2y^2 = 0$ and $2x^2 + 3xy - 2y^2 - 3x + y + 1 = 0$ form a square.

9. Show that the straight lines represented by $ax^2 + 2hxy + by^2 = 0$ and those represented by $ax^2 + 2hxy + by^2 + 2gx + 2fy + c = 0$ form a rhombus, if $(c - h) fg + h(f^2 - g^2) = 0$.

10. If the equation $ax^2 + 2hxy + by^2 + 2gx + 2fy + c = 0$ represents two straight lines and parallel lines to these two lines are drawn through the origin then show that the area of the parallelogram so formed is $\frac{|c|}{2\sqrt{h^2 - ab}}$.

11. If the straight lines given by $ax^2 + 2hxy + by^2 + 2gx + 2fy + c = 0$ intersects on the y-axis then show that $2fgh - hg^2 - ch^2 = 0$.

12. A parallelogram is such that two of its adjacent sides are along the lines $ax^2 + 2hxy + by^2 = 0$ and its centre is (a, b). Find the equation of the other two sides.

Ans.: $a(x - 2a) + 2h(x - 2a)(y - 2b) + b(y - 2b)^2 = 0$

Example 3.22

Show that the pair of lines given by $(a-b)(x^2 - y^2) + 4hxy = 0$ and the pair of lines given by $h(x^2 - y^2) = (a-b)xy$ are such that each pair bisects the angle between the other pairs.

Solution

$$(a-b)(x^2 - y^2) + 4hxy = 0 \tag{3.44}$$

$$h(x^2 - y^2) - (a-b)xy = 0 \tag{3.45}$$

The combined equation of the bisectors of the pair of lines given by (3.44) is

$$\frac{x^2-y^2}{(a-b)+(a-b)}=\frac{xy}{2h}$$

$$\text{(i.e.)}\quad h(x^2-y^2)=xy(a-b)$$

which is (3.45).

The combined equation of the bisectors of the angle between lines given by (3.45) is $\dfrac{x^2-y^2}{2h}=\dfrac{xy}{-\left(\dfrac{a-b}{2}\right)}$

$$\text{(i.e.)}\quad (x^2-y^2)(a-b)=-4hxy$$

$$\Rightarrow (x^2-y^2)(a-b)+4hxy=0$$

which is (3.44). Hence, each pair bisects the angle between the other.

Example 3.23

If the bisectors of the line $x^2-2pxy-y^2=0$ are $x^2-2qxy-y^2=0$ show that $pq+1=0$.

Solution

$$x^2-2pxy-y^2=0 \qquad (3.46)$$

$$x^2-2qxy-y^2=0 \qquad (3.47)$$

The combined equation of the bisectors of (3.46) is

$$\frac{x^2-y^2}{a-b}=\frac{xy}{h}\Rightarrow\frac{x^2-y^2}{1+1}=\frac{xy}{-p}$$

$$\Rightarrow -px^2+py^2=2xy$$

$$\Rightarrow px^2+2xy-py^2=0 \qquad (3.46)$$

But equation of the bisector is given by

$$x^2-2qxy-y^2=0 \qquad (3.47)$$

Comparing (3.46) and (3.47), we get $\dfrac{p}{1}=\dfrac{1}{-q}=\dfrac{p}{1}$.

$$\therefore\ pq+1=0$$

Example 3.24

Prove that if one of the lines given by the equation $ax^2+2hxy+by^2=0$ bisects the angle between the coordinate axes then $(a+b)^2=4h^2$.

Solution

The bisectors of the coordinate axes are given by $y = x$ and $y = -x$. If $y = x$ is one of the lines of $ax^2 + 2hxy + by^2 = 0$ then $ax^2 + 2hx^2 + bx^2 = 0$.

$$\text{(i.e.)} \quad a + b = -2h$$

If $y = -x$ is one of the lines of $ax^2 + 2hxy + by^2 = 0$, then $a + b = 2h$.

From these two equations, we get $(a + b)^2 = 4h^2$.

Example 3.25

Show that the line $y = mx$ bisects the angle between the lines $ax^2 + 2hxy + by^2 = 0$ if $h(1 - m^2) + m(a - b) = 0$.

Solution

The combined equation of the bisectors of the angles between the lines $ax^2 + 2hxy + by^2 = 0$ is

$$\frac{x^2 - y^2}{a - b} = \frac{xy}{-h}$$

If $y = mx$ is one of the bisectors, then it has to satisfy the above equation.

$$\text{(i.e.)} \quad \frac{x^2 - m^2x^2}{a - b} = \frac{mx^2}{-h}$$

$$\Rightarrow h(1 - m^2) + m(a - b) = 0$$

Example 3.26

Show that the pair of the lines given by $a^2x^2 + 2h(a + b)xy + b^2y^2 = 0$ is equally inclined to the pair given by $ax^2 + 2hxy + by^2 = 0$.

Solution

In order to show that the pair of lines given by $a^2x^2 + 2h(a + b)xy + b^2y^2 = 0$ is equally inclined to the pair of lines given by $ax^2 + 2hxy + by^2 = 0$, we have to show that both the pairs have the same bisectors. The combined equations of the bisectors of the first pair of lines is $\dfrac{x^2 - y^2}{a^2 - b^2} = \dfrac{xy}{h(a + b)} \Rightarrow \dfrac{x^2 - y^2}{a - b} = \dfrac{xy}{h}$, which is the combined equation of the second pair of lines.

Exercises

1. If the pair of lines $x^2 - 2axy - y^2 = 0$ bisects the angles between the lines $x^2 - 2pxy - y^2 = 0$ then show that the latter pair also bisects the angle between the former pair.

2. If one of the bisectors of $ax^2 + 2hxy + by^2 = 0$ passes through the point of intersection of the lines $ax^2 + 2hxy + by^2 + 2gx + 2fy + c = 0$ then show that $h(f^2 - g^2) + (a - b)fg = 0$.
3. If the pair of straight lines $ax^2 + 2hxy + by^2 = 0$ and $bx^2 + 2gxy + by^2 = 0$ be such that each bisects the angle between the other then prove that $hg - b = 0$.
4. Prove that the equations $6x^2 + xy - 12y^2 - 14x + 47y - 40 = 0$ and $14x^2 + xy - 4y^2 - 30x + 15y = 0$ represent two pairs of lines such that the lines of the first pair are equally inclined to those of the second pair.
5. Prove that two of the lines represented by the equation $ax^4 + bx^2y + cx^2y^2 + dxy^3 + ay^4 = 0$ will bisect the angle between the other two if $c + ba = 0$ and $b + d = 0$.

Example 3.27

If the straight lines joining the origin to the point of intersection of $3x^2 - xy + 3y^2 + 2x - 3y + 4 = 0$ and $2x + 3y = k$ are at right angles, prove that $6k^2 - 5k + 52 = 0$.

Solution

Let

$$3x^2 - xy + 3y^2 + 2x - 3y + 4 = 0 \tag{3.48}$$
$$2x + 3y = k \tag{3.49}$$

The combined equation of the lines joining the origin to the point of intersection of the lines given (3.48) and (3.49) is got by homogenising (3.48) with the help of (3.49). Hence, the combined equation of the lines joining the origin to the points of intersection of (3.48) and (3.49) is

$$3x^2 - xy + 3y^2 + (2x - 3y)\frac{(2x + 3y)}{k} + 4\left(\frac{2x + 3y}{k}\right)^2 = 0$$

(i.e.) $\quad k^2(3x^2 - xy + 3y^2) + k(4x^2 - 7y^2) + 4(4x^2 + 12xy + 9y^2) = 0$

Since the two straight lines are at right angles,

coefficient of x^2 + coefficient of $y^2 = 0$

$$\Rightarrow (3k^2 + 4k + 16) + (3k^2 - 9k + 36) = 0$$

$$\Rightarrow 6k^2 - 5k + 52 = 0$$

Example 3.28

Show that the pair of straight lines joining the origin to the point of intersection of the straight lines $y = mx + c$ and the circle $x^2 + y^2 = a^2$ are at right angles $2c^2 = a^2(1 + m^2)$.

Solution

It is given that $x^2 + y^2 = a^2$ and $y = mx + c$.

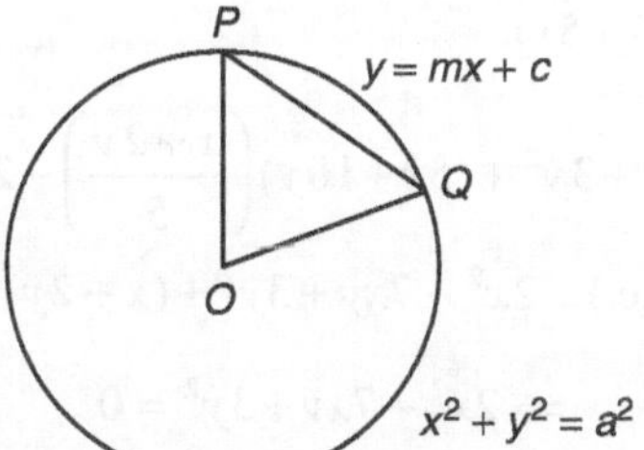

The combined equation of the lines OP and OQ is given by $x^2 + y^2 = a^2\left(\frac{y - mx}{c}\right)^2$.

$$\text{(i.e.)} \quad c^2(x^2 + y^2) = a^2(y^2 2xmy + m^2x^2)$$
$$\Rightarrow x^2(c^2 - m^2a^2) - 2ma^2xy + y^2(c^2 - a^2) = 0$$

Since OP and OQ are at right angles,
coefficient of x^2 + coefficient of $y^2 = 0$

$$c^2 - m^2a^2 + c^2 - a^2 = 0 \Rightarrow 2c^2 = a^2(1 + m^2)$$

Example 3.29

Show that the join of origin to the intersection of the lines $2x^2 - 7xy + 3y^2 + 5x + 10y - 25 = 0$ and the points at which these lines are cut by the line $x + 2y - 5 = 0$ are the vertices of a parallelogram.

Solution

$$2x^2 - 7xy + 3y^2 + 5x + 10y - 25 = 0 \quad (3.50)$$
$$x + 2y = 5 \quad (3.51)$$

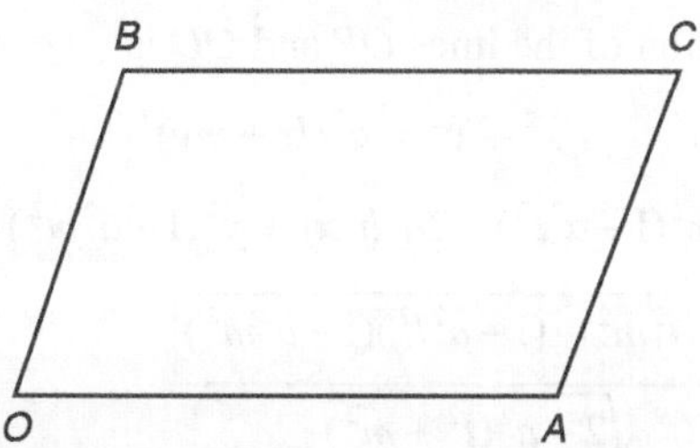

Let equation (3.50) represents the lines *CA* and *CB* and (3.51) represents the line *AB*.

The combined equation of the lines *OA* and *OB* is got by homogeniousing (3.50) with the help of (3.51).

$$\text{(i.e.)}\quad 2x^2-7xy+3y^2+(5x+10y)\left(\frac{x+2y}{5}\right)-25\left(\frac{x+2y}{5}\right)^2=0$$

$$\text{(i.e.)}\quad 2x^2-7xy+3y^2+(x+2y)^2-(x+2y)^2=0$$

$$\Rightarrow 2x^2-7xy+3y^2=0 \qquad (3.52)$$

Since the second degree terms in (3.50) and (3.52) are the same the two lines represented by (3.50) are parallel to the two lines represented by (3.52). Therefore, the four lines form a parallelogram.

Example 3.30

If the chord of the circle $x^2+y^2=a^2$ whose equation is $lx+my=1$ subtends an angle of 45° at the origin then show that $4[a^2(l^2+m^2)-1]=[a^2(l^2+m^2)-2]^2$.

Solution

It is given that,

$$x^2+y^2=a^2$$

$$lx+my=1$$

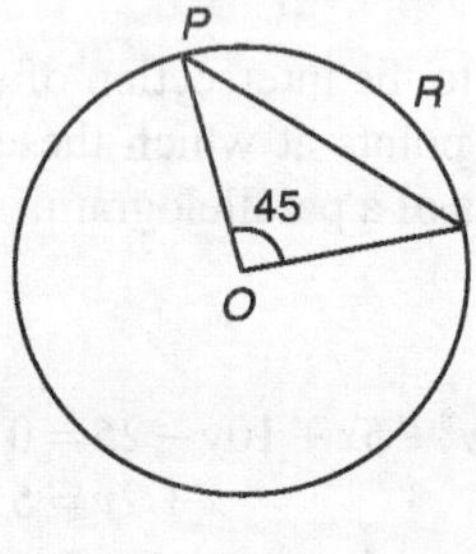

The combined equation of the lines *OP* and *OQ* is

$$x^2+y^2=a^2(lx+my)^2$$

$$\Rightarrow x^2(1-a^2l^2)-2a^2lmxy+y^2(1-a^2m^2)=0$$

$$\text{Then } \tan 45^\circ=\frac{\left|2\sqrt{a^4l^2m^2-(1-a^2l^2)(1-a^2m^2)}\right|}{\sqrt{2-a^2(l^2+m^2)}}$$

$$[a^2(l^2+m^2)-2]^2 = 4[a^4l^2m^2-(1-a^2l^2)(1-c^2m^2)]$$
$$= 4[a^4l^2m^2-1+a^2(l^2+m^2)-a^4l^2m^2]$$
$$= 4[a^2(l^2+m^2)-1]$$

Example 3.31

Find the equation to the straight lines joining the origin to the point of intersection of the straight line $\frac{x}{a}+\frac{y}{b}=1$ and the circle $5(x^2+y^2+ax+by)=9ab$ and find the conditions that the straight lines may be at right angles.

Solution

It is given that,

$$5(x^2+y^2)+5(ax+by)=9ab \quad (3.53)$$

$$\frac{x}{a}+\frac{y}{b}=1 \quad (3.54)$$

The combined equation of the lines joining the origin to the points of intersection of (3.53) and (3.54) is

$$5(x^2+y^2)+5(ax+by)\left(\frac{x}{a}+\frac{y}{b}\right)=9ab\left(\frac{x}{a}+\frac{y}{b}\right)^2$$

$$x^2\left[5+5-\frac{9b}{a}\right]+y^2\left[5+5-\frac{9a}{b}\right]+xy\left[\frac{5a}{b}+\frac{5b}{a}-18\right]=0$$

Since the lines are at right angles, coefficient of x^2 + coefficient of $y^2 = 0$

$$10-\frac{9b}{a}+10-\frac{9a}{b}=0 \Rightarrow 20ab-9b^2-9a^2=0$$
$$\Rightarrow 9(a^2+a^2)-20ab=0$$

Example 3.32

The line $lx+my=1$ meets the circle $x^2+y^2=a^2$ in P and Q. If O is the origin then show that $\underline{|POQ} = 2\cos^{-1}\left(\frac{1}{a\sqrt{l^2+m^2}}\right)$.

Solution

The perpendicular from the origin to the line $lx+my=1$ is $OR=\frac{1}{\sqrt{l^2+m^2}}$, $OP=a$

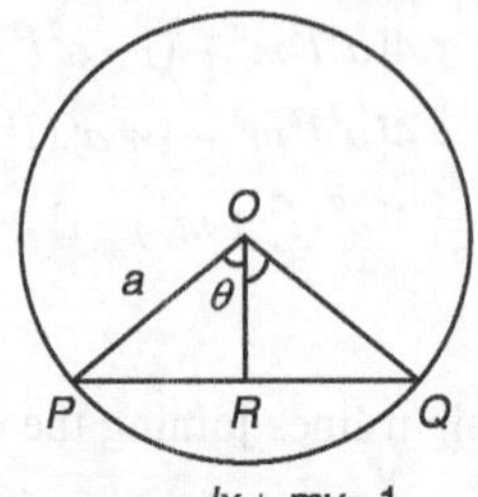

$$\angle POR = \theta \text{ then } \cos\theta = \frac{OR}{OP} = \frac{1}{a\sqrt{l^2+m^2}}$$

$$\angle POQ = 2\cos^{-1}\theta = 2\cos^{-1}\left(\frac{1}{a\sqrt{l^2+m^2}}\right)$$

Example 3.33

The straight line $y - k = m(x + 2a)$ intersects the curve $y^2 = 4a(x + a)$ in A and C. Show that the bisectors of angle $\angle AOC$, 'O' being the origin, are the same for all values of m.

Solution

$$y^2 = 4a(x + a)$$

Let

$$y^2 = 4ax + 4a^2$$

$$y - mx = 2am + k$$

The combined equation of the lines OA and OB is

$$y^2 = 4ax\frac{(y-mx)}{2am+k} + 4a^2\left(\frac{y-mx}{2am+k}\right)^2$$

(i.e.) $(2am+k)^2y^2 = (2am+k)4ax(y-mx) + 4a^2(y-mx)^2$

(i.e.) $x^2[4a^2m^2 - 4am(2am+k)] + xy[4a(2am+k) - 8a^2m] + y^2[4a^2 - (2am+k)]^2 = 0$

The combined equation of the bisectors is

$$\frac{x^2-y^2}{4a^2m^2 - 4am(2am+k) - 4a^2 + (2am+k)^2} = \frac{xy}{2a(2am+k) - 4a^2m}$$

(i.e.) $$\frac{x^2-y^2}{4a^2m^2-8a^2m^2-4amk-4a^2+4a^2m^2+4amk+k^2}=\frac{xy}{2ak}$$

$$\Rightarrow \frac{x^2y^2}{k^2-4a^2}=\frac{xy}{2ak} \text{ which is independent of } m.$$

Example 3.34

Prove that if all chords of $ax^2 + 2hxy + by^2 + 2gx + 2fy + c = 0$ subtend a right angle at the origin, then the equation must represent two straight lines at right angles through the origin.

Solution

$$ax^2 + 2hxy + by^2 + 2gx + 2fy + c = 0 \tag{3.55}$$

Let the equation of the chord be

$$lx + my = 1 \tag{3.56}$$

Let the lines (3.55) and (3.56) intersect at P and Q. The combined equation of OP and OQ is $ax^2 + 2hxy + by^2 + (2gx + 2fy)(lx + my) + c(lx + my)^2 = 0$.

Since $\underline{|OPQ} = 90°$, coefficient of x^2 + coefficient of $y^2 = 0$.

$$(a + 2gl + cl^2) + (b + 2fm + cm^2) = 0$$

Since l and m are arbitrary, coefficients of l^1, l^2, m^1, m^2 and the constant term vanish separately. Since $g = 0, f = 0, c = 0$ and $a + b = 0$.

Hence, equation (3.55) becomes $ax^2 + 2hxy + by^2 = 0$ which is a pair of perpendicular lines through the origin.

Exercises

1. Show that the line joining the origin to the points common to $3x^2 + 5xy + 3y^2 + 2x + 3y = 0$ and $3x - 2y = 1$ are at right angles.
2. If the straight lines joining the origin to the point of intersection $3x^2 - xy + 3y^2 + 2x - 3y + 4 = 0$ and $2x + 3y = k$ are at right angles then show that $6k^2 - 5k + 52 = 0$.
3. Show that all the chords of the curve $3x^2 - y^2 - 2x + y = 0$ which subtend a right angle at the origin pass through a fixed point.
4. If the curve $x^2 + y^2 + 2gx + 2fy + c = 0$ intercepts on the line $lx + my = 1$, which subtends a right angle at the origin then show that $a(l^2 + m^2) + 2(gl + fm + 1) = 0$.

5. If the straight lines joining the origin to the point of intersection of the line $kx + hy = 2hk$ with the curve $(x - h)^2 + (y - k)^2 = a^2$ are at right angles at the origin show that $h^2 + k^2 = a^2$.

6. Prove that the triangle formed by the lines $ax^2 + 2hxy + by^2 = 0$ and $lx + my = 1$ is isosceles if $(l^2 - m^2)h = (a - b)lm$.

7. Prove that the pair of lines joining the origin to the intersection of the curves $\frac{x^2}{a^2} + \frac{y^2}{b^2} = 1$ by the line $lx + my + n = 0$ are coincident if $a^2l^2 + b^2m^2 = n^2$.

8. Show that the straight lines joining the origin to the point of intersection of the curves $ax^2 + 2hxy + by^2 + 2gx = 0$ and $a_1x^2 + 2h_1xy + b_1y^2 + 2g_1x = 0$ will be at right angles if $g_1(a_1 + b_1) = g(h_1 + b_1)$.

9. Show that the angle between the lines drawn from the origin to the point of intersection of $x^2 + 2xy + y^2 + 2x + 2y - 5 = 0$ and $3x - y + 1 = 0$ is $\tan^{-1}\left(\frac{4\sqrt{6}}{13}\right)$.

CHAPTER 4

Circle

4.1 INTRODUCTION

Definition 4.1.1: A circle is the locus of a point in a plane such that its distance from a fixed point in the plane is a constant. The fixed point is called the centre of the circle and the constant distance is called the radius of the circle.

4.2 EQUATION OF A CIRCLE WHOSE CENTRE IS (*h*, *k*) AND RADIUS *r*

Let $C(h, k)$ be the centre of the circle and $P(x, y)$ be any point on the circle. $CP = r$ is the radius of the circle. $CP^2 = r^2$ (i.e.) $(x-h)^2 + (y-k)^2 = r^2$. This is the equation of the required circle.

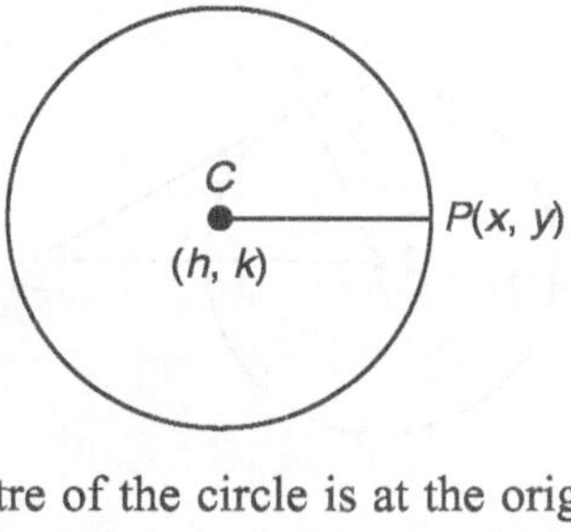

Note 4.2.1: If the centre of the circle is at the origin, then the equation of the circle is $x^2 + y^2 = r^2$.

4.3 CENTRE AND RADIUS OF A CIRCLE REPRESENTED BY THE EQUATION $x^2 + y^2 + 2gx + 2fy + c = 0$

$$x^2 + y^2 + 2gx + 2fy + c = 0$$

$$x^2 + y^2 + 2gx + 2fy = -c$$

Adding $g^2 + f^2$ to both sides, we get

$$x^2 + y^2 + 2gx + 2fy + g^2 + f^2 = g^2 + f^2 - c$$

$$\Rightarrow (x+g)^2 + (y+f)^2 = \left(\sqrt{g^2 + f^2 - c}\right)^2 \qquad (4.1)$$

This equation is of the form $(x-h)^2 + (y-k)^2 = r^2$, which is a circle with centre (h, k) and radius r. Thus, equation (4.1) represents a circle whose centre is $(-g, -f)$ and radius $\sqrt{g^2 + f^2 - c}$.

Note 4.3.1: A second degree equation in x and y will represent a circle if the coefficients of x^2 and y^2 are equal and the xy term is absent.

Note 4.3.2:

1. If $g^2+f^2 - c$ is positive, then the equation represents a real circle.
2. If $g^2+f^2 - c$ is zero, then the equation represents a point.
3. If $g^2+f^2 - c$ is negative, then the equation represents an imaginary circle.

4.4 LENGTH OF TANGENT FROM POINT $P(x_1, y_1)$ TO THE CIRCLE $x^2 + y^2 + 2gx + 2fy + c = 0$

The centre of the circle is $C(-g, -f)$ and radius $r = \sqrt{g^2 + f^2 - c}$. Let PT be the tangent from P to the circle.

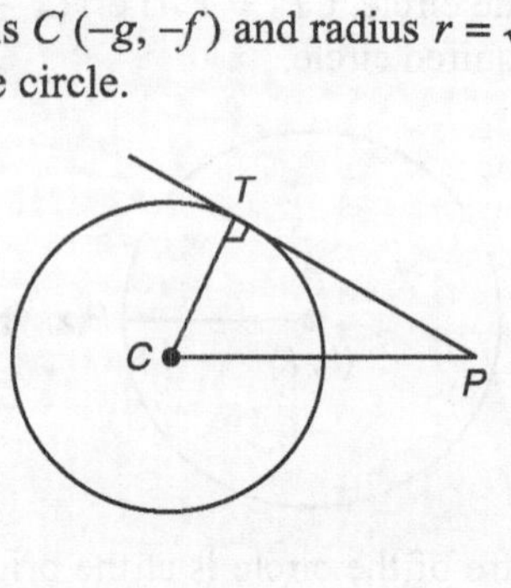

Then $PT^2 = PC^2 - r^2 = (x_1 + g)^2 + (y_1 + f)^2 - (g^2 + f^2 - c)$

$$= x_1^2 + 2gx_1 + g^2 + y_1^2 + 2fy_1 + f^2 - g^2 - f^2 + c$$

$$= x_1^2 + y_1^2 + 2gx_1 + 2fy_1 + c$$

$$PT = \sqrt{x_1^2 + y_1^2 + 2gx_1 + 2fy_1 + c}$$

Note 4.4.1:

(1) If $PT^2 > 0$ then point $P(x_1, y_1)$ lies outside the circle.

(2) If $PT^2 = 0$ then the point $P(x_1, y_1)$ lies on the circle.
(3) If $PT^2 < 0$ then point $P(x_1, y_1)$ lies inside the circle.

4.5 EQUATION OF TANGENT AT (x_1, y_1) TO THE CIRCLE $x^2 + y^2 + 2gx + 2fy + c = 0$

The centre of the circle is $(-g, -f)$. The slope of the radius $CP = \frac{y_1 + f}{x_1 + g}$.

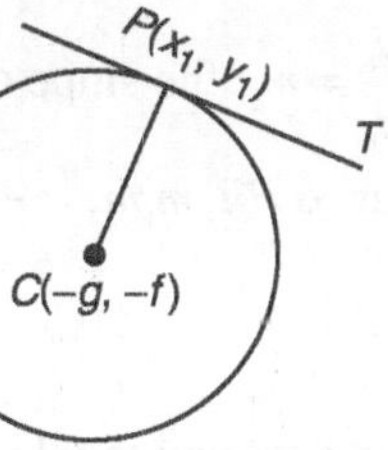

Hence, the equation of tangent at (x_1, y_1) is $(y - y_1) = m(x - x_1)$

(i.e.) $y - y_1 = \frac{-(x_1 + g)}{(y_1 + f)}(x - x_1) \Rightarrow (y - y_1)(y_1 + f) = -(x_1 + g)(x - x_1)$

(i.e. $yy_1 - y_1^2 + fy - fy_1 = -x_1x - gx + x_1^2 + gx_1$

(i.e.) $xx_1 + yy_1 + gx + fy = x_1^2 + y_1^2 + gx_1 + fy_1$

Adding $gx_1 + fy_1 + c$ to both sides,

$$xx_1 + yy_1 + g(x + x_1) + f(y + y_1) + c = x_1^2 + y_1^2 + 2gx_1 + 2fy_1 + c = 0$$

since the point (x_1, y_1) lies on the circle.

Hence, the equation of the tangent at (x_1, y_1) is $xx_1 + yy_1 + g(x + x_1) + f(y + y_1) + c = 0$.

4.6 EQUATION OF CIRCLE WITH THE LINE JOINING POINTS A (x_1, y_1) AND B (x_2, y_2) AS THE ENDS OF DIAMETER

$A(x_1, y_1)$ and $B(x_2, y_2)$ are the ends of a diameter. Let $P(x, y)$ be any point on the circumference of the circle. Then $\angle APB = 90°$ (i.e.) $AP \perp PB$.

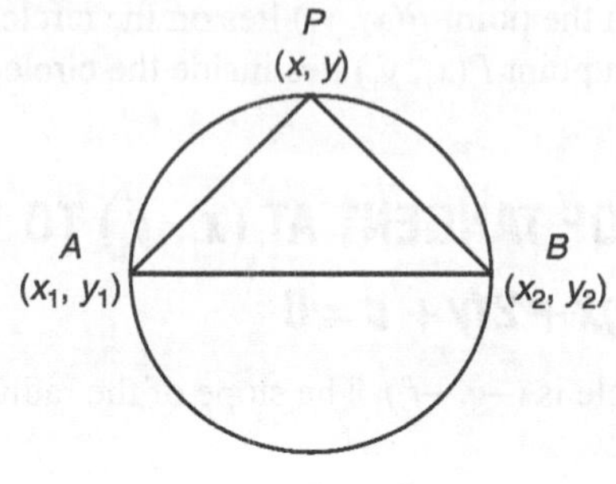

The slope of AP is $\dfrac{y-y_1}{x-x_1} = m_1$; the slope of BP is $\dfrac{y-y_2}{x-x_2} = m_2$.

Since AP is perpendicular to PB, $m_1 m_2 = -1$

(i.e.) $\dfrac{y-y_1}{x-x_1}, \dfrac{y-y_1}{x-x_2} = -1$

(i.e.) $(y-y_1)(y-y_2) = -(x-x_2)(x-x_1)$

(i.e.) $(x-x_1)(x-x_2)+(y-y_1)(y-y_2) = 0$

This is the required equation of the circle.

4.7 CONDITION FOR THE STRAIGHT LINE $y = mx + c$ TO BE A TANGENT TO THE CIRCLE $x^2 + y^2 = a^2$

Method 1:

The centre of the circle is (0, 0). The radius of the circle is a. If $y = mx + c$ is a tangent to the circle, the perpendicular distance from the centre on the straight line $y = mx + c$ is the radius of the circle.

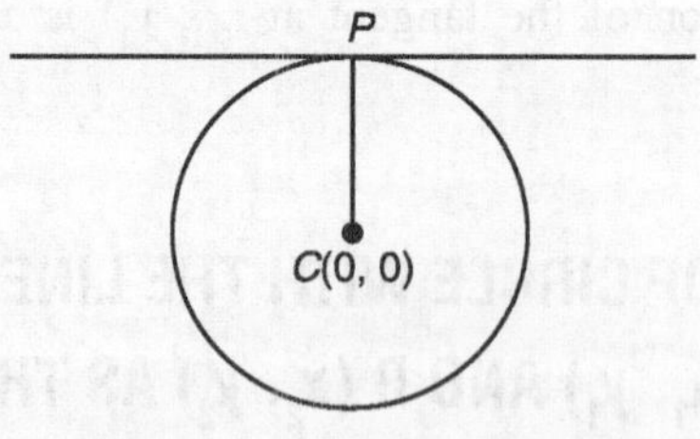

$$\therefore \frac{c}{\sqrt{1+m^2}} = \pm a. \Rightarrow c^2 = a^2\,(1+m^2)$$

This is the required condition.

Method 2:

The equation of the circle is

$$x^2 + y^2 = a^2 \tag{4.2}$$

The equation of the line is

$$y = mx + c \tag{4.3}$$

The x-coordinates of the point of intersection of circle (4.2) and line (4.3) are given by

$$x^2 + (mx + c)^2 = a^2 \text{ (i.e.) } (1 + m^2)x^2 + 2mcx + (c^2 - a^2) = 0 \tag{4.4}$$

If $y = mx + c$ is a tangent to the circle, then the two values of x given by equation (4.4) are equal. The condition for this is the discriminant of quadratic equation (4.4) is zero.

$$4m^2c^2 - 4c(1 + m^2)(c^2 - a^2) = 0$$

$$\text{(i.e.)} \quad m^2c^2 + c^2 - a^2 + m^2c^2 - ma^2 = 0 \tag{4.5}$$

$$\therefore c^2 = a^2(1 + m^2)$$

This is the required condition.

Note 4.7.1: Any tangent to the circle $x^2 + y^2 = a^2$ is of the form $y = mx + a\sqrt{1 + m^2}$.

4.8 EQUATION OF THE CHORD OF CONTACT OF TANGENTS FROM (x_1, y_1) TO THE CIRCLE $x^2 + y^2 + 2gx + 2fy + c = 0$

Let QR be the chord of contact of tangents from $P(x_1, y_1)$. Let Q and R be the points (x_2, y_2) and (x_3, y_3), respectively. The equations of tangents at Q and R are

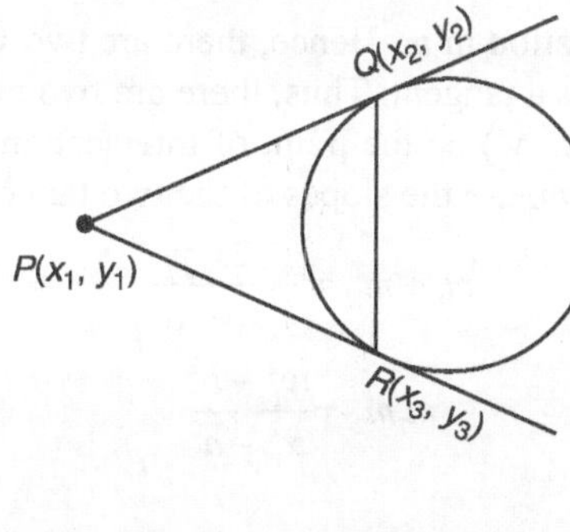

$$xx_2 + yy_2 + g(x+x_2) + f(y+y_2) + c = 0$$
$$xx_3 + yy_3 + g(x+x_3) + f(y+y_3) + c = 0$$

These two tangents pass through the point $P(x_1, y_1)$. Therefore, $x_1x_2 + y_1y_2 + g(x_1 + x_2) + f(y_1 + y_2) + c = 0$

and

$$x_1x_3 + y_1y_3 + g(x+x_3) + f(y+y_3) + c = 0$$

These two equations show that the points (x_2, y_2) and (x_3, y_3) lie on the straight line

$$xx_1 + yy_1 + g(x+x_1) + f(y+y_1) + c = 0$$

Hence, the equation of the chord of contact from (x_1, y_1) is

$$xx_1 + yy_1 + g(x+x_1) + f(y+y_1) + c = 0$$

4.9 TWO TANGENTS CAN ALWAYS BE DRAWN FROM A GIVEN POINT TO A CIRCLE AND THE LOCUS OF THE POINT OF INTERSECTION OF PERPENDICULAR TANGENTS IS A CIRCLE

Let the equation of the circle be

$$x^2 + y^2 = a^2 \tag{4.6}$$

Let (x_1, y_1) be a given point. Any tangent to the circle $x^2 + y^2 = a^2$ is $y = mx + a\sqrt{1+m^2}$. If this tangent points through (x_1, y_1), then

$$y_1 = mx_1 + a\sqrt{1+m^2} \Rightarrow (y_1 - mx_1)^2 = a^2(1+m^2)$$
$$m^2(x^2 - a^2) - 2mx_1y_1 + y_1^2 - a^2 = 0 \tag{4.7}$$

This is a quadratic equation in m. Hence, there are two values for m, and for each value of m there is a tangent. Thus, there are two tangents from a given point to a circle. Let (x_1, y_1) be the point of intersection of the two tangents from (x_1, y_1). If m_1 and m_2 are the slopes of the two tangents, then

$$m_1 + m_2 = -\frac{2x_1y_1}{x_1^2 - a^2} \tag{4.8}$$

$$m_1m_2 = \frac{y_1^2 - b^2}{x_1^2 - a^2} \tag{4.9}$$

If the two tangents are perpendicular, then $m_1m_2 = -1$.

$$\therefore \frac{y_1^2 - b^2}{x_1^2 - a^2} = -1 \Rightarrow y_1^2 - b^2 = -x_1^2 + a^2 \Rightarrow x_1^2 + y_1^2 = a^2 + b^2$$

The locus of (x_1, y_1) is $x^2 + y^2 = a^2 + b^2$, which is a circle.

4.10 POLE AND POLAR

Definition 4.10.1: The polar of a point with respect to a circle is defined to be the locus of the point of intersection of tangents at the extremities of a variable chord through that point. The point is called the pole.

4.10.1 Polar of the Point $P(x_1, y_1)$ with Respect to the Circle $x^2 + y^2 + 2gx + 2fy + c = 0$

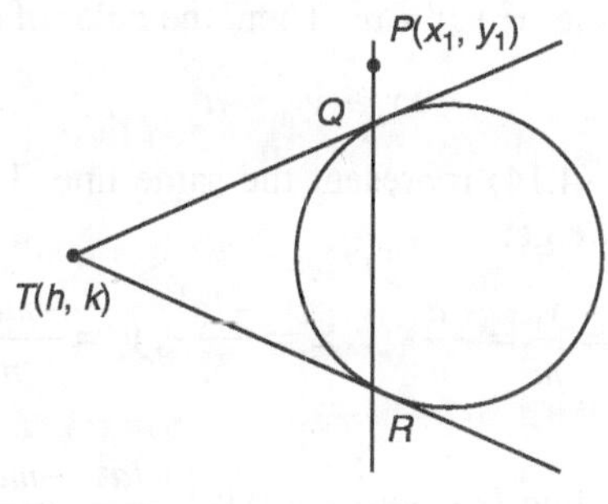

Let the equation of circle be

$$x^2 + y^2 + 2gx + 2fy + c = 0 \tag{4.10}$$

Let QR be a variable chord through the point $P(x_1, y_1)$. Let the tangents at Q and R to the circle intersect at $T(h, k)$. Then, QR is the chord of contact of the tangents from $T(h, k)$. Its equation is

$$xh + yk + g(x + h) + f(y + k) + c = 0$$

This chord passes through $P(x_1, y_1)$. Therefore,

$$x_1h + y_1k + g(x_1 + h) + f(y_1 + k) + c = 0 \tag{4.11}$$

The locus of (h, k) is

$$xx_1 + yy_1 + g(x + x_1) + f(y + y_1) + c = 0 \tag{4.12}$$

Hence, the polar of (x_1, y_1) is $xx_1 + yy_1 + g(x + x_1) + f(y + y_1) + c = 0$.

Note 4.10.1.1:

1. If the point (x_1, y_1) lies outside the circle, the polar of (x_1, y_1) is the same as the chord of contact from (x_1, y_1). If the point lies on the circle, then the tangent at (x_1, y_1) is the polar of the point $P(x_1, y_1)$.
2. The point (x_1, y_1) is called the pole of the line $xx_1 + yy_1 + g(x + x_1) + f(y + y_1) + c = 0$. Line (4.12) is called the polar of the point (x_1, y_1).
3. The polar of (x_1, y_1) with respect to the circle $x^2 + y^2 = a^2$ is $xx_1 + yy_1 = a^2$.

4.10.2 Pole of the Line $lx + my + n = 0$ with Respect to the Circle $x^2 + y^2 = a^2$

Let (x_1, y_1) be the pole of the line

$$lx + my + n = 0 \tag{4.13}$$

with respect to the circle $x^2 + y^2 = a^2$. Then, the polar of (x, y) is

$$xx_1 + yy_1 = a^2 \tag{4.14}$$

Equations (4.13) and (4.14) represent the same line. Therefore, identifying these two equations, we get

$$\frac{x_1}{l} = \frac{y_1}{m} = \frac{-a^2}{n} \quad \therefore x_1 = \frac{-la^2}{n}, y_1 = \frac{-ma^2}{n}$$

Hence, the pole of the line $lx + my + n = 0$ is $\left(\frac{-la^2}{n}, \frac{-ma^2}{n}\right)$

4.11 CONJUGATE LINES

Definition 4.11.1: Two lines are said to be conjugate with respect to the circle $x^2 + y^2 = a^2$ if the pole of either line lies on the other line.

4.11.1 Condition for the Lines $lx + my + n = 0$ and $l_1x + m_1y + n_1 = 0$ to be Conjugate Lines with Respect to the Circle $x^2 + y^2 = a^2$

The pole of the line $lx + my + n = 0$ is $\left(\frac{-la^2}{n}, \frac{-ma^2}{n}\right)$. Since the two given lines are conjugate to each other, this point lies on the line $l_1x + m_1y + n_1 = 0$.

$$\therefore l_1\left(\frac{-la^2}{n}\right) + m_1\left(\frac{-ma^2}{n}\right) + n_1 = 0 \text{ (i.e.) } ll_1a^2 + mm_1a^2 = nn_1$$

4.12 EQUATION OF A CHORD OF CIRCLE $x^2 + y^2 + 2gx + 2fy + c = 0$ IN TERMS OF ITS MIDDLE POINT

Let PQ be a chord of the circle $x^2 + y^2 + 2gx + 2fy + c = 0$ and $R(x_1, y_1)$ be its middle point.

The equation of any chord through (x_1, y_1) is

$$\frac{x - x_1}{\cos\theta} = \frac{y - y_1}{\sin\theta} = r \tag{4.15}$$

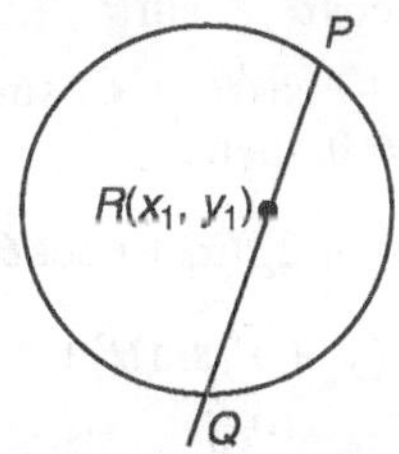

Any point on this line is $x = x_1 + r\cos\theta,\ y = y_1 + r\sin\theta$. When the chord PQ meets the circle this point lies on the circle. Therefore,

$$(x_1 + r\cos\theta)^2 + (y_1 + r\sin\theta)^2 + 2g(x_1 + r\cos\theta) + 2f(y_1 + r\sin\theta) + c = 0$$

$$\text{(i.e.) } r^2 + 2r[(x_1 + g)\cos\theta + (y_1 + f)\sin\theta] + x_1^2 + y_1^2 + 2gx_1 + 2fy_1 + c = 0$$

The values of r of this equation are the distances RP and RQ, which are equal in magnitude but opposite in sign. The condition for this is the coefficient of $r = 0$.

$$\therefore (x_1 + g)\cos\theta + (y_1 + f)\sin\theta = 0 \tag{4.16}$$

Eliminating $\cos\theta$ and $\sin\theta$, from (4.15) and (4.16), we get

$$(x_1 + g)\left(\frac{x - x_1}{r}\right) + (y_1 + f)\left(\frac{y - y_1}{r}\right) = 0$$

$$\text{(i.e.) } xx_1 + yy_1 + gx + fy = x_1^2 + y_1^2 + 2gx_1 + 2fy_1$$

Adding $gx_1 + fy_1 + c$ to both sides, we get

$$xx_1 + yy_1 + g(x + x_1) + f(y + y_1) + c = x_1^2 + y_1^2 + 2gx_1 + 2fy_1 + c$$

This is the required equation of the chord PQ in terms of its middle point (x_1, y_1). This equation can be expressed in the form $T = S_1$ where $T = xx_1 + yy_1 + g(x + x_1) + f(y + y_1) + c$ and $S_1 = x_1^2 + y_1^2 + 2gx_1 + 2fy_1 + c$.

Note 4.12.1: T is the expression we have in the equations of the tangent (x_1, y_1) to the circle S: $x^2+y^2+2gx+2fy+c=0$ and S_1 is the expression we get by substituting $x=x_1$ and $y=y_1$ in the left-hand side of $S=0$.

4.13 COMBINED EQUATION OF A PAIR OF TANGENTS FROM (x_1, y_1) TO THE CIRCLE $x^2+y^2+2gx+2fy+c=0$

Let the equation of a chord through (x_1, y_1) be

$$\frac{x-x_1}{\cos\theta}=\frac{y-y_1}{\sin\theta}=r \tag{4.17}$$

Any point on this line is $(x_1+r\cos\theta,\ y_1+r\sin\theta)$. If this point lies on the circle $x^2+y^2+2gx+2fy+c=0$, then

$$(x_1+r\cos\theta)^2+(y_1+r\sin\theta)^2+2g(x_1+r\cos\theta)+2f(y_1+r\sin\theta)+c=0$$

$$\text{(i.e.)}\quad r^2+2r[(x_1+g)\cos\theta+(y_1+f)\sin\theta]+x_1^2+y_1^2+2gx_1+2fy_1+c=0 \tag{4.18}$$

If chord (4.17) is a tangent to circle (4.18), then the two values of r of this equation are equal. The condition for this is $[(x_1+g)\cos\theta+(y_1+f)\sin\theta]^2 = [x_1^2+y_1^2+2gx_1+2fy_1+c]$.

$$\text{(i.e.)}\quad [(x_1+g)+(y_1+f)\tan\theta]^2=(x_1^2+y^2+2gx_1+2fy_1+c)\sec^2\theta$$

$$=(x_1^2+y_1^2+2gx_1+2fy_1+c)(1+\tan^2\theta) \tag{4.19}$$

But from (4.17) $\tan\theta=\dfrac{y-y_1}{x-x_1}$.

Substituting this in (4.19), we get

$$\left[(x_1+g)+(y_1+f)\left(\frac{y-y_1}{x-x_1}\right)\right]^2=(x_1^2+y_1^2+2gx_1+2fy_1+c)\left[1+\left(\frac{y-y_1}{x-x_1}\right)^2\right]$$

$$[(x_1+g)(x-x_1)+(y_1+f)(y-y_1)]^2=(x_1^2+y_1^2+2gx_1+2fy_1+c)\,[(x-x_1)^2+(y-y_1)^2]$$

$$[(xx_1+yy_1+gx+fy)-(x_1^2+y_1^2+gx_1+fy_1)]^2 = (x_1^2+y_1^2+2gx_1+2fy_1+c)[x^2+y^2-2(xx_1+yy_1)+(x_1^2+y_1^2)]$$

$$\text{(i.e.)}\quad (T-S_1)^2=S_1[S-2T+S_1]\quad\text{or}\quad T^2=SS_1$$

(i.e.) $[xx_1 + yy_1 + g(x+x_1) + f(y+y_1) + c]^2$

$$= (x^2 + y^2 + 2gx + 2gx_1 + 2fy_1 + c)(x_1^2 + y_1^2 + 2gx_1 + 2fy_1 + c)$$

This equation is the combined equation of the pair of tangents from (x_1, y_1).

4.14 PARAMETRIC FORM OF A CIRCLE

$x = a\cos\theta, y = a\sin\theta$ satisfy the equation $x^2 + y^2 = a^2$. This point is denoted by 'θ', which is called a parameter for the circle $x^2 + y^2 = a^2$.

4.14.1 Equation of the Chord Joining the Points 'θ' and 'ϕ' on the Circle and the Equation of the Tangent at θ

The two given points are $(a\cos\theta, a\sin\theta)$ and $(a\cos\phi, a\sin\phi)$. The equation of the chord joining these two points is

$$\frac{y - a\sin\theta}{x - a\cos\theta} = \frac{a(\sin\theta - \sin\phi)}{a(\cos\theta - \cos\phi)} = \frac{2\cos\frac{\theta+\phi}{2}\sin\frac{\theta-\phi}{2}}{-2\sin\frac{\theta+\phi}{2}\sin\frac{\theta-\phi}{2}}$$

$$x\cos\left(\frac{\theta+\phi}{2}\right) - a\cos\theta\cos\left(\frac{\theta+\phi}{2}\right) = -y\sin\left(\frac{\theta+\phi}{2}\right) + a\sin\theta\sin\left(\frac{\theta+\phi}{2}\right)$$

$$x\cos\left(\frac{\theta+\phi}{2}\right) + y\sin\left(\frac{\theta+\phi}{2}\right) = a\left(\cos\theta\cos\frac{\theta+\phi}{2} - \sin\theta\sin\frac{\theta+\phi}{2}\right)$$

$$= a\cos\left(\theta - \frac{\theta+\phi}{2}\right).$$

$$x\cos\left(\frac{\theta+\phi}{2}\right) + y\sin\left(\frac{\theta+\phi}{2}\right) = a\cos\left(\frac{\theta-\phi}{2}\right)$$

This chord becomes the tangent at 'θ' if $\phi = 0$. Therefore, the equation of the tangent at 'θ' is $x\cos\theta + y\sin\theta = a$.

ILLUSTRATIVE EXAMPLES

Example 4.1

Find the equation of the circle whose centre is (3, –2) and radius 3 units.

Solution

The equation of the circle is $$(x-h)^2+(y-k)^2=r^2$$

(i.e.) $$(x-3)^2+(y+2)^2=3^2$$

(i.e.) $x^2-6x+9+y^2+4y+4=9 \Rightarrow x^2+y^2-6x+4y+4=0$

Example 4.2

Find the equation of the circle whose centre is $(a, -a)$ and radius 'a'.

Solution

The centre of the circle is $(a, -a)$. The radius of the circle is a. The equation of the circle is $(x-a)^2+(y+a)^2=a^2$ (i.e.) $x^2-2ax+a^2+y^2+2ay+a^2=a^2$ (i.e.) $x^2+y^2-2ax+2ay+a^2=0$.

Example 4.3

Find the centre and radius of the following circles:

(i) $x^2+y^2-14x+6y+9=0$

(ii) $5x^2+5y^2+4x-8y-16=0$

Solution

(i) $x^2+y^2-14x+6y+9=0$

Centre is $(7, -3)$

Radius $=\sqrt{g^2+f^2=c}=\sqrt{49+9-9}=\sqrt{49}=7$ units

(ii) $5x^2+5y^2+4x-8y-16=0$

Dividing this by 5, we get

$$x^2+y^2+\frac{4}{5}x-\frac{8}{5}y-\frac{16}{5}=0$$

Centre is $\left(\frac{-2}{5}, \frac{4}{5}\right)$

Radius $=\sqrt{\frac{4}{25}+\frac{16}{25}+\frac{16}{5}}=\sqrt{\frac{100}{25}}=2$ units

Example 4.4

Find the equation of the circle whose centre is (2, –2) and which passes through the centre of the circle $x^2+y^2-6x-8y-5=0$

Solution

The centre of the required circle is (2, –2). The centre of the circle $x^2+y^2-6x-8y-5=0$ is (3, 4). The radius of the required circle is given by $r^2=(2-3)^2+(-2-4)^2=1+36=37$.

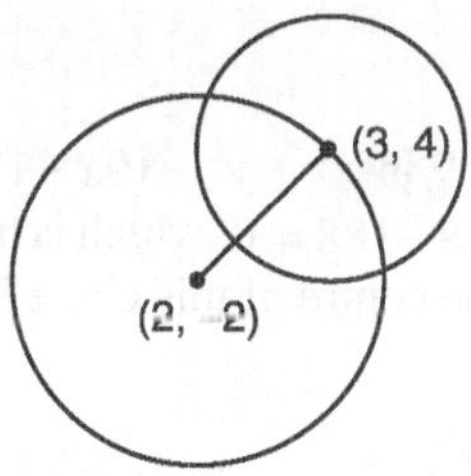

Therefore, the equation of the required circle is $(x-2)^2+(y+2)^2=37$ (i.e.) $x^2+y^2-4x+4y-29=0$

Example 4.5

Show that the line $4x-y=17$ is a diameter of the circle $x^2+y^2-8x+2y=0$.

Solution

The centre of the circle $x^2+y^2-8x+2y=0$ is (4, –1). Substituting $x=4$ and $y=-1$ in the equation $4x-y=17$, we get 16 + 1 = 17, which is true. Therefore, the line $4x-y=17$ passes through the centre of the given circle. Hence, the given line is a diameter of the circle.

Example 4.6

Prove that the centres of the circles $x^2+y^2+4y+3=0$, $x^2+y^2+6x+8y-17=0$ and $x^2+y^2-30x-16y-42=0$ are collinear.

Solution

The centres of the three given circles are $A(0,-2), B(-3,-4)$ and $C(15, 8)$.

The slope of AB is $\dfrac{-2+4}{0+3}=\dfrac{2}{3}$

The slope of BC is $\frac{-4-8}{-3-15}=\frac{-12}{-18}=\frac{2}{3}$.

Since the slopes AB and BC are equal and B is a common point, the points A, B and C are collinear.

Example 4.7

Show that the point (8, 9) lies on the circle $x^2+y^2-10x-12y+43=0$ and find the other end of the diameter through (8, 9).

Solution

Substituting $x=8$ and $y=9$ in $x^2+y^2-10x-12y+43=0$, we get $64+81-80-108+43=0$ (i.e.) $188-188=0$, which is true. Therefore, the point (8, 9) lies on the given circle. The centre of this circle is (5, 6). Let (x, y) be the other end of the diameter.

$$\text{Then } \frac{8+x}{2}=5,\quad \frac{9+y}{2}=6 \Rightarrow 8+x=10,\ 9+y=12$$

$$\Rightarrow x=2,\ y=3$$

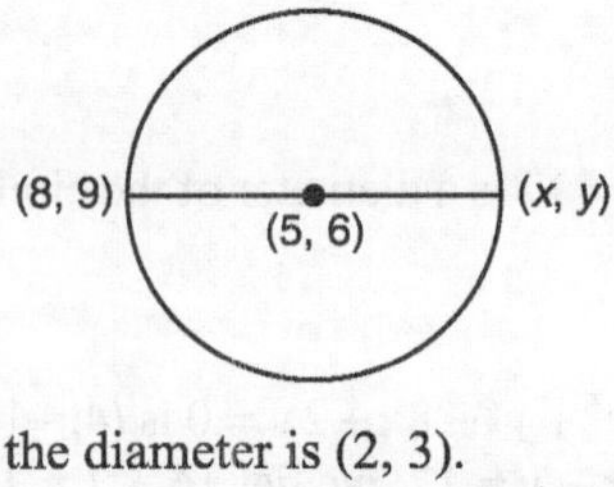

Hence, the other end of the diameter is (2, 3).

Example 4.8

Find the equation of the circle passing through the points (1, 1), (2, −1) and (3, 2).

Solution

Let the equation of the circle be $x^2+y^2+2gx+2fy+c=0$. The circle passes through the points (1, 1), (2, −1) and (3, 2).

$$\therefore 1+1+2g+2f+c=0$$

$$4+1+4g-2f+c=0$$

$$9+4+6g+4f+c=0$$

$$\text{(i.e.)}\quad 2g+2f+c=-2 \tag{4.20}$$
$$4g-2f+c=-5 \tag{4.21}$$
$$6g+4f+c=-13 \tag{4.22}$$

$$(4.20)-(4.21)\,\text{gives}-2g+4f=3 \tag{4.23}$$
$$(4.21)-(4.22)\,\text{gives}-2g-6f=8 \tag{4.24}$$
$$(4.23)-(4.24)\text{ gives } 10f=-5$$
$$\therefore f=\frac{-1}{2}$$

From equation (4.20), $-2g-2=3;\ -g=\dfrac{5}{2}$

From equation (4.20), $-5-1+c=-2 \Rightarrow c=4$

Therefore, the equation of the circle is $x^2+y^2-5x-y+4=0$.

Example 4.9

Show that the points (3, 4), (0, 5) (–3, –4) and (–5, 0) are concyclic and find the radius of the circle.

Solution

Let the equation of the circle be $x^2+y^2+2gx+2fy+c=0$. This passes through the points (3, 4), (0, 5) and (–3, –4). Therefore,

$$9+16+6g+8f+c=0,\ 0+25+0+10f+c=0,\ 9+16-6g-8f+c=0$$

(i.e.)
$$6g+8f+c=-25 \tag{4.25}$$
$$10f+c=-25 \tag{4.26}$$
$$-6g-8f+c=-25 \tag{4.27}$$

$$(4.25)-(4.26)\,\text{gives } 6g-2f=0\ \text{(ie)}\ f=3g \tag{4.28}$$
$$(4.26)-(4.27)\,\text{gives } 12g+16f=0\ \text{(ie)}\ 4f=-3g \tag{4.29}$$

$$\therefore 12g=-3g \Rightarrow 15g=0 \Rightarrow g=0$$

From equation (4.28), $f=0$
From equation (4.25), $c=25$
Hence, the equation of the circle is $x^2+y^2-25=0$ (4.30)

Substituting $x = -5$ and $y = 0$ in equation (4.30), we get $0 + 25 - 25 = 0$, which is true. Therefore, (–5, 0) also lies on the circle. Hence, the four given points are concyclic. The centre of the circle is (0, 0) and the radius is 5 units.

Example 4.10

Find the equation of the circle whose centre lies on the line $x = 2y$ and which passes through the points (–1, 2) and (3, –2).

Solution

Let the equation of the circle be $x^2 + y^2 + 2gx + 2fy + c = 0$. This passes through the points (–1, 2) and (3, –2). Therefore,

$$1 + 4 - 2g + 4f + c = 0$$
$$9 + 4 + 6g - 4f + c = 0$$

(i.e.)

$$-2g + 4f + c = -5 \quad (4.31)$$
$$6g - 4f + c = -13 \quad (4.32)$$

Subtracting, we get

$$-8g + 8f = 8 \text{ or } -g + f = 1 \quad (4.33)$$

$$(-g, -f) \text{ lies on } x = 2y$$
$$\therefore -g = -2f \text{ or } g = 2f \quad (4.34)$$

Substituting this in equation (4.33), we get

$$-2f + f = 1 \Rightarrow f = -1 \quad \therefore g = -2$$

From (4.31), $4 - 4 + c = -5 \Rightarrow c = \Rightarrow -5$

Hence, the equation of the circle is $x^2 + y^2 - 4x - 2y - 2y - 5 = 0$.

Example 4.11

Find the equation of the circle cutting off intercepts 4 and 6 on the coordinate axes and passing through the origin.

Solution

Let the equation of the circle be $x^2 + y^2 + 2gx + 2fy + c = 0$. This passes through the points (0, 0), (4, 0) and (0, 6).

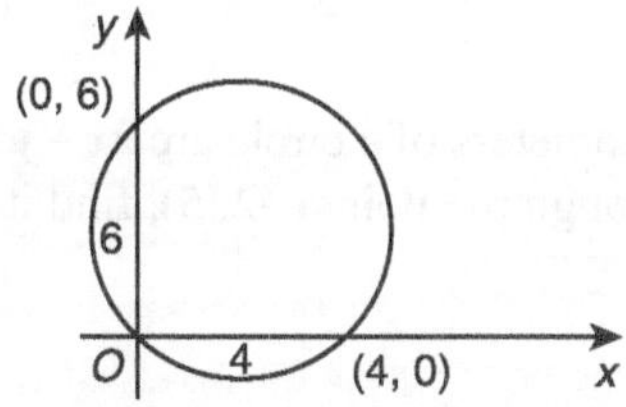

$$c = 0 \quad (4.35)$$

$$16 + 8g + c = 0 \quad (4.36)$$

$$36 + 12f + c = 0 \quad (4.37)$$

$$\therefore g = -2, \ f = -3$$

$$(4.35) - (4.36) \text{ gives } 16 + 8g = 0$$

$$(4.36) - (4.37) \text{ gives } 36 + 2f = 0$$

Thus, the equation of the circle is $x^2 + y^2 - 4x - 6y = 0$.

Example 4.12

Find the equation of the circle concentric with $x^2 + y^2 - 8x - 4y - 10 = 0$ and passing through the point (2, 3).

Solution

Two circles are said to be concentric if they have the same centre. Therefore, the equation of the concentric circle is $x^2 + y^2 - 8x - 4y + k = 0$. This circle passes through (2, 3).

$$\therefore 4 + 9 - 16 - 12 + k = 0 \quad \therefore k = 15$$

Hence, the equation of the concentric circle is $x^2 + y^2 - 8x - 4y + 15 = 0$.

Example 4.13

Find the equation of the circle on the joining the points (4, 7) and (–2, 5) as the extremities of a diameter.

Solution

The equation of the required circle is $(x - x_1)(x - x_2) + (y - y_1)(y - y_2) = 0$

$$\text{(i.e.)} \quad (x - 4)(x + 2) + (y - 7)(y - 5) = 0$$

$$\text{(i.e.)} \quad x^2 - 2x - 8 + y^2 - 12y + 35 = 0$$

$$\text{(i.e.)} \quad x^2 + y^2 - 2x - 2y + 27 = 0.$$

Example 4.14

The equation of two diameters of a circle are $2x+y-3=0$ and $x-3y+2=0$. If the circle passes through the point (–2, 5), find its equation.

Solution

The centre of the circle is the point of intersection of the diameter.

$$2x+y=3 \tag{4.38}$$

$$x-3y=-2 \tag{4.39}$$

(4.38)×3 gives $\quad 6x+3y=9$

(4.39)×1 gives $\quad x-3y=-2$

Adding these two equations, we get $7x=7$. $\quad \therefore x=1$

From (4.38), $y=1$.

Hence, the centre of the circle is (1, 1) and radius is $r=\sqrt{(1+2)^2+(1-5)^2}=\sqrt{9+16}=5$.

Therefore, the equation of the circle is

$$(x-1)^2+(y-1)^2=25$$

$$x^2+y^2-2x-2y-23=0$$

Example 4.15

Find the length of the tangent from the point (2, 3) to the circle $x^2+y^2+8x+4y+8=0$.

Solution

The length of the tangent from $P(x_1, y_1)$ to the circle $x^2+y^2+2gx+2fy+c=0$ is given by $PT^2=x_1^2+y_1^2+2gx_1+2fy_1+c$. Here, the length of the tangent from $P(2, 3)$ to the given circle is $PT=\sqrt{4+9+16+12+8}=\sqrt{49}=7$ units.

Example 4.16

Determine whether the following points lie outside, on or inside the circle $x^2+y^2-4x+4y-8=0$: $A(0,1), B(5,9), C(-2,3)$.

Solution

The equation of the circle is $x^2+y^2-4x+4y-8=0$.

$$AT_1^2=0+1-0+4-8=-3<0$$

$$BT_2^2 = 25+49-20+28-8 = 64 > 0$$
$$CT_3^2 = 4+9+8+12-8 = 25 > 0$$

Therefore, point A lies inside the circle. Points B and C lie outside the circle.

Example 4.17

Find the equation of the tangent at the point (2, –5) on the circle $x^2 + y^2 - 5x + y - 14 = 0$.

Solution

Given $x^2 + y^2 - 5x + y - 14 = 0$

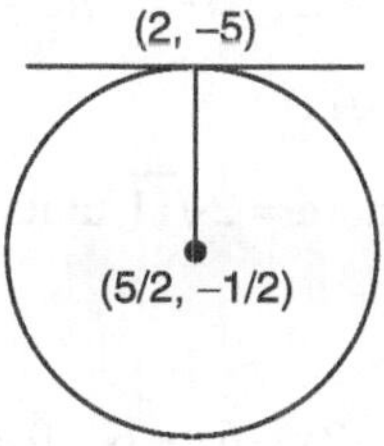

Centre is $\left(\frac{5}{2}, \frac{-1}{2}\right)$.

$$\text{Slope of the radius} = \frac{-5+\frac{1}{2}}{2-\frac{5}{2}} = 9$$

$$\text{Slope of the tangent} = \frac{-1}{9}$$

Therefore, the equation of the tangent is

$$y - y_1 = m(x - x_1) \text{ (i.e.) } y+5 = \frac{-1}{9}(x-2)$$
$$\Rightarrow 9y+45 = -x+2 \text{ or } x+9y+43 = 0$$

Example 4.18

Find the length of the chord of the circle $x^2 + y^2 - 4x - 6y - 3 = 0$ given that (1, 1) is the midpoint of a chord of the circle.

Solution

Centre of the circle is (2, 3) and radius $r = \sqrt{4+9+3} = 4$. Point M (1, 1) is the midpoint of the chord AB.

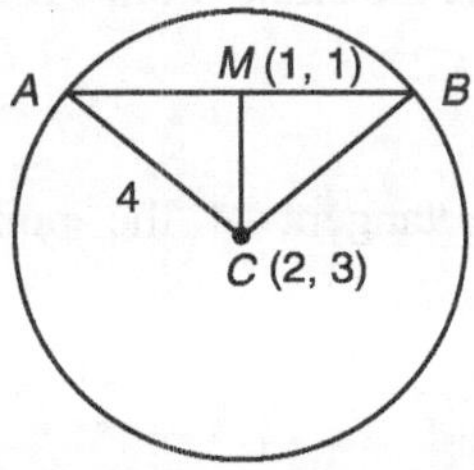

$$CM^2 = (2-1)^2 + (3-1)^2 = 1+4 = 5$$

$$\therefore AM^2 = r^2 - CM^2 = 16-5 = 11$$

$$\therefore AM = \sqrt{11}$$

Therefore, the length of the chord = $2\sqrt{11}$ units.

Example 4.19

Show that the circles $x^2 + y^2 - 2x + 6y + 6 = 0$ and $x^2 + y^2 - 5x + 6y + 15 = 0$ touch each other internally.

Solution

For the circle $x^2 + y^2 - 2x + 6y + 6 = 0$,

centre is $A(1, -3)$ and radius $r_1 = \sqrt{1+9-6} = 2$ units

For the circle $x^2 + y^2 - 5x + 6y + 15 = 0$,

centre is $B\left(\frac{5}{2}, -3\right)$ and radius $r_2 = \sqrt{\frac{25}{4} + 9 - 15} = \sqrt{\frac{25+36-60}{4}} = \frac{1}{2}$ units

Distance between the centres is $AB = \sqrt{\left(1-\frac{5}{2}\right)^2 + (-3+3)^2} = \frac{3}{2}$

$$\therefore r_1 - r_2 = 2 - \frac{1}{2} = \frac{3}{2}$$

Thus, the distance between the centres is equal to the difference in radii. Hence, the two circles touch each other internally.

Example 4.20

The abscissa of the two points A and B are the roots of the equation $x^2+2x-a^2=0$ and the ordinates are the roots of the equation $y^2+4y-b^2=0$. Find the equation of the circle with AB as its diameter. Also find the coordinates of the centre and the length of the radius of the circle.

Solution

Let the roots of the equation $x^2+2x-a^2=0$ be α and β.

Then
$$\alpha+\beta=-2 \tag{4.40}$$
$$\alpha\beta=-a^2 \tag{4.41}$$

Let γ, δ be the roots of the equation $x^2+4y-b^2=0$.

Then
$$\gamma+\delta=-4$$
$$\gamma\delta=-b^2$$

The coordinates of A and B are $(\alpha,\ \gamma)$ and $(\beta,\ \delta)$. The equation of the circle on the line joining the points A and B as the ends of a diameter is $(x-\alpha)(x-\beta)+(y-\gamma)(y-\delta)=0$.

$$\Rightarrow x^2-(\alpha+\beta)x+\alpha\beta+y^2-(\gamma+\delta)y+\gamma\delta=0$$
$$\Rightarrow x^2+y^2+2x+4y-a^2-b^2=0$$

The centre of the circle in $(-1,-2)$ and the radius $=\sqrt{1+4+a^2+b^2}=\sqrt{a^2+b^2+5}$.

Example 4.21

Find the equation of a circle that passes through the point (2, 0) and whose centre is the limit point of the intersection of the lines $3x + 5y = 1$ and $(2+c)x+5c^2y=1$ as $c\to1$.

Solution

The centre of the circle is the point of intersection of the lines
$$3x+5y=1 \tag{4.42}$$
$$(2+c)x+5c^2y=1 \tag{4.43}$$
$$\text{as}\quad c\to1$$

(4.42) $\times\, c^2-$ (4.43) gives $[3c^2-(2+c)]x=c^2-1\Rightarrow x=\dfrac{c^2-1}{3c^2-(2+c)}$

$$x=\frac{c^2-1}{3c^2-c-2}=\frac{\cancel{(c-1)}(c+1)}{\cancel{(c-1)}(3c-2)}=\frac{c+1}{3c+2}.$$

As $c \to 1$, the x-coordinate of the centre is $\frac{2}{5}$.

From (4.42), $y = \frac{1}{5}\left[1 - \frac{6}{5}\right] = \frac{-1}{25}$

Hence, the centre of the circle is $\left(\frac{2}{5}, \frac{-1}{25}\right)$.

Radius is the length of the line joining the points (2, 0) and $\left(\frac{2}{5}, \frac{-1}{25}\right)$.

$$r = \sqrt{\left(2 - \frac{2}{5}\right)^2 + \left(0 + \frac{1}{25}\right)^2} = \sqrt{\frac{1601}{625}}$$

Therefore, the equation of the circles is $\left(x - \frac{2}{5}\right)^2 + \left(y + \frac{1}{25}\right)^2 = \frac{1601}{625}$

$$x^2 + y^2 - \frac{4x}{5} - \frac{2}{25}y + \frac{4}{25} + \frac{1}{625} - \frac{1601}{625} = 0$$

$$\text{(i.e.) } 25x^2 + 25y^2 - 20x + 2y - 60 = 0$$

Example 4.22

Find the length intercepted on the y-axis by the chord of the circle joining the points (–4, 3) and (12, –1) as diameter.

Solution

The equation of the circle is

$$(x+4)(x-12) + (y-3)(y+1) = 0$$

(i.e.)

$$x^2 - 8x - 48 + y^2 - 2y - 3 = 0$$

$$x^2 + y^2 - 8x - 2y - 51 = 0$$

If y_1 and y_2 are the y-coordinates of the point of intersection of the circle and y-axis, then

$$y_1 + y_2 = 2$$

$$y_1 y_2 = -51$$

$$\therefore (y_1 - y_2)^2 = (y_1 + y_2)^2 - 4y_1 y_2 = 4 + 204 = 208$$

$$|y_1 - y_2| = \sqrt{208} = 4\sqrt{13} \text{ units}$$

Example 4.23

The rods whose lengths are a and b slide along the coordinate axes in such a way that their extremities are concyclic. Find the locus of the centre of the circle.

Solution

Let AB and CB be the portion of x-axis and y-axis, respectively, intercepted by the circle. Let $P(x_1, y_1)$ be the centre of the circle. Draw PL and PM perpendicular to x-axis and y-axis, respectively. Then, by second property

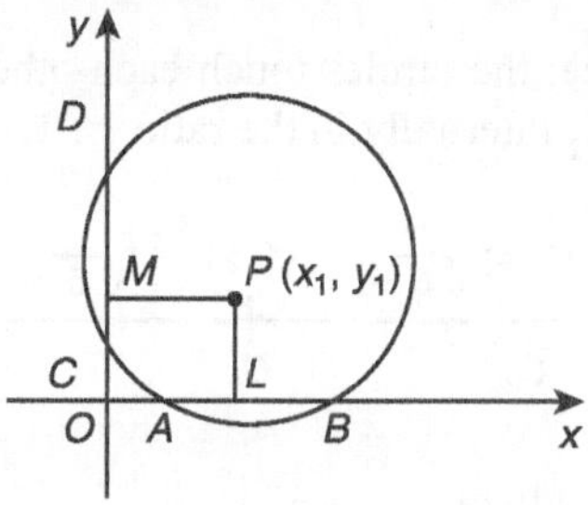

$$OA.OB = OC.OD = r^2 \quad OA = x_1 - \frac{a}{2}, OC = y_1 - \frac{b}{2}$$

$$OB = x_1 + \frac{a}{2}, OD = y_1 + \frac{b}{2}$$

$$\therefore \left(x_1 - \frac{a}{2}\right)\left(x_1 + \frac{a}{2}\right) = \left(y_1 - \frac{b}{2}\right)\left(y_1 + \frac{b}{2}\right)$$

$$\therefore x_1^2 - y_1^2 - \frac{(a^2 - b^2)}{4} = 0$$

The locus of $P(x_1, y_1)$ is $4(x^2 - y^2) = a^2 - b^2$.

Example 4.24

Show that the circles $x^2 + y^2 - 2x - 4y = 0$ and $x^2 + y^2 - 8y - 4 = 0$ touch each other. Find the coordinates of the point of contact and the equation of the common tangents.

Solution

$$x^2 + y^2 - 2x - 4y = 0 \tag{4.44}$$

$$x^2 + y^2 - 8y - 4 = 0 \tag{4.45}$$

The centres of these two circles are $C_1(1,2)$ and $C_2(0,4)$.
The radii of the two circles are

$$r_1 = \sqrt{1+4} = \sqrt{5}$$

$$r_2 = \sqrt{0+16+4} = \sqrt{20} = 2\sqrt{5}$$

The distance between the centres is

$$C_1C_2 = \sqrt{(1-0)^2+(2-4)^2} = \sqrt{5}$$

$\therefore r_1 - r_2 = C_1C_2$. Hence, the circles touch each other internally. The point of contact C divides C_1C_2 internally in the ratio 1 : 1.

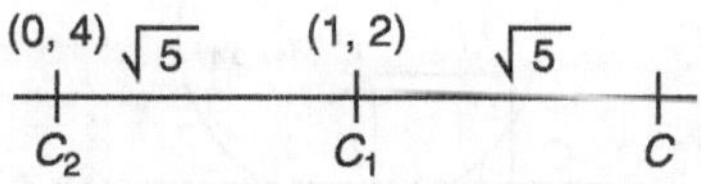

If C is the point (x_1, y_1) then

$$\frac{x_1+0}{2} = 1, \frac{y_1+4}{2} = 2 \quad \text{(i.e.)}\ x_1 = 2, y_1 = 0$$

$\therefore C(2,0)$ is the point of contact. The slope of $C_1C_2 = \dfrac{2-4}{1-0} = -2.$

Hence, the slope of the common tangent is 1/2. The equation of the common tangent is $y - 0 = \dfrac{1}{2}(x-2)$ or $x - 2y - 2 = 0$

Example 4.25

Show that the general equation of the circle that passes through the point $A(x_1, y_1)$ and $B(x_2, y_2)$ may be written as

$$(x-x_1)(x-x_2)+(y-y_1)(y-y_2)+k\begin{vmatrix} x & y & 1 \\ x_1 & y_1 & 1 \\ x_2 & y_2 & 1 \end{vmatrix} = 0.$$

Solution

Let $A(x_1, y_1)$ and $B(x_2, y_2)$ be the two points on the circumference of the circle and $A(x_1, y_1)$ be any point on the circumference.

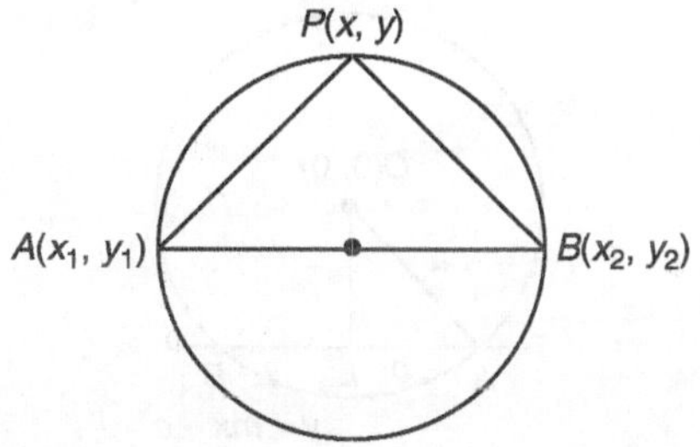

Let $\angle APB = \theta$. The slope of AP and BP are $m_1 = \dfrac{y-y_1}{x-x_1}$ and $m_2 = \dfrac{y-y_2}{x-x_2}$

$$\Rightarrow \tan\theta = \pm\frac{(m_1 - m_2)}{1+m_1m_2}$$

$$\Rightarrow \tan\theta = \pm\frac{\dfrac{y-y_1}{x-x_1} - \dfrac{y-y_2}{x-x_2}}{1+\dfrac{y-y_1}{x-x_1}\times\dfrac{y-y_2}{x-x_2}}$$

or

$$(x-x_1)(x-x_2)+(y-y_1)(y-y_2) \pm \cot\infty[(x-x_2)(y-y_1)-(x-x_1)(y-y_2)]$$

(i.e.) $$(x-x_1)(y-y_1)+(x-x_2)(y-y_2) \pm \cot\infty[xy - xy_2 - x_2y + x_2y_2 - xy + xy_1 + x_2y - x_2y_1] = 0$$

(i.e.) $$(x-x_1)(x-x_2)+(y-y_1)(y-y_2) \pm \cot\infty[x(y_1-y_2) - y(x_1-x_2) + x_1y_2 - x_2y_1] = 0$$

(i.e.) $$(x-x_1)(x-x_2)+(y-y_1)(y-y_2)+k\begin{vmatrix} x & y & 1 \\ x_1 & y_1 & 1 \\ x_2 & y_2 & 1 \end{vmatrix} = 0$$

Example 4.26

Show that if the circle $x^2 + y^2 = a^2$ cuts off a chord of length $2b$ on the line $y = mx + c$, then $c^2 = (1+m)^2(a^2 - b^2)$.

Solution

Given $x^2 + y^2 = a^2$. The centre of the circle is (0, 0). Radius $= r = a$. Draw OL perpendicular to AB. Then, L is the midpoint of AB.

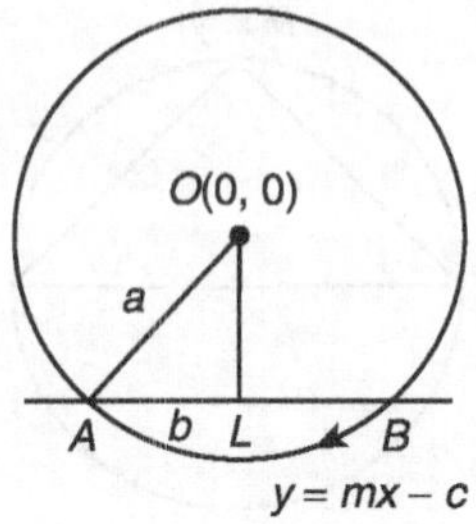

$$AL = b,\ OL = \frac{|c|}{\sqrt{1+m^2}},\ OA = a$$

Then

$$AL^2 = OA^2 - OL^2$$

$$b^2 = a^2 - \frac{c^2}{1+m^2}$$

$$\therefore c^2 = (1+m^2)(a^2 - b^2).$$

Example 4.27

A point moves such that the sum of the squares of the distances from the sides of a square of side unity is equal to *a*. Show that the locus is a circle whose centre coincides with the centre of the square.

Solution

Let the centre of the square be the origin. Let $P(x, y)$ be any point. Then, the equation of the sides are $x-\frac{1}{2}=0$, $x+\frac{1}{2}=0$, $y-\frac{1}{2}=0$ and $y+\frac{1}{2}=0$.

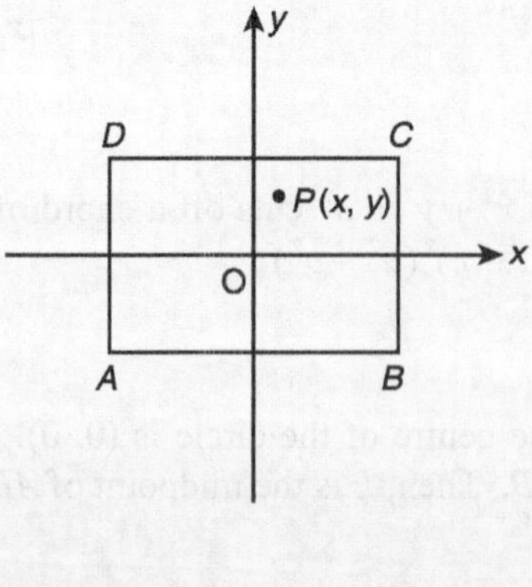

Sum of the perpendicular distances from P on the sides is equal to a

$$\therefore \left(x-\frac{1}{2}\right)^2+\left(x+\frac{1}{2}\right)^2+\left(y-\frac{1}{2}\right)^2+\left(y+\frac{1}{2}\right)^2=9$$

$$x^2+y^2-1=0$$

Hence, the locus of P is the circle $x^2+y^2-1=0$. The centre of the circle is (0, 0), which is the centre of the square.

Example 4.28

If the lines $l_1x+m_1y+n_1=0$ and $l_2x+m_2y+n_2=0$ cut the coordinate axes at concyclic points, prove that $l_1l_2=m_1m_2$.

Solution

Given $l_1x+m_1y+n_1=0$. The intercepts of the line on the axis are $\frac{-n_1}{l_1}$ and $\frac{-n_1}{m_1}$.

If the line meets the axes at L_1 and M_1, then $OL_1=\frac{-n_1}{l_1}$ and $OM_1=\frac{-n_1}{m_1}$. If the second line meets the axes at L_2 and M_2, then

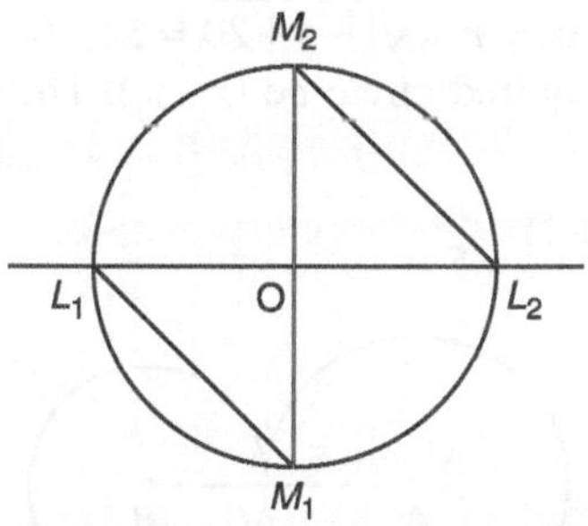

$OL_2=\frac{-n_2}{l_2}$ and $OM_2=\frac{-n_2}{m_2}$. Then $OL_1.OL_2=OM_1.OM_2$.

(i.e.) $$\left(\frac{-n_1}{l_1}\right)\left(\frac{-n_2}{l_2}\right)-\left(\frac{-n_1}{m_1}\right)\left(\frac{-n_2}{m_2}\right)=0$$

(i.e.) $$l_1l_2=m_1m_2$$

Example 4.29

Show that the locus of a point whose ratio of distances from two given points is constant is a circle. Hence, show that the circle cannot pass through the given points.

Solution

Let the two points A and B be chosen in the x-axis and the midpoint of AB be (0, 0). Then let $A(a, 0)$ and $B(-a, 0)$. Given that $PA = K \cdot PB \Rightarrow PA^2 = K^2 PB^2$ where k is a constant.

$$(x-a)^2 + (y-0)^2 = K^2[(x+a)^2 + y^2]$$

In this equation, the coefficients of x^2 and y^2 are the same and there is no xy term. Therefore, the locus of P is a circle. If $A(a, 0)$ lies on this circle, then $O = K^2[4a^2] \Rightarrow a = 0$ or $k = 0$, which are not possible. Therefore, the point A does not lie on the circle. Similarly, the point B $(-a, 0)$ also does not lie on the circle.

Example 4.30

Find the equation of the circle whose radius is 5 and which touches the circle $x^2 + y^2 - 2x - 4y - 20 = 0$ at the point (5, 5).

Solution

Given $x^2 + y^2 - 2x - 4y - 20 = 0$.

Centre is (1, 2) and radius $= r = \sqrt{1+4+20} = 5$

Let the centre of the required circle be (x_1, y_1). The point of contact is the midpoint of AB.

(i.e.) $\dfrac{(x+1)}{2} = 5, \ \dfrac{(y+2)}{2} = 5$

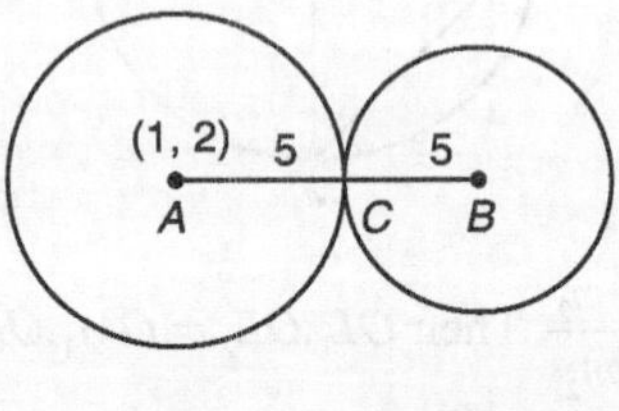

$\therefore x = 9$ and $y = 8$

Thus, B is (9, 8). Hence, the equation of the required circle is

$$(x-9)^2 + (y-8)^2 = 25$$

$$\text{(i.e.) } x^2 + y^2 - 18x - 16y + 120 = 0$$

Example 4.31

One of the diameters of the circle circumscribing the rectangle $ABCD$ is $4y = x + 7$. If A and B are the points (−3, 4) and (5, 4), respectively, find the area of the rectangle $ABCD$.

Solution

Let $P(x_1, y_1)$ be the centre of the circle and $4y = x + 7$ be the equation of the diameter of BD.

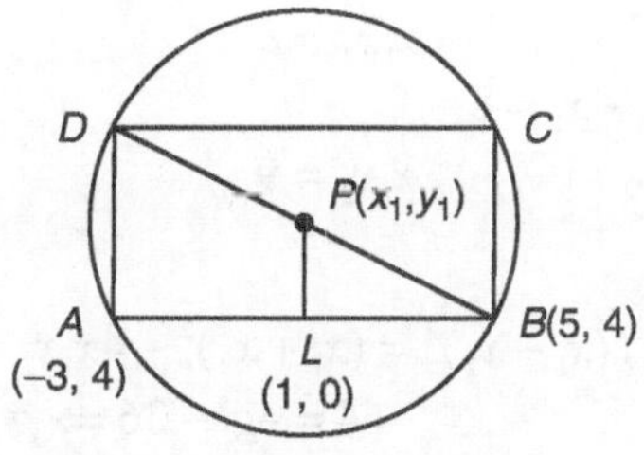

The midpoint of AC is (1, 1). The slope of AB is 0. Therefore, the slope of PL is ∞.

Hence $\frac{y_1 - 0}{x_1 - 1} = \infty \qquad \therefore x_1 = 1$

(x_1, y_1) lies on $4y = x + 7 \qquad \Rightarrow 4y_1 = 8$ or $y_1 = 2$

$\therefore P$ is (1, 2). Now $AB^2 = (-3-5)^2 + (4-4)^2 = 64$ or $AB = 8$

$$BC = 2 \cdot PL = 2\sqrt{(1-1)^2 + (2-0)^2} = 2 \times 2 = 4$$

Hence, the area of the rectangle $ABCD = 8 \times 4 = 32$ sq. cm.

Example 4.32

Find the equation of the circle touching the y-axis at (0, 3) and making an intercept of 8 cm on the x-axis.

Solution

Let the equation of the circle be $x^2 + y^2 + 2gx + 2fy + c = 0$. Centre is $(-g, -f)$, Thus, $-f = 3$ or $f = -3$.

When the circle meets the x-axis, $y = 0$.

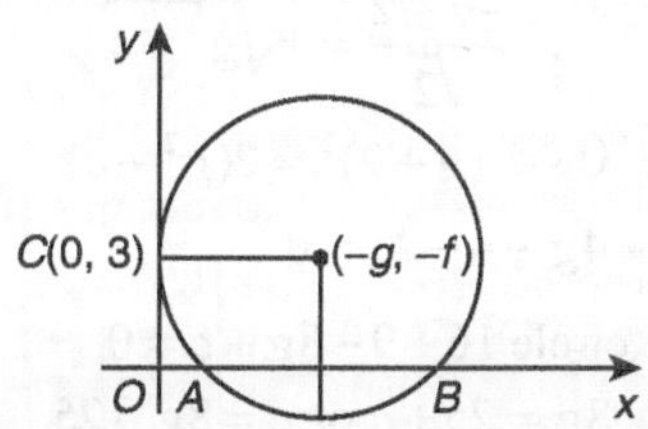

$$\therefore x^2 + 2gx + c = 0$$

If x_1, x_2 are the x coordinates of A and B then

$$x_1 + x_2 = -2g$$
$$x_1 x_2 = c$$

But $OA \cdot OB = 9$ and $x_2 - x_1 = 8$

$$x_1 x_2 = 9$$

Also $f = -3$

$$\therefore (x_1 - x_2)^2 = (x_1 + x_2)^2 - 4x_1 x_2$$
$$64 = 4g^2 - 36 \Rightarrow g = \pm 5$$

Hence, the equation of the circle is $x^2 + y^2 \pm 10x - 6y + 9 = 0$.

Example 4.33

Find the equation of the circle passing through the point (-4, 3) and touching the lines $x + y = 2$ and $x - y = 2$.

Solution

$x + y = 2$ and $x - y = 2$ intersect at the point (2, 0).

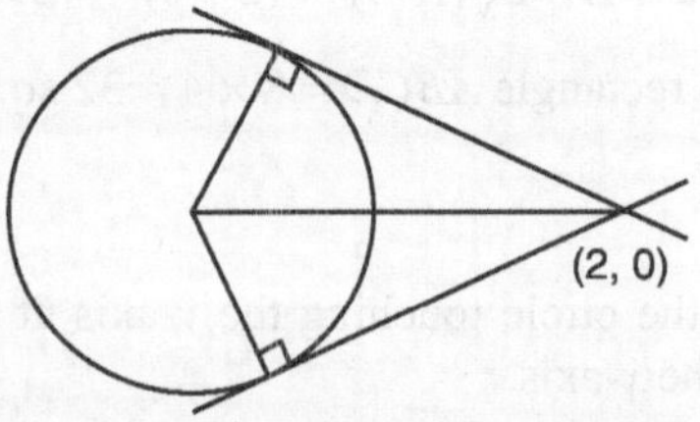

Moreover, these lines are perpendicular and their slopes are 1 and −1. So, they make 45° and 135° with the x-axis. Hence one of the bisectors is the x-axis and centre lies on one of the bisectors. If $x^2 + y^2 + 2gx + 2fy + c = 0$ is the equation of the circle, then $f = 0$. Also the perpendicular distance from $(-g, 0)$ to the tangents is equal to the radius.

$$\therefore \frac{-g-2}{\sqrt{2}} = \pm\sqrt{g^2 - c}$$

$$\text{(i.e.)} \; (g+2)^2 = 2(g^2 - c)$$

$$\therefore g^2 - 4g - 4 - 2c = 0 \qquad (4.46)$$

Since (−4, 3) lies on the circle $16 + 9 - 8g + c = 0$

$$8g = 25 + c \text{ or } c = 8g - 25 \qquad (4.47)$$

Hence, equation (4.46) becomes $g^2 - 4g - 4 - 16g + 50 = 0$ or $g^2 - 20g + 46 = 0$.

$$\therefore g = \frac{20 \pm \sqrt{400 - 4 \times 46}}{2}$$

$$= \frac{20 \pm \sqrt{216}}{2}$$

$$= 10 \pm 3\sqrt{6}$$

From (4.47), $c = 8g - 25 = 80 + 24\sqrt{6} - 25$

$$= 55 \pm 24\sqrt{6}$$

Thus, there are two circles whose equations are given by $x^2 + y^2 + 2(10 \pm 3\sqrt{6})x + 55 \pm 24\sqrt{6} = 0$

Example 4.34

A is the centre of the circle $x^2 + y^2 - 2x - 4y - 20 = 0$. If the tangents drawn at the points *B*(1, 7) and *D*(4, –2) on the circle meet at the point *C*, then find the area of the quadrilateral *ABCD*.

Solution

$$x^2 + y^2 - 2x - 4y - 20 = 0$$

Centre of this circle is (1, 2)

$$r = \sqrt{1 + 4 + 20} = 5$$

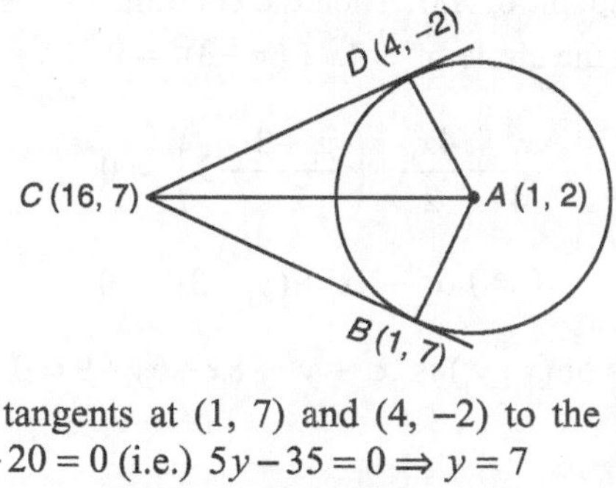

The equations of tangents at (1, 7) and (4, –2) to the circle are $x + 7y - (x+1) - 2(y+7) - 20 = 0$ (i.e.) $5y - 35 = 0 \Rightarrow y = 7$

and

$$4x - 2y - (x+4) - 2(y-2) - 20 = 0$$

$$\text{(i.e.) } 3x - 4y - 20 = 0$$

Since $y = 7$, $x = 16$. Hence, the point C is $(16, 7)$.
Area of the quadrilateral $ABCD = 2 \times$ area of ΔABC

$$= 2 \times \frac{1}{2}[1(7-7) + 16(7-2) + 1(2-7)]$$

$$= 0 + 80 - 5$$
$$= 75 \text{ sq. units}$$

Example 4.35

From the point $A(0, 3)$ on the circle $x^2 + 4x + (y-3)^2 = 0$, a chord AB is drawn and extended to a point M such that $AM = AB$. Find the equation of the locus of M.

Solution

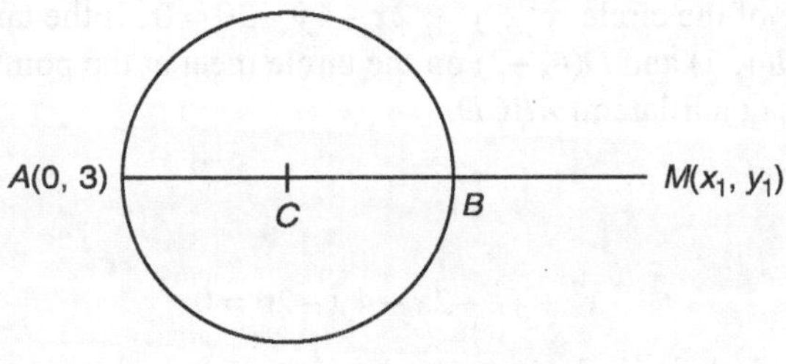

$$AM = 2.AB$$

Hence, B is the midpoint of AM. Then the coordinates of B are $\left(\frac{x_1}{2}, \frac{y_1+3}{2}\right)$. This point B lies on the circle $x^2 + 4x + (y-3)^2 = 0$.

$$\frac{x_1^2}{4} + \frac{4x_1}{2} + \left(\frac{y_1+3}{2} - 3\right)^2 = 0$$

$$\text{(i.e.) } x_1^2 + 8x_1 + (y_1 - 3)^2 = 0$$

Therefore, the locus of (x_1, y_1) is $x^2 + y^2 + 8x - 6y + 9 = 0$.

Example 4.36

AB is a diameter of a circle, CD is a chord parallel to AB and $2CD = AB$. The tangent at B meets the line AC produced at E. Prove that $AE = 2.AB$.

Solution

Let the equation of the circle be $x^2 + y^2 = a^2$ and PQ be the diameter along the x-axis. CD is parallel to AB.

Let $AB = 2a$ and points A and B be $(a, 0)$ and $(-a, 0)$, respectively. Also $CD = \frac{1}{2} AB$.

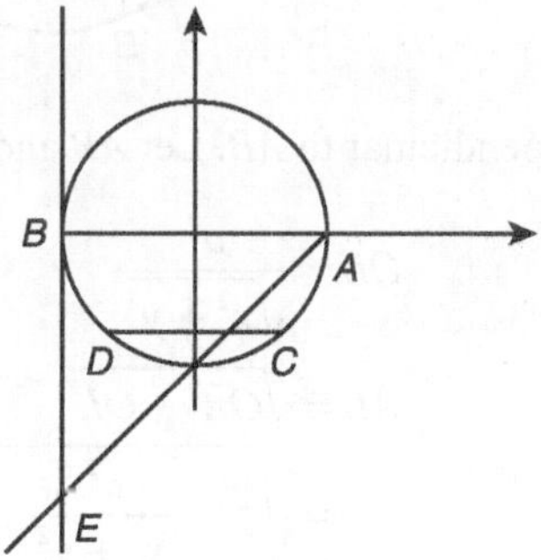

Hence, the coordinates of C are $\left(\frac{a}{2}, -h\right)$.

The point lies on the circle $x^2 + y^2 = a^2$.

$$\therefore \frac{a^2}{4} + h^2 = a^2 \text{ or } h = \frac{\sqrt{3}}{2}a \text{ and the point } C\left(\frac{a}{2}, \frac{\sqrt{3}}{2}a\right)$$

The equation of AE is $\frac{y-0}{x-a} = \frac{0+\frac{\sqrt{3}}{2}a}{a-\frac{a}{2}}$ or $y = \sqrt{3}(x-a)$.

Hence, the point of intersection E is $(-a, -2\sqrt{3}a)$.

$$AE^2 = (a+a)^2 + \left(0 + 2\sqrt{3}a\right)^2 = 16a^2$$

$$\therefore AE = 4a \text{ and } AB = 2a$$

$$\therefore AE = 2.AB$$

Example 4.37

Find the area of the triangle formed by the tangents from the point (h, k) to the circle $x^2 + y^2 = a^2$ and their chord of contact.

Solution

The equation of the circle is $x^2 + y^2 = a^2$. Let AB be the chord of contact of tangents from $C\,(x_1, y_1)$. Then the equation of AB is $xx_1 + yy_1 = a^2$.

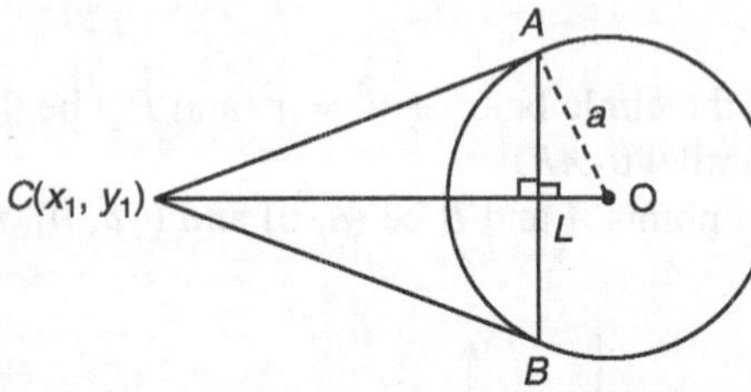

We know that OC is perpendicular to AB. Let AB and OC meet at L.

$$OL = \frac{a^2}{\sqrt{x^2 + y^2}}$$

$$AL = \sqrt{OA^2 - OL^2}$$

$$= \sqrt{a^2 - \frac{a^2}{x_1^2 + y_1^2}}$$

$$= a\sqrt{\frac{x_1^2 + y_1^2 - a^2}{x_1^2 + y_1^2}}$$

$$\therefore AB = 2AL = 2a\sqrt{\frac{x_1^2 + y_1^2 - a^2}{x_1^2 + y_1^2}}$$

The perpendicular distance from C on AB

$$CL = \frac{x_1^2 + y_1^2 - a^2}{\sqrt{x_1^2 + y_1^2}}$$

Hence, the area of $\Delta ABC = \frac{1}{2} \cdot AB \cdot CL$

$$= \frac{1}{2} \cdot 2a\sqrt{\frac{x_1^2 + y_1^2 - a^2}{x_1^2 + y_1^2}} \cdot \frac{(x_1^2 + y_1^2 - a^2)}{\sqrt{x_1^2 + y_1^2}}$$

$$= \frac{a(x_1^2 + y_1^2 - a^2)^{\frac{3}{2}}}{x_1^2 + y_1^2}$$

Example 4.38

Let a circle be given by $2x(x-a) + y(2y-b) = 0$, $(a, b \neq 0)$. Find the condition on a and b if two chords each intersected by the x-axis can be drawn to the circle from $\left(a, \frac{b}{2}\right)$.

Solution

$$2x(x-a)+y(2y-b)=0$$

$$x^2+y^2-ax-\frac{by}{2}=0$$

The chord is bisected by the x-axis. Let the midpoint of the chord be $(h, 0)$. The equation of the chord is

$$T=S_1$$

$$xh+y(0)-\frac{a}{2}(x+h)-\frac{b}{4}(y+0)=h^2-ah$$

This chord passes through $\left(a, \frac{b}{2}\right)$.

$$\therefore ah-\frac{a}{2}(a+h)-\frac{b}{4}\cdot\frac{h}{2}=h^2-ah$$

$$\text{(i.e.) } h^2-\frac{3ah}{2}+\frac{a^2}{2}+\frac{bh}{8}=0$$

Since the chord meets the x-axis at two reals,
Discriminant > 0

$$\therefore \frac{9a^2}{4}-4\left(\frac{a^2}{2}+\frac{b^2}{8}\right)>0$$

$$\text{(i.e.) } 9a^2-8a^2-2b^2>0$$

$$\Rightarrow a^2>2b^2$$

Example 4.39

Find the condition that the chord of contact from a point to the circle $x^2+y^2=a^2$ subtends a right angle at the centre of the circle.

Solution

The equation to the chord of contact from (x_1, y_1) to the circle

$$x^2+y^2=a^2 \tag{4.48}$$

is

$$xx_1+yy_1=a^2 \tag{4.49}$$

Then the combined equation to OA and OB is got by homogenizing equation (4.48) with the help of equation (4.49).

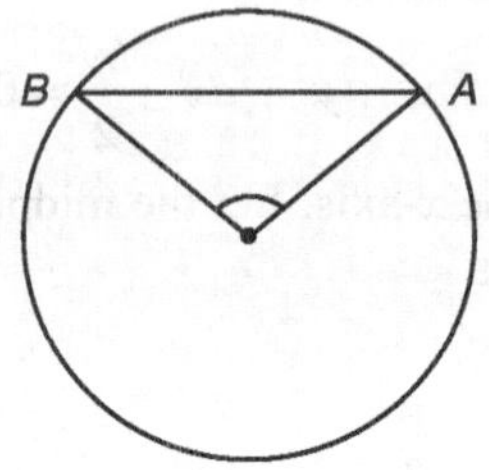

The combined equation of OA and OB is $x^2 + y^2 = a^2\left(\frac{xx_1 + yy_1}{a^2}\right)^2$

$$\text{(i.e.)}\quad x^2[a^2 - x_1^2] + y^2[a^2 - y_1^2] - 2x_1y_1xy = 0$$

Since $\lfloor AOB = 90°$, coefficient of x^2 + coefficient of $y^2 = 0$.

$$\therefore a^2 - x_1^2 + b^2 - y_1^2 = 0$$

$$x_1^2 + y_1^2 = a^2 + b^2$$

Example 4.40

If $y = mx$ be the equation of a chord of the circle whose radius is a, the origin being one of the extremities of the chord and the axis being a diameter of the circle, prove that the equation of a circle of which this chord is a diameter is $(1+m^2)(x^2+y^2) - 2a(x+my) = 0$.

Solution

Let a be the radius of the circle. Thus $(a, 0)$ is the centre of the circle. The equation of the circle is

$$(x-a)^2 + y^2 = a^2 \Rightarrow x^2 + y^2 - 2ax = 0$$

When $y = mx$ meets the circle $x^2 + m^2x^2 - 2ax = 0$.

$$\therefore x = 0 \text{ or } x = \frac{2a}{1+m^2}$$

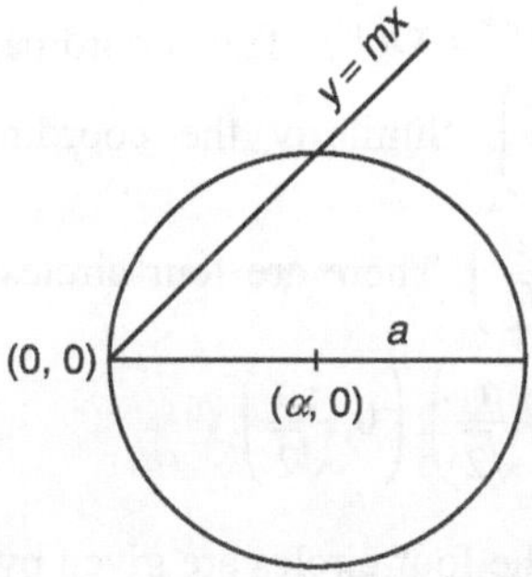

Therefore, the extremities of this chord are (0, 0) and $\left(\frac{2a}{1+m^2}, \frac{2am}{1+m^2}\right)$. Then, the equation of the circle with the chord as a diameter is

$$x\left(x-\frac{2a}{1+m^2}\right)+y\left(y-\frac{2am}{1+m^2}\right)=0$$

$$\text{(i.e.)}\ (x^2+y^2)(1+m^2)-2a(x+my)=0$$

Example 4.41

Find the equation to the circle that passes through the origin and cuts off equal chords of length a from the straight lines $y = x$ and $y = -x$.

Solution

Let the lines $y = x$ and $y = -x$ meet the circle at P, P' and Q, Q', respectively.

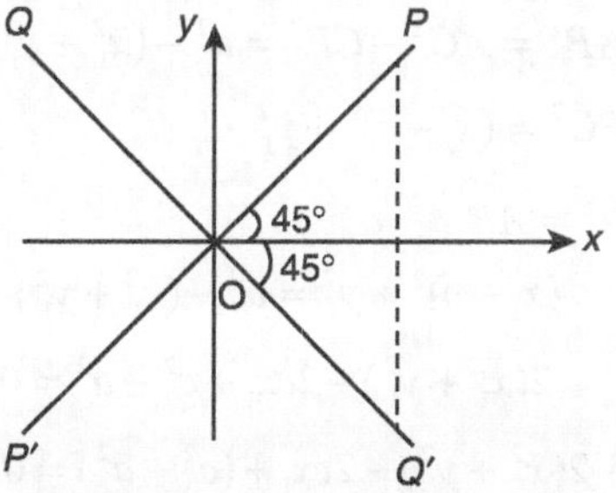

Then $OP = OQ = a = OP' = OQ'$. The coordinates of P and P' are $\left(\frac{a}{\sqrt{2}}, \frac{a}{\sqrt{2}}\right)$ and $\left(\frac{-a}{\sqrt{2}}, \frac{-a}{\sqrt{2}}\right)$. Similarly the coordinates of Q and Q' are $\left(\frac{-a}{\sqrt{2}}, \frac{a}{\sqrt{2}}\right)$ and $\left(\frac{a}{\sqrt{2}}, \frac{-a}{\sqrt{2}}\right)$. There are four circles possible having centres at $\left(\frac{a}{\sqrt{2}}, 0\right), \left(\frac{-a}{\sqrt{2}}, 0\right), \left(0, \frac{a}{\sqrt{2}}\right), \left(0, \frac{-a}{\sqrt{2}}\right)$.

Hence, the equations of the four circles are given by

$$x^2 + y^2 \pm \sqrt{2}ax = 0 \text{ and } x^2 + y^2 \pm \sqrt{2}ay = 0$$

Example 4.42

Find the locus of the midpoint of chords of the circle $x^2 + y^2 = a^2$, which subtends a right angle at the point $(c, 0)$.

Solution

Since AB subtends 90° at $C(c, 0)$, $PA = PB = PC$. Let P be the point (x_1, y_1).

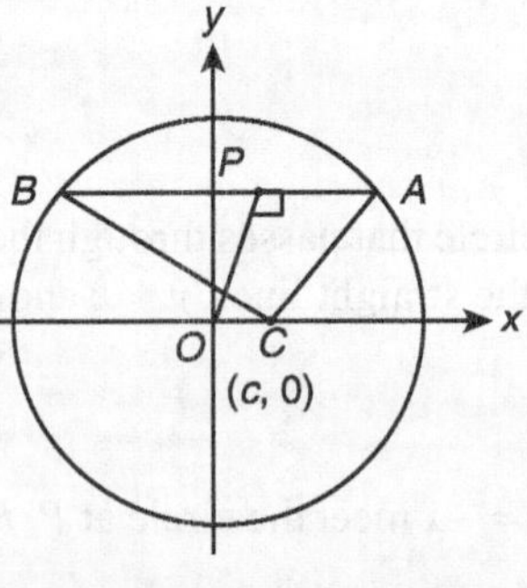

Since P is the midpoint of the chord AB, $CP \perp AP$

$$AP^2 = AC^2 - CP^2 = a^2 - (x_1^2 + y_1^2)$$

$$PC^2 = (x_1 - c)^2 + y_1^2$$

Since $\angle APC = 90°$, $PC = AP$.

$$\therefore (x_1 - c)^2 + y_1^2 = a^2 - (x_1^2 + y_1^2)$$

$$2(x_1^2 + y_1^2) - 2cx_1 + c^2 - a^2 = 0$$

The locus of (x_1, y_1) is $2(x^2 + y^2) - 2cx_1 + (c^2 - a^2) = 0$.

Example 4.43

Find the equations of the circles that touch the coordinate axes and the line $x = a$.

Solution

$y = 0, x = 0$ and $x = a$ are the tangents to the circle. There are two circles as shown in the figure.

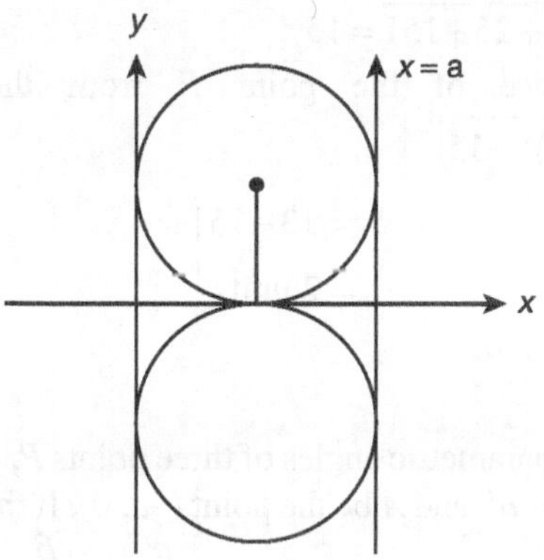

The centres are $\left(\frac{a}{2}, \pm\frac{a}{2}\right)$ and radius $\frac{a}{2}$. The equations of the circles are

$$\left(x-\frac{a}{2}\right)^2+\left(y\pm\frac{a}{2}\right)^2=\left(\frac{a}{2}\right)^2$$

$$\Rightarrow x^2+y^2-ax\pm ay+\frac{a^2}{4}=0$$

$$\Rightarrow 4(x^2+y^2)-4ax\pm 4ay+a^2=0$$

Example 4.44

Find the shortest distance from the point (2, –7) to the circle $x^2+y^2-14x-10y-151=0$.

Solution

$$x^2+y^2-14x-10y-151=0$$

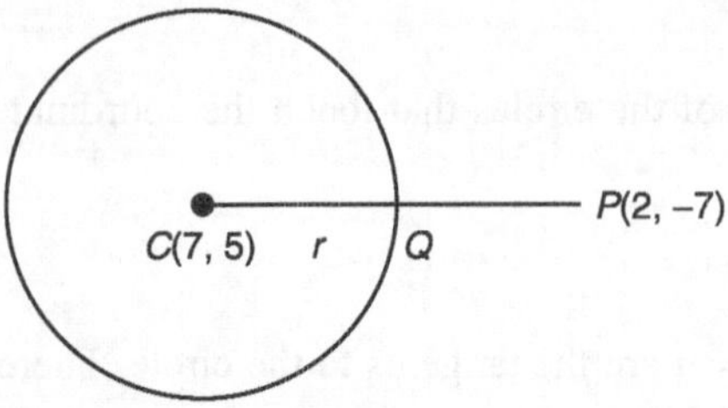

Centre is (7, 5)

Radius $= r = \sqrt{49+25+151} = 15$

The shortest distance of the point P from the circle $= |CP - r|$

$= \left|\sqrt{(2-7)^2 + (-7-5)^2} - 15\right|$

$= |13 - 15|$

$= 2$ units

Example 4.45

Let α, β and γ be the parametric angles of three points P, Q and R, respectively, on the circle $x^2 + y^2 = a^2$ and A be the point $(-a, 0)$. If the length of the chords AP, AQ and AR are in AP then show that $\cos\frac{\alpha}{2}, \cos\frac{\beta}{2}, \cos\frac{\gamma}{2}$ are also in AP.

Solution

Let $P(a\cos\alpha, a\sin\alpha)$, $Q(a\cos\beta, a\sin\beta)$, $R(a\cos r, a\sin r)$

A is $(a, 0)$

$$AP = \sqrt{(a\cos\alpha + a)^2 + a^2\sin^2\alpha} = \sqrt{4a^2\cos^4\frac{\alpha}{2} + 4a^2\sin^2\frac{\alpha}{2}\cos^2\frac{\alpha}{2}}$$

$$= 2a\sqrt{\cos^2\frac{\alpha}{2}\left(\cos^2\frac{\alpha}{2} + \sin^2\frac{\alpha}{2}\right)} = 2a\cos\frac{\alpha}{2}$$

Similarly $AQ = 2a\cos\frac{\beta}{2}$

$$AR = 2a\cos\frac{\gamma}{2}$$

The lengths of chords AP, AQ, AR are in AP.

$\therefore \cos\frac{\alpha}{2}, \cos\frac{\beta}{2}, \cos\frac{\upsilon}{2}$ are also in AP.

Example 4.46

Let $S = x^2 + y^2 + 2gx + 2fy + c = 0$. Find the locus of the foot of the perpendicular from the origin on any chord of the circle that subtends a right angle at the origin.

Solution

Let the equation of the line AB be

$$lx + my = 1 \tag{4.50}$$

Let (x_1, y_1) be the midpoint of AB.

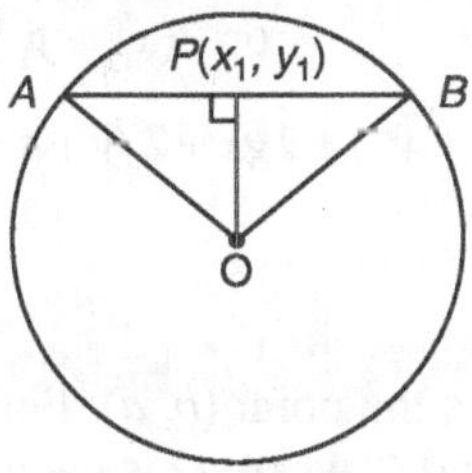

Let $P(x_1, y_1)$ be the foot of the perpendicular from thc origin on AB. Then, since OP is perpendicular to AP.

$$\frac{-l}{m} \times \frac{y_1}{x_1} = -1 \Rightarrow ly_1 = mx_1 \tag{4.51}$$

$$\frac{l}{x_1} = \frac{m}{y_1} = \frac{1}{x_1^2 + y_1^2}$$

Since (x_1, y_1) lies on the line $lx + my = 1$ we have

$$lx_1 + my_1 = 1 \tag{4.52}$$

The combined equation of lines OA and OB is got by homogenizing the equation of the circle $x^2 + y^2 + 2gx + 2fy + c = 0$ with the line $lx + my = 1$.

(i.e.) $x^2 + y^2 + (2gx + 2fy)(lx + my) + c(lx + my)^2 = 0$

(i.e.) $x^2(1 + 2gl + cl^2) + y^2(1 + 2fm + cm^2) + xy(2gm + 2fl + 2lmc) = 0$

Since $\lfloor AOB = 90°$, the condition is coefficient of x^2 + coefficient of $y^2 = 0$. Hence,

$$1 + 2gl + cl^2 + 1 + 2fm + cm^2 = 0$$

$$\text{(i.e.) } c(l^2 + m^2) + (2gl + 2fm) + 2 = 0$$

$$c\left[\frac{x_1^2}{(x_1^2 + y_1^2)^2} + \frac{y_1^2}{(x_1^2 + y_1^2)^2}\right] + 2g\left(\frac{x_1}{x_1^2 + y_1^2}\right) + 2f\left(\frac{y_1}{x_1^2 + y_1^2}\right) + 2 = 0$$

$$\text{(i.e.) } \frac{c}{x_1^2 + y_1^2} + \frac{2gx_1}{x_1^2 + y_1^2} + \frac{2fy_1}{x_1^2 + y_1^2} + 2 = 0$$

$$\text{or } 2(x_1^2 + y_1^2) + 2gx_1 + 2fy_1 + c = 0$$

The locus of (x_1, y_1) is $2(x^2 + y^2) + 2gx + 2fy + c = 0$.

Example 4.47

P is the point (a, b) and *Q* is the point (b, a). Find the equation of the circle touching *OP* and *OQ* at *P* and *Q* where *O* is the origin.

Solution

Let the equation of the circle be $x^2 + y^2 + 2gx + 2fy + c = 0$ (4.53)

Let $C(-g, -f)$ be the centre of the circle.

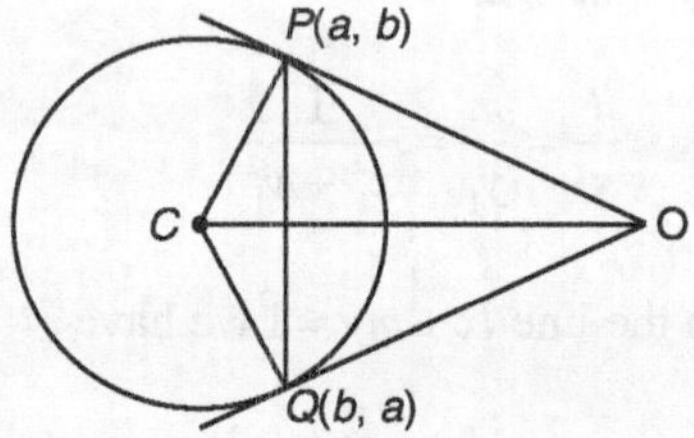

We know that *PQ* is the chord of the contact from *O* and *OC* is perpendicular to *PQ*.

$\therefore$ Slope of $PQ \times$ slope of $OC = -1$

$$\frac{f}{g} = -\frac{(b-a)}{(a-b)}$$

$$\therefore f = g \tag{4.54}$$

The equation of OP is $\frac{y}{x}=\frac{b}{a}$ or $bx-ay=0$.

Since CP is perpendicular to OP, r is the perpendicular distance from C on OP.

$$\sqrt{g^2+f^2-c}=\pm\frac{(-gb+af)}{\sqrt{a^2+b^2}}$$

(i.e.) $$(g^2+f^2-c)(a^2+b^2)=(bg-af)^2$$

or $$(2g^2-c)(a^2+b^2)=g^2(b-a)^2$$

$$g^2\left[2(a^2+b^2)-(b-a)^2\right]=c(a^2+b^2)$$

$$\Rightarrow c=\frac{g^2(a+b)^2}{a^2+b^2}$$

The point (a, b) lies on the circle (4.53).

$$a^2+b^2+2ga+2fb+c=0$$

$$a^2+b^2+2g(a+b)+\frac{g^2(a+b)^2}{a^2+b^2}=0$$

$$(a+b)^2g^2+2g(a+b)+(a^2+b^2)^2=0$$

$$\left[g(a+b)+(a^2+b^2)\right]^2=0 \quad \Rightarrow g=\frac{-(a^2+b^2)}{a+b}$$

Hence the equation of the circle is

$$x^2+y^2-2\frac{(a^2+b^2)}{a+b}x-2\frac{(a^2+b^2)}{a+b}y+\frac{(a+b)^2}{a^2+b^2}\cdot\frac{(a^2+b^2)^2}{(a+b)^2}=0$$

(i.e.) $$(a+b)(x^2+y^2)-2(a^2+b^2)(x+y)+(a+b)(a^2+b^2)=0$$

Example 4.48

A circle of circumradius $3k$ passes through the origin and meets the axes at A and B. Show that the locus of the centroid of ΔOAB is the circle $x^2+y^2=4K^2$.

Solution

Let A and B be the points $(a, 0)$ and $(0, b)$, respectively. Let (x_1, y_1) be the centroid of ΔOAB. Then since $\angle AOB = 90°$, AB is a diameter of the circle.

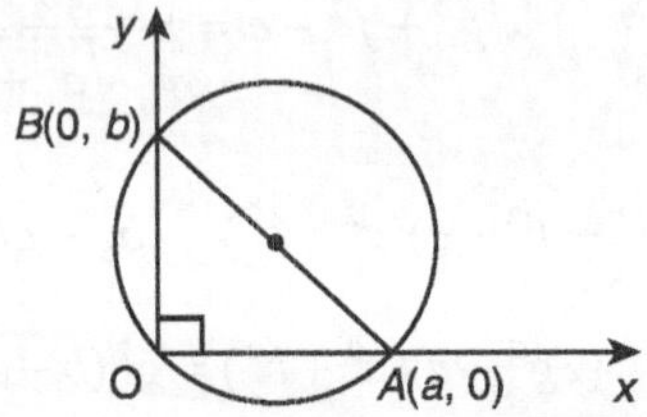

$$\therefore OA^2 + OB^2 = (6k)^2 \text{ or } a^2 + b^2 = 36k^2 \tag{4.55}$$

Let the centroid of ΔOAB be (x_1, y_1). Then $x_1 = \dfrac{a}{3}$ and $y_1 = \dfrac{b}{3}$ or $a = 3x_1$ and $b = 3y_1$. Substituting this in (4.55), we get $9x_1^2 + 9y_1^2 = 36k^2$ or $x_1^2 + y_1^2 = 4k^2$. The locus of (x_1, y_1) is $x^2 + y^2 = 4k^2$.

Example 4.49

A variable line passes through a fixed point (a, b) and cuts the coordinate axes at the points A and B. Show that the locus of the centre of the circle AB is $\dfrac{a}{x} + \dfrac{b}{y} = 2$.

Solution

Let AB be a variable line whose equation be $\dfrac{x}{\alpha} + \dfrac{y}{\beta} = 1$.

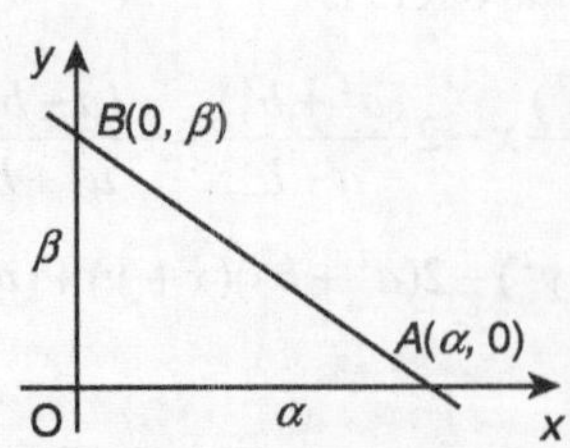

This passes through the point (a, b).

$$\therefore \frac{a}{\alpha} + \frac{b}{\beta} = 1$$

Since $\underline{|AOB} = 90°$, AB is a diameter of the circumcircle of ΔOAB. Its centre is $\left(\frac{\alpha}{2}, \frac{\beta}{2}\right)$. If (x_1, y_1) be the circumcentre, then $\frac{\alpha}{2} = x_1$ and $\frac{\beta}{2} = y_1$.

$$\therefore \alpha = 2x_1 \text{ and } \beta = 2y_1$$

Hence, from (4.55), we get

$$\frac{a}{2x_1} + \frac{b}{2y_1} = 1$$

The locus of (x_1, y_1) is $\frac{a}{x} + \frac{b}{y} = 2$.

Example 4.50

If $4l^2 - 5m^2 + 6l + 1 = 0$ then show that the line $lx + my + 1 = 0$ touches a fixed circle. Find the centre and radius of the circle.

Solution

Let the line

$$lx + my + 1 = 0 \tag{4.56}$$

touch the circle

$$(x - h)^2 + (y - k)^2 = r^2 \tag{4.57}$$

Then the perpendicular distance from (h, k) to line (4.56) is equal to the radius.

$$\therefore \frac{lh + mk + 1}{\sqrt{l^2 + m^2}} = \pm a$$

(i.e.) $a^2(l^2 + m^2) = (lh + mk + 1)^2$

or $l^2[a^2 - h^2] + m^2[a^2 - k^2] - 2lmhk - 2lh + 2mk - 1 = 0$ (4.58)

But the condition is given by

$$4l^2 - 5m^2 + 6l + 1 = 0 \tag{4.59}$$

Identifying (4.58) and (4.59), we get

$$\frac{a^2 - h^2}{4} = \frac{a^2 - k^2}{-5} = \frac{2hk}{0} = \frac{-2h}{6} = \frac{2k}{0} = \frac{-1}{1}$$

$$\therefore k = 0, h = 3 \text{ and } \frac{a^2 - h^2}{4} = -1$$

$$\Rightarrow a^2 - h^2 = -4$$

$$\Rightarrow a^2 = 9 - 4 = 5$$

Hence, the line touches the fixed circle $(x-3)^2 + y^2 = 5$ or $x^2 + y^2 - 6x + 4 = 0$ whose centre is (3, 0) and radius is $\sqrt{5}$.

Exercises

1. Find the equation of the following circles:
   (i) centre (2, –5) and radius 5 units **Ans.:** (i) $x^2 + y^2 - 4x + 10y + 4 = 0$
   (ii) centre (–2, –4) and radius 10 units **Ans.:** (ii) $x^2 + y^2 + 4x + 8y - 80 = 0$
   (iii) centre (a, b) and radius (a + b) **Ans.:** (iii) $x^2 + y^2 - 2ax - 2by = 0$
2. Find the centre and radius of the following circles:
   (i) $x^2 + y^2 - 22x - 4y + 25 = 0$ **Ans.:** (i) (11, 2), 10
   (ii) $4(x^2 + y^2) - 8(x - 2y) + 19 = 0$ **Ans.:** (ii) (1, –2), $\frac{1}{2}$
   (iii) $2x^2 + 2y^2 + 3x + y + 1 = 0$ **Ans.:** (iii) $\left(\frac{-3}{4}, -\frac{1}{4}\right), \frac{1}{2\sqrt{2}}$
3. Find the equation of the circle passing through the point (2, 4) and having its centre on the lines $x - y = 4$ and $2x + 3y = 8$.

   Ans.: $x^2 + y^2 - 8x - 4 = 0$
4. Find the equation of the circle whose centre is (–2, 3) and which passes through the point (2, –2).

   Ans.: $x^2 + y^2 + 4x - 6y - 28 = 0$
5. Show that the line $4x - y = 17$ is a diameter of the circle $x^2 + y^2 - 8x + 2y = 0$.
6. The equation of the circle is $x^2 + y^2 - 8x + 6y - 3 = 0$. Find the equation of its diameter parallel to $2x - 7y = 0$. Also find the equation of the diameter perpendicular to $3x - 4y + 1 = 0$.

   Ans.: $2x - 7y - 29 = 0$
   $4x + 3y - 7 = 0$
7. Find the equation of the circle passing through the following points:
   (i) (2, 1), (1, 2), (8, 9) **Ans.:** $x^2 + y^2 - 10x - 10y - 25 = 0$
   (ii) (0, 1), (2, 3), (–2, 5) **Ans.:** $3x^2 + 3y^2 + 2x - 20y + 17 = 0$
   (iii) (5, 2), (2, 1), (1, 4) **Ans.:** $x^2 + y^2 - 6x - 6y + 13 = 0$
8. Find the equation of the circle through the points (1, 0) and (0, 1) and having its centre on the line $x + y = 1$.

   Ans.: $x^2 + y^2 - x - y = 0$

9. Find the equation of the circle passing through the points (0, 1) and (4, 3) and having its centre on the line $4x-5y-5=0$.

Ans.: $x^2+y^2-5x-2y+1=0$

10. Two diameters of a circle are $5x-y=3$ and $2x+3y=8$. The circle passes through the point (–1, 7). Find its equation.

Ans.: $x^2+y^2-2x-4y=164$

11. Find the equation of the circle circumscribing the triangle formed by the axes and the straight line $3x+4y+12=0$.

Ans.: $x^2+y^2+4x+3y=0$

12. Show that the points (–1, 2), (–2, 4), (–1, 3) and (2, 0) are on a circle and find its equation.

13. If the coordinates of the extremities of the diameter of a circle are (3, 5) and (–7, –5), find the equation of the circle.

Ans.: $x^2+y^2+4x-3y=0$

14. Find the equation of the circle when the coordinates of the extremities of one of its diameters are (4, 1) and (–2, –7).

Ans.: $x^2+y^2-2x+6y-15=0$

15. If one end of the diameter of the circle $x^2+y^2-2x+6y-15=0$ is (4, 1), find the coordinates of the other end.

Ans.: (–2, –7)

16. Prove that the tangents from (0, 5) to the circles $x^2+y^2+2x-4=0$ and $x^2+y^2-y+1=0$ are equal.

17. Find the equation of the circle passing through the origin and having its centre at (3, 4). Also find the equation of the tangent to the circle at the origin.

Ans.: $x^2+y^2-6x-8y=0, 3x+4y=0$

18. Find the slope of the radius of the circle $x^2+y^2=25$ through the point (3, –4) and hence write down the equation of the tangent to the circle at the point. What are the intercepts made by this tangent on the x-axis and y-axis?

Ans.: $\dfrac{-4}{5}, 3x-4y=25, \dfrac{25}{3}, \dfrac{-25}{4}$

19. One vertex of a square is the origin and two others are (4, 0) and (0, 4). Find the equation of the circle circumscribing the square. Also find the equation of the tangent to this circle at the origin.

Ans.: $x^2+y^2-4x-4=0, x+y=0$

20. A circle passes through the origin and the points (6, 0) and (0, 8). Find its equation and also the equation of the tangent to the circle at the origin.

Ans.: $x^2 + y^2 - 6x - 8y = 0, 3x + 4y = 0$

21. A and B are two fixed points on a plane and the point P moves on the plane in such a way that $PA = 2PB$ always. Prove analytically that the locus of P is a circle.

22. Does the point (2, 1) lie (i) on, (ii) inside or (iii) outside the circle $x^2 + y^2 - 4x - 6y + 9 = 0$?

23. Show that the circles $x^2 + y^2 - 2x + 2y + 1 = 0$ and $x^2 + y^2 + 6x - 4y - 3 = 0$ touch each other externally.

24. Prove that the centres of the three circles $x^2 + y^2 - 2x + 6y + 1 = 0$, $x^2 + y^2 + 4x - 12y - 9 = 0$ and $x^2 + y^2 = 25$ lie on the same straight line. What is the equation of this line?

Ans.: $3x + 4y = 0$

25. Prove that the two circles $x^2 + y^2 + 2ax + c^2 = 0$ and $x^2 + y^2 + 2by + c^2 = 0$ touch each other if $\frac{1}{a^2} + \frac{1}{b^2} = \frac{1}{c^2}$.

26. Show that the circles $x^2 + y^2 - 4x + 2y + 1 = 0$ and $x^2 + y^2 - 12x + 8y + 43 = 0$ touch each other externally.

27. Show that the circles $x^2 + y^2 = 400$ and $x^2 + y^2 - 10x - 24y + 120 = 0$ touch one another. Find the co-ordinates of the point of contact.

Ans.: $\left(\frac{100}{13}, \frac{240}{13}\right)$

28. Find the length of the tangent from the origin to the circle $4x^2 + 4y^2 + 6x + 7y + 1 = 0$.

29. Show that the circles $x^2 + y^2 - 26x - 19 = 0$ and $x^2 + y^2 + 3x - 8y - 43 = 0$ touch externally. Find the point of contact and the common tangent.

30. A point moves so that the square of its distance from the base of an isosceles triangle is equal to the rectangle contained by its distances from the equal sides. Prove that the locus is a circle.

31. Prove that the centres of the circles $x^2 + y^2 = 1$, $x^2 + y^2 + 4x + 8y - 1 = 0$ and $x^2 + y^2 - 6x - 12y + 1 = 0$ are collinear.

32. Prove that the constant in the equation of the circle $x^2 + y^2 + 2gx + 2fy + c = 0$ is equal to the rectangle under the segments of the chords through the origin.

33. Find the equation of the locus of a point that moves in a plane so that the sum of the squares from the line $7x - 4y - 10 = 0$ and $4x + 7y + 5 = 0$ is always equal to 3.

Ans.: $13x^2 + 13y^2 - 20x + 30y - 14 = 0$

34. Show that the circles $x^2+y^2-10x+4y-20=0$ and $x^2+y^2+14x-6y+22=0$ touch each other. Find the equation of their common tangent at the point of contact and also the point of contact.

Ans.: $\left(\frac{-14}{13},\frac{9}{13}\right)$, $12x-5y+21=0$

35. L and M are the feet of the perpendicular from $(c, 0)$ on the lines $ax^2+2hxy+by^2=0$. Show that the equation of LM is $(a+b)x+2hy+bc=0$.

36. A circle has radius 3 units and its centre lies on the line $y=x-1$. Find the equation of the circle if it passes through $(-1, 3)$.

37. Find the equation of the circle on the line joining the points $(-4, 3)$ and $(12, -1)$. Find also the intercepts made by it on the y-axis.

Ans.: $x^2+y^2-8x-2y-51=0; 4\sqrt{13}$.

38. Show that the points $\left(1, \frac{\sqrt{11}+6}{6}\right)$ lies outside the circle $3x^2+3y^2-5x$

$-6y+4=0$.

39. Find the condition that the line $lx+my+n=0$ touches the circle $x^2+y^2=a^2$. Find also the point of contact.

Ans.: $\left(\frac{-a^2l}{n},\frac{-a^2m}{n}\right)$

40. Find the equation of the circle passing through the point $(3, 5)$ and $(5, 3)$ and having its centre on the line $2x+3y-1=0$.

Ans.: $5x^2+3y^2-14x-14y-50=0$

41. $ABCD$ is a square whose side is a. Taking line AO as the axis of coordinates, prove that the equation of the circumcircle of the square is $x^2+y^2-ax-ay=0$.

42. Find the equation of the circle with its centre on the line $2x+y=0$ and touching the lines $4x-3y+10=0$ and $4x-3y-3=0$.

Ans.: $x^2+y^2-2x+4y-11=0$

43. Find the equation of the circle that passes through the point $(1, 1)$ and touches the circle $x^2+y^2+4x-6y-3=0$ at the point $(2, 3)$ on it.

Ans.: $x^2+y^2+x-6y+3=0$

44. Prove that the tangent to the circle $x^2+y^2=5$ at the point $(1, -2)$ also touches the circle $x^2+y^2-8x+6y+20=0$ and find its point of contact.

Ans.: $(3, -1)$

45. A variable circle passes through the point $A(a, b)$ and touches the x-axis. Show that the locus of the other end of the diameter through A is $(x-c)^2=4by$.

46. Find the equation of the circle passing through the points $A(-5, 0)$, $B(1, 0)$, and $C(2, 1)$ and show that the line $4x-3y-5=0$ is a tangent to the line.

47. Find the equation of the circle through the origin and through the point of contact of the tangents from the origin to the circle.

Ans.: $2x^2 + 2y^2 - 11x - 13y = 0$

48. The circle $x^2 + y^2 - 4x - 4y + 4 = 0$ is inscribed in a triangle that has two of its sides along the coordinate axes. The locus of the circumference of the triangle is $x + y - xy + k\sqrt{x^2 + y^2} = 0$. Find k.

Ans.: $k = 1$

49. A circle of diameter 13 m with centre O coinciding with the origin of coordinate axes has diameter AB on the x-axis. If the length of the chord AC be 5 m, find the area of the smaller portion bounded between the circles and the chord AC.

Ans.: 1.9 m².

50. Find the radius of the smallest circle that touches the straight line $3x - y = 6$ at $(1, -3)$ and also touches the line $y = x$.

Ans.: $10\sqrt{2} - 4\sqrt{10}$

51. If $\left(m_i, \frac{1}{m_i}\right)$, $m_i > 0$, $i = 1, 2, 3, 4$ form distinct points on a circle show that $m_1, m_2, m_3, m_4 = 1$.

52. If the line $x\cos\alpha + y\sin\alpha = \rho$ cuts the circle $x^2 + y^2 = a^2$ in M and N, then show that the circle whose diameter is MN is $x^2 + y^2 - a^2 - 2\rho(x\cos\alpha + y\sin\alpha - \rho) = 0$.

53. Show that the tangents drawn from the point (8, 1) to the circle $x^2 + y^2 - 2x - 4y - 20 = 0$ are perpendicular to each other.

54. How many circles can be drawn each touching all the three lines $x + y = 1, y = x + 1$ and $7x - y = 6$? Find the centre and radius of all the circles.

Ans.: 4; $\left(\frac{7}{2}, 1\right), \frac{7\sqrt{2}}{4}$

55. Find the points on the line $x - y + 1 = 0$, the tangents from which to the circle $x^2 + y^2 - 3x = 0$ are of length 2 units.

Ans.: $\left(\frac{3}{2}, \frac{5}{2}\right)$; $(-1, 0)$

56. On the circle $16x^2 + 16y^2 + 48x - 3y - 43 = 0$, find the point nearest to the line $8x - 4y + 73 = 0$ and calculate the distance between this point and the line.

Ans.: $\left(\frac{-7}{2}, \frac{5}{4}\right)$; $2\sqrt{5}$

57. Find the equations of the lines touching the circle $x^2 + y^2 + 10x - 2y + 6 = 0$ and parallel to the line $2x + y - 7 = 0$.

Ans.: $2x + y - 1 = 0,\ 2x + y + 19 = 0$

58. Find the equation of the circle whose diameter is the chord of intersection of the line $x+3y=6$ and the curve $4x^2+9y^2=36$.

Ans.: $5(x^2+y^2)-12x-16y+12=0$

59. Find the equation for the circle concentric with the circle $x^2+y^2-8x+6y-5=0$ and passes through the point (–2, 7).

Ans.: $x^2+y^2-8x+6y-27=0$

60. Find the equation of the circle that cuts off intercepts –1 and –3 on the x-axis and touches the y-axis at the point $(0,\sqrt{3})$.

Ans.: $x^2+y^2+4x-2\sqrt{3}y+3=0$

61. Find the coordinates of the point of intersection of the line $5x-y+7=0$ and the circle $x^2+y^2+3x-4y-9=0$. Also find the length of the common segment.

Ans.: $\left(\frac{-1}{2},\frac{-1}{2}\right)$ and (1, 7), $\frac{\sqrt{117}}{2}$

62. The line $4x+3y+k=0$ is a tangent to the circle $x^2+y^2=4$. Find the value of k.

Ans.: $k=\pm 10$

63. Find the equations of tangents to the circle $x^2+y^2-6x+4y-17=0$ that are perpendicular to $3x-4y+5=0$.

Ans.: $4x+3y+19=0$, $4x+3y-31=0$

64. Find the equation of tangents to the circle $x^2+y^2-14x+y-5=0$ at the points whose abscissa is 10.

Ans.: $3x+7y-93=0, 3x-7y-64=0$

65. Show that the circles $x^2+y^2-4x+6y+8=0$ and $x^2+y^2-10x-6y+14=0$ touch each other. Find the point of contact.

Ans.: (3, –1)

66. Show that the tangent to the centre $x^2+y^2=0$ at the point (1, –2) also touches the circle $x^2+y^2-8x+6y+20=0$. Find the point of contact.

Ans.: (3, –1)

67. A straight line AB is divided at C so that $AB=3CB$. Circles are described on AC and CB as diameters and a common tangent meets AB produced at D. Show that BD is equal to the radius of the smaller circle.

68. The lines $3x-4y+4=0$ and $6x-8y-7=0$ are tangents to the same circle. Find the radius of this circle.

Ans.: $\frac{3}{4}$

69. From the origin, chords are drawn to the circle $(x-1)^2 + y^2 = 1$. Find the equation of the locus of the midpoint of these chords.

Ans.: $x^2 + y^2 - x = 0$

70. Find the equations of the pair of tangents to the circle $x^2 + y^2 - 2x + 4y = 0$ from (0,1).

Ans.: $2x^2 - 2y^2 + 3xy - 3x + 4y - 2 = 0$

71. If the polar of points on the circle $x^2 + y^2 = a^2$ with respect to the circle $x^2 + y^2 = b^2$ touch the circle $x^2 + y^2 = c^2$, show that a, b and c are in GP.

72. If the distances of origin to the centres of three circles $x^2 + y^2 - 2\lambda x = c^2$ where λ is a variable and c is a constant are in G. P, prove that the length of the tangent drawn to them from any point on the circle $x^2 + y^2 = c^2$ are in G. P.

73. A tangent is drawn to each of the circles $x^2 + y^2 = a^2$ and $x^2 + y^2 = b^2$. If the two tangents are mutually perpendicular, show that the locus of their point of intersection is a circle concentric with the given circles.

74. If the pole of any line with respect to the circle $x^2 + y^2 = a^2$ lies on the circle $x^2 + y^2 = 9a^2$, then show that the line will be a tangent to the circle $x^2 + y^2 = \frac{a^2}{9}$.

75. A triangle has two of its sides along the y-axis, and its third side touches the circle $x^2 + y^2 - 2ax - 2ay + a^2 = 0$. Prove that the locus of the circumcentre of the triangle is $2xy - 2a(x+y) + a^2 = 0$.

76. Lines $5x + 12y - 10 = 0$ and $6x - 11y - 40 = 0$ touch a circle C_1 of diameter 6. If the centre of C_1 lies in the first quadrant, find the equation of circle C_2. which is concentric with C_1 and cuts intercepts of length 8 on these lines.

Ans.: $\left(x - \frac{65}{5}\right)^2 + \left(y + \frac{5}{4}\right)^2 = 25$

77. Find the equation of the circle that touches the y-axis at a distance of 4 units from the origin and cuts off an intercept of 6 units from the x-axis.

Ans.: $x^2 + y^2 + 10x - 8y + 16 = 0$

78. Find the equation of the circle in which the line joining the points $(0, b)$ and $(b, -a)$ is a chord subtending an angle 45° at any point on its circumference

Ans.: $x^2 + y^2 - 2(a+b)x + 2(a-b)y + (a^2 + b^2)$

79. From any point on a given circle, tangents are drawn to another circle. Prove that the locus of the middle point of the chord of contact is a third circle; the distance between the centres of the given circle is greater than the sum of their radii.

80. A point moves so that the sum of the squares of the perpendiculars that fall from it on the sides of an equilateral triangle is constant. Prove that the locus is a circle.

81. A circle of constant radius passes through the origin O and cuts the axes in A and B. Show that the locus of the foot of the perpendicular from AB is $(x^2+y^2)^2(x^2+y^2)=4r^2$.

82. Find the equation of the image of the circle $(x-3)^2+(y-2)^2=1$ by the mirror $x+y=19$.

Ans.: $(x-17)^2+(y-16)^2=1$

83. Find the value of λ for which the circle $x^2+y^2+6x+5+\lambda(x^2+y^2-8x+7)=0$ dwindles into a point.

Ans.: $2\pm\frac{4\sqrt{2}}{3}$

84. A variable circle always touches the line $y=x$ and passes through the point (0, 0). Show that the common chords of this circle and $x^2+y^2+6x+8y-7=0$ will pass through a fixed point $\left(\frac{1}{2},\frac{1}{2}\right)$.

85. The equation of the circle that touches the axes of the coordinates and the line $\frac{x}{3}+\frac{y}{4}=1$ and whose centre lies in the first quadrant is $x^2+y^2-2cx-2cy+c^2=0$. Find the values of c.

Ans.: (1, 6)

86. A region in xy-plane is bounded by the curve $y=\sqrt{25-x^2}$ and the line $y=0$. If the point $(a, a+1)$ lies in the interior of the region, find the range of a.

Ans.: $a \subset (\ 1, 3)$

87. The points (4, –2) and (3, 6) are conjugate with respect to the circle $x^2+y^2=24$. Find the value of b.

Ans.: $b=-6$

88. If the two circles $x^2+y^2+2gx+2fy=0$ and $x^2+y^2+2g^1x+2f^1y=0$ touch each other, show that $f^1g=gf^1$.

89. Show that the locus of the points of chords of contact of tangents subtending a right angle at the centre is a concentric circle whose radius is $\sqrt{2}$ times the radius of the given circle. Also show that this is also the locus of the point of intersection of perpendicular tangents.

90. Show that the points (x_i, y_i), $i=1, 2, 3$ are collinear if and only if their poles with respect to the circles $x^2+y^2=a^2$ are concurrent.

91. The length of the tangents from two given points A and B to a circle are t_1 and t_2, respectively. If the points are conjugate points, show that $AO^2=t_1^2+t_2^2$.

92. Show that the equation to the pair of tangents drawn from the origin to the circle $x^2+y^2+2gx+2fy+c=0$ is $(gx+fy)^2=(f^2+g^2)$. Hence find the locus of the centre of the circle if these tangents are perpendicular.

Ans.: $x^2+y^2=2c$

93. Three sides of a triangle have the equations $L_i = y - m_r x - c_r = 0, r = 1, 2, 3.$ Then show that $\lambda L_2 L_3 + \mu L_3 L_1 + \upsilon L_1 L_2 = 0$ where $\lambda, \mu, \upsilon \neq 0$ is the equation of the circumcircle of the triangle if $\sum \lambda(m_2 + m_3) = 0$ and $\sum \lambda(m_2 m_3 - m_1) = 0$.

94. A triangle is formed by the lines whose combined equation is $c(x + y - 4)(xy - 2x - y + 2) = 0$. Show that the equation of its circumference is $x^2 + y^2 - 3x - 5y + 8 = 0$.

95. Two distinct chords drawn from the point (p, q) on the circle $x^2 + y^2 = px + qy$, where $pq \neq 0$, are bisected by the x-axis. Show that $p^2 > 8q^2$.

96. Show that the number of points with integral coordinates that are interior to the circle $x^2 + y^2 = 16$ is 45.

97. Find the number of common tangents to the circles $x^2 + y^2 - 6x - 14y + 48 = 0$ and $x^2 + y^2 - 6x = 0$.

Ans.: 4

98. The tangents to the circle $x^2 + y^2 = 4$ at the points A and B meet at $P(-4, 0)$. Find the area of the quadrilateral $PAOB$.

Ans.: $4\sqrt{3}$ m²

99. The equations of four circles are $(x \pm a)^2 + (y \pm a)^2 = a^2$. Find the radius of a circle touching all the four circles.

Ans.: $(\sqrt{2} \pm 1)a$

100. A circle of radius 2 touches the coordinate axes in the first quadrant. If the circle makes a complete rotation on the x-axis along the positive direction of the x-axis, then show that the equation of the circle in the new position is $x^2 + y^2 - 4(x + y) - 8\lambda x + (2 + 4\pi)^2 = 0$.

101. Two tangents are drawn from the origin to a circle with centre at $(2, -1)$. If the equation of one of the tangents is $3x + y = 0$, find the equation of the other tangent.

Ans.: $x - 3y = 0$

102. Find the equation of the chord of the circle $x^2 + y^2 = a^2$ passing through the point (2, 3) farther from the centre.

Ans.: $2x + 3y = 17$

103. An equilateral triangle is inscribed in the circle $x^2 + y^2 = a^2$, with the vertex at (a, 0). Find the equation of the side opposite to this vertex.

Ans.: $2x + a = 0$

104. A line is drawn through the point $P(3, 1)$ to cut the circle $x^2 + y^2 = 9$ at A and B. Find the value of $PA \cdot PB$.

Ans.: 121

105. C_1 and C_2 are circles of unit radius with their centres at (0, 0) and (1, 0), respectively. C_3 is a circle of unit radius, passing through the centres of the circles C_1 and C_2 and having its centre above the x-axis. Find the equation of the common tangent to C_1 and C_3 that passes through C_2.

Ans.: $\sqrt{3}x - y + 2 = 0$

106. A chord of the circle $x^2 + y^2 - 4x - 6y = 0$ passing through the origin subtends an angle $\tan^{-1}\left(\frac{7}{4}\right)$ at the point where the circle meets the positive y-axis. Find the equation of the chord.

Ans.: $x - 2y = 0$

107. A circle with its centre at the origin and radius equal to a meets the axis of x at A and B. P and Q are respectively the points $(a\cos\alpha, a\tan\alpha)$ and $(a\cos\beta, a\tan\beta)$ such that $\alpha - \beta = 2\gamma$. Show that the locus of the point of intersection of AP and BQ is $x^2 + y^2 - 2ay\tan\gamma = -2$.

108. A circle C_1 of radius touches the circle $x^2 + y^2 = a^2$ externally and has its centre on the positive x-axis. Another circle C_2 of radius c touches circle C_1 externally and has its centre on the positive x-axis. If $a < b < c$, show that the three circles have a common tangent if a, b, c are in GP.

109. Find the equations of common tangents to the circles $x^2 + y^2 + 14x - 14y + 28 = 0$ and $x^2 + y^2 - 14x + 4y - 28 = 0$

Ans.: $28y + 45y + 371 = 0$ and $y - 7 = 0$.

110. If a circle passes through the points of intersection of the coordinate axes with the line $x - \lambda y + 1 = 0 (\lambda \neq 0)$ and $x - 2y + 3 = 0$ then λ satisfies the equation $6\lambda^2 - 7\lambda + 2 = 0$.

111. OA and OB are equal chords of the circle $x^2 + y^2 - 2x + 4y = 0$ perpendicular to each other and passing through the origin. Show that the slopes of OA and OB satisfy the equation $3m^2 - 8m - 3 = 0$.

112. Find the equation of the circle passing through the points (1, 0) and (0, 1) and having the smallest possible radius.

Ans.: $x^2 + y^2 - x - y = 0$

113. Find the equation of the circle situated systematically opposite to the circle $x^2 + y^2 - 2x = 0$ with respect to the line $x + y = 2$.

Ans.: $x^2 + y^2 - 4x - 2y + 4 = 0$

114. O is a fixed point and R moves along a fixed line L not passing through O. If S is taken on OR such that $OR \cdot OS = K^2$, then show that the locus of S is a circle.

115. Show that the circumference of the triangle formed by the lines $ax + by + c = 0$, $bx + cy + a = 0$ and $cx + ay + b = 0$ passes through the origin if $(b^2 + c^2)(c^2 + a^2)(a^2 + b^2) = abc(b + c)(c + a)(a + b)$.

116. Two circles are drawn through the points $(a, 5a)$ and $(4a, a)$ to touch the y-axis. Prove that they intersect at an angle $\tan^{-1}\left(\frac{40}{9}\right)$.

117. Show that the locus of a point P that moves so that its distance from the given point O is always in a given ratio $n:1\cdot(n \neq -1)$ to its distance on the line joining the points that divides the line OA in the given ratio as diameter.

118. The lines $3x-4y+4=0$ and $6x-3y-7=0$ are tangents to the same circle. Find the radius of the circle.

Ans.: $r=\frac{3}{4}$

119. The line $y=x$ touches a circle at P so that $OP=4\sqrt{2}$ where O is the origin. The point $(-10, 2)$ is inside the circle and length of the chord on the line $x+y=0$ is $6\sqrt{2}$. Find the equation of the line.

Ans.: $x^2+y^2+18x-2y+32=0$

120. Find the intervals of values of a for which the line $y+x=0$ bisects two chords drawn from a point $\left(\frac{1+\sqrt{2}a}{2}, \frac{1-\sqrt{2}a}{2}\right)$ to the circle $2x^2+2y^2-\left(1+\sqrt{2}a\right)x-\left(1-\sqrt{2}a\right)y=0$.

121. Show that all chords of the circle $3x^2-y^2-2x+4y=0$ that subtend a right angle at the origin are concurrent. Does the result hold for the curve $3x^2+3y^2-2x+4y=0$? If yes, what is the point of concurrency, and if not, give the reason.

122. Find the equations of the common tangents to the circles $x^2+y^2-14x+6y+33=0$ and $x^2+y^2+30x-20y+1=0$.

Ans.: $4x-3y-12=0,\ 24x+7y-22=0$

123. Prove that the orthocentre of the triangle whose angular points are $(a\cos\alpha, a\sin\alpha)$, $(a\cos\beta, a\sin\beta)$ and $(a\cos\gamma, a\sin\gamma)$ is the point $[a(\cos\alpha+\cos\beta+\cos\gamma), a(\sin\alpha+\sin\beta+\sin\gamma)]$.

CHAPTER

5 System of Circles

5.1 RADICAL AXIS OF TWO CIRCLES

Definition 5.1.1: The radical axis of two circles is defined as the locus of a point such that the lengths of tangents from it to the two circles are equal.

Obtain the equation of the radical axis of the two circles $S \equiv x^2 + y^2 + 2gx + 2fy + c = 0$ and $S_1 \equiv x^2 + y^2 + 2g_1x + 2f_1y + c_1 = 0$.

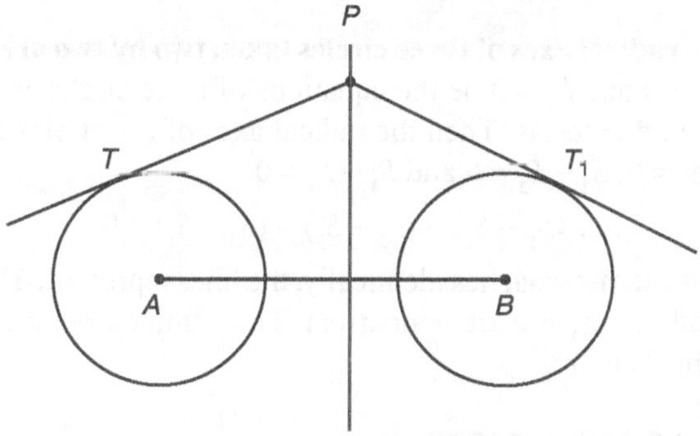

Let $P(x_1, y_1)$ be a point such that the lengths of tangents to the two circles are equal.

$$\text{(i.e.)} \quad PT = PT_1$$

$$\Rightarrow PT^2 = PT_1^2$$

$$\therefore x_1^2 + y_1^2 + 2gx_1 + 2fy_1 + c = x_1^2 + y_1^2 + 2g_1x_1 + 2f_1y_1 + c_1$$

$$2(g - g_1)x_1 + 2(f - f_1)y_1 + c - c_1 = 0$$

The locus of (x_1, y_1) is $2(g - g_1)x + 2(f - f_1)y + (c - c_1) = 0$ which is a straight line.
Therefore, the radical axis of two given circle is a straight line.

Note 5.1.1: If $S = 0$ and $S_1 = 0$ are the equations of two circles with unit coefficients for x^2 and y^2 terms then the equation of the radical axis is $S - S_1 = 0$.

Note 5.1.2: Radical axis of two circles is a straight line perpendicular to the line of centres.

The centres of the two circles are $A(-g, -f)$ and $B(-g_1, -f_1)$.

The slope of the line of centres is $m_1 = \dfrac{-f+f_1}{-g+g_1} = \dfrac{f-f_1}{g-g_1}$

The slope of the radical axis is $m_2 = \dfrac{-(g-g_1)}{f_1-f_1}$

$$\therefore\ m_1 m_2 = -1$$

Therefore, the radical axis is perpendicular to the line of centres.

Note 5.1.3: If the two circles $S = 0$ and $S_1 = 0$ intersect then the radical axis is the common chord of the two circles.

Note 5.1.4: If the two circles touch each other, then the radical axis is the common tangent to the circles.

Note 5.1.5: If a circle bisects the circumference of another circle then the radical axis passes through the centre of the second circle.

Show that the radical axes of three circles taken two by two are concurrent.
Let $S_1 = 0$, $S_2 = 0$ and $S_3 = 0$ be the equations of three circles with unit coefficients for x^2 and y^2 terms. Then the radical axes of the circles taken two by two are $S_1 - S_2 = 0$, $S_2 - S_3 = 0$ and $S_3 - S_1 = 0$.

$$\therefore\ (S_1 - S_2) + (S_2 - S_3) + (S_3 - S_1) \equiv 0$$

Since sum of the terms vanishes identically, the lines represented by $S_1 - S_2 = 0$, $S_2 - S_3 = 0$ and $S_3 - S_1 = 0$ are concurrent. The common point of the lines is called the radical centre.

5.2 ORTHOGONAL CIRCLES

Definition 5.2.1: Two circles are defined to be orthogonal if the tangents at their point of intersection are at right angles.

Find the condition for the circles $S \equiv x^2 + y^2 + 2gx + 2fy + c = 0$, $S_1 \equiv x^2 + y^2 + 2g_1x + 2f_1y + c_1 = 0$ to be orthogonal.

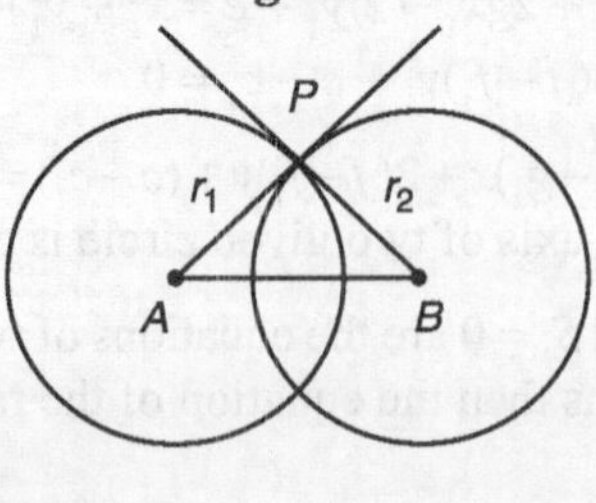

Let P be a point of intersection of the two circles $S = 0$ and $S_1 = 0$. The centres are $A(-g, -f)$, $B(-g_1, -f_1)$.

The radii are $r_1 = \sqrt{g^2 + f^2 - c}$ and $r_2 = \sqrt{g_2^2 + f_2^2 - c_1}$.

Since the two circles are orthogonal, PA is perpendicular to PB.
(i.e.) APB is a right triangle.

$$\therefore AB^2 = PA^2 + PB^2$$

$$\Rightarrow (-g + g_1)^2 + (-f + f_1)^2 = (g^2 + f^2 - c) + (g_1^2 + f_1^2 - c_1)$$

$$\Rightarrow g^2 + g_1^2 - 2gg_1 + f^2 + f_1^2 - 2ff_1 = g^2 + g_1^2 + f^2 + f_1^2 - c - c_1$$

$$\Rightarrow 2gg_1 + 2ff_1 = c + c_1$$

Show that if a circle cuts two given circles orthogonally then its centre lies on the radical axis of the two given circles.

Let $S_1 = x^2 + y^2 + 2g_1x + 2f_1y + c_1 = 0$ and $S_2 = x^2 + y^2 + 2g_2x + 2f_2y + c_1 = 0$ be the two given circles.

Let $S = x^2 + y^2 + 2gx + 2fy + c = 0$ cuts $S_1 = 0$ and $S_2 = 0$ orthogonally.
Since $S = 0$ cuts $S_1 = 0$ and $S_2 = 0$ orthogonally,

$$2gg_1 + 2ff_1 = c + c_1; \qquad 2gg_2 + 2ff_2 = c + c_2$$

By subtracting, we get

$$2g(g_1 - g_2) + 2f(f_1 - f_2) = c_1 - c_2$$

or
$$-2g(g_1 - g_2) - 2f(f_1 - f_2) + (c_1 - c_2) = 0$$

This shows that $(-g, -f)$ lies on the line, $2(g_1 - g_2)x + 2(f_1 - f_2)y + (c_1 - c_2) = 0$ which is the radical axis of the two circles.

Therefore, the centre of the circle $S = 0$ lies on the radical axis of the circles $S_1 = 0$ and $S_2 = 0$.

5.3 COAXAL SYSTEM

Definition 5.3.1: A system of circles is said to be coaxal if every pair of the system has the same radical axis.

Express the equation of a coaxal system of circles in the simplest form.

In a coaxal system of circles, every pair of the system has the same radical axis. Therefore, there is a common radical axis to a coaxal system of circles.

Hence, in a coaxal system the centres are all collinear and the common radical axis is perpendicular to the lines of centres. Therefore, let us choose the line of centres as x-axis and the common radical axis as y-axis.

Let us consider two circles of the coaxal system,

$$x^2 + y^2 + 2gx + 2fy + c = 0, x^2 + y^2 + 2g_1x + 2f_1y + c_1 = 0$$

Since the centres lie on the x-axis, $f_1 = 0$ and $f_2 = 0$.

Therefore, the equations of the circles are $x^2 + y^2 + 2gx + c = 0$ and $x^2 + y^2 + 2g_1x + c_1 = 0$.

The radical axis of these two circles is $2(g - g_1)x + (c - c_1) = 0$.

However, the common radical axis is the y-axis whose equation is $x = 0$.

$$\therefore c - c_1 = 0 \quad \text{or} \quad c = c_1.$$

Hence, the general equation to a coaxal system of circles is $x^2 + y^2 + 2gx + c = 0$ where g is a variable and c is a constant.

So the equation of a coaxal system can be expressed in the simplest form

$$x^2 + y^2 + 2\lambda x + c = 0$$

where λ is a variable and c is a constant.

5.4 LIMITING POINTS

Definition 5.4.1: Limiting points are defined to be the centres of point circles belonging to a coaxal system; that is, they are centres of circles of zero radii belonging to a coaxal system.

Obtain the limiting points of the coaxal system of circles $x^2 + y^2 + 2\lambda x + c = 0$. Centres are $(-\lambda, 0)$ and radii are $\sqrt{\lambda^2 - c}$.

For point circles radii are zero.

$$\therefore \sqrt{\lambda^2 - c} = 0 \text{ or } \lambda^2 = c$$

$$\therefore \lambda = (\pm\sqrt{c}, 0)$$

Therefore, limiting points are $(\sqrt{c}, 0)$ and $(-\sqrt{c}, 0)$.

Theorem 5.4.1: The polar of one limiting point of a coaxal system of circles with respect to any circle of the system passes through the other limiting point.

Proof: Let x-axis be the line of centres and y-axis be the common radical axis of a coaxal system of circles. Then any circle of the coaxal system is

$$x^2 + y^2 + 2\lambda x + c = 0 \tag{5.1}$$

where λ is a variable and c is a constant.

The limiting points of this coaxal system of circles are $(\sqrt{c},0)$ and $(-\sqrt{c},0)$. The polar of the point $(\sqrt{c},0)$ with respect to the circle (5.1) is

$$\sqrt{c}x+\lambda(x+\sqrt{c})+c=0$$

$$\text{(i.e.)}\quad (\lambda+\sqrt{c})(x+\sqrt{c})=0 \text{ since } \lambda+\sqrt{c}\neq 0,\ x+\sqrt{c}=0.$$

This linc passes through the other limiting point. For every coaxal system of circles there exists an orthogonal system of circles. Let x-axis be the line of centres and y-axis be the common radical axis. Then the equation to a coaxal system of circles is

$$x^2+y^2+2\lambda x+c=0 \tag{5.2}$$

Let us assume that the circle

$$x^2+y^2+2gx+2fy+k=0 \tag{5.3}$$

cut every circle of the coaxal system of circles given by (5.2) orthogonally. Then the condition for orthogonality is

$$2g\lambda=c+k \tag{5.4}$$

Let us now consider two circles of the coaxal system for the different values of λ, say λ_1 and λ_2. The condition (5.4) becomes $2g\lambda_1=c+k$, $2g\lambda_2=c+k$.

$$\therefore\ 2(\lambda_1-\lambda_2)g=0.$$

Since $\lambda_1-\lambda_2\neq 0, g=0$ and so $k=-c$.

Hence, from (5.3) the equation of the circle which cuts every member of the system (5.2) is $x^2+y^2+2fy-c=0$, where f is an arbitrary constant. Therefore, for every coaxal system of circles there exists an orthogonal system of circles given by $x^2+y^2+2fy-c=0$; where f is a variable and c is a constant. For this system of orthogonal circles y-axis is the line of centres and x-axis is the common radical axis.

Note 5.4.1: Every circle of the orthogonal coaxal system of circles passes through the limiting points $(\pm\sqrt{c},0)$.

Theorem 5.4.2: If $S=x^2+y^2+2gx+2fy+c=0$ and $S_1=x^2+y^2+2g_1x+2f_1y+c_1=0$ be any two circles of a coaxal system then any circle of coaxal system can be expressed in the form $S+\lambda S_1=0$.

Proof:

$$S=x^2+y^2+2gx+2fy+c=0 \tag{5.5}$$

$$S_1=x^2+y^2+2g_1x+2f_1y+c_1=0 \tag{5.6}$$

Consider,

$$S + \lambda S_1 = 0 \tag{5.7}$$

where λ is a variable. In this equation, $(1 + \lambda)$ is the coefficient of x^2 and y^2. Dividing by $(1 + \lambda)$ equation (5.7) becomes $\frac{S + \lambda S_1}{1+\lambda} = 0$ in which the coefficient of x^2 and y^2 are unity.

Now consider two different values of λ, that is, λ_1 and λ_2. Then, $\frac{S + \lambda_1 S_1}{1+\lambda_1} = 0$ and $\frac{S + \lambda_2 S_1}{1+\lambda_2} = 0$.

The radical axis of these two circles is $\frac{S + \lambda_1 S_1}{1+\lambda_1} - \frac{S + \lambda_2 S_1}{1+\lambda_2} = 0$

$$\text{(i.e.)} \quad (1 + \lambda_2)(S + \lambda_1 S_1) - (1 + \lambda_1)(S + \lambda_2 S_1) = 0$$

$$\text{(i.e.)} \quad (\lambda_1 - \lambda_2)(S - S_1) = 0.$$

Since $\lambda_1 - \lambda_2 \neq 0$ and therefore $S - S_1 = 0$ which is the common radical axis. Therefore, every member of the coaxal system can be expressed in the form $S + \lambda S_1 = 0$ where λ is a variable.

Theorem 5.4.3: If $S = x^2 + y^2 + 2gx + 2fy + c = 0$ is a circle of a coaxal system and $L = lx + my + n = 0$ is the common radical axis of the system then $S + \lambda L = 0$ is the equation of a circle of the coaxal system of circles.

Proof:

$$S = x^2 + y^2 + 2gx + 2fy + c = 0 \tag{5.8}$$

$$L = lx + my + n = 0 \tag{5.9}$$

$$S + \lambda L = 0 \tag{5.10}$$

$$\text{(i.e.)} \quad x^2 + y^2 + 2gx + 2fy + c + \lambda(lx + my + n) = 0$$

Consider two members of the system (5.10) for the different values of λ, that is, λ_1 and λ_2.

Then, $S + \lambda_1 L = 0$ and $S + \lambda_2 L = 0$

The radical axis of these two circles is $(\lambda_1 - \lambda_2)L = 0$.

Since $\lambda_1 - \lambda_2 \neq 0$, $L = 0$ which is the common radical axis. Therefore, $S + \lambda L = 0$ represents any circle of the coaxal system in which $S = 0$ is a circle and $L = 0$ is the common radical axis.

5.5 EXAMPLES (RADICAL AXIS)

Example 5.5.1

Find the radical axis of the two circles $x^2 + y^2 + 2x + 4y - 7 = 0$ and $x^2 + y^2 - 6x + 2y - 5 = 0$ and show that it is at right angles to the line of centres of the two circles.

Solution

$$x^2+y^2+2x+4y-7=0 \qquad (5.11)$$
$$x^2+y^2-6x+2y-5=0 \qquad (5.12)$$

The radical axis of the circles is $S-S_1=0$.

$$\text{(i.e.)}\quad 8x+2y-2=0$$
$$\text{(i.e.)}\quad 4x+y-1=0 \qquad (5.13)$$

The slope of the radical axis is $m_1=-4$.

The centres of the two circles are $(-1,-2)$ and $(3,-1)$.

The slope of the line of centres is $m_2=\dfrac{-2+1}{-1-3}=\dfrac{1}{4}$.

$$m_1m_2=(-4)\left(\frac{1}{4}\right)=-1$$

Therefore, the radical axis is perpendicular to the line of centres.

Example 5.5.2

Show that the circle $x^2+y^2+2gx+2fy+c=0$ will bisect the circumference of the circle $x^2+y^2+2g_1x+2f_1y+c_1=0$, if $2g_1(g-g_1)+2f_1(f-f_1)=c-c_1$.

Solution

Let

$$x^2+y^2+2gx+2fy+c=0 \qquad (5.14)$$
$$x^2+y^2+2g_1x+2f_1y+c_1=0 \qquad (5.15)$$

The radical axis of these two circles is $2(g-g_1)c+2(f-f_1)y+c-c_1=0$, Circle (5.14) bisects the circumference of the circle (5.15).

Therefore, radical axis passes through the centre of the second circle.

The radical axis of the two given circles be

$$2(g-g_1)(-g_1)+2(f-f_1)(-f_1)+c-c_1=0$$
$$\text{(i.e.)}\quad 2(g-g_1)\,g_1+2(f-f_1)f_1-(c-c_1)=0$$

Example 5.5.3

Show that the circles $x^2+y^2-4x+6y+8=0$ and $x^2+y^2-10x-6y+14=0$ touch each other and find the coordinates of the point of contact.

Solution

$$x^2+y^2-4x+6y+8=0 \qquad (5.16)$$
$$x^2+y^2-10x-6y+14=0 \qquad (5.17)$$

The radical axis of these two circles is $6x + 12y - 6 = 0$.

$$\text{(i.e.)}\quad x + 2y - 1 = 0 \tag{5.18}$$

The centres of the circles are $A\ (2, -3)$ and $B\ (5, 3)$.

The radii of the circles are $r_1 = \sqrt{4+9-8} = \sqrt{5}$ and $r_2 = \sqrt{25+9-14} = 2\sqrt{5}$.

The perpendicular distance from $A(2, -3)$ on the radical axis $x + 2y - 1 = 0$ is $\dfrac{12-6-11}{\sqrt{1+4}} = \dfrac{5}{\sqrt{5}} = \sqrt{5}$ = radius of the first circle.

Therefore, radical axis touches the first circle and hence the two circles touch each other.

The equation of the lines of centres is $\dfrac{y+3}{x-2} = \dfrac{-3-3}{2-5} = \dfrac{-6}{-3} = 2$

or

$$2x - 4 = y + 3 \Rightarrow 2x - y - 7 = 0 \tag{5.19}$$

Solving (5.18) and (5.19), we get the point of contact.

Therefore, the point of contact is $(3, -1)$.

Example 5.5.4

Show that the circles $x^2 + y^2 + 2ax + c = 0$ and $x^2 + y^2 + 2by + c = 0$ touch if $\dfrac{1}{a^2} + \dfrac{1}{b^2} = \dfrac{1}{c}$.

Solution

The radical axis of the two given circles is $2ax - 2by = 0$. The centre of the first circle is $(-a, 0)$. The radius of the first circle is $\sqrt{a^2 - c}$.

If the two circles touch each other, then the perpendicular distance from the centre $(-a, 0)$ to the radical axis is equal to the radius of the circle.

$$\therefore\ \frac{a^2}{\sqrt{a^2+b^2}} = \sqrt{a^2 - c} \Rightarrow a^4 = (a^2 - c)(a^2 + b^2)$$

$$\Rightarrow a^4 = a^4 - c(a^2 + b^2) + a^2b^2$$

$$\Rightarrow c(a^2 + b^2) = a^2b^2$$

On dividing by a^2b^2c, we get $\dfrac{1}{a^2} + \dfrac{1}{b^2} = \dfrac{1}{c}$.

Example 5.5.5

Find the radical centre of the circles $x^2 + y^2 + 4x + 7 = 0$, $2x^2 + 2y^2 + 3x + 5y + 9 = 0$ and $x^2 + y^2 + y = 0$.

Solution

Let

$$x^2 + y^2 + 4x + 7 = 0 \tag{5.20}$$

$$x^2 + y^2 + \frac{3}{2}x + \frac{5}{2}y + \frac{9}{2} = 0 \tag{5.21}$$

$$x^2 + y^2 + y = 0 \tag{5.22}$$

The radical axis of circles (5.20) and (5.22) is $\frac{5}{2}x - \frac{5}{2}y + \frac{5}{2} = 0$.

$$\Rightarrow x - y + 1 = 0 \tag{5.23}$$

The radical axis of the circles (5.20) and (5.22) is

$$4x - y + 7 = 0 \tag{5.24}$$

Solving (5.23) and (5.24) we get the radical centre as follows:

$$\begin{aligned} x - y &= -1 \\ 4x - y &= -7 \\ \hline 3x &= -6 \\ \Rightarrow x &= -2 \end{aligned}$$

$$-2 - y = -1$$

$$y = -2 + 1 = -1$$

Therefore, the radical centre is (–2, –1).

Example 5.5.6

Prove that if the points of intersection of the circles $x^2 + y^2 + ax + by + c = 0$ and $x^2 + y^2 + a_1x + b_1y + c_1 = 0$ by the lines $Ax + By + C = 0$ and $A_1x + B_1y + C_1 = 0$ are concyclic if

$$\begin{vmatrix} a - a_1 & b - b_1 & c - c_1 \\ A & B & C \\ A_1 & B_1 & C_1 \end{vmatrix} = 0.$$

Solution

Let

$$x^2+y^2+ax+by+c=0 \quad (5.25)$$

$$x^2+y^2+a_1x+b_1y+c_1=0 \quad (5.26)$$

$$Ax+By+C=0 \quad (5.27)$$

$$A_1x+B_1y+C_1=0 \quad (5.28)$$

$Ax + By + C = 0$ meets the circle (5.25) at P and Q and $A_1x + B_1y + C_1 = 0$ meets the circle (5.26) at R and S. Since P, Q, R and S are concyclic, the equation of this circle be

$$x^2+y^2+2gx+2fy+c=0 \quad (5.29)$$

The radical axis of the circles (5.25) and (5.26) is

$$(a-a_1)x+(b-b_1)y+(c-c_1)=0 \quad (5.30)$$

The radical axis of circles (5.25) and (5.29) is

$$Ax+By+C=0 \quad (5.31)$$

The radical axis of circles (5.26) and (5.29) is

$$A_1x+B_1y+C_1=0 \quad (5.32)$$

Since these three radical axes are concurrent we get from equations (5.30), (5.31) and (5.32),

$$\begin{vmatrix} a-a_1 & b-b_1 & c-c_1 \\ A & B & C \\ A_1 & B_1 & C_1 \end{vmatrix}=0.$$

Example 5.5.7

Prove that the difference of the square of the tangents to two circles from any point in their plane varies as the distance of the point from their radical axis.

Solution

Let $P(x_1, y_1)$ be any point and the two circles be

$$x^2+y^2+2gx+2fy+c=0 \quad (5.33)$$

$$x^2+y^2+2g_1x+2f_1y+c_1=0 \quad (5.34)$$

The equation to the radical axis of these two circles be

$$2(g-g_1)x+2(f-f_1)y+c-c_1=0 \tag{5.35}$$

$$PT^2-PT_1^2=(x_1^2+y_1^2+2gx_1+2fy_1+c)-(x^2+y_1^2+2g_1x_1+2f_1y_1+c_1)$$

$$PT^2-PT_1^2=2(g-g_1)x_1+2(f-f_1)y+c-c_1 \tag{5.36}$$

The perpendicular distance of the point from the radical axis is

$$PR=\frac{2(g-g_1)x_1+2(f-f_1)y_1+c-c_1}{2\sqrt{(g-g_1)^2+(f-f_1)^2}} \tag{5.37}$$

From equations (5.36) and (5.37), we get $PT^2-PT_1^2 \propto PR$.

Example 5.5.8

Prove that for all constants λ and μ, the circle $(x-a)(x-a+\lambda)+(y-b)(y-b+\mu)=r^2$ bisects the circumference of the circle $(x-a)^2+(y-b)^2=r^2$.

Solution

$$(x-a)^2+(y-b)^2+\lambda(x-a)+\mu(y-b)-r^2=0$$

(i.e.) $$x^2+y^2-x(\lambda-2a)+y(\mu-2b)+a^2+b^2-\lambda a+\mu b-r^2=0 \tag{5.38}$$

$$(x-a)^2+(y-b)^2=r^2$$

(i.e.) $$x^2+y^2-2ax-2by+a^2+b^2=0 \tag{5.39}$$

The radical axis of these two circles is

$$\lambda x+\mu y-\lambda a-\mu b=0 \tag{5.40}$$

The centre of the second circle is (a, b). Substituting $x=a$, $y=b$ in (5.40), we get

$$\lambda a+\mu b-\lambda a-\mu b=0.$$

$\therefore$ (a, b) lies on the radical axis.

Therefore, the radical axis bisects the circumference of the second circle.

Example 5.5.9

Prove that the length of common chord of the two circles $x^2+y^2+2\lambda x+c=0$ and $x^2+y^2+2\mu y-c=0$ is $2\sqrt{\dfrac{(\lambda^2-c)(\mu^2+c)}{\lambda^2+\mu^2}}$.

Solution

The two given circles are

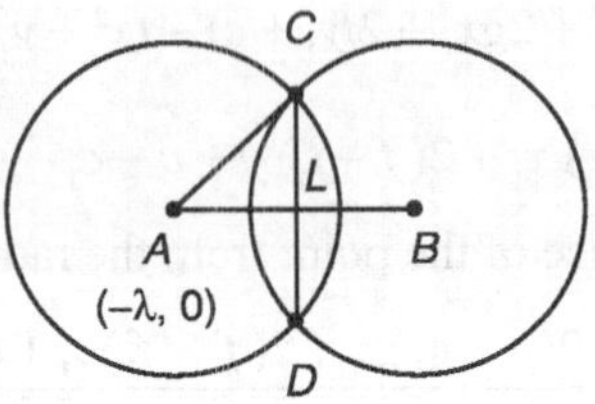

$$x^2 + y^2 + 2\lambda x + c = 0$$

$$x^2 + y^2 + 2\mu y - c = 0$$

Centres are $A(-\lambda, 0)$ and $B(-\mu, 0)$, radii are $AC = \sqrt{\lambda^2 - c}$ and $BC = \sqrt{\mu^2 + c}$.

The radical axis is $\lambda x - \mu y + c = 0$. The perpendicular distance from A on CD is $AL = \dfrac{|\lambda^2 - c|}{\sqrt{\lambda^2 + \mu^2}}$.

$$\therefore CL^2 = AC^2 - AL^2 = (\lambda^2 - c) - \frac{(\lambda^2 - c)^2}{\lambda^2 + \mu^2}$$

$$= \frac{(\lambda^2 - c)(\lambda^2 + \mu^2) - (\lambda^2 - c)^2}{\lambda^2 + \mu^2}$$

$$= \frac{(\lambda^2 - c)[\lambda^2 + \mu^2 - \lambda^2 + c]}{\lambda^2 + \mu^2}$$

$$= \frac{(\lambda^2 - c)(\mu^2 + c)}{\lambda^2 + \mu^2}$$

Therefore, the length of common chord $= CD = 2CL = 2\sqrt{\dfrac{(\lambda^2 - c)(\mu^2 + c)}{\lambda^2 + \mu^2}}$.

Example 5.5.10

Show that the circle $x^2 + y^2 - 8x - 6y + 21 = 0$ is orthogonal to the circle $x^2 + y^2 - 2y - 15 = 0$. Find the common chord and the equation of the circle passing through the centres and intersecting points of the circles.

Solution

$$x^2 + y^2 - 8x - 6y + 21 = 0$$
$$x^2 + y^2 - 2y - 15 = 0$$

The condition for orthogonality is $2gg_1 + 2ff_1 = c + c_1$.

$$\text{(i.e.)}\quad 2(-4)(0) + 2(-3)(-1) = 21 - 15$$

$$0 + 6 = 6 \text{ which is true.}$$

Therefore, the two circles cut each other orthogonally. The equation of the common chord is $S - S_1 = 0$.

$$\text{(i.e.)}\quad -8x - 4y + 36 = 0$$

$$\text{(i.e.)}\quad 2x + y - 9 = 0$$

Any circle passing through the intersection of the circles is $S + \lambda L = 0$.

$$\text{(i.e.)}\quad x^2 + y^2 - 8x - 6y + 21 + \lambda(2x + y - 9) = 0.$$

This passes through the centre (4, 3) of the first circle.

$$16 + 9 - 32 - 18 + 21 + \lambda(8 + 3 - 9) = 0$$

$$\Rightarrow -4 + \lambda(2) = 0$$

$$\Rightarrow \lambda = 2$$

Therefore, the equation of the required circle is $x^2 + y^2 - 8x - 6y + 21 + 2(2x + y - 9) = 0$.

$$\text{(i.e.)}\quad x^2 + y^2 - 4x - 4y + 3 = 0$$

Example 5.5.11

Find the equation to the circle which cuts orthogonally the three circles $x^2 + y^2 + 2x + 17y + 4 = 0$, $x^2 + y^2 + 7x + 6y + 11 = 0$ and $x^2 + y^2 - x + 22f + 33 = 0$.

Solution

Let the equation of the circle which cuts orthogonally the three given circles be $x^2 + y^2 + 2gx + 2fy + c = 0$.

Then the conditions for orthogonality are

$$2g + 17f = c + 4 \quad (5.41)$$

$$7g + 6f = c + 11 \quad (5.42)$$

$$-g + 22f = c + 3 \quad (5.43)$$

(5.41) – (5.42) gives $\quad -5g + 11f = -7 \quad$ (5.44)

$$(5.42) - (5.43) \text{ gives} \qquad 3g - 5f = 1 \qquad (5.45)$$

$$(5.44) \times 3 \text{ gives} \qquad -15g + 33f = -21$$

$$(5.45) \times 5 \text{ gives} \qquad \underline{15g - 25f = 5}$$

$$8f = 16$$

$$\therefore f = -2$$

From (5.45), we get $3g + 10 = 1$

$$g = \frac{-9}{3}$$

$$= -3$$

From (5.41), we get $-6 - 34 = c + 4$ or $c = -44$

Therefore, the equation of the circle which cuts orthogonally the three given circles is $x^2 + y^2 - 6x - 4y - 44 = 0$.

Aliter

The radical axis of circles (5.41) and (5.42) is

$$5x - 11y + 7 = 0 \qquad (5.46)$$

The radical axis of circles (5.41) and (5.43) is

$$3x - 5y + 1 = 0 \qquad (5.47)$$

$$(5.41) \times 3: \qquad 15x - 33y + 21 = 0$$

$$(5.47) \times 5: \qquad 15x - 25y + 5 = 0$$

$$- 8y + 16 = 0$$

$$\therefore y = 2; x = 3$$

Therefore, radical centre is (3, 2).

If R is the length of the tangent from points (3, 2) to the first circle then $R^2 = 9 + 4 + 6 + 34 + 4 = 57$.

Therefore, the equation of the required circle is $(x - 3)^2 + (y - 2)^2 = 57$.

$$\text{(i.e.)} \quad x^2 + y^2 - 6x - 4y - 44 = 0$$

Example 5.5.12

Find the equation of the circle which passes through the origin, has its centre on the line $x + y = 4$ and cuts orthogonally the circle $x^2 + y^2 - 4x + 2y + 4 = 0$.

Solution

Let the equation of the required circle passing through the origin be

$$x^2 + y^2 + 2gx + 2fy = 0 \tag{5.48}$$

This circle cuts orthogonally the circle

$$x^2 + y^2 - 4x + 2y + 4 = 0 \tag{5.49}$$

$$\text{(i.e.)} \quad -4g + 2f = 0 + 4 \tag{5.50}$$

The centre of the circle (5.48) lies on $x + y = 4$.

$$\therefore -g - f = 4 \tag{5.51}$$

$$\begin{array}{lr} (5.50) & -4g + 2f = 4 \\ (5.51)\ \times 2 & -2g - 2f = 8 \\ \hline & 6g = 12 \end{array}$$

Adding, we get

$$\Rightarrow g = -2$$

$$\therefore f = -2$$

Therefore, the equation of the required circle is $x^2 + y^2 - 4x - 4y = 0$.

Example 5.5.13

If the equation of the circles with radii r and R are $S = 0$ and $S_1 = 0$, respectively then show that the circles $\frac{S}{r} + \frac{S_1}{R} = 0$ will intersect orthogonally.

Solution

Without loss of generality, we can assume the line of centres of the two circles as x-axis and the distance between the centres as $2a$. Then the centres of the two circles are $(a, 0)$ and $(-a, 0)$. The equation of the two circles are $S = (x - a)^2 + y^2 - r^2 = 0$ and $S_1 = (x + a)^2 + y^2 - R^2 = 0$.

Consider $\frac{S}{r} \pm \frac{S_1}{R} = 0$

$\therefore RS \pm rS_1 = 0$

Clearly, the coefficients of the R and r in these equations are the same and so they represent circles.

Consider $RS + rS_1 = 0$

(i.e.) $$R[x^2+y^2-2ax+a^2-r^2]+r[x^2+y^2+2ax+a^2-R^2]=0$$

or $$(R+r)(x^2+y^2)-2a(R-r)x+a^2(R+r)-Rr(R+r)=0 \quad (5.52)$$

Also, $RS-rS=0$ has the equation

$$(R-r)(x^2+y^2)-2a(R+r)x+(R-r)a^2-Rr(R-r)=0 \quad (5.53)$$

Equations (5.52) and (5.53) can be written as $x^2+y^2-2a\left(\frac{R-r}{R+r}\right)x+a^2-rR=0$ and $x^2+y^2-2a\left(\frac{R+r}{R-r}\right)x+(R+r)a^2+Rr=0.$

The condition for orthogonality is $2gg_1+2ff_1=c+c_1$.

$$\text{(i.e.)} \quad 2a^2\frac{(R^2-r^2)}{(R^2-r^2)}+0=a^2-rR+a^2+rR$$

$$\text{(i.e.)} \quad 2a^2=2a^2 \text{ which is true.}$$

Therefore, the circles $\frac{S}{r}\pm\frac{S_1}{R}=0$ are orthogonal.

Exercises (Radical Axis)

1. Find the radical axis of the circles $x^2+y^2+2x+4y=0$ and $2x^2+2y^2-7x-8y+1=0$.

   Ans.: $11x+16y-1=0$

2. Find the radical axis of the circles $x^2+y^2-4x-2y-11=0$ and $x^2+y^2-2x-6y+1=0$ and show that the radical axis is perpendicular to the line of centres.

   Ans.: $x-2y+5=0$

3. Show that the circles $x^2+y^2-6x-9y+13=0$ and $x^2+y^2-2x-16y=0$ touch each other. Find the coordinates of point of contact.

   Ans.: (5, 1)

4. Find the equation of the common chord of the circles $x^2+y^2+2ax+2by+c=0$ and $x^2+y^2+2bx+2ay+c=0$ and also show that the circles touch if $(a+b)^2=2c$.

5. Show that the circles $x^2+y^2+2x-8y+8=0$ and $x^2+y^2+10x-2y+22=0$ touch each other and find the point of contact.

   Ans.: $\left(\frac{-17}{5},\frac{11}{5}\right)$

6. Find the equation of the circle passing through the intersection of the circles $x^2 + y^2 = 6$ and $x^2 + y^2 - 6x + 8 = 0$ and also through the point (1, 1).

Ans.: $x^2 + y^2 - x - y = 0$

7. Find the equation of the circle passing through the point of intersection of the circles $x^2 + y^2 - 6x + 2y + 4 = 0$ and $x^2 + y^2 + 2x - 4y - 6 = 0$ and whose radius is 3/2.

Ans.: $5x^2 + 5y^2 - 18x + y + 5 = 0$

8. If the circles $x^2 + y^2 + 2gx + 2fy = 0$ and $x^2 + y^2 + 2g_1x + 2f_1y = 0$ touch each other then show that $fg_1 = f_1g$.

9. Find the radical centre of the circles $x^2 + y^2 + a_ix + b_iy + c = 0$, $i = 1, 2, 3$.

Ans.: (0, 0)

10. Find the radical centre of the circles $x^2 + y^2 - x + 3y - 3 = 0$, $x^2 + y^2 - 2x + 2y + 2 = 0$ and $x^2 + y^2 + 2x + 2y - 9 = 0$.

Ans.: (2, 1)

11. The radical centre of three circles is at the origin. The equation of two of the circles are $x^2 + y^2 = 1$ and $x^2 + y^2 + 4x + 4y - 1 = 0$. Find the general form of the third circle. If it passes through (1, 1) and (–2, 1) then find its equation.

Ans.: $x^2 + y^2 + x - 2y - 1 = 0$

12. Find the radical centre of the circles $x^2 + y^2 + x + 2y + 3 = 0$, $x^2 + y^2 + 4x + 7 = 0$ and $2x^2 + 2y^2 + 3x + 5y + 9 = 0$.

Ans.: (–2, –1)

13. Find the equation of the circle whose radius is 3 and which touches the circle $x^2 + y^2 - 4x - 6y + 2 = 0$ internally at the point (–1, –1).

14. Show that the radical centres of three circles described on the sides of a triangle as diameter is the orthocentre of the triangle.

15. Find the equation of the circle which cuts orthogonally the three circles $x^2 + y^2 + y = 0$, $x^2 + 4y^2 + 4x + 7 = 0$, $21x^2 + y^2 + 3x + 5y + 9 = 0$.

Ans.: $x^2 + y^2 + 4x + 2y + 1 = 0$

16. *A* and *B* are two fixed points and *P* moves so that $PA = n{\cdot}PB$. Show that the locus of *P* is a circle and that for different values of *n*, all the circles have the same radical axis.

17. Find the equation of circle whose radius is 5 and which touches the circle $x^2 + y^2 - 2x - 4y - 20 = 0$ at the point (5, 5).

18. Prove that the length of the common chord of the two circles whose equations are $(x - a)^2 + (y - b)^2 = r^2$ and $(x - a)^2 + (y - b)^2 = c^2$ is $\sqrt{ac^2 - 2(a-b)^2}$.

19. Find the equation to two equal circles with centres (2, 3) and (5, 6) which cuts each other orthogonally.

20. If three circles with centres A, B and C cut each other orthogonally in pairs then prove that the polar of A with respect to the circle centre B passes through C.

21. Find the locus of centres of all the circles which touch the line $x = 2a$ and cut the circle $x^2 + y^2 = a^2$ orthogonally.

22. A, B are the points $(a, 0)$ and $(-a, 0)$. Show that if a variable circle S is orthogonal to the circle on AB as diameter, the polar of $(a, 0)$ with respect to the circle S passes through the fixed point $(-a, 0)$.

23. If a circle passes through the point (a, b) and cuts the circle $x^2 + y^2 = k^2$ orthogonally then prove that the locus of its centres is $2ax + 2by - (a^2 + b^2 + k^2) = 0$.

24. Show that the circles $x^2 + y^2 + 10x + 6y + 14 = 0$ and $x^2 + y^2 - 4x + 6y + 8 = 0$ touch each other at the point $(3, -1)$.

25. Show that the circles $x^2 + y^2 + 2ax + 4ay - 3a^2 = 0$ and $x^2 + y^2 - 8ax - 6ay + 7a^2 = 0$ touch each other at the point $(a, 0)$.

26. The equation of three circles are $x^2 + y^2 = 1$, $x^2 + y^2 + 8x + 15 = 0$ and $x^2 + y^2 + 10y + 24 = 0$. Determine the coordinate of the point such that the tangents drawn from it to the three circles are equal in length.

27. If P and Q be a pair of conjugate points with respect to a circle $S = 0$ then prove that the circle on PQ as diameter cuts the circle $S = 0$ orthogonally.

28. Find the equation of the circle whose diameter is the common chord of the circles $x^2 + y^2 + 2x + 3y + 1 = 0$ and $x^2 + y^2 + 4x + 3y + 2 = 0$.

5.6 EXAMPLES (LIMITING POINTS)

Example 5.6.1

If A, B and C are the centres of three coaxal circles and t_1, t_2 and t_3 are the lengths of tangents to them from any point then prove that $BC \cdot t_1^2 + CA \cdot t_2^2 + AB \cdot t_3^2 = 0$.

Solution

Let the three circles of coaxal system be

$$S_1 = x^2 + y^2 + 2g_1x + c = 0$$

$$S_2 = x^2 + y^2 + 2g_2x + c = 0$$

$$S_3 = x^2 + y^2 + 2g_3x + c = 0$$

The centres are $A(-g_1, 0)$, $B(-g_2, 0)$ and $C(-g_3, 0)$ and $BC = g_3 - g_2$, $CA = g_1 - g_3$,

$$AB = g_2 - g_1$$
$$t_1^2 = x_1^2 + y_1^2 + 2g_1x_1 + c$$
$$t_2^2 = x_1^2 + y_1^2 + 2g_2x_1 + c$$
$$t_3^2 = x_1^2 + y_1^2 + 2g_3x_1 + c$$

Then,

$$\begin{aligned}\Sigma BC \cdot t_1^2 &= -\Sigma \cdot (g_2 - g_3)(x_1^2 + y_1^2 + 2g_1x_1 + c)\\ &= -(x_1^2 + y_1^2 + c)\Sigma(g_2 - g_3) + 2x_1\Sigma g_1(g_2 - g_3)\\ &= 0\end{aligned}$$

since $\Sigma\,(g_2 - g_3) = 0$ and $\Sigma g_1(g_2 - g_3) = 0$.

Example 5.6.2

Find the equations of the circles which pass through the points of intersection of $x^2 + y^2 - 2x + 1 = 0$ and $x^2 + y^2 - 5x - 6y - 4 = 0$ and which touch the line $2x - y + 3 = 0$.

Solution

$$x^2 + y^2 - 2x + 1 = 0 \tag{5.54}$$

$$x^2 + y^2 - 5x - 6y + 4 = 0 \tag{5.55}$$

The radical axis of these two circles is

$$3x + 6y - 3 = 0 \text{ or } x + 2y - 1 = 0 \tag{5.56}$$

The equation of any circle passing through the intersection of these two circles is $x^2 + y^2 - 2x + 1 + \lambda(x + 2y - 1) = 0$.

The centre of this circle is $\left(1 - \frac{\lambda}{2}, -\lambda\right)$ and radius $= \sqrt{\left(1 - \frac{\lambda}{2}\right)^2 + \lambda^2 + \lambda - 1}$

The circle (5.56) touches the line $2x - y + 3 = 0$.

$$\therefore \frac{\left|2\left(1 - \frac{\lambda}{2}\right) + \lambda + 3\right|}{\sqrt{4+1}} = \sqrt{\left(1 - \frac{\lambda}{2}\right)^2 + \lambda^2 + \lambda - 1}$$

$$\therefore 5 = \left(1 - \frac{\lambda}{2}\right)^2 + \lambda^2 + \lambda - 1$$

$$= 1 + \frac{\lambda^2}{4} - \lambda + \lambda^2 + \lambda - 1$$

$$\Rightarrow 5 = \frac{\lambda^2}{4} + \lambda^2$$

$$\Rightarrow \frac{5\lambda^2}{4} = 5$$

$$\Rightarrow \lambda = \pm 2$$

The equation of the circles are $x^2 + y^2 - 2x + 1 \pm 2(x + 2y - 1) = 0$.

(i.e.) $x^2 + y^2 - 2x + 1 + 2x + 4y - 2 = 0$ and $x^2 + y^2 - 2x + 1 - 2x - 4y + 2 = 0$

(i.e.) $x^2 + y^2 + 4y - 1 = 0$ and $x^2 + y^2 - 4x + 4y + 3 = 0$

Example 5.6.3

Find the equation of the circle which passes through the intersection of the circles $x^2 + y^2 = 4$ and $x^2 + y^2 - 2x - 4y + 4 = 0$ and has a radius $2\sqrt{2}$.

Solution

$$x^2 + y^2 - 4 = 0 \tag{5.57}$$

$$x^2 + y^2 - 2x - 4y + 4 = 0 \tag{5.58}$$

The radical axis of these two circles is $2x + 4y - 8 = 0$. Any circle passing through the intersection of these two circles is

$$S + \lambda L = 0 \Rightarrow x^2 + y^2 - 4 + \lambda(2x + 4y - 8) = 0$$

Centre is $(-\lambda, -2\lambda)$.

$$\text{radius} = \sqrt{\lambda^2 + 4\lambda^2 + 4 + 8\lambda} = 2\sqrt{2}$$

$$\Rightarrow 5\lambda^2 + 8\lambda + 4 = 8$$

$$\Rightarrow 5\lambda^2 + 8\lambda - 4 = 0$$

$$\Rightarrow 5\lambda^2 + 10\lambda - 2\lambda - 4 = 0$$

$$\Rightarrow 5\lambda(\lambda + 2) - 2(\lambda + 2) = 0$$

$$\Rightarrow (\lambda + 2)(5\lambda - 2) = 0$$

$$\Rightarrow \lambda = -2, \frac{2}{5}$$

Therefore, the required circles are $x^2 + y^2 - 4 - 2(2x + 4y - 8) = 0$ and $x^2 + y^2 - 4 + \frac{2}{5}(2x + 4y - 8) = 0$.

(i.e.) $x^2 + y^2 - 4x - 8y + 12 = 0$ and $5x^2 + 5y^2 + 4x + 8y - 36 = 0$

Example 5.6.4

Find the equation of the circle whose diameter is the common chord of the circles $x^2 + y^2 + 2x + 3y + 1 = 0$ and $x^2 + y^2 + 4x + 3y + 2 = 0$.

Solution

$$x^2 + y^2 + 2x + 3y + 1 = 0$$
$$x^2 + y^2 + 4x + 3y + 2 = 0$$

The radical axis of these two circles is $2x + 1 = 0$.

Any circle of the system is $x^2 + y^2 + 2x + 3y + 1 + \lambda(2x + 1) = 0$.

Centre is $\left(-1-\lambda, \frac{-3}{2}\right)$.

Since the radical axis is a diameter, centre lies on the radical axis.

$$\therefore 2(-1 - \lambda) + 1 = 0$$
$$-2 - 2\lambda + 1 = 0$$
$$\Rightarrow 2\lambda = -1$$

or
$$\lambda = \frac{-1}{2}$$

Hence, the equation of the required circle is $x^2 + y^2 + 2x + 3y + 1 - \frac{1}{2}(2x + 1) = 0$.

$$\Rightarrow 2x^2 + 2y^2 + 4x + 6y + 2 - 2x - 1 = 0$$
$$\Rightarrow 2x^2 + 2y^2 + 2x + 6y + 1 = 0$$

Example 5.6.5

Find the equation of the circle which touches x-axis and is coaxal with the circles $x^2 + y^2 + 12x + 8y - 33 = 0$ and $x^2 + y^2 = 5$.

Solution

$$x^2 + y^2 + 12x + 8y - 33 = 0$$
$$x^2 + y^2 - 5 = 0$$

The radical axis of these two circles is

$$12x + 8y - 28 = 0$$
$$\text{(i.e.)} \quad 6x + 4y - 14 = 0 \qquad (5.59)$$

Any circle of the coaxal system is $x^2 + y^2 - 5 + \lambda(6x + 4y - 14) = 0$.
Centre is $(-3\lambda, -2\lambda)$.

$$\text{Radius} = \sqrt{9\lambda^2 + 4\lambda^2 + 5 + 14\lambda}$$
$$= \sqrt{13\lambda^2 + 14\lambda + 5}$$

The circles (5.69) touches x-axis (i.e.) $y = 0$.

$$\text{Radius} = \pm\frac{2\lambda}{\sqrt{1}} = \pm 2\lambda \Rightarrow 13\lambda^2 + 4\lambda + 5 = 4\lambda^2$$
$$\Rightarrow 9\lambda^2 + 14\lambda + 5 = 0$$
$$\Rightarrow (\lambda + 1)(9\lambda + 5) = 0 \quad \Rightarrow \lambda = -1 \quad \text{or} \quad \frac{-5}{9}$$

Therefore, the two circles of the system touching x-axis are

$$x^2 + y^2 - 5 - 1(6x + 4y - 14) = 0 \Rightarrow x^2 + y^2 - 6x - 4y + 9 = 0$$

and $9(x^2 + y^2 - 5) - 5(6x + 4y - 14) = 0 \Rightarrow 9x^2 + 9y^2 - 36x - 20y + 25 = 0$.

Example 5.6.6

The line $2x + 3y = 1$ cuts the circle $x^2 + y^2 = 4$ in A and B. Show that the equation of the circle on AB as diameter is $13(x^2 + y^2) - 4x - 6y - 50 = 0$.

Solution

Let

$$x^2 + y^2 - 4 = 0 \tag{5.60}$$
$$2x + 3y - 1 = 0 \tag{5.61}$$

Any circle passing through the intersection of the circle and the line is

$$x^2 + y^2 + \lambda(2x + 3y - 1) = 0 \tag{5.62}$$

Centre is $\left(-\lambda, -\frac{3\lambda}{2}\right)$ and radius $= \sqrt{\lambda^2 + \frac{9\lambda^2}{4} + \lambda}$.

If AB is a diameter of the circle (5.72), their centre should lie on AB.

$$-2\lambda + 3\left(\frac{-3\lambda}{2}\right) - 1 = 0$$
$$\Rightarrow -4\lambda - 9\lambda - 2 = 0 \Rightarrow \lambda = \frac{-2}{13}$$

Therefore, the equation of the circle on AB as diameter is $13(x^2 + y^2 - 4) - 2(2x + 3y - 1) = 0$.

$$\therefore 13(x^2 + y^2) - 4x - 6y + 50 = 0$$

Example 5.6.7

A point moves so that the ratio of the length of tangents to the circles $x^2 + y^2 + 4x + 3 = 0$ and $x^2 + y^2 - 6x + 5 = 0$ is 2:3. Show that the locus of the point is a circle coaxal with the given circles.

Solution

The lengths of tangents from a point $P(x_1, y_1)$ to the two circles are $PT_1^2 = x_1^2 + y_1^2 + 4x_1 + 3;\ PT_2^2 = x_1^2 + y_1^2 - 6x_1 + 5.$

Given that,

$$\frac{PT_1^2}{PT_2^2} = \frac{4}{9} \Rightarrow PT_1^2 = \frac{4}{9}PT_2^2$$

$$\Rightarrow x_1^2 + y_1^2 + 4x_1 + 3 = \frac{4}{9}(x_1^2 + y_1^2 - 6x_1 + 5)$$

The locus of (x_1, y_1) is $x^2 + y^2 + 4x_2 + 3 - \frac{4}{9}(x^2 + y^2 - 6x_2 + 6) = 0$

This is of the form $S_1 + \lambda S_2 = 0$

Hence the locus of circle is a circle coaxal with the two given circles.

Example 5.6.8

Find the limiting points of the coaxal system determined by the circle $x^2 + y^2 + 2x + 4y + 7 = 0$ and $x^2 + y^2 + 4x + 2y + 5 = 0$.

Solution

Given that,

$$x^2 + y^2 + 2x + 4y + 7 = 0$$
$$x^2 + y^2 + 4x + 2y + 5 = 0$$

The radical axis of these two circles is $2x - 2y - 2 = 0$. Any circle of the coaxal system is $x^2 + y^2 + 2x + 4y + 7 + \lambda(2x - 2y - 2) = 0$.

Centre is $(-1 - \lambda, -2 + \lambda)$.

Radius is $\sqrt{(1+\lambda)^2 + (2-\lambda)^2 + 2\lambda - 7} = \sqrt{2\lambda^2 - 2}$.

Limiting points are the centres of circles of radii zero.

Therefore, limiting points are $(-2, -1)$ and $(0, -3)$.

Example 5.6.9

The point (2, 1) is a limiting point of a system of coaxal circles of which $x^2 + y^2 - 6x - 4y - 3 = 0$ is a member. Find the equation to the radial axis and the coordinates of the other limiting point.

Solution

Given that

$$x^2 + y^2 - 6x - 4y - 3 = 0$$

Since (2, 1) is a limit point, the point circle corresponding to the coaxal system is

$$(x - 2)^2 + (y - 1)^2 = 0$$
$$\Rightarrow x^2 + y^2 - 4x - 2y + 5 = 0$$

The radical axis of the system is

$$S - S_1 = 0 \quad \text{or} \quad 2x + 2y + 8 = 0$$
$$\Rightarrow x + y + 4 = 0$$

Any circle of the coaxal system is $S + \lambda L = 0$.

$$x^2 + y^2 - 6x - 4y - 3 + \lambda(2x + 2y + 8) = 0$$

Centre is $(3 - \lambda, 2 - \lambda)$.

Radius $= \sqrt{(3-\lambda)^2 + (2-\lambda)^2 + (3-8\lambda)}$

For point circles, radius = 0.

$$\Rightarrow (3 - \lambda)^2 + (2 - \lambda)^2 + 3 - 8\lambda = 0$$
$$\Rightarrow 2\lambda^2 - 18\lambda + 16 = 0$$
$$\Rightarrow \lambda^2 - 9\lambda + 8 = 0$$
$$\therefore \lambda = 1, 8$$

Therefore, the limiting points are the centres of point circle of the coaxal system, that is, (2, 1) and (–5, –6).

Example 5.6.10

Find the equation of the circle which passes though the origin and belongs to the coaxal system of which limiting points are (1, 2) and (4, 3).

Solution

Since (1, 2) and (4, 3) are limiting points of two circles of the coaxal system and $(x - 1)^2 + (y - 2)^2 = 0$ and $(x - 4)^2 + (y - 3)^2 = 0$.

$$\text{(i.e.)} \quad x^2 + y^2 - 2x - 4y + 5 = 0$$
$$x^2 + y^2 - 8x - 6y + 25 = 0$$

Radical axis is $6x + 2y - 20 = 0$.

Any circle of the system is $x^2 + y^2 - 2x - 4y + 5 + \lambda(6x + 2y - 20) = 0$. This passes through the origin.

$$\therefore 5 - 20\lambda = 0 \Rightarrow \lambda = \frac{1}{4}$$

Hence, the equation of the system is

$$4(x^2 + y^2 - 2x - 4y + 5) + 6x + 2y - 20 = 0.$$

$$\Rightarrow 4x^2 + 4y^2 - 2x + 4y = 0$$

$$\Rightarrow 2x^2 + 2y^2 - x - 7y = 0$$

Example 5.6.11

A point P moves so that its distances from two fixed points are in a constant ratio λ. Prove that the locus of P is a circle. If λ varies then show that P generates a system of coaxal circles of which the fixed points are the limiting points.

Solution

Let $P(x_1, y_1)$ be a moving point and $A(c, 0)$ and $B(0, -c)$ be the two fixed points. Here, we have chosen the fixed points on the x-axis such that P is its midpoint. Given that

$$\frac{PA}{PB} = \lambda \Rightarrow PA^2 = \lambda^2 PB^2 \Rightarrow (x-a)^2 + y^2 - \lambda^2[(x+a)^2 + y^2] = 0$$

This equation is of the form $S + \lambda S' = 0$ which is the equation to a coaxal system of circles.

Therefore, for different values of λ, P generates a coaxal system of circles of which $(x - a)^2 + y^2 = 0$ and $(x + a)^2 + y^2 = 0$ are members. These equations are the equation of point circles whose centres are $(a, 0)$, $(-a, 0)$ which is the fixed points.

Example 5.6.12

Prove that the limiting point of the system $x^2 + y^2 + 2gx + c + \lambda(x^2 + y^2 + 2fy + k) = 0$ subtends a right angle at the origin if $\frac{c}{g^2} + \frac{k}{f^2} = 2$.

Solution

The two members of the system are $x^2 + y^2 + 2gx + c = 0$ and $x^2 + y^2 + 2fy + k = 0$.

Radical axis is $2gx - 2fy + c - k = 0$.

Any circle of the system is $x^2 + y^2 + 2gx + c + \lambda(2gx - 2fy + c - k) = 0$.

Centre is $(-g - g\lambda, f\lambda)$.

Radius $= \sqrt{(g + g\lambda)^2 + f^2\lambda^2 - c - c\lambda + k\lambda}$

For point circle, radius = 0.

$$(g + g\lambda)^2 + \lambda^2 f^2 - c - c\lambda + k\lambda = 0$$
$$g^2(\lambda^2 + 2\lambda + 1) + f^2\lambda^2 - c - c\lambda + k\lambda = 0$$
$$\lambda^2(g^2 + f^2) + \lambda(2g^2 - c + k) + g^2 - c = 0$$

Considering the two values of λ as λ_1, λ_2,

centres are $A(-g(1 + \lambda_1), f\lambda_1)$ and $B(-g(1 + \lambda_2), f\lambda_2)$

Since OAB is right angled at O, OA is perpendicular to OB.

$$\therefore \quad \frac{f^2\lambda_1\lambda_2}{g^2(1+\lambda_1)(1+\lambda_2)} = -1$$

$$\Rightarrow f^2\lambda_1\lambda_2 + g^2(1 + \lambda_1 + \lambda_2 + \lambda_1\lambda_2) = 0$$

$$\Rightarrow f^2\left(\frac{g^2 - c}{g^2 + f^2}\right) + g\left[1 - \frac{(2g^2 - c + k)}{g^2 + f^2} + \frac{g^2 - c}{g^2 + f^2}\right] = 0$$

$$\Rightarrow f^2(g^2 - c) + g^2(g^2 + f^2 - 2g^2 + c - k + g^2 - c) = 0$$

$$\Rightarrow f^2(g^2 - c) + g^2(f^2 - k) = 0$$

$$\Rightarrow 2f^2g^2 = cf^2 + kg^2$$

$$\Rightarrow \frac{c}{g^2} + \frac{k}{f^2} = 2$$

Exercises

1. Find the equation of the circle passing through the intersection of $x^2 + y^2 - 6 = 0$ and $x^2 + y^2 + 4y - 1 = 0$ through the point $(-1, 1)$.

   Ans.: $9x^2 + 9y^2 + 16y - 34 = 0$

2. Show that the circles $x^2 + y^2 = 480$ and $x^2 + y^2 - 10x - 24y + 120 = 0$ touch each other and find the equation, if a third circle which touches the circles at their point of intersection and the x-axis $x^2 + y^2 - 200x - 400y + 10000 = 0$.

   Ans.: $5x^2 + 5y^2 - 40x + 96y + 30 = 0$

3. Find the equation of the circle whose centre lies on the line $x+y-11=0$ and which passes through the intersection of the circle $x^2+y^2-3x+2y-4=0$ with the line $2x+5y-2=0$.

4. Find the length of the common chord of the circles $x^2+y^2+4x-22y=0$ and $x^2+y^2-10x+5y=0$.

Ans.: 40/7

5. Find the coordinates of the limiting points of the coaxal circles determined by the two circles $x^2+y^2-4x-6y-3=0$ and $x^2+y^2-24x-26y+277=0$.

Ans.: (1, 2),(3, 1)

6. Find the coordinates of the limiting points of the coaxal system of circles of which two members are $x^2+y^2+2x-6y=0$ and $2x^2+2y^2-10y+5=0$.

Ans.: (1, 2),(3, 1)

7. Find the coaxal system of circles if one of whose members is $x^2+y^2+2x-6y=0$ and *a* limiting point is (1, –2).

Ans.: $x^2+y^2+2x+3y-7-\lambda(4x-y-12)=0$

8. Find the limiting point of the coaxal system determined by the circles $x^2+y^2-6x-6y+4=0$ and $x^2+y^2-2x-4y+3=0$.

Ans.: $(-1,1),\left(\frac{2}{5}-\frac{8}{5}\right)$

9. Find the equation of the coaxal system of circles one of whose members is $x^2+y^2-4x-2y-5=0$ and the limiting point is (1, 2).

Ans.: $x^2+y^2-2x-4y+5+\lambda(x-y-5)=0$

10. If origin is a limiting point of a system of coaxal circles of which $x^2+y^2+2gx+2fy+c=0$ is a member then show that the other limiting points is $\left(\frac{-gc}{g^2+f^2},\frac{-fc}{f^2+g^2}\right)$.

11. Show that the equation of the coaxal system whose limiting points are (0, 0) and (*a*, *b*) is $x^2+y^2+k(2ax-2by-a^2-b^2)=0$.

12. The origin is a limiting point of a system of coaxal circles of which $x^2+y^2+2gx+2fy+c=0$ is a member. Show that the equation of circles of the orthogonal system is $(x^2+y^2)(g+\lambda f)+c(x-\lambda y)=0$ for different values of *x*.

13. Show that the circles $x^2+y^2+2ax+2by+2\lambda(ax-by)=0$ where λ is a parameter from a coaxal system and also show that the equation of the common radical axis and the equation of circles which are orthogonal to this system are $ax-by=0,\ 2(x^2+y^2)+c\left(\frac{x}{a}+\frac{y}{b}-2\right)=0$.

14. A point P moves such that the length of tangents to the circles $x^2+y^2-2x-4y+5=0$ and $x^2+y^2+4x+6y-7=0$ are in the ratio 3:4. Show that the locus is a circle.

15. Show that the limiting points of the circle $x^2+y^2=a^2$ and an equal circle with centre on the line $lx+my+n=0$ be on the line $(x^2+y^2)(lx+my+n)+a^2(ln+mn)=0$.

CHAPTER

6 Parabola

6.1 INTRODUCTION

If a point moves in a plane such that its distance from a fixed point bears a constant ratio to its perpendicular distance from a fixed straight line then the path described by the moving point is called a conic. In other words, if S is a fixed point, l is a fixed straight line and P is a moving point and PM is the perpendicular distance from P on l, such that $\frac{SP}{PM} = a$ constant, then the locus of P is called a conic. This constant is called the eccentricity of the conic and is denoted by e.

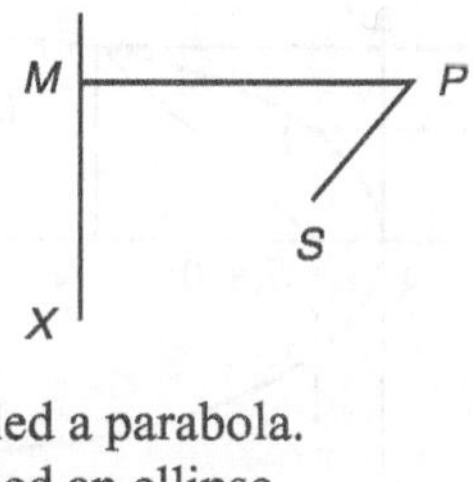

If $e = 1$, the conic is called a parabola.
If $e < 1$, the conic is called an ellipse.
If $e > 1$, the conic is called a hyperbola.

The fixed point S is called the focus of the conic. The fixed straight line is called the directrix of the conic. The property $\frac{SP}{PM} = e$ is called the focus-directrix property of the conic.

6.2 GENERAL EQUATION OF A CONIC

We can show that the equation of a conic is a second degree equation in x and y. This is derived from the focus-directrix property of a conic. Let $S(x_1, y_1)$ be the focus and $P(x, y)$ be any point on the conic and $lx + my + n = 0$ be the equation of the directrix. The focus-directrix property of the conic states $\frac{SP}{PM} = e$ (i.e.) $SP^2 = e^2 PM^2$

$$\text{(i.e.)} \qquad (x-x_1)^2+(y-y_1)^2=e^2\left(\frac{lx+my+n}{\sqrt{l^2+m^2}}\right)^2$$

This equation can be expressed in the form $ax^2+2hxy+by^2+2gx+2fy+c=0$ which is a second degree equation in x and y.

6.3 EQUATION OF A PARABOLA

Let S be the focus and the line l be the directrix. We have to find the locus of a point P such that its distance from the focus S is equal to its distance from the fixed line l.

(i.e.) $\frac{SP}{PM}=1$ where PM is perpendicular to the directrix.

Draw SX perpendicular to the directrix and bisect SX. Let A be the point of bisection and $SA=AX=a$. Then the point A is a point on the parabola since $\frac{SA}{AX}=1$. Take AS as the x-axis and AY perpendicular to AS as the y-axis. Then the coordinate of S are $(a, 0)$. Let (x, y) be the coordinates of the point P. Draw PN perpendicular to the x-axis.

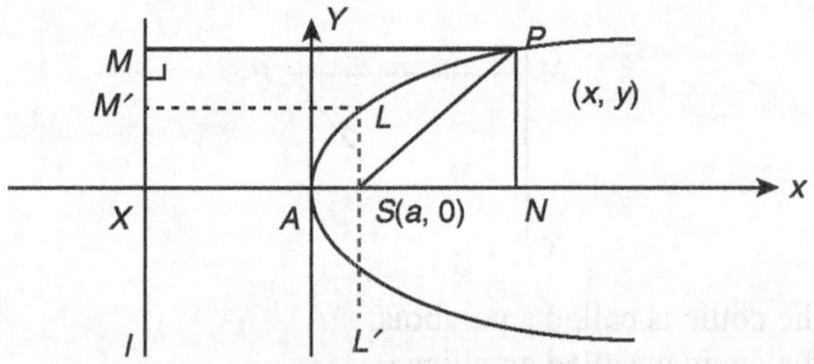

$$PM=NX=NA+AX=x+a$$

Since $$\frac{SP}{PM}=1$$

$$SP^2=PM^2$$

$$\text{(i.e.)} \quad (x-a)^2+y^2=(x+a)^2$$

$$y^2=(x+a)^2-(x-a)^2 \text{ or } y_2=4ax.$$

This, being the locus of the point P, is the equation of the parabola. This equation is the simplest possible equation to a parabola and is called the standard equation of the parabola.

Note 6.3.1:

1. The line AS(x-axis) is called the axis of the parabola.
2. The point A is called the vertex of the parabola.
3. AY(y-axis) is called the tangent at the vertex.
4. The perpendicular through the focus is called the latus rectum.
5. The double ordinate through the focus is called the length of the latus rectum.
6. The equation of the directrix is $x + a = 0$.
7. The equation of the latus rectum is $x - a = 0$.

6.4 LENGTH OF LATUS RECTUM

To find the length of the latus rectum,draw LM' perpendicular to the directrix. Then $\frac{SL}{LM'} = 1$

$$\therefore SL = LM' = SX = 2a$$

$$\therefore LL' = 2SX = 4a$$

6.4.1 Tracing of the curve $y^2 = 4ax$

1. If $x < 0$, y is imaginary. Therefore, the curve does not pass through the left side of y-axis.
2. When $y = 0$, we get $x = 0$. Therefore, the curve meets the y-axis at only one point, that is, (0, 0).
3. When $x = 0$, $y^2 = 0$, that is, $y = 0$. Hence the y-axis meets the curve at two coincident points (0, 0). Hence the y-axis is a tangent to the curve at (0, 0).
4. If (x, y) is a point on the parabola $y^2 = 4ax$, $(x, -y)$ is also a point. Therefore, the curve is symmetrical about the x-axis.
5. As x increases indefinitely, the values of y also increases indefinitely. Therefore the points of the curve lying on the opposite sides of x-axis extend to infinity towards the positive side of x-axis.

6.5 DIFFERENT FORMS OF PARABOLA

1. If the focus is taken at the point $(-a, 0)$ with the vertex at the origin and its axis as x-axis then its equation is $y^2 = -4ax$.

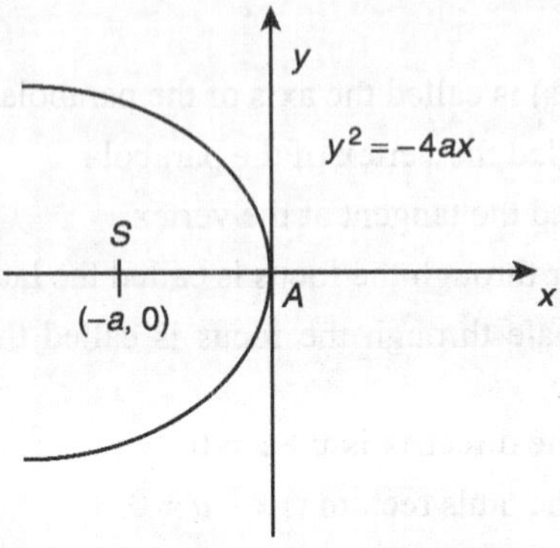

2. If the axis of the parabola is the y-axis, vertex at the origin and the focus at $(0, a)$, the equation of the parabola is $x^2 = 4ay$.

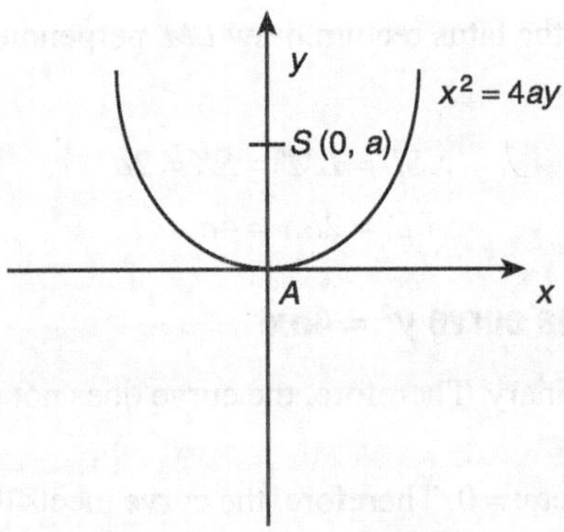

3. If the focus is at $(0, -a)$, vertex $(0, 0)$ and axis as y-axis, then the equation of the parabola is $x^2 = -4ay$.

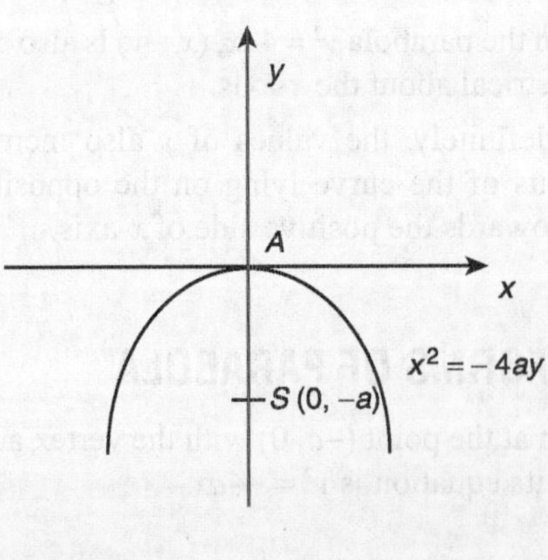

ILLUSTRATIVE EXAMPLES BASED ON FOCUS DIRECTRIX PROPERTY

Example 6.1

Find the equation of the parabola with the following foci and directrices:

(i) $(1, 2): x + y - 2 = 0$
(ii) $(1, -1): x - y = 0$
(iii) $(0, 0): x - 2y + 2 = 0$

Solution

(i) Let $P(x, y)$ be any point on the parabola. Draw PM perpendicular to the directrix. Then from the definition of the parabola, $\frac{SP}{PM} = 1$

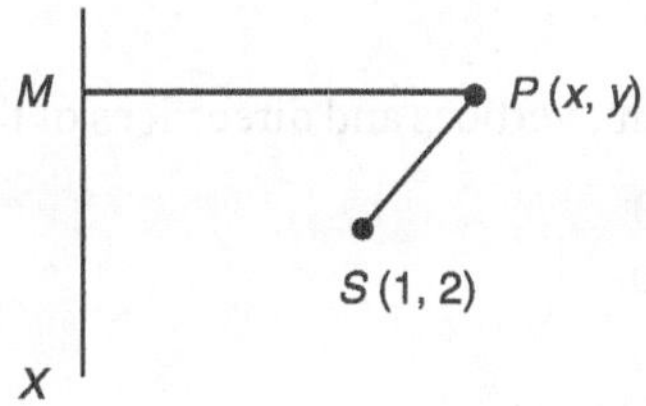

$\therefore SP^2 = (x-1)^2 + (y-2)^2$

$PM =$ perpendicular distance from (x, y) on $x + y - 2 = 0$

$$= \pm\frac{(x+y-2)}{\sqrt{2}}$$

$$\therefore (x-1)^2 + (y-2)^2 = \left(\frac{x+y-2}{\sqrt{2}}\right)^2$$

$\therefore 2(x^2 - 2x + 1) + 2(y^2 - 4y + 4) = x^2 + y^2 + 4 + 2xy - 4x - 4y$

(i.e.) $x^2 + y^2 - 2xy - 4y + 6 = 0$

This is the equation of the required parabola.

(ii) The point S is $(1, -1)$. Directrix is $x - y = 0$

$$SP^2 = (x-1)^2 + (y+1)^2 = x^2 + y^2 - 2x + 2y + 2$$

$$PM = \pm\frac{(x-y)}{\sqrt{2}}$$

From any point on the parabola, $\frac{SP}{PM} = 1$

$\therefore \quad SP^2 = PM^2 \Rightarrow 2(x^2 + y^2 - 2x + 2y + 2) = x^2 + y^2 - 2xy$

(i.e.) $x^2 + y^2 + 2xy - 4x + 4y + 4 = 0$

(iii) S is (0, 0). Directrix is $x - 2y + 2 = 0$

$$SP^2 = (x - 0)^2 + (y - 0)^2 = x^2 + y^2$$

$$PM = \pm\frac{(x-2y+2)}{\sqrt{1+4}} = \pm\frac{(x-2y+2)}{\sqrt{5}}$$

For any point P on the parabola, $\frac{SP}{PM} = 1$

$$\therefore SP^2 = PM^2$$

$$x^2 + y^2 = \left(\frac{x-2y+2}{\sqrt{5}}\right)^2$$

(i.e.) $5(x^2 + y^2) = x^2 + 4y^2 + 4 - 4xy + 4x - 8y$

(i.e.) $4x^2 + y^2 + 4xy - 4x + 8y - 4 = 0$

Example 6.2

Find the foci, latus rectum, vertices and directrices of the following parabolas:

(i) $y^2 + 4x - 2y + 3 = 0$
(ii) $y^2 - 4x + 2y - 3 = 0$
(iii) $y^2 - 8x - 9 = 0$

Solution

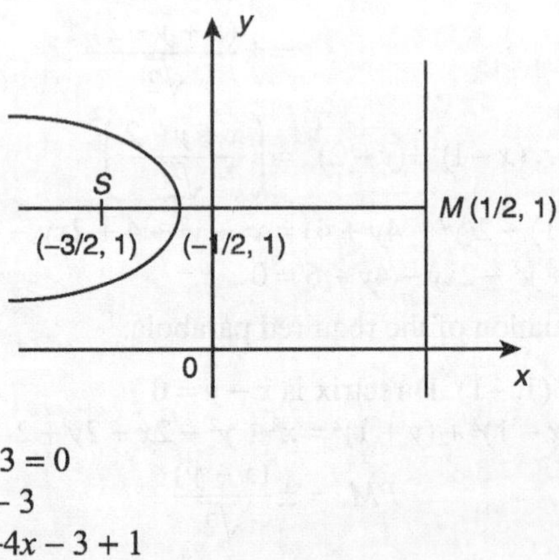

(i) $y^2 + 4x - 2y + 3 = 0$
$y^2 - 2y = -4x - 3$
$y^2 - 2y + 1 = -4x - 3 + 1$

$$\Rightarrow (y-1)^2 = -4\left(x+\frac{1}{2}\right)$$

Take $x+\frac{1}{2} = X$, $y - 1 = Y$. Shifting the origin to the point $\left(\frac{-1}{2}, 1\right)$ the equation of the parabola becomes $Y^2 = -4X$.

$\therefore$ Vertex is $\left(\frac{-1}{2}, 1\right)$, latus rectum is 4, focus is $\left(\frac{-3}{2}, 1\right)$ and foot of the directrix is $\left(\frac{1}{2}, 1\right)$. The equation of the directrix is $x = \frac{1}{2}$ or $2x - 1 = 0$.

(ii) $$y^2 - 4x + 2y - 3 = 0$$
$$y^2 + 2y = 4x + 3 \Rightarrow y^2 + 2y + 1 = 4x + 3 + 1$$
$$\Rightarrow (y + 1)^2 = 4(x + 1)$$

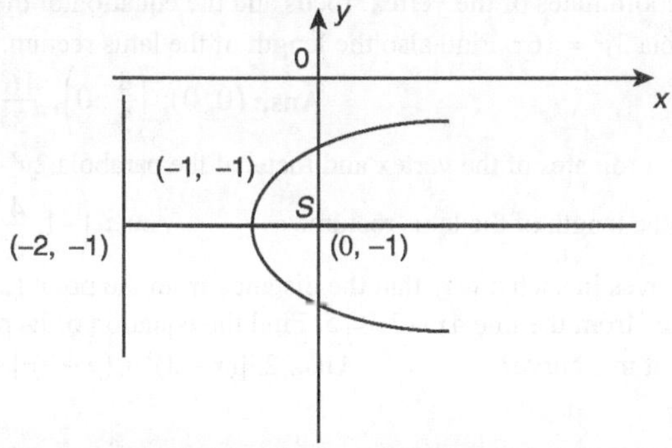

Shifting the origin to the point $(-1, -1)$ by taking $x + 1 = X$ and $y + 1 = Y$ the equation of the parabola becomes $Y^2 = 4X$.
$\therefore$ Vertex is $(-1, -1)$, latus rectum $= 4$, focus is $(0, -1)$ and foot of the directrix is $(-2, -1)$.
$\therefore$ The equation of the directrix is $x + 2 = 0$.

$$y^2 - 8x - 9 = 0 \Rightarrow y^2 = 8x + 9$$

Shift the origin to the point $\left(\frac{-9}{8}, 0\right)$ and take $x + \frac{9}{8} = X$ and $y = Y$.
$\therefore$ The equation of the parabola becomes $Y^2 = 8X$. Vertex is $\left(\frac{-9}{8}, 0\right)$, latus rectum $= 8$ andfocus is $\left(\frac{-9}{8} + 2, 0\right)$ (i.e.) $\left(\frac{7}{8}, 0\right)$.
The equation of the directrix is $x + \frac{9}{8} + 2 = 0$. (i.e.) $8x + 25 = 0$.

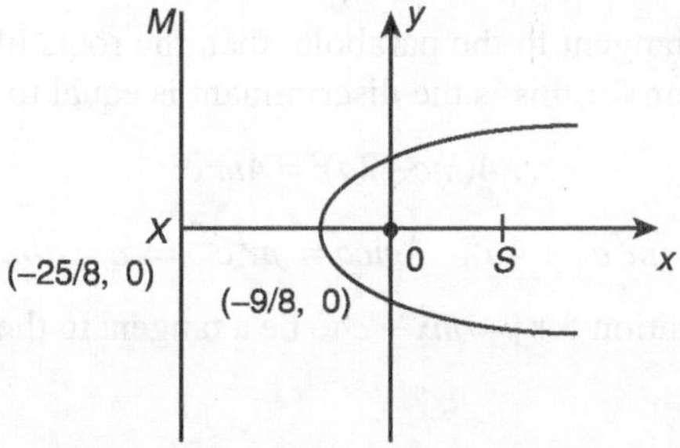

Exercises

1. Find the equation of the parabola whose focus is (2, 1) and directrix is $2x + y + 1 = 0$. **Ans.:** $x^2 - 4xy + 4y^2 - 24x - 12y + 24 = 0$
2. Find the equation of the parabola whose focus is (3, –4) and whose directrix is $x - y + 5 = 0$. **Ans.:** $x^2 + 2xy + y^2 - 16x - 26y + 25 = 0$
3. Find the coordinates of the vertex, focus and the equation of the directrix of the parabola $3y^2 = 16x$. Find also the length of the latus rectum.

   Ans.: $(0, 0); \left(\frac{4}{3}, 0\right); \frac{16}{3}, 3x + 4y = 0$
4. Find the coordinates of the vertex and focus of the parabola $2y^2 + 3y + 4x = 2$. Find also the length of the latus rectum. **Ans.:** $\left(-1, \frac{4}{3}\right); \left(-1, \frac{23}{24}\right); \frac{3}{2}$
5. A point moves in such a way that the distance from the point (2, 3) is equal to the distance from the line $4x + 3y = 5$. Find the equation of its path. What is the name of this curve? **Ans.:** $25[(x - 2)^2 + (y - 3)^2] - (4x + 3y - 5)^2$

6.6 CONDITION FOR TANGENCY

Find the condition for the straight line $y = mx + c$ to be a tangent to the parabola $y^2 = 4ax$ and find the point of contact.

Solution

The equation of the parabola is

$$y^2 = 4ax \tag{6.1}$$

The equation of the line is

$$y = mx + c \tag{6.2}$$

Solving equations (6.1) and (6.2), we get their points of intersection. The x-coordinates of the points of intersection are given by

$$(mx + c)^2 = 4ax \Rightarrow m^2 x^2 + 2(mc - 2a) x + c^2 = 0 \tag{6.3}$$

If $y = mx + c$ is a tangent to the parabola, then the roots of this equation are equal. The condition for this is the discriminant is equal to zero.

$$\therefore 4(mc - 2a)^2 = 4m^2c^2$$

$$\Rightarrow \cancel{m^2c^2} + 4a^2 - 4mca = \cancel{m^2c^2} \Rightarrow c = a/m$$

Hence, the condition for $y = mx + c$ to be a tangent to the parabola $y^2 = 4ax$ is $c = a/m$.

Substituting $c = a/m$ in equation (6.3), we get

$$m^2x^2 - 2ax + \frac{a^2}{m^2} = 0 \Rightarrow \left(mx - \frac{a}{m}\right)^2 = 0$$

Therefore, the point of contact is $\left(\frac{a}{m^2}, \frac{2a}{m}\right)$.

Note 6.6.1: Any tangent to the parabola is $y = mx + \frac{a}{m}$.

6.7 NUMBER OF TANGENTS

Show that two tangents can always be drawn from a point to a parabola.

Solution

Let the equation to the parabola be $y^2 = 4ax$. Let (x_1, y_1) be the given point. Any tangent to the parabola is $y = mx + \frac{a}{m}$. If this tangent passes through (x_1, y_1), then $y_1 = mx_1 + \frac{a}{m} \Rightarrow my_1 = m^2x_1 + a$ or $m^2x_1 - my_1 + a = 0$. (1) This is a quadratic equation in m.

Therefore, there are two values of m and for each value of m there is a tangent. Hence, there are two tangents from a given point to the parabola.

Note 6.7.1: If m_1, m_2 are the slopes of thc two tangcnts then they are the roots of equation (6.3).

$$m_1 + m_2 = -\frac{y_1}{x_1} \quad \text{and} \quad m_1m_2 = \frac{a}{x}$$

6.8 PERPENDICULAR TANGENTS

Show that the locus of the point of intersection of perpendicular tangents to a parabola is the directrix.

Solution

Let the equation of the parabola be $y^2 = 4ax$. Let (x_1, y_1) be the point of intersection of the two tangents to the parabola. Any tangent to the parabola is

$$y = mx + \frac{a}{m} \tag{6.4}$$

If this tangent passes through (x_1, y_1) then $y_1 = mx_1 + \frac{a}{m}$.

$$\text{(i.e.)} \quad m^2x_1 - my_1 + a = 0 \tag{6.5}$$

If m_1, m_2 are the slopes of the two tangents from (x_1, y_1), then they are the roots of equation (6.5). Since the tangents are perpendicular,

$$m_1 \cdot m_2 = -1 \Rightarrow \frac{a}{x_1} = -1 \text{ or } x_1 + a = 0$$

Therefore, the locus of (x_1, y_1) is $x + a = 0$, which is the directrix.

Show that the locus of the point of intersection of two tangents to the parabola that make complementary angles with the axis is a line through the focus.

Solution

Let (x_1, y_1) be the point of intersection of tangents to the parabola $y^2 = 4ax$. Any tangent to the parabola is $y = mx + \frac{a}{m}$. If this line passes through (x_1, y_1), then $y_1 = mx_1 + \frac{a}{m} \Rightarrow m^2 x_1 - my_1 + a = 0$. If m_1, m_2 are the slopes of the two tangents, then

$$m_1 + m_2 = \frac{y_1}{x_1}, m_1 m_2 = \frac{a}{x_1}$$

If the tangents make complementary angles with the axis of the parabola, then $m_1 = \tan\theta$ and $m_2 = \tan(90 - \theta)$.

$$m_1 m_2 = \tan\theta \times \tan(90 - \theta) = \tan\theta \times \cot\theta = 1$$

$$\Rightarrow \frac{a}{x_1} = 1 \quad \text{or} \quad x_1 - a = 0\,.$$

The locus of the point of intersection of the tangents is $x - a = 0$, which is a straight line through the origin.

6.9 EQUATION OF TANGENT

Find the equation of the tangent at (x_1, y_1) to the parabola $y^2 = 4ax$.

Let $P(x_1, y_1)$ and $Q(x_2, y_2)$ be two points on the parabola $y^2 = 4ax$. Then

$$y_1^2 = 4ax_1 \tag{6.6}$$

$$y_2^2 = 4ax_2 \tag{6.7}$$

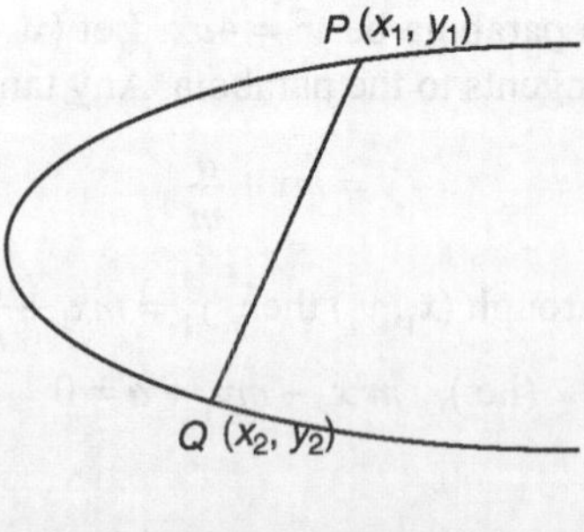

The equation of the chord joining the points (x_1, y_1) and (x_2, y_2) is

$$\frac{y-y_1}{x-x_1}=\frac{y_1-y_2}{x_1-x_2}$$

From equations (6.6) and (6.7), we get

$$y_1^2-y_2^2=4a(x_1-x_2) \quad \text{or} \quad \frac{y_1-y_2}{x_1-x_2}=\frac{4a}{y_1+y_2}$$

Hence, the equation of the chord PQ is

$$\frac{y-y_1}{x-x_1}=\frac{4a}{y_1+y_2} \quad \text{or} \quad y-y_1=\frac{4a(x-x_1)}{y_1+y_2}$$

When the point $Q(x_2, y_2)$ tends to coincide with $P(x_1, y_1)$, the chord PQ becomes the tangent at P.

Hence, the equation of the tangent at P is

$$y-y_1=\frac{2a}{y_1}(x-x_1) \quad \text{or} \quad yy_1-y_1^2=2ax-2ax_1$$

$$yy_1=2ax-2ax_1+y_1^2=2ax-2ax_2+4ax_1$$

$$\text{(i.e)} \quad yy_1=2a(x+x_1)$$

Aliter: The equation of the parabola is $y^2 = 4ax$.
Differentiating this equation with respect to x_1, we get $2y\frac{dy}{dx}=4a$

$$\Rightarrow \frac{dy}{dx}=\frac{2a}{y}$$

$$\left[\frac{dy}{dx}\right]_{\text{at }(x_1,y_1)}=\frac{2a}{y_1}=\text{slope of the tangent at }(x_1,y_1)$$

The equation of the tangent at (x_1, y_1) is

$$y-y_1=\frac{2a}{y_1}(x-x_1)\Rightarrow yy_1-y_1^2=2ax-2ax_1$$

$$\text{or} \quad yy_1=2ax-2ax_1+4ax_1=2a(x+x_1)$$

$$\therefore \quad yy_1=2a(x+x_1)$$

6.10 EQUATION OF NORMAL

Find the equation of the normal at (x_1, y_1) on the parabola $y^2 = 4ax$.

Solution

The slope of the tangent at (x_1, y_1) is $\frac{2a}{y_1}$.

Therefore, the slope of the normal at (x_1, y_1) is $\frac{-y_1}{2a}$.
The equation of the normal at (x_1, y_1) is

$$y - y_1 = -\frac{y_1}{2a}(x - x_1) \Rightarrow 2ay - 2ay_1 = -xy_1 + x_1y_1$$
$$\text{or} \quad xy_1 + 2ay = 2ay_1 + y_1x_1$$

6.11 EQUATION OF CHORD OF CONTACT

Find the equation of the chord of contact of tangents from (x_1, y_1) to the parabola $y^2 = 4ax$.

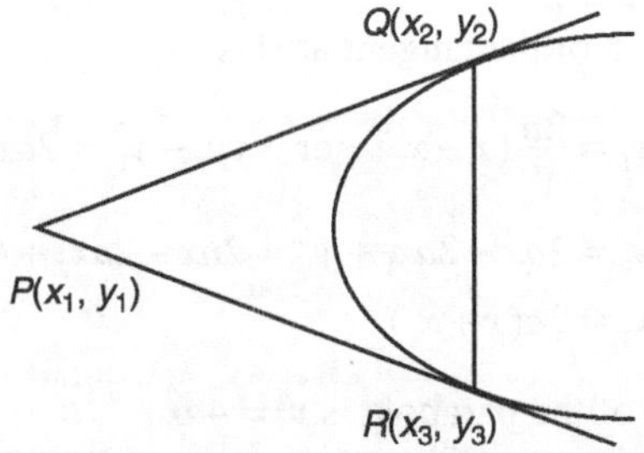

Solution

Let QR be the chord of contact of tangents from $P(x_1, y_1)$. Let Q and R be the points (x_2, y_2) and (x_3, y_3), respectively. Then, the equation of tangents at Q and R are

$$yy_2 = 2a(x + x_2) \tag{6.8}$$
$$yy_3 = 2a(x + x_3) \tag{6.9}$$

These two tangents pass through $P(x_1, y_1)$.

$$\therefore y_1y_2 = 2a(x_1 + x_2) \tag{6.10}$$
$$y_1y_3 = 2a(x_1 + x_3) \tag{6.11}$$

These two equations show that the points (x_2, y_2) and (x_3, y_3) lie on the line $yy_1 = 2a(x + x_1)$.

Therefore, the equation of the chord of contact of tangents from $P(x_1, y_1)$ is $yy_1 = 2a(x + x_1)$.

6.12 POLAR OF A POINT

Find the polar of the point with respect to the parabola $y^2 = 4ax$.

Definition 6.12.1 The polar of a point with respect to a parabola is defined as the locus of the point of intersection of the tangents at the extremities of a chord passing through that point.

Solution

Let $P(x_1, y_1)$ be the given point. Let QR be a variable chord passing through P. Let the tangents at Q and R intersect at (h, k). Then the equation of the chord of contact of tangents from (h, k) is $yk = 2a(x + h)$. This chord passes through $P(x_1, y_1)$.

$$\therefore\ y_1k = 2a\,(x_1 + h)$$

Then the locus of (h, k) is $yy_1 = 2a(x + x_1)$

Hence, the polar of (x_1, y_1) with respect to $y^2 = 4ax$ is

$$yy_1 = 2a(x + x_1)$$

Note 6.12.1: Point P is the pole of the line $yy_1 = 2a(x + x_1)$.

Note 6.12.2: Find the pole of the line $lx + my + n = 0$ with respect to the parabola $y^2 = 4ax$.

Let (x_1, y_1) be the pole. Then the polar of (x_1, y_1) is

$$yy_1 = 2a(x + x_1)$$

$$\text{(i.e.)}\quad 2ax - yy_1 + 2ax_1 = 0 \tag{6.12}$$

But the polar of (x_1, y_1) is given by $lx + my + n = 0$ (6.13)

Equations (6.12) and (6.13) represent the same line. Then, identifying these two equations, we get

$$\frac{2a}{l} = \frac{-y_1}{m} = \frac{-2ax_1}{n}$$

$$\therefore\ x_1 = \frac{-n}{l},\ y_1 = \frac{-2am}{l}$$

Hence, the pole of the line is $\left(\frac{-n}{l}, \frac{-2am}{l}\right)$.

6.13 CONJUGATE LINES

Definition 6.13.1 Two lines are said to be conjugate to each other if the pole of each lies on the other.

Find the condition for the lines $lx + my + n = 0$ and $l_1 x + m_1 y + n_1 = 0$ to be conjugate lines with respect to the parabola $y^2 = 4ax$.

Solution

Let (x_1, y_1) be the pole of the lines $l_1 x + m_1 y + n_1 = 0$ with respect to the parabola. The polar of (x_1, y_1) with respect to the polar $y^2 = 4ax$ is $lx + my + n = 0$.

The equation of the polar of (x_1, y_1) with respect to the parabola $y^2 = 4ax$ is

$$yy_1 = 2a(x + x_1) \Rightarrow 2ax - yy_1 + 2ax_1 = 0 \tag{6.14}$$

But the polar of (x_1, y_1) is given by $lx + my + n = 0$ (6.15)

Equations (6.14) and (6.15) represent the same line. Identifying these two equations, we get

$$\frac{2a}{l} = \frac{-y_1}{m} = \frac{2ax_1}{n}$$

The pole of the line $lx + my + n = 0$ is $\left(\frac{n}{l}, \frac{-2am}{l}\right)$.

Since the lines $lx + my + n = 0$ and $l_1 x + m_1 y + n_1 = 0$ are conjugate to each other, the pole of $lx + my + n = 0$ will lie on $l_1 x + m_1 y + n_1 = 0$.

$$l_1\left(\frac{n}{l}\right) + m_1\left(\frac{-2am}{l}\right) + n_1 = 0.$$

$$\text{(i.e.)} \quad l_1 n + l_1 n = 2amm_1.$$

This is the required condition.

6.14 PAIR OF TANGENTS

Find the equation of pair of tangents from (x_1, y_1) to the parabola $y^2 = 4ax$.

Solution

The equation of a line through (x_1, y_1) is $\dfrac{x - x_1}{\cos\theta} = \dfrac{y - y_1}{\sin\theta} = r$ (6.16)

Any point on this line is $(x_1 + r\cos\theta, y_1 + r\sin\theta)$. The points of intersection of the line and the parabola are given by

$$(y_1 + r\sin\theta)^2 = 4a(x_1 + \cos\theta)$$

$$\text{(i.e.)} \quad r^2 \sin^2\theta + 2r(y_1 \sin\theta - 2a\cos\theta) + y_1^2 - 4ax_1 = 0 \tag{6.17}$$

The two values of r of this equation are the distances of point (x, y) to the point (x_1, y_1). If line (6.16) is a tangent to the parabola, then the two values of r must be equal and the condition for this is the discriminant of quadratic (6.17) is zero.

$$\therefore 4(y_1 \sin\theta - 2a\cos\theta)^2 = 4\sin^2\theta(y_1^2 - 4ax_1)$$

Eliminating θ in this equation with the help of (6.16), we get

$$4\left[y_1\left(\frac{y-y_1}{r}\right) - 2a\left(\frac{x-x_1}{r}\right)\right]^2 = 4\left(\frac{y-y_1}{r}\right)^2(y_1^2 - 4ax_1)$$

$$\text{(i.e.) } (yy_1 - y_1^2 - 2ax + 2ax_1)^2 = (y-y_1)^2(y_1^2 - 4ax_1)$$

$$[yy_1 - 2a(x+x_1) - y_1^2 + 4ax_1]^2 = (y^2 + y_1^2 - 2yy_1)(y_1^2 - 4x_1^2)$$

$$= \{y^2 - 4ax + y_1^2 - 4ax_1 - 2yy_1 - 4a(x+x_1)\} \times (y_1^2 - 4ax_1)$$

$$\text{(i.e.) } (T - S_1)^2 = (S + S_1 - 2T)S_1$$

$$\text{or } T^2 + S_1^2 - 2TS_1 = SS_1 + S_1^2 - 2TS_1. \text{ (i.e.) } T^2 = SS_1$$

Therefore, the equation of pair of tangents from (x_1, y_1) is

$$[yy_1 - 2a(x+x_1)]^2 = (y^2 - 4ax)(y_1^2 - 4ax_1).$$

6.15 CHORD IN TERMS OF MID-POINT

Find the equation of a chord of the parabola in terms of its middle point (x_1, y_1).

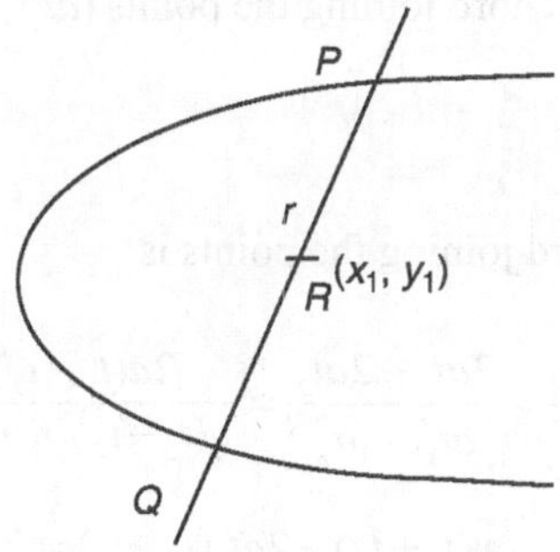

Solution

Let the equation of the chord be

$$\frac{x-x_1}{\cos\theta} = \frac{y-y_1}{\sin\theta} = r$$

Any point on this line is $(x_1 + r\cos\theta, y_1 + r\sin\theta)$. When the chord meets the parabola $y^2 = 4ax$, this point lies on the curve.

$$\therefore (y_1 + r\sin\theta)^2 = 4a(x_1 + r\cos\theta)$$

$$\text{(i.e.)} \quad r^2\sin^2\theta + 2r(y_1\sin\theta - 2a\cos\theta) + y_1^2 - 4ax_1 = 0$$

The two values of r are the distances RP and RQ, which are equal in magnitude but opposite in sign. The condition for this is the coefficient of r is equal to zero.

$$\therefore y_1\sin\theta - 2a\cos\theta = 0$$

$$\text{(i.e.)} \quad \frac{y_1(y-y_1)}{r} - \frac{2a(x-x_1)}{r} = 0$$

$$yy_1 - y_1^2 = 2ax - 2ax_1$$

$$\text{or} \quad yy_1 - 2a(x+x_1) = y_1^2 - 4ax_1$$

$$\text{(i.e.)} \quad T = S_1$$

This is the required equation of the chord.

6.16 PARAMETRIC REPRESENTATION

$x = at^2, y = 2at$ satisfy the equation $y^2 = 4ax$. This means $(at^2, 2at)$ is a point on the parabola. This point is denoted by 't' and t is called a parameter.

6.17 CHORD JOINING TWO POINTS

Find the equation of the chord joining the points $(at_1^2, 2at_1)$ and $(at_2^2, 2at_2)$ on the parabola $y^2 = 4ax$.

Solution

The equation of the chord joining the points is

$$\frac{y-2at_1}{x-at_1^2} = \frac{2at_1 - 2at_2}{at_1^2 - at_2^2} = \frac{2a(t_1-t_2)}{a(t_1-t_2)(t_1+t_2)} = \frac{2}{t_1+t_2}$$

$$\therefore y(t_1+t_2) - 2at_1(t_1+t_2) = 2x - 2at_1^2$$

$$\text{(i.e.)} \; y(t_1+t_2) = 2x - 2at_1^2 + 2at_1^2 + 2at_1t_2$$

$$y(t_1+t_2) = 2x + 2t_1t_2$$

Note 6.17.1: The chord becomes the tangent at 't' if $t_1 = t_2 = t$.
Therefore, the equation of the tangent at t is

$$y(2t) = 2x + 2at^2 \text{ or } yt = x + at^2$$

6.18 EQUATIONS OF TANGENT AND NORMAL

Find the equation of the tangent and normal at 't' on the parabola $y^2 = 4ax$.

Solution

The equation of the parabola is $y^2 = 4ax$. Differentiating with respect to x,

$$2y\frac{dy}{dx} = 4a \text{ (ie) } \frac{dy}{dx} = \frac{2a}{y}$$

$$\left(\frac{dy}{dx}\right) \text{ at } (at^2, 2at) = \frac{2a}{2at} = \frac{1}{t} = \text{slope of the tangent at } t$$

The equation of the tangent at t is

$$y - 2at = \frac{1}{t}(x - at^2)$$

$$yt - 2at^2 = x - at^2 \quad \text{(i.e.)} \quad yt = x + at^2$$

The slope of the normal at t is $-t$. The equation of the normal at 't' is

$$y - 2at = -t(x - at^2)$$

$$\text{(i.e.)} \quad y - 2at = -xt + at^3$$

$$\text{(i.e.)} \quad y + xy = 2at + at^3$$

6.19 POINT OF INTERSECTION OF TANGENTS

Find the point of intersection of tangents at t_1 and t_2 on the parabola $y^2 = 4ax$.

Solution

The equation of tangents at t_1 and t_2 are

$$yt_1 = x + at_1^2$$

$$yt_2 = x + at_2^2$$

Subtrating, $y(t_1 - t_2) = a(t_1^2 - t_2^2)$

$$\text{(i.e.)} \quad y = a\,(t_1 + t_2).$$

$$\therefore a(t_1 + t_2)\, t_1 = x + at_1^2$$
$$\therefore x = at_1t_2$$

Hence, the point of intersection is $[at_1t_2, a(t_1 + t_2)]$.

6.20 POINT OF INTERSECTION OF NORMALS

Find the point of intersection of normals at t_1 and t_2.

Solution

The equation of normals at t_1 and t_2 are

$$y + xt_1 = 2at_1 + at_1^3$$
$$y + xt_2 = 2at_2 + at_2^3$$

Subtracting, $x(t_1 - t_2) = 2a(t_1 - t_2) + a(t_1^3 - t_2^3)$

$$\Rightarrow x = 2a + a(t_1^2 + t_1t_2 + t_2^2)$$
$$\therefore y = -at_1t_2(t_1 + t_2)$$

Hence, the point of intersection of normals is

$$[2a + a(t_1^2 + t_1t_2 + t_2^2), -at_1t_2(t_1 + t_2)]$$

6.21 NUMBER OF NORMALS FROM A POINT

Show that three normals can always be drawn from a given point to a parabola.

Solution

Let the equation of the parabola be $y^2 = 4ax$.
The equation of the normal at t is

$$y + xt = 2at + at^3$$

If this passes through (x_1, y_1) then

$$y_1 + x_1t = 2at + at^3$$

$$\therefore at^3 - t(2a - x_1) - y_1 = 0 \qquad (6.18)$$

This being a cubic equation in t, there are three values for t. For each value of t there is a normal from(x_1, y_1) to the parabola $y^2 = 4ax$.

Note 6.21.1: If t_1, t_2, t_3 are the roots of equation (6.18), then

$$t_1 + t_2 + t_3 = 0 \tag{6.19}$$

$$t_1t_2 + t_2t_3 + t_3t_1 = \frac{2a - x_1}{a} \tag{6.20}$$

$$t_1t_2t_3 = \frac{y_1}{a} \tag{6.21}$$

Note 6.21.2: From (6.18), $2at_1 + 2at_2 + 2at_3 = 0$
Therefore, the sum of the coordinates of the feet of the normal is always zero.

6.22 INTERSECTION OF A PARABOLA AND A CIRCLE

Prove that a circle and a parabola meet at four points and show that the sum of the ordinates of the four points of intersection is zero.

Solution

Let the equation of the circle be

$$x^2 + y^2 + 2gx + 2fy + c = 0 \tag{6.22}$$

Let the equation of the parabola be

$$y^2 = 4ax \tag{6.23}$$

Any point on the parabola is $(at^2, 2at)$. When the circle and the parabola intersect, this point lies on the circle,

$$\therefore (a^2t^2)^2 + (2at)^2 + 2g(at^2) + 2f(2at) + c = 0$$

$$\text{(i.e.)} \quad a^2t^4 + (4a^2 + 2ga)t^2 + 4fat + c = 0 \tag{6.24}$$

This being a fourth degree equation in t, there are four values of t. For each value of t there is a point of intersection. Hence, there are four points of intersection of a circle and a parabola. If t_1, t_2, t_3, t_4 be the four roots of equation (6.24), then

$$t_1 + t_2 + t_3 + t_4 = 0 \tag{6.25}$$

Multiplying equation (6.25) by $2a$, we get

$$2at_1 + 2at_2 + 2at_3 + 2at_4 = 0$$

Therefore, the sum of the ordinates of the four points of intersection is zero.

ILLUSTRATIVE EXAMPLES BASED ON TANGENTS AND NORMALS

Example 6.3

Find the equations of the tangent and normal to the parabola $y^2 = 4(x - 1)$ at (5, 4).

Solution

$$y^2 = 4(x - 1)$$

Differentiating with respect to x,

$$2y\frac{dy}{dx} = 4 \Rightarrow \frac{dy}{dx} = \frac{4}{2y} = \frac{2}{y}$$

$$\left(\frac{dy}{dx}\right)_{\text{at}(5,\,4)} = \frac{2}{4} = \frac{1}{2} = \text{Slope of the tangent at } (5, 4)$$

∴ The equation of the tangent at (5, 4) is $y - 4 = \frac{1}{2}(x - 5)$.

$2y - 8 = x - 5$ or $x - 2y + 3 = 0$. The slope of the normal at (5, 4) is -2.

∴ The equation of normal at (5, 4) is $y - 4 = -2(x - 5)$ or $2x + y = 14$.

Example 6.4

Find the condition that the straight line $lx + my + n = 0$ is a tangent to the parabola.

Solution

Any straight line tangent to the parabola $y^2 = 4ax$ is of the form $y = mx + c$ if

$$c = \frac{a}{m}. \tag{6.26}$$

Consider the line $lx + my + n = 0$ (i.e.) $my = -lx - n$

$$\Rightarrow y = \frac{-l}{m}x - \frac{n}{m}$$

If this is a tangent to the parabola, $y^2 = 4ax$ then $\frac{-n}{m} = \frac{-a}{(l/m)}$

$$\Rightarrow \frac{-n}{m} = \frac{-am}{l} \text{ (i.e.) } am^2 = nl$$

Example 6.5

A common tangent is drawn to the circle $x^2 + y^2 = r^2$ and the parabola $y^2 = 4ax$. Show that the angle θ which it makes with the axis of the parabola is given by $\tan^2\theta = \frac{\sqrt{r^2 + 4a^2} - r}{2r}$.

Solution

Let $y = mx + c$ be a common tangent to the parabola

$$y^2 = 4ax \tag{6.27}$$

and the circle

$$x^2 + y^2 = r^2 \tag{6.28}$$

If $y = mx + c$ is tangent to the parabola (6.27) then

$$c = \frac{a}{m}$$

$$\text{(i.e.) } y = mx + \frac{a}{m} \tag{6.29}$$

If $y = mx + c$ is a tangent to the circle (6.29) then

$$y = mx + r\sqrt{1+m^2} \tag{6.30}$$

Equations (6.29) and (6.30) represent the same straight line. Identifying we get

$$\frac{a}{m} = r\sqrt{1+m^2} \Rightarrow a^2 = m^2 r^2 (1+m^2)$$

$$r^2 m^4 + r^2 m^2 - a^2 = 0$$

$$m^2 = \frac{-r^2 \pm \sqrt{r^4 + 4a^2 r^2}}{2r^2} = \frac{-r^2 \pm r\sqrt{r^2 + 4a^2}}{2r^2}$$

Since m^2 has to be positive,

$$m^2 = \frac{-r + \sqrt{r^2 + 4a^2}}{2r} \text{ or } \tan^2\theta = \frac{\sqrt{r^2 + 4a^2} - r}{2r} \quad \text{or}$$

Example 6.6

A straight line touches the circle $x^2 + y^2 = 2a^2$ and the parabola $y^2 = 8ax$. Show that its equation is $y = \pm(x + 2a)$.

Solution

The equation of the circle is

$$x^2 + y^2 = 2a^2 \tag{6.31}$$

The equation of the parabola is

$$y^2 = 4ax \tag{6.32}$$

A tangent to the parabola (6.32) is

$$y = mx + \frac{2a}{m} \tag{6.33}$$

A tangent to the circle (6.33) is

$$y = mx + \sqrt{2}a\sqrt{1+m^2} \tag{6.34}$$

Equations (6.33) and (6.34) represent the same straight line. Identifying we get,

$$\sqrt{2}a\sqrt{1+m^2} = \frac{2a}{m}$$

$$2a^2m^2(1+m^2) = 4a^2$$

$$\text{or } m^4 + m^2 - 2 = 0$$

$$(m^2 - 1)(m^2 + 2) = 0$$

$m^2 = 1$ or -2; $m^2 = -2$ is impossible.

$$\therefore m^2 = 1 \quad \text{or} \quad m = \pm 1$$

$\therefore$ The equation of the common tangent is $y = \pm\, x \pm 2a$.

$$\therefore y = \pm(x + 2a)$$

Example 6.7

Show that for all values of m, the line $y = m(x+a) + \frac{a}{m}$ will touch the parabola $y^2 = 4a(x+a)$. Hence show that the locus of a point,the two tangents form which to the parabolas $y^2 = 4a(x+a)$ and $y^2 = 4b(x+b)$ one to each are at right angles, is the line $x + a + b = 0$.

Solution

$$y^2 = 4a(x+a) \tag{6.35}$$

$$y = m(x+a) + \frac{a}{m} \tag{6.36}$$

Solving (6.35) and (6.36), we get their points of intersection. The x-coordinates of their points of intersection are given by,

$$\left[m(x+a) + \frac{a}{m}\right]^2 = 4a(x+a)$$

$$\therefore m^2(x+a)^2+\frac{a^2}{m^2}+2a(x+a)=4a(x+a)$$

$$\text{or } \left[m(x+a)-\frac{a}{m}\right]^2=0$$

$$\therefore x+a=\frac{a}{m^2}.$$

$\therefore$ The two values of x and hence of y of the points of intersection are the same. Hence, $y=m(x+a)+\frac{a}{m}$ is a tangent to the parabola $y^2=4a(x+a)$. Let (x_1, y_1) be the point of intersection of the two tangents to the parabola $y^2=4a(x+a), y^2=4a(y+b)$. The tangents are $y=m(x+a)+\frac{a}{m}$ and $y=m_1(x+b)+\frac{b}{m_1}$. Since they pass through (x_1, y_1), we have

$$y_1=m(x_1+a)+\frac{a}{m} \tag{6.37}$$

$$y_1=m_1(x_1+b)+\frac{b}{m_1} \tag{6.38}$$

Since the tangents are at right angles, $m_1m_2=-1$. Subtracting (6.38) from (6.37), we get

$$0=x_1(m-m_1)+a\left(m+\frac{1}{m}\right)-b\left(m_1+\frac{1}{m_1}\right)$$

$$0=x_1\left(m+\frac{1}{m}\right)+a\left(m+\frac{1}{m}\right)+b\left(m+\frac{1}{m}\right)$$

since $m_1=\frac{-1}{m}$

Cancelling $\left(m+\frac{1}{m}\right)$, $x_1+a+b=0$

The locus of (x_1, y_1) is $x+a+b=0$.

Example 6.8

Prove that the locus of the point of intersection of two tangents to the parabola $y^2=4ax$, which makes an angle of α with x-axis, is $y^2-4ax=(x+a)^2\tan^2\alpha$. Determine the locus of point of intersection of perpendicular tangents.

Solution

Let (x_1, y_1) be the point of intersection of tangents. Any tangent to the parabola is $y=mx+\frac{a}{m}$. If this passes through (x_1, y_1) then $y_1=mx_1+\frac{a}{m}$.

(i.e.) $m^2x_1 - my_1 + a = 0$.

$$m_1 + m_2 = \frac{y_1}{x_1}$$

$$m_1m_2 = \frac{a}{x_1} \quad \text{Now } \tan\alpha = \left|\frac{m_1 - m_2}{1 + m_1m_2}\right|$$

$$\tan^2\alpha = \frac{(m_1+m_2)^2 - 4m_1m_2}{(1+m_1m_2)^2} = \frac{\frac{y_1^2}{x_1^2} - 4\frac{a}{x_1}}{\left(1+\frac{a}{x_1}\right)^2} = \frac{y_1^2 - 4ax_1}{(x_1+a)^2}$$

$$\Rightarrow (x_1 + a)^2 \tan^2\alpha = y_1^2 - 4ax_1$$

$\therefore$ The locus of (x_1, y_1) is $y^2 - 4ax = (x+a)^2 \tan^2\alpha$.
If the tangents are perpendicular, $\tan\alpha = \tan 90° = \infty$
$\therefore$ The locus of perpendicular tangents is directrix.

Example 6.9

Prove that if two tangents to a parabola intersect on the latus rectum produced then they are inclined to the axis of the parabola at complementary angles.

Solution

Let (x_1, y_1) the equation of the parabola be $y^2 = 4ax$. Let $y = mx + \frac{a}{m}$ be any tangent to the parabola. Let the two tangents intersect at (a, y_1), a point on the latus rectum. Then (a_1, y_1) lies, on $y = mx + \frac{a}{m}$.

$$\therefore y_1 = ma + \frac{a}{m} \text{ or } m^2a - my_1 + a = 0 \tag{6.39}$$

If m_1 and m_2 are the slopes of the two tangents to the parabola then $m_1m_2 = 1$. (i.e.) $\tan\theta \cdot \tan(90 - \theta) = 1$.

(i.e.) The tangent makes complementary angles to the axis of the parabola.

Example 6.10

Prove that the locus of poles of the chords of the parabola $y^2 = 4ax$ which subtends a constant angle α at the vertex is the curve $(x+4a)^2 \tan^2\alpha = 4(y^2 - 4ax)$.

Solution

Let (x_1, y_1) be the pole of a chord of the parabola. Then the polar of (x_1, y_1) is

$$yy_1 = 2a(x + x_1) \tag{6.40}$$

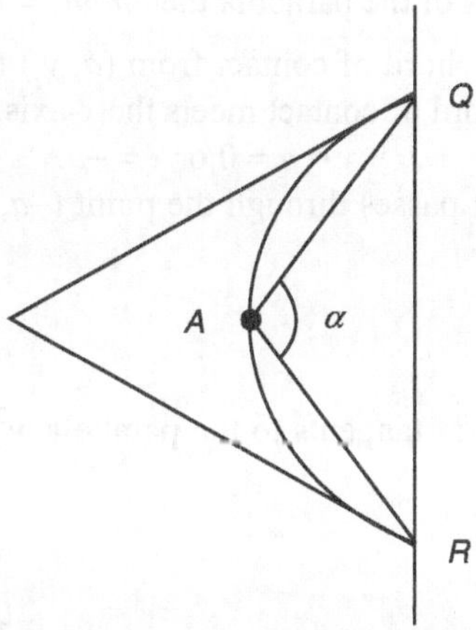

which is the chord of contact from (x_1, y_1). The combined equation of the lines AQ and AR is got by homogenization of the equation of the parabola $y^2 = 4ax$ with the help of (6.40).

∴ The combined equation of the lines is

$$y^2 = 4ax\left[\frac{yy_1 - 2ax}{2ax_1}\right] \text{ (i.e.) } x_1y^2 = 2xyy_1 - 4ax^2 \text{ or } 4ax^2 - 2x_1xy + x_1y^2 = 0$$

$$\text{Since } \angle QAR = \alpha, \tan\alpha = \left|\frac{2\sqrt{h^2 - ab}}{a+b}\right| \Rightarrow \tan^2\alpha = \frac{4(h^2 - ab)}{(a+b)^2},$$

$$\Rightarrow \tan^2\alpha = \frac{4(y_1^2 - 4ax_1)}{(x_1 + 4a)^2}$$

The locus of (x_1, y_1) is $(x + 4a)^2 \tan^2\alpha = 4(y^2 - 4ax)$.

Note 6.10.1: If $\alpha = 90°$, the locus of (x_1, y_1) is $x + 4a = 0$.

Example 6.11

If two tangents are drawn to a parabola making complementary angles with the axis of the parabola, prove that the chord of contact passes through the point where the axis cuts the directrix.

Solution

Let $y = mx + \frac{a}{m}$ be a tangent from a point (x_1, y_1) to the parabola $y^2 = 4ax$. Then $y_1 = mx_1 + \frac{a}{m}$ or $m^2x_1 - my_1 + a = 0$. If the tangents make complementary angles with the axis of the parabola then $m_1m_2 = 1$ or $\frac{a}{x_1} = 1 \Rightarrow x_1 = a$.

The equation of the chord of contact from (a, y_1) to the parabola is $yy_1 = 2a(x + a)$. When the chord of contact meets the x-axis, $y = 0$.

$$\therefore \quad x + a = 0 \text{ or } x = -a.$$

∴ The chord of contact passes through the point $(-a, 0)$ where the axis cuts the directrix.

Example 6.12

Find the locus of poles of tangents to the parabola $y^2 = 4ax$ with respect to the parabola $x^2 = 4by$.

Solution

Let (x_1, y_1) be the pole with respect to the parabola $x^2 = 4by$. Then the polar of (x_1, y_1) is $xx_1 = 2b(y + y_1)$, $y = \frac{xx_1}{2b} - y_1$. This is a tangent to the parabola $y^2 = 4ax$. The condition for tangent is $c = \frac{a}{m}$ (i.e.) $-y_1 = \frac{2ab}{x_1}$ or $x_1y_1 + 2ab = 0$.

The locus of (x_1, y_1) is the straight line $xy + 2ab = 0$.

Example 6.13

From a variable point on the tangent at the vertex of a parabola, the perpendicular is drawn to its polar. Show that the perpendicular passes through a fixed point on the axis of the parabola.

Solution

The equation of the tangent at the vertex is $x = 0$. Any point on this line is $(0, y_1)$. The polar of $(0, y_1)$ with respect to the parabola $y^2 = 4ax$ is $yy_1 = 2ax$.

The equation of the perpendicular to the polar of (x_1, y_1) is $y_1x + 2ay = k$. This passes through $(0, y_1)$.

$$\therefore k = 2ay_1.$$

∴ The equation of the perpendicular to the polar from $(0, y_1)$ is $y_1x + 2ay = 2ay_1$, when this line meets the x-axis, $y_1x = 2ay_1$ or $x = 2a$

Hence, the perpendicular passes through the point $(2a, 0)$, a fixed point on the axis of the parabola.

Example 6.14

The polar of any point with respect to the circle $x^2 + y^2 = a^2$ touches the parabola $y^2 = 4ax$. Show that the point lies on the parabola $y^2 = -ax$.

Solution

The polar of the point (x_1, y_1) with respect to the circle $x^2 + y^2 = a^2$ is $xx_1 + yy_1 = a^2$

(i.e.) $$y = -\frac{x_1}{y_1}x + \frac{a^2}{y_1}$$

This is a tangent to the parabola $y^2 = 4ax$. The condition for that is $c = \frac{a}{m}$.

(i.e.) $$\frac{a^2}{y_1} = a\left(\frac{-y_1}{x_1}\right) \text{ or } ax_1 = -y_1^2$$

The locus of (x_1, y_1) is $y^2 = -ax$ which is a parabola.

Example 6.15

Find the locus of poles with respect to the parabola $y^2 = 4ax$ of tangents to the circle $x^2 + y^2 = c^2$.

Solution

Let the pole with respect to the parabola $y^2 = 4ax$ be (x_1, y_1). Then the polar of (x_1, y_1) is

$$yy_1 = 2a(x + x_1)$$

(i.e.) $y = \frac{2a}{y_1}x + \frac{2ax_1}{y_1}$. This is a tangent to the circle $x^2 + y^2 = a^2$. The condition for this is '$c^2 = a^2(1 + m^2)$.

(i.e.) $$\frac{4a^2x_1^2}{y^2} = c^2\left(1 + \frac{4a^2}{y_1^2}\right)$$

$$\Rightarrow 4a^2x_1^2 = c^2(y_1^2 + 4a^2).$$

The locus of (x_1, y_1) is $4a^2x^2 = c^2(y^2 + 4a^2)$.

Example 6.16

A point P moves such that the line through it perpendicular to its polar with respect to the parabola $y^2 = 4ax$ touches the parabola $x^2 = 4by$. Show that the locus of P is $2ax + by + 4a^2 = 0$.

Solution

Let P be the point (x_1, y_1). The polar of P with respect to $y^2 = 4ax$ is

$$yy_1 = 2a(x + x_1) \text{ or } 2ax - yy_1 + 2ax_1 = 0 \tag{6.41}$$

The equation of the perpendicular to (6.41) is $y_1x + 2ay + k = 0$. This passes through (x_1, y_1).

$$\therefore x_1y_1 + 2ay_1 + k = 0. \Rightarrow k = -(x_1y_1 + 2ay_1)$$

Hence, the equation of the perpendicular is $y_1x + 2ay - (x_1y_1 + 2ay_1) = 0$.

(i.e.) $$x = \frac{-2a}{y_1}y + \frac{x_1y_1 + 2ay_1}{y_1} \Rightarrow x = \frac{-2a}{y_1}y + (x_1 + 2a)$$

This is a tangent to the parabola $x^2 = 4by$.

$\therefore$ The condition $x_1 + 2a = \dfrac{-by_1}{2a}$

$\therefore$ The locus of (x_1, y_1) is $2ax + by + 4a^2 = 0$.

Example 6.17

If the polar of the point P with respect to a parabola passes through Q then show that the polar of Q passes through P.

Solution

Let the equation of the parabola be $y^2 = 4ax$. Let P and Q be the points (x_1, y_1) and (x_2, y_2). Then the polar of P is $yy_1 = 2a(x + x_1)$.

Since this passes through $Q(x_2, y_2)$, we get $y_1y_2 = 2a(x_1 + x_2)$. This condition shows that the point (x_1, y_1) lies on the line $yy_2 = 2a(x + x_2)$.

$\therefore$ The polar of Q passes through the point $P(x_1, y_1)$.

Example 6.18

P is a variable point on the tangent at the vertex of the parabola $y^2 = 4ax$. Prove that the locus of the foot of the perpendicular from P on its polar with respect to the parabola is the circle $x^2 + y^2 - 2ax = 0$.

Solution

P is the variable point on the tangent at the vertex of the parabola $y^2 = 4ax$. The equation of the tangent at the vertex is $x = 0$. Any point on the tangent at the vertex is $P(0, y_1)$. The polar of $(0, y)$ is

$$yy_1 = 2ax \tag{6.42}$$

The equation of the perpendicular to this polar is

$$y_1x + 2ay = k \tag{6.43}$$

This passes through $(0, y_1)$.

$$\therefore\ 2ay_1 = k.$$

$\therefore$ The equation of the perpendicular from P to its polar is

$$y_1x + 2ay = 2ay_1 \tag{6.44}$$

Let (l, m) be the point of intersection of (6.42) and (6.43). Then $2al - my_1 = 0$.

$$y_1l + 2am - 2ay_1 = 0$$

Solving $\quad \dfrac{l}{2ay_1^2} = \dfrac{m}{4a^2y_1} = \dfrac{1}{4a^2 + y_1^2}$

$$\therefore l = \frac{2ay_1^2}{y_1^2 + 4a^2}, m = \frac{4a^2y_1}{4a^2 + y_1^2}$$

Now, $\quad l^2 + m^2 - 2am = \dfrac{4a^2y_1^4 + 16a^4y_1^2}{(y_1^2 + 4a^2)^2} - \dfrac{4a^2y_1^2}{y_1^2 + 4a^2}$

$$= \frac{4a^2y_1^2(y_1^2 + 4a^2)}{(y_1^2 + 4a^2)^2} - \frac{4a^2y_1^2}{y_1^2 + 4a^2}$$

$$= \frac{4a^2y_1^2}{y_1^2 + 4a^2} - \frac{4a^2y_1^2}{y_1^2 + 4a^2} = 0$$

$\therefore$ The locus of (l, m) is $x^2 + y^2 - 2ax = 0$.

Example 6.19

If from the vertex of the parabola $y^2 = 4ax$, a pair of chords can be drawn at right angles to one another and with these chords as adjacent sides a rectangle be made, prove that the locus of further angle of the rectangle is the parabola $y^2 = 4a(x - 8a)$.

Solution

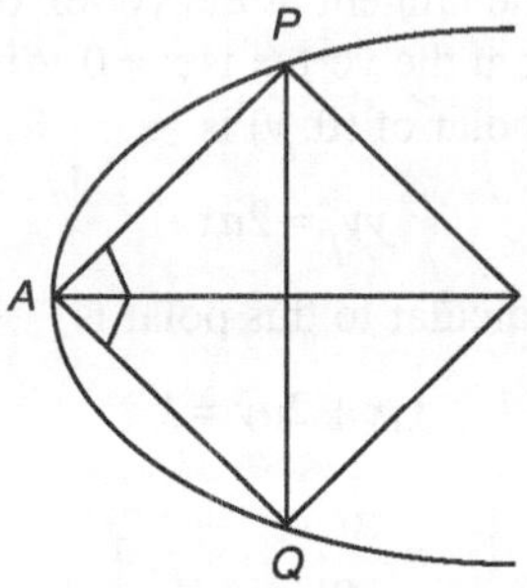

Let AP and AQ be the chords of the parabola such that $\underline{|PAQ} = 90°$. Complete the rectangle $APRQ$. Then the midpoints of AR and PQ are the same. Let the equations of AP be $y = mx$. Solving $y = mx$ and $y^2 = 4ax$, we get $m^2x^2 = 4ax$ or $x = \frac{4a}{m^2}$.

$\therefore$ The point P is $\left(\frac{4a}{m^2}, \frac{4a}{m}\right)$. Since AQ is perpendicular to AP, slope of AQ is $\frac{-1}{m}$.

Hence, the point Q is $(4am^2, -4am)$.

Let (x_1, y_1) be the point R. The midpoint of AR is $\left(\frac{x_1}{2}, \frac{y_1}{2}\right)$. The midpoint of PQ is $\left[2a\left(m^2 + \frac{1}{m^2}\right); -2a\left(m - \frac{1}{m}\right)\right]$. Since the midpoint of AR is the same as that of PQ,

$$\frac{x_1}{2} = 2a\left(m^2 + \frac{1}{m^2}\right), \quad \frac{y_1}{2} = -2a\left(m - \frac{1}{m}\right)$$

$$\therefore \frac{x_1}{2} = 2a\left[\left(m - \frac{1}{m}\right)^2 + 2\right] \quad \text{or} \quad x_1 = 4a\left[\frac{y_1^2}{16a^2} + 2\right] = \frac{y_1^2}{4a} + 8a$$

Hence, the locus of (x_1, y_1) is $y^2 = 4a(x - 8a)$.

Example 6.20

Show that if r_1 and r_2 be the lengths of perpendicular chords of a parabola drawn through the vertex then $(r_1r_2)^2 = 16a^2\left(r_1^{2/3} + r_2^{2/3}\right)$.

Solution

$$AP = r_1, AQ = r_2, \underline{|PAX} = \theta, \underline{|QAX} = 90 - \theta$$

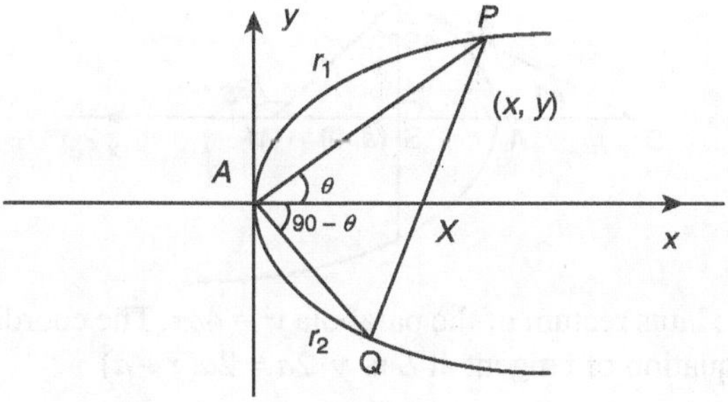

The coordinates of P are $(r_1\cos\theta, r_1\sin\theta)$.
The coordinates of Q are $(r_2\sin\theta, r_2\cos\theta)$.
Since P lies on the parabola $y^2 = 4ax$,

$$r_1^2 \sin^2\theta = 4a \cdot r_1 \cos\theta \quad \therefore r_1 = \frac{4a\cos\theta}{\sin^2\theta}$$

Similarly, $$r_2 = \frac{4a\sin\theta}{\cos^2\theta};\ r_1r_2 = \frac{16a^2}{\cos\theta\sin\theta} = \frac{(4a)^2}{(\cos\theta\sin\theta)}$$

$$(r_1r_2)^{4/3} = \frac{(4a)^{8/3}}{(\cos\theta\sin\theta)^{4/3}} \tag{6.45}$$

Also $$\text{RHS} = 16a^2(r_1^{2/3} + r_2^{2/3}) = 16a^2\left[\frac{(4a\cos\theta)^{2/3}}{(\sin^2\theta)^{2/3}} + \frac{(4a\sin\theta)^{2/3}}{(\cos^2\theta)^{2/3}}\right]$$

$$= 16a^2\left[\frac{(4a)^{2/3}(\cos^2\theta + \sin^2\theta)}{(\cos\theta\sin\theta)^{4/3}}\right] = \frac{(4a)^{8/3}}{(\cos\theta\sin\theta)^{4/3}} \tag{6.46}$$

From equations (6.45) and(6.46), $(r_1r_2)^{4/3} = 16a^2(r_1^{2/3} + r_2^{2/3})$.

Example 6.21

Show that the latus rectum of a parabola bisects the angle between the tangents and normal at either extremity.

Solution

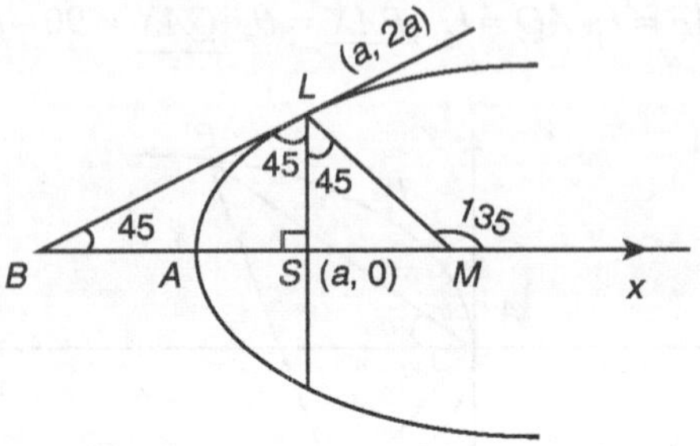

Let LSL' be the latus rectum of the parabola $y^2 = 4ax$. The coordinates of L are $(a, 2a)$. The equation of tangent at L is $y \cdot 2a = 2a(x + a)$

$$y = x + a \tag{6.47}$$

The slope of the tangent is 1.

$\therefore$ The slope of the normal at L is -1. LS is perpendicular to x-axis.

$$\therefore \underline{|BLS} = \underline{|SLM} = 45°.$$

$\therefore$ Latus rectum bisects the angle between the tangents and normal at L.

Example 6.22

Show that the locus of the points of intersection of tangents to $y^2 = 4ax$ which intercept a constant length d on the directrix is$(y^2 - 4ax)(x + a)^2 = d^2x^2$.

Solution

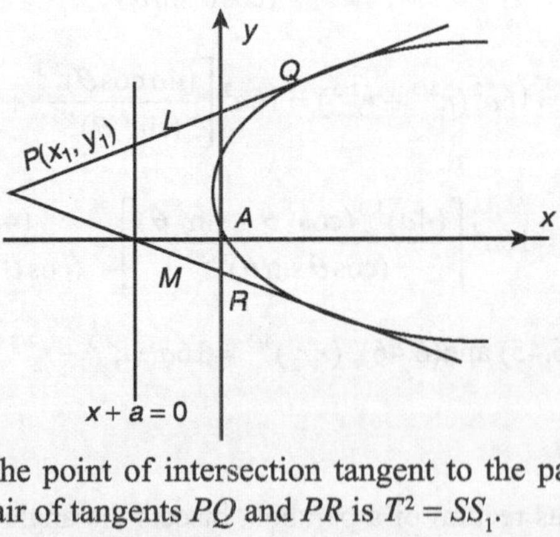

Let $P(x_1, y_1)$ be the point of intersection tangent to the parabola. Then the equation of the pair of tangents PQ and PR is $T^2 = SS_1$.

(i.e.) $[yy_1 - 2a(x + x_1)]^2 = (y^2 - 4ax)(y_1^2 - 4ax_1)$. When these lines meet the directrix $x = -a$, we have

$$[yy_1 - 2a(-a + x_1)]^2 = (y^2 + 4a^2)(y_1^2 - 4ax_1)$$

$$\text{(i.e.)} \quad y^2[y_1^2 - y_1^2 + 4ax_1] - y[4a(-a + x_1)y_1]$$

$$+[4a^2(-a + x_1)^2 - 4a^2(y_1^2 - 4ax_1)]y = 0.$$

$$x_1y^2 - y_1(x_1 - a)y + a[(x_1 - a)^2 - (y_1^2 - 4ax_1)] = 0$$

$$x_1y^2 - y_1(x_1 - a)y + a[(x_1 + a)^2 - y_1^2] = 0.$$

If y_1 and y_2 are the ordinates of the point of intersection of tangents with the directrix $x + a = 0$, then

$$y_1 + y_2 = \frac{(x_1 - a)y_1}{x_1} \text{ or } y_1y_2 = \frac{a\left[(x_1 + a)^2 - y_1^2\right]}{x_1}$$

Then $d^2 = (y_1 - y_2)^2 = (y_1 + y_2)^2 - 4y_1y_2 = \dfrac{(x_1 - a)^2 y_1^2 - 4ax_1\left[(x_1 + a)^2 - y_1^2\right]}{x_1^2}$

$$\therefore d^2x_1^2 = (x_1 - a)^2 y_1^2 - 4ax_1\left[(x_1 + a)^2 - y_1^2\right]$$

$$= y_1^2\left[(x_1 - a)^2 + 4ax_1\right] - 4ax_1(x_1 + a)^2$$

$$= y_1^2(x_1 + a)^2 - 4ax_1(x_1 + a)^2$$

$$= (y_1^2 - 4ax_1)(x_1 + a)^2$$

$\therefore$ The locus of (x_1, y_1) is $(y^2 - 4ax)(x + a)^2 = d^2x^2$.

Example 6.23

Show that the locus of midpoints of chords of a parabola which subtend a right angle at the vertex is another parabola whose latus rectum is half the latus rectum of the parabola.

Solution

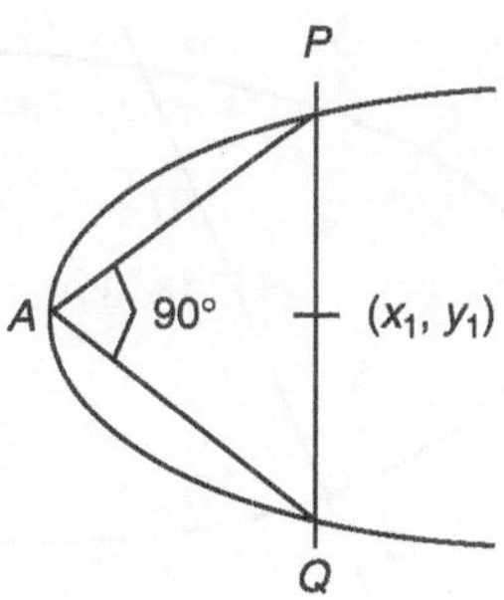

Let the equation of the parabola be

$$y^2 = 4ax \tag{6.48}$$

Let (x_1, y_1) be the midpoint of the chord PQ. Then the equation of PQ is $T = S_1$

$$\text{(i.e.)} \quad yy_1 - 2a(x + x_1) = y_1^2 - 4ax_1$$

$$yy_1 - 2ax = y_1^2 - 2ax_1 \qquad (6.49)$$

The combined equation of the lines AP and AQ is got by homogenization of equation (6.48) with the help of (6.49).

$\therefore$ The combined equation of OP and OQ is $y^2 = 4ax\left(\dfrac{y_1 y - 2ax}{y_1^2 - 2ax_1}\right)$.

$$(y_1^2 - 2ax_1)y^2 - 4ax(yy_1 - 2ax) = 0$$

Since $\angle PAQ = 90°$, coefficient of x^2 + coefficient of $y^2 = 0$

$$\text{(i.e.)} \quad y_1^2 - 2ax_1 + 8a^2 = 0 \quad \Rightarrow y_1^2 = 2a(x_1 - 4a).$$

The locus of (x_1, y_1) is $y^2 = 2a(x - 4a)$ which is a parabola whose latus rectum is half the latus rectum of the given parabola.

Example 6.24

Show that the locus of midpoints of chords of the parabola of constant length $2l$ is $(y^2 - 4ax)(y^2 + 4a^2) + 4a^2l^2 = 0$.

Solution

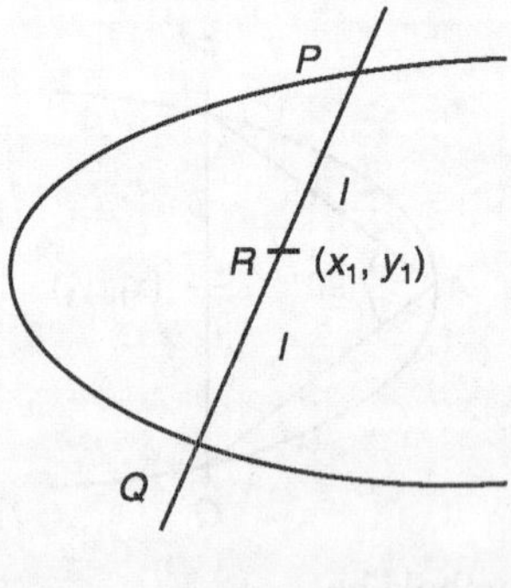

Let (x_1, y_1) be the midpoint of a chord of the parabola

$$y^2 = 4ax \qquad (6.50)$$

Let the equation of the chord be

$$\frac{x-x_1}{\cos\theta}=\frac{y-y_1}{\sin\theta}=r \tag{6.51}$$

Any point on this line is $(x_1 + r\cos\theta, y_1 + r\sin\theta)$. This point lies on the parabola $y^2 = 4ax$.

(i.e.) $$(y_1 + r\sin\theta)^2 = 4a(x_1 + r\cos\theta).$$

$$r^2\sin^2\theta + 2r(y_1\sin\theta - 2a\cos\theta) + y_1^2 - 4ax_1 = 0 \tag{6.52}$$

The two values of r are the distances RP and RQ which are equal in magnitude but opposite in sign. The condition for this is the coefficient of $r = 0$. (i.e.) $y_1\sin\theta - 2a\cos\theta = 0$.

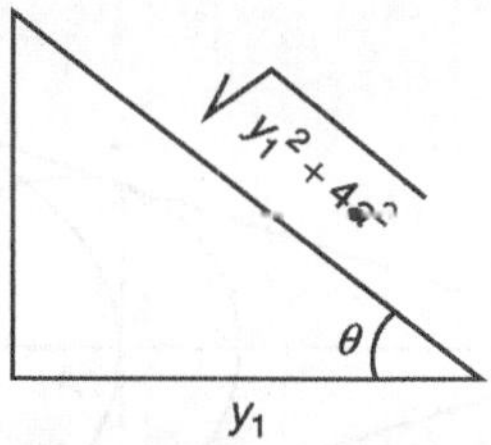

Then from Equation (6.50), $r^2\sin^2\theta + y_1^2 - 4ax_1 = 0$

$$r^2\left(\frac{4a^2}{y_1^2+4a^2}\right)+(y_1^2-4ax_1)=0$$

The locus of (x_1, y_1) is $(y^2 - 4ax)(y^2 + 4a^2) + 4a^2l^2 = 0$.
(since $r = l$)

Example 6.25

Show that the locus of the midpoints of focal chords of a parabola is another parabola whose vertex is at the focus of the given parabola.

Solution

Let the given parabola be

$$y^2 = 4ax \tag{6.53}$$

Let (x_1, y_1) be the midpoint of a chord of this parabola. Then its equation is $yy_1 - 2ax = y_1^2 - 2ax_1$.

If this is a focal chord then this passes through $(a, 0)$.

$$\therefore -2a^2 = y_1^2 - 2ax_1.$$

The locus of (x_1, y_1) is $y^2 = 2a(x - a)$ which is a parabola whose vertex is at the focus of the given parabola.

Example 6.26

From a point common tangents are drawn to the circle $x^2 + y^2 = \frac{a^2}{2}$ and the parabola $y^2 = 4ax$. Find the area of the quadrilateral formed by the common tangents, the chord of contact of the circle and the chord of contact of the parabola.

Solution

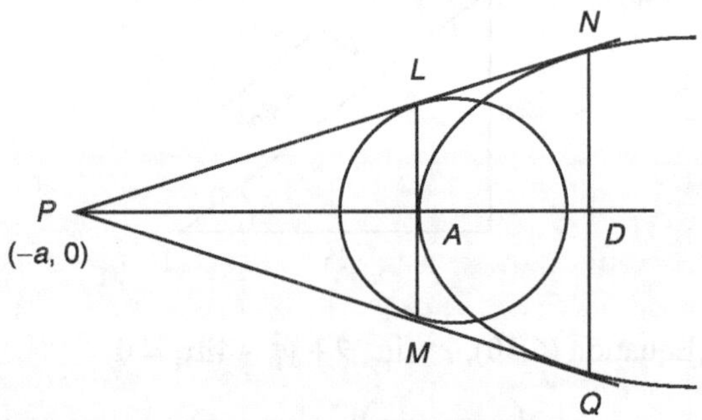

Any tangent to the parabola $y^2 = 4ax$ is

$$y = mx + \frac{a}{m} \tag{6.54}$$

If this is also a tangent to the circle $x^2 + y^2 = \frac{a^2}{2}$ then $\frac{a}{m} = \frac{a}{\sqrt{2}}\sqrt{(1+m^2)}$

$\therefore m^2(1 + m^2) = 2$ or $m^4 + m^2 - 2 = 0$ or $(m^2 - 1)(m^2 + 2) = 0$

$\Rightarrow m^2 = 1$ or -2. But $m^2 = -2$ is inadmissible since m^2 has to be positive. $\therefore m = \pm 1$.

Hence the common tangents are $y = \pm (x + a)$. The two tangents meet at $P(-a, 0)$.

The equation of the chord of contact from $(-a, 0)$ to the circle $x^2 + y^2 = \frac{a^2}{2}$ is $-xa = \frac{a^2}{2}$ or $x = \frac{-a}{2}$.

The equation of the chord of contact from $(-a, 0)$ to the parabola $y^2 = 4ax$ is $0 = 2a(x - a)$ or $x - a = 0$.

When $x = a$, $y = \pm 2a$. Hence N and Q are $(a, 2a)$ and $(a, -2a)$.

When $x = -a$, $y = \pm\frac{a}{2}$,

$\therefore L$ and M are $\left(\frac{-a}{2}, \frac{a}{2}\right)$ and $\left(\frac{-a}{2}, \frac{-a}{2}\right)$.

$\therefore$ Area of quadrilateral $LMQN$ = Area of trapezium $LMQN$

$$= \frac{1}{2}(LM + NQ) \cdot AD$$

Since $PD = 2a$, $PA = \frac{a}{2}$,

$$\therefore AD = 2a - \frac{a}{2} = \frac{3a}{2}.$$

$$= \frac{1}{2}(a + 4a)\frac{3a}{2}$$

$$= \frac{15a^2}{4}.$$

Example 6.27

The polar of a point P with respect to the parabola $y^2 = 4ax$ meets the curve in Q and R. Show that if P lies on the line $lx + my + n = 0$ then the locus of the middle point of the QR is $l(y^2 - 4ax) + 2a(lx + my + n) = 0$.

Solution

Let P be the point (h, k). The polar of $P(h, k)$ with respect to the parabola $y^2 = 4ax$ is

$$yk = 2a(x + h) \tag{6.55}$$

The polar of P meets the parabola $y^2 = 4ax$ at Q and R. Let $P(x_1, y_1)$ be the midpoint of QR. Its equation is

$$yy_1 - 2ax = y_1^2 - 2ax_1 \tag{6.56}$$

Equations (6.55) and (6.56) represent the same line.

$\therefore$ Identifying equations (6.55) and (6.56) we get,

$$\frac{k}{y_1} = \frac{1}{1} = \frac{2ah}{y_1^2 - 2ax_1}$$

$$\therefore y_1 = k \text{ and } y_1^2 - 2ax_1 = 2ah \tag{6.57}$$

Since the point (h, k) lies on $lx + my + n = 0$, $lh + mk + n = 0$.

Using equation (6.56), $\frac{l(y_1^2 - 2ax_1)}{2a} + my_1 + n = 0$

(i.e.) $l(y_1^2 - 2ax_1) + 2a(my_1 + n) = 0$

(i.e.) $l(y_1^2 - 4ax_1) + 2a(lx_1 + my_1 + n) = 0$

$\therefore$ The locus of (x_1, y_1) is $l(y^2 - 4ax) + 2a(lx + my + n) = 0$.

Example 6.28

Prove that area of the triangle inscribed in the parabola $y^2 = 4ax$ is $\frac{1}{8a}|(y_1 - y_2)(y_2 - y_3)(y_3 - y_1)|$ where y_1, y_2 and y_3 are the ordinates of the vertices of the triangle.

Solution

Let (x_1, y_1), (x_2, y_2) and (x_3, y_3) be the vertices of the triangle inscribed in the parabola $y^2 = 4ax$. Then the vertices are $\left(\frac{y_1^2}{4a}, y_1\right)$, $\left(\frac{y_2^2}{4a}, y_2\right)$ and $\left(\frac{y_3^2}{4a}, y_3\right)$. The area of the triangle is

$$= \frac{1}{2}\left|\sum x_1(y_2 - y_3)\right| = \frac{1}{2}\left|\sum \frac{y_1^2}{4a}(y_2 - y_3)\right| = \frac{1}{8a}\left|\sum y_1^2(y_2 - y_3)\right|$$

$$= \frac{1}{8a}\left|y_1(y_2 - y_3) \cdot y_2(y_3 - y_1) \cdot y_3(y_1 - y_2)\right|$$

Example 6.29

An equilateral triangle is inscribed in the parabola $y^2 = 4ax$ one of whose vertices is at the vertex of the parabola. Find its side.

Solution

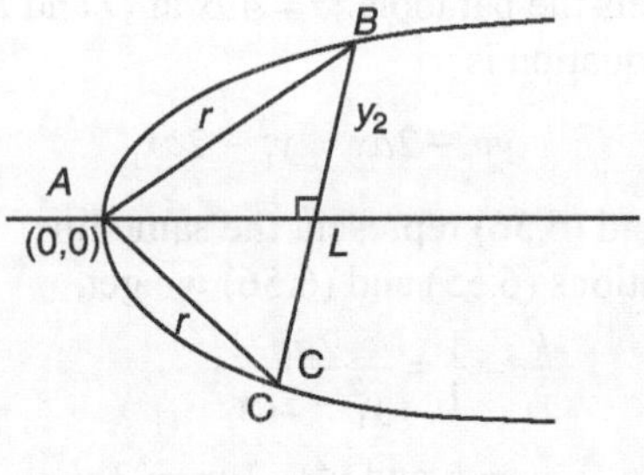

The coordinates of B are $B(r \cos 30°, r\sin 30°)$, $\left(\frac{\sqrt{3}}{2}r, \frac{r}{2}\right)$.

Since this point lies on the parabola $y^2 = 4ax$, then

$$\frac{r^2}{4} = 4a \cdot \frac{r}{2}\sqrt{3} \qquad \therefore r = 8a\sqrt{3}$$

Exercises

1. Show that two tangents can be drawn from a given point to a parabola. If the tangents make angles θ_1 and θ_2 with x axis such that
   (i) $\tan\theta_1 + \tan\theta_2$ is a constant show that the locus of point of intersection of tangents is a straight line through the vertex of a parabola.
   (ii) if $\tan\theta_1 \cdot \tan\theta_2$ is a constant show that the locus of the point of intersecting is a straight line.
   (iii) if $\theta_1 + \theta_2$ is a constant show that the locus of the point of intersection of tangents is a straight line through the focus.
   (iv) if θ_1 and θ_2 are complementary angles then the locus of point of intersection is the straight line $x = a$.

2. Find the locus of point of intersection of tangents to the parabola $y^2 = 4ax$ which includes an angle of $\frac{\pi}{3}$.

   Ans.: $3(x + a)^2 = y^2 - 4ax$

3. Show that the locus of the poles of chords of the parabola $y^2 = 4ax$ which subtends an angle of 45° at the vertex is the curve $(x + a)^2 = 4(y^2 - 4ax)$.

4. Show that the locus of poles of all tangents to the parabola $y^2 = 4ax$ with respect to the parabola $y^2 = 4bx$ is the parabola $ay^2 = 4b^2x$.

5. Show that the locus of poles of chords of the parabola which subtends a right angle at the vertex is $x + 4a = 0$.

6. Show that if tangents be drawn to the parabola $y^2 = 4ax$ from any point on the straight line $x + 4a = 0$, the chord of contact subtends a right angle at the vertex of the parabola.

7. Perpendiculars are drawn from points on the tangent at the vertex on their polars with respect to the parabola $y^2 = 4ax$. Show that the locus of the foot of the perpendicular is a circle centre at $(a, 0)$ and radius a.

8. Show that the locus of poles with respect to the parabola $y^2 = 4ax$ of tangents to the circle $x^2 + y^2 = 4a^2$ is $x^2 - y^2 = 4a^2$.

9. A point P moves such that the line through the perpendicular to its polar with respect to the parabola $y^2 = 4ax$ touches the parabola $x^2 = 4by$. Show that the locus of P is $2ax + by + 4a^2x = 0$.

10. If a chord of the parabola $y^2 = 4ax$ subtends a right angle at its focus, show that the locus of the pole of this chord with respect to the given parabola is $x^2 - y^2 + 6ax + a^2 = 0$.

11. Show that the locus of poles of all chords of the parabola $y^2 = 4ax$ which are at a constant distance d from the vertex is $d^2y^2 + 4a^2\,(d^2 - x^2) = 0$.

12. Show that the locus of poles of the focal chords of the parabola $y^2 = 4ax$ is $x + a = 0$.

13. If two tangents to the parabola $y^2 = 4ax$ make equal angles with a fixed line show that the chord of contact passes through a fixed point.

14. Prove that the polar of any point on the circle $x^2 + y^2 - 2ax - 3a^2 = 0$ with respect to the circle $x^2 + y^2 + 2ax - 3a^2 = 0$ touches the parabola $y^2 = 4ax$.

15. Show that the locus of the poles with respect to the parabola $y^2 = 4ax$ of the tangents to the curve $x^2 - y^2 = a$ is the ellipse $4x^2 + y^2 = 4ax$.

16. P is a variable point on the line $y = b$, prove that the polar of P with respect to the parabola $y^2 = 4ax$ is a fixed directrix.

17. The perpendicular from a point O on its polar with respect to a parabola meet the polar in the points M and cuts the axis in G. The polar meets x-axis in T and the ordinate through O intersects the curve in P and P'. Show that the points G, M, P, P' and T lie on a circle whose centre is at the focus S.

18. Tangents are drawn to the parabola $y^2 = 4ax$ from a point (h, k). Show that the area of the triangle formed by the tangents and the chord of contact is $\dfrac{(k^2 - 4ah)^{3/2}}{a}$.

19. Prove that the length of the chord of contact of the tangents drawn from the point (x_1, y_1) to the parabola $y^2 = 4ax$ is $\frac{1}{a}\sqrt{y_1^2 + 4a^2}\,(y_1^2 - 4ax_1)$. Hence show that one of the triangles formed by these tangents and their chord of contact is $\frac{1}{2a}(y_1^2 - 4ax_1)^{3/2}$.

20. Tangents are drawn from a variable point P to the parabola $y^2 = 4ax$ such that they form a triangle of constant area with the tangent at the vertex. Show that the locus of P is $(y^2 - 4ax)x^2 = 4c^2$.

21. Prove that the tangent to a parabola and the perpendicular to it from its focus meet on the tangent at the vertex.

22. Show that a portion of a tangent to a parabola intercepted between directrix and the curve subtends a right angle at the focus.

23. The tangent to the parabola $y^2 = 4ax$ make angles θ_1 and θ_2 with the axis. Show that the locus of the point of intersection such that $\cot\theta_1 + \cot\theta_2 = c$ is $y = ac$.

24. If perpendiculars be drawn from any two fixed points on the axis of a parabola equidistant from the focus on any tangent to it, show that the difference of their squares is a constant.

25. Prove that the equation of the parabola whose vertex and focus on x-axis at distances $4a$ and $5a$ from the origin respectively $(a > 0)$ is $y^2 = 4a(x - 4a)$. Also obtain the equation to the tangent to this curve at the end of latus rectum in the first quadrant. **Ans.:** $y = x - a$

26. Chords of a parabola are drawn through a fixed point. Show that the locus of the middle points is another parabola.

27. Find the locus of the middle points of chords of the parabola $y^2 = 2x$ which touches the circle $x^2 + y^2 - 2x - 4 = 0$.
28. A tangent to the parabola $y^2 + 4bx = 0$ meets the parabola $y^2 = 4ax$ at P and Q. Show that the locus of the middle point of PQ is $y^2(2a + b) = 4a^2x$.
29. Through each point of the straight line $x - my = h$ is drawn a chord of the parabola $y^2 = 4ax$ which is bisected at the point. Prove that it always touches the parabola $(y + 2am)^2 = 8axh$.
30. Two lines are drawn at right angles, one being a tangent to the parabola $y^2 = 4ax$ and the other to $y^2 = 4by$. Show that the locus of their point of intersection is the curve $(ax + by)(x^2 + y^2) = (bx - ay)^2$.
31. A circle cuts the parabola $y^2 = 4ax$ at right angles and passes through the focus. Show that the centre of the circle lies on the curve $y^2(a + x) = a(a + bx)^2$.
32. Two tangents drawn from a point to the parabola make angles θ_1 and θ_2 with the x-axis. Show that the locus of their point of intersection if $\tan^2\theta_1 + \tan^2\theta_2 = c$ is $y^2 - cx^2 = 2ax$.
33. If a triangle PQR is inscribed in a parabola so that the focus S is the orthocentre and the sides meet the axes in points K, L and M then prove that $SK \cdot SL \cdot SM - 4SA^2 = 0$ where A is the vertex of the parabola.
34. Chords of the parabola $y^2 = 4ax$ are drawn through a fixed point (h, k). Show that the locus of the midpoint is a parabola whose vertex is $\left(h - \frac{k^2}{8a}, \frac{k}{2}\right)$ and latus rectum is $2a$.
35. Show that the locus of the middle points of a system of parallel chords of a parabola is a line which is parallel to the axis of the parabola.
36. Show that the locus of the midpoints of chords of the parabola which subtends a constant angle α at the vertex is $(y^2 - 2ax - 8a^2)^2 \tan^2\alpha = 16a^2(4ax - y^2)$.

ILLUSTRATIVE EXAMPLES BASED ON PARAMETERS

Example 6.30

Prove that perpendicular tangents to the parabola will intersect on the directrix.

Solution

Let the tangents at t_1 and t_2 intersect at P. The equation of tangents at t_1 and t_2 are $yt_1 = x + at_1^2$ and $yt_2 = x + at_2^2$.

The slopes of the tangents are $\frac{1}{t_1}$ and $\frac{1}{t_2}$. Since the tangents are perpendicular, $\frac{1}{t_1} \cdot \frac{1}{t_2} = -1$

$$\therefore t_1 t_2 = -1$$

The point of intersection of the tangents at t_1 and t_2 is $P(at_1t_2, a(t_1+t_2))$ (i.e.) $(-a, a(t_1+t_2))$. This point lies on the line $x + a = 0$.

$\therefore$ Perpendicular tangents intersect on the directrix.

Example 6.31

Prove that the tangents at the extremities of a focal chord intersect at right angles on the directrix.

Solution

Let t_1 and t_2 be the extremities of a focal chord. Then the equation of the chord is $y(t_1+t_2) = 2x + 2at_1t_2$.

This passes through the focus $(a, 0)$.

$$\therefore 0 = 2a + 2at_1t_2 \text{ or } t_1t_2 = -1 \tag{6.58}$$

$\therefore$ Tangents at t_1 and t_2 are perpendicular.

The point of intersection of tangents at t_1 and t_2 is $[at_1t_2, a(t_1+t_2)]$ (i.e.) $(-a, a(t_1+t_2))$. This point lies on the directrix. Hence the tangents at the extremities of a focal chord intersect at right angles on the directrix.

Example 6.32

Prove that any tangent to a parabola and perpendicular on it from the focus meet on the tangent at the vertex.

Solution

Let the equation of the parabola $y^2 = 4ax$. The equation of the tangent at t is

$$yt = x + at^2 \tag{6.59}$$

The slope of the tangent is $\frac{1}{t}$. The slope of the perpendicular to it is $-t$. Hence the equation of the perpendicular line passing through focus $(a, 0)$ is

$$y - 0 = -t(x - a) \text{ or } y + xt = at \tag{6.60}$$

Multiplying equation (6.60) by t, we get

$$yt + xt^2 = at^2 \tag{6.61}$$

Equation (6.59) – equation (6.61) gives $x(1 + t^2) = 0$ or $x = 0$.

$$\therefore y = at$$

Hence, the point of intersection of (6.59) and (6.60) is $(0, at)$ and this point lies on y-axis.

Example 6.33

Show that the orthocentre of the triangle formed by the tangents at three points on a parabola lies on the directrix.

Solution

Let t_1, t_2 and t_3 be points of contact of the tangents at the points A, B and C, respectively on the parabola $y^2 = 4ax$, forming a triangle PQR. The equation of QR is

$$yt_3 = xt_3 + at_3^2 \tag{6.62}$$

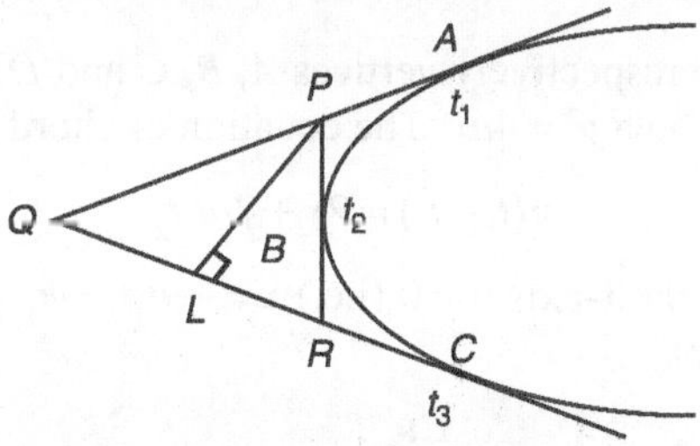

P is the point of intersection of tangents at t_1 and t_2. This point is $[at_1t_2, a(t_1 + t_2)]$. The slope of PL, perpendicular to QR is $-t_3$.

∴ The equation of PL is $y - a(t_1 + t_2) = -t_3[x - at_1t_2]$

(i.e.) $$y + xt_3 = a(t_1 + t_2) + at_1t_2t_3 \tag{6.63}$$

Then the equation of QM perpendicular from Q on PR is

$$y + xt_2 = a(t_2 + t_3) + at_1t_2t_3 \tag{6.64}$$

Equation (6.63) – equation (3) gives $x(t_3 - t_2) = a(t_2 - t_3)$ or $x = -a$.

This point lies on the directrix $x + a = 0$. Hence the orthocentre lies on the directrix.

Example 6.34

The coordinates of the ends of a focal chord of the parabola $y^2 = 4ax$ are (x_1, y_1) and (x_2, y_2). Prove that $x_1x_2 = a^2$ and $y_1y_2 = -4a^2$.

Solution

Let t_1 and t_2 be the ends of a focal chord. Then the equation of the focal chord is $y(t_1 + t_2) = 2x + at_1t_2$. Since this passes through the focus $(a, 0)$, $0 = 2a + at_1t_2$ or $t_1t_2 = -1$.

$$x_1x_2 = at_1^2 \cdot at_2^2 = at_1^2t_2^2 = a \text{ since } t_1t_2 = -1$$
$$y_1y_2 = 2at_1 \cdot 2at_2 = -4a^2$$

Example 6.35

A quadrilateral is inscribed in a parabola and three of its sides pass through fixed points on the axis. Show that the fourth side also passes through a fixed point on the axis of the parabola.

Solution

Let t_1, t_2, t_3 and t_4 be respectively vertices A, B, C and D of the quadrilateral inscribed in the parabola $y^2 = 4ax$. The equation of chord AB is

$$y(t_1 + t_2) = 2x + 2at_1t_2 \tag{6.65}$$

When this meets the x-axis $y = 0$ (i.e.) $x = -at_1t_2 = k_1$. Since AB meets the x-axis at a fixed point,

$$\therefore t_1t_2 = \frac{-k_1}{a}.$$

Similarly, $$t_2t_3 = \frac{-k_2}{a},\ t_3t_4 = \frac{-k_3}{a}.$$

Multiplying these, we get

$$t_1t_2^2t_3^2t_4 = \frac{-k_1k_2k_3}{a^3}$$

$$t_1t_4\left(\frac{k_2^2}{a^2}\right) = \frac{-k_1k_2k_3}{a^3}$$

$$\therefore t_1t_2 = \frac{-k_1k_3}{k_2a} = a \text{ constant.}$$

Hence, the fourth side of the quadrilateral also passes through a fixed point.

Example 6.36

Tangents to the parabola $y^2 = 4ax$ are drawn at points whose abscissae are in the ratio k:1. Prove that the locus of their point of intersection is the curve $y^2 = (k^{1/4} + k^{-1/4})^2x^2$.

Solution

Let the tangents at t_1 and t_2 intersect at $P(x_1, y_1)$

Given that $\frac{at_1^2}{at_2^2} = k$

$$\Rightarrow t_1 = \sqrt{k}t_2 \tag{6.66}$$

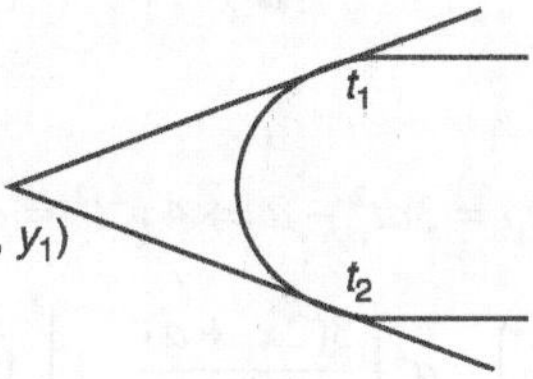

The point of intersection of the tangents at t_1 and t_2 is $x_1 = at_1t_2$ and $y_1 = a(t_1 + t_2)$.

$$x_1 = \sqrt{k}at_2^2 \text{ and } y_1 = a(\sqrt{k}+1)t_2$$

$$x_1 = \sqrt{k}\cdot a \cdot \frac{y_1^2}{a^2(\sqrt{k}+1)^2} \Rightarrow y_1^2 = \left(\frac{\sqrt{k}+1}{k^{1/4}}\right)^2 x_1$$

$\therefore$ The locus of (x_1, y_1) is $y^2 = (k^{1/4} + k^{-1/4})^2 x$.

Example 6.37

Show that the locus of the middle point of all tangents from points on the directrix to the parabola $y^2 = 4ax$ is $y^2(2x + a) = a(x + 3a)^2$.

Solution

Let $(-a, y_1)$ be a point on the directrix. Let t be the point of contact of tangents from $(-a, y_1)$ to the parabola $y^2 = 4ax$. The equation of the tangent at t is

$$yt = x + at^2 \tag{6.67}$$

Since this passes through $(-a, y_1)$, $y_1 = \frac{-a+at^2}{t}$

$\therefore$ The point on the directrix is $\left(-a, \frac{-a+at^2}{t}\right)$.

Let (x_1, y_1) be the midpoint of the portion of tangent between the directrix and the point of contact. Then

$$x_1 = \frac{at^2 - a}{2},\ y_1 = \frac{2at + \dfrac{-a + at^2}{t}}{2}$$

$$\frac{2x_1 + a}{a} = t^2 \tag{6.68}$$

and $2y_1 t = 2at^2 - a + at^2$

(i.e.) $$2y_1 t = 3at^2 - a \Rightarrow 4y_1^2 t^2 = a^2(3t^2 - 1)^2 \tag{6.69}$$

$$\therefore 4y_1^2\left(\frac{2x_1 + a}{a}\right) = a^2\left[\frac{3(2x_1 + a)}{a} - 1\right]^2 \text{ using equation (6.68)}$$

$$\therefore 4y_1^2(2x_1 + a) = a[6x_1 + 2a]^2$$

$$y_1^2(2x_1 + a) = a[3x_1 + a]^2$$

The locus of (x_1, y_1) is $y^2(2x + a) = a(3x + a)^2$.

Example 6.38

Tangents are drawn from a variable point P to the parabola $y^2 = 4ax$, such that they form a triangle of constant area c^2 with the tangent at the vertex. Show that the locus of P is $(y^2 - 4ax)x^2 = 4c^4$.

Solution

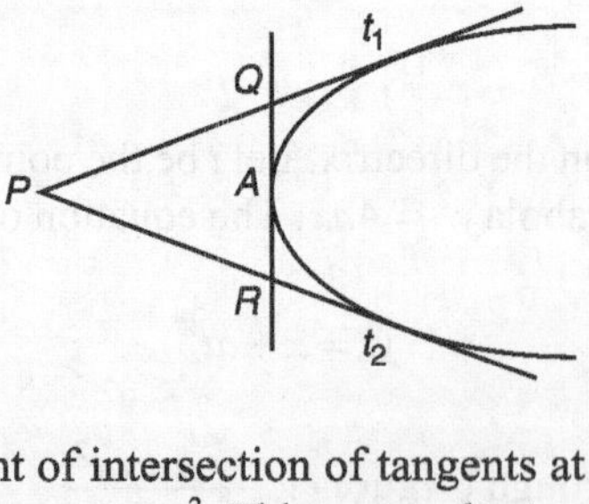

Let $P(x_1, y_1)$ be the point of intersection of tangents at t_1 and t_2. The equation of the tangent at t_1 is $yt_1 = x + at_1^2$. This meets the tangent at the vertex at Q. $\therefore Q$ is $(0, at_1)$. Similarly, R is $(0, at_2)$. P is the point of intersection of tangents at t_1 and t_2 and the point is $P(at_1t_2, a(t_1 + t_2))$. The area of ΔPQR is given as c^2.

$$\frac{1}{2}[0+0+at_1t_2(at_1-at_2)]=c^2$$

$$\Rightarrow a^2t_1t_2(t_1-t_2)=2c^2 \qquad \text{(i.e.) } a^4t_1^2t_2^2(t_1-t_2)^2=4c^4$$

$$\Rightarrow a^4t_1^2t_2^2\left[(t_1+t_2)^2-4t_1t_2\right]=4c^4$$

$$\Rightarrow x_1^2(y_1^2-4ax_1)=4c^4$$

Therefore, the locus of (x_1, y_1) is $x^2(y^2-4ax)=4c^4$.

Example 6.39

Prove that the distance of the focus from the intersection of two tangents to a parabola is a mean proportional to the focal radii of the point of constant.

Solution

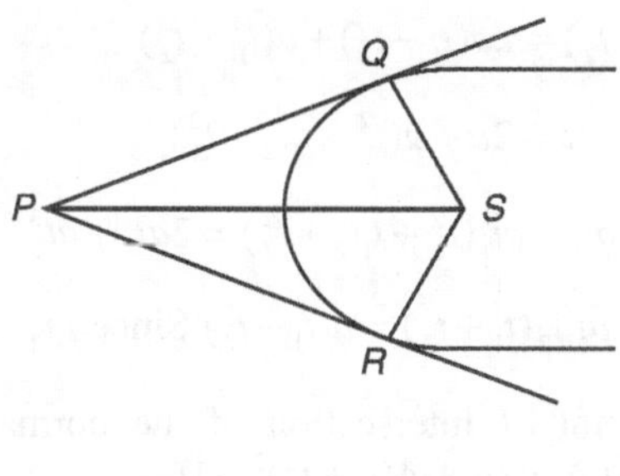

Let the tangents at $Q(at_1^2, 2at_1)$ and $R(at_2^2, 2at_2)$ intersect at P. Then the coordinates of the point P are $(at_1t_2, a(t_1+t_2))$. S is the point $(a, 0)$.

$$SQ^2=(a-at_1^2)^2+(2at_1-0)^2=a^2(1-t_1^2)^2+4a^2t_1^2=a^2(1+t_1^2)^2$$

$$\therefore\ SQ=a(1+t_1^2).$$

Similarly, $SR=a(1+t_2^2)$

$$SP^2=(a-at_1t_2)^2+[0-a(t_1+t_2)]^2 \Rightarrow a^2(1-t_1t_2)^2+a^2(t_1+t_2)^2$$

$$=a^2\left[(t_1+t_2)^2+(1-t_1t_2)^2\right]=a^2\left[t_2^2+t_1^2+1+t_1^2t_2^2\right]$$

$$=a^2(1+t_1^2)(1+t_2^2)$$

$\therefore SQ\cdot SR=SP^2$

$\therefore SP$ is the mean proportional between SQ and SR.

Example 6.40

Prove that the locus of the point of intersection of normals at the ends of a focal chord of a parabola is another parabola whose latus rectum is one fourth of that of the given parabola.

Solution

Let the equation of the parabola be

$$y^2 = 4ax \tag{6.70}$$

Let t_1 and t_2 bethe ends of a focal chord of the parabola.
For a focal chord $t_1t_2 = -1$.
The equation of the normal at t_1 and t_2 are

$$y + xt_1 = 2at_1 + at_1^3 \tag{6.71}$$

$$y + xt_2 = 2at_2 + at_2^3 \tag{6.72}$$

Subtracting, $x(t_1 - t_2) = 2a(t_1 - t_2) + a(t_1^3 - t_2^3)$

$$x = 2a + a(t_1^2 + t_1t_2 + t_2^2)$$

From (6.71), $y + 2at_1 + at_1(t_1^2 + t_1t_2 + t_2^2) = 2at_1 + at_1^3$

$$y = -at_1t_2(t_1 + t_2) = a(t_1 + t_2) \text{ Since } t_1t_2 = -1.$$

If (x_1, y_1) is a point of intersection of the normals at t_1 and t_2 then $x_1 = 2a + a(t_1^2 + t_2^2 + t_1t_2) = 2a + a[(t_1 + t_2)^2 + 1]$

$$y_1 = a(t_1 + t_2)$$

$$\therefore x_1 = 2a + a\left[\frac{y_1^2}{a^2} + 1\right] \Rightarrow x_1 - 2a = a\frac{(y_1^2 + a^2)}{a^2}$$

$$\Rightarrow a(x_1 - 2a) = y_1^2 + a^2 \qquad \text{or} \qquad y_1^2 = a_1(x_1 - 3a)$$

The locus of (x_1, y_1) is $y^2 = a(x - 3a)$ which is a parabola whose latus rectum is one fourth of the latus rectum of the original parabola.

Example 6.41

If the normal at the point t_1 on the parabola $y^2 = 4ax$ meets the curve again at t_2 prove that $t_2 = -t_1 - \dfrac{2}{t_1}$.

Solution

The equation of the normal at t_1 is

$$y + xt_1 = 2at_1 + at_1^3 \qquad (6.73)$$

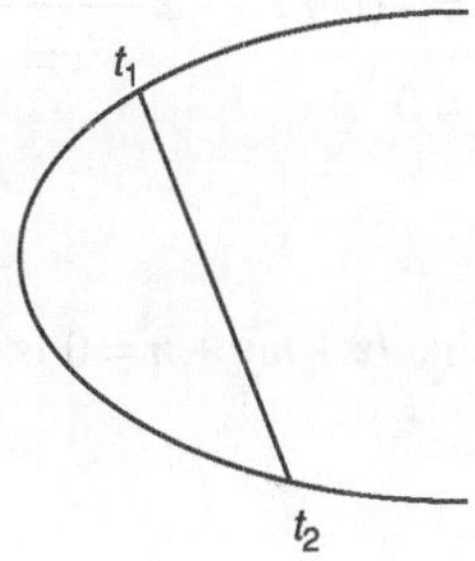

The equation of the chord joining the points t_1 and t_2 is

$$y(t_1 + t_2) = 2x + 2at_1t_2 \qquad (6.74)$$

Equations(6.73) and (6.74) represent the same lines. Therefore identifying we get $\frac{t_1 + t_2}{1} = \frac{-2}{t_1} = \frac{2at_1t_2}{2at_1 + at_1^3}$

$$\therefore t_1 + t_2 = \frac{-2}{t_1} \Rightarrow t_2 = -t_1 - \frac{2}{t_1}$$

Example 6.42

If the normals at two points t_1, t_2 on the parabola $y^2 = 4ax$ intersect again at a point on the curve show that $t_1 + t_2 + t_3 = 0$ and $t_1t_2 = 2$ and the product of ordinates of the two points is $8a^2$.

Solution

The normals t_1 and t_2 meet at t_3.

$$\therefore t_3 = -t_1 - \frac{2}{t_1} \qquad (6.75)$$

$$t_3 = -t_2 - \frac{2}{t_2} \qquad (6.76)$$

Subtracting $0 = -(t_1 - t_2) + 2\frac{(t_1 - t_2)}{t_1t_2}$. Since $t_1 - t_2 \neq 0$, $t_1t_2 = 2$.

Solving equations (6.75) and (6.76), we get

$$2t_3 = -(t_1 + t_2) - 2\left(\frac{1}{t_1} + \frac{1}{t_2}\right)$$

$$= -(t_1 + t_2) - 2\frac{(t_1 + t_2)}{t_1 t_2} = -(t_1 + t_2) - (t_1 + t_2)$$

$$\therefore t_1 + t_2 + t_3 = 0$$

Example 6.43

Find the condition that the line $lx + my + n = 0$ is a normal to the parabola is $y^2 = 4ax$.

Solution

Let the line $lx + my + n = 0$ be a normal at 't'. The parabola is $y^2 = 4ax$. The equation of the normal at t is

$$y + xt = 2at + at^3 \tag{6.77}$$

But the equation of the normal is given as

$$lx + my = -n \tag{6.78}$$

Identifying equations (6.77) and (6.78), we get $\frac{t}{l} = \frac{1}{m} = \frac{2at + at^3}{-n} \therefore t = \frac{l}{m}$

and $(2a + at^2)l = 2\left(2a + \frac{al^2}{m^2}\right)l = -n$

$$\text{(i.e.)} \quad al^3 + 2alm^2 + m^2 n = 0.$$

Example 6.44

Show that the locus of poles of normal chords of the parabola is $y^2 = 4ax$ is $(x + 2a)y^2 + 4a^3 = 0$.

Solution

Let (x_1, y_1) be the pole of a normal chord normal at t. The equation of the polar of (x_1, y_1) is

$$yy_1 = 2a(x + x_1) \tag{6.79}$$

The equation of the normal at t is

$$y + xt = 2at + at^3 \quad (6.80)$$

Equations (6.79) and (6.80) represent the same line.
∴ Identifying equations (6.79) and (6.80), we get

$$\frac{y_1}{t} = \frac{-2a}{t} = \frac{2ax_1}{2at + at^3}$$

$$t = \frac{-2a}{y_1}; \text{ also } y_1(2t + t^3) = 2x_1$$

$$y_1\left[\frac{-4a}{y_1} - \frac{8a^3}{y_1^3}\right] = 2x_1 \quad \therefore 4a + \frac{8a^3}{y_1^2} = -2x_1$$

(i.e.)

$$x_1 + 2a + \frac{4a^3}{y_1^2} = 0.$$

The locus of (x_1, y_1) is $(x + 2a)y^2 + 4a^3 = 0$.

Example 6.45

In the parabola $y^2 = 4ax$ the tangent at the point P whose abscissa is equal to the latus rectum meets the axis on T and the normal at P cuts the curve again in Q. Prove that $PT{:}TQ = 4{:}5$.

Solution

Let P and Q be the points t_1 and t_2 respectively. Given that $at_1^2 = 4a$ or $t_1^2 = 4$. The equation of the tangent at t_1 is

$$yt_1 = x + at_1^2 \quad (6.81)$$

when this meets the x-axis, $y = 0$.

$$\therefore x = -at_1^2 = -4a.$$

Hence T is the point $(-4a, 0)$. Also as the normal at t_1 meets the curve at t_2,

$$t_2 = -t_1 - \frac{2}{t_1} = -\frac{(t_1^2 + 2)}{t_1}$$

$$t_2 = \frac{-6}{t_1} \quad (6.82)$$

$$\text{Now, } PT^2 = (at_1^2 + 4a)^2 + (2at_1 - 0)^2 = 64a^2 + 16a^2 = 80a^2$$

$$PQ^2 = (at_1^2 - at_2^2)^2 + (2at_1 - 2at_2)^2 = (4a - 9a)^2 + (2at_1 - 2at_2)^2$$

$$= 25a^2 + \left(4at_1 + \frac{4a}{t_1}\right)^2 = 25a^2 + 64a^2 + 4a^2 + 32a^2$$

$$= 125a^2$$

$$\therefore PT : PQ = \sqrt{80a^2} : \sqrt{125a^2} = 4:5$$

Example 6.46

Show that the locus of a point such that two of the three normal drawn from it to the parabola $y^2 = 4ax$ coincide is $27ay^2 = 4(x-2a)^3$.

Solution

Let (x_1, y_1) be a given point and t be foot of the normal from (x_1, y_1) to the parabola $y^2 = 4ax$. The equation of the normal at 't' is

$$y + xt = 2at + at^3 \tag{6.83}$$

Since this passes through (x_1, y_1) we have

$$y_1 + x_1 = 2at + at^3 \text{ or } at^3 + t(2a - x_1) - y_1 = 0 \tag{6.84}$$

If t_1, t_2 and t_3 be feet of the normals from (x_1, y_1) to the parabola then t_1, t_2 and t_3 are the roots of equation (6.84).

$$\therefore t_1 + t_2 + t_3 = 0 \tag{6.85}$$

$$t_1t_2 + t_2t_3 + t_3t_1 = \frac{2a - x_1}{a} \tag{6.86}$$

$$t_1t_2t_3 = \frac{y_1}{a} \tag{6.87}$$

If two of the three normals coincide then $t_1 = t_2$.

$$(6.86) \Rightarrow 2t_1 - t_3 = 0$$

$$\text{or} \quad t_3 = 2t_1 \tag{6.88}$$

$$t_1^2 + 2t_3t_1 = \frac{2a - x_1}{a} \tag{6.89}$$

$$t_1^2 t_3 = \frac{y_1}{a} \tag{6.90}$$

From equations (6.88) and (6.89), $-2t_1^3 = \frac{y_1}{a}$ or $t_1^3 = \frac{-y_1}{2a}$. Since t_1 is a root of equation (6.84) $at_1^3 - (2a - x_1)t_1 - y_1 = 0$.

$$a\left(\frac{-y_1}{2a}\right) + (2a - x_1)t_1 - y_1 = 0$$

$$\frac{-3y_1}{2} = -t_1(2a - x_1) \Rightarrow \frac{27y_1^3}{8} = t_1^3(2a - x_1)^3$$

$$\Rightarrow 27ay_1^3 - 4(x_1 - 2a)^3 = 0 \Rightarrow \text{Locus of } (x_1, y_1) \text{ is } 27ay^2 = 4(x_1 - 2a)^3.$$

Example 6.47

If the normals from a point to the parabola $y^2 = 4ax$ cut the axis in points whose distances from the vertex are in AP then show that the point lies on the curve $27ay^2 = 2(x - 2a)^3$.

Solution

Let (x_1, y_1) be a given point and t be the foot of a normal from (x_1, y_1). The equation of the normal at t is

$$y + xt = 2at + at^3 \tag{6.91}$$

Since this passes through (x_1, y_1), $y_1 + x_1 t = 2at + at^3$.

or
$$at^3 + t(2a - x_1) - y_1 = 0 \tag{6.92}$$

If t_1, t_2 and t_3 be the feet of the normals from (x_1, y_1) then

$$t_1 + t_2 + t_3 = 0 \tag{6.93}$$

$$t_1 t_2 + t_2 t_3 + t_3 t_4 = \frac{2a - x_1}{y_1} \tag{6.94}$$

$$t_1 t_2 t_3 = \frac{y_1}{a} \tag{6.95}$$

When the normal at t meets the x-axis, $y = 0$, from (6.91) we get $xt = 2at + at^3$ or $x = 2a + at^2$.

Then the x-coordinates of the points where the normal meets the x-axis are given by $(2a+at_1^2, 2a+at_2^2, 2a+at_3^2)$. Given these are in AP.

$\therefore t_1^2, t_2^2$ and t_3^2 are in AP.

(i.e.) $$2t_2^2 = t_1^2 + t_3^2 \tag{6.96}$$

$$2t_2^2 = (t_1+t_3)^2 - 2t_1t_3 = t_2^2 - 2t_1t_3 \text{ or } 2t_1t_3 = -t_2^2$$

From equation (6.95), or $t_2^3 = \dfrac{-2y_1}{a}$

Since t_2 is a root of equation (6.91), $at_2^3 + t_2(2a - x_1) - y_1 = 0$.

$$a\left(\frac{-2y_1}{a}\right) - y_1 = -t_2(2a - x_1) \text{ or } 3y_1 = t_2(2a - x_1),$$

$$27y_1^3 = t_2^3(2a - x_1)^3 \Rightarrow 27ay_1^3 = 2(x_1 - 2a)^3$$

$\therefore$ The locus of (x_1, y_1) is $27ay^3 = 2(x - 2a)^3$.

Example 6.48

Show that the locus of the point of intersection of two normals to the parabola which are at right angles is $y^2 = a(x - 3a)$.

Solution

If (x_1, y_1) is the point of intersection of two normals to the parabola $y^2 = 4ax$ then

$$at^3 + t(2a - x_1) - y_1 = 0 \tag{6.97}$$

If t_1, t_2 and t_3 be the feet of the three normals from (x_1, y_1) then

$$t_1 + t_2 + t_3 = 0 \tag{6.98}$$

$$t_1t_2 + t_2t_3 + t_3t_1 = \frac{2a - x_1}{a} \tag{6.99}$$

$$t_1t_2t_3 = \frac{y_1}{a} \tag{6.100}$$

Since two of the normals are perpendicular then $t_1t_2 = -1$

$$\therefore t_3 = \frac{-y_1}{a} \tag{6.101}$$

Since t_3 is a root of equation (6.97), $at_3^3+(2a-x_1)t_3-y_1=0$.

$$a\left(\frac{-y_1}{a}\right)^3-(2a-x_1)\left(\frac{-y_1}{a}\right)-y_1=0.$$

or $-y_1^3-ay_1(2a-x_1)-a^2y_1=0$ or $y_1^2=a(x_1-3a)$
The locus of (x_1, y_1) is $y^2=a(x-3a)$.

Example 6.49

Prove that a normal chord of a parabola which subtends a right angle at the vertex makes an angle $\tan^{-1}(\sqrt{2})$ with the x-axis.

Solution

Let the equation of the parabola be

$$y^2=4ax \tag{6.102}$$

The equation of the normal at t is

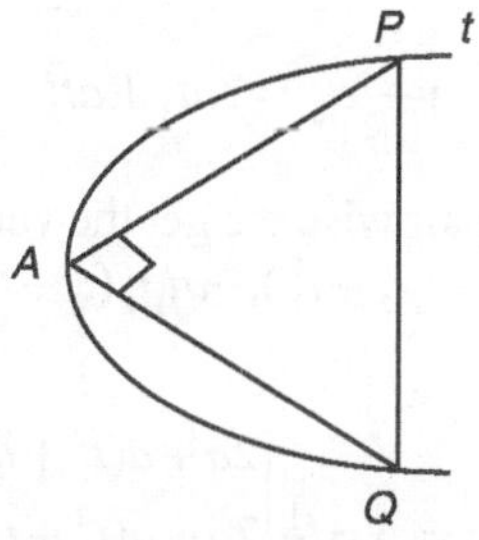

$$y+xt=2at+t^3 \tag{6.103}$$

The combined equation of the lines AP and AQ is

$$y^2=\frac{4ax(y+xt)}{2at+at^3} \text{ or } y^2(2at+at^3)=4ax(y+xt)$$

$$4atx^2+4axy-(2at+at^3)y^2=0.$$

Since the two lines are at right angles, coefficient of x^2 + coefficient of $y^2=0$.

$$\therefore 4at-2at-at^3=0 \qquad \Rightarrow at(2-t^2)=0$$

$$\therefore t = 0 \quad \text{or} \quad t^2 = 2.$$

$t = 0$ corresponds to the normal joining through the vertex.

$$t^2 = 2 \Rightarrow (-t)^2 = 2 \qquad \text{or} \qquad \tan^2\theta = 2$$

$$\Rightarrow \theta = \tan^{-1}(\pm\sqrt{2})$$

$\therefore$ The normals make an angle $\tan^{-1}(\sqrt{2})$ with the x-axis.

Example 6.50

Prove that the area of the triangle formed by the normals to the parabola $y^2 = 4ax$ at the points t_1, t_2 and t_3 is $\frac{1}{2}a^2(t_1 - t_2)(t_2 - t_3)(t_3 - t_1)(t_1 + t_2 + t_3)^2$.

Solution

The equations of the normals at t_1, t_2, t_3 are

$$y + xt_1 = 2at_1 + at_1^3 \tag{6.104}$$

$$y + xt_2 = 2at_2 + at_2^3 \tag{6.105}$$

$$y + xt_3 = 2at_3 + at_3^3 \tag{6.106}$$

Solving these equations pair wise we get the vertices of the triangle. Hence the vertices are $[2a + a(t_1^2 + t_1t_2 + t_2^2), -at_1t_2(t_1 + t_2)]$ and two other similar points.

$$\therefore \text{The area of the triangle} = \Delta = \frac{1}{2}\begin{vmatrix} 2a + a(t_1^2 + t_1t_2 + t_2^2) & -at_1t_2(t_1 + t_2) & 1 \\ 2a + a(t_2^2 + t_2t_3 + t_3^2) & -at_2t_3(t_2 + t_3) & 1 \\ 2a + a(t_3^2 + t_3t_1 + t_1^2) & -at_3t_1(t_3 + t_1) & 1 \end{vmatrix}$$

$$\Delta = \frac{a^2}{2}\begin{vmatrix} t_1^2 + t_1t_2 + t_2^2 & -t_1t_2(t_1 + t_2) & 1 \\ t_2^2 + t_2t_3 + t_3^2 & -t_2t_3(t_2 + t_3) & 1 \\ t_3^2 + t_3t_1 + t_1^2 & -t_3t_1(t_3 + t_1) & 1 \end{vmatrix}$$

$$= \frac{a^2}{2}\begin{vmatrix} t_1^2 + t_1t_2 + t_2^2 & t_1t_2(t_1 + t_2) & 1 \\ (t_3 - t_1)(t_1 + t_2 + t_3) & -t_2(t_3 - t_1)(t_1 + t_2 + t_3) & 0 \\ (t_1 - t_2)(t_1 + t_2 + t_3) & -t_3(t_1 - t_2)(t_1 + t_2 + t_3) & 0 \end{vmatrix}$$

$$= \frac{a^2}{2}(t_1 + t_2 + t_3)^2(t_1 - t_2)(t_2 - t_3)(t_3 - t_1)$$

Example 6.51

Prove that the length of the intercepts on the normal at the point $P(at^2, 2at)$ to the parabola $y^2 = 4ax$ made by the circle described on the line joining the focus and P as diameter is $a\sqrt{1+t^2}$.

Solution

The equation of the normal at P is $y + xt = 2at + at^3$.

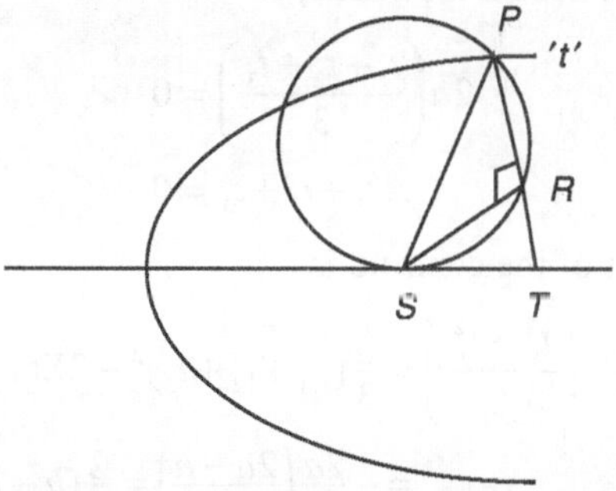

Let the circle on PS as diameter cut the normal at P at R and the x-axis at T.

$$SP = \sqrt{(at^2 - a)^2 + (2at)^2} = a(1+t^2)$$

$$SR = \frac{at + at^3}{\sqrt{1+t^2}} = at\sqrt{1+t^2}$$

$$PR = \sqrt{SP^2 - SR^2} = \sqrt{a^2(1+t^2)^2 - a^2t^2(1+t^2)}$$

$$= \sqrt{a^2(1+t^2)(1+t^2-t^2)}$$

$$PR = a\sqrt{1+t^2}$$

Example 6.52

Normals at three points P, Q and R of the parabola $y^2 = 4ax$ meet in (h, k). Prove that the centroid of ΔPQR lies on the axis at a distance $\frac{2(h-2a)}{3}$ from the vertex.

Solution

Let t be a foot of a normal from (h, k). The equation of the normal at t is

$$y + xt = 2at + at^3 \quad (6.107)$$

This passes through (h, k).

$$\therefore k + ht = 2at + at^3 \qquad (6.108)$$

$\Rightarrow at^3 + t(2a - h) - k = 0.$

If t_1, t_2 and t_3 are the feet of the normals from (h, k) then $t_1 + t_2 + t_3 = 0$, $\Sigma t_1 t_2 = \frac{2a-h}{a}$, $t_1 t_2 t_3 = \frac{k}{a}$.

The centroid of the ΔPQR is $\left(\frac{at_1^2 + at_2^2 + at_3^2}{3}, \frac{2at_1 + 2at_2 + 2at_3}{3}\right)$

Since the centroid lies on the x-axis,

$$2a\left(\frac{t_1 + t_2 + t_3}{3}\right) = 0$$

$$\therefore t_1 + t_2 + t_3 = 0$$

The x-coordinates of the centroid is

$$a\left(\frac{t_1^2 + t_2^2 + t_3^2}{3}\right) = \frac{a}{3}[(t_1 + t_2 + t_3)^2 - 2\Sigma t_1 t_2]$$

$$= -\frac{2a}{3}\left(\frac{2a-h}{a}\right) = \frac{2}{3}(h - 2a)$$

$\therefore$ Centroid is at a distance $\frac{2}{3}(h - 2a)$ from the vertex of the parabola.

Example 6.53

The normals at three points P, Q and R on a parabola meet at T and S be the focus of the parabola. Prove that $SP \cdot SQ \cdot SR = aTS^2$.

Solution

Let T be the point (h, k). Then P, Q and R are the feet of the normals from $T(h, k)$. The equation of the normal at t is $y + xt = 2at + at^3$. If t_1, t_2 and t_3 be the feet of the normals from T then $t_1 + t_2 + t_3 = 0$.

$$\Sigma t_1 t_2 = \frac{2a-h}{a}, \quad t_1 t_2 t_3 = \frac{k}{a}$$

S is the point $(a, 0)$.

$$SP^2 = (a - at_1^2)^2 + (2at_1)^2 = a^2(1 + t_1^2)^2$$

$$SP = a(1 + t_1^2)$$

$$SQ = a(1 + t_2^2)$$

$$SR = a(1 + t_3^2)$$

$$\therefore SP\cdot SQ\cdot SR = a^3(1+t_1^2)(1+t_2^2)(1+t_3^2)$$
$$= a^3[1+\Sigma t_1^2 + \Sigma t_1^2t_2^2 + (t_1t_2t_3)^2]$$
$$= a^3[1+(t_1+t_2+t_3)^2 - 2\Sigma t_1t_2 + (t_1t_2+t_2t_3+t_3t_1)^2$$
$$-2t_1t_2t_3(t_1+t_2+t_3)+(t_1t_2t_3)^2]$$
$$= a^3\left[1-\frac{2(2a-h)}{a}+\frac{(2a-h)^2}{a^2}+\frac{k^3}{a^3}\right]$$
$$= a\left[a^2-2a(2a-h)+(2a-h)^2+k^3\right] = a\left[a^2-2ah+h^2+k^2\right]$$
$$= a\left[(h-a)^2+k^2\right] = a\cdot ST^2$$

Example 6.54

The equation of a chord PQ of the parabola $y^2 = 4ax$ is $lx + my = 1$. Show that the normals at P, Q meet on the normal at $\left(\frac{4am^2}{l^2}, \frac{4am}{l}\right)$.

Solution

Let P and Q be the points t_1 and t_2. The normals at P and Q meet at R. If t_3 is the foot of the normal of the 3rd point then

$$t_1 + t_2 + t_3 = 0 \tag{6.109}$$

The equation of the chord PQ is $lx + my = 1$. Since P and Q are the points t_1 and t_2,

$$lat_1^2 + 2amt_1 = 1$$
$$lat_2^2 + 2amt_2 = 1$$

$$\therefore la(t_1 - t_2)(t_1 + t_2) + 2am(t_1 - t_2) = 0$$

Since $t_1 - t_2 \neq 0$, $la + 2am(-t_3) = 0$ or $t_3 = \frac{2m}{n}$

Hence, the 3rd point $(at_3^2, 2at_3)$ is $\left(\frac{4am^2}{l^2}, \frac{4am}{l}\right)$.

Example 6.55

If the normal at P to the parabola $y^2 = 4ax$ meets the curve at Q and make an angle θ with the axis show that

(i) it will cut the parabola at θ at an angle $\tan^{-1}\left(\frac{1}{2}\tan\theta\right)$ and
(ii) $PQ = 4a\sec\theta\,\mathrm{cosec}^2\theta$.

Solution

Let P be the point $(at^2, 2at)$. The equation of the normal at t is $y + xt = 2at + at^3$. The normal at t meets the curve at $Q(at_1^2, 2at_1)$.

$$\therefore t_1 = -t - \frac{2}{t}$$

Let ϕ be the angle between the normal and the tangent at Q. The slope of the tangent at Q is $\frac{1}{t_1} = \frac{1}{-t-\frac{2}{t}} = \frac{-t}{t^2+2}$

Slope of the normal at t is $-t$.

$$\therefore \tan\phi = \frac{\left|-t + \frac{t}{t^2+2}\right|}{1+\frac{t^2}{t^2+2}} = \frac{\left|-t(t^2+2)\right|}{2(t^2+2)} = \left|\frac{t}{2}\right|$$

$$= \left|\frac{1}{2}\tan\theta\right| \text{ since } \tan\theta = -t$$

Since $\tan\theta = -t$

$$\therefore \phi = \tan^{-1}\left(\frac{1}{2}\tan\theta\right)$$

$$PQ^2 = (at^2 - at_1^2)^2 + (2at - 2at_1)^2$$

$$= a^2(t+t_1)^2(t-t_1)^2 + 4a^2(t-t_1)^2$$

$$= a^2(t-t_1)^2[(t_1+t)^2+4] = a^2\left[t+t+\frac{2}{t}\right]^2\left[\left(t-t-\frac{2}{t}\right)^2+4\right]$$

$$= a^2\frac{4(t^2+1)^2}{t^2}\left(\frac{4}{t^2}+4\right) = \frac{16a^2(1+t^2)^2(1+t^2)}{t^4}$$

$$= \frac{16a^2(1+t^2)^3}{t^4} = \frac{16a^2\sec^6\theta}{\tan^4\theta} = 16a^2\sec^2\theta\,\text{cosec}^4\theta$$

Example 6.56

Prove that the circle passing through the feet of the three normals to a parabola drawn from any point in the plane passes through the vertex of the parabola. Also find the equation of the circle passing through the feet of the normals.

Solution

Let the equation of the parabola be

$$y^2 = 4ax \tag{6.110}$$

Let the equation of the circle be

$$x^2 + y^2 + 2gx + 2fy + c = 0 \tag{6.111}$$

Let P, Q and R be the feet of the normals to $y^2 = 4ax$ from a given point (h, k). Then we have $at^3 + (2a - h)t - k = 0$.

If t_1, t_2 and t_3 be the feet of the normals at P, Q and R then $t_1 + t_2 + t_3 = 0$. We know that the circle (6.111) and the parabola (6.110) cut at four points and if t_1, t_2, t_3 and t_4 are the four points of intersection of the circle and the parabola then they are the roots of the equation,

$$a^2t^4 + (4a^2 + 2ga)t^2 + 4fat + c = 0 \tag{6.112}$$

$$\therefore t_1 + t_2 + t_3 + t_4 = 0 \tag{6.113}$$

If t_1, t_2 and t_3 correspond to the feet of the normals from (h, k) then

$$t_1 + t_2 + t_3 = 0 \tag{6.114}$$

From equations (6.113) and (6.114), $t_4 = 1$.

But t_4 is the point $(at_4^2, 2at_4)$ (i.e.) (0, 0) which is the vertex of the parabola. Hence the circle passing through the feet of the normals from a given point also passes through the vertex of the parabola. Hence equation (6.112) becomes

$$at^3 + (4a + 2g)t + 4f = 0 \tag{6.115}$$

since $c = 0$

Equations (6.111) and (6.115) are the same. By comparing the coefficients,

we get $1 = \dfrac{4a+2g}{2a-h} = \dfrac{4f}{-k}$

$$\therefore 2a - h = 4a + 2g \text{ and } 4f = -k$$

$$\text{or } -2a - h = 2g \text{ and } 2f = \frac{-k}{2}$$

$\therefore$ The equation of the circle passing through the feet of the normal is $x^2 + y^2 - (2a+h)x - \dfrac{k}{2}y = 0$ or $2(x^2 + y^2) - 2(2a + h)x - ky = 0$.

Exercises

1. Show that the portion of the tangent intercepted between the point of contact and the directrix subtends a right angle at the focus.

2. If the tangent at a point P on the parabola meets the axis at T and PN is the ordinate at P then show that $AN = AT$.

3. If the tangent at P meets the tangent at the vertex in Y then show that SY is perpendicular to TP and $SY^2 = AS\ SP$.

4. If A, B and C, are three points on a parabola whose ordinates are in GP then prove that the tangents at A and C meet on the ordinates of B.

5. Prove that the middle point of the intercepts made on a tangent to a parabola by the tangents at two points P and Q lies on the tangent which is parallel to PQ.

6. If points $(at^2, 2at)$ is one extremity of a focal chord of the parabola $y^2 = 4ax$, show that the length of the focal chord is $a\left(t+\frac{1}{t}\right)^2$.

7. Show that the tangents at one extremity of a focal chord of a parabola is parallel to the normal at the other extremity.

8. If the tangents at three points on the parabola $y^2 = 4ax$ make angles 60°, 45° and 30° with the axis of the parabola, show that the abscissae and ordinates of the three points are in GP.

9. Show that the circle described on the focal chord of a parabola as diameter touches the directrix.

10. Show that the tangent at one extremity of a focal chord of a parabola is parallel to the normal at the other extremity.

11. Prove that the semilatus rectum of a parabola is the harmonic mean of the segments of a focal chord.

12. Prove that the circle described on focal radii as diameter touches the tangents at the vertex of a parabola.

13. Three normals to a parabola $y^2 = 4x$ are drawn through the point (15, 12). Show that the equations are $3x - y - 33 = 0$, $4x + y - 72 = 0$ and $x - y - 3 = 0$.

14. The normals at two points P and Q of a parabola $y^2 = 4ax$ meet at the point (x_1, y_1) on the parabola. Show that $PQ = (x_1 + 4a)(x_1 - 8a)$.

15. Show that the coordinates of the feet of the normals of the parabola $y^2 = 4ax$ drawn from the point $(6a, 0)$ are $(0, 0)$, $(4a, 4a)$ and $(4a, -4a)$.

16. The normal at P to the parabola $y^2 = 4ax$ makes an angle a with the axis. Show that the area of the triangle, formed by it is the tangents at its extremities is a constant.

17. If P, Q and R are the points t_1, t_2 and t_3 on the parabola $y^2 = 4ax$, such that the normal at Q and R meet at P then show that:
    (i) the line PQ is passes through a fixed point on the axis.
    (ii) the locus of the pole of PQ is $x = a$.
    (iii) the locus of the midpoint of PQ is $y^2 = 2a(x + 2a)$.
    (iv) the ordinates of P and Q are the roots of the equation $y^2 + xy + 8a^2 = 0$ where t_3 is the ordinate of the point of intersection of the normals at P and Q.

18. If a circle cuts a parabola at P, Q, R and S show that PQ and RS are equally inclined to the axis.

19. The normals at the points P and R on the parabola $y^2 = 4ax$ meet on the parabola at the point P. Show that the locus of the orthocentre of ΔPQR is $y^2 = a(x + 6a)$ and the locus of the circumcentre of ΔPQR is the parabola $2y = x(x - a)$.

20. Prove that the area of the triangle inscribed in a parabola is twice the area of the triangle formed by the tangents at the vertices.

21. Prove that any three tangents to a parabola whose slopes are in HP encloses a triangle of constant area.

22. Prove that the circumcircle of a triangle circumscribing a parabola passes through the focus.

23. If the normals at any point P of the parabola $y^2 = 4ax$ meet the axis at G and the tangent at vertex at H and if A be the vertex of the parabola and the rectangle $AGQH$ be completed, prove that the equation to the locus of Q is $x^2 = 2ax + ay^2$.

24. The normal at a point P of a parabola meets the curve again at Q and T is the pole of PQ. Show that T lies on the directrix passing through P and that PT is bisected by the directrix.

25. If from the vertex of the parabola $y^2 = 4ax$, a pair of chords be drawn at right angles to one another and with these chords as adjacent sides a rectangle be made then show that the locus of further angle of the rectangle is the parabola $y^2 = 4a(x - 8a)$.

26. The normal to the parabola $y^2 = 4ax$ at a point P on it meets the axis in G. Show that P and G are equidistant from the focus of the parabola.

27. Two perpendicular straight lines through the focus of the parabola $y^2 = 4ax$ meet its directrix in T and T' respectively. Show that the tangents to the parabola to the perpendicular lines intersect at the midpoint of TT'.

28. If the normals at any point $P(18, 12)$ to the parabola $y^2 = 8x$ cuts the curve again at Q show that $9 \cdot PQ = 80\sqrt{10}$.

29. If the normal at P to the parabola $y^2 = 4ax$ meets the curve again at Q and if PQ and the normal at Q make angles θ and ϕ, respectively with the axis, prove that $\tan\theta \tan\phi + \tan^2\theta + 2 = 0$.

30. PQ is a focal chord of a parabola. PP' and QQ' are the normals at P and Q cutting the curve again at P' and Q'. Show that $P'Q'$ is parallel to PQ and is three times PQ.

31. If PQ be a normal chord of the parabola. $y^2 = 4ax$ and if S be the focus, show that the locus of the centroid of the triangle SPQ is $y^2(ay^2 + 180a^2 - 108ax) + 128\ a^4 = 0$.

32. If the tangents at P and Q meet at T and the orthocenter of the ΔPTQ lies on the parabola, show that either the orthocentre is at the vertex or the chord PQ is normal to the parabola.

33. If three normals from a point to the parabola $y^2 = 4ax$ cuts the axis in points, whose distances from the vertex are in AP, show that the point on the curve $27ay^2 = 2(x-a)^3$.

34. Tangents are drawn to a parabola from any point on the directrix. Show that the normals at the points of contact are perpendicular to each other and that they intersect on another parabola.

35. Show that if two tangents to a parabola $y^2 = 4ax$ intercept a constant length on any fixed tangent, the locus of their point of intersection is another equal parabola.

36. Show that the equation of the circle described on the chord intercepted by the parabola $y^2 = 4ax$ on the line $y = mx + c$ as diameter is $m^2(x^2 + y^2) + 2(mc - 2a)x - 4ay + c(4am + c) = 0$.

37. Circles are described on any two common chords of a parabola as diameter. Prove that their common chord passes through the vertex of the parabola.

38. If $P(h, k)$ is a fixed point in the plane of a parabola $y^2 = 4ax$. Through P a variable secant is drawn to cut the parabola in Q and R. T is a point on QR such that
   (i) $PQ \cdot PR = PT^2$. Show that the locus of T is $(y-k)^2 = k^2 - 4ah$.
   (ii) $PQ + PR = PT$. Show that the locus of T is $y^2 - k^2 = 4a\ (x - h)$.

39. Show that the locus of the point of intersection of tangents, to the parabola $y^2 = 4ax$ at points whose ordinates are in the ratio $\alpha^2:\beta^2$ is $y^2 = \left(\dfrac{\alpha}{\beta} + \dfrac{\beta}{\alpha}\right)^2 ax$.

40. Show that the locus of the middle points of a system of parallel chords of a parabola is a line which is parallel to the axis of the parabola.

41. P, Q and R are three points on a parabola and the chord PQ meets the diameter through R in T. Ordinates PM and QN are drawn to this diameter. Show that $RMRN = RT^2$.

CHAPTER

7 Ellipse

7.1 STANDARD EQUATION

A conic is defined as the locus of a point such that its distance from a fixed point bears a constant ratio to its distance from a fixed line. The fixed point is called the focus and the fixed straight line is called the directrix. The constant ratio is called the eccentricity of the conic. If the eccentricity is less than unity the conic is called an ellipse. Let us now derive the standard equation of an ellipse using the above property called focus-directrix property.

7.2 STANDARD EQUATION OF AN ELLIPSE

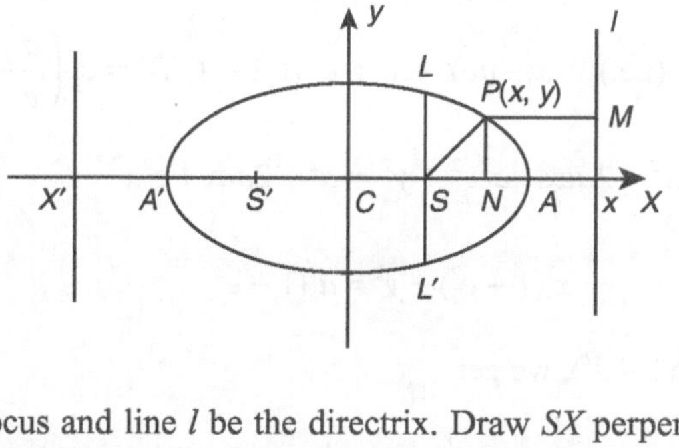

Let S be the focus and line l be the directrix. Draw SX perpendicular to the directrix. Divide SX internally and externally in the ratio $e:1(e<1)$. Let A and A' be the points of division. Since $\frac{SA}{AX} = e$ and $\frac{SA'}{A'X} = e$, from the definition of ellipse, the points A and A' lie on the ellipse. Let $AA' = 2a$ and C be its middle point.

$$SA = eAX \tag{7.1}$$

$$SA' = eA'X \tag{7.2}$$

Adding equations (7.1) and (7.2), we get $SA + SA' = e(AX + A'X)$.

$$\text{(i.e.)} \quad AA' = e(AX + A'X) = e(CX - CA + CX + CA')$$

$$= e \cdot 2CX \text{ Since } CA = CA'$$

$$\therefore CX = \frac{a}{e} \tag{7.3}$$

Subtracting equations (7.1) from (7.2), we get $SA' - SA = e(CX' - CX)$

$$CS + CA + (CS - CA) = e(AA')$$

$$2CS = e \cdot 2a \Rightarrow CS = ae \tag{7.4}$$

Take CS as the x-axis and CM perpendicular to CS, as y-axis.

Let $P(x, y)$ be any point on the ellipse. Draw PM perpendicular to the directrix. Then the coordinates of S are $(ae, 0)$. From the focus-directrix property of the ellipse, $\frac{SP}{PM} = e$.

$$\therefore SP^2 = e^2 PM^2 = e^2 NX^2 = e^2(CX - CN)^2$$

$$\text{(i.e.)} \quad (x - ae)^2 + y^2 = e^2 (CX - CN)^2 = e^2\left(\frac{a}{e} - x\right)^2$$

$$x^2 - 2aex + a^2e^2 + y^2 = a^2 - 2aex + e^2x^2$$

$$x^2(1 - e^2) + y^2 = a^2(1 - e^2)$$

Dividing by $a^2(1 - e^2)$, we get

$$\frac{x^2}{a^2} + \frac{y^2}{a^2(1-e^2)} = 1$$

$$\text{(i.e.)} \quad \frac{x^2}{a^2} + \frac{y^2}{b^2} = 1 \tag{7.5}$$

where $b^2 = a^2(1 - e^2)$

This is called the standard equation of an ellipse.

Note 7.2.1:

1. Equation (7.5) can be written as:

$$\frac{y^2}{b^2} = 1 - \frac{x^2}{a^2} = \frac{a^2 - x^2}{a^2} = \frac{(a+x)(a-x)}{a^2}$$

$$\frac{PN^2}{b^2} = \frac{AN \cdot NX'}{a^2} \Rightarrow \frac{PN^2}{AN \cdot NA'} = \frac{b^2}{a^2} = \frac{BC^2}{AC^2}$$

2. AA' is called the major axis of the ellipse.
3. BB' is called the minor axis of the ellipse.
4. C is called the centre of the ellipse.
5. The curve meets the x-axis at the point $A(a, 0)$ and $A'(-a, 0)$.
6. The curve meets the y-axis at the points $B(0, b)$ and $B'(0, -b)$.
7. The curve is symmetrical about both the axes. If (x, y) is a point on the curve, then $(x, -y)$ and $(-x, y)$ are also the points on the curve.
8. From the equation of the ellipse, we get

$$x = \pm\frac{a\sqrt{b^2 - y^2}}{b}, y = \pm\frac{b\sqrt{a^2 - x^2}}{a}$$

Therefore, for any point (x, y) on the curve, $-a \le x \le a$ and $-b \le y \le b$.

9. The double ordinate through the focus is called the latus rectum of the ellipse.
(i.e.) LSL' is the latus rectum.

$$\text{Length of latus rectum} = LL' = 2SL = 2e(CX - CS) = 2e\left(\frac{a}{e} - ae\right)$$

$$= 2a(1 - e^2).$$

$$= 2a \cdot \frac{b^2}{a^2} = \frac{2b^2}{a}$$

10. **Second focus and second directrix:** On the negative side of the origin, take a point S' such that $CS = CS'$ and another point X' such that $CX = CX' = a$.
Draw $X'M'$ perpendicular to AA' and PM' perpendicular to $X'M'$.
Then we can show that $\frac{S'P}{PM'} = e$ gives the locus of P as $\frac{x^2}{a^2} + \frac{y^2}{b^2} = 1$.
Here $S'$is called the second focus and $X'M'$ is the second directrix.

11. (i) Shifting the origin to the focus S, the equation of the ellipse is $\dfrac{(x-ae)^2}{a^2}+\dfrac{y^2}{b^2}=1$.

(ii) Shifting the origin to A, the equation of the ellipse is $\dfrac{(x-a)^2}{a^2}+\dfrac{y^2}{b^2}=1$.

(iii) Shifting the origin to X, the equation of the focus is $\dfrac{\left(x-\dfrac{a}{2}\right)^2}{a^2}+\dfrac{y^2}{b^2}=1$.

12. The equation of an ellipse is easily determined if we are given the focus and the equation of the directrix.

7.3 FOCAL DISTANCE

The sum of the focal distances of any point on the ellipse is equal to the length of the major axis.

In the above figure, (section 2.2)

$$SP = ePM$$

$$S'P = ePM^2.$$

Adding $\quad SP + S'P = e(PM + PM') = e(NX + NX')$

$$= e[CX - CN + CX' + CN] = e\left[\frac{a}{e} - x_1 + \frac{a}{e} + x_1\right]$$

$$= 2a \text{ where } P \text{ is } (x_1, y_1).$$

$$= \text{length of the major axis}$$

Note 7.3.1:

$$SP = ePM = a - ex_1$$

$$S'P = ePM' = a + ex_1$$

7.4 POSITION OF A POINT

A point (x_1, y_1) lies inside, on or outside of the ellipse according as $\dfrac{x^2}{a^2}+\dfrac{y^2}{b^2}-1$ is negative, zero or positive.

Let $Q(x_1, y_1)$ be a point on the ordinate PN where P is a point on the ellipse $\frac{x^2}{a^2}+\frac{y^2}{b^2}-1=0$.

Then, $QN < PN \Rightarrow \frac{QN^2}{b^2} < \frac{PN^2}{b^2} \Rightarrow \frac{y'^2}{b^2} < 1-\frac{x'^2}{a^2}$

$$\therefore \frac{x'^2}{a^2}+\frac{y'^2}{b^2} < 1$$

Similarly, if the $Q'(x', y')$ is a point outside the ellipse, $\frac{x'^2}{a^2}+\frac{y'^2}{b^2} > 1$. Evidently if $Q(x', y')$ is a point on the ellipse, $\frac{x'^2}{a^2}+\frac{y'^2}{b^2}-1=0$.

7.5 AUXILIARY CIRCLE

The circle described on the major axis as diameter is called the auxiliary circle.

Let P be any point on the ellipse. Let the ordinate through P meet the auxiliary circle at P'. Since $\underline{|A'P'A} = 90°$, we have the geometrical relation, $P'N^2 = AN \cdot A'N$.

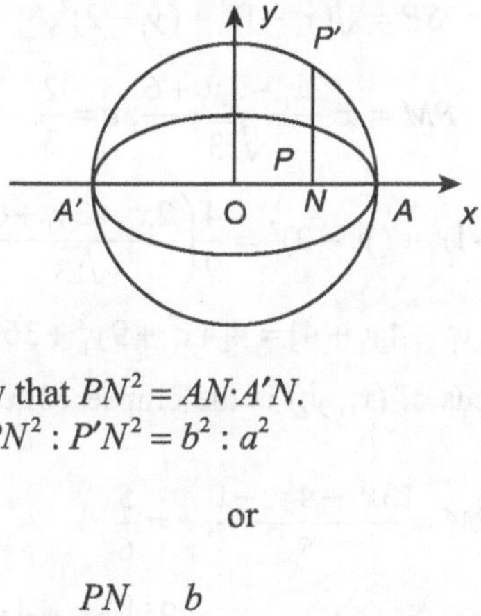

However, we know that $PN^2 = AN \cdot A'N$.

$$\therefore PN^2 : P'N^2 = b^2 : a^2$$

or

$$\frac{PN}{P'N} = \frac{b}{a}$$

The point P' where the ordinate PN meets the auxiliary circle is called the corresponding point of P. Therefore, the ordinate of any point on the ellipse to that of corresponding point on the ellipse are in the ratios of lengths of semi-minor axis and semi-major axis. This ratio gives another definition to an ellipse. Consider a circle and from each point on it, draw perpendicular to a diameter.

The locus of these points dividing these perpendiculars in a given ratio is an ellipse and for this ellipse the given circle is the auxiliary circle.

ILLUSTRATIVE EXAMPLES BASED ON FOCUS-DIRECTRIX PROPERTY

Example 7.1

Find the equation of the ellipse whose foci, directrix and eccentricity are given below:

(i) Focus is (1, 2), directrix is $2x - 3y + 6 = 0$ and eccentricity is $2/3$
(ii) Focus is (0, 0), directrix is $3x + 4y - 1 = 0$ and eccentricity is $5/6$
(iii) Focus is (1, –2), directrix is $3x - 2y + 1 = 0$ and eccentricity is $1/\sqrt{2}$

Solution

(i) Let $P(x_1, y_1)$ be a point on the ellipse. Then $\frac{SP}{PM} = e \Rightarrow \frac{SP^2}{PM^2} = e^2$

$$\therefore SP^2 = e^2 PM^2$$

$$SP = \sqrt{(x_1 - 1)^2 + (y_1 - 2)^2},$$

$$PM = \pm\frac{2x_1 - 3y_1 + 6}{\sqrt{13}};\ e = \frac{2}{3}.$$

Hence, $$(x_1 - 1)^2 + (y_1 - 2)^2 = \frac{4}{9}\left(\frac{2x_1 - 3y_1 + 6}{\sqrt{13}}\right)^2$$

$$117[x_1^2 - 2x_1 + 1 + y_1^2 - 4y_1 + 4] = 4[4x_1^2 + 9y_1^2 + 36 - 12x_1y_1 + 24x_1 - 36y_1]$$

Therefore, the locus of (x_1, y_1) is the ellipse $101x^2 + 81y^2 + 48x - 330x - 324y + 441 = 0$.

(ii) $SP = \sqrt{x_1^2 + y_1^2}$; $PM = \frac{3x_1 + 4y_1 - 1}{5}$, $e = \frac{5}{6}$

$$SP^2 = e^2 PM^2 \Rightarrow x_1^2 + y_1^2 = \frac{25}{36}\left[\frac{3x_1 + 4y_1 - 1}{5}\right]^2$$

$$\Rightarrow 36(x_1^2 + y_1^2) = 9x_1^2 + 16y_1^2 + 1 + 24x_1y_1 - 6x_1 - 8y_1$$

Therefore, the locus of (x_1, y_1) is the ellipse $27x^2 + 20y^2 - 24xy + 6x + 8y - 1 = 0$.

(iii) $$SP = \sqrt{(x_1 - 1)^2 + (y_1 + 2)^2}$$

$$\Rightarrow SP^2 = x_1^2 + y_1^2 - 2x_1 + 4y_1 + 5$$

$$PM = \pm\frac{3x_1 - 2y_1 + 1}{\sqrt{13}};\ e = \frac{1}{\sqrt{2}},\ SP^2 = e^2 PM^2$$

$$x_1^2 + y_1^2 - 2x_1 + 4y_1 + 5 = \frac{1}{2}\left(\frac{3x_1 - 2y_1 + 1}{\sqrt{13}}\right)^2$$

$$26(x_1^2 + y_1^2 - 2x_1 + 4y_1 + 5) = 9x_1^2 + 4y_1^2 + 1 - 12x_1y_1 + 6x - 4y_1$$

Therefore, the locus of (x_1, y_1) is the ellipse $17x^2 + 22y^2 + 12xy - 58x + 108y + 129 = 0$.

Example 7.2

Find the equation of the ellipse whose

(i) Foci are (4, 0) and (–4, 0) and $e = \frac{1}{3}$

(ii) Foci are (3, 0) and (–3, 0) and $e = \sqrt{\frac{3}{8}}$

Solution

(i) If the foci are $(ae, 0)$ and $(-ae, 0)$ then the equation of the ellipse is $\frac{x^2}{a^2} + \frac{y^2}{b^2} = 1$. Here, $ae = 4$ and $e = \frac{1}{3}$.

$$a = \frac{4}{e} = 4 \times 3 = 12$$

$$b^2 = a^2(1 - e^2) = 144\left(1 - \frac{1}{9}\right) = 144 \times \frac{8}{9} = 128$$

$\therefore$ The equation of the ellipse is $\frac{x^2}{144} + \frac{y^2}{128} = 1$.

(ii) If the foci are $(ae, 0)$ and $(-ae, 0)$ the equation of the ellipse is $\frac{x^2}{a^2} + \frac{y^2}{b^2} = 1$.

Here, $ae = 3$ and $e = \sqrt{\frac{3}{8}}$; $a^2e^2 = 9$ and $e^2 = \frac{3}{8}$

$$\therefore a^2 = 9\times\frac{8}{3} = 24,\ b^2 = a^2(1-e^2) = 24\left(\frac{5}{8}\right) = 15$$

Therefore, the equation of the ellipse is $\frac{x^2}{24}+\frac{y^2}{15}=1.$

Example 7.3

Find the eccentricity, foci and the length of the latus rectum of the ellipse.

(i) $9x^2+4y^2=36$
(ii) $3x^2+4y^2-12x-8y+4=0$
(iii) $25x^2+9y^2-150x-90y+225=0.$

Solution

(i) $9x^2+4y^2=36$
Dividing by 36, we get

$$\frac{9x^2}{36}+\frac{4y^2}{36}=1$$

$$\text{(i.e.)}\quad \frac{x^2}{4}+\frac{y^2}{5}=1$$

$$\therefore a^2=4,\ b^2=9.$$

This is an ellipse whose major axis is the y-axis and minor axis is the x-axis and centre at the origin.

$$\therefore a^2=b^2(1-e^2)\Rightarrow 4=9(1-e^2)$$

$$\therefore 9e^2=5$$

Therefore, eccentricity $= e = \frac{\sqrt{5}}{3}$

Therefore, foci are $\left(0, \pm\frac{be}{1}\right)$ (i.e.) $(0, \pm\sqrt{5})$.

Therefore, latus rectum $= \frac{2a^2}{b} = 2\times\frac{4}{3} = \frac{8}{3}.$

(ii)
$$3x^2+4y^2-12x-8y+4=0$$
$$(3x^2-12x)+(4y^2-8y)+4=0$$

$$3(x^2 - 4x) + 4(y^2 - 2y) + 4 = 0$$
$$3(x^2 - 4x + 4) - 12 + 4(y^2 - 2y + 1) - 4 + 4 = 0$$
$$\Rightarrow 3(x - 2)^2 + 4(y - 1)^2 = 12$$
$$\Rightarrow \frac{(x-2)^2}{4} + \frac{(y-1)^2}{3} = 1$$

Shift the origin to the point (2, 1).
Therefore, centre is (2, 1).

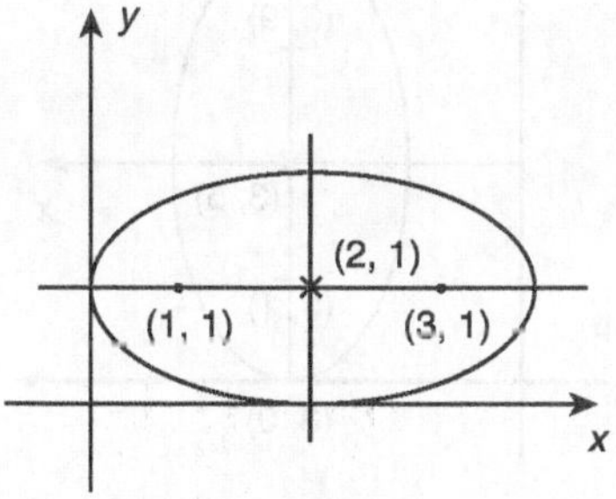

Therefore, the equation of the ellipse is

$$\frac{X^2}{4} + \frac{Y^2}{3} = 1$$
$$a^2 = 4,\ b^2 = 3$$
$$a^2e^2 = a^2 - b^2 = 4 - 3 = 1$$
$$4e^2 = 1 \text{ or } e^2 = \frac{1}{4} \text{ or } e = \frac{1}{2} \quad \therefore ae = 2 \times \frac{1}{2} = 1$$

Therefore, foci are (3, 1) and (1, 1) with respect to old axes.

$$\text{Length of the latus rectum} = \frac{2b^2}{a} = 2 \times \frac{3}{2} = 3.$$

(iii)
$$25x^2 + 9y^2 - 150x - 90y + 225 = 0$$
$$(25x^2 - 150x) + (9y^2 - 90y) + 225 = 0$$
$$25(x^2 - 6x) + 9(y^2 - 10y) + 225 = 0$$
$$25[(x-3)^2 - 9] + 9[(y-5)^5 - 25] + 225 = 0$$
$$25(x - 3)^2 + 9(y - 5)^2 = 225$$
$$\Rightarrow \frac{(x-3)^2}{9} + \frac{(y-5)^2}{25} = 1.$$

Shift the origin to the point (3, 5).
Therefore, the equation of the ellipse is $\frac{X^2}{9}+\frac{Y^2}{25}=1$.
Therefore, centre is (3, 5). This is an ellipse with y-axis on the major axis and x-axis as the minor axis.

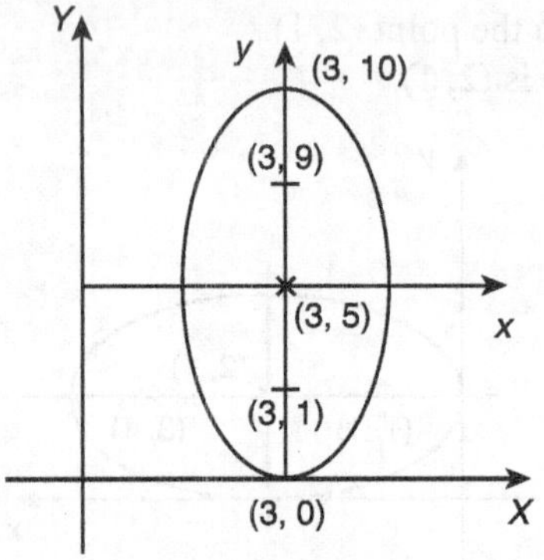

$$a^2=9, b^2=25$$
$$a=3, b=5$$
$$b^2e^2=b^2-a^2$$
$$25e^2=25-9=16$$
$$e^2=\frac{16}{25}\Rightarrow e=\frac{4}{5}$$

Therefore, foci lie on the line $x=3$.

$$be=5\times\frac{4}{5}=4$$

Therefore, foci are (3, 9) and (3, 1) and Leangth of the latus rectum $=\frac{2a^2}{b}=\frac{18}{5}$.

Exercises

1. Find the centre, foci and latus rectum of the ellipse:

   (i) $3x^2+4y^2+12x+8y-32=0$ **Ans.:** (–2, –1); (0, –1); (–4, –1); 6

   (ii) $9x^2+25y^2=225$ **Ans.:** (0, 0); $(\pm\sqrt{7}, 0)$; $\frac{9}{8}$

(iii) $x^2+9y^2=9$ **Ans.:** $(0, 0); (\pm\sqrt{6}, 0); \frac{2}{3}$

(iv) $2x^2+3y^2-4x+6y+4=0$ **Ans.:** $(-1,-1); (1\pm\frac{1}{\sqrt{6}}, -1); \frac{2\sqrt{2}}{3}$

2. Find the equation of the ellipse whose foci are $(0, \pm 2)$ and the length of major axis is $2\sqrt{5}$.
Ans.: $5x^2+y^2=5$

3. Find the equation of the ellipse whose foci is (3, 1), eccentricity $\frac{1}{2}$ and directrix is $x-y+6=0$.
Ans.: $7x^2+2xy+7y^2-60x-20y+44=0$

4. Find the equation of ellipse whose centre is at the origin, one focus is (0, 3) and the length of semi-major axis is 5.
Ans.: $\frac{x^2}{12}+\frac{y^2}{25}=1$

5. Find the equation of ellipse whose focus is (1, –1), eccentricity is $\frac{1}{2}$ and directrix is $x-y+3=0$.
Ans.: $7x^2+2xy+7y^2-22x+22y+7=0$

6. Find the equation of the ellipse whose centre is (2, –3), one focus at (3, –3) and one vertex at (4, –3).
Ans.: $3x^2+4y^2-12x+24y+36=0$

7. Find the coordinates of the centre, eccentricity and foci of the ellipse $8x^2+6y^2-6x+12y+13=0$
Ans.: $(1,-1),1,(-2,\sqrt{6}\pm 1)$

8. Find the equation of the ellipse with foci at (0, 1) and (0, –1) and minor axis of length 1.
Ans.: $2x^2+4y^2=5$

9. An ellipse is described by using one endless string which is passed through two points. If the axes are 6 and 4 units find the necessary length and the distance between the points.
Ans.: $6+2\sqrt{5}, 2\sqrt{5}$

7.6 CONDITION FOR TANGENCY

To find the condition that the straight line $y = mx + c$ may be a tangent to the ellipse:

Let the equation of the ellipse be

$$\frac{x^2}{a^2}+\frac{y^2}{b^2}=1 \tag{7.6}$$

Let the equation of the straight line be

$$y = mx + c \tag{7.7}$$

Solving equations (7.6) and (7.7), we get their points of intersection; the x-coordinates of the points of intersection are given by $\frac{x^2}{a^2} + \frac{(mx+c)^2}{b^2} = 1.$

$$\text{(i.e.)} \quad b^2x^2 + a^2(mx+c)^2 = a^2b^2$$

$$\text{(i.e.)} \quad x^2(b^2 + a^2m^2) + 2cma^2x + a^2(c^2 - b^2) = 0 \tag{7.8}$$

If $y = mx + c$ is a tangent to the ellipse then the two values of x of this equation are equal. The condition for that is the discriminant of the quadratic equation is zero.

$$\therefore 4a^4m^2c^2 - 4(a^2m^2 + b^2)\cdot a^2(c^2 - b^2) = 0$$

$$a^2m^2c^2 - (a^2m^2 + b^2)(c^2 - b^2) = 0$$

$$a^2m^2c^2 - (a^2m^2c^2 - a^2m^2b^2 + b^2c^2 - b^4) = 0$$

or

$$b^2c^2 = a^2m^2b^2 + b^4$$

$$\therefore c^2 = a^2m^2 + b^2$$

This is the required condition for the line $y = mx + c$ to be a tangent to the given ellipse.

Note 7.6.1: The equation of any tangent to the ellipse is given by $y = mx + \sqrt{a^2m^2 + b^2}$.

7.7 DIRECTOR CIRCLE OF AN ELLIPSE

To show that always two tangents can be drawn from a given point to an ellipse and the locus of point of intersection of perpendicular tangents is a circle:

Let the equation of the ellipse be

$$\frac{x^2}{a^2} + \frac{y^2}{b^2} = 1 \tag{7.9}$$

Any tangent to this ellipse is

$$y = mx + \sqrt{a^2m^2 + b^2} \tag{7.10}$$

If this tangent passes through the point (x_1, y_1) then $y_1 = mx_1 + \sqrt{a^2m^2 + b^2}$.

$$\text{(i.e.)} \quad (y_1 - mx_1)^2 = a^2m^2 + b^2$$

$$\therefore m^2(x_1^2 - a^2) - 2mx_1y_1 + y_1^2 - b^2 = 0 \tag{7.11}$$

This is a quadratic equation in m and hence there are two values for m. For each value of m, there is a tangent (real or imaginary) and hence there are two tangents from a given point to an ellipse. If m_1 and m_2 are the roots of the equation (7.11), then $m_1 + m_2 = \dfrac{-2x_1y_1}{x_1^2 - a^2}$, $m_1m_2 = \dfrac{y_1^2 - b^2}{x_1^2 - a^2}$.

If the two tangents are perpendicular then $m_1m_2 = -1$.

$$\therefore \frac{y_1^2 - b^2}{x_1^2 - a^2} = -1 \text{ or } y_1^2 - b^2 = -x_1^2 + a^2$$

$$\Rightarrow x_1^2 + y_1^2 = a^2 + b^2$$

The locus of (x_1, y_1) is $x^2 + y^2 = a^2 + b^2$ which is a circle, centre at (0, 0) and radius $\sqrt{a^2 + b^2}$.

Note 7.7.1: This circle is called the director circle of the ellipse.

7.8 EQUATION OF THE TANGENT

To find the equation of the chord joining the points (x_1, y_1) and (x_2, y_2) and find the equation of the tangent at (x_1, y_1) to the ellipse:

Let $P(x_1, y_1)$ and $Q(x_2, y_2)$ be two points on the ellipse. Let the equation of ellipse be

$$\frac{x^2}{a^2} + \frac{y^2}{b^2} = 1 \tag{7.12}$$

Then

$$\frac{x_1^2}{a^2} + \frac{y_1^2}{b^2} = 1 \tag{7.13}$$

and

$$\frac{x_2^2}{a^2} + \frac{y_2^2}{b^2} = 1 \tag{7.14}$$

Subtracting, $$\frac{x_1^2 - x_2^2}{a^2} + \frac{y_1^2 - y_2^2}{b^2} = 0$$

$$\frac{(x_1 - x_2)(x_1 + x_2)}{a^2} + \frac{(y_1 - y_2)(y_1 + y_2)}{b^2} = 0 \tag{7.15}$$

From equation (7.15), we get the equation of the chord joining the points (x_1, y_1) and (x_2, y_2) as:

$$\frac{y - y_1}{x - x_1} = \frac{y_1 - y_2}{x_1 - x_2} = \frac{-b^2}{a^2}\left(\frac{x_1 + x_2}{y_1 + y_2}\right) \tag{7.16}$$

This chord becomes the tangent at (x_1, y_1) if Q tends to P and coincides with P. Hence, by putting $x_2 = x_1$ and $y_2 = y_1$ in equation (7.16), we get the equation of the tangent at (x_1, y_1).

Therefore, the equation of the tangent at (x_1, y_1) is:

$$\frac{y - y_1}{x - x_1} = \frac{-b^2 x_1}{a^2 y_1}$$

$$\text{or} \quad a^2 yy_1 - a^2 y_1^2 = -b^2 xx_1 + b^2 x_1^2$$

$$b^2 xx_1 + a^2 yy_1 = b^2 x_1^2 + a^2 y_1^2$$

Dividing by a^2b^2, we get

$$\frac{xx_1}{a^2} + \frac{yy_1}{b^2} = \frac{x_1^2}{a^2} + \frac{y_1^2}{b^2} \tag{7.17}$$

However, $\frac{x_1^2}{a^2} + \frac{y_1^2}{b^2} = 1$ since (x_1, y_1) lies on the ellipse.

Therefore, from equation (7.17), the equation of the tangent at (x_1, y_1) is $\frac{xx_1}{a^2} + \frac{yy_1}{b^2} = 1$.

7.9 EQUATION OF TANGENT AND NORMAL

To find the equation of tangent and normal at (x_1, y_1) to the ellipse $\frac{x^2}{a^2} + \frac{y^2}{b^2} = 1$:

The equation of the ellipse is

$$\frac{x^2}{a^2} + \frac{y^2}{b^2} = 1 \tag{7.18}$$

Differentiating with respect to x, we get

$$\frac{2x}{a^2}+\frac{2y}{b^2}\frac{dy}{dx}=0 \Rightarrow \frac{dy}{dx}=\frac{-b^2x}{a^2y}$$

$$\left(\frac{dy}{dx}\right)_{(x_1,\,y_1)}=\frac{-b^2x_1}{a^2y_1}$$

However, $\frac{dy}{dx}$ at (x_1, y_1) = slope of the tangent at (x_1, y_1). Therefore, the equation of the tangent at (x_1, y_1) is,

$$y-y_1=\frac{-b^2x_1}{a^2y_1}(x-x_1)$$

$$\text{or } a^2yy_1-a^2y_1^2=-b^2xx_1+b^2x_1^2$$

$$\text{or } a^2yy_1+b^2xx_1=b^2x_1^2+a^2y_1^2$$

Dividing by a^2b^2, we get

$$\frac{xx_1}{a^2}+\frac{yy_1}{b^2}=\frac{x_1^2}{a^2}+\frac{y_1^2}{b^2}=1 \quad \text{since } \frac{x_1^2}{a^2}+\frac{y_1^2}{b^2}=1.$$

Slope of the normal at (x_1, y_1) is $\frac{a^2y_1}{b^2x_1}$.

The equation of the normal at (x_1, y_1) to the ellipse is

$$y-y_1=\frac{a^2y_1}{b^2x_1}(x-x_1)$$

$$b^2x_1y-b^2x_1y_1=a^2y_1x-a^2x_1y_1$$

or

$$b^2x_1y-a^2xy_1=(b^2-a^2)x_1y_1$$

Dividing by x_1, y_1, we get,

$$\frac{a^2x}{x_1}-\frac{b^2y}{y_1}=a^2-b^2$$

Therefore, the equation of normal at (x_1, y_1) to the ellipse $\frac{x^2}{a^2}+\frac{y^2}{b^2}=1$ is $\frac{a^2x}{x_1}-\frac{b^2y}{y_1}=a^2-b^2$.

7.10 EQUATION TO THE CHORD OF CONTACT

To find the equation to the chord of contact of tangents drawn from (x_1, y_1) to the ellipse $\frac{x^2}{a^2}+\frac{y^2}{b^2}=1$:

The equation of the ellipse is

$$\frac{x^2}{a^2}+\frac{y^2}{b^2}=1 \tag{7.19}$$

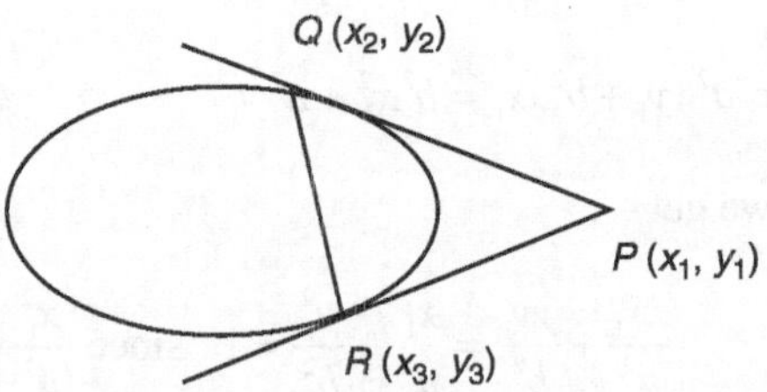

Let QR be the chord of contact of tangents from $P(x_1, y_1)$. Let Q and R be the points (x_2, y_2) and (x_3, y_3), respectively. Then the equation of tangents at Q and R are:

$$\frac{xx_2}{a^2}+\frac{yy_2}{b^2}=1 \tag{7.20}$$

$$\frac{xx_3}{a^2}+\frac{yy_3}{b^2}=1 \tag{7.21}$$

These two tangents pass through $P(x_1, y_1)$.

Therefore, $\frac{x_1x_2}{a^2}+\frac{y_1y_2}{b^2}=1$ and $\frac{x_1x_3}{a^2}+\frac{y_1y_3}{b^2}=1$.

The above two equations show that the points (x_2, y_2) and (x_3, y_3) lie on the line $\frac{xx_1}{a^2}+\frac{yy_1}{b^2}=1$. Hence, the equation of the chord of contact is $\frac{xx_1}{a^2}+\frac{yy_1}{b^2}=1$.

7.11 EQUATION OF THE POLAR

To find the equation of the polar of the point $P(x_1, y_1)$ on the ellipse $\frac{x^2}{a^2}+\frac{y^2}{b^2}=1$:

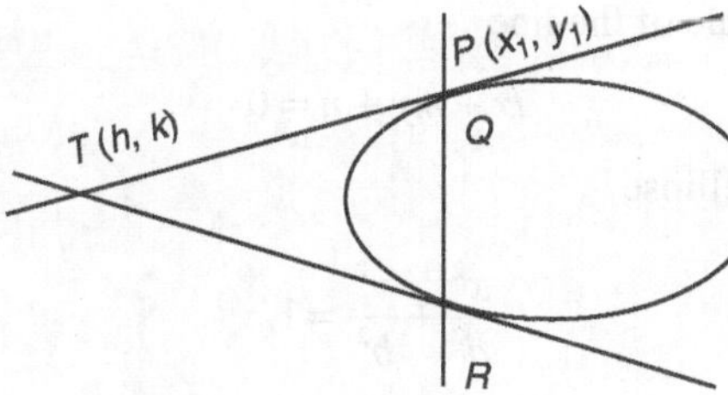

Let $P(x_1, y_1)$ be the given point. Let QR be a variable chord through the point $P(x_1, y_1)$. Let the tangents at Q and R meet at $T(h, k)$. The equation of the chord contact from $T(h, k)$ is:

$$\frac{xh}{a^2}+\frac{yk}{b^2}=1 \tag{7.22}$$

This chord of contact passes through (x_1, y_1).

$$\therefore \frac{x_1h}{a^2}+\frac{y_1k}{b^2}=1 \tag{7.23}$$

The locus of $T(h, k)$ is the polar of the point (x_1, y_1).

Therefore, the polar of (x_1, y_1) is $\frac{xx_1}{a^2}+\frac{yy_1}{b^2}=1$.

Note 7.11.1:

1. When the point (x_1, y_1) lies on the ellipse, the polar of (x_1, y_1) is the tangent at (x_1, y_1). When the point (x_1, y_1) lies inside the ellipse the polar of (x_1, y_1) is the chord of contact of tangents from (x_1, y_1).

2. The line $\frac{xx_1}{a^2}+\frac{yy_1}{b^2}=1$ is called the polar of the point (x_1, y_1) and (x_1, y_1) is called the pole of the line $\frac{xx_1}{a^2}+\frac{yy_1}{b^2}=1$.

7.12 CONDITION FOR CONJUGATE LINES

To find the pole of the line $lx + my + n = 0$ with respect to the ellipse $\frac{x^2}{a^2}+\frac{y^2}{b^2}=1$ and deduce the condition for the lines $lx + my + n = 0$ and $l_1x+m_1y+n_1=0$ to be conjugate lines:

Let (x_1, y_1) be the pole of the line

$$lx + my + n = 0 \tag{7.24}$$

with respect to the ellipse

$$\frac{x^2}{a^2}+\frac{y^2}{b^2}=1 \tag{7.25}$$

Then the polar of (x_1, y_1) is:

$$\frac{xx_1}{a^2}+\frac{yy_1}{b^2}=1. \tag{7.26}$$

Then the equations (7.24) and (7.26) represent the same line.

∴ Identifying equations (7.24) and (7.26), we get

$$\frac{x_1/a^2}{l}=\frac{y_1/b^2}{m}=\frac{-1}{n}$$

$$\therefore x_1=\frac{-la^2}{n}, y_1=\frac{-mb^2}{n}$$

Hence, the pole of the line $lx + my + n = 0$ is $\left(\frac{-la^2}{n},\frac{-mb^2}{n}\right)$. Two lines are said to be conjugate if the pole of the each lies on the other.

∴ The point $\left(\frac{-la^2}{n},\frac{-mb^2}{n}\right)$ lies on the line $l_1x+m_1y+n_1=0$.

$$l_1\left(\frac{-la^2}{n}\right)+m_1\left(\frac{-mb^2}{n}\right)+n_1=0$$

$$\text{(i.e.)}\quad ll_1a^2+mm_1b^2=nn_1$$

This is the required condition for the lines $lx + my + n = 0$ and $l_1x + m_1y + n_1 = 0$ to be conjugate lines.

ILLUSTRATIVE EXAMPLES BASED ON TANGENTS, NORMALS, POLE-POLAR AND CHORD

Example 7.4

Find the equation of the tangent to the ellipse $x^2 + 2y^2 = 6$ at $(2, -1)$.

Solution

The equation of the ellipse is $x^2 + 2y^2 = 6$.

$$\text{(i.e.)} \quad \frac{x^2}{6} + \frac{y^2}{3} = 1$$

The equation of the tangent at (x_1, y_1) is $\frac{xx_1}{6} + \frac{yy_1}{3} = 1$.

Therefore, the equation of the tangent at $(2, -1)$ is $\frac{2x}{6} - \frac{y}{3} = 1$.

$$\text{(i.e.)} \quad 2x - 2y = 6 \Rightarrow x - y = 3$$

Example 7.5

Find the equation of the normal to the ellipse $3x^2 + 2y^2 = 5$ at $(-1, 1)$.

Solution

The equation of the ellipse is $3x^2 + 2y^2 = 5$.

$$\text{(i.e.)} \quad \frac{x^2}{\frac{5}{3}} + \frac{y^2}{\frac{5}{2}} = 1$$

The equation of the normal at (x_1, y_1) is $\frac{ax^2}{x_1} - \frac{by^2}{y_1} = a^2 - b^2$.

$$\frac{5}{3} \cdot \frac{x}{(-1)} - \frac{5}{2} \cdot \frac{y}{1} = \frac{5}{3} - \frac{5}{2}.$$

$$\frac{-5x}{3} - \frac{5y}{2} = \frac{5}{3} - \frac{5}{2} \Rightarrow \frac{-5x}{3} - \frac{5y}{2} = \frac{10-15}{6}$$

$$\Rightarrow \frac{-5x}{3} - \frac{5y}{2} = \frac{-5}{6} \Rightarrow 2x + 3y = 1$$

Therefore, the equation of the normal to the ellipse $3x^2 + 2y^2 = 5$ is $2x + 3y = 1$.

Example 7.6

If B and B' are the ends of the minor axis of an ellipse then prove that $SB = S'B' = a$ where S and S' are the foci and a' is the semi-major axis. Show also that $SBS'B'$ is a rhombus whose area is $2abe$.

Solution

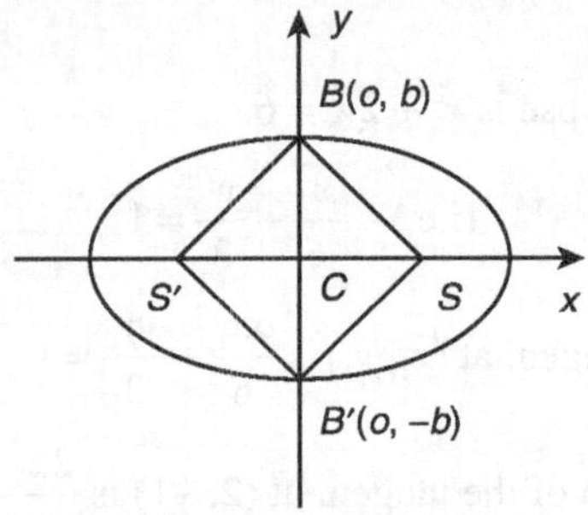

S is $(ae, 0)$; S' is $(-ae, 0)$
B is $(0, b)$; B' is $(0, -b)$

$$SB = \sqrt{(ae-0)^2+(0-b)^2} = \sqrt{a^2e^2+b^2}$$
$$= \sqrt{a^2-b^2+b^2} = a$$

$$S'B' = \sqrt{(-ae-0)^2+(0+b)^2} = \sqrt{a^2e^2+b^2}$$
$$= \sqrt{a^2-b^2+b^2} = a$$

$$\therefore\ SB = S'B' = a.$$

In the figure, $SBS'B'$ the diagonals SS' and $BB'$are at right angles. Therefore, $SBS'B'$ is a rhombus.

$$\text{Area of the rhombus} = 4(\Delta CSB)$$
$$= 4\times\frac{1}{2}\times CS\cdot CB$$
$$= 2ae\cdot b$$
$$= 2abe$$

Example 7.7

If the tangent at P of the ellipse $\frac{x^2}{a^2}+\frac{y^2}{b^2}=1$ meets the major axis at T and PN is the ordinate of P, then prove that $CN\cdot CT = a^2$ where C is the centre of the ellipse.

Solution

Let P be the point (x_1, y_1)

The equation of tangent at (x_1, y_1) is $\frac{xx_1}{a^2} + \frac{yy_1}{b^2} = 1$.

When the tangent meets the x-axis, $y = 0$

$$\therefore \frac{xx^2}{a^2} = 1 \text{ or } x = \frac{a^2}{x_1}$$

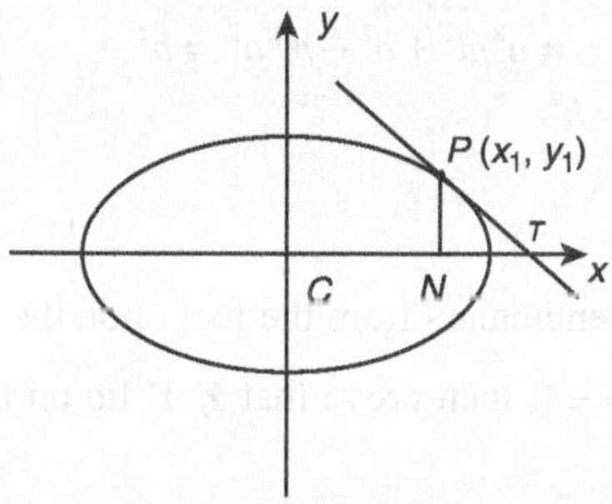

$$\therefore CT = \frac{a^2}{x_1}$$

However, $CN = x_1$

$$\therefore CN \cdot CT = a^2$$

Example 7.8

The tangent at any point P on the ellipse $\frac{x^2}{a^2} + \frac{y^2}{b^2} = 1$ meets the tangents at A and A' (extremities of major axis) in L and M, respectively. Prove that $AL \cdot A'M = b^2$.

Solution

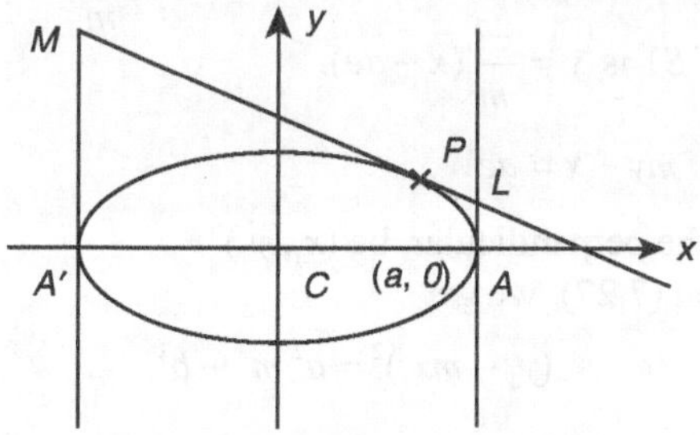

Let the equation of the tangent at P be $y = mx + \sqrt{a^2m^2 + b^2}$. The equation of the tangent at the point A is $x = a$. Solving these two equations, we get $y = ma + \sqrt{a^2m^2 + b^2}$.

$$\therefore AL = \sqrt{a^2m^2 + b^2} + ma$$

Similarly, $A'M = \sqrt{a^2m^2 + b^2} - ma$

$$\therefore AL \cdot A'M = \left(\sqrt{a^2m^2 + b^2} - ma\right)\left(\sqrt{a^2m^2 + b^2} + ma\right)$$

$$= a^2m^2 + b^2 - m^2a^2 = b^2$$

Therefore, $AL \cdot A'M = b^2$.

Example 7.9

If SY and $S'Y'$ be perpendiculars from the foci upon the tangents at any point of the ellipse $\frac{x^2}{a^2} + \frac{y^2}{b^2} = 1$, then prove that Y, Y' lie on the circle $x^2 + y^2 = a^2$ and that $SY \cdot S'Y' = b^2$.

Solution

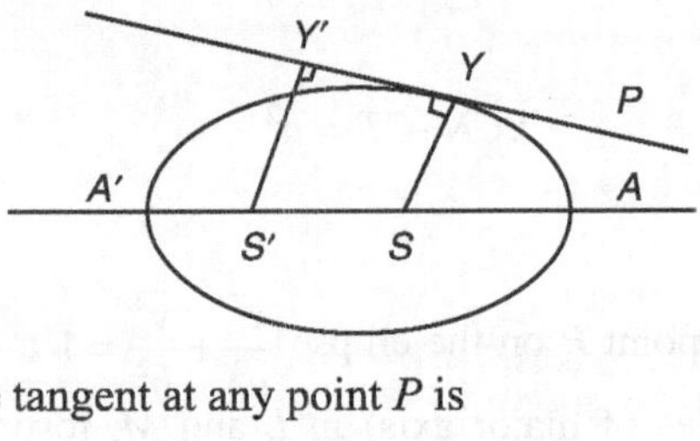

The equation of the tangent at any point P is

$$y = mx + \sqrt{a^2m^2 + b^2} \tag{7.27}$$

The slope of the tangent is m.
Therefore, the slope of the perpendicular line SY is $\frac{-1}{m}$. S is the point $(ae, 0)$.

$\therefore$ The equation of SY is $y = \frac{-1}{m}(x - ae)$.

$$my + x = ae \tag{7.28}$$

Let Y, the foot of the perpendicular, be (x_1, y_1)
Then from equation (7.27), we get

$$(y_1 - mx_1)^2 = a^2\, m^2 + b^2 \tag{7.29}$$

From equation (7.28), we get

$$(my + x_1)^2 = a^2e^2 = a^2 - b^2 \tag{7.30}$$

Adding equations (7.29) and (7.30), we get

$$x_1^2(1+m^2) + y_1^2(1+m^2) = a^2(1+m^2).$$

Cancelling, $1+m^2, x_1^2 + y_1^2 = a^2$

The locus of (x_1, y_1) is $x^2 + y^2 = a^2$. Similarly, we can prove that the locus of Y' is also this circle. Hence, Y and Y' lie on this circle.

$$SY = \frac{mae + \sqrt{a^2m^2 + b^2}}{\sqrt{1+m^2}}, S'Y' = \frac{-mae + \sqrt{a^2m^2 + b^2}}{\sqrt{1+m^2}}$$

$$SY \cdot S'Y' = \frac{(a^2m^2 + b^2) - m^2a^2e^2}{1+m^2} = \frac{a^2m^2 + b^2 - m^2(a^2 - b^2)}{1+m^2}$$

Since $(a^2 e^2 = a^2 - b^2)$

$$= \frac{b^2(1+m^2)}{1+m^2} = b^2$$

Note 7.12.1: This circle is called the auxiliary circle $(x^2 + y^2 = a^2)$. This is the circle described on the major axis as diameter.

Example 7.10

If normal at a point P on the ellipse $\frac{x^2}{a^2} + \frac{y^2}{b^2} = 1$ meets the major axis at G then prove that:

(i) $CG = e^2CN$, where C is the centre of the ellipse and N is the foot of the perpendicular from P to the major axis.

(ii) $SG = eSP$ where S is the focus of the ellipse.

Solution

(i) Let P be the point (x_1, y_1).
The equation of the normal at (x_1, y_1) is $\frac{a^2x}{x_1} - \frac{b^2y}{y_1} = a^2e^2$.

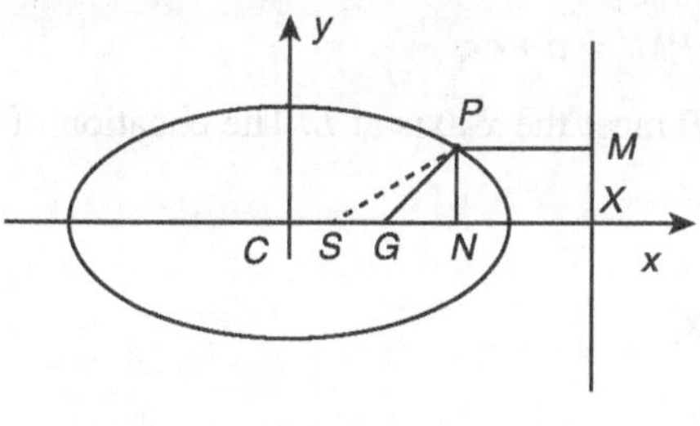

When this meets the x-axis, $y = 0$.

$$\therefore x = \frac{a^2 e^2 x_1}{a^2}$$

$$\therefore CG = e^2 x_1 \quad \text{(i.e.)} \quad CG = e^2 CN$$

(ii) $\dfrac{SP}{PM} = e$

$$\therefore SP = ePM = e.NX = e(CX - CN) = e\left(\frac{a}{e} - x_1\right) = a - ex$$

$$\therefore eSP = ae - e^2 x_1 = CS - CG = SG$$

$$\therefore SG = eSP$$

Example 7.11

In an ellipse, prove that the tangent and normal at any point P are the external and internal bisectors of the angle SPS' where S and S' are the foci.

Solution

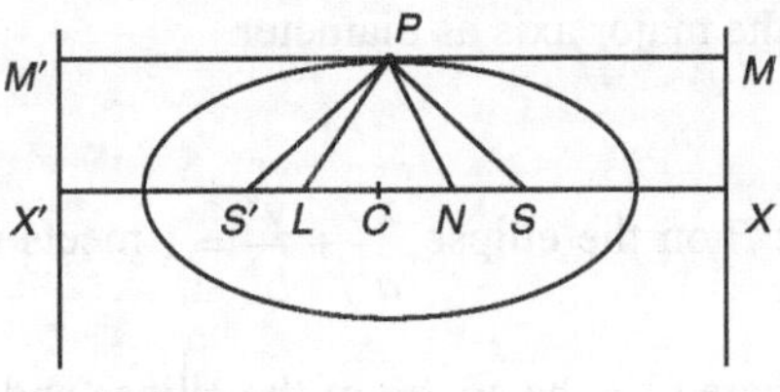

Let $P(x_1, y_1)$ be any point on the ellipse.

$$\frac{SP}{PM} = e$$

$$\therefore SP = ePM = e \cdot NX = e(CX - CN) = e\left(\frac{a}{e} - x_1\right) = a - ex_1 \qquad (7.31)$$

$$\frac{S'P}{PM'} = e$$

$$\therefore S'P = ePM' = a + ex_1 \qquad (7.32)$$

Let the normal at P meet the x-axis at L. The equation of the normal at P is $\dfrac{a^2 x}{x_1} - \dfrac{b^2 y}{y_1} = a^2 e^2$.

When $y = 0$, $x = e^2 x_1$

$\therefore L$ is $(e^2 x_1, 0)$.

$$S'L = S'C + CL = ae + e^2x_1 \quad (7.33)$$

$$SL = CS - CL = ae + e^2x_1 \quad (7.34)$$

From equations (7.33) and (7.34), we get

$$\frac{S'P}{SP} = \frac{a+ex_1}{a-ex_1} \quad (7.35)$$

From equations (7.33) and (7.34), we get

$$\frac{S'L}{LS} = \frac{ae+e^2x_1}{ae-e^2x_1} = \frac{a+ex_1}{a-ex_1}$$

$$= \frac{S'P}{SP} \text{ from equation} \quad (7.35)$$

Therefore, the normal PL is the internal bisector of $\underline{|S'PS}$. Since the tangent at P is perpendicular to the normal at P, the tangent P is the external bisector.

Example 7.12

Find the angle subtended by a focal chord of the ellipse $\frac{x^2}{a^2}+\frac{y^2}{b^2}=1$ passing through an end of the minor axis at the centre of the ellipse.

Solution

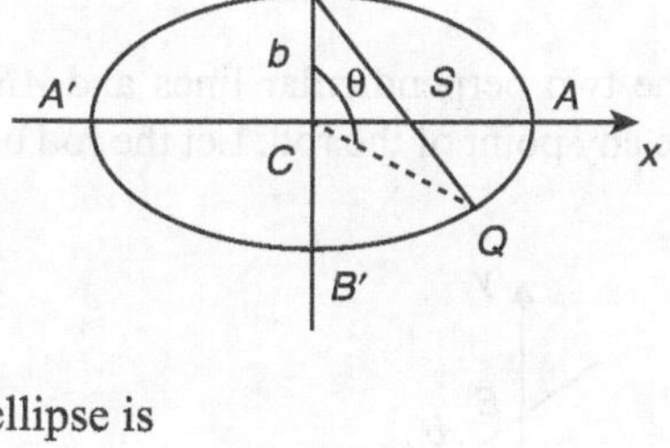

The equation of the ellipse is

$$\frac{x^2}{a^2}+\frac{y^2}{b^2}=1 \quad (7.36)$$

The equation of the focal chord is

$$\frac{x}{ae}+\frac{y}{b}=1 \quad (7.37)$$

The combined equation of the lines *CB* and *CQ* is got by homogenization of the equation of the ellipse with the help of straight line (7.37).

$$\text{(i.e.)} \quad \frac{x^2}{a^2}+\frac{y^2}{b^2}=\left(\frac{x}{ae}+\frac{y}{b}\right)^2$$

$$\frac{x^2}{a^2}\left(1-\frac{1}{e^2}\right)-\frac{2xy}{abe}=0$$

$$\text{(i.e.)} \quad \frac{x^2(e^2-1)}{ae}-\frac{2xy}{b}=0$$

$$\text{(i.e.)} \quad b(1-e^2)\,x^2+2aexy=0$$

The angle between the lines *CB* and *CQ* is given by

$$\tan\theta=\pm\frac{2\sqrt{h^2-ab}}{a+b}=\frac{\pm2\sqrt{a^2e^2-0}}{b(1-e^2)}=\pm\frac{2ae}{b(1-e^2)}.$$

Since the angle *BCQ* is obtuse, $\theta=\tan^{-1}\left(\frac{-2ae}{b(1-e^2)}\right)$.

Example 7.13

A bar of given length moves with its extremities on two fixed straight lines at right angles. Prove that any point of the rod describes an ellipse.

Solution

Let *OA* and *OB* be the two perpendicular lines and *AB* be the rod of fixed length. Let $P(x_1, y_1)$ be any point of the rod. Let the rod be inclined at an angle θ with *OX*.

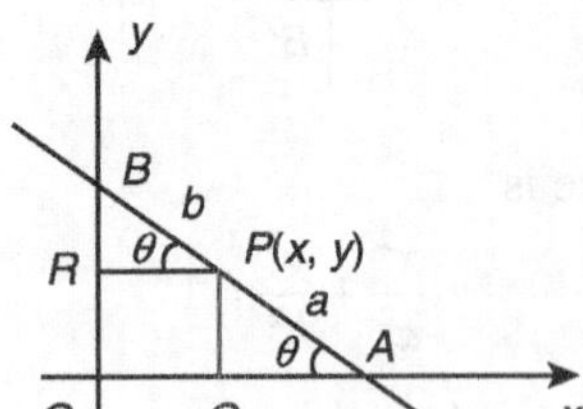

(i.e.) $\angle OAB=\theta$

Take $PA=a$ and $PB=b$.

Then $x = OQ = RP = b\cos\theta$, $y = QP = b\sin\theta$,

$\therefore\ \dfrac{x}{a} = \cos\theta$ and $\dfrac{y}{b} = \sin\theta$

Hence, $\dfrac{x^2}{a^2} + \dfrac{y^2}{b^2} = \cos^2\theta + \sin^2\theta = 1$

Therefore, the locus of P is an ellipse.

Example 7.14

The equation $25(x^2 - 6x + 9) + 16y^2 = 400$ represents an ellipse. Find the centre and foci of the ellipse. How should the axis be transformed so that the ellipse is represented by the equation $\dfrac{x^2}{25} + \dfrac{y^2}{16} = 1$?

Solution

$$25(x^2 - 6x + 9) + 16y^2 = 400$$

$$25(x-3)^2 + 16y^2 = 400$$

Dividing by 400, $\dfrac{(x-3)^2}{16} + \dfrac{y^2}{25} = 1$; Take $x - 3 = X, y = Y$.

Then $\dfrac{X^2}{16} + \dfrac{Y^2}{25} = 1$.

The major axis of this ellipse is the Y-axis.

$$\therefore\ 16 = 25(1 - e^2) \Rightarrow 1 - e^2 = \frac{16}{25} \Rightarrow e^2 = 1 - \frac{16}{25} = \frac{9}{25}$$

$$\therefore\ e = \frac{3}{5}.$$

Centre is (3, 0). Foci are $(3, \pm ae)$ (i.e.) $\left(3, \pm 5 \times \dfrac{3}{5}\right)$ (i.e.) $(3, \pm 3)$. Now shift origin to the point (3, 0) and then rotate the axes through right angles. Then the equation of the ellipse becomes $\dfrac{x^2}{25} + \dfrac{y^2}{16} = 1$.

Example 7.15

Show that if s, s' are the lengths of the perpendicular on a tangent from the foci, a, a' those from the verlices and e that from the centre then $s, s' - e^2 = e^2(aa' - c^2)$ where e is the eccentricity.

Solution

Let the equation of the ellipse be

$$\frac{x^2}{a^2}+\frac{y^2}{b^2}=1 \tag{7.38}$$

Foci are $S(ae, 0)$ and $S'(-ae, 0)$. Vertices are $A(a, 0)$ and $A'(-a, 0)$, centre is $(0, 0)$. Any tangent to the ellipse (7.38) is $y = mx+\sqrt{a^2m^2+b^2}$.

The perpendicular distance from $S(ae, 0)$ is $= \dfrac{ae+\sqrt{a^2m^2+b^2}}{\sqrt{1+m^2}}$.

Similarly, $s' = \dfrac{-ae+\sqrt{a^2m^2+b^2}}{\sqrt{1+m^2}}, \quad a = \dfrac{a+\sqrt{a^2m^2+b^2}}{\sqrt{1+m^2}}$

$$a' = \frac{-ae+\sqrt{a^2m^2+b^2}}{\sqrt{1+m^2}}, \qquad c = \frac{\sqrt{a^2m^2+b^2}}{\sqrt{1+m^2}}$$

$$SS'-c^2 = \left|\frac{a^2m^2+b^2-a^2e^2}{1+m^2}-\frac{a^2m^2+b^2}{1+m^2}\right| = \frac{a^2e^2}{1+m^2} \tag{7.39}$$

$$aa'-c^2 = \left|\frac{a^2m^2+b^2-a^2}{1+m^2}-\frac{a^2m^2+b^2}{1+m^2}\right| = \frac{a^2}{1+m^2}$$

$$\therefore e^2(aa'-c^2) = \frac{a^2e^2}{1+m^2} \tag{7.40}$$

From equations (7.39) and (7.40), we get $ss' - c^2 = e^2(aa' - c^2)$.

Example 7.16

A circle of radius r is concentric with the ellipse $\dfrac{x^2}{a^2}+\dfrac{y^2}{b^2}=1$. Prove that each common tangent is inclined to the axis at an angle $\tan^{-1}\sqrt{\dfrac{r^2-b^2}{a^2-r^2}}$ and towards its length.

Solution

The equation of the ellipse is

$$\frac{x^2}{a^2}+\frac{y^2}{b^2}=1 \tag{7.41}$$

The equation of the circle concentric with the ellipse is

$$x^2 + y^2 = r^2 \tag{7.42}$$

Any tangent to the ellipse is

$$y = mx + \sqrt{m^2a^2 + b^2} \tag{7.43}$$

Any tangent to the circle is

$$y = mx + r\sqrt{1+m^2} \tag{7.44}$$

If the tangent is a common tangent then

$$r^2(1+m^2) = a^2m^2 + b^2 \Rightarrow m^2(a^2 - r^2) = r^2 - b^2, m^2 = \frac{r^2 - b^2}{a^2 - r^2}$$

(i.e.) $\tan^2\theta = \dfrac{r^2 - b^2}{a^2 - r^2}$ or $\tan\theta = \sqrt{\dfrac{r^2 - b^2}{a^2 - r^2}}$.

Therefore, the inclination to the major axis is $\theta = \tan^{-1}\sqrt{\dfrac{r^2 - b^2}{a^2 - r^2}}$.

Example 7.17

Prove that the sum of the squares of the perpendiculars of any tangent of an ellipse $\dfrac{x^2}{a^2} + \dfrac{y^2}{b^2} = 1$ from two points on the minor axis, each distance $\sqrt{a^2 - b^2}$ from the centre is $2a^2$.

Solution

The equation of the ellipse is $\dfrac{x^2}{a^2} + \dfrac{y^2}{b^2} = 1$.

Any tangent to the ellipse is $y = mx + \sqrt{a^2m^2 + b^2}$. The perpendicular distance from $\left(0, \sqrt{a^2 - b^2}\right)$ to the tangent is $p_1 = \pm\dfrac{\sqrt{a^2m^2 + b^2} - \sqrt{a^2 - b^2}}{\sqrt{1+m^2}}$.

The perpendicular distance from $\left(0, -\sqrt{a^2 - b^2}\right)$ is given by

$$p_2 = \pm\frac{\sqrt{a^2m^2 + b^2} + \sqrt{a^2 - b^2}}{\sqrt{1+m^2}}$$

$$p_1^2 + p_2^2 = \frac{\left(\sqrt{a^2m^2+b^2} - \sqrt{a^2-b^2}\right)^2 + \left(\sqrt{a^2m^2+b^2} + \sqrt{a^2-b^2}\right)^2}{1+m^2}$$

$$= \frac{2(a^2m^2+b^2+a^2-b^2)}{1+m^2}$$

$$= \frac{2a^2(1+m^2)}{(1+m^2)}$$

$$= 2a^2$$

Example 7.18

Let d be the perpendicular distance from the centre of the ellipse $\frac{x^2}{a^2}+\frac{y^2}{b^2}=1$ to the tangent drawn at a point P on the ellipse. If F_1 and F_2 are the two foci of the ellipse then show that $(PF_1 - PF_2)^2 = 4a^2\left(1-\frac{b^2}{d^2}\right)$.

Solution

The equation of the ellipse is $\frac{x^2}{a^2}+\frac{y^2}{b^2}=1$. Let $P(x_1, y_1)$ be any point on it.

The equation of the tangent at (x_1, y_1) is $\frac{x^2}{a^2}+\frac{y^2}{b^2}=1$.

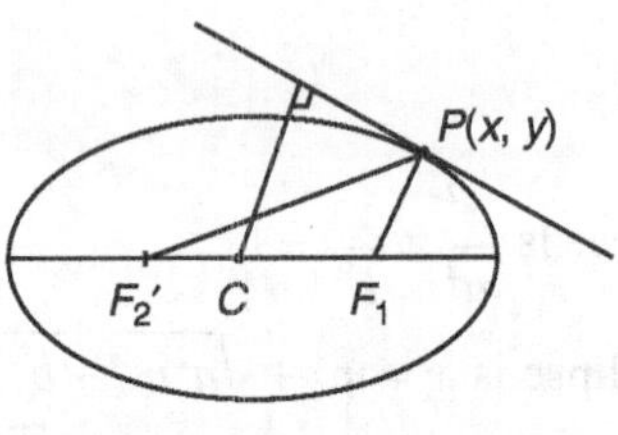

The perpendicular distance from C on this tangent is

$$d = \frac{1}{\sqrt{\frac{x_1^2}{a^4}+\frac{y_1^2}{b^4}}}$$

$$\Rightarrow \frac{1}{d} = \sqrt{\frac{x_1^2}{a^4}+\frac{y_1^2}{b^4}}$$

We know that $PF_1 = a - ex_1$, $PF_2 = a + ex_1$

$$(PF_1 - PF_2)^2 = [(a - ex_1) - (a + ex_1)]^2 = 4e^2 x_1^2 \tag{7.45}$$

$$4a^2\left(1 - \frac{b^2}{d^2}\right) = 4a^2\left[1 - b^2\left(\frac{x_1^2}{a^4} + \frac{y_1^2}{b^4}\right)\right]$$

$$4a^2\left(1 - \frac{b^2}{d^2}\right) = 4a^2\left[\frac{x_1^2}{a^2} - \frac{b^2 x_1^2}{a^4}\right] = 4a^2\left[\frac{x_1^2}{a^2} - \frac{a^2(1-e^2)x_1^2}{a^4}\right]$$

$$= 4a^2\left[\frac{x_1^2}{a^2} - \frac{(1-e^2)x_1^2}{a^2}\right]$$

$$= 4a^2\left(\frac{e^2 x_1^2}{a^2}\right) = 4e^2 x_1^2 \tag{7.46}$$

From equations (7.45) and (7.46), we get

$$(PF_1 - PF_2)^2 = 4a^2\left(1 - \frac{b^2}{d^2}\right).$$

Example 7.19

Show that the locus of the middle points of the portion of a tangent to the ellipse $\frac{x^2}{a^2} + \frac{y^2}{b^2} = 1$ included between the axes is the curve $\frac{a^2}{x^2} + \frac{b^2}{y^2} = 4$.

Solution

Any tangent to the ellipse $\frac{x^2}{a^2} + \frac{y^2}{b^2} = 1$ is

$$y = mx + \sqrt{a^2 m^2 + b^2} \tag{7.47}$$

When the tangent meets the x-axis, $y = 0$.

$$\therefore x = \frac{-\sqrt{a^2 m^2 + b^2}}{m}$$

When it meets the y-axis, $x = 0$.

$$\therefore y = \sqrt{a^2 m^2 + b^2}$$

Therefore, the points of intersection of the tangents with the axes are $A\left(\frac{-\sqrt{a^2m^2+b^2}}{m},0\right)$ and $B\left(0,\sqrt{a^2m^2+b^2}\right)$. Let (x_1, y_1) be the midpoint of line AB.

$$2x_1 = \frac{-\sqrt{a^2m^2+b^2}}{m},\ 2y_1 = \sqrt{a^2m^2+b^2}$$

$$\therefore \frac{a^2}{4x_1^2}+\frac{b^2}{4y_1^2} = \frac{a^2m^2}{a^2m^2+b^2}+\frac{b^2}{a^2m^2+b^2}$$

$$= \frac{a^2m^2+b^2}{a^2m^2+b^2} = 1$$

Therefore, the locus of $P(x_1, y_1)$ is $\frac{a^2}{x^2}+\frac{b^2}{y^2} = 4.$

Example 7.20

Prove that the tangent to the ellipse $\frac{x^2}{a^2}+\frac{y^2}{b^2}=1$ meets the ellipse $\frac{x^2}{a^2}+\frac{y^2}{b^2}=a+b$ in points, tangents at which are at right angles.

Solution

Any tangent to the ellipse $\frac{x^2}{a^2}+\frac{y^2}{b^2}=1$ is

$$y = mx+\sqrt{a^2m^2+b^2} \tag{7.48}$$

At Q and R let the tangents meet the ellipse

$$\frac{x^2}{a}+\frac{y^2}{b} = a+b \tag{7.49}$$

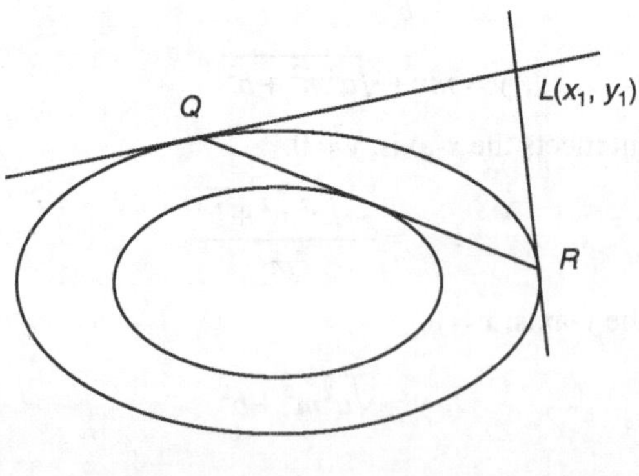

Let $L(x_1, y_1)$ be the point of intersection of tangents at Q and R. Then QR is the chord of contact form L. Its equation is

$$\frac{xx_1}{a}+\frac{yy_1}{b}=a+b \tag{7.50}$$

Equations (7.48) and (7.49) represent the same line. Identifying equations (7.48) and (7.49), we get $\frac{m}{\left(\frac{x_1}{a}\right)}=\frac{-1}{\left(\frac{y_1}{b}\right)}=\frac{\sqrt{a^2m^2+b^2}}{a+b}$.

$$x_1=\frac{am(a+b)}{\sqrt{a^2m^2+b^2}}$$

$$y_1=-\frac{b(a+b)}{\sqrt{a^2m^2+b^2}}$$

Therefore, $x_1^2+y_1^2=(a+b)^2$. The locus of (x_1, y_1) is the equation of the director circle of the ellipse (7.49). However, director circle is the intersection of perpendicular tangents. Hence, the tangents at Q and R are at right angles.

Example 7.21

A chord PQ of an ellipse subtends a right angle at the centre of the ellipse $\frac{x^2}{a^2}+\frac{y^2}{b^2}+1$. Show that the locus of the intersection of the tangents at Q and R is the ellipse $\frac{x^2}{a^4}+\frac{y^2}{b^4}=\frac{1}{a^2}+\frac{1}{b^2}$.

Solution

The equation of ellipse is $\frac{x^2}{a^2}+\frac{y^2}{b^2}+1$ (7.51)

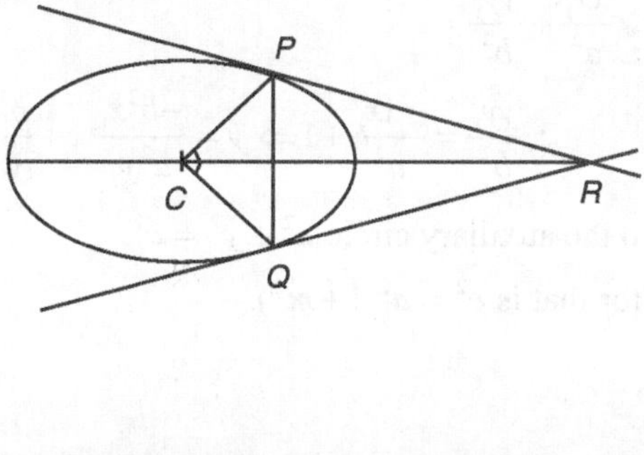

Let $R(x_1, y_1)$ be the point of intersection of tangents at P and Q. The equation of the chord of contact of PQ is

$$\frac{xx_1}{a^2}+\frac{yy_1}{b^2}=1 \tag{7.52}$$

The combined equation of CP and CQ is got by homogenization of equation (7.51) with the help of equation (7.52).

$$\frac{x^2}{a^2}+\frac{y^2}{b^2}=\left(\frac{xx_1}{a^2}+\frac{yy_1}{b^2}\right)^2$$

$$x^2\left[\frac{x_1^2}{a^4}-\frac{1}{a^2}\right]+y^2\left[\frac{y_1^2}{b^4}-\frac{1}{b^2}\right]-\frac{2x_1y_1xy}{a^2b^2}=0.$$

Since $\underline{|PLQ} = 90°$, coefficient of x^2 + coefficient of $y^2 = 0$.

$$\therefore \frac{x_1^2}{a^4}-\frac{1}{a^2}+\frac{y_1^2}{b^4}-\frac{1}{b^2}=0$$

$$\Rightarrow \frac{x_1^2}{a^4}+\frac{y_1^2}{b^4}=\frac{1}{a^2}+\frac{1}{b^2}$$

The locus of $P(x_1, y_1)$ is $\dfrac{x^2}{a^4}+\dfrac{y^2}{b^4}=\dfrac{1}{a^2}+\dfrac{1}{b^2}$

Example 7.22

Show that the locus of poles with respect to the ellipse $\frac{x^2}{a^2}+\frac{y^2}{b^2}+1$ of any tangent to the auxiliary circle is $\frac{x^2}{a^2}+\frac{y^2}{b^2}=\frac{1}{b^2}$.

Solution

Let (x_1, y_1) be the pole with respect to the ellipse $\frac{x^2}{a^2}+\frac{y^2}{b^2}+1$.
The polar of (x_1, y_1) is

$$\frac{xx_1}{a^2}+\frac{yy_1}{b^2}=1 \tag{7.53}$$

$$\therefore \frac{yy_1}{b^2}=\frac{-xx_1}{a^2}+1 \Rightarrow y=\frac{-b^2x_1}{a^2y_1}x+\frac{b^2}{y_1}$$

This is a tangent to the auxiliary circle $x^2 + y^2 = a^2$.

The condition for that is $c^2 = a^2(1+m^2)$.

(i.e.) $\left(\frac{b^2}{y_1}\right)^2 = a^2\left[1+\frac{b^4x_1^2}{a^4y_1^2}\right] \Rightarrow \frac{b^4}{\cancel{y_1^2}} = \frac{a^4y_1^2+b^4x_1^2}{a^2\cancel{y_1^2}}$

(i.e.) $b^4a^2 = a^4y_1^2 + b^4x_1^2$

Dividing by a^2b^4, $\frac{x_1^2}{a^4}+\frac{y_1^2}{b^4}=\frac{1}{a^2}$.

The locus of (x_1, y_1) is $\frac{x^2}{a^4}+\frac{y^2}{b^4}=\frac{1}{a^2}$.

Example 7.23

Show that the locus of poles of tangents to the circle $(x-h)^2+(y-k)^2=r^2$ with respect to the ellipse $\frac{x^2}{a^2}+\frac{y^2}{b^2}=1$ is $r^2\left(\frac{x^2}{a^2}+\frac{y^2}{b^2}\right)=\left(\frac{hx}{a^2}+\frac{ky}{b^2}-1\right)$.

Solution

Let (x_1, y_1) be the pole with respect to the ellipse $\frac{x^2}{a^2}+\frac{y^2}{b^2}=1$. Then the polar of (x_1, y_1) is $\frac{xx_1}{a^2}+\frac{yy_1}{b^2}=1$. This line is a tangent to the circle $(x-h)^2+(y-k)^2=r^2$. The condition for this is that the radius of the circle should be equal to the perpendicular distance from the centre on the tangents.

$$\therefore r = \pm\frac{\left(\frac{hx_1}{a^2}+\frac{ky_1}{b^2}-1\right)}{\sqrt{\frac{x_1^2}{a^4}+\frac{y_1^2}{b^2}}}$$

$$r^2\left(\frac{x_1^2}{a^4}+\frac{y_1^2}{b^4}\right)=\left(\frac{hx_1}{a^2}+\frac{ky_1}{b^2}-1\right)^2$$

Therefore, the locus of (x_1, y_1) is $r^2\left(\frac{x^2}{a^4}+\frac{y^2}{b^4}\right)=\left(\frac{hx}{a^2}+\frac{ky}{b^2}-1\right)^2$.

Example 7.24

Find the locus of the poles with respect to the ellipse of the tangents to the parabola $y^2 = 4px$.

Solution

Let (x_1, y_1) be the pole with respect to the ellipse $\frac{x^2}{a^2}+\frac{y^2}{b^2}+1$. The polar of (x_1, y_1) is $\frac{xx_1}{a^2}+\frac{yy_1}{b^2}=1$.

$$\text{(i.e.)}\quad y=\frac{-b^2x_1}{a^2y_1}x+\frac{b^2}{y_1}.$$

This is a tangent to the parabola $y^2 = 4px$.

$\therefore$ The condition is $\frac{b^2}{y_1}=\frac{p}{-\left(\frac{b^2x_1}{a^2y_1}\right)}$.

$$\text{(i.e.)}\quad \frac{b^2}{y_1}=-\frac{a^2py_1}{b^2x_1}$$

$$\text{(i.e.)}\quad b^4x_1+a^2py_1^2=0$$

Therefore, the locus of (x_1, y_1) is $a^2py^2+b^4x=0$.

Example 7.25

Any tangent to an ellipse is cut by the tangents at the extremities of the major axis in the point T and T'. Prove that the circle drawn on TT' as diameter passes through the foci.

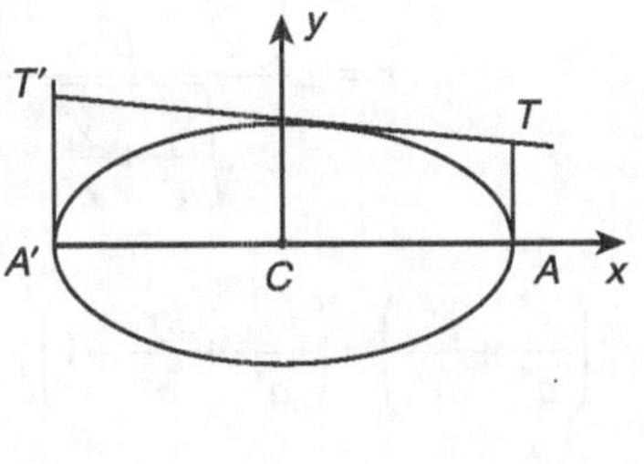

Solution

Let the equation of the ellipse be

$$\frac{x^2}{a^2}+\frac{y^2}{b^2}=1 \tag{7.54}$$

The ends of major axis are $A(a, 0)$ and $A'(-a, 0)$. Any tangent to the ellipse is

$$y=mx+\sqrt{a^2+m^2+b^2} \tag{7.55}$$

This tangent meets the tangents at A, A' at T and T', respectively. Then the co-ordinates of T and T' are $T\left(a,\ ma+\sqrt{a^2m^2+b^2}\right)$, $T'\left(-a,\ ma+\sqrt{a^2m^2+b^2}\right)$. The equation of the circle on TT' as diameter is

$$(x-a)(x+a)+\left(y-ma-\sqrt{a^2m^2+b^2}\right)\left(y+ma-\sqrt{a^2m^2+b^2}\right)=0.$$

$$x^2-a^2+(y-\sqrt{a^2m^2+b^2})^2-a^2m^2=0.$$

$$x^2-a^2+y^2-2y\sqrt{a^2m^2+b^2}+m^2a^2-a^2m^2+b^2=0.$$

or

$$x^2+y^2-2(\sqrt{a^2m^2+b^2})y=a^2-b^2$$

or

$$x^2+y^2-2(\sqrt{a^2m^2+b^2})y=a^2e^2.$$

This circle passes through the foci $S(ae, 0)$ and $S'(-ae, 0)$.

Example 7.26

The ordinate NP of a point P on the ellipse is produced to meet the tangent at one end of the latus rectum through the focus S in Q. Prove that $QN = SP$.

Solution

Let LSL' be the latus rectum through the focus S. The equation of tangent at L is

$$\frac{x(ae)}{a^2}+\frac{y(b^2/a)}{b^2}=1 \tag{7.56}$$

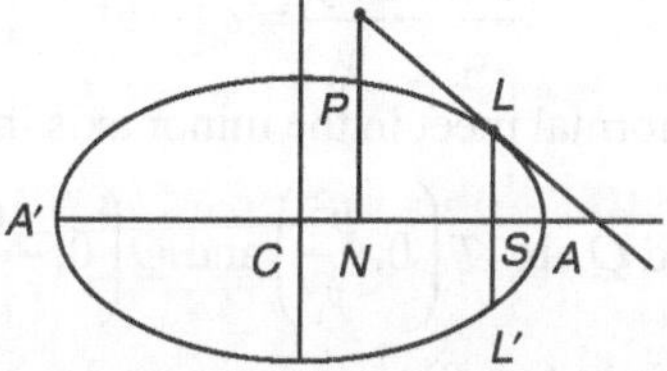

Let P be the point (x_1, y_1). The equation of the ordinate at P is

$$x=x_1 \tag{7.57}$$

When the tangent at L meets the ordinate at P in Q, the coordinates of Q are given by solving equations (7.56) and (7.57).

$$\frac{x_1}{a}(e)+\frac{y_1}{a}=1$$

or

$$y_1=a-ex_1$$

$$\therefore QN = a - ex_1$$

We know that $SP = a - ex_1$.
Therefore, $QN = SP$.

Example 7.27

The tangent and normal at a point P on the ellipse meet the minor axis in T and Q. Prove that TQ subtends a right angle at each of the foci.

Solution

The equation of ellipse is

$$\frac{x^2}{a^2} + \frac{y^2}{b^2} = 1 \quad (7.58)$$

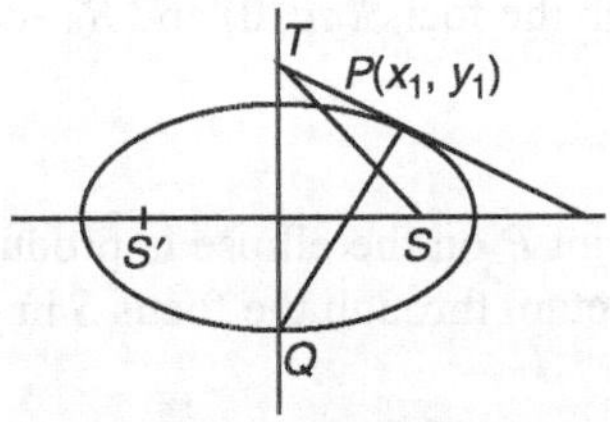

The equation of the tangent and normal at $P(x_1, y_1)$ is

$$\frac{xx_1}{a^2} + \frac{yy_1}{b_2} = 1 \quad (7.59)$$

$$\frac{a^2x}{x_1} - \frac{b^2y}{y_1} = a^2 - b^2 \quad (7.60)$$

When the tangent and normal meet in the minor axis in T and Q, respectively, the coordinates of T and Q are $T\left(0, \frac{b^2}{y_1}\right)$ and $Q\left(0, -\frac{(a^2-b^2)y_1}{b^2}\right)$.

The coordinates of S are $(ae, 0)$.

Slope of $\quad TS = \left(\frac{b^2}{y_1}\right) \Big/ (-ae) = \left(\frac{-b^2}{aey_1}\right) = m_1$ (say)

Slope of $\quad QS = -\frac{(a^2-b^2)y_1}{b^2} \Big/ (-ae) = \frac{a^2e^2y_1}{bae} = \frac{aey_1}{b^2} = m_2$ (say)

Now, $m_1 m_2 = -1$.
Therefore, TQ subtends a right angle at the focus S. The coordinates of S' are $(-ae, 0)$. Hence it is proved that TQ subtends a right angle at S'.

Example 7.28

If S and S' be the foci of the ellipse $\frac{x^2}{a^2}+\frac{y^2}{b^2}=1$ and e be its eccentricity then prove that $\tan\left(\frac{\underline{|PSS'}}{2}\right)\times\tan\left(\frac{\underline{|PS'S}}{2}\right)=\frac{1-e}{1+e}$ where P is any point on the ellipse.

Solution

The equation of ellipse is $\frac{x^2}{a^2}+\frac{y^2}{b^2}=1$.

The coordinates of S are $(ae, 0)$ and S' are $(-ae, 0)$.

$\therefore SS' = 2ae$

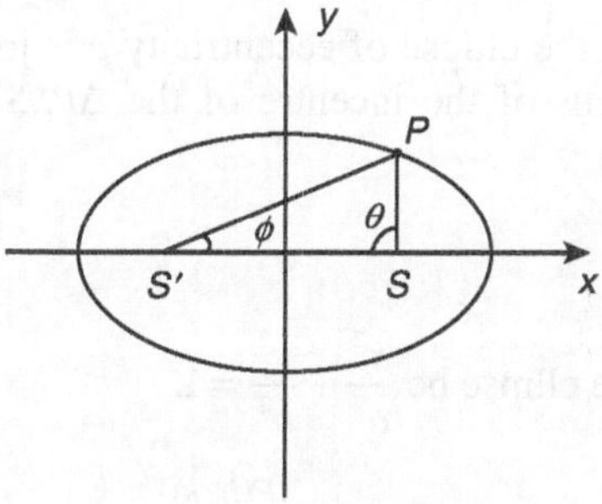

In any ΔABC, we know that $\tan\frac{A}{2}=\sqrt{\frac{(s-b)(s-c)}{s(s-a)}}$ where s is the semi perimeter of ΔABC.

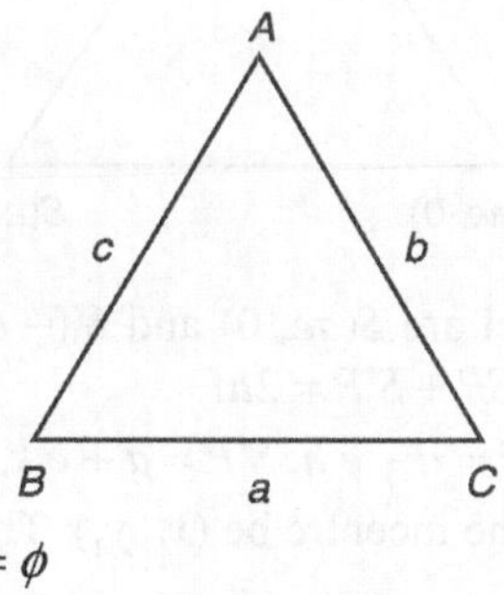

Let $\underline{|PSS'} = \theta$ and $\underline{|PS'S} = \phi$

Then
$$s = \frac{SP+S'P+SS'}{2}$$
$$= \frac{2a+ae}{2} = a+ae$$

$$\tan\frac{\theta}{2}=\sqrt{\frac{(a+ae-SS')(a+ae-SP)}{(a+ae)(a+ae-SP')}}$$

$$\tan\frac{\phi}{2}=\sqrt{\frac{(a+ae-SS')(a+ae-S'P)}{(a+ae)(a+ae-SP)}}$$

$$\therefore \tan\frac{\theta}{2}\cdot\tan\frac{\phi}{2}=\frac{a+ae-2ae}{a+ae}=\frac{a(1-e)}{a(1+e)}=\frac{1-e}{1+e}$$

Hence, $\tan\left(\frac{\angle PSS'}{2}\right)\cdot\tan\left(\frac{\angle PS'S}{2}\right)=\frac{1-e}{1+e}$

Example 7.29

A variable point P on the ellipse of eccentricity e is joined to its foci S and S'. Prove that the locus of the incentre of the $\Delta PSS'$ is an ellipse whose eccentricity is $\sqrt{\frac{2e}{1+e}}$.

Solution

Let the equation of the ellipse be $\frac{x^2}{a^2}+\frac{y^2}{b^2}=1$.

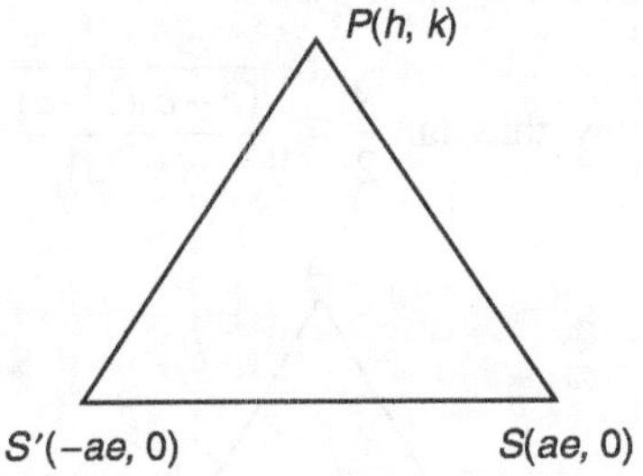

The coordinates of the foci are $S(ae, 0)$ and $S'(-ae, 0)$. Let $P(h, k)$ be any point on the ellipse. Then $SP + S'P = 2a$.

Also $SS' = 2ae$. Also $SP = a - e\,h$, $S'P = a + e\,k$.

Let the coordinates of the incentre be (x_1, y_1). Then

$$x_1=\frac{2ae.h+(a-eh)(-ae)+(a+eh)ae}{2ae+(a-eh)+(a+eh)}=\frac{2aeh+2ae^2h}{2a(1+e)}$$

$$x_1=2\frac{aeh\,\cancel{(1+e)}}{2a\,\cancel{(1+e)}}=eh \qquad (7.61)$$

$$y_1 = \frac{2aek + (a - eh) + (a + eh)}{2a(1+e)} = \frac{ek}{1+e} \quad (7.62)$$

Since (h, k) lies on the ellipse $\frac{x^2}{a^2} + \frac{y^2}{b^2} = 1 \Rightarrow \frac{h^2}{a^2} + \frac{k^2}{b^2} = 1$

$$\text{(i.e.)} \quad \frac{x_1^2}{a^2e^2} + \frac{y_1^2}{\left(\frac{be}{1+e}\right)^2} = 1.$$

The locus of $P(x_1, y_1)$ is $\frac{x^2}{a^2e^2} + \frac{y^2}{\left(\frac{be}{1+e}\right)^2} = 1$ which is an ellipse whose eccentricity e_1 is given by,

$$\left(\frac{be}{1+e}\right)^2 = a^2e^2(1 - e_1^2)$$

$$\text{or} \quad \frac{b^2}{(1+e)^2} = a^2(1 - e_1^2)$$

$$\text{or} \quad (1 - e_1^2) = \frac{b^2}{a^2(1+e)^2}$$

$$e_1^2 = \frac{2e}{1+e}$$

$$e_1 = \sqrt{\frac{2e}{1+e}}$$

Therefore, the locus of the incentre of the $\Delta PSS'$ is an ellipse whose eccentricity e_1 is $\sqrt{\frac{2e}{1+e}}$.

Exercises

1. **Find the equation of the tangent to the ellipse which makes equal intercepts on the axes.**

   Ans.: $x + y = \sqrt{a^2 + b^2}$

2. **Find the length of latus rectum, eccentricity, equation of the directrix and foci of the ellipse $25x^2 + 16y^2 = 400$.**

   Ans.: $\frac{32}{5}, \frac{3}{5}, (0, \pm 3), y = \pm\frac{25}{3}$

3. The equation to the ellipse is $2x^2 + y^2 - 8x - 2y + 1 = 0$. Find the length of its semi axes coordinates of the foci, length of latus rectum and equation of the directrix.

Ans.: $2, 2\sqrt{3}, (2,-1), (2,3), 2\sqrt{2}, x-2=0, y-1=0$

4. Prove that $x + y = \sqrt{a^2 + b^2}$ touches the ellipse $\frac{x^2}{a^2} + \frac{y^2}{b^2} = 1$ and find the coordinates of the point of contact.

Ans.: $\left(\frac{a^2}{\sqrt{a^2+b^2}}, \frac{b^2}{\sqrt{a^2+b^2}}\right)$

5. If p be the length of the perpendicular from the focus S of the ellipse $\frac{x^2}{a^2} + \frac{y^2}{b^2} = 1$ on the tangents at P then show that $\frac{b^2}{p^2} = \frac{2a}{SP} - 1$.

6. If ST be the perpendicular from the focus S on the tangent at any point P on the ellipse $\frac{x^2}{a^2} + \frac{y^2}{b^2} = 1$ then show that T lies on the auxiliary circle of the ellipse.

7. The line $x \cos \alpha + y \sin \alpha = p$ intercepted by the ellipse $\frac{x^2}{a^2} + \frac{y^2}{b^2} = 1$ subtends a right angle at its centre prove that the value of p is $\frac{ab}{\sqrt{a^2 + b^2}}$.

8. If the chord of contact of the tangents drawn from the point (α, β) to the ellipse $\frac{x^2}{a^2} + \frac{x^2}{b^2} = 1$ touches the circle $x^2 + y^2 = c^2$ prove that the point (α, β) lies on the ellipse $\frac{x^2}{a^2} + \frac{x^2}{b^2} = \frac{1}{c^2}$.

9. P is a point on the ellipse $\frac{x^2}{a^2} + \frac{x^2}{b^2} = 1$ and Q, the corresponding point on the auxiliary circle. If the tangent at P to the ellipse cuts the minor axis in T, then prove that the line QT touches the auxiliary circle.

10. Tangents to the ellipse $\frac{x^2}{a^2} + \frac{x^2}{b^2} = 1$ make angles θ_1 and θ_2 with the major axis. Find the equation of the locus of their intersection when $\tan(\theta_1 + \theta_2)$ is a constant.

11. Show that the locus of the point of intersection of two perpendicular tangents to an ellipse is a circle.

12. Prove that a chord of an ellipse is divided harmonically by any point on it and its pole with respect to the ellipse.

13. If the polar of P with respect to an ellipse passes through the point Q, show that polar of Q passes through P.

14. Find the condition for the pole of the straight line $lx + my = 1$ with respect to the ellipse $\frac{x^2}{a^2} + \frac{y^2}{b^2} = 1$ may lie on the ellipse $\frac{x^2}{a^2} + \frac{y^2}{b^2} = 4$.

Ans.: $a^2l^2 + b^2m^2 = 4$

15. Chords of the ellipse $\frac{x^2}{a^2}+\frac{y^2}{b^2}=1$ touch the circle $x^2+y^2=r^2$. Find the locus of their poles.

16. Chords of the ellipse $\frac{x^2}{a^2}+\frac{y^2}{b^2}=1$ always touch the ellipse $\frac{x^2}{a^2}+\frac{y^2}{b^2}=1$. Show that the locus of the poles is $\frac{b^2x^2}{a^4}+\frac{a^2y^2}{b^4}=1$.

17. Prove that the perpendicular from the focus of an ellipse whose centre is C on any polar of P will meet CP on the directrix.

18. Show that the focus of an ellipse is the pole of the corresponding directrix.

19. A tangent to the ellipse $\frac{x^2}{a^2}+\frac{y^2}{b^2}=1$ meets the ellipse $\frac{x^2}{a}+\frac{y^2}{b}=a+b$ at Q and R. Show that the locus of the pole of QR with respect to the latter is $x^2+y^2=a^2+b^2$.

20. If the midpoint of a chord lies on a fixed line $lx+my+n=0$, show that the locus of pole of the chord is the ellipse $n\left(\frac{x^2}{a^2}+\frac{y^2}{b^2}-1\right)+lx+my+n=0$.

21. Find the locus of the poles of chords of the ellipse $\frac{x^2}{a^2}+\frac{y^2}{b^2}=1$ which touch the parabola $ay^2=-2b^2x$.

22. The perpendicular from the centre of the ellipse $\frac{x^2}{a^2}+\frac{y^2}{b^2}=1$ on the polar of a point with respect to the ellipse is equal to c. Prove that the locus of the point is the ellipse, $\frac{x^2}{a^4}+\frac{y^2}{b^4}=\frac{1}{c^2}$.

23. Show that the locus of the poles with respect to an ellipse of a straight line which touches the circle described on the minor axis of the ellipse as diameter.

24. Show that the locus of poles of tangents to the ellipse $\frac{x^2}{a^2}+\frac{y^2}{b^2}=1$ with respect to $x^2+y^2=ab$ is an equal ellipse.

25. Prove that the tangents at the extremities of latus rectum of an ellipse intersect on the corresponding directrix.

26. Find the coordinates of all points of intersection of the ellipse $\frac{x^2}{4}+\frac{y^2}{9}=1$ and the circle $x^2+y^2=6$. Write down the equation of the tangents to the ellipse and circle at the point of intersection and find the angle between them.

Ans.: $\left(\pm\sqrt{\frac{12}{5}},\pm\sqrt{\frac{18}{5}}\right), \frac{1}{2}\sqrt{\frac{3}{5}}x+\frac{1}{3}\sqrt{\frac{2}{5}}y=1$

$2\sqrt{\frac{3}{5}}x+3\sqrt{\frac{3}{5}}y=6; \tan^{-1}\left(\frac{1}{\sqrt{6}}\right)$

27. Tangents are drawn from any point on the ellipse $\frac{x^2}{a^2}+\frac{y^2}{b^2}=1$ to the circle $x^2+y^2=a^2$. Prove that their chord of contact touches the ellipse $a^2x^2+b^2y^2=r^4$.

28. Prove that the angle between the tangents to the ellipse $\frac{x^2}{a^2}+\frac{y^2}{b^2}=1$ and the circle $x^2+y^2=ab$ at their point of intersection is $\tan^{-1}\left(\frac{a-b}{\sqrt{ab}}\right)$.

29. Prove that the sum of the reciprocals of the squares of any two diameters of an ellipse which are at right angles to one another is a constant.

30. An ellipse slides between two straight lines at right angles to each other. Show that the locus of its centre is a circle.

31. Show that the locus of the foot of perpendiculars drawn from the centre of the ellipse $\frac{x^2}{a^2}+\frac{y^2}{b^2}=1$ on any tangent to it is $(x^2+y^2)^2=a^2x^2+b^2y^2$.

32. Two tangents to an ellipse interest at right angles. Prove that the sum of the squares of the chords which the auxiliary circle intercepts on them is constant and equal to the square of the line joining the foci.

33. Show that the conjugate lines through a focus of an ellipse are at right angles.

34. An archway is in the form of a semi-ellipse, the major axis of which coincides with the road level. If the breadth of the road is 34 feet and a man who is 6 feet high, just reaches the top when 2 feet from a side of the road, find the greatest height of the arch.

35. If the pole of the normal to an ellipse at P lies on the normal at Q then show that the pole of the normal at Q lies on the normal at P.

36. PQ, PR is a pair of perpendicular tangents to the ellipse $\frac{x^2}{a^2}+\frac{y^2}{b^2}=1$. Prove that QR always touches the ellipse $\frac{x^2}{a^4}+\frac{y^2}{b^4}=\frac{1}{a^2+b^2}$.

37. Show that the points (x_r, y_r), $r = 1, 2, 3$ are collinear if their polars with respect to the ellipse $\frac{x^2}{a^2}+\frac{y^2}{b^2}=1$ are concurrent.

38. If l_1 and l_2 be the length of two tangents to the ellipse $\frac{x^2}{a^2}+\frac{y^2}{b^2}=1$ at right angles to one another, prove that $4(a^2+b^2)^3=(l_1^2+l_2^2)\left[a^2+b^2+a^2b^2\left(\frac{1}{l_1^2}+\frac{1}{l_2^2}\right)\right]^2$.

39. If RP and RQ are tangents from an external point $R(x_1, y_1)$ to the ellipse $\frac{x^2}{a^2}+\frac{y^2}{b^2}=1$ and S be the focus then show that $\frac{RS^2}{SP\cdot SQ}=\frac{x_1^2}{a^2}+\frac{y_1^2}{b^2}$.

7.13 ECCENTRIC ANGLE

Let P be a point on the ellipse and P' be the corresponding point on the auxiliary circle. The angle CP' makes with the positive direction of x-axis is called the eccentric angle of the point P on the ellipse. If this angle is denoted by θ, then $CN = a\cos\theta$ and $NP' = a\sin\theta$.

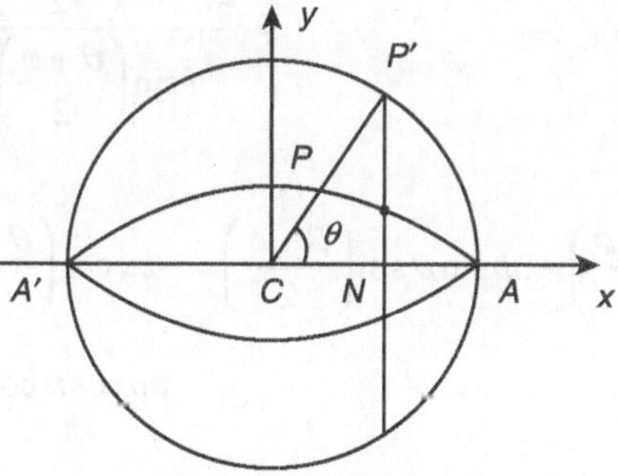

We know that

$$\frac{NP}{NP'} = \frac{b}{a}$$

or

$$NP = \frac{b}{a} NP'$$

$$= \frac{b}{a} \times a\sin\theta$$

$$= b\sin\theta$$

Then the coordinates of any point P are (CN, NP).

(i.e.) $(a\cos\theta, b\sin\theta)$

$\therefore$ 'θ' is called the eccentric angle and it is also called the parameter of the point P.

7.14 EQUATION OF THE CHORD JOINING THE POINTS

To find the equation of the chord joining the points whose eccentric angles are 'θ' and 'ϕ':

The two given points are $(a\cos\theta, b\sin\theta)$ and $(a\cos\phi, b\sin\phi)$. The equation of the chord joining the two points is $\frac{y - y_1}{x - x_1} = \frac{y_1 - y_2}{x_1 - x_2}$.

$$\text{(i.e.)}\quad \frac{y-b\sin\theta}{x-a\cos\theta}=\frac{b(\sin\theta-\sin\phi)}{a(\cos\theta-\cos\phi)}=\frac{b}{a}\left(\frac{2\cos\frac{\theta+\phi}{2}\sin\frac{\theta-\phi}{2}}{-2\sin\frac{\theta+\phi}{2}\sin\frac{\theta-\phi}{2}}\right)$$

$$=\frac{-b}{a}\frac{\cos\left(\frac{\theta+\phi}{2}\right)}{\sin\left(\frac{\theta+\phi}{2}\right)}$$

$$\text{(i.e.)}\quad ay\sin\left(\frac{\theta+\phi}{2}\right)-ab\sin\theta\sin\left(\frac{\theta+\phi}{2}\right)=-bx\cos\left(\frac{\theta+\phi}{2}\right)+ ab\cos\theta\cos\left(\frac{\theta+\phi}{2}\right)$$

$$\text{(i.e.)}\quad bx\cos\left(\frac{\theta+\phi}{2}\right)+ay\sin\left(\frac{\theta+\phi}{2}\right)=ab\left(\cos\theta\cos\frac{\theta+\phi}{2}+\sin\theta\sin\frac{\theta+\phi}{2}\right)$$

$$=ab\cos\left(\theta-\left(\frac{\theta+\phi}{2}\right)\right)$$

$$=ab\cos\frac{\theta+\phi}{2}$$

Dividing by ab,

$$\frac{x}{a}\cos\left(\frac{\theta+\phi}{2}\right)+\frac{y}{b}\sin\left(\frac{\theta+\phi}{2}\right)=\cos\frac{\theta+\phi}{2}$$

Therefore, the equation of the chord joining the points whose eccentric angles are 'θ' and 'ϕ' is $\frac{x}{a}\cos\left(\frac{\theta+\phi}{2}\right)+\frac{y}{b}\sin\left(\frac{\theta+\phi}{2}\right)=\cos\frac{\theta+\phi}{2}$.

Note 7.14.1: This chord becomes the tangent at 'θ' if $\phi=\theta$

$\therefore$ The equation of the tangent at 'θ' is $\frac{x}{a}\cos\theta+\frac{y}{b}\sin\theta=1$.

7.15 EQUATION OF TANGENT AT 'θ' ON THE ELLIPSE $\frac{x^2}{a^2}+\frac{y^2}{b^2}=1$

The equation of the ellipse is $\frac{x^2}{a^2}+\frac{y^2}{b^2}=1$.

Differentiating with respect to x, we get,

$$\frac{2x}{a^2}+\frac{2y}{b^2}\frac{dy}{dx}=0$$

$$\frac{dy}{dx}=-\frac{b^2x}{a^2y}$$

$$\frac{dy}{dx} \text{ at } (a\cos\theta, b\sin\theta)=\frac{-b^2a\cos\theta}{a^2b\sin\theta}=\frac{-b\cos\theta}{a\sin\theta}$$

The equation of the tangent at 'θ' is,

$$y-b\sin\theta=\frac{-b\cos\theta}{a\sin\theta}(x-a\cos\theta)$$

$$ay\sin\theta+bx\cos\theta=ab(\sin^2\theta+\cos^2\theta)=ab$$

Dividing by ab,

$$\frac{x}{a}\cos\theta+\frac{y}{b}\sin\theta=1$$

The slope of the normal at θ is $\frac{a\sin\theta}{b\cos\theta}$.

Therefore, the equation of the normal at θ is:

$$y-b\sin\theta=\frac{a\sin\theta}{b\cos\theta}(x-a\cos\theta)$$

$$by\cos\theta-b^2\sin\theta\cos\theta=ax\sin\theta-a^2\sin\theta\cos\theta$$

$$ax\sin\theta-by\cos\theta=(a^2-b^2)\sin\theta\cos\theta$$

Dividing by sin θ cos θ, we get,

$$\frac{ax}{\cos\theta} - \frac{by}{\sin\theta} = a^2 - b^2$$

Therefore, equation of normal at 'θ' on the ellipse $\frac{x^2}{a^2} + \frac{y^2}{b^2} = 1$ is $\frac{ax}{\cos\theta} - \frac{bk}{\sin\theta} = a^2 - b^2$.

7.16 CONORMAL POINTS

In general, four normals can be drawn from a given point to an ellipse. If α, β, γ, and δ be the eccentric angles of these four conormal points then $\alpha + \beta + \gamma + \delta$ is an odd multiple of π.

Let (h, k) be a given point. Let $P(a\cos\theta, b\sin\theta)$ be any point on the ellipse $\frac{x^2}{a^2} + \frac{y^2}{b^2} = 1$.

The equation of the normal at θ is $\frac{ax}{\cos\theta} - \frac{bk}{\sin\theta} = a^2 - b^2$.

If the normal passes through (h, k) then

$$\frac{ah}{\cos\theta} - \frac{bk}{\sin\theta} = a^2 - b^2$$

$$\text{(i.e.)} \quad ah\left(\frac{1+t^2}{1-t^2}\right) - bk\left(\frac{1+t^2}{2t}\right) = a^2 - b^2 \quad \text{where } t = \tan\frac{\theta}{2}$$

$$\text{(i.e.)} \quad ah(1+t^2)2t - bk(1-t^4) - (a^2 - b^2)2t(1-t^2) = 0$$

$$\text{(i.e.)} \quad bkt^4 + 2[ah + (a^2 - b^2)]\, t^3 + 2[ah - (a^2 - b^2)]\, t - bk = 0 \tag{7.63}$$

This is a fourth degree equation in t and hence there are four values for t. For each value of t, there is a value of θ and hence there are four values of θ say α, β, γ and δ. Hence, there are four normals from a given point to an ellipse.

Hence, $\tan\frac{\alpha}{2}, \tan\frac{\beta}{2}, \tan\frac{\gamma}{2}$ and $\tan\frac{\delta}{2}$ are the roots of the equation (7.63).

$$\sum \tan\frac{\alpha}{2} = \frac{-2[ah+(a^2-b^2)]}{bk}; \sum \tan\frac{\alpha}{2}\tan\frac{\beta}{2} = 0$$

$$\sum \tan\frac{\alpha}{2}\tan\frac{\beta}{2}\tan\frac{\gamma}{2} = \frac{-2[ah-(a^2-b^2)]}{bk}$$

$$\tan\frac{\alpha}{2}\tan\frac{\beta}{2}\tan\frac{\gamma}{2}\tan\frac{\delta}{2} = -1 \tag{7.64}$$

We know that, $\tan\left(\frac{\alpha}{2}+\frac{\beta}{2}+\frac{\gamma}{2}+\frac{\delta}{2}\right) = \frac{S_1 - S_3}{1 - S_2 + S_4}$

$$= \frac{\frac{-2(ab+a^2-b^2)}{bk} + \frac{2(ab-a^2+b^2)}{bk}}{1-0-1}$$

$$= \infty$$

$\therefore \frac{\alpha+\beta+\gamma+\delta}{2}$ = an odd multiples of $\frac{\pi}{2}$.

$\therefore \alpha+\beta+\gamma+\delta$ = an odd multiple of π.

7.17 CONCYCLIC POINTS

A circle and an ellipse will cut four points and that the sum of the eccentric angles of the four points of intersection is an even multiple of π.

Let the equation of the ellipse be

$$\frac{x^2}{a^2} + \frac{y^2}{b^2} = 1 \tag{7.65}$$

Let the equation of the circle be

$$x^2 + y^2 + 2gx + 2fy + c = 0 \tag{7.66}$$

Any point on the ellipse is $(a\cos b, a\sin\theta)$. When the circle and the ellipse intersect, this point lies on the circle.

$$\therefore a^2\cos^2\theta + b^2\sin^2\theta + 2ga\cos\theta + 2fb\sin\theta + c = 0 \tag{7.67}$$

We know that, $\cos\theta = \dfrac{1-\tan^2\dfrac{\theta}{2}}{1+\tan^2\dfrac{\theta}{2}} = \dfrac{1-t^2}{1+t^2}$

$$\sin\theta = \frac{2\tan\frac{\theta}{2}}{1+\tan^2\frac{\theta}{2}} = \frac{2t}{1+t^2} \text{ where } t = \tan\frac{\theta}{2}$$

Substituting these values in equation (7.67), we get

$$a^2\left(\frac{1-t^2}{1+t^2}\right)^2 + b^2\left(\frac{2t}{1+t^2}\right)^2 + 2ga\left(\frac{1-t^2}{1+t^2}\right) + 2fb\left(\frac{2t}{1+t^2}\right) + c = 0.$$

(i.e.) $(a^2 - 2ga + c)t^4 + 4bft^3 + (4b^2 - 2a^2 + 2c)t^2 + 4bft$
$+ (a^2 + 2ga + c) = 0$ (7.68)

where $\sum t_1 = \dfrac{-4bf}{a^2 - 2ga + c}; \sum t_1t_2 = \dfrac{4b^2 - 2a^2 + 2c}{a^2 - 2ga + c}$

$\sum t_1t_2t_3 = \dfrac{-4bf}{a^2 - 2ga + c}; t_1t_2t_3t_4 = \dfrac{a^2 + 2ga + c}{a^2 - 2ga + c}.$

Equation (7.68) is a fourth degree equation in t and hence there are four values for t, real or imaginary.

For each value of t there corresponds a value of θ. Hence in general there are four points of intersection of a circle and an ellipse with eccentric angles θ_1, θ_2, θ_3, and θ_4.

We know that,

$$\tan\left(\frac{\theta_1 + \theta_2 + \theta_3 + \theta_4}{2}\right) = \frac{S_1 - S_3}{1 - S_2 + S_4} \text{ with usual notation.}$$

$$= \frac{\Sigma t_1 - \Sigma t_1t_2t_3}{1 - \Sigma t_1t_2 + t_1t_2t_3t_4} = 0$$

$$\therefore \frac{\theta_1 + \theta_2 + \theta_3 + \theta_4}{2} = n\pi$$

$$\Rightarrow \theta_1 + \theta_2 + \theta_3 + \theta_4 = 2n\pi$$

$$= \text{an even multiple of } \pi.$$

7.18 EQUATION OF A CHORD IN TERMS OF ITS MIDDLE POINT

To find the equation of a chord in term of its middle point:

Let the equation of the ellipse be

$$\frac{x^2}{a^2}+\frac{y^2}{b^2}=1 \tag{7.69}$$

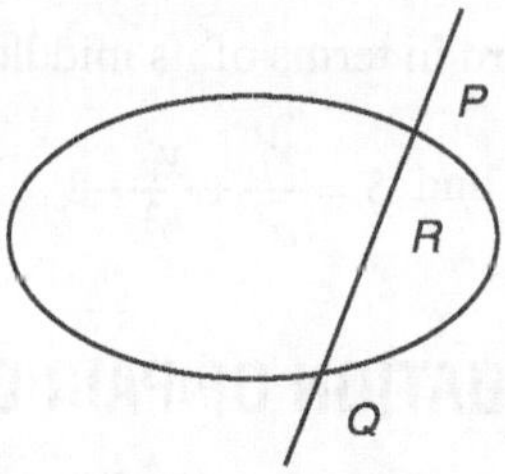

Let $R(x_1, y_1)$ be the midpoint of a chord PQ of this ellipse. Let the equation of chord PQ be

$$\frac{x-x_1}{\cos\theta}=\frac{y-y_1}{\sin\theta}=r \tag{7.70}$$

Any point on this line is $(x_1+r\cos\theta, y_1+r\sin\theta)$. When the chord meets the ellipse this point lies on the ellipse (7.69).

$$\therefore \frac{(x_1+r\cos\theta)^2}{a^2}+\frac{(y_1+r\sin\theta)^2}{b^2}=1$$

$$\text{(i.e.) } r^2\left(\frac{\cos^2\theta}{a^2}+\frac{\sin^2\theta}{b^2}\right)+2r\left(\frac{x_1\cos\theta}{a^2}+\frac{y_1\sin\theta}{b^2}\right)+\frac{x_1^2}{a^2}+\frac{y_1^2}{b^2}-1=0 \tag{7.71}$$

If $R(x_1, y_1)$ is the midpoint of the chord PQ then the two values of r are the distances PR and RQ which are equal in magnitude but opposite in sign.

The condition for this is the coefficient of $r = 0$.

$$\therefore \frac{x_1}{a^2}\cos\theta+\frac{y_1}{b^2}\sin\theta=0 \tag{7.72}$$

Substituting $\cos\theta = \dfrac{x - x_1}{r}$ and $\sin\theta = \dfrac{y - y_1}{r}$ in equation (7.72), we get

$$\frac{x_1}{a^2}\left(\frac{x - x_1}{r}\right) + \frac{y_1}{b^2}\left(\frac{y - y_1}{r}\right) = 0$$

$$\text{(i.e.)} \quad \frac{xx_1}{a^2} + \frac{yy_1}{b^2} = \frac{x_1^2}{a^2} + \frac{y_1^2}{b^2}$$

$$\text{or} \quad \frac{xx_1}{a^2} + \frac{yy_1}{b^2} - 1 = \frac{x_1^2}{a^2} + \frac{y_1^2}{b^2} - 1$$

Hence, the equation of chord in terms of its middle point is $T = S_1$

where $\quad T = \dfrac{xx_1}{a^2} + \dfrac{yy_1}{b^2} - 1$ and $S = \dfrac{x_1^2}{a^2} + \dfrac{y_1^2}{b^2} - 1$.

7.19 COMBINED EQUATION OF PAIR OF TANGENTS

To find the combined equation of pair of tangents from (x_1, y_1) to the ellipse $\dfrac{x^2}{a^2} + \dfrac{y^2}{b^2} = 1$:

Let the equation of the chord through (x_1, y_1) be

$$\frac{x - x_1}{\cos\theta} = \frac{y - y_1}{\sin\theta} = r \tag{7.70a}$$

Any point on this line is $(x_1 + r\cos\theta, y_1 + r\sin\theta)$

If this point lies on the ellipse $\dfrac{x^2}{a^2} + \dfrac{y^2}{b^2} = 1$,

$$\frac{(x_1 + r\cos\theta)^2}{a^2} + \frac{(y_1 + r\sin\theta)^2}{b^2} = 1$$

$$\text{(i.e.)} \quad r^2\left(\frac{\cos^2\theta}{a^2} + \frac{\sin^2\theta}{b^2}\right) + 2r\left(\frac{x_1\cos\theta}{a^2} + \frac{y_1\sin\theta}{b^2}\right) + \frac{x_1^2}{a^2} + \frac{y_1^2}{b^2} - 1 = 0$$

The two values of r are the distances of the point of intersection of the chord and the ellipse from (x_1, y_1). The line will become a tangent if the two values of r are equal. The condition for this is the discriminant of the quadratic equation is zero.

$$\therefore \left(\frac{x_1}{a^2}\cos\theta + \frac{y_1}{b^2}\sin\theta\right)^2 = \left(\frac{x_1^2}{a^2} + \frac{y_1^2}{b^2} - 1\right)\left(\frac{\cos^2\theta}{a^2} + \frac{\sin^2\theta}{b^2}\right)$$

Using the values of $\cos\theta$ and $\sin\theta$ from equation (7.70a),

$$\left[\frac{x_1(x-x_1)}{a^2} + \frac{y_1(y-y_1)}{b^2}\right]^2 = \left(\frac{x_1^2}{a^2} + \frac{y_1^2}{b^2} - 1\right)\left[\frac{(x-x_1)^2}{a^2} + \frac{(y-y_1)^2}{b^2}\right]$$

(i.e.)

$$\left[\frac{xx_1}{a^2} + \frac{yy_1}{b^2} - \left(\frac{x_1^2}{a^2} + \frac{y_1^2}{b^2}\right)\right] = \left(\frac{x_1^2}{a^2} + \frac{y_1^2}{b^2} - 1\right)\left[\frac{x^2}{a^2} + \frac{y^2}{b^2} - 2\left(\frac{xx_1}{a^2} + \frac{yy_1}{b^2}\right) + \frac{x_1^2}{a^2} + \frac{y_1^2}{b^2}\right]$$

(i.e.) $$\left[\left(\frac{xx_1}{a^2} + \frac{yy_1}{b^2} - 1\right) - \left(\frac{x_1^2}{a^2} + \frac{y_1^2}{b^2} - 1\right)\right]^2 = \left[\frac{x_1^2}{a^2} + \frac{y_1^2}{b^2} - 1\right]\left[\left(\frac{x^2}{a^2} + \frac{y^2}{b^2} - 1\right)\right.$$

$$\left. -2\left(\frac{xx_1}{a^2} + \frac{yy_1}{b^2} - 1\right) + \left(\frac{x_1^2}{a^2} + \frac{y_1^2}{b^2} - 1\right)\right]$$

(i.e.) $(T - S_1)^2 = S_1(S - 2T + S_1)$

$$T^2 - 2TS_1 + S_1^2 = SS_1 - 2TS_1 + S_1^2$$

(i.e.) $T^2 = SS_1$ where $T = \dfrac{xx_1}{a^2} + \dfrac{yy_1}{b^2} - 1$

$$S = \frac{x^2}{a^2} + \frac{y^2}{b^2} - 1$$

$$S_1 = \frac{x_1^2}{a^2} + \frac{y_1^2}{b^2} - 1$$

This is the combined equation of the pair of tangents from (x_1, y_1).

Note 7.19.1: The combined equation of the pair of tangents from the point (x_1, y_1) is

$$\left(\frac{xx_1}{a^2} + \frac{yy_1}{b^2} - 1\right)^2 = \left(\frac{x^2}{a^2} + \frac{y^2}{b^2} - 1\right)\left(\frac{x_1^2}{a^2} + \frac{y_1^2}{b^2} - 1\right).$$

If the two tangents are perpendicular then coefficient of x^2 + coefficient of $y^2 = 0$.

$$\left(\begin{array}{l}\dfrac{x_1^2}{a^4}-\dfrac{1}{a^2}\left(\dfrac{x_1^2}{a^2}+\dfrac{y_1^2}{b^2}-1\right)+\dfrac{y_1^2}{b^4}-\dfrac{1}{b^2}\left(\dfrac{x_1^2}{a^2}+\dfrac{y_1^2}{b^2}-1\right)\\ \text{(i.e.)}\ \dfrac{-y_1^2}{a^2b^2}+\dfrac{1}{a^2}-\dfrac{x_1^2}{a_1^2b^2}+\dfrac{1}{b^2}=0\\ \text{(i.e.)}\ \dfrac{x_1^2+y_1^2}{a^2b^2}=\dfrac{1}{a^2}+\dfrac{1}{b^2}\Rightarrow x_1^2+y_1^2=a^2+b^2\end{array}\right)$$

The locus of (x_1, y_1) is $x^2 + y^2 = a^2 + b^2$.
Therefore, the locus of the point of intersection of perpendicular tangents is a circle. This equation is called the directrix of the circle.

7.20 CONJUGATE DIAMETERS

Example 7.30

Find the condition that the line $lx + my + n = 0$ may be a tangent to the ellipse $\dfrac{x^2}{a^2}+\dfrac{y^2}{b^2}=1$.

Solution

Let $lx + my + n = 0$ be a tangent to the ellipse $\dfrac{x^2}{a^2}+\dfrac{y^2}{b^2}=1$.

Let the line be tangent at 'θ'. The equation of the tangent at θ is

$$\frac{x}{a}\cos\theta+\frac{y}{b}\sin\theta=1 \tag{7.71a}$$

However, the equation of tangent is given as

$$lx + my = -n \tag{7.72a}$$

Identifying equations (7.71a) and (7.72a), we get

$$\frac{l}{\left(\dfrac{\cos\theta}{a}\right)}=\frac{m}{\left(\dfrac{\sin\theta}{b}\right)}=\frac{1}{-n}$$

$$\therefore\ \cos\theta=\frac{-al}{n},\quad \sin\theta=\frac{-bm}{n}.$$

Squaring and adding, we get

$$\cos^2\theta+\sin^2\theta=\frac{a^2l^2}{n^2}+\frac{b^2m^2}{n^2}\Rightarrow 1=\frac{a^2l^2}{n^2}+\frac{b^2m^2}{n^2}$$

(i.e.) $a^2l^2+b^2m^2=n^2$.

This is the required condition.

Example 7.31

Find the condition for the line $lx + my + n = 0$ to be a normal to the ellipse $\frac{x^2}{a^2}+\frac{y^2}{b^2}=1$.

Solution

The equation of the ellipse is $\frac{x^2}{a^2}+\frac{y^2}{b^2}=1$.

The equation of normal is

$$lx+my=-n \tag{7.73}$$

Let this equation be normal at 'θ'. The equation of the normal at 'θ' is

$$\frac{ax}{\cos\theta}-\frac{by}{\sin\theta}=a^2-b^2 \tag{7.74}$$

The equations (7.73) and (7.74) represent the same line.

Therefore, identifying equations (7.73) and (7.74), we get

$$\frac{l}{\left(\frac{a}{\cos\theta}\right)}=\frac{m}{\left(\frac{b}{\sin\theta}\right)}=\frac{-n}{a^2-b^2}$$

$$\therefore\quad \cos\theta=\frac{-an}{l(a^2-b^2)},\quad \sin\theta=\frac{-bn}{m(a^2-b^2)}$$

Squaring and adding equations (7.78) and (7.79), we get

$$\frac{a^2n^2}{l^2(a^2-b^2)^2}+\frac{b^2n^2}{m^2(a^2-b^2)^2}=1$$

$$\therefore\frac{a^2}{l^2}+\frac{b^2}{m^2}=\frac{(a^2-b^2)^2}{n^2}$$

This is required condition.

Example 7.32

Show that the locus of the point of intersection of tangents to an ellipse at the points whose eccentric angles differ by a constant is an ellipse.

Solution

Let the eccentric angles of P and Q be $\alpha+\beta$ and $\alpha-\beta$.
$\therefore (\alpha+\beta)-(\alpha-\beta)=2\beta=2k$; a constant $\therefore \beta=k$.

The equation of tangents at P and Q are

$$\frac{x}{a}\cos(\alpha+\beta)+\frac{y}{b}\sin(\alpha+\beta)=1 \tag{7.75}$$

$$\frac{x}{a}\cos(\alpha-\beta)+\frac{y}{b}\sin(\alpha-\beta)=1 \tag{7.76}$$

Let (x_1, y_1) be their point of intersection. Then

$$\frac{x_1}{a}\cos(\alpha+\beta)+\frac{y_1}{b}\sin(\alpha+\beta)=1 \tag{7.77}$$

$$\frac{x_1}{a}\cos(\alpha-\beta)+\frac{y_1}{b}\sin(\alpha-\beta)=-1 \tag{7.78}$$

Solving equations (7.77) and (7.78), we get

$$\frac{x_1/a}{-[\sin(\alpha-\beta)-\sin(\alpha-\beta)]}=\frac{y_1/b}{\cos(\alpha+\beta)-\cos(\alpha-\beta)}$$

$$=\frac{1}{\cos(\alpha+\beta)\sin(\alpha-\beta)-\sin(\alpha+\beta)\cos(\alpha-\beta)}$$

$$\Rightarrow \frac{x_1}{-2a\cos\alpha\sin\beta}=\frac{y_1}{-2b\sin\alpha\sin\beta}=\frac{1}{-\sin 2\beta}=\frac{-1}{2\sin\beta\cos\beta}$$

$$\therefore x_1=\frac{a\cos\alpha}{\cos\beta},\quad y_1=\frac{b\sin\alpha}{\cos\beta}$$

$$\therefore \cos\alpha=\frac{x_1\cos\beta}{a},\quad \sin\alpha=\frac{y_1\cos\beta}{b}$$

Squaring and adding, we get

$$1=\frac{x_1^2\cos^2\beta}{a^2}+\frac{y_1^2\sin^2\beta}{b^2}$$

Since $\beta = k$, the locus of (x_1, y_1) is $\frac{x\cos^2 k}{a^2}+\frac{y\sin^2 k}{b^2}=1$ which is an ellipse.

Example 7.33

Show that the locus of poles of normal chords of the ellipse $\frac{x^2}{a^2}+\frac{y^2}{a^2}=1$ is $\frac{a^6}{x^2}+\frac{b^6}{y^2}=(a^2-b^2)^2$.

Solution

Let (x_1, y_1) be the pole of the normal chord of the ellipse

$$\frac{x^2}{a^2}+\frac{y^2}{b^2}=1 \tag{7.79}$$

Then the polar of (x_1, y_1) with respect to ellipse is

$$\frac{xx_1}{a^2}+\frac{yy_1}{b^2}=1 \tag{7.80}$$

Let this be normal at 'θ' on the ellipse of equation (7.84). Then the equation of the normal at 'θ' is

$$\frac{ax}{\cos\theta}-\frac{by}{\sin\theta}=a^2-b^2 \tag{7.81}$$

Equations (7.80) and (7.81) represent the same line.

Therefore, identifying equations (7.80) and (7.81), we get

$$\frac{\frac{x}{a^2}}{\frac{a}{\cos\theta}}=\frac{\frac{y}{b^2}}{\frac{-b}{\sin\theta}}=\frac{1}{a^2-b^2}$$

$$\text{(i.e.)}\quad \frac{x_1\cos\theta}{a^3}=\frac{-y_1\sin\theta}{b^3}=\frac{1}{a^2-b^2}$$

$$\therefore \cos\theta=\frac{a^3}{x_1(a^2-b^2)}$$

$$\sin\theta=\frac{-b^3}{y_1(a^2-b^2)}$$

Squaring and adding we get,

$$1=\frac{a^6}{x_1^2(a^2-b^2)^3}+\frac{b^6}{y_1^2(a^2-b^2)^2}$$

Therefore, the locus of (x_1, y_1) is $\frac{a^6}{x^2}+\frac{b^6}{y^2}=(a^2-b^2)^2$.

Example 7.34

Find the locus of midpoints of the normal chords of the ellipse $\frac{x^2}{a^2}+\frac{y^2}{b^2}=1.$

Solution

Let (x_1, y_1) be the midpoint of a chord of the ellipse which is normal at θ. The equation of the chord in terms of its middle point is

$$\frac{xx_1}{a^2}+\frac{yy_1}{b^2}=\frac{x_1^2}{a^2}+\frac{y_1^2}{b^2} \tag{7.82}$$

The equation of the normal at 'θ' is

$$\frac{ax}{\cos\theta}-\frac{by}{\sin\theta}=a^2-b^2 \tag{7.83}$$

Equations (7.82) and (7.83) represent the same line.

Therefore, identifying equations (7.87) and (7.88) we get,

$$\frac{\frac{x_1}{a_2}}{\frac{a}{\cos\theta}}=\frac{\frac{y_1}{b_2}}{\frac{b}{\sin\theta}}=\frac{\frac{x_1^2}{a^2}+\frac{y_1^2}{b^2}}{a^2-b^2}$$

$$\frac{x_1\cos\theta}{a^3}=\frac{y_1\sin\theta}{b^3}=\frac{\frac{x_1}{a_2}+\frac{x_1^2}{b^2}}{a^2-b^2}$$

$$\cos\theta=\frac{a^3\left(\frac{x_1^2}{a_2}+\frac{y_1^2}{b^2}\right)}{x_1(a^2-b^2)},\ \sin\theta=\frac{b^3\left(\frac{x_1^2}{a_2}+\frac{y_1^2}{b^2}\right)}{y_1(a^2-b^2)}$$

Squaring and adding, we get

$$\frac{a^6}{x_1^2}\frac{\left(\frac{x_1^2}{a_2}+\frac{y_1^2}{b^2}\right)^2}{(a^2-b^2)^2}+\frac{b^3}{y_1^2}\frac{\left(\frac{x_1^2}{a_2}+\frac{y_1^2}{b^2}\right)^2}{(a^2-b^2)^2}=1$$

$$\text{(i.e)}\quad \frac{a^6}{x_1^2}+\frac{b^6}{y_1^2}=\frac{(a^2-b^2)^2}{\left(\frac{x_1^2}{a_2}+\frac{y_1^2}{b^2}\right)^2}$$

$$\left(\frac{a^6}{x_1^2}+\frac{b^6}{y_1^2}\right)\left(\frac{x_1^2}{a^2}+\frac{y_1^2}{b^2}\right)^2=(a^2-b^2)^2$$

Therefore, the locus of (x_1, y_1) is $\left(\frac{a^6}{x^2}+\frac{b^6}{y^2}\right)\left(\frac{x^2}{a^2}+\frac{y^2}{b^2}\right)^2 = (a^2-b^2)^2.$

Example 7.35

If the chord joining two points, whose eccentric angles are α and β on the ellipse $\frac{x^2}{a^2}+\frac{y^2}{b^2}=1$ cuts the major axis at a distance d from the centre, show that $\tan\frac{\alpha}{\beta}\cdot\tan\frac{\beta}{\alpha}=\frac{d-a}{d+a}$.

Solution

The equation of the chord joining the points whose eccentric angles are α and β is $\frac{x}{a}\cos\left(\frac{\alpha+\beta}{2}\right)+\frac{y}{b}\sin\frac{\alpha+\beta}{2}=\cos\frac{\alpha-\beta}{2}$.

This line meets the major axis at the point $(d, 0)$.

$$\therefore \quad \frac{d}{a}\cos\left(\frac{\alpha+\beta}{2}\right)=\cos\frac{(\alpha-\beta)}{2}$$

$$\therefore \quad \frac{d}{a}=\frac{\cos\left(\frac{\alpha-\beta}{2}\right)}{\cos\left(\frac{\alpha+\beta}{2}\right)}$$

$$\Rightarrow \frac{d-a}{d+a}=\frac{\cos\frac{\alpha-\beta}{2}-\cos\frac{\alpha+\beta}{2}}{\cos\frac{\alpha-\beta}{2}+\cos\frac{\alpha+\beta}{2}}$$

$$=\frac{2\sin\frac{\alpha}{2}\sin\frac{\beta}{2}}{2\cos\frac{\alpha}{2}\cos\frac{\beta}{2}}$$

$$=\tan\frac{\alpha}{2}\tan\frac{\beta}{2}$$

$$\therefore \tan\frac{\alpha}{2}\tan\frac{\beta}{2}=\frac{d-a}{d+a}$$

Example 7.36

The tangent at the point α on the ellipse meet auxiliary circle on two points which subtend a right angle at the centre. Show that the eccentricity of the ellipse is $(1+\sin^2\alpha)^{-1/2}$.

Solution

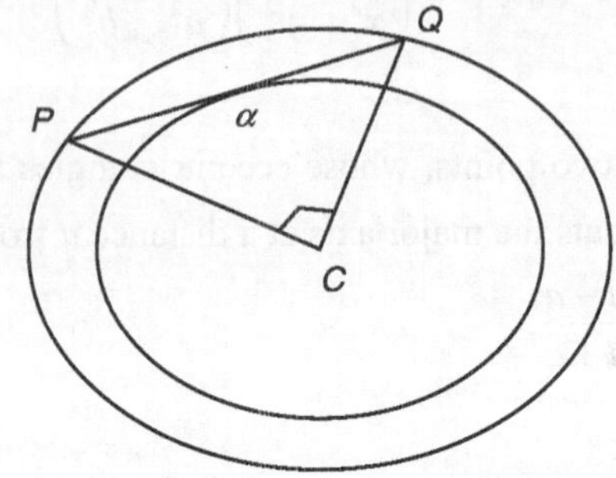

Let the equation of the ellipse be

$$\frac{x^2}{a^2}+\frac{y^2}{b^2}=1 \tag{7.84}$$

The equation of the auxiliary circle is

$$x^2+y^2=a^2 \tag{7.85}$$

The equation of the tangent at α is $\frac{x}{a}\cos\alpha+\frac{y}{b}\sin\alpha=1$. This line meets the auxiliary circle at P and Q. Then the combined equation of the lines CQ and CR is $x^2+y^2=a^2\left(\frac{x\cos\alpha}{a}+\frac{y}{b}\sin\alpha\right)^2$.

(i.e.) $$x^2[1-\cos^2\alpha]-\frac{2xya\sin\alpha\cos\alpha}{b}+y^2\left[1-\frac{a^2}{b^2}\sin^2\alpha\right]=0$$

Since $\underline{|QCR}=90°$, coefficient of x^2 + coefficient of $y^2=0$

$$\therefore \quad 1-\cos^2\alpha+1-\frac{a^2}{b^2}\sin^2\alpha=0$$

$$\sin^2\alpha+1-\frac{\sin^2\alpha}{1-e^2}=0$$

$$(1-e^2)\sin^2\alpha+(1-e^2)-\sin^2\alpha=0$$

$$e^2(1+\sin^2\alpha)=1\Rightarrow e^2=\frac{1}{1+\sin^2\alpha}$$

$$\Rightarrow e=(1+\sin^2\alpha)^{-1/2}$$

Example 7.37

If the normal at the end of a latus rectum of an ellipse passes through one extremity of the minor axis, show that the eccentricity of the curve is given by $e^4 + e^2 - 1 = 0$.

Solution

Let the equation of the ellipse be $\frac{x^2}{a^2} + \frac{y^2}{b^2} = 1$.

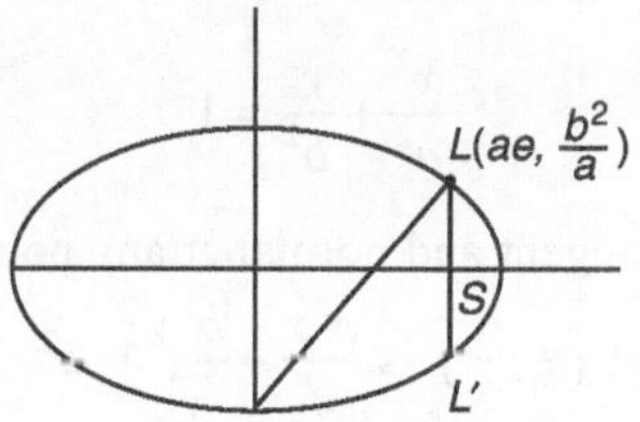

The coordinates of the end L of the latus rectum are $\left(ae, \frac{b^2}{a}\right)$. The equation of the normal at L is $\frac{a^2x}{x_1} - \frac{b^2y}{y_1} = a^2 - b^2$.

This normal passes at the point $\left(ae, \frac{b^2}{a}\right)$.

$$\therefore \frac{a^2x}{ae} - \frac{b^2}{\frac{b^2}{a}} = a^2 - b^2$$

This line passes through the point $(0, -b)$.

$$\therefore \frac{b^3}{\frac{b^2}{a}} = a^2e^2 \Rightarrow ab = a^2e^2$$

Squaring $a^2b^2 = a^4e^4 \Rightarrow a^2(a^2(1-e^2)) = a^4e^4$

$$\Rightarrow 1 - e^2 = e^4$$

$$\Rightarrow e^4 + e^2 - 1 = 0$$

Example 7.38

Prove that the tangent and normal at a point on the ellipse bisect the angle between the focal radii of that point.

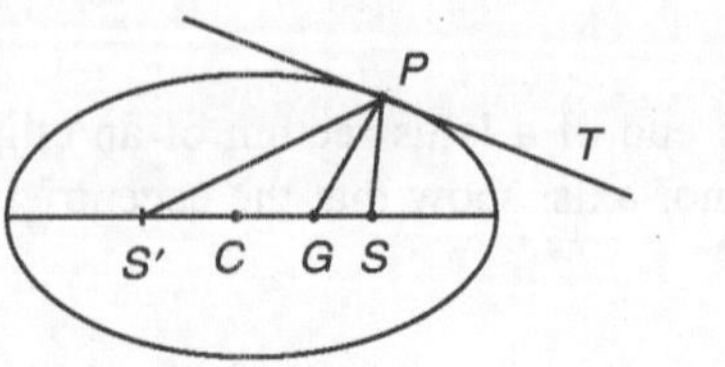

Solution

Let the equation of the ellipse be

$$\frac{x^2}{a^2}+\frac{y^2}{b^2}=1 \tag{7.86}$$

Let PT and PQ be the tangent and normal at any point P on the ellipse. The equation of the normal at (x_1, y_1) is $\frac{a^2x}{x^1}-\frac{b^2y}{y^1}=a^2-b^2$.

When this meet the major axis, $y = 0$.

$$\therefore \frac{a^2x}{x_1}=a^2e^2 \Rightarrow x=e^2x_1$$

$$\therefore G \text{ is } (e^2x_1, 0)$$

$$\frac{S'G}{GS}=\frac{SC+CG}{CS-CG}=\frac{ae+e^2x_1}{ae-e^2x_1}=\frac{a+ex_1}{a-ex_1}=\frac{S'P}{SP}$$

Since $SP' = a+ex_1$ and $SP = a-ex_1$.

$$\therefore \frac{S'G}{GS}=\frac{S'P}{PS}$$

Hence, PG bisects internally $\underline{|S'PS}$. Since the tangent PT is perpendicular to SG, PT is the external bisector of $\underline{|S'PS}$. Therefore, the tangent and normal at P are the bisectors of the angles between the focal radii through that point.

Example 7.39

Show that the locus of the middle point of chord of the ellipse $\frac{x^2}{a^2}+\frac{y^2}{b^2}=1$ which subtends a right angle at the centre is $\frac{x^2}{a^4}+\frac{y^2}{b^4}=\frac{a^2+b^2}{a^2b^2}\left(\frac{x^2}{a^2}+\frac{y^2}{b^2}\right)^2$.

Solution

The equation of the ellipse is

$$\frac{x^2}{a^2}+\frac{y^2}{b^2}=1 \tag{7.87}$$

Let (x_1, y_1) be the midpoint of a chord of the ellipse of equation (7.87).

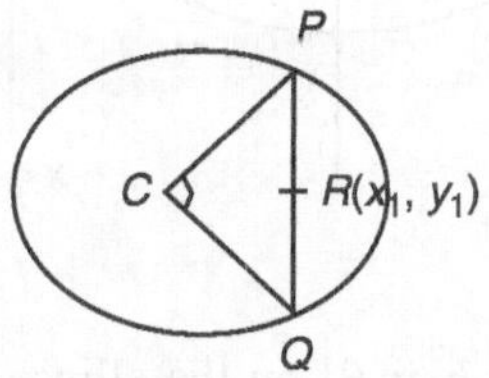

Then its equation is $\frac{xx_1}{a^2}+\frac{yy_1}{b^2}=\frac{x_1^2}{a^2}+\frac{y_1^2}{b^2}$.

If C is the centre of the ellipse, the combined equation of the lines CP and CQ is

$$\frac{x^2}{a^2}+\frac{y^2}{b^2}=\left(\frac{\frac{xx_1}{a^2}+\frac{yy_1}{b^2}}{\frac{x_1^2}{a^2}+\frac{y_1^2}{b^2}}\right)^2.$$

$$\Rightarrow\left(\frac{x^2}{a^2}+\frac{y^2}{b^2}\right)\left(\frac{x_1^2}{a^2}+\frac{y_1^2}{b^2}\right)^2=\left(\frac{xx_1}{a^2}+\frac{yy_1}{b^2}\right)^2$$

Since $\underline{|PCQ} = 90°$, coefficient of x^2 + coefficient of $y^2 = 0$.

$$\frac{1}{a^2}\left(\frac{x_1^2}{a^2}+\frac{y_1^2}{b^2}\right)^2-\frac{x_1^2}{a^4}+\frac{1}{b^2}\left(\frac{x_1^2}{a^2}+\frac{y_1^2}{b^2}\right)^2-\frac{y_1^2}{b^4}=0.$$

The locus of (x_1, y_1) is $\frac{x^2}{a^4}+\frac{y^2}{b^4}=\left(\frac{1}{a^2}+\frac{1}{b^2}\right)\left(\frac{x^2}{a^2}+\frac{y^2}{b^2}\right)^2$

$$\text{(i.e)}\quad \frac{x^2}{a^4}+\frac{y^2}{b^4}=\frac{a^2+b^2}{a^2b^2}\left(\frac{x^2}{a^2}+\frac{y^2}{b^2}\right)^2$$

Example 7.40

Prove that the portion of the tangent to the ellipse intercepted between the curve and the directrix subtends a right angle at the corresponding focus.

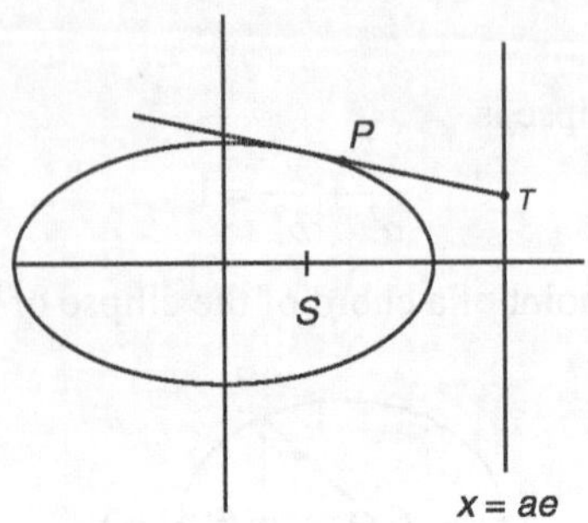

Solution

Let P be the point $(a\cos\theta, b\sin\theta)$ on the ellipse $\frac{x^2}{a^2}+\frac{y^2}{b^2}=1$. The equation of the tangent at θ is

$$\frac{x}{a}\cos\theta+\frac{y}{b}\sin\theta=1 \tag{7.88}$$

The equation of the corresponding directrix is

$$x=\frac{a}{e} \tag{7.89}$$

Solving equations (7.88) and (7.89), we get T, the point of intersection.

$$\frac{\frac{a}{e}\cos\theta}{a}+\frac{y}{b}\sin\theta=1 \Rightarrow \frac{y}{b}\sin\theta=1-\frac{1}{e}\cos\theta$$

$$\therefore \frac{y}{b}\sin\theta=\frac{e-\cos\theta}{e} \Rightarrow y=\frac{b(e-\cos\theta)}{e\sin\theta}$$

$$\therefore T \text{ is } \left(\frac{a}{e}, \frac{b(e-\cos\theta)}{e\sin\theta}\right)$$

The slope of SP is $\frac{b\sin\theta}{a(\cos\theta-e)}=m_1$.

The slope of ST is $\frac{\left(\frac{b(e-\cos\theta)}{e\sin\theta}\right)}{\frac{a}{e}-ae}=m_2$.

$$\therefore m_2=\frac{b(e-\cos\theta)}{a\sin\theta(1-e^2)}$$

$$m_1m_2 = \frac{b\sin\theta}{a(\cos\theta - e)} \times \frac{b(e-\cos\theta)}{a\sin\theta(1-e^2)} = \frac{a^2(1-e^2)\sin\theta(e-\cos\theta)}{a(\cos\theta-e)a\sin\theta(1-e^2)}$$

$$= -1$$

$$\therefore \underline{|PST} = \frac{\pi}{2}$$

Example 7.41

A normal inclined at 45° to the x-axis of the ellipse $\frac{x^2}{a^2}+\frac{y^2}{b^2}=1$ is drawn. It meets the major and minor axis in P and Q respectively. If C is the centre of the ellipse, show that the area of ΔCPQ is $\frac{(a^2-b^2)^2}{2(a^2+b^2)}$ sq. units.

Solution

The equation of the normal at 'θ' is

$$\frac{ax}{\cos\theta} - \frac{by}{\sin\theta} = a^2 - b^2$$

When this meets x-axis, $y = 0$.

$$x = \frac{\left(a^2-b^2\right)\cos\theta}{a}$$

Therefore, P is $\left[\frac{(a^2-b^2)\cos\theta}{a}, 0\right]$.

When it meets y-axis, $x = 0$.

$$y = \frac{-(a^2-b^2)\sin\theta}{b}$$

Therefore, Q is $\left[0, \frac{-(a^2-b^2)\sin\theta}{b}\right]$.

C is (0, 0).

Slope of the normal $= \frac{a\sin\theta}{b\cos\theta} = \tan 45° \Rightarrow \frac{a}{b}\tan\theta = 1$

$$\therefore \tan\theta = \frac{b}{a}.$$

Area of the $\Delta CPY = \frac{1}{2}(x_1y_2 - x_2y_1)$

$$= \frac{1}{2}\frac{(a^2-b^2)^2 \sin\theta\cos\theta}{ab}$$

$$= \frac{1}{2}\frac{(a^2-b^2)^2}{ab}\cdot\frac{b}{\sqrt{a^2+b^2}}\cdot\frac{a}{\sqrt{a^2+b^2}}$$

$$= \frac{(a^2-b^2)^2}{2(a^2+b^2)} \text{sq.units.}$$

Example 7.42

If $\alpha - \beta$ is a constant, prove that the chord joining the points, 'α' and 'β' touches a fixed ellipse.

Solution

The equation of the chord joining the points α and β is

$$\frac{x}{a}\cos\left(\frac{\alpha+\beta}{2}\right)+\frac{y}{b}\sin\left(\frac{\alpha+\beta}{2}\right)=\cos\frac{\alpha-\beta}{2}.$$

(i.e.) $\frac{x}{a}\cos\left(\frac{\alpha+\beta}{2}\right)+\frac{y}{b}\sin\left(\frac{\alpha+\beta}{2}\right)=\cos k$ where $\alpha-\beta=2k$.

Take $\frac{\alpha+\beta}{2}=\phi$, then the above equation becomes $\frac{x}{a}\cos\theta+\frac{y}{b}\sin\theta=\cos k$.

$$\text{(i.e.)} \quad \frac{x\cos\theta}{a\cos k}+\frac{y\sin\theta}{b\cos k}=1.$$

This line is a tangent to the ellipse $\frac{x^2}{(a\cos k)^2}+\frac{y^2}{(b\cos k)^2}=1.$

Example 7.43

If the chord joining the variable points at θ and ϕ on the ellipse $\frac{x^2}{a^2}+\frac{y^2}{b^2}=1$ subtends a right angle at the point $(a, 0)$ then show that $\tan\frac{\theta}{2}\cdot\tan\frac{\phi}{2}=\frac{-b^2}{a^2}$.

Solution

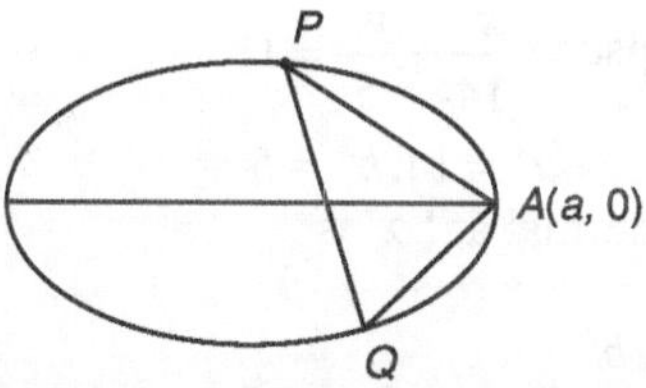

P is the point $(a\cos\theta, b\sin\theta)$.

Q is the point $(a\cos\phi, b\sin\phi)$.

Slope of AP is $\dfrac{b\sin\theta}{a(\cos\theta-1)}$.

Slope of AQ is $\dfrac{b\sin\phi}{a(\cos\phi-1)}$.

Since AP is perpendicular to AQ,

$$\frac{b\sin\theta}{a(\cos\theta-1)}\cdot\frac{b\sin\phi}{a(\cos\phi-1)}=-1$$

$$\text{(i.e.)}\quad \frac{b^2}{a^2}\frac{\sin\theta\sin\phi}{(1-\cos\theta)(1-\cos\phi)}=-1$$

$$\therefore \frac{b^2}{a^2}\cdot\frac{4\cos\frac{\theta}{2}\sin\frac{\theta}{2}\sin\frac{\phi}{2}\cos\frac{\phi}{2}}{2\sin^2\frac{\theta}{2}\cdot 2\sin^2\frac{\phi}{2}}=-1$$

$$\Rightarrow \frac{b^2}{a^2}\frac{\cos\frac{\theta}{2}\cos\frac{\phi}{2}}{\sin\frac{\theta}{2}\sin\frac{\phi}{2}}=-1$$

$$\therefore \tan\frac{\theta}{2}\tan\frac{\phi}{2}=\frac{-a^2}{b^2}$$

Example 7.44

If the normal to the ellipse $\dfrac{x^2}{14}+\dfrac{y^2}{5}=1$ at the point α cuts the curve join in 2α show that $\cos\alpha=\dfrac{-2}{3}$.

Solution

The equation of the ellipse is $\frac{x^2}{14}+\frac{y^2}{5}=1$.

$$a^2=14, b^2=5$$

The equation of the normal at 'α' is

$$\frac{ax}{\cos\alpha}-\frac{by}{\sin\alpha}=a^2-b^2$$

This passes through the point 2α.

$$\therefore \frac{a}{\cos\alpha}\cdot a\cos 2\alpha-\frac{b}{\sin\alpha}\cdot b\sin 2\alpha=a^2-b^2$$

$$\frac{14\cos 2\alpha}{\cos\alpha}-\frac{5\sin 2\alpha}{\sin\alpha}=a^2-b^2$$

$$\frac{14(2\cos^2\alpha-1)(\cos^2\alpha-1)}{\cos\alpha}-\frac{5(2\sin\alpha\cos\alpha)}{\sin\alpha}=14-5=9$$

(i.e.) $28\cos^2\alpha-14-10\cos^2\alpha=9\cos\alpha$.

$$18\cos^2\alpha-9\cos\alpha-14=0.$$

$$(6\cos\alpha-7)(3\cos\alpha+2)=0 \text{ or } \cos\alpha=\frac{7}{6} \text{ or } \frac{-2}{3}$$

$$\text{As } |\cos\alpha|<1, \cos\alpha=\frac{-2}{3}$$

Example 7.45

If the normal at any point P to the ellipse $\frac{x^2}{a^2}+\frac{y^2}{b^2}=1$ meets the major and minor axes in G and g and if CF be the perpendicular upon this normal, where C is the centre of the ellipse, then prove that $PF\cdot Pg=a^2$ and $PF\cdot PG=b^2$.

Solution

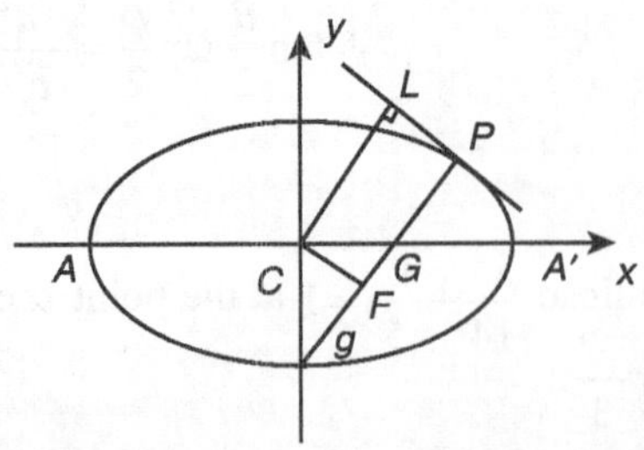

Let $P(a\cos\theta, b\sin\theta)$ be any point on the ellipse. Let the normal at P meet the major axis in G and minor axis in g. Let CF be the perpendicular from C to the normal at P. The equations of the tangent and normal at P are

$$\frac{x}{a}\cos\theta + \frac{y}{b}\sin\theta = 1 \tag{7.90}$$

$$\frac{ax}{\cos\theta} - \frac{by}{\sin\theta} = a^2 - b^2 \tag{7.91}$$

Then the coordinates of G and g are $\left[\frac{(a^2-b^2)\cos\theta}{a}, 0\right]$ and $\left[0, \frac{-(a^2-b^2)\sin\theta}{n}\right]$

$PF = CL$ where CL is the perpendicular on the tangent.

$$= \frac{1}{\sqrt{\frac{\cos^2\theta}{a^2} + \frac{\sin^2\theta}{b^2}}}$$

$$= \frac{ab}{\sqrt{a^2\sin^2\theta + b^2\cos^2\theta}} \tag{7.92}$$

$$PG^2 = \left(a\cos\theta - \frac{(a^2-b^2)\cos\theta}{a}\right)^2 + b^2\sin^2\theta = \frac{b^4\cos^2\theta}{a^2} + b^2\sin^2\theta.$$

$$= \frac{b^2(b^2\cos^2\theta + a^2\sin^2\theta)}{a^2}$$

$$\therefore PF \cdot Pg = \frac{ab}{\sqrt{a^2\sin^2\theta + b^2\cos^2\theta}} \cdot \frac{b}{a}\sqrt{a^2\sin^2\theta + b^2\cos^2\theta} = b^2$$

$$Pg^2 = a^2\cos^2\theta + \left(b\sin\theta + \frac{a^2-b^2}{b}\sin\theta\right)^2$$

$$= a^2\cos^2\theta + \frac{a^4\sin^2\theta}{b^2} = \frac{a^2(b^2\cos^2\theta + a^2\sin^2\theta)}{b^2}$$

$$PF \cdot Pg = \frac{ab}{\sqrt{a^2\sin^2\theta + b^2\cos^2\theta}} \cdot \frac{a}{b}\sqrt{a^2\sin^2\theta + b^2\cos^2\theta}$$

$$= a^2$$

Example 7.46

Show that the condition for the normals at the points (x_i, y_i), $i = 1, 2, 3$ on the ellipse $\frac{x^2}{a^2}+\frac{y^2}{b^2}=1$ to be concurrent is $\begin{vmatrix} x_1 & y_1 & x_1y_1 \\ x_2 & y_2 & x_2y_2 \\ x_3 & y_3 & x_3y_3 \end{vmatrix} = 0.$

Solution

Let (h, k) be the point of concurrence of the normal. The equation of the normal at (x_i, y_i) is $\frac{a^2x}{x_1}-\frac{b^2y}{y_1}=a^2-b^2$ (7.93)

Since this normal passes through (h, k),

$$\frac{a^2h}{x_1}-\frac{b^2k}{y_1}=a^2-b^2$$

(i.e.) $b^2kx_1 - a^2hy_1 + (a^2-b^2)x_1y_1 = 0.$ (7.94)

Similarly,

$$b^2kx_2 - a^2hy_2 + (a^2-b^2)x_2y_2 = 0 \quad (7.95)$$

$$b^2kx_3 - a^2hy_3 + (a^2-b^2)x_3y_3 = 0. \quad (7.96)$$

Eliminating h and k from equations (7.94), (7.95) and (7.96), we get

$$\begin{vmatrix} b^2x_1 & -a^2y_1 & (a^2-b^2)x_1y_1 \\ b^2x_2 & -a^2y_2 & (a^2-b^2)x_2y_2 \\ b^3x_3 & -a^2y_3 & (a^2-b^2)x_3y_3 \end{vmatrix} = 0$$

(i.e.) $b^2(-a^2)(a^2-b^2)\begin{vmatrix} x_1 & y_1 & x_1y_1 \\ x_2 & y_2 & x_2y_2 \\ x_3 & y_3 & x_3y_3 \end{vmatrix} = 0.$

(i.e.) $\begin{vmatrix} x_1 & y_1 & x_1y_1 \\ x_2 & y_2 & x_2y_2 \\ x_3 & y_3 & x_3y_3 \end{vmatrix} = 0.$

Example 7.47

Show that the area of the triangle inscribed in an ellipse is $2ab \sin \frac{\beta-\gamma}{2} \sin \frac{\gamma-\alpha}{2} \sin \frac{\alpha-\beta}{2}$ where α, β and γ are the eccentric angles of the vertices and hence find the condition that the area of the triangle inscribed in an ellipse is maximum.

Solution

Let ΔABC be inscribed in the ellipse

$$\frac{x^2}{a^2}+\frac{y^2}{b^2}=1 \qquad (7.97)$$

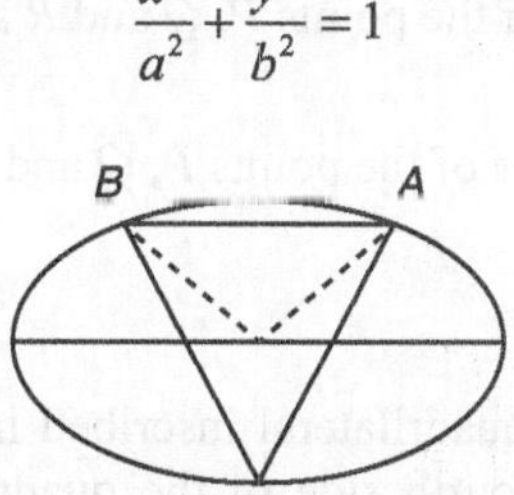

Let A, B and C be the points $(a\cos\alpha, b\sin\alpha)$, $(a\cos\beta, b\sin\beta)$ and $(a\cos\gamma, b\sin\gamma)$, respectively. Then the area of the ΔABC is given by,

$$\Delta ABC = \frac{1}{2}\begin{vmatrix} a\cos\alpha & b\sin\alpha & 1 \\ a\cos\beta & b\sin\beta & 1 \\ a\cos\theta & b\sin\theta & 1 \end{vmatrix}$$

$$= \frac{ab}{2}\begin{vmatrix} \cos\alpha & \sin\alpha & 1 \\ \cos\beta & \sin\beta & 1 \\ \cos\gamma & \sin\gamma & 1 \end{vmatrix}$$

$$= \frac{ab}{2}[1(\cos\alpha\sin\beta - \sin\alpha\cos\beta) - 1(\cos\alpha\sin\gamma - \sin\alpha\cos\gamma) + 1(\cos\beta\sin\gamma - \sin\beta\cos\gamma)]$$

$$= \frac{ab}{2}[\sin(\beta-\gamma)+\sin(\gamma-\alpha)+\sin(\alpha-\beta)]$$

$$= \frac{ab}{2}\left|4\sin\frac{\beta-\gamma}{2}\sin\frac{\gamma-\alpha}{2}\sin\frac{\alpha-\beta}{2}\right| \quad \text{since } \beta-\gamma+\gamma-\alpha+\alpha-\beta=\pi$$

If A', B' and C' are the corresponding points on the auxiliary circle then

$$\Delta A'B'C' = 2a^2 \sin\frac{\beta-\gamma}{2}\sin\frac{\gamma-\alpha}{2}\sin\frac{\alpha-\beta}{2}$$

$$\therefore \frac{\Delta ABC}{\Delta A'B'C'} = \frac{b}{a}$$

$$\text{(i.e.)}\quad \Delta ABC = \frac{b}{a}\Delta A'B'C'$$

Area of ΔABC is the greatest when the area of $\Delta A'B'C'$ is the greatest. However, the area of $A'B'C'$ is the greatest when the triangle is equilateral. In this case the eccentric angles of the points P, Q and R are $\alpha, \alpha+\frac{2\pi}{3}, \alpha+\frac{4\pi}{3}$.

(i.e.) The eccentric angles of the points P, Q and R differ by $\frac{2\pi}{3}$.

Example 7.48

If three of the sides of a quadrilateral inscribed in an ellipse are in a fixed direction, show that the fourth side of the quadrilateral is also in a fixed direction.

Solution

Let α, β, γ and δ be the eccentric angles of the vertices of the quadrilateral $ABCD$ inscribed in the ellipse

$$\frac{x^2}{a^2}+\frac{y^2}{b^2}=1 \qquad (7.98)$$

Then the equation of the chord PQ is $\frac{x}{a}\cos\frac{\alpha+\beta}{2}+\frac{y}{b}\sin\frac{\alpha+\beta}{2}=\cos\frac{\alpha-\beta}{2}$.

The slope of the chord PQ is $\frac{-b}{a}\cot\frac{\alpha+\beta}{2}$.

Since the direction of PQ is fixed, $\frac{-b}{a}\cot\frac{\alpha+\beta}{2}$ = a constant.

$$\therefore \frac{\alpha+\beta}{2} = k, \text{ a constant.}$$

Similarly, $\frac{\beta+\gamma}{2}=k_2$ and $\frac{\gamma+\delta}{2}=k_3$.

Adding $\dfrac{\alpha+2\beta+2\gamma+\delta}{2}=k_1+k_2+k_3$

(i.e.) $\dfrac{\alpha+\delta}{2}+\beta+\gamma=k_1+k_2+k_3$

(i.e.) $$\frac{\alpha+\delta}{2}=k_1+k_2+k_3-2k_2$$
$$=k_1+k_2-k_3$$
$$=k \text{ (a constant)}$$

Therefore, the direction of PS is also fixed.

Example 7.49

Prove that the area of the triangle formed by the tangents at the points α, β and γ is

$$ab\tan\frac{\beta-\gamma}{2}\tan\frac{\gamma-\alpha}{2}\cdot\tan\frac{\alpha-\beta}{2}.$$

Solution

The equation of tangents at α and β are

$$\frac{x}{a}\cos\alpha+\frac{y}{b}\sin\alpha=1$$

$$\frac{x}{a}\cos\beta+\frac{y}{b}\sin\beta=1$$

(i.e.) $$bx\cos\alpha+ay\sin\alpha-ab=0 \tag{7.99}$$
$$bx\cos\beta+ay\sin\beta-ab=0 \tag{7.100}$$

Solving equations (7.104) and (7.105) we get,

$$\frac{x}{a^2b(\sin\beta-\sin\alpha)}=\frac{y}{ab^2(\cos\beta-\cos\alpha)}=\frac{1}{ab(\cos\alpha\sin\beta-\sin\alpha\cos\beta)}$$

(i.e.) $$\frac{x}{a(\sin\alpha-\sin\beta)}=\frac{y}{b(\cos\alpha-\cos\beta)}=\frac{1}{\sin(\alpha-\beta)}$$

$$\therefore\ x=\frac{a(\sin\alpha-\sin\beta)}{\sin(\alpha-\beta)}=\frac{a\cdot 2\cos\frac{\alpha+\beta}{2}\sin\frac{\alpha-\beta}{2}}{2\cdot\sin\frac{\alpha-\beta}{2}\cdot\cos\frac{\alpha-\beta}{2}}$$

$$x=\frac{a\cos\left(\frac{\alpha+\beta}{2}\right)}{\cos\left(\frac{\alpha-\beta}{2}\right)}$$

$$y=\frac{b\cdot\sin\left(\frac{\alpha+\beta}{2}\right)}{\cos\left(\frac{\alpha-\beta}{2}\right)}$$

Therefore, the point of intersection of tangents at P is

$$\left[\frac{a\cos\left(\frac{\alpha+\beta}{2}\right)}{\cos\left(\frac{\alpha-\beta}{2}\right)},\frac{b\sin\frac{\alpha+\beta}{2}}{\cos\frac{\alpha-\beta}{2}}\right]$$

Hence, the area of the triangle formed by the tangents at α, β and γ is

$$\Delta=\frac{1}{2}\frac{1}{\cos\frac{\beta-\gamma}{2}\cos\frac{\gamma-\alpha}{2}\cos\frac{\alpha-\beta}{2}}\begin{vmatrix} a\cos\frac{\alpha+\beta}{2} & b\sin\frac{\alpha+\beta}{2} & \cos\frac{\alpha-\beta}{2} \\ a\cos\frac{\beta+\gamma}{2} & b\sin\frac{\beta+\gamma}{2} & \cos\frac{\beta-\gamma}{2} \\ a\cos\frac{\gamma+\alpha}{2} & b\sin\frac{\gamma+\alpha}{2} & \cos\frac{\gamma-\alpha}{2} \end{vmatrix}$$

$$\Delta=\frac{1}{2}\left|\frac{ab}{\cos\frac{\alpha-\beta}{2}\cos\frac{\beta-\gamma}{2}\cos\frac{\gamma-\alpha}{2}}\left[\sum\cos\frac{\alpha+\beta}{2}\sin\frac{\alpha-\beta}{2}\right]\right|$$

$$=\frac{ab}{4}\left[\frac{\sin(\alpha-\beta)+\sin(\beta-\gamma)+\sin(\gamma-\alpha)}{\cos\frac{\alpha-\beta}{2}\cos\frac{\beta-\gamma}{2}\cos\frac{\gamma-\alpha}{2}}\right]$$

$$=\frac{ab}{4}\left[\frac{4\sin\frac{\alpha-\beta}{2}\sin\frac{\beta-\gamma}{2}\sin\frac{\gamma-\alpha}{2}}{\cos\frac{\alpha-\beta}{2}\cos\frac{\beta-\gamma}{2}\cos\frac{\gamma-\alpha}{2}}\right]$$

$$=ab\tan\frac{\alpha-\beta}{2}\tan\frac{\beta-\gamma}{2}\tan\frac{\gamma-\alpha}{2}$$

Exercises

1. If α and β be the eccentric angles at the extremities of a chord of an ellipse of eccentricity *e*, prove that $\cos\left(\dfrac{\alpha-\beta}{2}\right)=e\cdot\cos\dfrac{\alpha+\beta}{2}$.

2. Let *P* and *Q* be two points on the major axis of an ellipse $\dfrac{x^2}{a^2}+\dfrac{y^2}{b^2}=1$ equidistant from the centre. Chords are drawn through *P* and *Q* meeting the ellipse at points whose eccentric angles are α, β, gand δ. Then prove that $\tan\dfrac{\alpha}{2}\tan\dfrac{\beta}{2}\tan\dfrac{\gamma}{2}\tan\dfrac{\delta}{2}=1$.

3. Prove that the chord joining the points on the ellipse $\dfrac{x^2}{a^2}+\dfrac{y^2}{b^2}=1$ whose eccentric angles differ by $\dfrac{2\pi}{3}$ touches another ellipse whose semi-axes are half those of the first.

4. *PSP′* and *QSQ′* are two focal chords of ellipse $\dfrac{x^2}{a^2}+\dfrac{y^2}{b^2}=1$ such that *PQ* is a diameter. Prove that *P′Q′* passes through a fixed point on the major axis of the ellipse. Find also its equation. **Ans.:** $\left(\dfrac{2ae}{e^2+1},0\right)$

5. *P* and *P′* are the corresponding points on an ellipse and its auxiliary circle. Prove that the tangents at *P* and *P′* intersect on the major axis.

6. The tangent at one end *P* of a diameter *PP′* of an ellipse and any chord *P′Q* through the other end meet at *R*. Prove that the tangent at *Q* bisects *PR*.

7. Prove that the three ellipses $\dfrac{x^2}{a_1^2}+\dfrac{y^2}{b_1^2}=1$, $\dfrac{x^2}{a_2^2}+\dfrac{y^2}{b_2^2}=1$ and $\dfrac{x^2}{a_3^2}+\dfrac{y^2}{b_3^2}=1$ will have a common tangent if $\begin{vmatrix} a_1^2 & b_1^2 & 1\\ a_2^2 & b_2^2 & 1\\ a_3^2 & b_3^2 & 1\end{vmatrix}=0$.

8. Any tangent to the ellipse is cut by the tangents at the ends of the major axis in *T* and *T′*. Prove that the circle on *TT′* as diameter will pass through the foci.

9. Find the coordinates of the points on the ellipse $\dfrac{x^2}{a^2}+\dfrac{y^2}{b^2}=1$, the tangents at which will make equal angles with the axis. Also prove that the length of the perpendicular from the centre on either of these is $\sqrt{\dfrac{a^2+b^2}{2}}$.

Ans.: $\left(\pm\dfrac{a^2}{\sqrt{a^2+b^2}},\pm\dfrac{b^2}{\sqrt{a^2+b^2}}\right)$

10. Find the condition for the line $x\cos\alpha + y\sin\alpha = p$ is a tangent to the ellipse $\frac{x^2}{a^2}+\frac{y^2}{b^2}=1.$ **Ans.:** $a\cos^2\alpha + b\sin^2\alpha = p^2$

11. If the tangent to the ellipse $\frac{x^2}{a^2}+\frac{y^2}{b^2}=1$, intercepts lengths α and β on the coordinate axes then show that $\frac{a^1}{\alpha^2}+\frac{b^1}{\beta^2}=1.$

12. If $x\cos\alpha + y\sin\alpha - p = 0$ be a tangent to the ellipse $\frac{x^2}{a^2}+\frac{y^2}{b^2}=1$, prove that $p^2 = a^2\cos^2\alpha + b^2\sin\alpha$. If P be the point of contact of the tangent $x\cos\alpha + y\sin\alpha = p$ and N, the foot of the perpendicular on it, from the centre of the ellipse, prove that $pNP = \frac{a^2-b^2}{2}\sin 2\alpha.$

13. The tangent at one end of P of a diameter OP' of an ellipse and any chord $P'Q$ through the other end meet in R. Prove that the tangent at Q bisects OR.

14. P and P' are corresponding points on an ellipse and the auxiliary circle. Prove that the tangents at P and P' intersect on the major axis.

15. If the normal at a point P on the ellipse of semi-axes a, b and centre C cuts the major and minor axes at G and g, show that $a^2Cg^2 + b^2CG^2 = (a^2 - b^2)^2$. Also prove that $PG = e\cdot GN$, where PN is the ordinate of P.

16. The tangents and normal at a point P on the ellipse $\frac{x^2}{a^2}+\frac{y^2}{b^2}=1$ meet the major axis in T and T' so that $TT' = a$. Prove that the eccentric angle of P is given by $e^2\cos^2\theta + \cos\theta - 1 = 0.$

17. Prove that, the line joining the extremities of any two perpendicular diameters of an ellipse always touches a concentric circle.

18. Show that the locus of the foot of the perpendicular drawn from the centre of the ellipse $\frac{x^2}{a^2}+\frac{y^2}{b^2}=1$ on any tangent to it is $(x^2+y^2)^2 = (a^2x^2+b^2y^2)^2$.

19. If P is any point on the ellipse $\frac{x^2}{a^2}+\frac{y^2}{b^2}=1$ whose ordinate is y', prove that the angle between the tangent at P and the fixed distance of P is $\tan^{-1}\left(\frac{b^2}{a\cdot ey'}\right)$.

20. Show that the feet of the normals that can be drawn from the point (h, k) to the ellipse $\frac{x^2}{a^2}+\frac{y^2}{b^2}=1$ lie on the curve $b^2(k-y) + a^2y\,(x-h) = 0$.

21. If the normals at the four points (x_i, y_i), $i = 1, 2, 3, 4$ on the ellipse $\frac{x^2}{a^2}+\frac{y^2}{b^2}=1$ are concurrent show that: (i) $(\Sigma x_1)\left(\Sigma\frac{1}{x_1}\right)=4$ and (ii) $(\Sigma y_1)\left(\Sigma\frac{1}{y_1}\right)=4$.

22. If the normals at the four points θ_i, $i = 1, 2, 3, 4$ are concurrent, prove that $(\Sigma \cos \theta_i)(\Sigma \sec\theta_i) = 4$. Show that the mean position of these four points is $\left(\frac{1}{2}\frac{a^2h}{a^2-b^2}, \frac{1}{2}\cdot\frac{b^2k}{a^2-b^2}\right)$ where (h, k) is the point of concurrency.

23. If the normals at the points α, β and γ on the ellipse $\frac{x^2}{a^2}+\frac{y^2}{b^2}=1$ are concurrent then prove that $\begin{vmatrix} \sec\alpha & \operatorname{cosec}\alpha & 1 \\ \sec\beta & \operatorname{cosec}\beta & 1 \\ \sec\gamma & \operatorname{cosec}\gamma & 1 \end{vmatrix} = 0$.

24. If α, β, γ and δ are the eccentric angles of the four corner points on the ellipse $\frac{x^2}{a^2}+\frac{y^2}{b^2}=1$ then prove that: (i) $\Sigma\cos(\alpha+\beta)=0$ and (ii) $\Sigma\sin(\alpha+\beta)=0$.

25. If the pole of the normal to an ellipse at P lies on the normal at Q, show that the pole of the normal at Q lies on the normal at P.

26. Find the locus of the middle points of the chords of ellipse whose distance from the centre C is constant c. **Ans.:** $\left(\frac{x^2}{a^2}+\frac{y^2}{b^2}\right)^2 = c^2\left(\frac{x^2}{a^2}+\frac{y^2}{b^2}\right)$

27. Find the locus of the midpoint of chords of the ellipse of constant length $2l$.

Ans.: $\left(\frac{x^2}{a^2}+\frac{y^2}{b^2}-1\right)\left(\frac{x^2}{a^4}+\frac{y^2}{b^4}\right)=\frac{l^2}{a^2b^2}\left(\frac{x^2}{a^2}+\frac{y^2}{b^2}\right)$

28. Show that the locus of midpoints of chords of the ellipse $\frac{x^2}{a^2}+\frac{y^2}{b^2}=1$, tangents at the ends of which intersect on the circle $x^2+y^2=a^2$ is $a^2\left(\frac{x^2}{a^2}+\frac{y^2}{b^2}\right)^2 = x^2+y^2$.

29. If the midpoint of a chord lies on a fixed line $lx+my+n=0$ show that the locus of the pole of the chord is the ellipse $n\left(\frac{x^2}{a^2}+\frac{y^2}{b^2}-1\right)+lx+my+n=0$.

30. Show that the locus of middle points of the chords of the ellipse that pass through a fixed point (h, k) is the ellipse $\frac{\left(x-\frac{h}{2}\right)^2}{a^2}+\frac{\left(y-\frac{k}{2}\right)^2}{b^2}=\frac{1}{h}\left(\frac{x^2}{a^2}+\frac{y^2}{b^2}\right)$.

31. Prove that the locus of the point of intersection of tangents to an ellipse at two points whose eccentric angles differ by a constant is an ellipse. If the sum of the eccentric angles be constant then prove that the locus is a straight line.

32. TP and TQ are the tangents drawn to an ellipse from a point T and C is its centre. Prove that the area of the quadrilateral $CPTQ$ is $ab\tan\frac{\theta-\phi}{2}$ where θ and ϕ are the eccentric angles of P and Q.

33. The eccentric angles of two points P and Q on the ellipse $\frac{x^2}{a^2}+\frac{y^2}{b^2}=1$ are θ and ϕ. Prove that the area of this parallelogram formed by the tangents at the ends of the diameters through P and Q is $4ab\,\text{cosec}(\theta-\phi)$.

34. Chords of the ellipse $\frac{x^2}{a^2}+\frac{y^2}{b^2}=1$ pass through a fixed point (h, k). Show that the locus of their middle points is the ellipse $\frac{x^2}{a^2}+\frac{y^2}{b^2}=\frac{xh}{a^2}+\frac{yk}{b^2}$.

35. If P is any point on the director circle, show that the locus of the middle points of the chord in which the polar of P cuts the ellipse is $\left(\frac{x^2}{a^2}+\frac{y^2}{b^2}\right)^2=\frac{x^2+y^2}{a^2+b^2}$.

36. Show that the locus of midpoints of the chords of the ellipse $\frac{x^2}{a^2}+\frac{y^2}{b^2}=1$ touching the ellipse $\frac{x^2}{\alpha^2}+\frac{y^2}{\beta^2}=1$ is $\alpha^2\beta^4x^2+\beta^2\alpha^4y^2=a^4b^4\left(\frac{x^2}{a^2}+\frac{y^2}{b^2}\right)^2$.

37. If the normals to an ellipse at P_i, $i = 1, 2, 3, 4$ are concurrent then the circle through P_1, P_2 and P_3 meets the ellipse again in a point P_4 which is the other end of the diameter through P_4.

38. Find the centre of the circle passing through the three points, on the ellipse whose eccentric angles are α, β and γ.

Ans.: $$\left\{\frac{b^2-a^2}{4a}[\cos\alpha+\cos\beta+\cos\gamma+\cos(\alpha+\beta+\gamma)],\ \frac{a^2-b^2}{4b}[\cos\alpha+\cos\beta+\cos\gamma+\cos(\alpha+\beta+\gamma)]\right\}.$$

39. If ABC be a maximum triangle inscribed in an ellipse then show that the eccentric angles of the vertices differ by $\frac{2\pi}{3}$ and the normals A, B and C are concurrent.

40. The tangent and normal to the ellipse $x^2+4y^2=2$, at the point P meet the major axis in Q and R, respectively and $QR = 2$. Show that the eccentric angle of P is $\cos^{-1}\left(\frac{2}{3}\right)$.

41. If two concentric ellipses be such that the foci of one lie on the other then prove that the angle between their axes is $\sqrt{\frac{e_1^2+e_2^2-1}{e_1e_2}}$ where e_1 and e_2 are their eccentricities.

42. Show that the length of the focal chord of the ellipse $\frac{x^2}{a^2}+\frac{y^2}{b^2}=1$ which makes an angle θ with the major axis is $\frac{2ab^2}{a^2\sin^2\theta+b^2\cos^2\theta}$.

43. If the normals are drawn at the extremities of a focal chord of an ellipse, prove that a line through their point of intersection parallel to the major axis will bisect the chord.

44. If tangents from the point to the ellipse $\frac{x^2}{a^2}+\frac{y^2}{b^2}=1$ cut off a length equal to the major axis from the tangent at $(a, 0)$, prove that T lies on a parabola.

45. If the normal at any point P on an ellipse cuts the major axis at G, prove that the locus of the middle point of PQ is an ellipse.

46. Show that the locus of the intersection of two normals to the ellipse $\frac{x^2}{a^2}+\frac{y^2}{b^2}=1$ which are perpendicular to each other is $(a^2-b^2)^2\cdot\left(\frac{x^2}{a^2}+\frac{y^2}{b^2}\right)^2 = (a^2+b^2)(x^2+y^2)\left(\frac{x^2}{a^2}+\frac{y^2}{b^2}\right)^2$.

47. If the angle between the diameter of any point of the ellipse $\frac{x^2}{a^2}+\frac{y^2}{b^2}=1$ and the normal at that point is θ, prove that the greatest value of $\tan\theta$ is $\frac{a^2-b^2}{2ab}$.

48. P is any point on an ellipse. Prove that the locus of the centroid G of the point P and the two foci of the ellipse is a concentric ellipse of the same eccentricity.

49. If P, Q, R and S are conormal points on an ellipse, show that the circle passing through P and R will cut the ellipse at a point S' where S and S' are the ends of a diameter of the ellipse.

50. Show that the locus of pole of any tangent to the ellipse with respect to the auxiliary circle is a similar concentric ellipse whose major axis is at right angles to that of the original ellipse.

51. The normals of four points of an ellipse meet at (h, k). If two of the points lie on $\frac{lx}{a}+\frac{my}{b}+1=0$, prove that the other two points lie on $\frac{x}{al}+\frac{y}{bm}-1=0$.

52. If the normals to the ellipse $\frac{x^2}{a^2}+\frac{y^2}{b^2}=1$ at the ends of the chords $lx+my=1$ and $l_1x+m_1y=1$ be concurrent then show that $a^2ll_1=b^2mm_1=-1$.

53. Prove that two straight lines through the points of intersection of an ellipse with any circle make equal angles with the axes of the ellipse.

54. Show that the equation of a pair of straight lines which are at right angles and each of which passes through the pole of the other may be written as $lx+my+n=0$ and $n(mx-ny)+lm(a^2-b^2)=0$. Also prove that the product of the distances of such pair of lines from the centre commonly exceeds $\frac{a^2-b^2}{2}$.

55. Show that the rectangle under the perpendicular drawn to the normal at a point of an ellipse from the centre and from the pole of the normal is equal to the rectangle under the focal distances of P.

56. Prove that if P, Q, R and S are the feet of the normals to the ellipse $\frac{x^2}{a^2}+\frac{y^2}{b^2}=1$ and the coordinates (x_1, y_1), (x_2, y_2), are the poles of PQ and RS then they are connected by the relations $x_1x_2^2=-a^2$ and $y_1y_2^2=-b^2$.

57. If the normals at four points of the ellipse $\frac{x^2}{a^2}+\frac{y^2}{b^2}=1$ are concurrent and if two points lie on the line $lx+my=1$, show that the other two points lie on the line $\frac{x}{a^2l}+\frac{y}{b^2m}=0$. Hence show that if the feet of the two normals from a point P to this ellipse are coincident then the locus of the midpoints of the chords joining the feet of the other normals is $\left(\frac{xy}{ab}\right)^2=\left(\frac{x^2}{a^2}+\frac{y^2}{b^2}\right)^2$.

7.20.1 Locus of Midpoint

Locus of midpoint of a series of parallel chords of the ellipse:
Let (x_1, y_1) be the midpoint of a chord parallel to the line $y=mx$.
Then the equation of the chord is $T=S_1$.

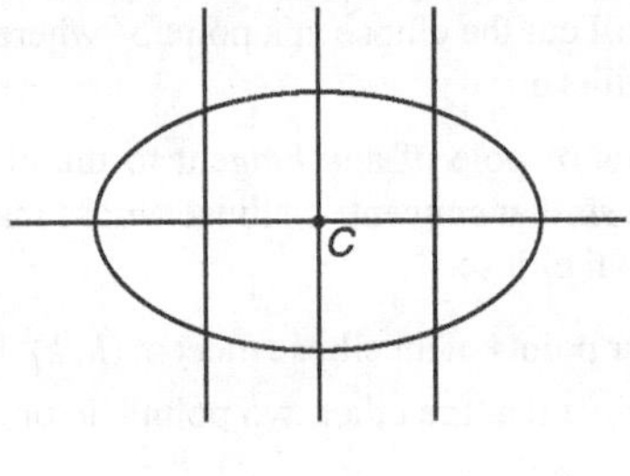

$$\text{(i.e.)}\quad \frac{xx_1}{a^2}+\frac{yy_1}{b^2}-1=\frac{x_1^2}{a^2}+\frac{y_1^2}{b^2}-1$$

Its slope is $\frac{-b^2x_1}{a^2b_1}$

Since this chord is parallel to $y=mx$, $m=\frac{-b^2x_1}{a^2y_1}$ (i.e.) $y_1=\frac{-b^2x_1}{a^2m}$.

The locus of (x_1, y_1) is $y=\frac{-b^2x}{a^2m}$ which is a straight line passing through the centre of the ellipse. If $y=m_1x$ bisect all chords parallel to $y=mx$ then $m_1=\frac{-b^2}{a^2m}$.

$$\therefore mm_1 = \frac{-b^2}{a^2}$$

By symmetry of this result, we see that the diameter $y = mx$ bisect all the chords parallel to $y = m_1 x$.

Definition 7.20.1 Two diameters are said to be conjugate to each other if chords parallel to one is bisected by the other. Therefore, the condition for the diameter $y = mx$ and $y = m_1 x$ to be conjugate diameters is $mm_1 = \frac{-b^2}{a^2}$.

7.20.2 Property: The Eccentric Angles of the Extremities of a Pair of Semi-conjugate Diameter Differ by a Right Angle

Let PCP' and DCD' be a pair of conjugate diameters. Let P be the points $(a\cos\theta, b\sin\theta)$ and D be the points $(a\cos\phi, b\sin\phi)$. Then the slope of CP is $m = \frac{b\sin\theta}{a\cos\theta}$.

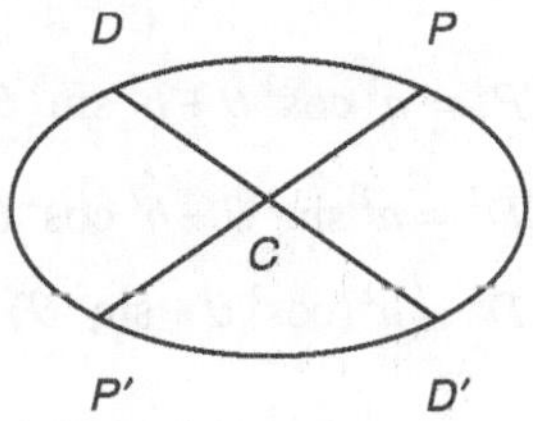

The slope of CD is $m_1 = \frac{b\sin\phi}{a\cos\phi}$.

Since CP and CD are semi-conjugate diameters

$$mm_1 = \frac{-b^2}{a} \Rightarrow \frac{b\sin\theta}{a\cos\theta}\left(\frac{b\sin\phi}{a\cos\phi}\right) = \frac{-b^2}{a^2}.$$

$$\therefore \cos\theta\cos\phi = -\sin\theta\sin\phi$$

$$\text{(i.e.)} \quad \cos\theta\cos\phi + \sin\theta\sin\phi = 0 \Rightarrow \cos(\phi - \theta) = 0$$

$$\therefore \phi - \theta = \frac{\pi}{2} \Rightarrow \phi = \theta + \frac{\pi}{2}$$

Therefore, the eccentric angles of a pair of semi-conjugate diameters differ by a right angle.

Note 7.20.1: The coordinates of D are $\left(a\cos\left(\frac{\pi}{2}+\theta\right), a\sin\left(\frac{\pi}{2}+\theta\right)\right)$.

$$\text{(i.e.)} \quad (-a\sin\theta, b\cos\theta)$$

Therefore, if the coordinates of P are $(a\cos\theta, b\sin\theta)$ then the coordinates of D are $(-a\sin\theta, b\cos\theta)$. The coordinates of P' are $(-a\cos\theta, -b\sin\theta)$. The coordinates of D' are $(a\sin\theta, -b\cos\theta)$.

7.20.3 Property: If *CP* and *CD* are a Pair of Semi-conjugate Diameters then $CD^2 + CP^2$ is a Constant

The coordinates of C, P and D are $C(0, 0)$

$$P(a\cos\theta, b\sin\theta) \text{ and } D(-a\sin\theta, b\cos\theta).$$

Then

$$CP^2 = a^2\cos^2\theta + b^2\sin^2\theta$$

$$CD^2 = a^2\sin^2\theta + b^2\cos^2\theta$$

$$\therefore CP^2 + CD^2 = a^2(\cos^2\theta + \sin^2\theta) + b^2(\sin^2\theta + \cos^2\theta)$$

$$= a^2 + b^2 = \text{constant}$$

7.20.4 Property: The Tangents at the Extremities of a Pair of Conjugate Diameters of an Ellipse Encloses a Parallelogram Whose Area Is Constant

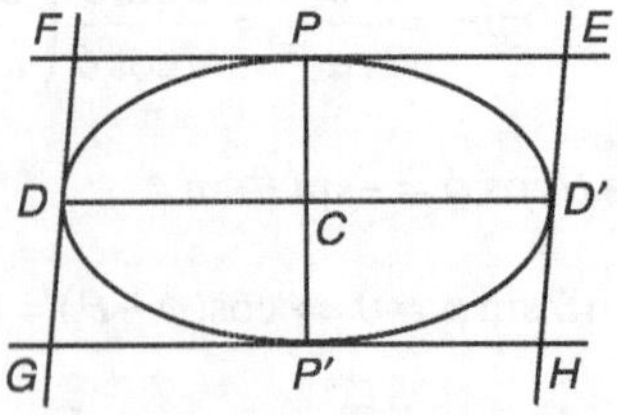

Let PCP' and DCD' be a pair of conjugate diameters. Let P be the point $(a\cos\theta, b\sin\theta)$. Then D is the point $\left[a\cos\left(\frac{\pi}{2}+\theta\right), b\sin\left(\frac{\pi}{2}+\theta\right)\right]$.

$$\text{(i.e.)} \quad (-a\sin\theta,\, b\cos\theta).$$

The equation of the tangent at P is $\frac{x}{a}\cos\theta + \frac{y}{b}\sin\theta = 1$.

The slope of the tangent at P is $\frac{-b\cos\theta}{a\sin\theta}$.

The slope of CD is $\frac{-b\cos\theta}{a\sin\theta}$.

Since the two slopes are equal, the tangents at P is parallel to DCD'. Similarly, we can show that the tangent at P' is parallel to DCD'. Therefore, the tangent at P and P' are parallel. Similarly, the tangent D and D' are parallel. Hence, the tangents at P, P', D, D' from a parallelogram $EFGH$.
The area of the parallelogram $EFGH$

$$= 4 \times \text{Area of parallelogram } CPFD$$
$$= 4 \times 2 \text{ area of } \Delta CPD = 8 \text{ area of } \Delta CPD$$
$$= 8 \times \frac{1}{2}[a\cos\theta(b\cos\theta) - a\sin\theta(-b\sin\theta)]$$
$$= 4ab(\cos^2\theta + \sin^2\theta)$$
$$= 4ab \text{ which is a constant.}$$

7.20.5 Property: The Product of the Focal Distances of a Point on an Ellipse Is Equal to the Square of the Semi-diameter Which Is Conjugate to the Diameter Through the Point

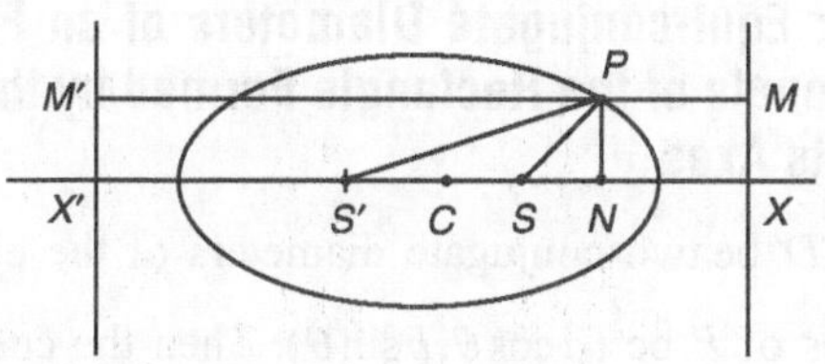

Let S and S' be the foci of ellipse $\frac{x^2}{a^2} + \frac{y^2}{b^2} = 1$. Let P be any point on the ellipse and draw MPM' perpendicular to the directrix.

Then, $$\frac{SP}{PM} = e \Rightarrow SP = e\text{PM} = e \cdot NX = e \cdot [CX - CN]$$

$$= e\left[\frac{a}{e} - a\cos\theta\right] = a - ae\cos\theta$$

$$S'P = e \cdot PM' = e \cdot NX' = e[CX' + CN]$$

$$= e\left[\frac{a}{e} + a\cos\theta\right] = a + ae\cos\theta$$

$$SP \cdot S'P = (a - ae\cos\theta)(a + ae\cos\theta) = a^2 - a^2e^2\cos^2\theta = a^2 - (a^2 - b^2)\cos^2\theta$$

$$= a^2(1 - \cos^2\theta) + b^2\cos^2\theta = a^2\sin^2\theta + b^2\cos^2\theta = CD^2$$

7.20.6 Property: If *PCP′* and *DCD′* are Conjugate Diameter then They are also Conjugate Lines

We know that the polar of a point and the chord of contact of tangents from it to the ellipse are the same. Therefore, the pole of the diameter *PCP′* will be the point of intersection of the tangents at *P* and *P′* which are parallel. Therefore, the pole of *PCP′* lies at infinity on the conjugate diameter *DCD′*. Hence, *PCP′* and *DCD′* are conjugate lines.

Note 7.20.2: Conjugate diameter is a special case of conjugate lines.

7.21 EQUI-CONJUGATE DIAMETERS

Definition 7.21.1 Two diameters of an ellipse are said to be equi conjugate diameters if they are of equal length.

7.21.1 Property: Equi-conjugate Diameters of an Ellipse Lie along the Diagonals of the Rectangle Formed by the Tangent at the Ends of its Axes

Let *PCP′* and *DCD′* be two conjugate diameters of the ellipse $\frac{x^2}{a^2} + \frac{y^2}{b^2} = 1.$ Let the coordinates of *P* be $(a\cos\theta, b\sin\theta)$. Then the coordinates of *D* are $(-a\sin\theta, b\cos\theta)$. *C* is the point $C(0, 0)$.

$$CP^2 = a^2\cos^2\theta + b^2\sin^2\theta$$
$$CD^2 = a^2\sin^2\theta + b^2\cos^2\theta$$

If *CP* and *CD* are equi-conjugate diameters then $CP^2 = CD^2$.

$$a^2\cos^2\theta + b^2\sin^2\theta = a^2\sin^2\theta + b^2\cos^2\theta.$$

$$(a^2 - b^2)\sin^2\theta = (a^2 - b^2)\cos^2\theta$$

$$\therefore \tan^2\theta = 1 \text{ or } \tan\theta = \pm 1$$

$$\therefore \theta = \frac{\pi}{4} \text{ or } \frac{3\pi}{4}$$

When $\theta = \frac{\pi}{4}$, the equation of the diameter is

$$y = \frac{b\sin\theta}{a\cos\theta} \quad \text{(i.e.)} \quad y = \frac{b}{a}x \tag{7.101}$$

The equation of CD is $y = \frac{b\cos\theta}{-a\sin\theta}x$

$$\text{(i.e.)} \quad y = \frac{-b}{a}x \tag{7.102}$$

When $\theta = \frac{3\pi}{4}$, equations of these two conjugate diameters are $y = \pm\frac{b}{a}x$. Therefore, the equi-conjugate diameters are $y = \pm\frac{b}{a}x$ which are the equations of the diagonals formed by the tangents at the four vertices of the ellipse.

ILLUSTRATIVE EXAMPLES BASED ON CONJUGATE DIAMETERS

Example 7.50

Show that the locus of the point of intersection of tangents at the extremities of a pair of conjugate diameters of the ellipse $\frac{x^2}{a^2} + \frac{y^2}{b^2} = 1$ is the ellipse $\frac{x^2}{a^2} + \frac{y^2}{b^2} = 2$.

Solution

Let PCP' and DCD' be a pair of conjugate diameters of the ellipse $\frac{x^2}{a^2} + \frac{y^2}{b^2} = 1$. Let P be the point $(a\cos\theta\ b\sin\theta)$.

Then D' is the point

$$\left(a\cos\left(\theta + \frac{\pi}{2}\right),\ b\sin\left(\theta + \frac{\pi}{2}\right)\right)$$

$$\text{(i.e.)} \quad (-a\sin\theta, b\cos\theta)$$

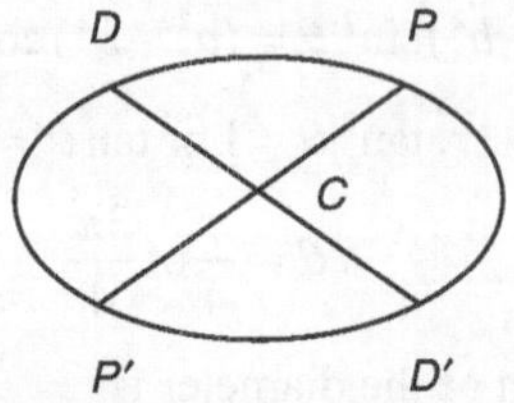

Let (x_1, y_1) be the point of intersection of the tangents at P and D. The equations of the tangents at P and D are

$$\frac{x}{a}\cos\theta + \frac{y}{b}\sin\theta = 1$$

$$-\frac{x}{a}\sin\theta + \frac{y}{b}\cos\theta = 1$$

Since these two tangents meet at (x_1, y_1),

$$\frac{x_1}{a}\cos\theta + \frac{y_1}{b}\sin\theta = 1 \qquad (7.103)$$

and

$$\frac{-x_1}{a}\sin\theta + \frac{y_1}{b}\cos\theta = 1 \qquad (7.104)$$

Squaring and adding from equations (7.103) and (7.104), we get $\frac{x_1^2}{a^2} + \frac{y_1^2}{b^2} = 2.$

Therefore, the locus of (x_1, y_1) is $\frac{x^2}{a^2} + \frac{y^2}{b^2} = 2.$

Example 7.51

If P and D are the extremities of a pair of conjugate diameter of the ellipse $\frac{x^2}{a^2} + \frac{y^2}{b^2} = 1$ show that the locus of the midpoint of PD is $\frac{x^2}{a^2} + \frac{y^2}{b^2} = \frac{1}{2}.$

Solution

Let P be the point $(a\cos\theta, b\sin\theta)$. Then the coordinates of D are $(-a\sin\theta, b\cos\theta)$. Let (x_1, y_1) be the midpoint of PD.

Then $x_1 = \frac{a\cos\theta - a\sin\theta}{2}$, $y_1 = \frac{b\sin\theta + b\cos\theta}{2}$

$$\text{(i.e.)} \quad \frac{2x_1}{a} = \cos\theta - \sin\theta \qquad (7.105)$$

$$\frac{2y_1}{b} = \sin\theta + \cos\theta \qquad (7.106)$$

Squaring and adding from equations (7.105) and (7.106), we get

$$\frac{4x_1^2}{a^2}+\frac{4y_1^2}{b^2}=2$$

$$\text{(i.e.)}\quad \frac{x_1^2}{a^2}+\frac{y_1^2}{b^2}=\frac{1}{2}$$

Therefore, the locus of (x_1, y_1) is $\frac{x^2}{a^2}+\frac{y^2}{b^2}=\frac{1}{2}$ which is a concentric ellipse.

Example 7.52

If CP and CD are two conjugate semi-diameters of an ellipse $\frac{x^2}{a^2}+\frac{y^2}{b^2}=1$ then prove that the line PD touches the ellipse $\frac{x^2}{a^2}+\frac{y^2}{b^2}=\frac{1}{2}$.

Solution

Let the eccentric angle of P be θ. Then the eccentric angle of D is $\theta+\frac{\pi}{2}$. The equation of the chord PD is

$$\frac{x}{a}\cos\left(\frac{\theta+\theta+\frac{\pi}{2}}{2}\right)+\frac{y}{b}\sin\left(\frac{\theta+\theta+\frac{\theta}{2}}{2}\right)=\cos\left(\frac{\theta-\left(\theta+\frac{\theta}{2}\right)}{2}\right)$$

$$\text{(i.e)}\quad \frac{x}{a}\cos\left(\theta+\frac{\pi}{4}\right)+\frac{y}{b}\sin\left(\theta+\frac{\pi}{4}\right)=\cos\frac{\pi}{4}=\frac{1}{\sqrt{2}}$$

$$\text{(i.e)}\quad \frac{x}{\left(\frac{a}{\sqrt{2}}\right)}\cos\left(\theta+\frac{\pi}{4}\right)+\frac{y}{\frac{b}{\sqrt{2}}}\sin\left(\theta+\frac{\pi}{4}\right)=1$$

$$\text{(i.e.)}\quad \frac{x}{A}\cos\phi+\frac{y}{B}\sin\phi=1$$

where $A=\frac{a}{\sqrt{2}}$, $B=\frac{b}{\sqrt{2}}$ and $\theta+\frac{\pi}{4}=\phi$.

This straight line touches the ellipse $\frac{x^2}{A^2}+\frac{y^2}{B^2}=1$.

$$\text{(i.e.)}\quad \frac{2x^2}{a^2}+\frac{2y^2}{b^2}=1 \text{ or } \frac{x^2}{a^2}+\frac{y^2}{b^2}=\frac{1}{2}$$

Example 7.53

Find the condition that the two straight lines represented by $Ax^2 + 2Hxy + By^2 = 0$ may be a pair of conjugate diameters of the ellipse $\frac{x^2}{a^2}+\frac{y^2}{b^2}=1$.

Solution

Let the two straight lines represented by $Ax^2 + 2Hxy + By^2 = 0$ be $y = m_1x$ and $y = m_2x$. Then $m_1 + m_2 = \frac{-2H}{B}$ and $m_1m_2 = \frac{A}{B}$.

The condition for the lines to be conjugate diameters is $m_1m_2 = -\frac{b^2}{a^2}$.

$$\text{(i.e.)}\quad \frac{A}{B}=\frac{-b^2}{a^2} \Rightarrow a^2A+b^2B=0$$

This is the required condition.

Example 7.54

If P and D be the ends of conjugate semi-diameters of the ellipse then show that the locus of the foot of the perpendicular from the centre on the line PD is $2(x^2+y^2)^2 = a^2x^2+b^2y^2$.

Solution

Let the eccentric angle of P be θ. Then the eccentric angle of D is $\theta+\frac{\pi}{2}$. The equation of PD is

$$\frac{x}{a}\cos\left(\theta+\frac{\pi}{4}\right)+\frac{y}{b}\sin\left(\theta+\frac{\pi}{4}\right)=\frac{1}{\sqrt{2}} \tag{7.107}$$

The equation of the line perpendicular to this and passing through the centre (0, 0) is

$$\frac{x}{b}\sin\left(\theta+\frac{\pi}{4}\right)-\frac{y}{a}\cos\left(\theta+\frac{\pi}{4}\right)=0 \tag{7.108}$$

Let (x_1, y_1) be the foot of the perpendicular from (0, 0) on PD. Then (x_1, y_1) lies on the above two lines.

$$\frac{x_1}{a}\cos\left(\theta+\frac{\pi}{4}\right)+\frac{y_1}{b}\sin\left(\theta+\frac{\pi}{4}\right)=\frac{1}{\sqrt{2}} \tag{7.109}$$

$$\frac{x_1}{b}\sin\left(\theta+\frac{\pi}{4}\right)-\frac{y_1}{a}\cos\left(\theta+\frac{\pi}{4}\right)=0 \tag{7.110}$$

Solving for $\cos\left(\theta+\frac{\pi}{4}\right)$ and $\sin\left(\theta+\frac{\pi}{4}\right)$, we get

$$\frac{\cos\left(\theta+\frac{\pi}{4}\right)}{ax_1}=\frac{\sin\left(\theta+\frac{\pi}{4}\right)}{by_1}=\frac{1}{\sqrt{a^2x_1^2+b^2y_1^2}}$$

Substituting for $\cos\left(\theta+\frac{\pi}{4}\right)$ and $\sin\left(\theta+\frac{\pi}{4}\right)$ in equation (7.107), we get

$$\frac{x_1^2+y_1^2}{\sqrt{x_1^2a^2+b^2y_1^2}}=\frac{1}{\sqrt{2}}$$

$$\text{(i.e.)}\quad 2(x_1^2+y_1^2)^2=a^2x_1^2+b^2y_1^2$$

Therefore, the locus of (x_1, y_1) is $2(x^2+y^2)=a^2x^2+b^2y^2$.

Example 7.55

CP and *CD* are semi-conjugate diameters of the ellipse $\frac{x^2}{a^2}+\frac{y^2}{b^2}=1$. If the circles on *CP* and *CD* as diameters intersect in *R* then prove that the locus of the point *R* is $2(x^2+y^2)^2=a^2x^2+b^2y^2$.

Solution

Let *P* be the point $(a\cos\theta, b\sin\theta)$. Then *D* is the point $(-a\sin\theta, b\cos\theta)$. *C* is the point (0, 0).

The equations of the circles on *CP* and *CD* as diameters are $x(x-a\cos\theta)+y(y-b\sin\theta)=0$ and $x(x+a\sin\theta)+y(y-b\cos\theta)=0$.

(i.e.) $x^2+y^2=ax\cos\theta+by\sin\theta$ and $x^2+y^2=-ax\sin\theta+by\cos\theta$.

Let (x_1, y_1) be a point of intersection of these two circles. Then

$$x_1^2+y_1^2=ax_1\cos\theta+by_1\sin\theta \tag{7.111}$$

$$x_1^2+y_1^2=-ax_1\sin\theta+by_1\cos\theta \tag{7.112}$$

By squaring and adding equations (7.111) and (7.112), we get $(x_1^2+y_1^2)^2=a^2x_1^2+b^2y_1^2$.

Therefore, the locus of (x_1, y_1) is $(x^2+y^2)^2=a^2x^2+b^2y^2$.

Example 7.56

If the points of intersection of the ellipses $\frac{x^2}{a^2}+\frac{y^2}{b^2}=1$ and $\frac{x^2}{a^2}+\frac{y^2}{b^2}=1$ be the points of conjugate diameters of the former prove that $\frac{a^2}{A^2}+\frac{b^2}{B^2}=2.$

Solution

Any conic passing through the point of intersection of the ellipses

$$\frac{x^2}{a^2}+\frac{y^2}{b^2}=1 \tag{7.113}$$

and

$$\frac{x^2}{A^2}+\frac{y^2}{B^2}=1 \tag{7.114}$$

is

$$\left(\frac{x^2}{a^2}+\frac{y^2}{b^2}-1\right)+\lambda\left(\frac{x^2}{A^2}+\frac{y^2}{B^2}-1\right)=0 \tag{7.115}$$

where $\lambda=-1$, equation (7.118) reduces to

$$\left(\frac{x^2}{a^2}+\frac{y^2}{b^2}\right)-\left(\frac{x^2}{A^2}+\frac{y^2}{B^2}\right)=0$$

$$\text{(i.e.)}\quad x^2\left(\frac{1}{a^2}-\frac{1}{A^2}\right)+y^2\left(\frac{1}{b^2}+\frac{1}{B^2}\right)=0 \tag{7.116}$$

This being a homogeneous equation of second degree in x and y represents a pair of straight lines, that is, equation (7.116) represents a pair of straight lines passing through the origin. $m_1m_2=-\frac{b^2}{a^2}$.

$$\text{(i.e.)}\quad \frac{\frac{1}{a^2}-\frac{1}{A^2}}{\frac{1}{b^2}-\frac{1}{B^2}}=-\frac{b^2}{a^2}\Rightarrow 1-\frac{a^2}{A^2}=1+\frac{b^2}{B^2}$$

or

$$\frac{a^2}{A^2}+\frac{b^2}{B^2}=2 \text{ which is required condition.}$$

Example 7.57

If α and β be the angles subtended by the major axis of an ellipse at the extremities of a pair of conjugate diameters then show that $\cos^2\alpha+\cos^2\beta$ is a constant.

Solution

Let equation of the ellipse be $\frac{x^2}{a^2}+\frac{y^2}{b^2}=1$.

Let P be the point $(a\cos\alpha,\ b\sin\beta)$. Then D is the point $\left[a\cos(\alpha+\pi),\ b\sin\left(\alpha+\frac{\pi}{2}\right)\right]$.

The slope of AP is $m_1=\frac{b\sin\alpha}{a(\cos\alpha-1)}$.

The slope of $A'P$ is $m_2=\frac{b\sin\alpha}{a(\cos\alpha+1)}$.

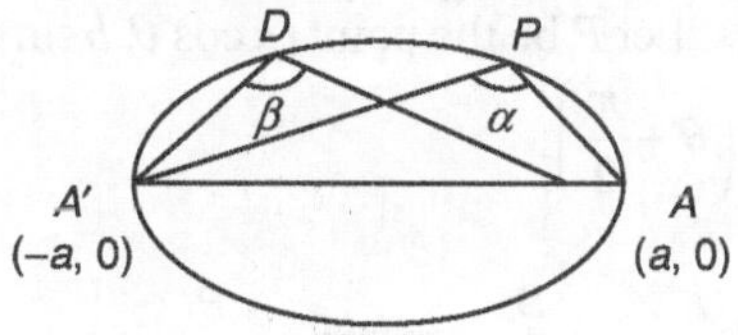

$$\tan\alpha=\pm\frac{(m_1-m_2)}{1+m_1m_2}$$

$$=\pm\frac{\frac{b\sin\alpha}{a(\cos\alpha-1)}-\frac{b\sin\alpha}{a(\cos\alpha+1)}}{1+\frac{b^2\sin^2\alpha}{a^2(\cos^2\alpha-1)}}$$

$$=\pm\frac{ab\sin\alpha(\cos\alpha+1-\cos\alpha+1)}{a^2(\cos^2\alpha-1)+b^2\sin^2\alpha}$$

$$\tan\alpha=\pm\frac{2ab\sin\alpha}{(a^2-b^2)\sin^2\alpha}$$

$$\therefore\ \cot^2\alpha=\frac{(a^2-b^2)^2\sin^2\alpha}{4a^2b^2} \qquad (7.117)$$

Changing a into $\alpha+\dfrac{\pi}{2}$,

$$\therefore \cot^2\beta=\frac{(a^2-b^2)^2\cos^2\alpha}{4a^2b^2} \tag{7.118}$$

Adding equations (7.117) and (7.118), we get

$$\cot^2\alpha+\cot^2\beta=\frac{(a^2-b^2)^2}{4a^2b^2}=a \text{ constant.}$$

Example 7.58

If $x\cos\alpha+y\sin\alpha=p$ is a chord joining the ends P and D of conjugate semi-diameters of the ellipse then prove that $a^2\cos^2\alpha+b^2\sin^2\alpha=2p^2$.

Solution

Let the equation of the ellipse be $\dfrac{x^2}{a^2}+\dfrac{y^2}{b^2}=1$. Let PCP' and DCD' be a pair of conjugate diameters. Let P be the point $(a\cos\theta, b\sin\theta)$ then D is the point $\left[a\cos\left(\theta+\dfrac{\pi}{2}\right), b\sin\left(\theta+\dfrac{\pi}{2}\right)\right]$.

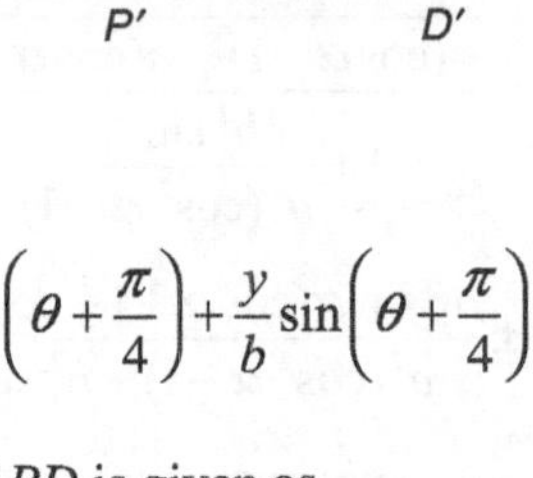

The equation of PD is

$$\frac{x}{a}\cos\left(\theta+\frac{\pi}{4}\right)+\frac{y}{b}\sin\left(\theta+\frac{\pi}{4}\right)=\frac{1}{\sqrt{2}} \tag{7.119}$$

However, the equation of PD is given as

$$x\cos\alpha+y\sin\alpha=\pi \tag{7.120}$$

Equations (7.119) and (7.120) represent the same line. Identifying equations (7.119) and (7.120), we get

$$\frac{\cos\left(\theta+\frac{\pi}{4}\right)}{a\cos\alpha}=\frac{\sin\left(\theta+\frac{\pi}{4}\right)}{b\sin\alpha}=\frac{1}{\sqrt{2}p}$$

$$\cos\left(\theta+\frac{\pi}{4}\right)=\frac{a\cos\alpha}{\sqrt{2}p}$$

$$\sin\left(\theta+\frac{\pi}{4}\right)=\frac{b\sin\alpha}{\sqrt{2}p}$$

(i.e.) $a^2\cos^2\alpha+b^2\sin^2\alpha=2p^2$

Example 7.59

CP and *CD* are conjugate diameters of the ellipse $\frac{x^2}{a^2}+\frac{y^2}{b^2}=1$. A tangent is drawn parallel to *PD* meeting *CP* and *CD* in *R* and *S* respectively. Prove that *R* and *S* lie on the ellipse $\frac{x^2}{a^2}+\frac{y^2}{b^2}=2$.

Solution

Let *CP* and *CD* be a pair of conjugate diameters of the ellipse

$$\frac{x^2}{a^2}+\frac{y^2}{b^2}=1 \tag{7.121}$$

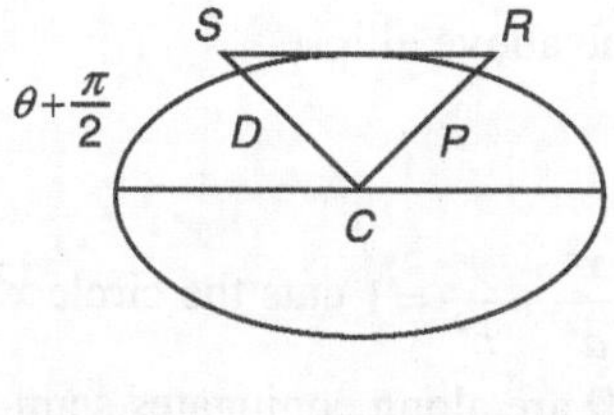

Let *P* be the point $(a\cos\theta, b\sin\theta)$. Then *D* is the point $(-a\sin\theta, b\cos\theta)$. Slope of *PD* is

$$m=\frac{b(\sin\theta-\cos\theta)}{a(\cos\theta+\sin\theta)}=\frac{b(\tan\theta-1)}{a(\tan\theta+1)} \tag{7.122}$$

Let the equation of the tangent parallel to *PD* be $y=mx+\sqrt{a^2m^2+b^2}$.

Let R be the point (h, k). Since (h, k) lies on this tangent,

$$(k - mh) = \sqrt{a^2m^2 + b^2} \tag{7.123}$$

In addition, the equation of CP is $y = \dfrac{b\sin\theta}{a\cos\theta}x$.

Since this passes through (h, k), $k = \dfrac{b\sin\theta}{a\cos\theta}h$.

$$\therefore \quad \frac{ak}{bh} = \tan\theta$$

Substituting in equation (7.122), we get

$$m = \frac{b(ak - bh)}{a(ak + bh)} \tag{7.124}$$

Eliminating m from equations (7.123) and (7.124),

$$k - h\frac{b}{a}\cdot\frac{ak - bh}{ak + bh} = \sqrt{\frac{a^2b^2}{a^2}\left(\frac{ak - bh}{ak + bh}\right)^2 + b^2}$$

$$a^2k^2 + b^2h^2 = ab\sqrt{2(a^2k^2 + b^2h^2)}$$

$$\sqrt{(a^2k^2 + b^2h^2)} = ab\sqrt{2}$$

$$\text{(i.e.)} \quad (a^2k^2 + b^2h^2) = 2a^2b^2$$

Dividing by a^2b^2, $\dfrac{k^2}{b^2} + \dfrac{h^2}{a^2} = 2$. The locus of (h, k) is $\dfrac{x^2}{a^2} + \dfrac{y^2}{b^2} = 2$. Similarly, the point S also lies on the above ellipse.

Example 7.60

A tangent to the ellipse $\dfrac{x^2}{a^2} + \dfrac{y^2}{b^2} = 1$ cuts the circle $x^2 + y^2 = a^2 + b^2$ in P and Q. Prove that CP and CQ are along conjugates semi-diameters of the ellipse where C is the centre of the circle.

Solution

The equation of the ellipse is

$$\frac{x^2}{a^2} + \frac{y^2}{b^2} = 1$$

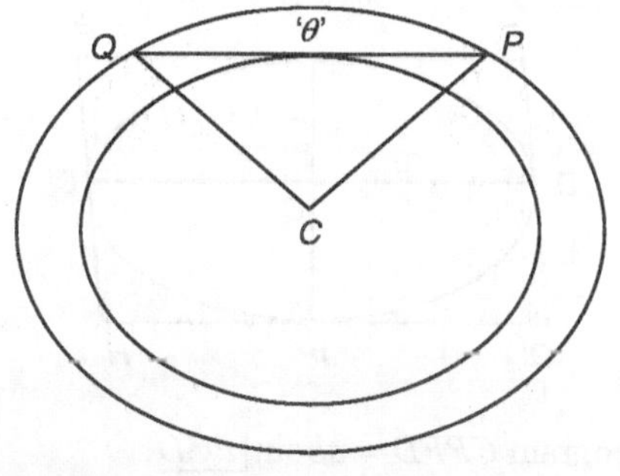

The equation of the circle is $x^2 + y^2 = a^2 + b^2$. (7.125)

The equation of the tangent at θ on the ellipse is

$$\frac{x}{a}\cos\theta + \frac{y}{b}\sin\theta = 1 \quad (7.126)$$

This meets the circle in P and Q.

The combined equation CP and CQ is got by *homogenization* of equation (7.125) with the help of equation (7.126),

$$x^2 + y^2 = (a^2+b^2)\left(\frac{x\cos\theta}{a} + \frac{y\sin\theta}{b}\right)^2$$

$$x^2\left[1 - \frac{a^2+b^2}{a^2}\cos^2\theta\right] + xy\left[\frac{2(a^2+b^2)}{ab}\cos\theta\sin\theta\right] + y^2\left[1 - \frac{a^2-b^2}{b^2}\sin^2\theta\right] = 0$$

$$m_1m_2 = \left[\frac{a^2-(a^2+b^2)\cos^2\theta}{a^2}\cdot\frac{b^2}{b^2-(a^2+b^2)\sin^2\theta}\right]$$

$\therefore$ CP and CQ are conjugate semi-diameters of the ellipse.

Example 7.61

Prove that the acute angle between two conjugate diameters is least when they are of equal length.

Solution

Let PCP' and DCD' be the conjugate diameters.

$$(\overrightarrow{CP} - \overrightarrow{CD})^2 = \overrightarrow{CP}^2 + \overrightarrow{CD}^2 - 2\overrightarrow{CP}\cdot\overrightarrow{CD}$$

$$2\overrightarrow{CP}\cdot\overrightarrow{CD} = (a^2+b^2) - (\overrightarrow{CP} - \overrightarrow{CD})^2 \quad (7.127)$$

$$\because \overrightarrow{CP}^2 + \overrightarrow{CD}^2 = a^2 + b^2$$

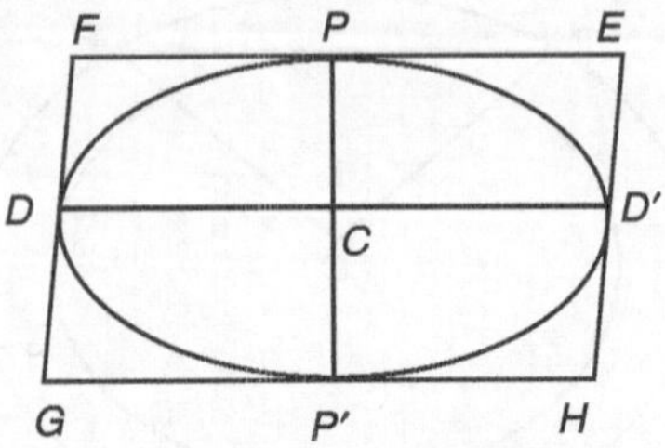

Now area of paralleogram $CPFD = ab\sin\underline{|PCD}$.

$$ab\sin\underline{|PCD} = \overrightarrow{CP}\cdot\overrightarrow{CD} \tag{7.128}$$

From equations (7.127) and (7.128),

$$\frac{2ab}{\sin\underline{|PCD}} = a^2 + b^2 - (\overrightarrow{CP} - \overrightarrow{CD})^2$$

$$\sin\underline{|PCD} = \frac{2ab}{a^2 + b^2 - (\overrightarrow{CP} - \overrightarrow{CD})^2}$$

RHS is least when the denominator is the largest. This happens when $\overrightarrow{CP} = \overrightarrow{CD}$. Therefore, the acute angle between the diameters is minimum when the conjugate diameters are of equal length and the least acute angle is given by $\theta = \sin^{-1}\left(\dfrac{2ab}{a^2 + b^2}\right)$.

Example 7.62

Find the locus of the point of intersection of normals at two points on an ellipse which are extremities of conjugate diameters.

Solution

Let the equation of the ellipse be

$$\frac{x^2}{a^2} + \frac{y^2}{b^2} = 1 \tag{7.129}$$

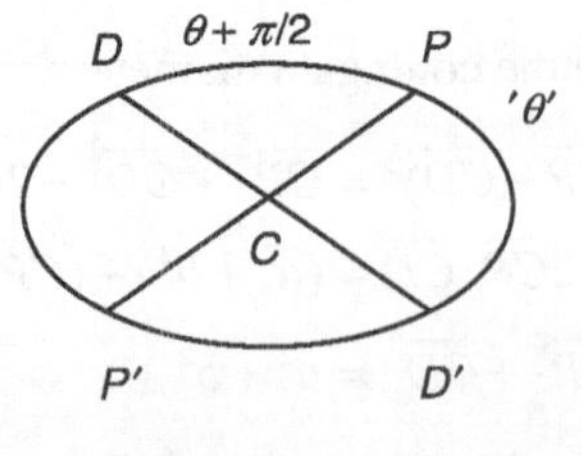

Let P and D be the extremities of a pair of conjugate diameters of the ellipse (7.129). Let P and D be the points $P(a\cos\theta, b\sin\theta)$ and $D(-a\sin\theta, b\cos\theta)$. The equations of the normal at P and D are

$$\frac{ax}{\cos\theta} - \frac{by}{\sin\theta} = a^2 - b^2$$

$$-\frac{ax}{\sin\theta} - \frac{by}{\cos\theta} = a^2 - b^2 \qquad (7.130)$$

$$\text{(i.e.)} \quad \frac{ax}{\sin\theta} + \frac{by}{\cos\theta} = -(a^2 - b^2) \qquad (7.131)$$

Solving equations (7.130) and (7.131), we get,

$$\cos\theta = \frac{a^2x^2 + b^2y^2}{(ax - by)(a^2 - b^2)} \qquad (7.132)$$

$$\sin\theta = \frac{a^2x^2 + b^2y^2}{-(ax + by)(a^2 - b^2)} \qquad (7.133)$$

Squaring and adding, we get

$$1 = \frac{(a^2x^2 + b^2y^2)^2}{(a^2 - b^2)^2}\left[\frac{1}{(ax - by)^2} + \frac{1}{(ax + by)^2}\right]$$

$$1 = \left(\frac{a^2x^2 + b^2y^2}{a^2 - b^2}\right)^2 = \frac{2(a^2x^2 + b^2y^2)}{a^2x^2 - b^2y^2}$$

Therefore, the locus of the point of intersection of these two normals is $(a^2x^2 + b^2y^2)^3 = (a^2 - b^2)^2 (a^2x^2 - b^2y^2)^2$.

Example 7.63

If the point of intersection of the ellipses $\frac{x^2}{a^2} + \frac{y^2}{b^2} = 1$ and $\frac{a^2}{\alpha^2} + \frac{b^2}{\beta^2} = 1$ be at the extremities of the conjugate diameters of the former then prove that $\frac{x^2}{a^2} + \frac{y^2}{b^2} = 2$.

Solution

The given ellipses are

$$\frac{x^2}{a^2} + \frac{y^2}{b^2} = 1 \qquad (7.134)$$

$$\frac{x^2}{\alpha^2}+\frac{y^2}{\beta^2}=1 \tag{7.135}$$

Solving equations (7.134) and (7.135) we get their point of intersections.

Equation (7.134) – (7.135) gives $x^2\left(\frac{1}{a^2}-\frac{1}{\alpha^2}\right)+y^2\left(\frac{1}{b^2}-\frac{1}{\beta^2}\right)=0.$

This is a pair of straight lines passing through the origin. If $y = mx$ is one of the lines then $x^2\left(\frac{1}{a^2}-\frac{1}{\alpha^2}\right)+m^2x^2\left(\frac{1}{a^2}-\frac{1}{\beta^2}\right)=0$

$$\text{(i.e.)}\quad \frac{1}{a^2}-\frac{1}{\alpha^2}+m^2\left(\frac{1}{b^2}-\frac{1}{\beta^2}\right)=0 \tag{7.136}$$

This is a quadratic equation in m. If m_1 and m_2 are the slopes of the two straight lines through the origin then

$$m_1m_2=\frac{\frac{1}{a^2}-\frac{1}{\alpha^2}}{\frac{1}{b^2}-\frac{1}{\beta^2}} \tag{7.137}$$

If m_1 and m_2 are the slopes of the pair of conjugate diameters then

$$m_1m_2=\frac{-b^2}{a^2} \tag{7.138}$$

From equations (7.137) and (7.138), we get

$$\frac{\frac{1}{a^2}-\frac{1}{\alpha^2}}{\frac{1}{b^2}-\frac{1}{\beta^2}}=\frac{-b}{a^2}$$

$$\text{(i.e.)}\quad a^2\left(\frac{1}{a^2}-\frac{1}{\alpha^2}\right)=-b^2\left(\frac{1}{b^2}-\frac{1}{\beta^2}\right)$$

$$1-\frac{a^2}{\alpha^2}=-1+\frac{b^2}{\beta^2}$$

$$\therefore \frac{a^2}{\alpha^2}+\frac{b^2}{\beta^2}=2$$

Example 7.64

Let P and Q be the extremities of two conjugate diameters of the ellipse $\frac{x^2}{a^2}+\frac{y^2}{b^2}=1$ and S be the focus. Then prove that $PQ^2-(SP-SQ)^2=2b^2$.

Solution

Let S be $(ae, 0)$ and P be $(a\cos\theta, b\sin\theta)$. Then $SP = a - ae\cos\theta$, $SQ = a + ae\sin\theta$.

$$\begin{aligned} PQ^2 &= (a\cos\theta + a\sin\theta)^2 + (b\sin\theta - b\cos\theta)^2 \\ &= a^2 + 2a^2\cos\theta\sin\theta - 2b^2\cos\theta\sin\theta + b^2 \\ &= a^2 + b^2 + 2\cos\theta\sin\theta(a^2-b^2) \\ &= a^2 + b^2 + 2(a^2-b^2)\cos\theta\sin\theta \\ (SP-SQ)^2 &= [(a-ae\cos\theta)-(a+ae\sin\theta)]^2 \\ &= a^2e^2(\cos\theta+\sin\theta)^2 \\ &= a^2e^2(1+2\sin\theta\cos\theta) \\ PQ^2-(SP-SQ)^2 &= a^2+b^2+2(a^2-b^2)\sin\theta\cos\theta-(a^2-b^2) \\ &\quad -(a^2-b^2)\cos\theta\sin\theta \\ &= a^2+b^2-(a^2-b^2) \\ &= 2b^2 \end{aligned}$$

Example 7.65

If CP and CD are semi-conjugate diameters of the ellipse $\frac{x^2}{a^2}+\frac{y^2}{b^2}=1$, prove that the locus of the orthocentre of ΔPCD is $2(b^2y^2+a^2x^2)^3 = (a^2-b^2)^2(a^2x^2-b^2y^2)^2$.

Solution

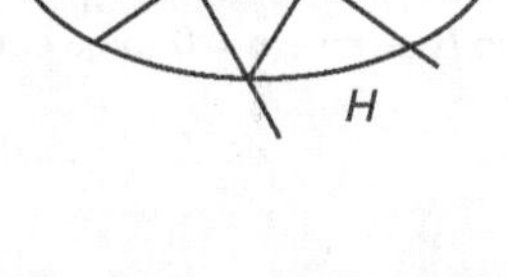

Let P be the point $(a\sin\theta, b\cos\theta)$. Then D is $(-a\sin\theta, b\cos\theta)$. Tangent at P is parallel to CD. Tangent at D is parallel to CP.

Therefore, the altitudes through P and D are the normals at P and D, respectively. Let (x_1, y_1) be the orthocentre.

The equation of the normal at P is

$$\frac{ax}{\cos\theta} - \frac{by}{\sin\theta} = a^2 - b^2 \tag{7.139}$$

The equation of the normal at Q is

$$\frac{-ax}{\sin\theta} - \frac{by}{\sin\theta} = a^2 - b^2 \tag{7.140}$$

Solving equations (7.139) and (7.140) we get the coordinates of the orthocentre.

$$\frac{1}{(ax-by)\cos\theta} = \frac{1}{-(ax+by)\sin\theta} = \frac{a^2-b^2}{a^2x^2+b^2y^2}$$

$$\therefore \cos\theta = \frac{a^2x^2+b^2y^2}{+(a^2-b^2)(ax-by)},\ \sin\theta = \frac{a^2x^2+b^2y^2}{-(a^2-b^2)(ax+by)}$$

Squaring and adding, we get

$$1 = \frac{(a^2x^2+b^2y^2)^2}{(a^2-b^2)^2}\left[\frac{1}{(ax-by)^2} + \frac{1}{(ax+by)^2}\right]$$

$$\therefore \frac{(a^2-b^2)^2}{(a^2x^2+b^2y^2)^2} = \frac{2(a^2x^2+b^2y^2)}{(a^2x^2-b^2y^2)^2}$$

Therefore, the locus of the orthocentre is

$$2(a^2x^2+b^2y^2)^3 = (a^2-b^2)^2(a^2x^2-b^2y^2)^2.$$

Exercises

1. Let CP and CQ be a pair of conjugate diameters of an ellipse and let the tangents at P and Q meet at R. Show that CR and PQ bisect each other.
2. Find the condition that for the diameters of $\frac{x^2}{a^2} + \frac{y^2}{b^2} = 1$ through its points of intersection with the line $lx + my + n = 0$ to be conjugate.

Ans.: $l^2 + m^2 = a^2l^2 + b^2m^2$

3. Prove that $b^2x^2 + 2hxy - a^2y^2 = 0$ represents conjugate diameters of the ellipse $\frac{x^2}{a^2} + \frac{y^2}{b^2} = 1$ for all values of h.
4. Prove that $a^2x^2 + 2hxy - b^2y^2 = 0$ represents conjugate diameters of the ellipse $ax^2 + by^2 = 1$ for all values of h.
5. Find the coordinates of the ends of the diameter of the ellipse $16x^2 + 25y^2 = 400$ which is conjugate to $5y = 4x$. **Ans.:** $(2\sqrt{2}, -\frac{5}{2}\sqrt{2}), (-2\sqrt{2}, \frac{5\sqrt{2}}{2})$
6. Find the length of semi-diameter conjugate to the diameter whose equation is $y = 3x$.
7. Through the foci of an ellipse, perpendiculars are drawn to a pair of conjugate diameters. Prove that they meet on a concentric ellipse.
8. A diameter of the ellipse $\frac{x^2}{a^2} + \frac{y^2}{b^2} = 1$ meets one latus rectum in P and the conjugate diameter meets the other latus rectum in Q. Prove that PQ touches $\frac{x^2}{a^2} + \frac{y^2}{b^2} = e^2$.
9. If PP' is a diameter and Q is any point on the ellipse, prove that QP and QP' are parallel to a pair of conjugate diameters of the ellipse.
10. If $\alpha + \beta = \gamma$ (a constant) then prove that the tangents at a and b on the ellipse $\frac{x^2}{a^2} + \frac{y^2}{b^2} = 1$. intersect on the diameter through γ.
11. Show that the line joining the extremities of any two diameters of an ellipse which are at right angles to one another will always touch a fixed circle.
12. Show that the sum of the reciprocals of the square of any two diameters of an ellipse which are at right angles to one another is a constant.
13. P and Q are extremities of two conjugate diameters of the ellipse $\frac{x^2}{a^2} + \frac{y^2}{b^2} = 1$. and S is a focus. Prove that $PQ^2 + (SP - SQ)^2 = 2b^2$.
14. If the distance between the two foci of an ellipse subtends angles 2α and 2β at the ends of a pair of conjugate diameters. Show that $\tan 2\alpha + \tan 2\beta$ is a constant.
15. Show that the sum of the squares of the normal at the extremities of conjugate semi-diameters and terminated by major axis is $a^2(1 - e^2)(2 - e^2)$.
16. If P and Q are two points on an ellipse such that CP is conjugate to the normal at Q, prove that CQ is conjugate to the normal at P.
17. Two conjugate diameters of the ellipse $\frac{x^2}{a^2} + \frac{y^2}{b^2} = 1$ centre at C meet the tangent at any point P is E and F. Prove that $PE \cdot PF = CD^2$.
18. If CP and CD are conjugate semi-diameters of the ellipse $\frac{x^2}{a^2} + \frac{y^2}{b^2} = 1$, the normal at P cuts the major axis at G and the line DC in F then prove that $PG : CD = b : a$.

19. The normal at a variable point P of an ellipse $\frac{x^2}{a^2}+\frac{y^2}{b^2}=1$ cuts the diameter CD conjugate to P in Q. Prove that the equation of the locus of Q is $\frac{a^2}{x^2}+\frac{b^2}{y^2}=\left(\frac{a^2-b^2}{x^2+y^2}\right)^2$.

20. Show that for a parallelogram inscribed in an ellipse, the sum of the squares of the sides is constant.

21. Show that the maximum value of the smaller of two angles between two conjugate diameters of an ellipse is $\frac{\pi}{2}$ and the minimum value of this angle is $\tan^{-1}\left(\frac{2ab}{a^2-b^2}\right)$ where a and b are its semi-major and semi-minor axes, respectively.

22. If PCP' and DCD' are two conjugate diameters of the ellipse $\frac{x^2}{a^2}+\frac{y^2}{b^2}=1$. and Q is any point on the circle $x^2+y^2=c^2$ then prove that $PQ^2+DQ^2+P'Q^2+PQ^2=2(a^2+b^2+2c^2)$.

23. Two conjugate diameters of the ellipse $\frac{x^2}{a^2}+\frac{y^2}{b^2}=1$ cut the circle $x^2+y^2=r^2$ at P and Q. Show that the locus of the midpoint of PQ is $a^2[(x^2+y^2)^2-r^2x^2]+b^2[(x^2+y^2)^2-r^2y^2]=0$.

24. In an ellipse whose semi-axes are a and b, prove that the acute-angle between two conjugate diameters cannot be less than $\sin^{-1}\left(\frac{2ab}{a^2+b^2}\right)$.

25. If CP and CD are conjugate diameters of an ellipse show that $4(CP^2-CD^2)=(SP-S'P)^2-(SD-S'D)^2$.

26. Two conjugate semi-diameters of an ellipse are inclined at angles α and β to the major axis. Show that their lengths c and d are connecting the relation $c^2\sin 2\alpha+d^2\sin 2\beta=0$.

27. Find the condition for the lines $l_1x+m_1y=0$ and $l_2x+m_2y=0$ to be conjugate diameters of $\frac{x^2}{a^2}+\frac{y^2}{b^2}=1$. **Ans.:** $a^2ll_1+b^2mm_1=0$

28. Show that $ax^2+2hxy-by^2=0$ represents conjugate diameters of the ellipse $ax^2+by^2=1$ for all values of a.

29. Prove that $ax^2+2hxy-by^2=0$ represents conjugate diameters of the ellipse $ax^2+by^2=1$ for all values of h.

30. CP and CQ are conjugate semi-diameters of the ellipse $\frac{x^2}{a^2}+\frac{y^2}{b^2}=1$. A tangent parallel to PQ meets CP and CQ in R and S, respectively. Show that R and S lie on the ellipse $\frac{x^2}{a^2}+\frac{y^2}{b^2}=2$.

31. If two conjugate diameters CP and CQ of an ellipse cut the director circle in L and M, prove that LM touches the ellipse.

32. Two conjugate diameters of the ellipse $\frac{x^2}{a^2}+\frac{y^2}{b^2}=1$ cuts the circle $x^2+y^2=r$ at P and Q. Show that the locus of the midpoint of PQ is $a^2[(x^2+y^2)^2-r^2x^2]+b^2[(x^2+y^2)^2-r^2y^2]=0$.

33. The eccentric angles of two points P and Q on the ellipse $\frac{x^2}{a^2}+\frac{y^2}{b^2}-1$ are α and β. Prove that the area of the parallelogram formed by the tangents at the ends of the diameters through P and Q is $\frac{4ab}{\sin(\alpha-\beta)}$ and hence show that it is least when P and Q are the extremities of a pair of conjugate diameters.

34. Let PCP' be a diameter of the ellipse $\frac{x^2}{a^2}+\frac{y^2}{b^2}=1$. If the normal at P meets the ordinate at P' in T, show that the locus of T is $\frac{x^2}{a^2}+\frac{h^2y^2}{(2a^2-b^2)}=1$.

35. If two conjugate diameters CP and CQ of an ellipse cut the director circle in L and M, prove that LM touches the ellipse.

36. In an ellipse, a pair of conjugate diameters is produced to meet a directrix. Show that the orthocentre of the triangles so formed is a focus.

37. Through a fixed point P, a pair of lines is drawn parallel to a variable pair of conjugate diameters of a given ellipse. The lines meet the principal axes in Q and R, respectively. Show that the midpoint of QR lies on a fixed line.

38. Perpendiculars PM and PN are drawn from any point P of an ellipse on the equi-conjugate diameter of the ellipse. Prove that the perpendiculars from P to its polar bisect MN.

39. In the ellipse $3x^2+7y^2=21$, find the equations of the equi-conjugate diameters and their lengths. **Ans.:** $\sqrt{3}x \pm\sqrt{7}y=0$.

40. Prove that the tangents to the ellipse $\frac{x^2}{a^2}+\frac{y^2}{b^2}=1$ at the points whose eccentric angles are θ and $\frac{\pi}{2}-\theta$ meet on one of the equi-conjugate diameters.

41. From a point on one of the equi-conjugate diameters of an ellipse tangents are drawn to the ellipse. Show that the sum of the eccentric angles of the point of contact is an odd multiple of $\frac{\pi}{2}$.

42. Tangents are drawn from any point on the ellipse $\frac{x^2}{a^2}+\frac{y^2}{b^2}=1$ to the circle $x^2+y^2=r^2$. Prove that the chords of contact are tangents to the ellipse $a^2x^2+b^2y^2=r^4$. If $\frac{1}{r^4}=\frac{1}{a^2}-\frac{1}{b^2}$, prove that the line and the centre to the points of contact with the circle are conjugate diameters of the second ellipse.

43. Any tangent to an ellipse meets the director circle in P and D. Prove that CP and CD are in the directions of conjugate diameters of the ellipse.

44. If CP is conjugate to the normal at Q, prove that CQ is conjugate to the normal at P.

45. Prove that the straight lines joining the centre to the intersection of the straight line $y = mx + \frac{\sqrt{a^2m^2 + b^2}}{2}$ with the ellipse are conjugate diameters.

CHAPTER

8 Hyperbola

8.1 DEFINITION

A hyperbola is defined as the locus of a point that moves in a plane such that its distance from a fixed point is always e times ($e > 1$) its distance from a fixed line. The fixed point is called the focus of the hyperbola. The fixed straight line is called the directrix and the constant e is called the eccentricity of the hyperbola.

8.2 STANDARD EQUATION

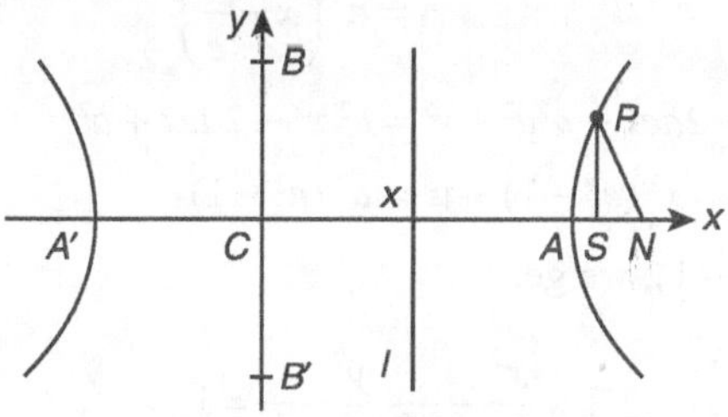

Let S be the focus and the line l be the directrix. Draw SX perpendicular to the directrix. Divide SX internally and externally in the ratio $e : 1$ ($e > 1$). Let A and A' be the point of division. Since $\frac{SA}{AX} = e$ and $\frac{SA'}{A'X} = e$, the points A and A' lie on the curve.

Let $AA' = 2a$ and C be its middle point.

$$SA = e \cdot AX \tag{8.1}$$

$$SA' = e \cdot A'X \tag{8.2}$$

Adding equations (8.1) and (8.2), we get

$$SA + SA' = e(AX + A'X)$$

$$(CS - CA) + (CS + CA') = eAA'$$
$$2CS = e \cdot 2a$$
$$\therefore\ CS = ae$$

Subtracting equation (8.1) from equation (8.2), we get

$$SA' - SA = e(A'X - AX)$$
$$AA' = e \cdot 2CX$$
$$2a = e \cdot 2CX$$
$$\therefore\ CX = \frac{a}{e}$$

Take CS as the x-axis and CY perpendicular to CX as the y-axis. Then, the coordinates of S are $(ae, 0)$. Let $P(x, y)$ be any point on the curve. Draw PM perpendicular to the directrix and PN perpendicular to x-axis.

From the focus directrix property of hyperbola, $\frac{SP}{PM} = e.$

$$SP^2 = e^2 \cdot PM^2$$
$$\Rightarrow (x - ae)^2 + y^2 = e^2 \cdot NX^2 = e^2(CN - CX)^2$$
$$= e^2\left(x - \frac{a}{e}\right)^2$$
$$x^2 - 2aex + a^2e^2 + y^2 = e^2 x^2 - 2aex + a^2$$
$$x^2 (e^2 - 1) - y^2 = a^2 (e^2 - 1)$$

Dividing by $a^2 (e^2 - 1)$, we get

$$\frac{x^2}{a^2} - \frac{y^2}{a^2(e^2 - 1)} = 1$$

$$\text{(i.e.)}\quad \frac{x^2}{a^2} - \frac{y^2}{b^2} = 1 \text{ where } b^2 = a^2 (e^2 - 1)$$

This is called the standard equation of hyperbola.

Note 8.2.1:

1. The curve meets the x-axis at points $(a, 0)$ and $(-a, 0)$.
2. When $x = 0$, $y^2 = -a^2$. Therefore, the curve meets the y-axis only at imaginary points, that is, there are no real points of intersection of the curve and y-axis.
3. If (x, y) is a point on the curve, $(x, -y)$ and $(-x, y)$ are also points on the curve. This shows that the curve is symmetrical about both the axes.

4. For any value of y, there are two values of x; as y increases, x increases and when $y \to \infty$, x also $\to \infty$. The curve consists of two symmetrical branches, each extending to infinity in both the directions.
5. AA' is called the **transverse axis** and its **length** is **$2a$.**
6. BB' is called the **conjugate axis** and its **length** is **$2b$.**
7. A hyperbola in which $a = b$ is called a **rectangular hyperbola**. Its equation is $x^2 - y^2 = a^2$. Its eccentricity is $e = \sqrt{\frac{a^2 + b^2}{a^2}} = \sqrt{2}$.
8. The double ordinate through the focus S is called **latus rectum** and its length is $\frac{2b^2}{a}$.
9. There is a second focus S' and a second directrix l' to the hyperbola.

8.3 IMPORTANT PROPERTY OF HYPERBOLA

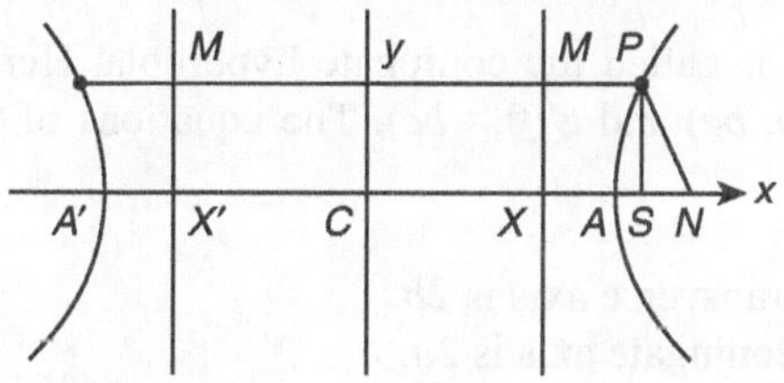

The difference of the focal distances of any point on the hyperbola is equal to the length of transverse axis.

$$SP = ePM;\ S'P = ePM'$$

$$S'P - SP = e(PM' - PM) = e \cdot MM' = eXX' = e(CX + CX') = e\left(2 \cdot \frac{a}{e}\right) = 2a.$$

8.4 EQUATION OF HYPERBOLA IN PARAMETRIC FORM

$(a \sec \theta, b \tan \theta)$ is a point on the hyperbola $\frac{x^2}{a^2} - \frac{y^2}{b^2} = 1$ for all values of θ, θ is called a parameter and is denoted by 'θ'. The parametric equations of hyperbola are $x = a \sec \theta$, $y = b \tan \theta$.

8.5 RECTANGULAR HYPERBOLA

A hyperbola in which $b = a$ is called a rectangular hyperbola. The standard equation of the rectangular hyperbola is $x^2 - y^2 = a^2$.

8.6 CONJUGATE HYPERBOLA

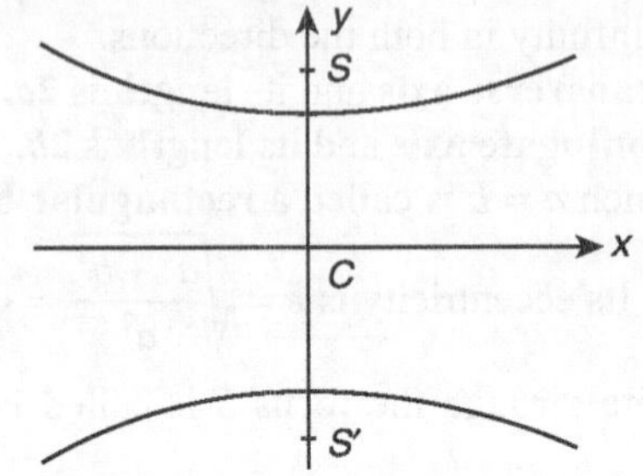

The foci are $S(ae, 0)$ and $S'(-ae, 0)$ and the equations of the directrices are $x = \pm\frac{a}{e}$. By the symmetry of the hyperbola, if we take the transverse axis as the y-axis and the conjugate axis as x-axis, then the equation of the hyperbola is $\frac{y^2}{b^2} - \frac{x^2}{a^2} = 1$ (i.e.) $\frac{x^2}{a^2} - \frac{y^2}{b^2} = -1$.

This hyperbola is called the conjugate hyperbola. Here, the coordinates of the foci are $S(0, be)$ and $S'(0, -be)$. The equations of the directrices are $y = \pm\frac{b}{e}$.

The length of the transverse axis is $2b$.
The length of the conjugate axis is $2a$.
The length of the latus rectum is $\frac{2a^2}{b}$.

The following are some of the standard results of the hyperbola whose equation is $\frac{x^2}{a^2} - \frac{y^2}{b^2} = 1$:

1. The equation of the tangent at (x_1, y_1) is $\frac{xx_1}{a^2} - \frac{yy_1}{b^2} = 1$.
2. The equation of the normal at (x_1, y_1) is $\frac{a^2x}{x_1} + \frac{b^2y}{y_1} = a^2 + b^2$.
3. The equation of the chord of contact of tangents from (x_1, y_1) is $\frac{xx_1}{a^2} - \frac{yy_1}{b^2} = 1$.
4. The polar of (x_1, y_1) is $\frac{xx_1}{a^2} - \frac{yy_1}{b^2} = 1$.
5. The condition that the straight line $y = mx + c$ is a tangent to the hyperbola is $c^2 = a^2m^2 - b^2$ and $y = mx + \sqrt{a^2m^2 - b^2}$ is the equation of a tangent.
6. The equation of the chord of the hyperbola having (x_1, y_1) as the midpoint is $T = S_1$ (i.e.) $\frac{xx_1}{a^2} - \frac{yy_1}{b^2} - 1 = \frac{x_1^2}{a^2} - \frac{y_1^2}{b^2} - 1$ or $\frac{xx_1}{a^2} - \frac{yy_1}{b^2} = \frac{x_1^2}{a^2} - \frac{y_1^2}{b^2}$.

7. The equation of the pair of tangents from (x_1, y_1) is $T^2 = SS_1$

(i.e.) $\left(\frac{xx_1}{a^2}-\frac{yy_1}{b^2}-1\right)^2 = \left(\frac{x^2}{a^2}-\frac{y^2}{b^2}-1\right)\left(\frac{x_1^2}{a^2}-\frac{y_1^2}{b^2}-1\right)$

8. **Parametric representation:** $x = a\sec\theta, y = b\tan\theta$ is a point on the hyperbola and this point is denoted by θ. θ is called a parameter of the hyperbola.

The equation of the tangent at θ is $\frac{x}{a}\sec\theta - \frac{y}{b}\tan\theta = 1$.

The equation of the normal at θ is $\frac{ax}{\sec\theta}+\frac{by}{\tan\theta} = a^2+b^2$.

9. The circle described on the transverse axis as diameter is called the auxiliary circle and its equation is $x^2 + y^2 = a^2$.

10. The equation of the director circle (the locus of the point of intersection of perpendicular tangents) is $x^2 + y^2 = a^2 - b^2$.

Example 8.6.1

Find the equation of the hyperbola whose focus is (1, 2), directrix $2x + y = 1$ and eccentricity $\sqrt{3}$.

Solution

Let $P(x_1, y_1)$ be any point on the hyperbola. Then $\frac{SP}{PM} = e$. S is (1,2).

$$\therefore SP^2 = (x_1 - 1)^2 + (y_1 - 2)^2 = x_1^2 + y_1^2 - 2x_1 - 4y_1 + 5$$

$$PM = \pm\frac{2x_1 + y_1 - 1}{\sqrt{5}}$$

Since $$SP^2 = e^2 PM^2,\ x_1^2 + y_1^2 - 2x_1 - 4y_1 + 5 = 3\left(\frac{2x_1 + y_1 - 1}{\sqrt{5}}\right)^2$$

$$5(x_1^2 + y_1^2 - 2x_1 - 4y_1 + 5) = 3(4x_1^2 + y_1^2 + 1 + 4x_1y_1 - 4x_1 - 2y_1)$$

$$\text{(i.e.)}\quad 7x_1^2 - 2y_1^2 + 12x_1y_1 - 2x_1 + 14y_1 - 22 = 0$$

Hence, the equation of the hyperbola which is the locus of

(x_1, y_1) is $7x^2 + 12xy - 2y^2 - 2x + 14y - 22 = 0$.

Example 8.6.2

Show that the equation of the hyperbola having focus (2,0), eccentricity 2 and directrix $x - y = 0$ is $x^2 + y^2 - 4xy + 4x - 4 = 0$.

Solution

S is (2, 0) : $e = 2$ and equation of the directrix is $x - y = 0$. Let P (x, y) be any point on the hyperbola. Then, $\frac{SP}{PM} = e$.

$$SP^2 = e^2 PM^2 \Rightarrow (x-2)^2 + y^2 = 4\left(\frac{x-y}{\sqrt{2}}\right)^2$$

$$x^2 + 4 - 4x + y^2 = 2\,(x^2 + y^2 - 2xy)$$

Hence, the equation of the hyperbola is $x^2 + y^2 - 4xy + 4x - 4 = 0$.

Example 8.6.3

Find the equation of the hyperbola whose focus is (2, 2), eccentricity $\frac{3}{2}$ and directrix

$$3x - 4y = 1$$

Solution

S is $(2,2) : e = \frac{3}{2}$ and directrix $3x - 4y = 1$. Let P (x, y) be any point on the hyperbola.

$$\frac{SP}{PM} = e \quad \text{or} \quad SP^2 = e^2 PM^2$$

$$(x-2)^2 + (y-2)^2 = \frac{9}{4}\left(\frac{3x-4y-1}{5}\right)^2$$

$$\Rightarrow 100\,(x^2 - 4x + 4 + y^2 - 4y + 4) = 9\,(9x^2 + 16y^2 + 1 - 24xy - 6x + 8y)$$

Hence, the equation of the hyperbola is $19x^2 + 216xy - 44y^2 - 346x - 472y - 791 = 0$.

Example 8.6.4

Find the equation of the hyperbola whose focus is (0, 0), eccentricity $\frac{5}{4}$ and directrix

$$x\cos\alpha + y\sin\alpha = p$$

Solution

$$SP^2 = x^2 + y^2 : \; e = \frac{5}{4}$$

$$PM = \left|\frac{x\cos\alpha + y\sin\alpha - p}{\sqrt{\cos^2\alpha + \sin^2\alpha}}\right| = |x\cos\alpha + y\sin\alpha - p|$$

For any point on the hyperbola,

$$SP^2 = e^2 PM^2 \Rightarrow x^2 + y^2 = \frac{25}{16}(x\cos\alpha + y\sin\alpha - p)^2$$

Hence, the equation of the hyperbola is $16(x^2 + y^2) - 25(x\cos\alpha + y\sin\alpha - p)^2 = 0$.

Example 8.6.5

Find the equation of the hyperbola whose foci are (6, 4) and (–4, 4) and eccentricity 2.

Solution

S is (6, 4) and S' (–4, 4), and C is the midpoint of SS'

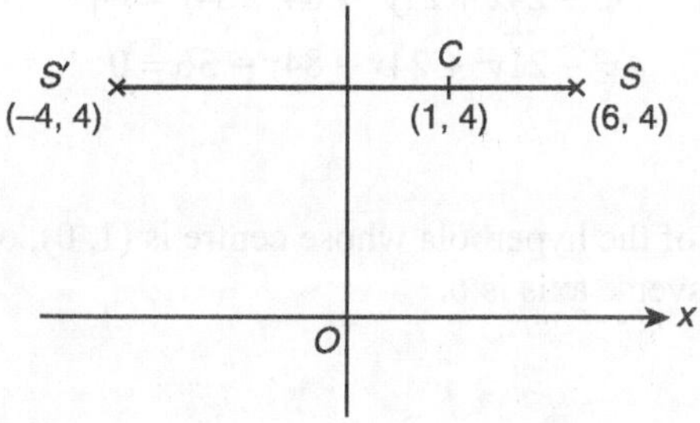

$\therefore$ C is (1, 4)

$$CS = 6 - 1 = 5. \text{ But } CS = ae = 2a.$$

$$\therefore 2a = 5 \quad \text{or} \quad a = \frac{5}{2}$$

$$b^2 = a^2(e^2 - 1) = \frac{25}{4} \times 3 = \frac{75}{4}$$

Hence, the equation of the hyperbola is

$$\frac{(x-1)^2}{\frac{25}{4}} - \frac{(y-4)^2}{\frac{75}{4}} = 1$$

$$\frac{4(x-1)^2}{25} - \frac{4(y-4)^2}{75} = 1$$

Example 8.6.6

Find the equation of the hyperbola whose center is (–3, 2), one end of the transverse axis is (–3, 4) and eccentricity is $\frac{5}{2}$.

Solution

Centre is (--3, 4)
A is (−3, 4) ∴ A' is (−3, 6); $a = 2$

$$b^2 = a^2(e^2 - 1) = 4\left(\frac{25}{4} - 1\right) = 21$$

Hence, the equation of the hyperbola is

$\frac{(y-2)^2}{4} - \frac{(x+3)^2}{21} = 1$ (since the line parallel to y-axis is the transverse axis)

$$\frac{y^2 + 4 - 4y}{4} - \frac{x^2 + 9 + 6x}{21} = 1$$

$$-4x^2 - 36 - 24x + 21y^2 + 84 - 84y = 84$$

$$4x^2 - 21y^2 + 24x + 84y + 36 = 0$$

Example 8.6.7

Find the equation of the hyperbola whose centre is (1, 0), one focus is (6, 0), and length of transverse axis is 6.

Solution

$$2a = 6 \therefore a = 3$$

$$b^2 = a^2(e^2 - 1) = 9\left(\frac{25}{9} - 1\right) = 16$$

$$ae = 5 \quad \therefore e = \frac{5}{3}$$

Hence, the equation of the hyperbola is $\frac{(x-1)^2}{9} - \frac{y^2}{16} = 1$ (i.e.) $16x^2 - 9y^2 - 32x - 128 = 0$.

Example 8.6.8

Find the equation of the hyperbola whose centre is (3, 2), one focus is (5, 2) and one vertex is (4, 2).

Solution

C is (3, 2), A is (4, 2) and S is (5, 2).
Hence, $CA = 1$and the transverse axis is parallel to x-axis.

$$\therefore a = 1$$

Also $ae = 2$. Since $a = 1$ and $e = 2$, $b^2 = a^2 (e^2 - 1) = 1 (4 - 1) = 3$.

Hence, the equation of the hyperbola is $\dfrac{(x-3)^2}{1} - \dfrac{(y-2)^2}{3} = 1$

$$3x^2 - 18x + 27 - y^2 + 4y - 4 = 3$$

$$3x^2 - y^2 - 18x + 4y + 20 = 0$$

Example 8.6.9

Find the equation of the hyperbola whose centre is (6, 2), one focus is (4, 2) and $e = 2$.

Solution

Transverse axis is parallel to x-axis and $CS = 2$ units in magnitude.

$$ae = 2 \quad \therefore a = 1$$

$$b^2 = a^2 (e^2 - 1) = 1 (4 - 1) = 3$$

Hence, the equation of the hyperbola is $\dfrac{(x-6)^2}{1} - \dfrac{(y-2)^2}{3} = 1.$

Example 8.6.10

Find the centre, eccentricity and foci of hyperbola $9x^2 - 16y^2 = 144$.

Solution

Dividing by 144, we get

$$\frac{x^2}{16} - \frac{y^2}{9} = 1$$

Hence, the centre of the hyperbola is (0, 0)

$$\therefore a^2 = 16, b^2 = 9 \quad \therefore a = 4, b = 3$$

$$b^2 = a^2(e^2 - 1) \Rightarrow 9 = 16(e^2 - 1) \Rightarrow e^2 - 1 = \frac{9}{16}$$

$$\therefore e = \frac{5}{4}$$

Hence, the foci are (5, 0) and (–5, 0).

Example 8.6.11

Find the centre, foci and eccentricity of $12x^2 - 4y^2 - 24x + 32y - 127 = 0$

Solution

$$(12x^2 - 24x) - (4y^2 - 32y) - 127 = 0$$
$$12(x^2 - 2x) - 4(y^2 - 8y) - 127 = 0$$
$$12(x-1)^2 - 12 - 4(y-4)^2 + 64 - 127 = 0$$
$$12(x-1)^2 - 4(y-4)^2 = 75$$
$$\frac{(x-1)^2}{\left(\frac{75}{12}\right)} - \frac{(y-4)^2}{\left(\frac{75}{4}\right)} = 1$$

Hence, centre is (1, 4).

$$a^2 = \frac{75}{12}, b^2 = \frac{75}{4}$$
$$b^2 = a^2(e^2 - 1)$$
$$\frac{75}{4} = \frac{75}{12}(e^2 - 1)$$
$$\Rightarrow e^2 - 1 = \frac{12}{4} = 3$$
$$\Rightarrow e^2 = 4 \quad \text{or} \quad e = 2$$
$$CS = ae = \frac{5\sqrt{3}}{2\sqrt{3}} \times 2 = \frac{5}{2} \times 2 = 5$$

Hence, the foci are (6, 4) and (– 4, 4).

Example 8.6.12

Find the centre and eccentricity of the hyperbola $9x^2 - 4y^2 + 18x + 16y - 43 = 0$.

Solution

$$9(x^2 + 2x) - 4(y^2 - 4y) - 43 = 0$$
$$9(x+1)^2 - 9 - 4(y-2)^2 + 19 - 43 = 0$$

$$9(x+1)^2 - 4(y-2)^2 = 36$$

Hence, centre is $(-1, 2)$, $a^2 = 4$ and $b^2 = 9$.

$$\frac{(x+1)^2}{4} - \frac{(y-2)^2}{9} = 1$$

$$b^2 = a^2(e^2 - 1) \text{ gives } e = \frac{\sqrt{13}}{2}.$$

Example 8.6.13

If from the centre C of the hyperbola $x^2 - y^2 = a^2$, CM is drawn perpendicular to the tangent at any point of the curve meeting the tangent at M and the curve at N, show that $CM \cdot CN = a^2$.

Solution

The equation of the tangent at $P(x_1, y_1)$ in $x^2 - y^2 = a^2$ is $xx_1 - yy_1 = a^2$.

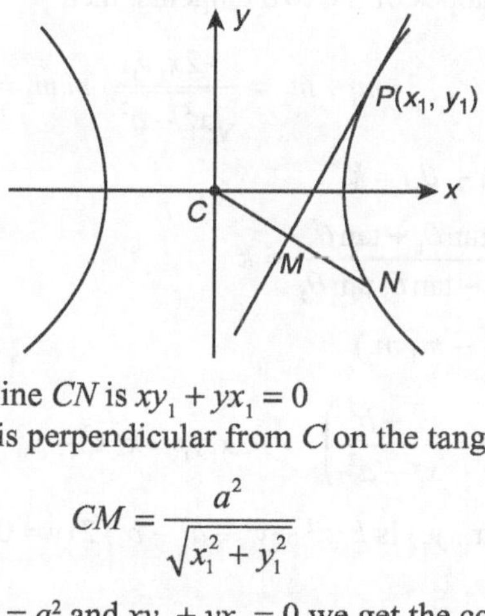

The equation of the line CN is $xy_1 + yx_1 = 0$

Then CM, which is perpendicular from C on the tangent, is given by

$$CM = \frac{a^2}{\sqrt{x_1^2 + y_1^2}}$$

Solving $x^2 - y^2 = a^2$ and $xy_1 + yx_1 = 0$ we get the coordinates of N

$$x^2 - \frac{x^2 y_1^2}{x_1^2} = a^2 \Rightarrow x^2(x_1^2 - y_1^2) = a^2, x_1^2$$

$$\therefore x = \pm \frac{ax_1}{\sqrt{x_1^2 - y_1^2}} = \pm x_1 \qquad \therefore y = \pm y_1$$

$$\therefore N \text{ is } (\pm x_1, \pm y_1)$$

$$\therefore CM \times CN = \frac{a^2}{\sqrt{x_1^2 + y_1^2}} \cdot \sqrt{x_1^2 + y_1^2} = a^2$$

Example 8.6.14

Tangents to the hyperbola make angles θ_1, θ_2 with the transverse axis. Find the equation of the locus of point of intersection such that tan $(\theta_1 + \theta_2)$ is a constant.

Solution

Let the equation of the hyperbola be $\frac{x^2}{a^2} - \frac{y^2}{b^2} = 1$. Then, the equation of the tangent to the hyperbola is $y = mx + \sqrt{a^2m^2 - b^2}$.

If this tangent passes through (x_1, y_1), then $y_1 = mx_1 + \sqrt{a^2m^2 - b^2}$.

$$(y_1 - mx_1)^2 = a^2 m^2 - b^2$$

$$m^2 (x_1{}^2 - a^2) - 2mx_1 y_1 + y_1^2 + b^2 = 0$$

If m_1 and m_2 are the slopes of the two tangents, then

$$m_1 + m_2 = \frac{-2x_1y_1}{\sqrt{x_1^2 - a^2}},\ m_1m_2 = \frac{y_1^2 + b^2}{x_1^2 - a^2}.$$

It is given that tan $(\theta_1 + \theta_2) = k$.

$$\frac{\tan\theta_1 + \tan\theta_2}{1 - \tan\theta_1 \tan\theta_2} = k$$

(i.e.) $m_1 + m_2 = k\,(1 - m_1\, m_2)$

(i.e.) $\frac{-2x_1y_1}{x_1^2 - a^2} = k\left(1 - \frac{y_1^2 + b^2}{x_1^2 - a^2}\right) \Rightarrow -2x_1y_1 = k(x_1^2 - y_1^2 - a^2 - b^2)$

Hence, the locus of (x_1, y_1) is $k\,(x^2 + y^2 - a^2 - b^2)\, 2xy = 0$.

Example 8.6.15

Prove that two tangents that can be drawn from any point on the hyperbola $x^2 - y^2 = a^2 - b^2$ to the ellipse $\frac{x^2}{a^2} + \frac{y^2}{b^2} = 1$ which make complementary angles with the axes.

Solution

The tangent drawn from any point to the ellipse $\frac{x^2}{a^2} + \frac{y^2}{b^2} = 1$ is

$$y = mx + \sqrt{a^2m^2 - b^2}$$

Since this passes through (x_1, y_1)

$$y_1 = mx_1 + \sqrt{a^2m^2 - b^2} \Rightarrow (y_1 - mx_1)^2 = a^2 m^2 - b^2$$

$$\text{(i.e.)} \quad m^2 (x_1^2 - a^2) - 2mx_1y_1^2 + y_1^2 - b^2 = 0.$$

If m_1 and m_2 are the slopes of the tangents, then

$$m_1m_2 = \frac{y_1^2 - b^2}{x_1^2 - a^2}$$

Since (x_1, y_1) lies on $x^2 - y^2 = a^2 - b^2$, we have $x_1^2 - y_1^2 = a^2 - b^2$.

$$\therefore x_1^2 - a^2 = y_1^2 - b^2$$

$$\frac{y_1^2 - b^2}{x_1^2 - a^2} = 1$$

$$\therefore m_1m_2 = 1$$

Hence, the two tangents make complementary angles with the axes.

Example 8.6.16

Chords of the hyperbola $\frac{x^2}{a^2} - \frac{y^2}{b^2} = 1$ are at a constant distance from the centre. Find the locus of their poles.

Solution

Let (x_1, y_1) be the pole with respect to hyperbola $\frac{x^2}{a^2} - \frac{y^2}{b^2} = 1$. The polar of (x_1, y_1) is $\frac{xx_1}{a^2} - \frac{yy_1}{b^2} = 1$. The perpendicular distance from the centre on the polar is $\frac{1}{\sqrt{\frac{x_1^2}{a^2} + \frac{y_1^2}{b^2}}} = k$ (a constant) (i.e.) $\frac{x_1^2}{a^4} + \frac{y_1^2}{b^4} = \frac{1}{k}$.

Hence, the locus of (x_1, y_1) is $\frac{x^2}{a^4} + \frac{y^2}{b^4} = \frac{1}{k}$.

Example 8.6.17

Find the equation of common tangents to the hyperbolas $\frac{x^2}{a^2} - \frac{y^2}{b^2} = 1$ and $\frac{y^2}{a^2} - \frac{x^2}{b^2} = 1$.

Solution

The two given hyperbolas are $\frac{x^2}{a^2} - \frac{y^2}{b^2} = 1$ and $\frac{x^2}{-b^2} - \frac{y^2}{-a^2} = 1$.

The conditions for $y = mx + c$ to be a tangent to the hyperbolas are

$$c^2 = a^2 m^2 - b^2$$

and

$$c^2 = -b^2 m^2 + a^2$$

$$\therefore a^2 m^2 - b^2 = -b^2 m^2 + a^2$$

$$m^2 (a^2 + b^2) = a^2 + b^2$$

$$\text{(i.e.) } (m^2 - 1)(a^2 + b^2) = 0$$

$$\therefore m = \pm 1$$

Hence, there are two common tangents whose equations are $y = \pm x \pm \sqrt{a^2 - b^2}$.
(i.e.) $y = \pm (x + \sqrt{a^2 - b^2})$.

Example 8.6.18

Show that the locus of midpoints of normal chords of the hyperbola $x^2 - y^2 = a^2$ is

$$(y^2 - x^2)^3 = 4a^2 x^2 y^2.$$

Solution

The equation of the hyperbola is $x^2 - y^2 = a^2$.

Let (x_1, y_1) be the midpoint of a normal chord of the hyperbola. The equation of the normal is $\dfrac{ax}{\sec\theta} + \dfrac{ay}{\tan\theta} = 2a^2$ and the equation of the chord in terms of the middle point is $xx_1 - yy_1 = x_1^2 - y_1^2$. Both these equations represent the same line. Hence, identifying them, we get

$$\frac{a/\sec\theta}{x_1} = \frac{a/\tan\theta}{-y_1} = \frac{2a^2}{x_1^2 - y_1^2}$$

$$\therefore \sec\theta = \frac{a(x_1^2 - y_1^2)}{2a^2 x_1} = \frac{x_1^2 - y_1^2}{2ax_1} \text{ and } \tan\theta = -\frac{x_1^2 - y_1^2}{2ay_1}$$

Squaring and subtracting, we get

$$\frac{(x_1^2 - y_1^2)}{4a^2 x_1^2} - \frac{(x_1^2 - y_1^2)^2}{4a^2 y_1^2} = 1$$

$$\frac{(x_1^2 - y_1^2)^2}{4a^2}\left(\frac{1}{x_1^2} - \frac{1}{y_1^2}\right) = 1$$

$$\Rightarrow (y_1^2 - x_1^2)^3 = 4a^2 x_1^2 y_1^2$$

The focus of (x_1, y_1) is $(y^2 - x^2)^3 = 4a^2x^2 y^2$.

Example 8.6.19

Prove that the locus of middle points of chords of the hyperbola $\frac{x^2}{a^2} - \frac{y^2}{b^2} = 1$ passing through a fixed point (h, k) is a hyperbola whose centre is $\left(\frac{h}{2}, \frac{k}{2}\right)$.

Solution

The equation of the hyperbola is $\frac{x^2}{a^2} - \frac{y^2}{b^2} = 1$.

The equation of the chord of the hyperbola in terms of its middle point is $\frac{xx_1}{a^2} - \frac{yy_1}{b^2} = \frac{x_1^2}{a^2} - \frac{y_1^2}{b^2}$

Since this chord passes through the fixed point (h, k),

$$\frac{xh}{a^2} - \frac{yk}{b^2} = \frac{x_1^2}{a^2} - \frac{y_1^2}{b^2}$$

(i.e.) $$\left(\frac{x_1}{a} - \frac{h}{2}\right)^2 - \left(\frac{y_1}{b} - \frac{k}{2}\right)^2 = \frac{1}{4}\left(\frac{h^2}{a^2} - \frac{k^2}{b^2}\right)$$

The locus of (x_1, y_1) is $\left(\frac{x}{a} - \frac{h}{2}\right)^2 - \left(\frac{y}{b} - \frac{k}{2}\right)^2 = \frac{1}{4}\left(\frac{h^2}{a^2} - \frac{k^2}{b^2}\right)$, which is a hyperbola whose centre is $\left(\frac{h}{2}, \frac{k}{2}\right)$.

Example 8.6.20

Show that the locus of the foot of the perpendicular from the centre upon any normal to the hyperbola $\frac{x^2}{a^2} - \frac{y^2}{b^2} = 1$ is $(x^2 + y^2)^2\left(\frac{a^2}{x^2} - \frac{b^2}{y^2}\right) = \frac{(a^2 + b^2)^2}{y}$.

Solution

Let P $(a \sec \theta, b \tan \theta)$ be a point on the hyperbola. Let m (x_1, y_1) be the foot of the perpendicular from the centre with normal at P. The equation of the normal at P is $\frac{ax}{\sec\theta} + \frac{by}{\tan\theta} = a^2 + b^2$.

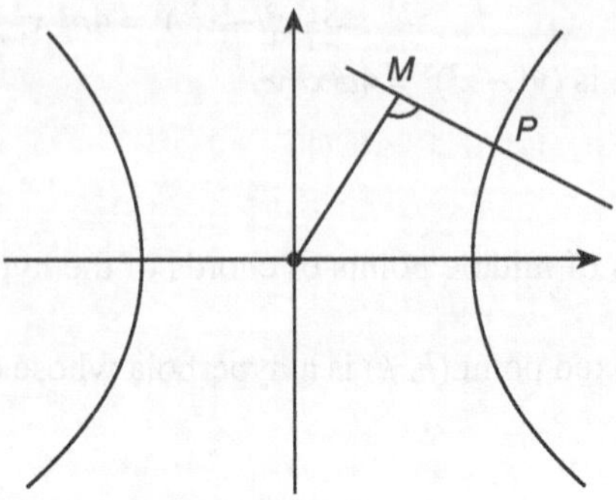

The equation of the perpendicular from C (0, 0) on this normal is

$$\frac{b}{\tan\theta}x - \frac{a}{\sec\theta}y = 0$$

These two lines intersect at (x_1, y_1)

$$\therefore \quad \frac{ax_1}{\sec\theta} + \frac{by_1}{\tan\theta} = a^2 + b^2 \text{ and}$$

$$\frac{bx_1}{\tan\theta} - \frac{ay_1}{\sec\theta} = 0$$

Solving these two equation for x_1 and y_1 we get

$$\frac{x_1}{\dfrac{-(a^2+b^2)a}{\sec\theta}} = \frac{y_1}{\dfrac{-b(a^2+b^2)}{\tan\theta}} = \frac{1}{-\dfrac{a^2}{\sec^2\theta} - \dfrac{b^2}{\tan^2\theta}}$$

$$\frac{x_1\sec\theta}{a(a^2+b^2)} = \frac{y_1\tan\theta}{b(a^2+b^2)} = \frac{\sec^2\theta\tan^2\theta}{a^2\tan^2\theta + b^2\sec^2\theta}$$

$$x_1 = \frac{a(a^2+b^2)\sec\theta\tan^2\theta}{a^2\tan^2\theta + b^2\sec^2\theta},\ y_1 = \frac{b(a^2+b^2)\sec^2\theta\tan\theta}{a^2\tan^2\theta + b^2\sec^2\theta}$$

$$\frac{a^2}{x_1^2} = \frac{(a^2\tan^2\theta + b^2\sec^2\theta)^2}{(a^2+b^2)^2\sec^2\theta\tan^4\theta},\ \frac{b^2}{y_1^2} = \frac{(a^2\tan^2\theta + b^2\sec^2\theta)^2}{(a^2+b^2)^2\sec^4\theta\tan^2\theta}$$

$$\left(\frac{a^2}{x_1^2} - \frac{b^2}{y_1^2}\right) = \frac{(a^2\tan^2\theta + b^2\sec^2\theta)^2}{(a^2+b^2)^2}\left(\frac{\sec^2\theta - \tan^2\theta}{\sec^4\theta\tan^4\theta}\right)$$

$$x_1^2 + y_1^2 = \frac{(a^2+b^2)^2}{(a^2\tan^2\theta + b^2\sec^2\theta)^2}\left(\begin{matrix} a^2\sec^2\theta\tan^4\theta + \\ b^2\sec^4\theta\tan^2\theta \end{matrix}\right)$$

$$= \frac{(a^2+b^2)^2 \sec^2\theta \tan^2\theta}{(a^2\tan^2\theta + b^2\sec^2\theta)^2}\left(a^2\tan^2\theta + b^2\sec^2\theta\right)$$

$$(x_1^2+y_1^2)^2 = \frac{(a^2+b^2)^4 \sec^4\theta \tan^4\theta}{(a^2\tan^2\theta + b^2\sec^2\theta)^4}\left(a^2\tan^2\theta + b^2\sec^2\theta\right)^2$$

$$(x_1^2+y_1^2)^2\left(\frac{a^2}{x_1^2}-\frac{b^2}{y_1^2}\right) = \frac{(a^2+b^2)^2 \sec^4\theta \tan^4\theta}{(a^2\tan^2\theta + b^2\sec^2\theta)^2}\left(\frac{1}{\sec^4\theta\tan^4\theta}\right)$$

$$(a^2\tan^2\theta + b^2\sec^2\theta)^2$$

$$= (a^2+b^2)^2$$

Hence, the locus of (x_1, y_1) is $(x^2+y^2)^2\left(\frac{a^2}{x^2}-\frac{b^2}{y^2}\right) = (a^2+b^2)^2$.

Example 8.6.21

Chords of the curve $x^2 + y^2 = a^2$ touch the hyperbola $\frac{x^2}{a^2}-\frac{y^2}{b^2}=1$. Prove that their middle points lie on the curve $(x^2 + y^2)^2 = a^2x^2 - b^2y^2$.

Solution

Let (x_1, y_1) be the midpoint of the chord of the circle. Its equation is

$$xx_1 + yy_1 = x_1^2 + y_1^2$$

$$\text{(i.e.)} \quad yy_1 = -xx_1 + x_1^2 + y_1^2$$

$$\text{(i.e.)} \quad y = \frac{-xx_1}{y_1} + \frac{x_1^2+y_1^2}{y_1}$$

This is a tangent to the hyperbola $\frac{x^2}{a^2}-\frac{y^2}{b^2}=1$.

Hence, the condition is

$$\left(\frac{x_1^2+y_1^2}{y_1}\right)^2 = a^2\frac{x_1^2}{y_1^2} - b^2$$

(i.e.) $(x_1^2 + y_1^2)^2 = (a^2x_1^2 - b^2y_1^2)$

Hence, the locus of (x_1, y_1) is $(x^2 + y^2)^2 = (a^2x^2 - b^2y^2)$.

Example 8.6.22

Show that the locus of midpoints of normal chords of the hyperbola $x^2 - y^2 = a^2$ is $(y^2 - x^2)^2 = 4a^2 xy$.

Solution

Let (x_1, y_1) be the midpoint of the normal chord of the hyperbola $x^2 - y^2 = a^2$. Then, the equation of the chord is

$$xx_1 - yy_1 = x_1^2 - y_1^2$$

The equation of the normal at 'θ' is

$$\frac{ax}{\sec\theta} + \frac{ay}{\tan\theta} = 2a^2$$

These two equations represent the same line. Identifying, we get

$$\frac{x_1 \sec\theta}{a} = \frac{-y_1 \tan\theta}{a} = \frac{x_1^2 - y_1^2}{2a^2}$$

$$\therefore \sec\theta = \frac{x_1^2 - y_1^2}{2ax_1},\ \tan\theta = \frac{x_1^2 - y_1^2}{2ay_1},$$

$$\sec^2\theta - \tan^2\theta = 1 \Rightarrow \frac{(x_1^2 - y_1^2)^2}{4a^2x_1^2} - \frac{(x_1^2 - y_1^2)^2}{4a^2y_1^2} = 1$$

$$\text{(i.e.)}\quad 1 = \frac{(x_1^2 - y_1^2)^2}{4a^2}\left[\frac{1}{x_1^2} - \frac{1}{y_1^2}\right] \Rightarrow (y_1^2 - x_1^2)^3 = 4a^2x_1^2y_1^2$$

The locus of (x_1, y_1) is $(y^2 - x^2)^3 = 4a^2 x^2 y^2$.

Example 8.6.23

A normal to the hyperbola $\frac{x^2}{a^2} - \frac{y^2}{b^2} = 1$ meets the axes at Q and R and lines QL and RL are drawn at right angles to the axes and meet at L. Prove that the locus of the point L is the hyperbola $(a^2 x^2 - b^2 y^2) = (a^2 + b^2)^2$. Prove further that the locus of the middle point of QR is $4\,(a^2 x^2 - b^2 y^2) = (a^2 + b^2)^2$.

Solution

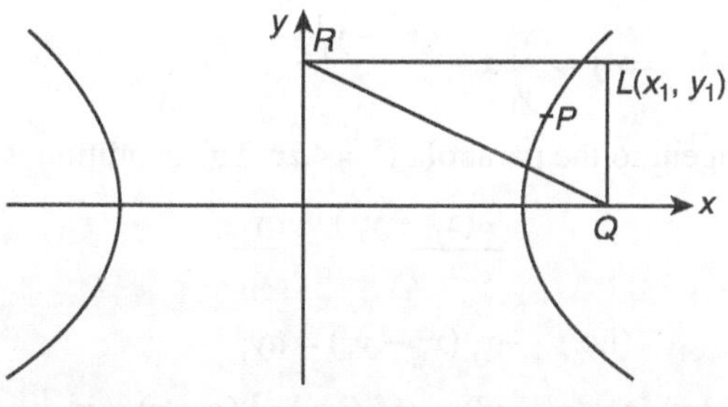

Let $P(h, k)$ be the point on the hyperbola $\frac{x^2}{a^2}-\frac{y^2}{b^2}=1$. The equation of the normal at (h, k) is $\frac{a^2x}{h}+\frac{b^2y}{k}=a^2+b^2$. When this line meets the x-axis $y = 0$

Therefore, the coordinates of Q are $\left(\frac{(a^2+b^2)h}{a^2}, 0\right)$. The coordinates of R are $\left(0, \frac{(a^2+b^2)k}{b^2}\right)$. Let (x_1, y_1) be the coordinates of L. Then, $x_1=(a^2+b^2)\frac{h}{a^2}$ and $y_1=(a^2+b^2)\frac{k}{b^2}$.

$$\therefore ax_1=(a^2+b^2)\frac{h}{a} \text{ and } by_1=(a^2+b^2)\frac{k}{b}$$

$\therefore a^2x_1^2-b^2y_1^2=(a^2+b^2)^2\left(\frac{h^2}{a^2}-\frac{k^2}{b^2}\right)=(a^2+b^2)^2$ since (h, k) lies on the hyperbola. The locus of (x_1, y_1) is $a^2x^2-b^2y^2=(a^2+b^2)^2$.

Let (α, β) be the midpoint of QR. Then, $\alpha=\frac{(a^2+b^2)h}{2a^2}$ and $\beta=\frac{(a^2+b^2)k}{2b^2}$.

$\therefore 4(a^2\alpha^2-b^2\beta^2)=(a^2+b^2)^2\left(\frac{x^2}{a^2}-\frac{y^2}{b^2}\right)=(a^2+b^2)^2$ since (h, k) lies on the hyperbola.

The locus of (α, β) is $4(a^2x^2-b^2y^2)=(a^2+b^2)^2$.

Example 8.6.24

The chords of the hyperbola $x^2-y^2=a^2$ touch the parabola $y^2=4ax$. Prove that the locus of their midpoint is the curve $y^2(a-y)=x^3$.

Solution

Let (x_1, y_1) be the midpoint of the chord of the hyperbola $\frac{x^2}{a^2}-\frac{y^2}{a^2}=1$. Its equation is $\frac{xx_1}{a^2}-\frac{yy_1}{a^2}=\frac{x_1^2}{a^2}-\frac{y_1^2}{a^2}$.

$$xx_1 - yy_1 = x_1^2 - y_1^2 \Rightarrow yy_1 = -x_1 x - (x_1^2 - y_1^2)$$

$$\Rightarrow y = \frac{x_1}{y_1}x - \frac{(x_1^2 - y_1^2)}{y_1}$$

The line is a tangent to the parabola $y^2 = 4ax$. The condition is

$$\frac{-(x_1^2 - y_1^2)}{y_1} = \frac{ay_1}{x_1}$$

$$\text{(i.e.)} \quad -x_1 (x_1^2 - y_1^2) = ay_1^2$$

The locus of (x_1, y_1) is $x (x^2 - y^2) = ay^2$ (i.e.) $y^2 (a - y) = x^8$.

Example 8.6.25

A variable tangent to the hyperbola $\frac{x^2}{a^2} - \frac{y^2}{b^2} = 1$ meets the transverse axis at Q and the tangent at the vertex at R. Show that the locus of the midpoint QR is $x (4y^2 + b^2) = ab^2$.

Solution

The equation of the tangent at 'θ' is $\frac{x}{a}\sec\theta - \frac{y}{b}\tan\theta = 1$.

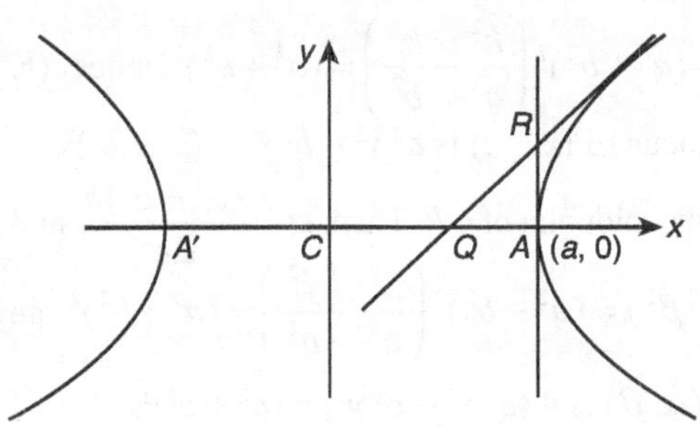

When this line meets the transverse axis, $y = 0$ and $x = a \cos\theta$. Here Q is $(a \cos \theta, 0)$. When it meets the line $x = a$,

$$\sec\theta - 1 = \frac{y}{b}\tan\theta \quad \therefore y = \frac{b(1-\cos\theta)}{\sin\theta} \quad \therefore R \text{ is } \left(a, \frac{b(1-\cos\theta)}{\sin\theta}\right)$$

Let (h, k) be the midpoint of QR. Then,

$$h = \frac{a\cos\theta + a}{2}$$

$$\therefore 2h = a(1+\cos\theta) \text{ or } \cos\theta = \frac{2h-a}{a}$$

$$k = \frac{b(1-\cos\theta)}{2\sin\theta}$$

$$\therefore 2h = a(1+\cos\theta), 2k = \frac{b(1-\cos\theta)}{\sin\theta}$$

$$4hk = ab\sin\theta \Rightarrow \sin\theta = \frac{4hk}{ab}, \cos\theta = \frac{2h-a}{a}$$

Squaring and adding, $1 = \frac{(2h-a)^2}{a^2} + \frac{16h^2k^2}{a^2b^2}$

$$a^2b^2 = b^2(2h-a)^2 + 16h^2k^2 \Rightarrow 4h^2b^2a^2 - 4b^2ha + 16h^2k^2 = 0$$

$\therefore\ hb^2 - ab^2 + 4hk^2 = 0$

Hence, the locus of (h,k) is $b^2x + 4xy^2 = ab^2$ or $x(b^2+4y^2) = ab^2$.

Example 8.6.26

Show that the locus of the midpoints of the chords of the hyperbola $\frac{x^2}{a^2} - \frac{y^2}{b^2} = 1$ that subtends a right angle at the centre is $\left(\frac{x^2}{a^2} - \frac{y^2}{b^2}\right)^2 \left(\frac{1}{a^2} - \frac{1}{b^2}\right) = \frac{x^2}{a^4} + \frac{y^2}{b^4}$.

Solution

Let $P(x_1, y_1)$ be the midpoint of a chord of the hyperbola $\frac{x^2}{a^2} - \frac{y^2}{b^2} = 1$.

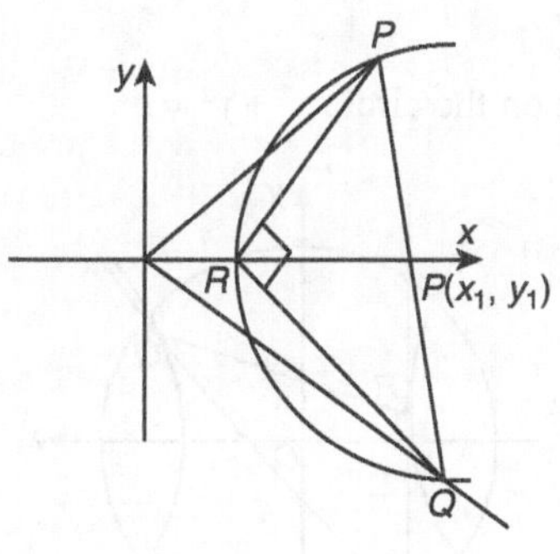

Then, the equation of the chord is $T = S_1$

(i.e.) $\frac{xx_1}{a^2} - \frac{yy_1}{b^2} = \frac{x_1^2}{a^2} - \frac{y_1^2}{b^2}$

The chord subtends a right angle at the centre of the hyperbola.
Hence, the combined equation of the lines CP and CQ is

$$\frac{x^2}{a^2}-\frac{y^2}{b^2}=\left(\frac{\frac{xx_1}{a^2}-\frac{yy_1}{b^2}}{\frac{x_1^2}{a^2}-\frac{y_1^2}{b^2}}\right)^2$$

$$\text{(i.e.)}\quad \left(\frac{x^2}{a^2}-\frac{y^2}{b^2}\right)\left(\frac{x_1^2}{a^2}-\frac{y_1^2}{b^2}\right)^2=\left(\frac{xx_1}{a^2}-\frac{yy_1}{b^2}\right)^2$$

Since $\angle QCR = 90°$, coefficient of x^2 + coefficient of $y^2 = 0$.

$$\text{(i.e.)}\quad \left[\left[\frac{1}{a^2}\left(\frac{x_1^2}{a^2}-\frac{y_1^2}{b^2}\right)^2-\frac{x_1^2}{a^4}\right]-\frac{1}{b^2}\left(\frac{x_1^2}{a^2}-\frac{y_1^2}{b^2}\right)^2+\frac{y_1^2}{b^4}\right]=0$$

$$\text{(i.e.)}\quad \left(\frac{x_1^2}{a^2}-\frac{y_1^2}{b^2}\right)^2\left(\frac{1}{a^2}-\frac{1}{b^2}\right)=\frac{x_1^2}{a^4}-\frac{y_1^2}{b^4}$$

The locus of (x_1, y_1) is $\left(\frac{x^2}{a^2}-\frac{y^2}{b^2}\right)^2\left(\frac{1}{a^2}-\frac{1}{b^2}\right)=\frac{x^2}{a^4}-\frac{y^2}{b^4}.$

Example 8.6.27

From points on the circle $x^2 - y^2 = a^2$ tangents are drawn to the hyperbola $x^2 - y^2 = a^2$. Prove that the locus of the middle points of the chords of contact is the curve $(x^2 - y^2) = a^2 (x^2 + y^2)$.

Solution

Let $P (x_1, y_1)$ be a point on the circle $x^2 + y^2 = a^2$.

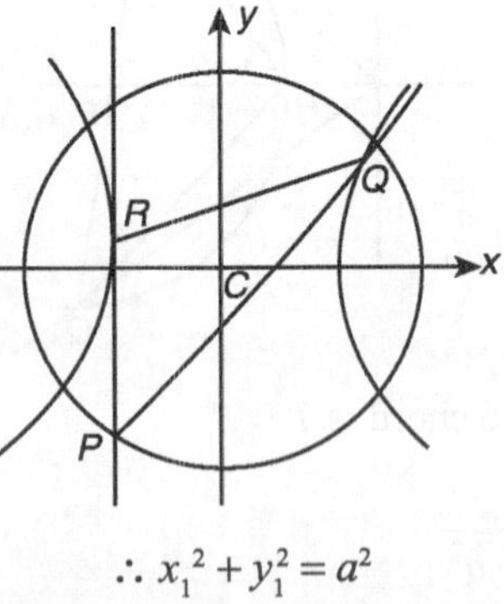

$$\therefore x_1^2 + y_1^2 = a^2$$

Let (h, k) be the midpoint of the chord of contact QR of the tangents from P to the hyperbola $x^2 - y^2 = a^2$. Then the equation of chord of contact to the hyperbola is

$$xx_1 - yy_1 = a^2$$

The equation of the chord in terms of the middle point (h, k) is

$$xh - yk = h^2 - k$$

These two equations represent the same line. Identifying them, we get

$$\frac{x_1}{h} = \frac{y_1}{k} = \frac{a^2}{h^2 - k^2}$$

$$\therefore\ x_1 = \frac{a^2 h}{h^2 - k^2} \quad \text{and} \quad y_1 = \frac{a^2 k}{h^2 - k^2}$$

$$x_1^2 + y_1^2 = \frac{a^2(h^2 + k^2)}{(h^2 - k^2)^2} \quad \text{(i.e.)} \quad a^2 = \frac{a^4(h^2 + k^2)}{(h^2 - k^2)^2} \ (\text{since } x_1^2 + y_1^2 = a^2)$$

Hence, the locus of (h, k) is $(x^2 - y^2)^2 = a^2 (x^2 + y^2)$.

Example 8.6.28

If the tangent and normal at any point of the hyperbola $\frac{x^2}{a^2} - \frac{y^2}{b^2} = 1$ meet on the conjugate axis at Q and R, show that the circle described with QR as the diameter passes through the foci of the hyperbola.

Solution

The equation of the tangent and normal at (x_1, y_1) on the hyperbola $\frac{x^2}{a^2} - \frac{y^2}{b^2} = 1$ are

$$\frac{xx_1}{a^2} - \frac{yy_1}{b^2} = 1 \tag{8.3}$$

$$\frac{a^2 x}{x_1} + \frac{b^2 y}{y_1} = a^2 + b^2 \tag{8.4}$$

These two lines meet the conjugate axis at Q and R. Therefore substitute $x = 0$ in equations (8.3) and (8.4). The coordinates of Q are $\left(0, \frac{-b^2}{y_1}\right)$. The coordinates of R are $\left(0, (a^2 + b^2)\frac{y_1}{b^2}\right)$.

The equation of the circle with QR as diameter is

$$x^2 + \left(y + \frac{b^2}{y_1}\right)\left(y - \frac{(a^2+b^2)^2 y_1}{b^2}\right) = 0$$

Substituting $x = \pm ae$ and $y = 0$

$$a^2e^2 - \frac{b^2}{y_1}\frac{(a^2+b^2)y_1}{b^2} = 0$$

(i.e.) $a^2e^2 - (a^2+b^2) = 0$ (i.e.) $a^2e^2 - a^2e^2 = 0$ which is true.

Hence, the circle with QR as diameter passes through the foci.

Exercises

1. Find the equation of the hyperbola whose focus is (1, 2), directrix $2x + y = 1$ and eccentricity $\sqrt{3}$.

   Ans.: $7x^2 + 12xy - 2y^2 - 2x + 14y - 22 = 0$

2. Show that the equation of the hyperbola having focus (2, 0), eccentricity 2 and directrix $x - y = 0$ is $x^2 + y^2 - 4xy + 4 = 0$.

3. Find the equation of the hyperbola whose focus is (2, 2), eccentricity $\frac{3}{2}$ and directrix $3x - 4y = 1$.

   Ans.: $19x^2 + 44y^2 - 216xy - 346x + 472y - 791 = 0$

4. Find the equation of the hyperbola whose focus is (0, 0), eccentricity $\frac{5}{4}$ and directrix $x\cos\alpha + y\sin\alpha = p$.

   Ans.: $16(x^2 + y^2) - 25(x\cos\alpha + y\sin\alpha - p)^2 = 0$

5. Find the equation of the hyperbola whose centre is (–3, 2) and one end of the transverse axis is (–3, 4) and eccentricity is $\frac{5}{2}$.

   Ans.: $4x^2 - 21y^2 + 24x + 84y + 36 = 0$

6. Find the equation of the hyperbola whose foci are (6, 4) and (–4, 4) and eccentricity 2. **Ans.:** $\frac{4(x-1)^2}{25} - \frac{4(y-4)^2}{75} = 1$

7. Find the equation of the hyperbola whose centre is (1, 0), one focus is (6, 0) and length of transverse axis is 6.

   Ans.: $16x^2 - 9y^2 - 32x - 128 = 0$

8. Find the equation of the hyperbola whose centre is (3, 2), one focus is (5, 2) and one vertex is (4, 2). **Ans.:** $3x^2 - y^2 - 18x - 4y + 20 = 0$

9. Find the equation of the hyperbola whose centre is (6, 2), one focus is (4, 2) and eccentricity 2. **Ans.:** $\frac{(x-6)^2}{1}-\frac{(y-2)^2}{3}=1$

10. Find the centre, eccentricity and foci of hyperbola $9x^2-16y^2=144$.

Ans.: $(0,0), \frac{5}{4}, (5,0)$ and $(-5,0)$

11. Find the centre, foci and eccentricity of $12x^2-4y^2-24x+32y-127=0$.

Ans.: (1, 4), (6, 4) and (–4, 4)

12. Find the centre, foci and eccentricity of the hyperbola $9x^2-4y^2-18x+16y-43=0$.

Ans.: $(-1,2), (\sqrt{13},0)$ and $(-\sqrt{13},0), \frac{\sqrt{13}}{2}$

13. If S and S' are the foci of a hyperbola and p is any point on the hyperbola, show that $S'P-SP=2a$.

14. Find the latus of the hyperbola $\frac{x^2}{a^2}-\frac{y^2}{b^2}=1$. **Ans.:** $\frac{2b^2}{a}$

15. Find the equation of the hyperbola referred to its axis as the axis of coordinate if length of transverse axis is 5 and conjugate axis is 4.

Ans.: $\frac{x^2}{25}-\frac{y^2}{16}=1$

16. Find the latus rectum of the hyperbola $4x-9y^2=36$. **Ans.:** $\frac{8}{3}$

17. Find the centre, eccentricity and foci of the hyperbola $x^2-2y^2-2x+8y-1=0$.

Ans.: $(1,2), \sqrt{3}, (1,5), (1,-1), y=5, x=-1$

18. Find the centre, eccentricity, foci and directrix of the hyperbola $16x^2-9y^2+32x+36y-164=0$.

Ans.: $(-1,2), \frac{5}{3}, (-6,2), (4,1), 5x=4, 5x=-14$

19. The hyperbola $\frac{x^2}{a^2}-\frac{y^2}{b^2}=1$ passes through the intersection of the lines $7x+13y-87=0$ and $5x-8y+7=0$ and its latus rectum is $\frac{32\sqrt{2}}{5}$. Find a and b. **Ans.:** $\frac{5\sqrt{2}}{2}, 4$

20. Tangents are drawn to the hyperbola $3x^2-2y^2=6$ from the point P and make θ_1, θ_2 with x-axis. If the $\tan\theta_1 \tan\theta_2$ is a constant, prove that locus of P is $2x^2-y^2=7$.

21. Find the equation of tangents to the hyperbola $3x^2 - 4y^2 = 15$ which are parallel to $y = 2x + k$. Find the coordinates of the point of contact.

Ans.: $y = 2x \pm \frac{\sqrt{65}}{2}, \left(\frac{20}{\sqrt{65}}, \frac{5}{2\sqrt{65}}\right), \left(\frac{-20}{\sqrt{65}}, \frac{-5}{2\sqrt{65}}\right)$

22. Tangents are drawn to the hyperbola $x^2 - y^2 = c^2$ are inclined at an angle of 45°, show that the locus of their intersection is $(x^2 + y^2)^2 + 4a^2(x^2 - y^2) = 4a^4$.

23. Prove that the polar of any point on the ellipse $\frac{x^2}{a^2} + \frac{y^2}{b^2} = 1$ with respect to the hyperbola $\frac{x^2}{a^2} - \frac{y^2}{b^2} = 1$ will touch the ellipse at the other end of the ordinate through the point.

24. If the polar of points (x_1, y_1) and (x_2, y_2) with respect to hyperbola are at right angles then show that $b^4 x_1 x_2 + a^4 y_1 y_2 = 0$.

25. Find the locus of poles of normal chords of the hyperbola $\frac{x^2}{a^2} - \frac{y^2}{b^2} = 1$.

26. Chords of the hyperbola $\frac{x^2}{a^2} - \frac{y^2}{b^2} = 1$ subtend a right angle at one of the vertices. Show that the locus of poles of all such chords is the straight line $x(a^2 + b^2) = a(a^2 - b^2)$.

27. If chords of the hyperbola are at a constant distance k from the centre, find the locus of their poles. **Ans.:** $\frac{x^2}{a^4} + \frac{y^2}{b^4} = \frac{1}{k^2}$

28. Obtain the locus of the point of intersection of tangents to the hyperbola $\frac{x^2}{a^2} - \frac{y^2}{b^2} = 1$ which includes an angle β.

Ans.: $4(a^2y^2 - b^2x^2 + a^2b^2) = (x^2 + y^2 - a^2 + b^2)\tan^2\beta$

29. If a variable chord of the hyperbola $\frac{x^2}{a^2} - \frac{y^2}{b^2} = 1$ is a tangent to the circle $x^2 + y^2 = c^2$ then prove that the locus of its middle point is

$$\left(\frac{x^2}{a^2} - \frac{y^2}{b^2}\right)^2 = c^2\left(\frac{x^2}{a^4} + \frac{y^2}{b^4}\right).$$

30. Show that the condition for the line $x\cos\alpha + y\sin\alpha = \beta$ touches the hyperbola $\frac{x^2}{a^2} - \frac{y^2}{b^2} = 1$ is $a^2\cos^2\alpha - b^2\sin^2\alpha = p^2$.

31. Prove that the tangent at any point bisects the angle between focal distances of the point.

32. Prove that the midpoints of the chords of the hyperbola $\frac{x^2}{a^2}-\frac{y^2}{b^2}=1$ parallel to the diameter $y = mx$ be on the diameter $a^2my = b^2x$.

33. If the polar of the point *A* with respect to a hyperbola passes through another point *B*, then show that the polar *B* passes through *A*.

34. If the polars of (x_1, y_1) and (x_2, y_2) with respect to the hyperbola $\frac{x^2}{a^2}-\frac{y^2}{b^2}=1$ are at right angles, then prove that $\frac{x_1x_2}{y_1-y_2}+\frac{a^4}{b^4}=0$.

35. Prove that the polar of any point on $\frac{x^2}{a^2}+\frac{y^2}{b^2}=1$ with respect to the hyperbola $\frac{x^2}{a^2}-\frac{y^2}{b^2}=1$ touches $\frac{x^2}{a^2}+\frac{y^2}{b^2}=1$.

36. Obtain the equation of the chord joining the points θ and ϕ on the hyperbola in the form $\frac{x}{a}\cos\left(\frac{\theta-\phi}{2}\right)-\frac{y}{b}\sin\left(\frac{\theta+\phi}{2}\right)=\cos\left(\frac{\theta+\phi}{2}\right)$. If $\theta-\phi$ is a constant and equal to 2α, show that *PQ* touches the hyperbola $\frac{x^2\cos^2\alpha}{a^2}-\frac{y^2}{b^2}=1$.

37. If a circle with centre $(3\alpha, 3\beta)$ and of variable radius cuts the hyperbola $x^2 - y^2 = 9a^2$ at the points P, Q, R and S then prove that the locus of the centroid of the triangle PQR is $(x-2\alpha)^2-(y-2\beta)^2=a^2$.

38. If the normal at *P* meets the transverse axis in *r* and the conjugate axis in g and *CF* be perpendicular to the normal from the centre then prove that $PF \cdot Pr = CB^2$ *and* $PF \cdot Pg = CF^2$.

39. Show that the locus of the points of intersection of tangents at the extremities of normal chords of the hyperbola $\frac{x^2}{a^2}-\frac{y^2}{b^2}=1$ is $\frac{a^6}{x^2}-\frac{b^6}{y^2}=(a^2+b^2)^2$.

40. Find the equation and length of the common tangents to hyperbolas $\frac{x^2}{a^2}-\frac{y^2}{b^2}=1$ and $\frac{y^2}{a^2}-\frac{x^2}{b^2}=1$.

Ans.: $\left[\frac{1}{\sqrt{a^2-b^2}};\ x-y=\sqrt{a^2-b^2}\right]$

41. Tangents are drawn from any point on hyperbola $x^2 - y^2 = a^2 + b^2$ to the hyperbola $\frac{x^2}{a^2}-\frac{y^2}{b^2}=1$. Prove that they meet the axes in conjugate points.

42. Prove that the part of the tangent at any point of a hyperbola intercepted between the point of contact and the transverse axis is a harmonic mean between the lengths of the perpendiculars drawn from the foci on the normal at the same point.

43. If the chord joining the points α and β on the hyperbola $\frac{x^2}{a^2}-\frac{y^2}{b^2}=1$ is a focal chord then prove that $\tan\frac{\alpha}{2}\tan\frac{\beta}{2}+\frac{ke-1}{ke+1}=0$ where $k \neq 1$.

44. Let the tangent and normal at a point P on the hyperbola meet the transverse axis in T and G respectively, prove that $CT \cdot CG = a^2+b^2$.

45. If the tangent at the point (h,k) to the hyperbola cuts the auxiliary circle in points whose ordinates are y_1 and y_2 then show that $\frac{1}{y_1}+\frac{1}{y_2}=\frac{2}{k}$.

46. If a line is drawn parallel to the conjugate axis of a hyperbola to meet it and the conjugate hyperbola in the points P and Q then show that the tangents at P and Q meet on the curve $\frac{y^2}{b^2}-\frac{x^2}{a^2}=\frac{4b^4x^2}{x^2y^n}$.

47. If an ellipse and a hyperbola have the same principal axes then show that the polar of any point on either curve with respect to the other touches the first curve.

48. If the tangent at any point P on the hyperbola $\frac{x^2}{a^2}-\frac{y^2}{b^2}=1$ whose centre is C, meets the transverse and conjugate axes in T_1 and T_2, then prove that (i) $CN \cdot CT_1 = a^2$ and (ii) $CM \cdot CT_2 = -b^2$ where PM and PN are perpendiculars in the transverse and conjugate axes, respectively.

49. If P is the length of the perpendicular from C, the centre of the hyperbola $\frac{x^2}{a^2}-\frac{y^2}{b^2}=1$ on the tangent at a point P on it and $CP=r$, prove that

$$\frac{a^2b^2}{p^2}=r^2+b^2-a^2.$$

8.7 ASYMPTOTES

Definition 8.7.1 An asymptote of a hyperbola is a straight line that touches the hyperbola at infinity but does not lie altogether at infinity.

8.7.1 Equations of Asymptotes of the Hyperbola

Let the equation of the hyperbola be $\frac{x^2}{a^2}-\frac{y^2}{b^2}=1$. Let $y=mx+c$ be an asymptote of the hyperbola. Solving these two equations, we get their points of intersection. The x coordinates of the points of intersection are given by

$$\frac{x^2}{a^2}-\frac{(mx+c)^2}{b^2}=1$$

$$\text{(i.e.)} \quad b^2x^2-a^2(mx+c)^2=a^2b^2$$

$$\text{(i.e.)} \quad x^2(b^2-a^2m^2)-2mca^2x-a^2(b^2+c^2)=0$$

If $y=mx+c$ is an asymptote, then the roots of the above equation are infinite. The conditions for these are the coefficient of $x^2=0$ and the coefficient of $x=0$, $b^2-a^2m^2=0$ and $mca^2=0$.

$$\text{(i.e.)} \quad m=\pm\frac{b}{a} \text{ and } c=0$$

The equations of the asymptotes are $y=\pm\frac{b}{a}x$.

$$\text{(i.e.)} \quad \frac{x}{a}-\frac{y}{b}=0 \text{ and } \frac{x}{a}+\frac{y}{b}=0$$

The combined equation of the asymptotes is $\frac{x^2}{a^2}-\frac{y^2}{b^2}=0$.

Note 8.7.1.1:

1. The asymptotes of the conjugate hyperbola $\frac{x^2}{a^2}-\frac{y^2}{b^2}=-1$ are also given by $\frac{x^2}{a^2}-\frac{y^2}{b^2}=0$. Therefore, the hyperbola and the conjugate hyperbola have the same asymptotes.
2. The equation of the hyperbola is $H:\frac{x^2}{a^2}-\frac{y^2}{b^2}-1=0$.

   The equation of the asymptotes is $A:\frac{x^2}{a^2}-\frac{y^2}{b^2}=0$.

   The equation of the conjugate hyperbola is $C:\frac{x^2}{a^2}-\frac{y^2}{b^2}+1=0$.
3. The equation of the asymptotes differs from that of the hyperbola by a constant and the equation of the conjugate hyperbola differs from that of the asymptotes by the same constant term. This result holds good even when the equations of the hyperbola and its asymptotes are in the most general form.
4. The asymptotes pass through the centre $(0,0)$ of the hyperbola.
5. The slopes of the asymptotes are $\frac{b}{a}$ and $\frac{-b}{a}$.

Hence, they are equally inclined to the coordinate axes, which are the transverse and conjugate axes.

8.7.2 Angle between the Asymptotes

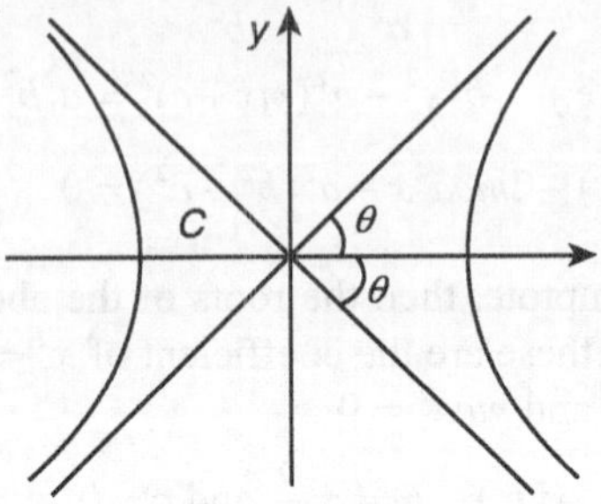

Let 2θ be the angle between the asymptotes. Then,

$$m = \tan\theta = \frac{b}{a}$$

$$\sec^2\theta = 1 + \tan^2\theta = 1 + \frac{b^2}{a^2} = \frac{a^2+b^2}{a^2} = e^2$$

$$\therefore \sec\theta = e$$

Hence, the angle between the asymptotes is $2\sec^{-1}(e)$.

Example 8.7.1

Find the equation of the asymptotes of the hyperbola $3x^2 - 5xy - 2y^2 + 17x + y + 14 = 0$.

Solution

The combined equation of the asymptotes should differ from that of the hyperbola only by a constant term.

$\therefore$ The combined equation of the asymptotes is

$$3x^2 - 5xy - 2y^2 + 17x + y + k = 0$$

$$\begin{aligned} 3x^2 - 5xy - 2y^2 &= 3x^2 - 6xy + xy - 2y^2 \\ &= 3x(x-2y) + y(x-2y) \\ &= (3x+y)(x-2y) \end{aligned}$$

Hence, the asymptotes are $3x + y + l = 0$ and $x - 2y + m = 0$.

$$(3x+y+l)(x-2y+m) = 3x^2 - 5yx - 2y^2 + 17x + y + k$$

Equating the coefficients of the terms x and y and the constant terms, we get

$$l+3m=17$$

$$-2l+m=1$$

Solving these two equations, we get $l=2$ and $m=5$.

$$lm=k$$

$\therefore\ k=10$. The combined equation of the asymptotes is $(3x+y+2)(x-2y+5)=0$.

Example 8.7.2

Find the equation of the asymptotes of the hyperbola $xy=xh+yk$.

Solution

The combined equation of the asymptotes is $xy=xh+yk+n$ or $xy-xh-yk-n=0$.

The asymptotes are $x+l=0$ and $y+m=0$.

$$(x+l)(y+m)=xy-xh-yk-n$$

Equating the coefficients of the terms x and y and the constant terms, we get

$$m=-h,\ l=-k \quad \text{and} \quad lm=-n$$

$$\therefore \quad n=-hk$$

Hence, the equation of the asymptotes is $(x-h)(y-k)=0$.

Example 8.7.3

Find the equation to the hyperbola that passes through $(2,3)$ and has for its asymptotes the lines $4x+3y-7=0$ and $x-2y=1$.

Solution

The combined equation of the asymptotes is $(4x+3y-7)(x-2y-1)=0$.

Hence, the equation of the hyperbola is $(4x+3y-7)(x-2y-1)+k=0$.

This pass through $(2,3)$.

$$(8+9-7)(2-6-1)+k=0$$

$$\therefore\ k=50$$

Hence, the equation of the hyperbola is

$$(4x+3y-7)(x-2y-1)+50=0$$

$$\text{(i.e.)}\quad 4x^2-5xy-6y^2-11x+11y+57=0$$

Example 8.7.4

Find the equation of the hyperbola that has $3x-4y+7=0$ and $4x+3y+1=0$ as asymptotes and passes through the origin.

Solution

The combined equation of the asymptotes is

$$(3x-4y+7)(4x+3y+1)=0$$

Hence, the equation of the hyperbola is $(3x-4y+7)(4x+3y+1)+k=0$.
This passes through the origin $(0,0)$. $\therefore\ 7+k=0$ or $k=-7$
Hence, the equation of the hyperbola is

$$(3x-4y+7)(4x+3y+1)-7=0$$

$$12x^2-12y^2-7xy+31x+17y=0$$

Example 8.7.5

Find the equations of the asymptotes and the conjugate hyperbola given that the hyperbola has eccentricity $\sqrt{2}$, focus at the origin and the directrix along $x+y+1=0$.

Solution

From the focus directrix property, the equation of the hyperbola is

$$x^2+y^2=2\left(\frac{x+y+1}{\sqrt{2}}\right)^2$$

$$\text{(i.e.)}\quad 2xy+2x+2y+1=0$$

The combined equation of the asymptotes is $2xy+2x+2y+k=0$, where k is a constant. Let the asymptotes be $2x+l=0$ and $y+m=0$. Then,

$$2xy+2x+2y+k=(2x+l)(y+m)$$

Equating like terms, we get $2m=2$. $\therefore\ m=1$. Similarly, $l=2$. As $lm=k$, we get $k=2$.

Therefore, the asymphtes of the combined equation of the asymptotes is $2xy+2x+2y+2=0$.

The equation of the asymptotes of the conjugate hyperbola should differ by the same constant. The equation of the asymptotes of the conjugate hyperbola is $2xy+2x+2y+1=0$.

Example 8.7.6

Derive the equations of asymptotes.

Solution

The equation of the hyperbola is $\frac{x^2}{a^2}-\frac{y^2}{b^2}=1$ (i.e.) $f(x_1,y_1)=b^2x^2-a^2y^2-a^2b^2=0$. This being a second-degree equation, it can have maximum two asymptotes. As the coefficients of the highest degree terms in x and y are constants, there is no asymptote parallel to the axes of coordinates. Take $x=1$ and $y=m$ in the highest degree terms $\phi(m)=b^2-a^2m^2$. Similarly $\phi(m)=0$. The slopes of the oblique asymptotes are given by $\phi_2(m)=0$. (i.e.) $m=\pm\frac{b}{a}$.

Also, $$c=-\frac{\phi_1(m)}{\phi_2'(m)}=\frac{0}{-2a^2m}=0$$

The equations of the asymptotes are given by $y=\pm\frac{b}{a}x$.

(i.e.) $\frac{x}{a}-\frac{y}{b}=0$ and $\frac{x}{a}+\frac{y}{b}=0$. Therefore, the combined equation is $\frac{x^2}{a^2}-\frac{y^2}{b^2}=0$.

Exercises

1. Prove that the tangent to the hyperbola $x^2-3y^2=3$ at $(\sqrt{3},0)$ when associated with the two asymptotes form an equilateral triangle whose area is $\sqrt{3}$ square units.
2. Prove that the polar of any point on any asymptote of a hyperbola with respect to the hyperbola is parallel to the asymptote.
3. Prove that the rectangle contained by the intercepts made by any tangent to a hyperbola on its asymptotes is constant.

4. From any point of the hyperbola tangents are drawn to another which has the same asymptotes. Show that the chord of contact cuts off a constant area from the asymptotes.

5. Find the equation of the hyperbola whose asymptotes are $x+2y+3=0$ and $3x+4y+5=0$ and which passes through the point $(1,-1)$

Ans.: $(x+2y-13)(3x+4y+3)-8=0$

6. Find the asymptotes of the hyperbola $3x^2-5xy-2y^2+5x+11y-8=0$.

Ans.: $x-2y+3=0 \quad 3x+y-4=0$

7. Prove that the locus of the centre of the circle circumscribing the triangle formed by the asymptotes of the hyperbola $\frac{x^2}{a^2}-\frac{y^2}{b^2}=1$ and a variable tangent is $4\left(\frac{x^2}{b^2}-\frac{y^2}{a^2}\right)=\frac{1}{a^2}+\frac{1}{b^2}$.

8. Find the equation of the asymptotes of the hyperbola $9y^2-4x^2=36$ and obtain the product of the perpendicular distance of any point on the hyperbola from the asymptotes. $ay^2-4x^2=0$; $\frac{36}{13}$

9. Show that the locus of the point of intersection of the asymptotes with the directrices of the hyperbola $\frac{x^2}{a^2}-\frac{y^2}{b^2}=1$ is the circle $x^2+y^2=a^2$.

10. Let C be the centre of a hyperbola. The tangent at P meets the axes in Q and R and the asymptotes in L and M. The normal at P meets the axes in A and B. Prove that L and M lie on the circle OAB and Q and R are conjugate with respect to the circle.

11. If a line through the focus S drawn parallel to the asymptotes $\frac{x}{a}-\frac{y}{b}=0$ of the hyperbola $\frac{x^2}{a^2}-\frac{y^2}{b^2}=1$ meets the hyperbola and the corresponding directrix at P and Q then show that $SQ=2\cdot SP$.

12. Find the asymptotes of the hyperbola $\frac{x^2}{a^2}-\frac{y^2}{b^2}=1$ and show that the straight line parallel to an asymptote will meet the curve in one point at infinity.

13. Prove that the product of the intercepts made by any tangent to a hyperbola on its asymptotes is a constant.

14. If a series of hyperbolas is drawn having a common transverse axis of length $2a$ then prove that the locus of a point P on each hyperbola, such that its distance from one asymptote is the curve $(x^2-b^2)^2=4x^2(x^2-a^2)$.

8.8 CONJUGATE DIAMETERS

Locus of mid points of parallel chords of the hyperbola is $\frac{x^2}{a^2}-\frac{y^2}{b^2}=1$.

Let (x_1, y_1) be the mid point of a chord of the hyperbola $\frac{x^2}{a^2}-\frac{y^2}{b^2}=1$.

Then its equation is $\frac{xx_1}{a^2}-\frac{yy_1}{b^2}=\frac{x_1^2}{a^2}-\frac{y_1^2}{b^2}$

The slope of this chord is $\frac{b^2x_1}{a^2y_1}$

Let this chord be parallel to $y = mx$.

Then $m=\frac{b^2x_1}{a^2y_1}$

$$\text{(i.e.)} \quad y_1=\frac{b^2x_1}{a^2m}$$

The locus of (x_1, y_1) is $y=\frac{b^2x}{a^2m}$, which is a straight line passing through the origin.

If $y = m'x$ bisects all chords parallel to $y = mx$ then $m'=\frac{b^2}{a^2m}$ or $mm'=\frac{b^2}{a^2}$.

By symmetry, we note that $y = mx$ will bisect all chords parallel to $y = m'x$.

Definition 8.8.1 Two diameters are said to be conjugate if each bisects chords parallel to the other. The condition of the diameters $y = mx$ and $y = m'x$ to be conjugate diameters is $mm'=\frac{b^2}{a^2}$.

Note 8.8.2 These diameters are also conjugate diameters of the conjugate hyperbola $\frac{x^2}{-a^2}-\frac{y^2}{-b^2}=1$ since $mm'=\frac{-b^2}{-a^2}=\frac{b^2}{a^2}$.

Property 8.8.1

If a diameter meets a hyperbola in real points, it will meet the conjugate hyperbola in imaginary points and its conjugate diameter will meet the hyperbola in imaginary points and the conjugate hyperbola in real points and vice versa.

Proof

Let the equation of the hyperbola be

$$\frac{x^2}{a^2}-\frac{y^2}{b^2}=1 \tag{8.5}$$

Then the equation of the conjugate hyperbola is

$$\frac{x^2}{a^2}-\frac{y^2}{b^2}=-1 \tag{8.6}$$

Let $y = mx$ and $y = m'x$ be a pair of conjugate diameters of the hyperbola (8.5).
Then

$$mm' = \frac{b^2}{a^2} \tag{8.7}$$

The points of intersection of $y = mx$ and the hyperbola (8.5) are given by $\frac{x^2}{a^2}-\frac{m^2x^2}{b^2}=1.$

$$\therefore \quad x^2 = \frac{a^2b^2}{b^2-a^2m^2} \quad \text{or} \quad x = \pm\frac{ab}{\sqrt{b^2-a^2m^2}} \tag{8.8}$$

Since the hyperbola meets $y = mx$ in real points from (8.8) $b^2 - a^2m^2 > 0$.
The points of intersection of (8.6) with $y = mx$ are given by

$$x^2 = -\frac{a^2b^2}{b^2-a^2} < 0 \text{ since } b^2 - a^2m^2 > 0$$

Therefore, $y = mx$ meets the conjugate hyperbola in imaginary points.
The points of intersection of $y = m'x$ with the hyperbola (8.5) are given by

$$x^2 = -\frac{a^2b^2}{b^2-a^2m'^2}$$

$$= \frac{a^2b^2}{b^2-a^2\left(\frac{b^2}{a^4m^2}\right)} \left(\because mm' = \frac{b^2}{a^2}\right)$$

$$= \frac{a^4m^2}{b^2-a^2m^2} < 0$$

The conjugate diameter meets the hyperbola in imaginary points. Also its intersection with the conjugate hyperbola is given by

$$x^2 = \frac{-a^4m^2}{b^2 - a^2m^2} > 0$$

$y = m'x$ meets the conjugate hyperbola in real points.

Property 8.8.2

If a pair of conjugate diameters meet the hyperbola and its conjugate hyperbola in P and D, respectively then $CP^2 - CD^2 = a^2 - b^2$.

Proof

Let P be the point $(a \sec \theta, b \tan \theta)$

Then D will have coordinates $(-a \tan \theta, -b \sec \theta)$.

Then $CP^2 = a^2 \sec^2 \theta + b^2 \tan^2 \theta$

$CD^2 = a^2 \tan^2 \theta + b^2 \sec^2 \theta$

$CP^2 - CD^2 = a^2 (\sec^2 \theta - \tan^2 \theta) - b^2 (\sec^2 \theta - \tan^2 \theta)$

$= a^2 - b^2$

Property 8.8.3

The parallelogram formed by the tangents at the extremities of conjugate diameters of hyperbola has its vertices lying on the asymptotes and is of constant area.

Proof

Let P and D be points $(a \sec \theta, b \tan \theta)$ and $(a \tan \theta, b \sec \theta)$ on the hyperbola and its conjugate.

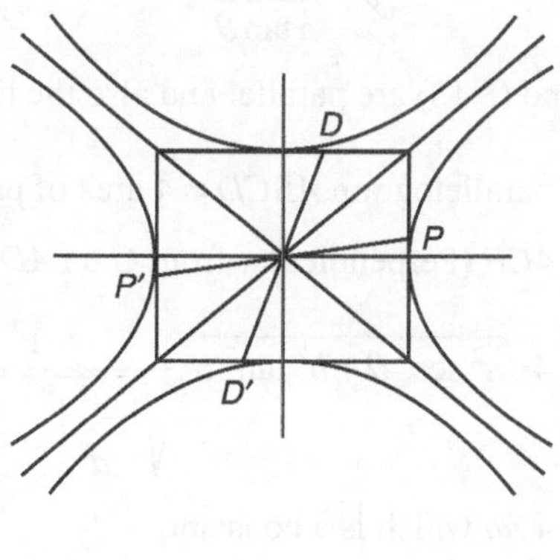

Then D' and P' are $(-a\tan\theta, -b\sec\theta)$ and $(-a\sec\theta, -b\tan\theta)$, respectively.

The equations of the asymptotes are

$$\frac{x}{a}-\frac{y}{b}=0 \text{ and } \frac{x}{a}+\frac{y}{b}=0.$$

The equations of the tangents at P, P', D, D' are

$$\frac{x\sec\theta}{a}-\frac{y\tan\theta}{b}=1 \tag{8.9}$$

$$-\frac{x\sec\theta}{a}+\frac{y\tan\theta}{b}=1 \tag{8.10}$$

$$\frac{x\tan\theta}{a}-\frac{y\sec\theta}{b}=-1 \tag{8.11}$$

$$-\frac{x\tan\theta}{a}+\frac{y\sec\theta}{b}=-1 \tag{8.12}$$

respectively. Clearly the tangents at P and P' are parallel and also the tangents at D and D' are parallel. Solving (8.9) and (8.11) we get the coordinates of D are $[a(\sec\theta+\tan\theta), b[\sec\theta+\tan\theta)]$.

This lies on the asymptote $\frac{x}{a}-\frac{y}{b}=0$.

Similarly the other points of intersection also lie on the asymptotes.

The equations of PCP' and DCD' are

$$y=\frac{b\tan\theta}{a\sec\theta}x \tag{8.13}$$

$$y=\frac{b\sec\theta}{a\tan\theta}x \tag{8.14}$$

Lines (8.11), (8.12) and (8.13) are parallel and also the lines (8.9), (8.10) and (8.11) are parallel.

Therefore, area of parallelogram $ABCD = 4$ area of parallelogram $CPAD$.

$$= 4CP \text{ (Perpendicular from } C \text{ on } AD)$$

$$= 4\sqrt{a^2\sec^2\theta+b^2\tan^2\theta}\cdot\frac{1}{\sqrt{\frac{\tan^2\theta}{a^2}+\frac{\sec^2\theta}{b^2}}}$$

$$= 4\,ab \text{ which is a constant.}$$

Example 8.8.1

If a pair of conjugate diameters meet hyperbola and its conjugate, respectively in *P* and *D* then prove that *PD* is parallel to one of the asymptotes and is bisected by the other asymptote.

Solution

Let the equation of the hyperbola be

$$\frac{x^2}{a^2}-\frac{y^2}{b^2}=1 \tag{8.15}$$

The equation of the conjugate hyperbola is

$$\frac{x^2}{a^2}-\frac{y^2}{b^2}=-1 \tag{8.16}$$

The asymptotes of the hyperbola (1) are

$$\frac{x}{a}-\frac{y}{b}=0 \tag{8.17}$$

$$\frac{x}{a}+\frac{y}{b}=0 \tag{8.18}$$

Let *P* be the point $(a\sec\theta, b\tan\theta)$.

Then *D* is the point $(a\tan\theta, b\sec\theta)$.

The slope of the chord *PD* is $\dfrac{b(\sec\theta-\tan\theta)}{a(\tan\theta-\sec\theta)}=-\dfrac{b}{a}$ = The slope of the asymptote (8.18)

PD is parallel to the asymptote (8.18).

The midpoint of *PD* is a $\left(\dfrac{a}{2}(\sec\theta+\tan\theta),\dfrac{b}{2}(\tan\theta+\sec\theta)\right)$. This point lies on the asymptotes given by (8.17).

Therefore, *PD* is bisected by the other asymptote.

Example 8.8.2

In the hyperbola $16x^2-9y^2=144$ find the equation of the diameter conjugate to the diameter $x=2y$.

Solution

The equation of the hyperbola is $16x^2-9y^2=144$

$$\text{(i.e.)}\quad \frac{x^2}{9}-\frac{y^2}{16}=1$$

$$a^2=9,\quad b^2=16$$

The slope of the line $x = 2y$ is $m = \frac{1}{2}$

If m and m' are the slopes of the conjugate diameters then $mm' = \frac{b^2}{a^2}$

$$\frac{1}{2}m' = \frac{16}{9} \text{ or } m' = \frac{32}{9}$$

Therefore, the equation of the conjugate diameter is $y = \frac{32}{9}x$ or $32x - 9y = 0$.

Example 8.8.3

Find the condition that the pair of lines $Ax^2 + 2Hxy + By^2 = 0$ to be conjugate diameters of the hyperbola $\frac{x^2}{a^2} - \frac{y^2}{b^2} = 1$.

Solution

Let the two straight lines represented by $Ax^2 + 2Hxy + By^2 = 0$ be $y = m_1x$ and $y = m_2x$. Then

$$m_1 + m_2 = -\frac{2H}{B} \quad \text{and} \quad m_1m_2 = \frac{A}{B} \tag{8.19}$$

If these lines are the conjugate diameters of the hyperbola then

$$m_1m_2 = \frac{b^2}{a^2} \tag{8.20}$$

From (8.19) and (8.20)

$$\frac{A}{B} = \frac{b^2}{a^2}$$

or

$$a^2A = b^2B.$$

Property 8.8.4

Any two conjugate diameters of a rectangular hyperbola are equally inclined to the asymptotes.

Proof

Let the equation of the rectangular hyperbola be $x^2 - y^2 = a^2$. The equation of the asymptotes is $x^2 - y^2 = 0$.

Let $y = mx$ and $y = \frac{1}{m}x$ be a pair of conjugate diameters of the rectangular hyperbola be $x^2 - y^2 = a^3$.

Then the combined equation of the conjugate diameters is

$$(y - mx)(x - my) = 0$$

$$\text{(i.e.)} \quad mx^2 - (m^2 + 1)xy + my^2 = 0.$$

The combined equation of the bisectors of the angles between these two lines is

$$\frac{x^2 - y^2}{0} = \frac{-xy}{\frac{1}{2}(m^2 + 1)}$$

$$\text{(i.e.)} \quad x^2 - y^2 = 0.$$

This is the combined equation of the asymptotes.

Therefore, the asymptotes bisect the angle between the conjugate diameter.

8.9 RECTANGULAR HYPERBOLA

Definition 8.9.1 If in a hyperbola the length of the semi-transverse axis is equal to the length of the semi-conjugate axis, then the hyperbola is said to be a rectangle hyperbola.

8.9.1 Equation of Rectangular Hyperbola with Reference to Asymptotes as Axes

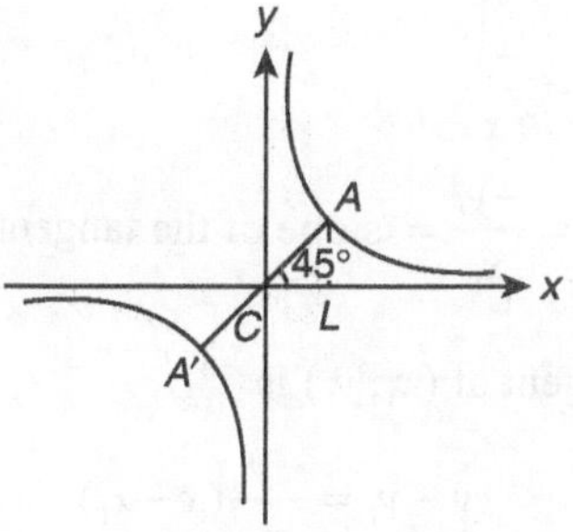

In a rectangular hyperbola, the asymptotes are perpendicular to each other. Since the axes of coordinates are also perpendicular to each other, we can take the asymptotes as the x- and y-axes. Then the equations of the asymptotes are $x = 0$ and $y = 0$. The combined equation of the asymptotes is $xy = 0$.

The equation of the hyperbola will differ from that of asymptotes only by a constant. Hence, the equation of the rectangular hyperbola is $xy = k$ where k is a constant to be determined. Let AA' be the transverse axis and its length be $2a$. Then, $AC = CA' = a$. Draw AL perpendicular to x-axis. Since the asymptotes bisect the angle between the axes, $\underline{|ACL} = 45°$.

$$CL = CA\cos 45° = \frac{a}{\sqrt{2}} \text{ and } AL = CA\sin 45° = \frac{a}{\sqrt{2}}$$

The coordinates of A are $\left[\frac{a}{\sqrt{2}}, \frac{a}{\sqrt{2}}\right]$. Since it lies on the rectangular hyperbola $xy = k$, we get $\frac{a^2}{2} = k$. Hence, the equation of the rectangular hyperbola is $xy = \frac{a^2}{2}$ or $xy = c^2$ where $c^2 = \frac{a^2}{2}$.

Note 8.9.1.1: The parametric equations of the rectangular hyperbola $xy = c^2$ are $x = ct$ and $y = \frac{c}{t}$.

8.9.2 Equations of Tangent and Normal at (x_1, y_1) on the Rectangular Hyperbola $xy = c^2$

The equation of rectangular hyperbola is $xy = c^2$. Differentiating with respect to x, we get

$$x\frac{dy}{dx} + y = 0$$

$$\frac{dy}{dx} = \frac{-y}{x}$$

$$\left|\frac{dy}{dx}\right|_{at(x_1, y_1)} = \frac{-y_1}{x_1} = \text{slope of the tangent at}(x_1, y_1).$$

The equation of the tangent at (x_1, y_1) is

$$y - y_1 = \frac{-y_1}{x_1}(x - x_1)$$

$$\Rightarrow x_1 y - x_1 y_1 = -y_1 x + x_1 y_1$$

$$\Rightarrow y_1 x + x_1 y = 2x_1 y_1$$

$$\text{(i.e.)} \quad y_1 x + x_1 y = 2c^2$$

since $\qquad x_1 y_1 = c^2.$

The slope of the normal at (x_1, y_1) is $\frac{x_1}{y_1}$.

The equation of the normal at (x_1, y_1) is

$$y - y_1 = \frac{x_1}{y_1}(x - x_1)$$

$$yy_1 - y_1^2 = xx_1 - x_1^2$$

$$\text{(i.e.)} \quad xx_1 - yy_1 = x_1^2 - y_1^2$$

8.9.3 Equation of Tangent and Normal at $\left(ct, \frac{c}{t}\right)$ on the Rectangular Hyperbola $xy = c^2$

The equation of the rectangular hyperbola is $xy = c^2$. Differentiating with respect to x, we get

$$x\frac{dy}{dx} + y = 0$$

$$\therefore \quad \frac{dy}{dx} = \frac{-y}{x}$$

$\frac{dy}{dx}\left(ct, \frac{c}{t}\right) = \frac{-1}{t^2}$ = slope of the tangent at $\left(ct, \frac{c}{t}\right)$

The equation of the tangent at is $\left(ct, \frac{c}{t}\right)$ is

$$y - \frac{c}{t} = \frac{-1}{t^2}(x - ct)$$

$$\text{(i.e.)} \quad yt^2 - ct = -x + ct \text{ or } x + yt^2 = 2ct$$

The slope of the normal at 't' is $-t^2$. The equation of the normal at 't' is

$$y - \frac{c}{t} = -t^2(x - ct)$$

$$y - \frac{c}{t} = -xt^2 + ct^3$$

Dividing by t, we get

$$\frac{y}{t} - \frac{c}{t^2} = -xt + ct^2 \quad \text{or} \quad xt - \frac{y}{t} = c\left(t^2 - \frac{1}{t^2}\right)$$

8.9.4 Equation of the Chord Joining the Points 't_1' and 't_2' on the Rectangular Hyperbola $xy = c^2$ and the Equation of the Tangent at t

The two points are $\left(ct_1, \frac{c}{t_1}\right)$ and $\left(ct_2, \frac{c}{t_2}\right)$. The equations of the chord joining the two points are

$$\frac{y-y_1}{x-x_1} = \frac{y_1-y_2}{x_1-x_2}$$

$$\frac{y-\frac{c}{t_1}}{x-ct_1} = \frac{\frac{c}{t_1}-\frac{c}{t_2}}{ct_1-ct_2} = \frac{-c(t_1-t_2)}{ct_1t_2(t_1-t_2)} = \frac{-1}{t_1t_2}$$

Cross multiplying, we get

$$yt_1t_2 - ct_2 = -x + ct_1$$
$$x + yt_1t_2 = c(t_1+t_2)$$

This chord becomes the tangent at 't' if $t_1 = t_2 = t$.
Hence, the equation of the tangent at 't' is $x + yt^2 = 2ct$.

8.9.5 Properties

Any two conjugate diameters of a rectangular hyperbola are equally inclined to the asymptotes.

Proof Let the equation of the rectangular hyperbola be $x^2 - y^2 = a^2$. The equation of the asymptotes is $x^2 - y^2 = 0$. Let $y = mx$ and $y = \frac{1}{m}x$ be a pair of conjugate diameters of the rectangular hyperbola $x^2 - y^2 = a^2$. Then, the combined equation of the conjugate diameters is

$$(y - mx)(x - my) = 0$$

$$\text{(i.e.)}\quad mx^2 - (m^2+1)xy + my^2 = 0$$

The combined equation of the bisectors of the angles between these two lines is $\frac{x^2-y^2}{0} = \frac{-xy}{\frac{1}{2}(m^2+1)}$ (i.e.) $x^2 - y^2 = 0$ which is the combined equation of the asymptotes. Therefore, the asymptotes bisect the angle between the conjugate diameter.

8.9.6 Results Concerning the Rectangular Hyperbola

1. The equation of the tangent at (x_1, y_1) on the rectangular hyperbola $xy = c^2$ is $\frac{1}{2}(xy_1 + yx_1) = c^2$.

2. The equation of the normal at (x_1, y_1) is $xx_1 - yy_1 = x_1^2 - y_1^2$.
3. The equation of the pair of tangents from (x_1, y_1) is $(xy_1 + yx_1 - 2c^2)^2 = 4(xy - c^2)(x_1y_1 - c^2)$.
4. The equations of the chord having (x_1, y_1) as its midpoint is $xy_1 + yx_1 = 2x_1y_1$.
5. The equation of the chord of contact from (x_1, y_1) is $xy_1 + yx_1 = 2c^2$.

8.9.7 Conormal Points—Four Normal from a Point to a Rectangular Hyperbola

Let (x_1, y_1) be a given point and t be the foot of the normal from (x_1, y_1) on the rectangular hyperbola $xy = c^2$.

The equation of the normal at t is $xt - \dfrac{y}{t} = c\left(t^2 - \dfrac{1}{t^2}\right)$.

Since this normal passes through (x_1, y_1),

$$x_1t - \frac{y_1}{t} = c\left(t^2 - \frac{1}{t^2}\right)$$

$$\text{(i.e.)}\quad x_1t^3 - y_1t = c(t^4 - 1) \text{ or } ct^4 - x_1t^3 + y_1t - c = 0$$

This is a fourth-degree equation in t and there are four values of t(real or imaginary). Corresponding to each value of t there is a normal, and hence there are four normals from a given point to the rectangular hyperbola.

Note 8.9.7.1: If t_1, t_2, t_3 and t_4 are the four points of intersection, then $\Sigma t_1 = \dfrac{x_1}{c}$, $\Sigma t_1t_2 = 0$, $\Sigma t_1t_2t_3 = \dfrac{-y}{c}$ and $t_1t_2t_3t_4 = -1$.

8.9.8 Concyclic Points on the Rectangular Hyperbola

Let the equation of the rectangular hyperbola be $xy = c^2$.

Let the equation of the circle be $x^2 + y^2 + 2gx + 2fy + k = 0$.

Let $\left(ct, \dfrac{c}{t}\right)$ be a point of intersection of rectangular hyperbola and the circle. Then, the point $\left(ct, \dfrac{c}{t}\right)$ also lies on the circle. Substituting $x = ct$, $y = \dfrac{c}{t}$ in the equation of the circle we get

$$c^2t^2 + \frac{c^2}{t^2} + 2gct + 2f \cdot \frac{c}{t} + k = 0$$

$$\text{(i.e.)} \quad c^2t^4 + 2gct^3 + kt^2 + 2fct + c^2 = 0$$

This is a fourth degree equation in t. For each value of t, there is a point of intersection (real or imaginary). Hence, there are four points of intersection for a rectangular hyperbola with the circle.

Note 8.9.8.1: If t_1, t_2, t_3 and t_4 are the four points of intersection, then $\Sigma t_1 = \frac{-2a}{c}$, $\Sigma t_1 t_2 = \frac{k}{c^2}$, $\Sigma t_1 t_2 t_3 = \frac{-2f}{c}$ and $t_1 t_2 t_3 t_4 = 1$.

Example 8.9.1

Show that the locus of poles with respect to the parabola $y^2 = 4ax$ of tangents to the hyperbola $x^2 - y^2 = a^2$ is the ellipse $4x^2 + y^2 = 4a^2$.

Solution

Let (x_1, y_1) be the pole with respect to the parabola $y^2 = 4ax$. Then, the polar of (x_1, y_1) is $yy_1 = 2a(x + x_1)$ (i.e.) $y = \frac{2ax}{y_1} + \frac{2ax_1}{y_1}$.
This is a tangent to the rectangular hypherbola $x^2 - y^2 = a^2$.
The condition for tangency is

$$c^2 = a^2(m^2 - 1)$$

$$\text{(i.e.)} \quad \left(\frac{2ax_1}{y_1}\right)^2 = a^2\left[\left(\frac{2a}{y_1}\right)^2 - 1\right]$$

$$\text{(i.e.)} \quad 4a^2x_1^2 = a^2(4a^2 - y_1^2)$$

$$4x_1^2 + y_1^2 = 4a^2$$

The locus of (x_1, y_1) is $4x^2 + y^2 = 4a^2$ which is an ellipse.

Example 8.9.2

P is a point on the circle $x^2 + y^2 = a^2$ and PQ and PR are tangents to the hyperbola $x^2 - y^2 = a^2$. Prove that the locus of the middle point of QR is the curve $(x^2 - y^2)^2 = a^2(x^2 + y^2)$.

Solution

Let $P(x_1, y_1)$ be a point on the circle $x^2 + y^2 = a^2$.

$$\therefore\ x_1^2 + y_1^2 = a^2$$

Since PQ and PR are tangents from P to the rectangular hyperbola $x^2 - y^2 = a^2$, QR is the chord of contacts of tangents from $P(x_1, y_1)$. Therefore, its equation is $xx_1 + yy_1 = a^2$. Let (h, k) be the midpoint of QR. Its equation is $xh - yk = h^2 - k^2$. These two equations represent the same line. Therefore, identifying them, we get

$$\frac{x_1}{h} = \frac{y_1}{k} = \frac{a^2}{h^2 - k^2}$$

$$x_1 = \frac{a^2 h}{h^2 - k^2} \text{ and } y_1 = \frac{a^2 k}{h^2 - k^2}$$

Since $x_1^2 + y_1^2 = a^2$, $\dfrac{a^4}{(h^2 - k^2)^2}(h^2 + k^2) = a^2$ (i.e.) $a^2(h^2 + k^2) = (h^2 - k^2)^2$.

The locus of (h, k) is $(x^2 - y^2)^2 = a^2(x^2 + y^2)$.

Example 8.9.3

Prove that the locus of poles of all normal chords of the rectangular hyperbola $xy = c^2$ is the curve $(x^2 - y^2) + 4c^2xy = 0$.

Solution

Let (x_1, y_1) be the pole of the normal chord of rectangular hyperbola $xy = c^2$.
The poles of (x_1, y_1) is $xy_1 + yx_1 = 2c^2$. Let the chord be normal at t.

The equation of the normal at t is $xt - \dfrac{y}{t} = c\left(t^2 - \dfrac{1}{t^2}\right)$. These two equations represent the same straight line. Identifying them, we get

$$\frac{y_1}{t} = -tx_1 = \frac{2c^2}{c\left(t^2 - \dfrac{1}{t^2}\right)}$$

$$\therefore t^2 = \frac{-y_1}{x_1}$$

Also $$x_1^2t_1^2 = \frac{4c^2}{\left(t^2 - \frac{1}{t^2}\right)^2}$$

$$x_1^2t_1^2\left(t^2 - \frac{1}{t^2}\right)^2 = 4c^2 \text{ (i.e.) } x_1^2\left(\frac{-y_1}{x_1}\right)\left(\frac{-y_1}{x_1} + \frac{x_1}{y_1}\right)^2 = 4c^2$$

or $(x_1^2 - y_1^2)^2 + 4c^2x_1y_1 = 0$

The locus of (x_1, y_1) is $(x^2 - y^2)^2 + 4c^2xy = 0$.

Example 8.9.4

If P is any point on the parabola $x^2 + 16ay = 0$, prove that the poles of P with respect to rectangular hyperbola $xy = 2a^2$ will touch the parabola $y^2 = ax$.

Solution

Let (x_1, y_1) be any point. The polar of P with respect to the hyperbola is $xy_1 + x_1y = 4a^2$ (i.e.) $y = \frac{-y_1}{x_1}x + \frac{4a^2}{x_1}$. This is a tangent to the parabola $y^2 = ax$. The condition is $c = \frac{a}{4m}$.

$$\text{(i.e.) } \frac{4a^2}{x_1} = \frac{a}{4\left(\frac{-y_1}{x_1}\right)} \quad \text{(i.e.) } 16ay_1 + x_1^2 = 0$$

The locus of (x_1, y_1) is $x^2 + 16ay = 0$.

Example 8.9.5

A tangent to the parabola $x^2 = 4ay$ meets the hyperbola $xy = c^2$ at P and Q. Prove that the middle point of PQ lies on a fixed parabola.

Solution

Let (x_1, y_1) be the midpoint of the chord PQ of the rectangular hyperbola $xy = c^2$. The equation of chord PQ is

$$xy_1 + yx_1 = 2x_1y_1$$

$$\text{(i.e.) } xy_1 = -yx_1 + 2x_1y_1 \Rightarrow x = \frac{-yx_1}{y_1} + 2x_1$$

This is a tangent to the parabola $x^2 = 4ay$. Therefore, the condition is $2x_1 = \dfrac{a}{\left(\dfrac{-x_1}{y_1}\right)}$.

(i.e.) $2x_1^2 + ay_1 = 0$. The locus of (x_1, y_1) is $2x^2 + ay = 0$, which is a fixed parabola.

Example 8.9.6

Find the locus of midpoints of chords of constant length $2l$ of the rectangular hyperbola $xy = c^2$.

Solution

Let $R(x_1, y_1)$ be the midpoint of the chord PQ. Let the equation of the chord PQ be $\dfrac{x - x_1}{\cos\theta} = \dfrac{y - y_1}{\sin\theta} = r$.

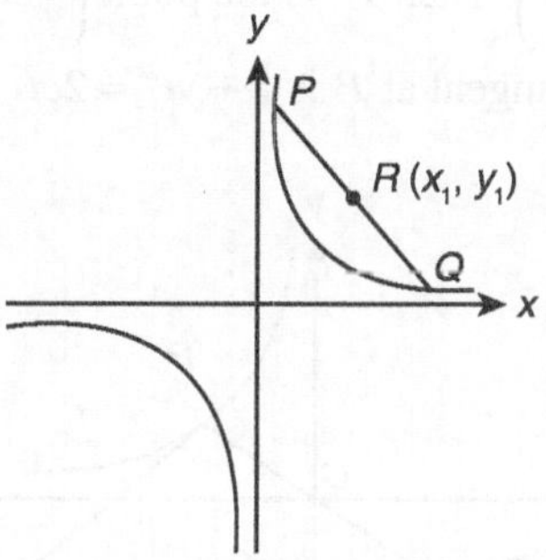

Any point on this line is $x = x_1 + r\cos\theta,\ y = y_1 + r\sin\theta$. If this point lies on the rectangular hyperbola $xy = c^2$, we get $(x_1 + r\cos\theta)(y_1 + r\sin\theta) = c^2$.

$$r^2 \sin\theta\cos\theta + r(x_1\sin\theta + y_1\cos\theta) + x_1y_1 - c^2 = 0 \qquad (8.21)$$

This is a quadratic equation in r. The two values of r are the distances RP and RQ which are equal in magnitude but opposite in sign. The condition for this is the coefficient of r is equal to zero.

$$x_1\sin\theta + y_1\cos\theta = 0 \qquad (8.22)$$

Then, equation (8.21) becomes

$$r^2 \sin\theta\cos\theta + x_1y_1 - c^2 = 0 \qquad (8.23)$$

From equation (8.22), $\tan\theta = \dfrac{-y_1}{x_1}, \sin\theta = \dfrac{-y}{\sqrt{x_1^2 + y_1^2}}$ and $\cos\theta = \dfrac{x_1}{\sqrt{x_1^2 + y_1^2}}$

Substituting these in equation (8.23), we get $r^2\left(\dfrac{-x_1 y_1}{x_1^2 + y_1^2}\right) + x_1 \cdot y_1 - c^2 = 0.$

$(x_1^2 + y_1^2)(x_1 y_1 - c^2) - r^2 x_1 y_1 = 0$; but $r = l$.

Therefore, the locus of (x_1, y_1) is $(x^2 + y^2)(xy - c^2) - l^2 xy = 0$.

Example 8.9.7

If PP' is a diameter of the rectangular hyperbola $xy = c^2$ show that the locus of the intersection of tangents at P with the straight line through P' parallel to either asymptote is $xy + 3c^2 = 0$.

Solution

Let P be the point $\left(ct, \dfrac{c}{t}\right)$. Then P' is the point $\left(-ct, \dfrac{-c}{t}\right)$.

The equation of the tangent at P is $x + yt^2 = 2ct$.

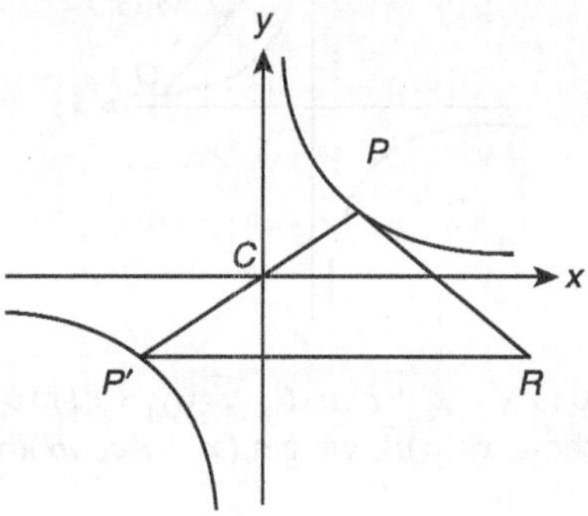

The equation of the straight line $P'R$ parallel to x-axis is $y = \dfrac{-c}{t}$.

Let (x_1, y_1) be the point of intersection of these two lines. Then

$$x_1 + y_1 t^2 = 2ct \tag{8.24}$$

$$y_1 = \frac{-c}{t} \tag{8.25}$$

$$\text{(i.e.)} \quad t = \frac{-c}{y_1}$$

Substituting in equation (8.24), $x_1 + y_1 \cdot \frac{c^2}{y_1^2} = 2c\left(\frac{-c}{y_1}\right)$

$$x_1 + \frac{c^2}{y_1} = \frac{-2c^2}{y_1}$$

$$xy_1 + c^2 = -2c^2 \quad \text{or} \quad x_1y_1 + 3c^2 = 0$$

The locus of (x_1, y_1) is $xy + 3c^2 = 0$.

Example 8.9.8

The tangents to the rectangular hyperbola $xy = c^2$ and the parabola $y^2 = 4ax$ at their point of intersections are inclined at angles α and β, respectively, to the x-axis. Show that $\tan\alpha + 2\tan\beta = 0$.

Solution

Let (x_1, y_1) be the point of intersection of the rectangular hyperbola $xy = c^2$ and the parabola $y^2 = 4ax$. The equation of tangent at (x_1, y_1) to the parabola is $yy_1 = 2a(x + x_1)$. The equation of tangent to the rectangular hyperbola is $xy_1 + yx_1 = 2c^2$.

The slope of the tangent to the parabola is $\tan\beta = \frac{2a}{y_1}$.

The slope of the tangent to the tangent to the rectangular hyperbola is $\tan\alpha = \frac{-y_1}{x_1}$.

$$\tan\alpha + 2\tan\beta = \frac{-y_1}{x_1} + 2\cdot\frac{2a}{y_1} = \frac{-y_1^2 + 4ax_1}{x_1y_1} = 0$$

since (x_1, y_1) lies on the parabola $y^2 = 4ax$.

$\therefore \tan\alpha + 2\tan\beta = 0$

Example 8.9.9

If the normal to the rectangular hyperbola $xy = c^2$ at the point t as it intersect the rectangular hyperbola at t_1 then show that $t^3t_1 = -1$.

Solution

The equation of the normal at t is $xt - \frac{y}{t} = c\left(t^2 - \frac{1}{t^2}\right)$.

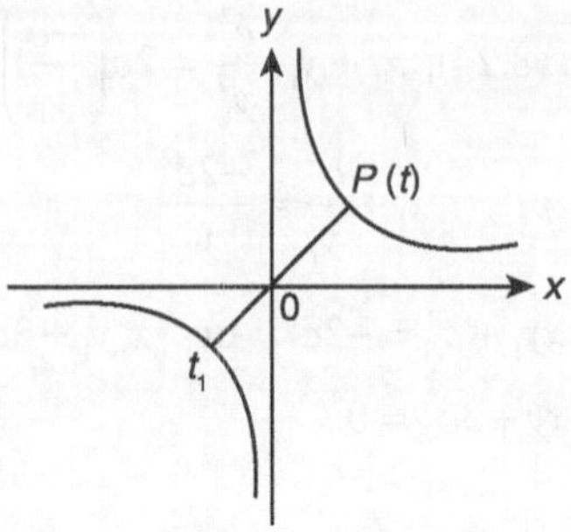

The equation of the chord joining the points t and t_1 is $x + ytt_1 = c(t + t_1)$.

These two equations represent the same straight line. Identifying them, we get

$$\frac{t}{1} = \frac{-1/t}{tt_1} = \frac{c\left(t^2 - \frac{1}{t^2}\right)}{c(t + t_1)}$$

$$\therefore t = \frac{-1}{t^2 t_1} \text{ or } t^3 t_1 = -1$$

Example 8.9.10

Show that the area of the triangle formed by the two asymptotes of the rectangular hyperbola $xy = c^2$ and the normal at (x_1, y_1) on the hyperbola is $\frac{1}{2}\left[\frac{(x_1^2 - y_1^2)}{c}\right]^2$.

Solution

The equation of the normal at (x_1, y_1) is $xx_1 - yy_1 = x_1^2 - y_1^2$.

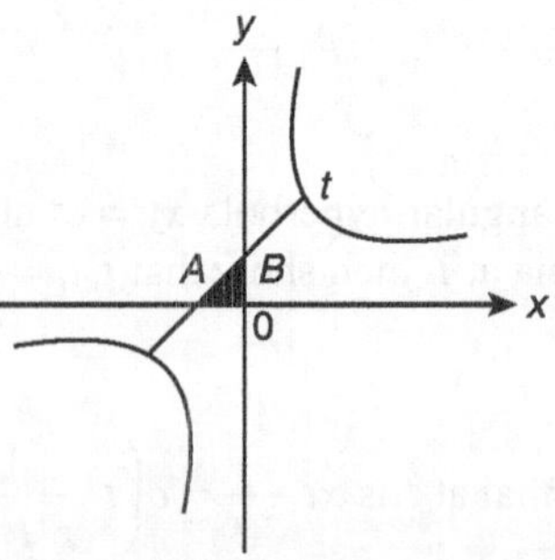

When the normal meets the x-axis, $y = 0$.

$$\therefore x = \frac{x_1^2 - y_1^2}{x_1} = OA$$

When the normal meets y-axis, $x = 0$

$$\therefore y = \frac{-(x_1^2 - y_1^2)}{y_1} = OB$$

The area of the triangle $OAB = \frac{1}{2} \cdot OA \cdot OB$

$$\Delta = \frac{-1}{2} \cdot \frac{x_1^2 - y_1^2}{x_1} \cdot \frac{x_1^2 - y_1^2}{y_1}$$

$$= \frac{-1}{2} \cdot \frac{(x_1^2 - y_1^2)^2}{x_1 y_1}$$

(i.e.) $\Delta = \frac{1}{2}\left[\frac{(x_1^2 - y_1^2)}{c}\right]^2$ since $x_1 y_1 = c^2$ and ignoring the negative sign.

Example 8.9.11

If four points be taken on a rectangular hyperbola such that the chord joining any two is perpendicular to the chord joining the other two and $\alpha, \beta, \gamma, \delta$ are the inclinations of the straight lines joining these points to the centre. prove that $\tan\alpha \tan\beta \tan\gamma \tan\delta = 1$.

Solution

Let $t_1, t_2, t_3,$ and t_4 be four points $P, Q, R,$ and S on the rectangular hyperbola $xy = c^2$. The equation of the chord joining t_1 and t_2 is $x + y t_1 t_2 = c(t_1 + t_2)$. The slope of this chord is $\frac{-1}{t_1 t_2}$.

Similarly, the slope of the chord joining t_3 and t_4 is $\frac{-1}{t_3 t_4}$.

Since these two chords are perpendicular, $\left(\frac{-1}{t_1 t_2}\right)\left(\frac{-1}{t_3 t_4}\right) = -1$

$$\text{(i.e.)} \quad t_1\, t_2\, t_3\, t_4 = -1 \qquad (8.26)$$

The slope of the line CP is $\tan\alpha = \frac{(c/t_1)}{ct_1}$ (i.e.) $\tan\alpha = \frac{1}{t_1^2}$

Similarly, $\tan\beta = \frac{1}{t_2^2}$, $\tan\gamma = \frac{1}{t_3^2}$, $\tan\delta = \frac{1}{t_4^2}$

$\therefore \tan\alpha\cdot\tan\beta\cdot\tan\gamma\cdot\tan\delta = \frac{1}{t_1^2\cdot t_2^2\cdot t_3^2\cdot t_4^2} = 1$ from equation (8.26)

Example 8.9.12

If the normals at three parts P, Q and R on a rectangular hyperbola intersect at a point S on the curve then prove that the centre of the hyperbola is the centroid of the triangle PQR.

Solution

If the normal at t meets the curve at t' then $t^2t' = -1$.

$$\text{(i.e.)}\quad t^2t' + 1 = 0 \tag{8.27}$$

This is a cubic equation in t. If t_1, t_2 and t_3 are the roots of this equation they can be regarded as the parameters of the points P, Q and R, the normals at these points meet at t'which is S. From equation (8.27), we get $t_1 + t_2 + t_3 = 0$ and $t_1t_2 + t_2t_3 + t_3t_1 = 0$. Let (h, k) be the centroid of ΔPQR.

Then $h = \frac{x_1 + x_2 + x_3}{3} = \frac{c(t_1 + t_2 + t_3)}{3} = 0.$

$$k = \frac{y_1 + y_2 + y_3}{3} = \frac{c}{3}\left[\frac{1}{t_1} + \frac{1}{t_2} + \frac{1}{t_3}\right] = \frac{c}{3}\left(\frac{t_1t_2 + t_2t_3 + t_3t_1}{t_1t_2t_3}\right) = 0.$$

The centroid is the centre of the rectangular hyperbola.

Example 8.9.13

Show that four normals can be drawn from a point (h, k) to the rectangular hyperbola $xy = c^2$ and that its feet form a triangle and its orthocentre.

Solution

The equation of the normal at t is $xt - \frac{y}{t} = c\left(t^2 - \frac{1}{t^2}\right)$.

$$\text{(i.e.)}\quad ct^4 - xt^3 + yt - c = 0$$

Since this passes through (h,k), $ct^4 - ht^3 + kt - c = 0$.

This is a fourth degree equation in t. Its roots are t_1, t_2, t_3 and t_4 which are the feet of the four normals from (h,k).

$$\therefore t_1 + t_2 + t_3 + t_4 = \frac{h}{c}, \sum t_1t_2 = 0, \sum t_1t_2t_3 = \frac{-k}{c}, t_1t_2t_3t_4 = -1$$

If t_1, t_2, t_3 and t_4 are the points P, Q, R and S on the rectangular hyperbola $xy = c^2$, it can be shown that the orthocentre of the triangle is $\left(\frac{-c}{t_1t_2t_3}, -ct_1\,t_2\,t_3\right)$.

This point is $\left(t_4, \frac{c}{t_4}\right)$ since $t_1t_2t_3t_4 = -1$.

$\therefore$ The four points P, Q, R and S form a triangle and its orthocentre.

Example 8.9.14

Prove that from any point (h,k) four normals can be drawn to the rectangular hyperbola $xy = c^2$ and that if the coordinates of the four feet of the normals P, Q, R and S be $(x_r, y_r), r = 1,2,3,4$. Then (i) $x_1 + x_2 + x_3 + x_4 = h$, $y_1 + y_2 + y_3 + y_4 = k$ and (ii) $x_1x_2x_3x_4 = y_1y_2y_3y_4 = -c^4$.

Solution

The equation of the normal at t is

$$xt - \frac{y}{t} = c\left(t^2 - \frac{1}{t^2}\right)$$

Since this passes through (h,k)

$$ht - \frac{k}{t} = c\left(t^2 - \frac{1}{t^2}\right) \text{ (i.e.) } ct^4 - ht^3 + kt - c = 0.$$

The form values of t correspond to the feet of the four normals from the point (h,k). If t_1, t_2, t_3 and t_4 are the four feet of the normals then they are the roots of the above equation.

$$t_1 + t_2 + t_3 + t_4 = \frac{h}{c} \tag{8.28}$$

$$\sum t_1t_2 = 0 \tag{8.29}$$

$$\sum t_1t_2t_3 = \frac{-k}{c} \tag{8.30}$$

$$t_1t_2t_3t_4 = -1 \tag{8.31}$$

From equation (8.28), $c(t_1 + t_2 + t_3 + t_4) = h$

$$\text{(i.e.)} \quad x_1 + x_2 + x_3 + x_4 = h$$

Dividing equation (8.30) by equation (8.31), we get $\frac{1}{t_1} + \frac{1}{t_2} + \frac{1}{t_3} + \frac{1}{t_4} = \frac{k}{c}$.

$$\frac{c}{t_1} + \frac{c}{t_2} + \frac{c}{t_3} + \frac{c}{t_4} = k$$

$$\text{(i.e.)} \quad y_1 + y_2 + y_3 + y_4 = k$$

Also, $\qquad x_1x_2x_3x_4 = c^4 \ t_1t_2t_3t_4 = -c'$

$$y_1y_2y_3y_4 = \frac{c^4}{t_1t_2t_3t_4} = -c^4,\ x_1x_2x_3x_4 = y_1y_2y_3y_4 = -c^4$$

Example 8.9.15

Prove that the feet of the concurrent normals on the rectangular hyperbola $xy = c^2$ which meets at (h, k) lie on another rectangular hyperbola which passes through $(0, 0)$ and (h, k).

Solution

The equation of the normal at (x_1, y_1) is $xx_1 - yy_1 = x_1^2 - y_1^2$.

Since this passes through (h, k), $hx_1 - ky_1 = x_1^2 - y_1^2$

The locus of (x_1, y_1) is $x^2 - y^2 - hx + ky = 0$. Clearly this is a rectangular hyperbola passing through $(0, 0)$ and (h, k).

Example 8.9.16

If a rectangular hyperbola whose centre is c is cut by any circle of radius r in four points P, Q, R, S then prove that $CP^2 + CQ^2 + CR^2 + CS^2 = 4r^2$.

Solution

Let the equation of the rectangular hyperbola be

$$xy = c^2 \tag{8.32}$$

Let the equation of the circle be

$$x^2 + y^2 + 2gx + 2fy + k = 0 \tag{8.33}$$

Solving these two equations, we get their points of intersections $y = \frac{c^2}{x}$.

Substituting in equation (8.33), $x^2 + \frac{c^4}{x^4} + 2gx + 2f\frac{c^2}{x} + k = 0$.

$$\text{(i.e.)} \quad x^4 + 2gx^3 + kx^2 + 2fc^2x + c^4 = 0.$$

If x_1, x_2, x_3, x_4 are the abscissae of the four points of intersection $x_1 + x_2 + x_3 + x_4 = -2g$.

$$\sum x_1x_2 = k,\ x_1^2 + x_2^2 + x_3^2 + x_4^2 = (x_1 + x_2 + x_3 + x_4)^2 - 2\sum x_1x_2 = 4g^2 - 2k.$$

Similarly, $y_1^2 + y_2^2 + y_3^2 + y_4^2 = 4f^2 - 2k = CP^2 + CQ^2 + CR^2 + CS^2$

$$= \sum (x_1^2 + y_1^2) = \sum x_1^2 + y_1^2$$

$$= 4(g^2 + f^2 - k) = 4r^2$$

Example 8.9.17

A, B, C and D are four points of intersection of a circle and a rectangular hyperbola. If AB passes through the centre of the hyperbola, show that CD passes through the centre of the circle.

Solution

Let the equation of the rectangular hyperbola be $xy = c^2$. Let the equation of the circle be $x^2 + y^2 + 2gx + 2fy + k = 0$. Let A, B, C and D be the points t_1, t_2, t_3 and t_4, respectively. When the circle and rectangular hyperbola intersect we know that

$$t_1 + t_2 + t_3 + t_4 = \frac{-2g}{c} \qquad (8.34)$$

$$\sum t_1t_2 = \frac{k}{c^2} \qquad (8.35)$$

$$\sum t_1t_2t_3 = \frac{-2f}{c} \qquad (8.36)$$

$$t_1t_2t_3t_4 = 1 \qquad (8.37)$$

The equation of the chord AB is $x + yt_1t_2 = c(t_1 + t_2)$. Since AB passes through $(0,0)$, we get

$$t_1 + t_2 = 0 \qquad (8.38)$$

$\therefore$ From equation (8.34),

$$t_3 + t_4 = -\frac{2g}{c} \tag{8.39}$$

However, $\sum t_1t_2t_3 = \frac{-2f}{c}$

$$t_1t_2t_3 + t_1t_2t_4 + t_1t_3t_4 + t_2t_3t_4 = \frac{-2f}{c}$$

$$t_1t_2(t_3 + t_4) + (t_1 + t_2)t_3t_4 = \frac{-2f}{c}.$$

$\frac{1}{t_3t_4}(t_3 + t_4) + 0 = \frac{-2f}{c}$ using equation (8.38)

$$t_3 + t_4 = \frac{-2f}{c}(t_3t_4) \tag{8.40}$$

$t_3t_4 = \left(\frac{-2g}{c}\right)\left(\frac{-c}{2f}\right) = \frac{g}{h}$ [From equations (8.39) and (8.40)]

The equation of the chord CD is $x + yt_3t_4 = c(t_3 + t_4)$.

$$x + y\left(\frac{g}{f}\right) = -2g$$

This straight line passes through the point $(-g, -f)$.

Therefore, CD passes through the centre of the circle.

Example 8.9.18

Show that through any given point P in the plane of $xy = c^2$, four normals can be drawn to it. If P_1, P_2, P_3 and P_4 are feet of these normals and C is centre then show that $CP_1^2 + CP_2^2 + CP_3^2 + CP_4^2 = CP^2$.

Solution

The equation of the normal at t is $xt - \frac{y}{t} = c\left(t^2 - \frac{1}{t^2}\right)$.

Let p be the point (h, k). Since the normal passes through (h, k),

$ht - \frac{k}{t} = c\left(t^2 - \frac{1}{t^2}\right)$ (i.e.) $ct^4 - ht^3 + kt - c = 0.$

This brings a fourth degree equation, there are four normals from P.

If t_1, t_2, t_3 and t_4 are the feet of the normals then $t_1+t_2+t_3+t_4=\frac{h}{c}$.

$\sum t_1t_2=0, \sum t_1t_2t_3=\frac{-k}{c}, t_1t_2t_3t_4=-1.$

Also, $\frac{1}{t_1},\frac{1}{t_2},\frac{1}{t_3},\frac{1}{t_4}$ are the roots of $ct^4-kt^3+ht-c=0.$ $\sum t_1=\frac{k}{c}, \sum\frac{1}{t_1t_2}=0.$

$$\sum CP^2=\sum c^2\left(t_1^2+\frac{1}{t_1^2}\right)=c^2\left[\sum t_1^2+\sum\frac{1}{t_1^2}\right]=c^2\left[\begin{matrix}\left(\sum t_1\right)^2-2\sum t_1t_2+\\ \left(\sum\frac{1}{t_1}\right)^2-2\sum\frac{1}{t_1t_2}\end{matrix}\right]$$

$$=c^2\left[\left(\frac{h}{c}\right)^2+\left(\frac{k}{c}\right)^2\right]=h^2+k^2$$

$$=CP^2$$

Example 8.9.19

The slopes of the sides of triangle ABC inscribed in a rectangular hyperbola $xy=c^2$ are $\tan\alpha, \tan\beta$ and $\tan\gamma$. If the normals at A, B and C are concurrent show that $\cot 2\alpha+\cot 2\beta+\cot 2\gamma=0$.

Solution

Let A, B, C be the points t_1, t_2 and t_3 respectively.

The slope of AB is $-\frac{1}{t_1t_2}$.

$$\therefore \tan\alpha=-\frac{1}{t_1t_2}$$

$$\sum\cot 2\alpha=\sum\frac{1-\tan^2\alpha}{2\tan\alpha}=\sum\frac{1-t_1^2t_2^2}{2t_1t_2}=\frac{1-t_1^2t_2^2}{2t_1t_2}+\frac{1-t_2^2t_3^2}{2t_2t_3}+\frac{1-t_3^2t_1^2}{2t_3t_1}$$

$$=\frac{t_1+t_2+t_3-t_1t_2t_3(t_1t_2+t_2t_3+t_3t_1)}{2t_1t_2t_3}$$

$$=\frac{t_4(t_1+t_2+t_3)-t_1t_2t_3t_4(t_1t_2+t_2t_3+t_3t_1)}{2t_1t_2t_3t_4}$$

$$=\frac{t_4(t_1+t_2+t_3)+(t_1t_2+t_2t_3+t_3t_1)}{-2}$$

$$= \sum \frac{t_1 t_2}{-2} = 0$$

$$\therefore \cot 2\alpha + \cot 2\beta + \cot 2\gamma = 0$$

Example 8.9.20

Show that an infinite number of triangles can be inscribed in a rectangular hyperbola $xy = c^2$ whose sides touch the parabola $y^2 = 4ax$.

Solution

Let ABC be a triangle inscribed in the rectangular hyperbola $xy = c^2$. Let A, B and C be the points $t_1, t_2,$ and t_3, respectively. Suppose the sides AB and AC touch the parabola $y^2 = 4ax$. The equation of the chord AB is $x + yt_1t_2 = c(t_1 + t_2)$.

This touches the parabola $y^2 = 4ax$. (i.e.) $\frac{c(t_1 + t_2)}{t_1 t_2} = -at_1t_2$

(i.e.) $c(t_1 + t_2) + a(t_1t_2)^2 = 0$

(i.e.) $at_1^2t_2^2 + c(t_1 + t_2) = 0$. Since AC also touches the parabola,

$$at_1^2t_3^2 + c(t_1 + t_3) = 0$$

From these equations, we note that t_2, t_3 are the roots of the equation $at_1^2t^2 + ct + ct_1 = 0$.

$$t_2 + t_3 = -\frac{c}{at_1^2}$$

$$t_2t_3 = \frac{c}{at_1}$$

The equation of the chord BC is $x + yt_2t_3 = c(t_2 + t_3)$.

$$\text{(i.e.)} \quad x + \frac{c}{at_1} y = \frac{-c^2}{at_1^2}$$

$$\text{(i.e.)} \quad y = \frac{-at_1}{c} x - \frac{c}{t_1}$$

This equation shows that BC touches the parabola $y^2 = 4ax$. Since ABC is an arbitrary triangle inscribed in the rectangular hyperbola $xy = c^2$ there are infinite number of such triangles touching the parabola $y^2 = 4ax$.

Exercises

1. Prove that the portion of the tangent intercepted between by its asymptotes is bisected at the point of contact and form a triangle of contact area.
2. If the tangent and normal to a rectangular hyperbola make intercepts a_1 and a_2 on one asymptote and b_1 and b_2 on the other then show that $a_1a_2 + b_1b_2 = 0$.
3. P and Q are variable points on the rectangular hyperbola $xy = c^2$ such that the tangent at Q passes through the foot of the ordinate of P. Show that the locus of the intersection of the tangents at P and Q is a hyperbola with the same asymptotes as the given hyperbola.
4. If the lines $x - \alpha = 0$ and $y - \beta = 0$ are conjugate lines with respect to the hyperbola $xy = c^2$ then prove that the point (α, β) is on the hyperbola $xy - 2c^2 = 0$.
5. If the chords of the hyperbola $x^2 - y^2 = a^2$ touch the parabola $y^2 = 4ax$ then prove that the locus of their middle points is the curve $y^2(x-a) = x^3$.
6. If PQ and PR are two perpendicular chords of the rectangular hyperbola $xy = c^2$ then show that QR is parallel to the normal at P.
7. If the polar of a point with respect to the parabola $y^2 = 4ax$ touches the parabola $x^2 = 4by$, show that the point should lie on a rectangular hyperbola.
8. Show that the normal at the rectangular hyperbola $xy = c^2$ at the point $\left(ct, \frac{c}{t}\right)$ meets the curve again at the point $\left(\frac{-c}{t^3}, -ct^3\right)$. Show that PQ varies as CP^2 where C is the centre.
9. If PQ is a chord of the rectangular hyperbola $xy = c^2$ which is the normal at P show that $3CP^2 + CQ^2 = PQ^2$ where C is the centre of the conic.
10. Two rectangular hyperbolas are such that the axes of one are along the asymptotes of the other. Find the distance between the point of contact of a common tangent to them.
11. Prove that any line parallel to either of the asymptotes of a hyperbola should meet it in one point at infinity.
12. The tangent at any point of the hyperbola meets the asymptotes at Q and R. Show that $CQ \cdot CR$ is a constant.
13. Prove that the locus of the centre of the circle circumscribing the triangle formed by the asymptotes of the hyperbola $\frac{x^2}{a^2} - \frac{y^2}{b^2} = 1$ and a variable tangent is $4(a^2x^2 - b^2y^2) = a^2 + b^2$.
14. Show that the coordinates of the point of intersection of two tangents to a rectangular hyperbola are harmonic means between the coordinates of the point of contact.

15. If the normals at A, B, C and D to the rectangular hyperbola $xy = c^2$ meet in $P(h,k)$ then prove that $PA^2 + PB^2 + PC^2 + PD^2 = 3(h^2 + k^2)$.

16. If $(c\tan\phi, c\cot\phi)$ be a point on the rectangular hyperbola $xy = c^2$ then show that the chords through the points ϕ and ϕ' where $\phi + \phi'$ is a constant passes through a fixed point on the conjugate axis of the hyperbola.

17. Prove that the poles with respect to the circle $x^2 + y^2 = a^2$ of any tangent to the rectangular hyperbola $xy = c^2$ lies on rectangular hyperbola $4c^2xy = c^2$.

18. If a normal to a rectangular hyperbola makes an acute angle θ with its transverse axis then prove that the acute angle at which it cuts the curve again is $\cot^{-1}(2\tan^2\theta)$.

19. If a circle cuts the rectangular hyperbola $xy = c^2$ in four points then prove that the product of the abscissae of the points is c^4.

20. Let the rectangular hyperbola $xy = c^2$ is cut by a circle passing through its centre C in four points P, Q, R and S. If p, q be the perpendiculars from c on PQ, RS then show that $pq = c^2$.

21. If a triangle is inscribed in a rectangular hyperbola $xy = c^2$ and two of its sides are parallel to $y = m_1x$ and $y = m_2x$ then prove that the third side touches the hyperbola $4m_1m_2xy = c^2(m_1 + m_2)^2$.

22. If a circle cuts the rectangular hyperbola $xy = c^2$ in P, Q, R and S and the parameters of these four points be t_1, t_2, t_3 and t_4, respectively then prove that the centre of the mean position of these points bisect the distance between the centres of the two curves.

23. If three tangents are drawn to the rectangular hyperbola $xy = c^2$ at the points $(x_i, y_i), i = 1, 2, 3$ and form a triangle whose circumcircle passes through the centre of the hyperbola then show that $\dfrac{\sum x_1}{x_1x_2x_3} + \dfrac{\sum y_1}{y_1y_2y_2} = 0$ and that the centre of the circle lies on the hyperbola.

24. If a circle with fixed centre $(3p, 3q)$ and of variable radius cuts the rectangular hyperbola $x^2 - y^2 = 9c^2$ at the points P, Q, R and S then show that the locus of the centroid of the triangle PQR is given by $(x - 2p)^2 - (y - 2q)^2 = a^2$.

25. Show that the sum of the eccentric angles of the four points of intersection of an ellipse and a rectangular hyperbola whose asymptotes are parallel to the axes of the ellipse is an odd multiple of π.

26. If from any point on the line $lx + my + 1 = 0$ tangents PQ, PR are drawn to the rectangular hyperbola $2xy = c^2$ and the circle PQR cuts the hyperbola again in T and T' then prove that TT' touches the parabola $(l^2 + m^2)(x^2 + y^2) = (lx + my + 1)^2$.

27. If a circle cuts two fixed perpendicular lines so that each intercept is of given length then prove that the locus of the centre of the circle is a rectangular hyperbola.

28. If A and B are points on the opposite branches of a rectangular hyperbola. The circle on AB as diameter cuts the hyperbola again at C and D then prove that CD is a diameter of the hyperbola.

29. If A, B and C are three points on the rectangular hyperbola $xy = d^2$ whose abscissae are a, b and c respectively then prove that the area is $\frac{1}{2}\cdot\frac{d^2(a-b)(b-c)(c-a)}{abc}$ and the area of the triangle enclosed by tangents at these points is $\frac{2d^2(a-b)(b-c)(c-a)}{(a+b)(b+c)(c+a)}$.

30. If four points on a rectangular hyperbola $xy = c^2$ lie on a circle, then prove that the product of their abscissae is c^4.

31. If x_1, x_2, x_3 and x_4 be the abscissae of the angular points and the orthocentre of a triangle inscribed in $xy = c^2$ then prove that $x_1x_2x_3x_4 = -c^4$.

32. Show that the length of the chord of the rectangular hyperbola $xy = c^2$ which is bisected at the point (h,k) is $2\sqrt{\frac{(h^2+k^2)(hk-c^2)}{hk}}$.

33. Prove that the point of intersection of the asymptotes of a rectangular hyperbola with the tangent at any point P and of the axes with the normal at P are equidistant from P.

34. If P is any point on a rectangular hyperbola whose vertices are A and A' then prove that the bisectors of angle APA' are parallel to the asymptotes of the curve.

35. Let QCQ' is a diameter of a rectangular hyperbola and P is any point on the curve. Prove that PQ, PQ' are equally inclined to the asymptotes of the hyperbola.

36. Through the point $P(0,b)$ a line is drawn cutting the same branch of the rectangular hyperbola $xy = c^2$ in Q and R such that $PQ = QR$. Show that its equation is $9c^2y + 2b^2x = 9bc^2$.

37. If a rectangular hyperbola $xy = c^2$ is cut by a circle passing through its centre O in points A, B, C and D whose parameters are t_1, t_2, t_3 and t_4 then show that $(t_1+t_2)(t_3+t_4) + t_1t_2 + t_3t_4 = 0$ and deduce that the product of the perpendicular from O on AB and CD is c^2.

CHAPTER

9 Polar Coordinates

9.1 INTRODUCTION

A coordinate system represents a point in a plane by an ordered pair of numbers called coordinates. Earlier we used Cartesian coordinates which are directed distances from two perpendicular axes. Now we describe another coordinate system introduced by Newton called polar coordinates which is more convenient for some special purposes.

9.2 DEFINITION OF POLAR COORDINATES

We choose a point in the plane and it is called the pole (or origin) and is denoted by O. Then we draw a ray (half linc) starting at O called polar axis. This is usually drawn horizontally to the right and corresponds to positive x-axis in Cartesian coordinates.

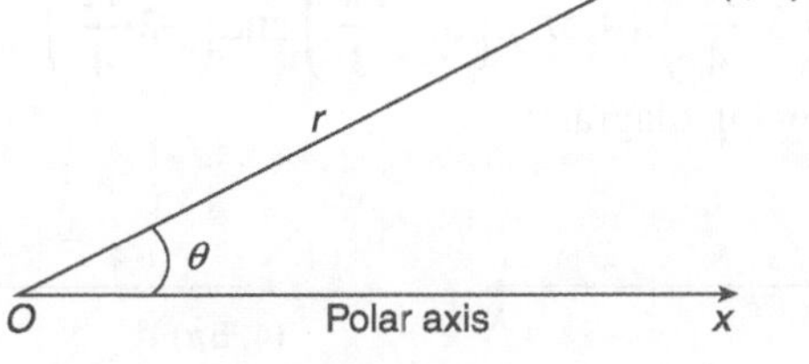

Let P be any point in the plane and r be the distance from O to P. Let θ be the angle (usually measured in radians) between the polar axis and the line OP. Then the point P is represented by the ordered pair (r, θ) and (r, θ) are called the polar coordinates of the point P. We use the convention that an angle is positive if measured in the anti-clockwise direction from the polar axis and negative in the clockwise direction.

If P coincides with O then $r = \theta$. Then (r, θ) represent the coordinates of the pole for any value of θ. Let us now extend the meaning of polar coordinates (r, θ) when r is negative, agreeing that the points $(-r, \theta)$ and (r, θ) lie on the same line through O and at the same distance $|r|$ from O but on opposite sides of O. If $r > 0$, the point (r, θ) lies on the same quadrant as θ. If $r < 0$, then it lies in the quadrant of the opposite side of the pole. We note that the point (r, θ) represents the same point as $(r, \theta + \pi)$

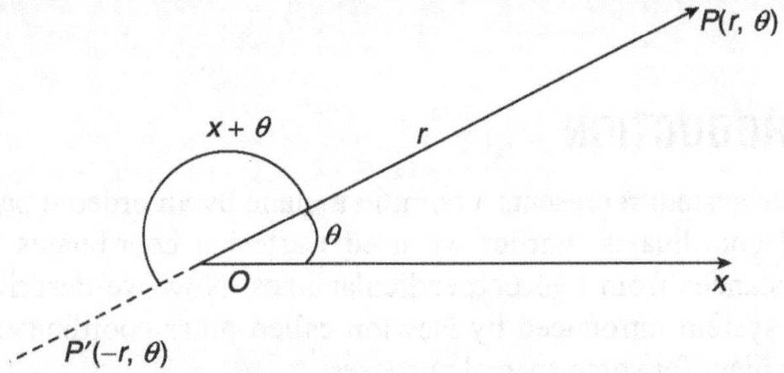

Example 9.2.1

Represent the following polar coordinates in the polar plane:

$$\left(3, \frac{5\pi}{4}\right), (4, 5\pi), \left(3, -\frac{2\pi}{3}\right) \text{ and } \left(-2\frac{3\pi}{4}\right)$$

Solution

The coordinates $\left(3, \frac{5\pi}{4}\right), (4, 5\pi), \left(3, -\frac{2\pi}{3}\right)$ and $\left(-2\frac{3\pi}{4}\right)$ are represented by points in the following diagram:

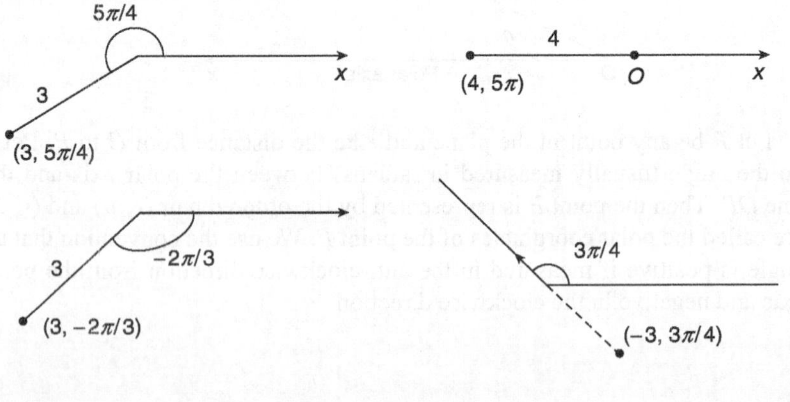

In Cartesian system of coordinates, every point has only one representation. But in polar coordinates system each point has many representations, for example, point $\left(3, \frac{5\pi}{4}\right)$ is also represented by $\left(3, \frac{-3\pi}{4}\right), \left(3, \frac{13\pi}{4}\right), \left(-3, \frac{\pi}{4}\right)$, etc.

In general, the point (r, θ) is also represented by $(r, \theta + 2n\pi)$ or $(-r, \theta + \overline{2n+1}\pi)$ where n is any integer.

9.3 RELATION BETWEEN CARTESIAN COORDINATES AND POLAR COORDINATES

If (x, y) is the Cartesian coordinates and (r, θ) are the polar coordinates of the point P, then $\cos\theta = \frac{x}{r}$ and $\sin\theta = \frac{y}{r}$.

Therefore, the transformations from one system to another are given by $x = r\cos\theta$, $y = r\sin\theta$. To find r from x and y, we use the relation $r^2 = x^2 + y^2$ and θ is given by

$$\cos\theta = \frac{x}{\sqrt{x^2 + y^2}}, \ \sin\theta = \frac{y}{\sqrt{x^2 + y^2}}$$

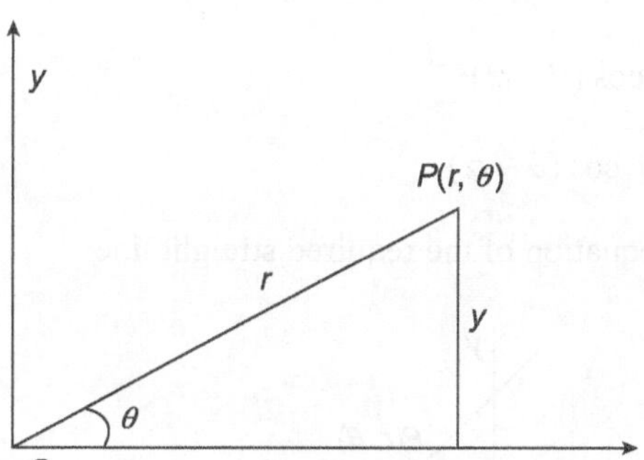

We have already studied the distance between two points, area of a triangle, equations of a straight line, equations to a circle and equation of conics in Cartesian coordinates system. Let us now derive the results in polar coordinate system.

9.4 POLAR EQUATION OF A STRAIGHT LINE

The general equation of a straight line in Cartesian coordinates is $Ax + By + C = 0$, where A, B and C are constants. Let (r, θ) be polar coordinates of a point and the x-axis be the initial line. Then for any point (x, y) on the straight line $x = r\cos\theta, y = r\sin\theta$. Substituting these in the equation of straight line, we get

$$A(r\cos\theta) + B(r\sin\theta) + C = 0$$

$$\text{(i.e.)} \quad A\cos\theta + B\sin\theta + C/r = 0$$

This can be written in the form

$$l/r = A\cos\theta + B\sin\theta \tag{9.1}$$

where A, B and l are constants. Therefore, equation (9.1) is the general equation of a straight line in polar coordinates.

9.5 POLAR EQUATION OF A STRAIGHT LINE IN NORMAL FORM

Let the origin be the pole and the x-axis be the initial line. Draw ON perpendicular to the straight line. Let $ON = p$ and $\angle XON = \alpha$.

In ΔOPN, $\dfrac{ON}{OP} = \cos(\theta - \alpha)$

(i.e.) $\dfrac{p}{r} = \cos(\theta - \alpha)$

(i.e.) $p = r\cos(\theta - \alpha)$

This is the polar equation of the required straight line.

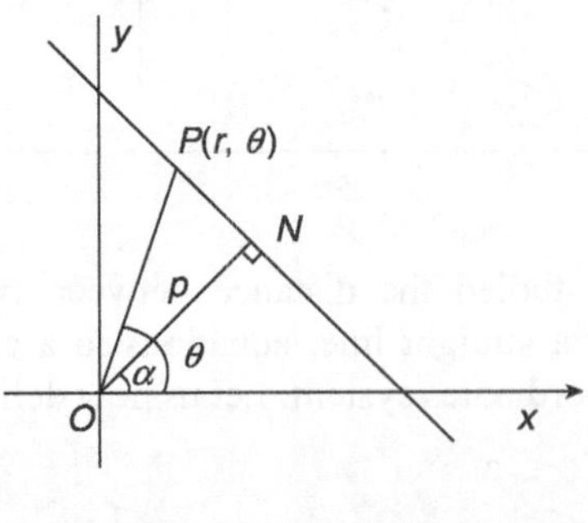

Note 9.5.1: Polar equation of the straight line perpendicular to $A\cos\theta + B\sin\theta = \frac{l}{r}$ is of the form $A\cos\left(\theta+\frac{\pi}{2}\right)+B\sin\left(\theta+\frac{\pi}{2}\right)=k\frac{l}{r}$.

$$\text{(i.e.)}\quad -A\sin\theta + B\cos\theta = \frac{kl}{r}$$

or $B\cos\theta - A\sin\theta = \frac{kl}{r}$, where k is a constant.

Note 9.5.2: The polar equation of the straight line parallel to $A\cos\theta + B\sin\theta = \frac{l}{r}$ is $A\cos\theta + B\sin\theta = \frac{kl}{r}$, where k is a constant.

Note 9.5.3: The condition for the straight lines $A\cos\theta + B\sin\theta = \frac{l}{r}$ and $A_1\cos\theta + B_1\sin\theta = \frac{l}{r}$ to be perpendicular to each other is $AA_1 + BB_1 = 0$. This result can be easily seen from their cartesian equations.

Note 9.5.4: If the line is perpendicular to the initial line then $\alpha = 0$ or π. Therefore, the equation of the straight line is $r\cos\theta = p$ or $r\cos\theta = -p$.

Note 9.5.5: If the line is parallel to the initial line then $\alpha = \frac{\pi}{2}$ or $\frac{3\pi}{2}$. In this case the equation of the line is

$r\cos\left(\theta-\frac{\pi}{2}\right) = p$ or $r\cos\left(\theta-\frac{3\pi}{2}\right) = p$ (i.e.) $r\sin\theta = p$ or $r\sin\theta = -p$.

Example 9.5.6

Find the equation of the straight line joining the two points $P(r_1,\theta_1)$ and $Q(r_2,\theta_2)$.

Solution

Let $R(r,\theta)$ be any point on the line joining the points P and Q.
The area of the triangle formed by the points $P(r_1,\theta_1), Q(r_2,\theta_2)$ and (r_3,θ_3) is

$$\Delta = \frac{1}{2}[r_1r_2\sin(\theta_2-\theta_1) + r_2r_3\sin(\theta_3-\theta_2) + r_3r_1\sin(\theta_1-\theta_3)]$$

Taking $r_3 = r$ and $\theta_3 = \theta$, we get

$$\Delta = \frac{1}{2}[r_1r_2\sin(\theta_2-\theta_1) + r_2r\sin(\theta-\theta_2) + rr_1\sin(\theta_1-\theta)]$$

Since the points P, Q and R are collinear, $\Delta = 0$.

$$r_1 r_2 \sin(\theta_2 - \theta_1) + r r_2 \sin(\theta - \theta_2) + r r_1 \sin(\theta_1 - \theta) = 0$$

Dividing by $r\, r_1\, r_2$, we get

$$\frac{1}{r}\sin(\theta_2 - \theta_1) = \frac{1}{r_1}\sin(\theta_2 - \theta) + \frac{1}{r_2}\sin(\theta - \theta_1)$$

This is the equation of the required straight line.

Example 9.5.7

Find the slope of the straight line $\frac{l}{r} = \cos(\theta - \alpha) + e\cos\theta$.

Solution

The equation of the straight line is $\frac{l}{r} = \cos(\theta - \alpha) + e\cos\theta$.

$$l = r(\cos\theta\cos\alpha + \sin\theta\sin\alpha) + re\cos\alpha$$

$$l = r(e + \cos\alpha)\cos\theta + r\sin\theta\sin\alpha$$

$$\text{(i.e.)}\quad l = (e + \cos\alpha)\,x + y\sin\alpha$$

Therefore, the slope of the straight line is $-\frac{(e + \cos\alpha)}{\sin\alpha}$.

Example 9.5.8

Find the point of intersection of the straight lines $\frac{l}{r} = \cos(\theta - \theta_1) + e\cos\theta$ and $\frac{i}{r} = \cos(\theta - \theta_2) + e\cos\theta$.

Solution

The equations of the straight lines are

$$\frac{l}{r} = \cos(\theta - \theta_1) + e\cos\theta \tag{9.2}$$

$$\frac{i}{r} = \cos(\theta - \theta_2) + e\cos\theta \tag{9.3}$$

Solving equations (9.2) and (9.3), we get

$$\cos(\theta - \theta_1) = \cos(\theta - \theta_2)$$

$$\theta - \theta_1 = \pm(\theta - \theta_2)$$

Therefore, the only possibility is $\theta = \frac{\theta_1 + \theta_2}{2}$.

Then from the equation of the first straight line, we get

$$\frac{l}{r} = \cos\left(\frac{\theta_2 - \theta_1}{2}\right) + e\cos\left(\frac{\theta_1 + \theta_2}{2}\right)$$

$$\therefore r = \frac{l}{\cos\left(\frac{\theta_2 - \theta_1}{2}\right) + e\cos\left(\frac{\theta_1 + \theta_2}{2}\right)}$$

Hence, the point of intersection of the two given lines is

$$\left[\frac{l}{\cos\left(\frac{\theta_2 - \theta_1}{2}\right) + e\cos\left(\frac{\theta_1 + \theta_2}{2}\right)}, \frac{\theta_2 + \theta_1}{2}\right].$$

Example 9.5.9

Find the equation of the line joining the points $\left(2, \frac{\pi}{3}\right)$ and $\left(3, \frac{\pi}{6}\right)$ and deduce that this line also passes through the point $\left(\frac{6}{3\sqrt{3} - 2}, \frac{\pi}{2}\right)$.

Solution

The equation of the line joining the points (r_1, θ_1) and (r_2, θ_2) is

$$\frac{1}{r}\sin(\theta_2 - \theta_1) = \frac{1}{r_1}\sin(\theta_2 - \theta) + \frac{1}{r_2}\sin(\theta - \theta_1).$$

Therefore, the equation of the line joining the points $\left(2, \frac{\pi}{3}\right)$ and $\left(3, \frac{\pi}{6}\right)$ is

$$\frac{1}{r}\sin\left(\frac{\pi}{6} - \frac{\pi}{3}\right) = \frac{1}{2}\sin\left(\frac{\pi}{6} - \theta\right) + \frac{1}{3}\sin\left(\theta - \frac{\pi}{3}\right).$$

$$-\frac{1}{r}\sin\left(\frac{\pi}{6}\right) = \frac{-3\sin\left(\theta - \frac{\pi}{6}\right) + 2\sin\left(\theta - \frac{\pi}{3}\right)}{6}$$

$$\text{when } \theta = \frac{\pi}{2}$$

$$\therefore \frac{1}{r} = \frac{-3\cos\frac{\pi}{6} + 2\cos\frac{\pi}{3}}{-6\sin\frac{\pi}{6}} = \frac{-3\frac{\sqrt{3}}{2} + 2\frac{1}{2}}{-6\times\frac{1}{2}} = \frac{3\sqrt{3}-2}{6}$$

$$r = \frac{6}{3\sqrt{3}-2}$$

Hence, the point $\left(\frac{6}{3\sqrt{3}-2}, \frac{\pi}{2}\right)$ lies on the straight line.

Example 9.5.10

Show that the straight lines $r(\cos\theta + \sin\theta) = \pm 1$ and $r(\cos\theta - \sin\theta) = \pm 1$ enclose a square and calculate the length of the sides of this square.

Solution

Converting into Cartesian form the four lines are

$$x + y = \pm 1$$

$$x - y = \pm 1$$

These four lines form a parallelogram and in $x + y = \pm 1, x - y = \pm 1$ the adjacent lines are perpendicular and hence $ABCD$ is a rectangle.

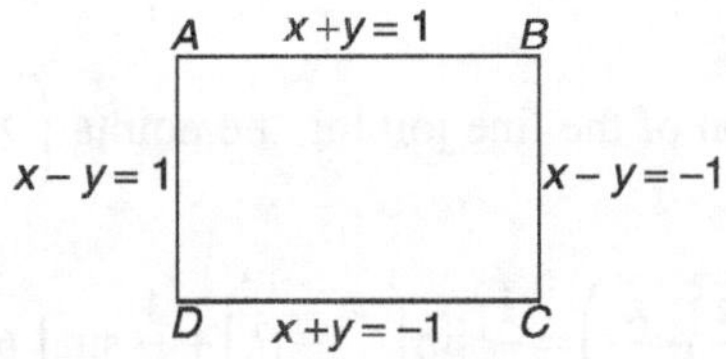

Also the distance between AB and $CD = \sqrt{2}$.

The distance between AD and $BC = \sqrt{2}$.

Therefore, these four lines form a square.

Example 9.5.11

Find the angle between the lines $r\cos\left(\theta-\frac{\pi}{2}\right)=p$ and $r\cos\left(\theta-\frac{\pi}{3}\right)=p'$.

Solution

The angle between the lines $=|\theta_2-\theta_1|$

$$=\left|\frac{\pi}{3}-\frac{\pi}{6}\right|$$

$$=\frac{\pi}{6}$$

Exercises

1. Find the angle between the lines

   (i) $r\cos\theta=p,\ r\sin\theta=p_1$

   (ii) $r\cos\theta=2,\ r\cos\left(\theta-\frac{\pi}{6}\right)=3$

   Ans.: (a) $\frac{\pi}{2}$ (b) $\frac{\pi}{6}$

2. Show that the points $\left(2,\frac{\pi}{3}\right),\left(3,\frac{\pi}{6}\right)$ and $\left(\frac{6\left(3\sqrt{3}+2\right)}{25},\frac{\pi}{2}\right)$ are collinear.

3. Show that the equation of any line parallel to $\frac{l}{r}=\cos(\theta-\alpha)+e\cos\theta$ through the pole is $\theta=\tan^{-1}\left(\frac{\cos\alpha+e}{\sin\alpha}\right)$.

4. Find the equation of the line perpendicular to $\frac{l}{r}=\cos(\theta-\alpha)+e\cos\theta$ and passing through the point (r_1,θ_1).

   Ans.: $\frac{r_1\sin(\theta_1-\alpha)+e\sin\theta_1}{r}=\sin(\theta-\alpha)+e\sin\theta$

9.6 CIRCLE

9.6.1 Polar Equation of a Circle

Let O be the pole and OX be the initial line. Let $C(c,\alpha)$ be the polar coordinates of the centre of the circle. Let $P(r,\theta)$ be any point on the circle. Then $\angle COP=\theta-\alpha$. Let a be the radius of the circle.

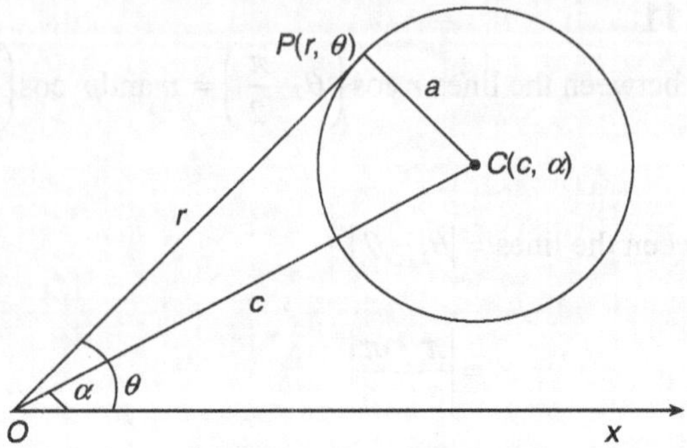

$$\text{In } \Delta\, COP,\ CP^2 = OC^2 + OP^2 - 2OC \cdot OP\cos(\theta - \alpha)$$

$$\text{(i.e.)}\quad a^2 = c^2 + r^2 - 2cr\cos(\theta - \alpha) \tag{9.4}$$

This is the polar equation of the required circle.

Note 9.6.1.1: If the pole lies on the circumference of the circle then $c = a$. Then the equation of the circle becomes,

$$a^2 = a^2 + r^2 - 2ar\cos(\theta - \alpha)$$

$$\text{(i.e.)}\quad r = 2a\cos(\theta - \alpha)$$

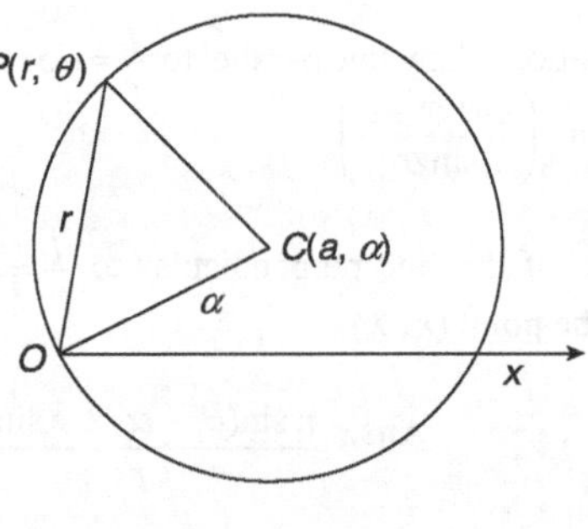

Note 9.6.1.2: The equation of the circle $r = 2a\cos(\theta - \alpha)$ can be written in the form $r = A\cos\theta + B\sin\theta$ where A and B are constants.

Note 9.6.1.3: If the pole lies on the circumference of the circle and the initial line passes through the centre of the circle then the equation of the circle becomes, $r = 2a\cos\theta$ since $\alpha = 0$.

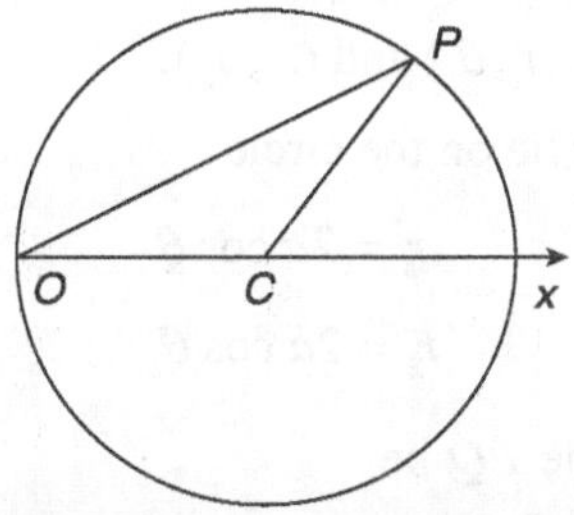

Note 9.6.1.4: Suppose the initial line is a tangent to the circle. Then $c = a \operatorname{cosec} \alpha$. Therefore, from equation (9.4) the equation of the circle becomes, $a^2 = a^2 \operatorname{cosec}^2\alpha + r^2 - 2ar \operatorname{cosec} \alpha \cos(\theta - \alpha)$

$$\text{(i.e.) } r^2 - 2ra \operatorname{cosec} \alpha \cos(\theta - \alpha) + a^2 \cot^2 \alpha = 0$$

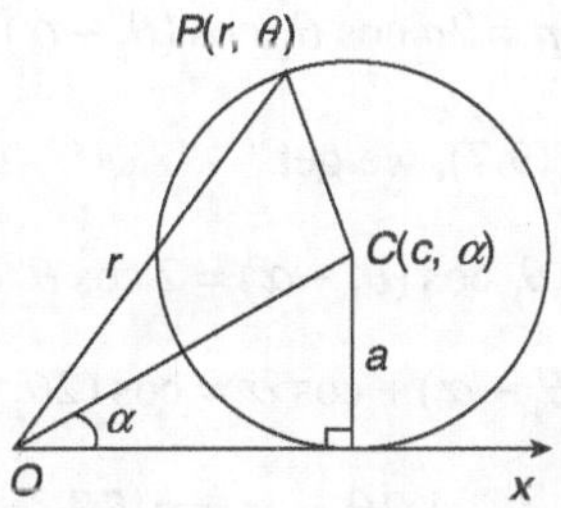

Note 9.6.1.5: Suppose the initial line is a tangent and the pole is at the point of contact. In this case $\alpha = 90°$. The equation of the circle becomes, $r^2 - 2ra \sin\theta = 0$ (*or*) $r = 2a \sin\theta$.

9.6.2 Equation of the Chord of the Circle $r = 2a \cos\theta$ on the Line Joining the Points (r_1, θ_1) and (r_2, θ_2).

Let PQ be the chord of the circle $r = 2a \cos\theta$.

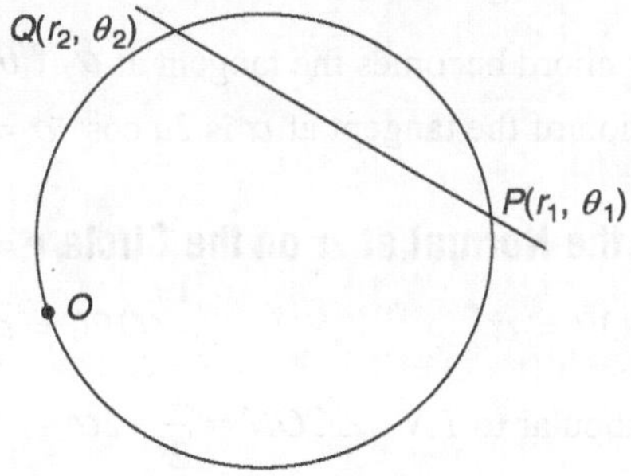

Let P and Q be the points (r_1, θ_1) and (r_2, θ_2).

Since the points P and Q lie on the circle

$$r_1 = 2a \cos \theta_1$$

$$r_2 = 2a \cos \theta_2$$

Let the equation of the line PQ be

$$p = r \cos(\theta - \alpha) \tag{9.5}$$

Since the points P and Q lie on this line

$$p = 2a \cos \theta_1 \cos(\theta_1 - \alpha) \tag{9.6}$$

$$p = 2a \cos \theta_2 \cos(\theta_2 - \alpha) \tag{9.7}$$

From equations (9.6) and (9.7), we get

$$2\cos\theta_1 \cos(\theta_1 - \alpha) = 2\cos\theta_2 \cos(\theta_2 - \alpha)$$

$$\text{(i.e.)} \quad \cos(2\theta_1 - \alpha) + \cos\alpha = \cos(2\theta_2 - \alpha) + \cos\alpha$$

$$\therefore 2\theta_1 - \alpha = \pm(2\theta_2 - \alpha)$$

$$\therefore \alpha = \theta_1 + \theta_2 \text{ or } \theta_1 = \theta_2$$

$$\text{But} \quad \theta_1 \neq \theta_2$$

$$\therefore \alpha = \theta_1 + \theta_2$$

Hence, from equation (9.6), we get $p = 2a \cos\theta_1 \cos\theta_2$.
Hence, from equation (9.5) the equation of the chord is $2a \cos\theta_1 \cos\theta_2 = r\cos(\theta_1 - \overline{\theta_1 + \theta_2})$.

Note 9.6.2.1: This chord becomes the tangent at α if $\theta_1 = \theta_2 = \alpha$.
Therefore, the equation of the tangent at α is $2a\cos^2\alpha = r\cos(\theta - 2\alpha)$.

9.6.3 Equation of the Normal at α on the Circle $r = 2\alpha\cos\theta$

$$\angle XOP = \alpha \qquad \therefore \angle OPC = \alpha$$

Since ON is perpendicular to PN, $\angle CON = \frac{\pi}{2} - 2\alpha$

$$ON = a\cos\left(\frac{\pi}{2} - 2\alpha\right)$$
$$= a\sin 2\alpha$$

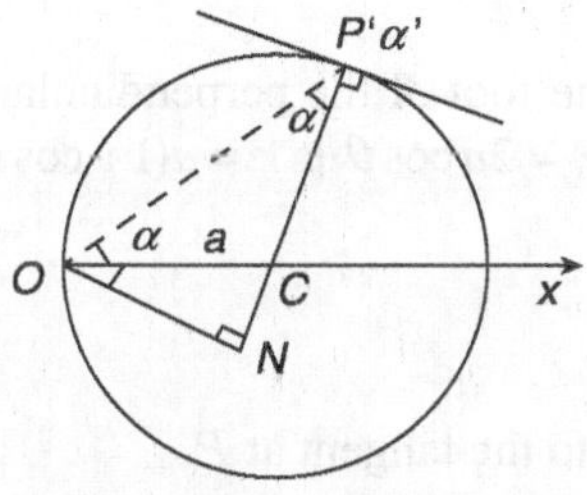

The equation of the normal is $p = r\cos(\theta - \alpha)$.

$$\text{(i.e.)} \quad a\sin 2\alpha = r\cos\left(\theta - \frac{\pi}{2} - 2\alpha\right)$$
$$a\sin 2\alpha = r\sin(\theta - 2\alpha)$$

9.6.4 Equation of the Circle on the Line Joining the Points (a, α) and (b, β) as the ends of a Diameter

$$AP^2 = a^2 + r^2 - 2ar\cos(\theta - \alpha)$$
$$BP^2 = r^2 + b^2 - 2br\cos(\theta - \beta)$$
$$AB^2 = a^2 + b^2 - 2ab\cos(\alpha - \beta)$$

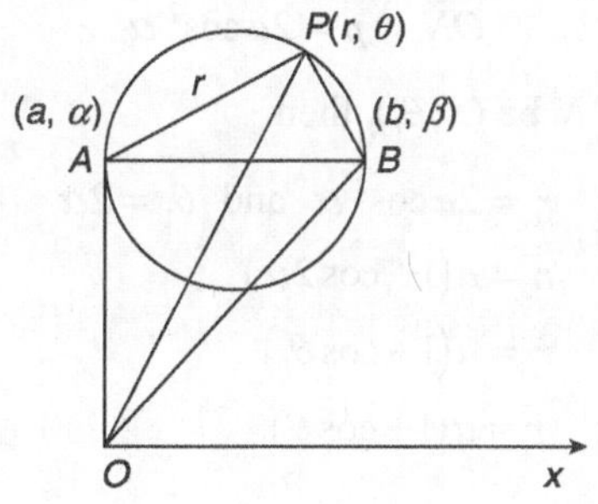

Since $\angle APB = 90°$

$$AB^2 = AP^2 + BP^2$$

$$a^2+b^2-2ab\cos(\theta-\beta)=a^2+r^2-2ar\cos(\theta-\alpha)+b^2+r^2-2br\cos(\theta-\beta)$$

$$\text{(i.e.)}\quad r^2-r\left[a\cos(\theta-\alpha)+b\cos(\theta-\beta)\right]+ab\cos(\theta-\beta)=0$$

Example 9.6.1

Show that the locus of the foot of the perpendicular drawn from the pole to the tangent to the circle $r=2a\cos\theta$ is $r=a(1+\cos\theta)$.

Solution

Let P be the point (r,α).
Draw ON perpendicular to the tangent at P.

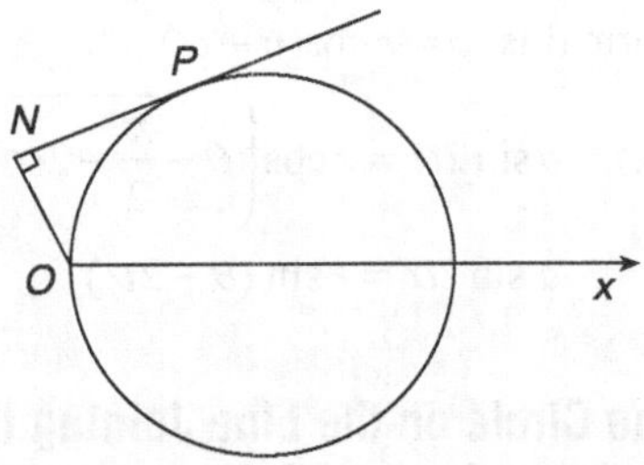

The equation of the tangent at P is

$$r\cos(\theta-2\alpha)=2a\cos^2\alpha$$

Since ON is the perpendicular distance from O on the line PN, from the normal form of the straight line, we get

$$ON=p=2a\cos^2\alpha$$

Let the coordinates of N be (r_1,θ_1), then

$$r_1=2a\cos^2\alpha \text{ and } \theta_1=2\alpha$$
$$r_1=a(1+\cos 2\alpha)$$
$$r_1=a(1+\cos\theta_1)$$

Locus of (r_1,θ_1) is $\quad r=a(1+\cos\theta)$.

Example 9.6.2

Show that the feet of the perpendiculars from the origin on the sides of the triangle formed by the points with vectorial angles α,β,γ

and which lie on the circle $r = 2a\cos\theta$ lie on the straight line $2a\cos\alpha\cos\beta\cos\gamma = r\cos(\pi-\alpha-\beta-\gamma)$.

Solution

The equation of the circle is $r = 2a\cos\theta$.
Let the vectorial angles of P, Q, R be α, β, γ respectively.
The equations of the chord PQ, QR and RP are

$$2a\cos\alpha\cos\beta = r\cos(\theta - \overline{\alpha+\beta})$$

$$2a\cos\beta\cos\gamma = r\cos(\theta - \overline{\beta+\gamma})$$

$$2a\cos\gamma\cos\alpha = r\cos(\theta - \overline{\gamma+\alpha})$$

Let L, M and N be the feet of the perpendiculars from O on the lines PQ, QR and RP

Then from the above equations, we infer that the coordinates of L, M and N are

$$(2a\cos\alpha\cos\beta, \alpha+\beta)$$
$$(2a\cos\beta\cos\gamma, \beta+\gamma)$$
$$(2a\cos\gamma\cos\alpha, \gamma+\alpha)$$

These three points satisfy the equation

$$2a\cos\alpha\cos\beta\cos\gamma = r\cos(\theta-\alpha-\beta-\gamma)$$

Hence L, M and N lies on the above line.

Example 9.6.3

Show that the straight line $\frac{l}{r} = A\cos\theta + B\sin\theta$ touches the circle $r = 2a\cos\theta$ if $a^2B^2 + 2alA = l^2$.

Solution

The equation of the circle is

$$r = 2a\cos\theta \tag{9.8}$$

The equation of the straight line is

$$\frac{l}{r} = A\cos\theta + B\sin\theta \tag{9.9}$$

Solving these two equations we get their point of intersection.

$$\frac{l}{2a\cos\theta} = A\cos\theta + B\sin\theta$$

Dividing by $\cos\theta$, we get

$$\frac{l}{2a\cos^2\theta} = A + B\tan\theta$$

$$\frac{l}{2a}\left(1+\tan^2\theta\right) = A + B\tan\theta$$

$$l\tan^2\theta + l = 2aA + 2aB\tan\theta$$

$$l\tan^2\theta - 2aB\tan\theta + l - 2aA = 0 \qquad (9.10)$$

If the line (9.9) is a tangent to (9.8) then the two values of $\tan\theta$ of the equation (9.10) are equal. The condition for that is the discriminant is equal to zero.

$$4a^2B^2 - 4l\left(l - 2aA\right) = 0$$

$$a^2B^2 - l^2 + 2aAl = 0$$

Exercises

1. Show that $r = A\cos\theta + B\sin\theta$ represents a circle and find the polar coordinates of the centre.

2. Show that the equation of the circle of radius a which touches the lines $\theta = 0, \theta = \frac{\pi}{2}$ is $r^2 - 2ar(\cos\theta + \sin\theta) + a^2 = 0$. Show that locus of the equation $r^2 - 2ra\cos 2\theta\sec\theta - 2a^2 = 0$ consists of a straight line and a circle.

3. Find the polar equations of circles passing through the points whose polar coordinates are $\left(a, \frac{\pi}{2}\right), \left(b, \frac{\pi}{2}\right)$ and touching the straight line $\theta = 0$.

   Ans.: $r^2 - r[(a+b)\sin\theta \pm 2b\cos\theta] + c^2 = 0$

4. A circle passes through the point (r, θ) and touches the initial line at a distance c from the pole. Show that its polar equation is $r_1\sin\theta_1(r^2 - 2ar\cos\theta + c^2) = r\sin\theta(r_1^2 - 2cr_1\cos\theta 1_1 + c)$.

5. Show that $r^2 - kr\cos(\theta - \alpha) + kd = 0$ represents a system of general circles for different values of k. Find the coordinates of the limiting points and the equation of the common radical axis.

6. Find the equation of the circle whole centre is $\left(4, \frac{\pi}{4}\right)$ and radius is 2.

Ans.: $r^2 - 8\cos\left(q - \frac{p}{4}\right) + 12 = 0$

7. Find the centre and radius of the circle $r^2 - 10r\cos\theta + 9 = 0$.

Ans.: $(5,0); 4$

8. Prove that the equation to the circle described on the line joining the points $\left(1, \frac{\pi}{3}\right)$ and $\left(2, \frac{\pi}{6}\right)$ as diameter is $r^2 - r\left[\cos\left(\theta - \frac{\pi}{3}\right)\right] + 2\cos\left(\cos - \frac{\pi}{6}\right) + \sqrt{3} = 0$.

9. Find the condition that the line $\frac{l}{r} = a\cos\theta + b\sin\theta$ may be a

   (i) tangent

   (ii) a normal to the circle $r = 2\cos\theta$.

10. Find the equation of the circle which touches the initial line, the vectorial angle of the centre being α and the radius of the circle a.

11. A circle passes through the point (r_1, θ_1) and touches the initial line at a distance c from the pole. Show that its polar equation is

$$\frac{r^2 - 2cr\cos\theta + c^2}{r\sin\theta} = \frac{r_1^2 - 2cr_1\cos\theta + c^2}{r_1\sin\theta_1}.$$

9.7 POLAR EQUATION OF A CONIC

Earlier we defined parabola, ellipse and hyperbola in terms of focus directrix. Now let us show that it is possible to give a more unified treatment of all these three types of conic using polar coordinates. Furthermore, if we place the focus at the origin then a conic section has simple polar equation.

Let S be a fixed point (called the focus) and XM, a fixed straight line (called the directrix) in a plane. Let e be a fixed positive number (called the eccentricity). Then the set of all points P in the plane such that $\frac{SP}{PM} = e$ is called a conic section. The conic is

(i) an ellipse if $e < 1$.

(ii) a parabola if $e = 1$.

(iii) a hyperbola if $e > 1$.

9.7.1 Polar Equation of a Conic

Let S be focus and XM be the directrix. Draw SX perpendicular to the directrix. Let S be the pole and SX be the initial line. Let $P(r, \theta)$ be any point on the conic; then $SP = r, \angle XSP = \theta$. Draw PM perpendicular to the directrix and PN perpendicular to the initial line.

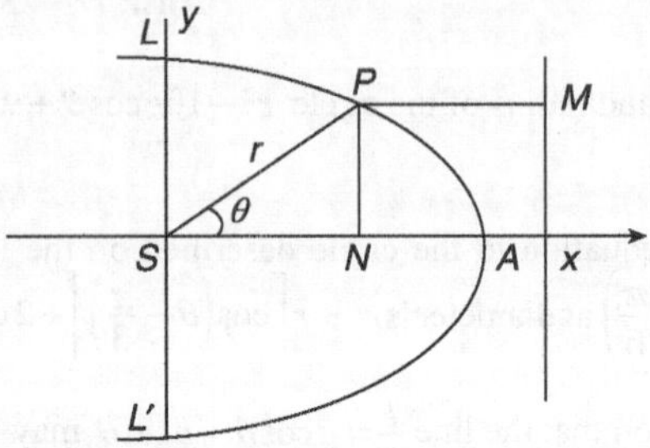

Let LSL' be the double ordinate through the focus (latus rectum).

The focus directrix property is

$$\frac{SP}{PM} = e$$

$$\begin{aligned}
\text{(i.e.)}\quad SP &= ePM \\
&= eNX \\
&= e(SX - SN)
\end{aligned}$$

$$\text{(i.e.)}\quad r = e\left(\frac{l}{e} - r\cos\theta\right)$$

$$r = l - er\cos\theta$$

$$r(1 + e\cos\theta) = l$$

$$\text{or}\quad \frac{l}{r} = 1 + e\cos\theta$$

This is the required polar equation of the conic.

Note 9.7.1.1: If the axis of the conic is inclined at an angle α to the initial line then the polar equation of conic is $\frac{l}{r} = 1 + e\cos(\theta - \alpha)$.

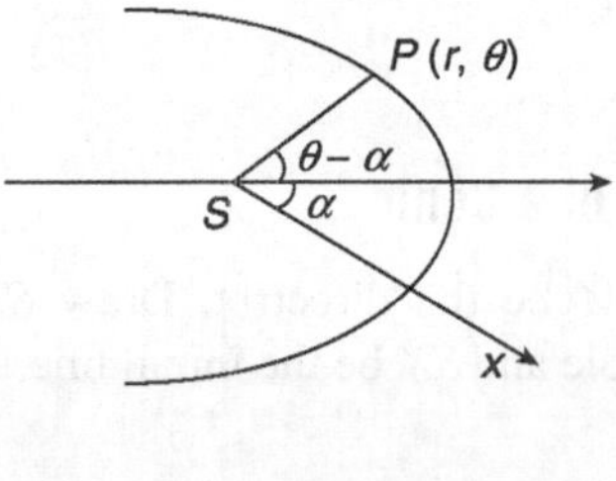

To trace the conic, $\frac{l}{r} = 1 + e\cos\theta$.

$\cos\theta$ is a periodic function of period 2π.

Therefore, to trace the conic it is enough if we consider the variation of θ from $-\pi$ to π. Since $\cos(-\theta) = \cos\theta$ the curve is symmetrical about the initial line. Hence it is enough if we study the variation of θ from 0 to π. Let us discuss the various cases for different values of θ.

Case 1: Let $e = 0$. In this case, the conic becomes $r = l$ which is a circle of radius l with its centre at the pole.

Case 2: Let $e = 1$. In this case, the equation of the conic becomes, $\frac{l}{r} = 1 + \cos\theta$ or $r = \frac{l}{1+\cos\theta}$. When θ varies from 0 to π,

$1 + \cos\theta$ varies from 2 to 0.

and $\frac{1}{1+\cos\theta}$ varies from $\frac{1}{2}$ to ∞.

The conic in this case is a parabola and is shown below.

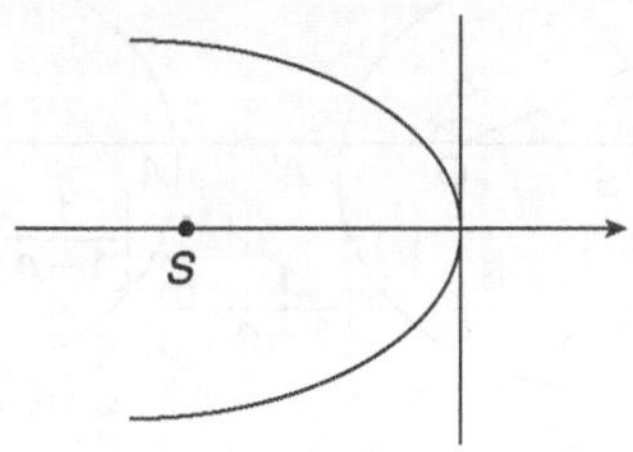

Case 3: Let $e < 1$. $r = \frac{l}{1+e\cos\theta}$

As θ varies from 0 to π, $1 + e\cos\theta$ decreases from $1+e$ to $1-e$. r increases from $\frac{l}{1+e}$ to $\frac{l}{1-e}$

The curve is clearly closed and is symmetrical about the initial line. The conic is an ellipse.

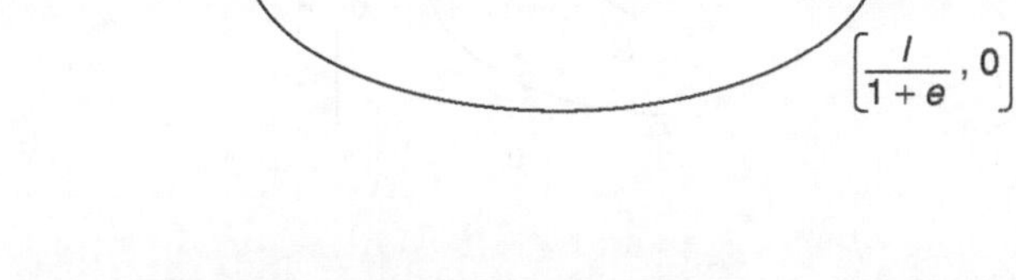

Case 4: Let $e>1$. $r=\dfrac{l}{1+e\cos\theta}$.

As θ varies from 0 to $\dfrac{\pi}{2}$, $1+e\cos\theta$ decreases from $(1+e)$ to 1 and hence r increases from $\dfrac{l}{1+e}$ to l.

As θ varies from $\dfrac{\pi}{2}$ to π, $1+e\cos\theta$ decreases from 1 to $(1-e)$.

Therefore, there exists an angle α such that $\dfrac{\pi}{2}<\alpha<\pi$ at which $1+e\cos\theta>0$. (i.e.) $\cos\alpha>-\dfrac{1}{e}$.

Hence, as θ varies from $\dfrac{\pi}{2}$ to α, r increases from 1 to ∞. As θ varies from α to π, $1+e\cos\theta$ remains negative and varies from 0 to $(1-e)$.

r varies from $-\infty$ to $\dfrac{1}{1-e}$.

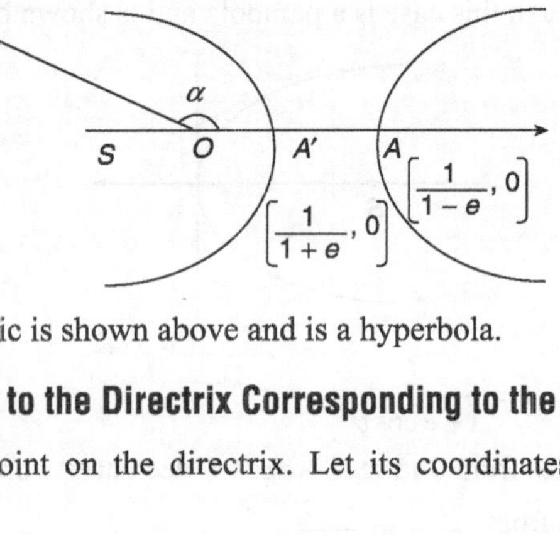

The conic is shown above and is a hyperbola.

9.7.2 Equation to the Directrix Corresponding to the Pole

Let Q be any point on the directrix. Let its coordinates be (r,θ). Then

$SX = r\cos\theta$

$\dfrac{l}{e}=r\cos\theta$ or $\dfrac{l}{r}=e\cos\theta$. Since this is true for all points (r,θ) on the directrix, the polar equation of the directrix is $\dfrac{l}{r}=e\cos\theta$.

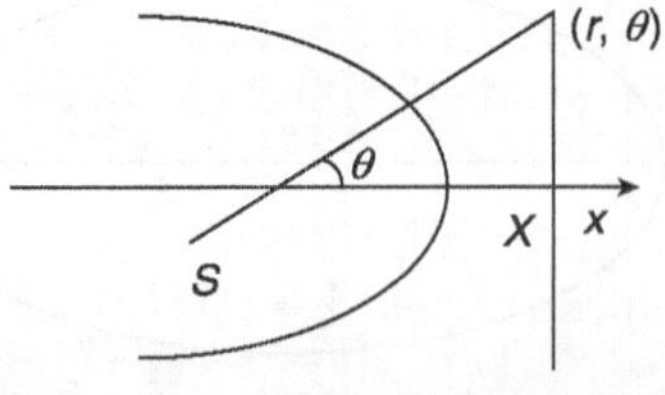

Note 9.7.2.1: The equation of the directrix of the conic $\frac{l}{r} = 1 + e\cos(\theta - \alpha)$ is $\frac{l}{r} = e\cos(\theta - \alpha)$.

The polar equation of the conic for different form of directrix is given below.

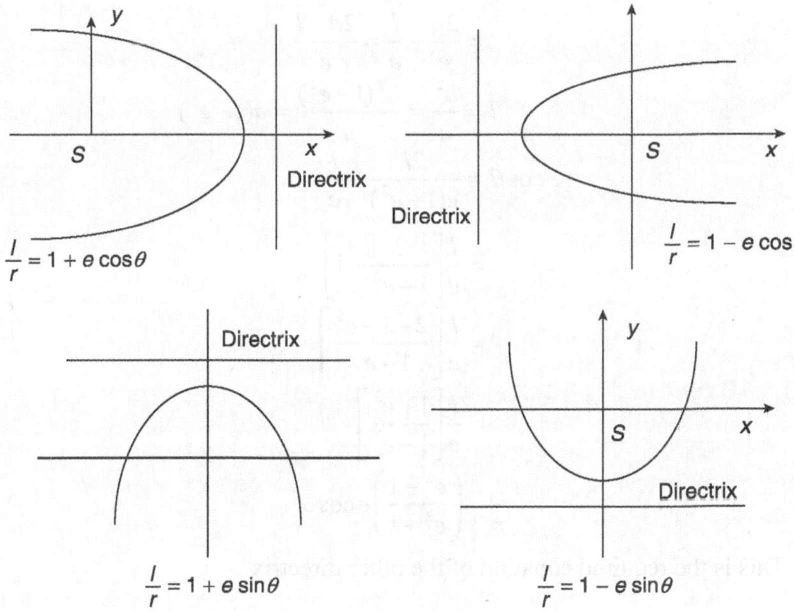

Note 9.7.2.2: The above conic is an ellipse if $e < 1$, parabola if $e = 1$ and hyperbola if $e > 1$.

9.7.3 Equation to the Directrix Corresponding to Focus Other than the Pole

Let (r, θ) be the coordinates of a point on the directrix $X'M'$.

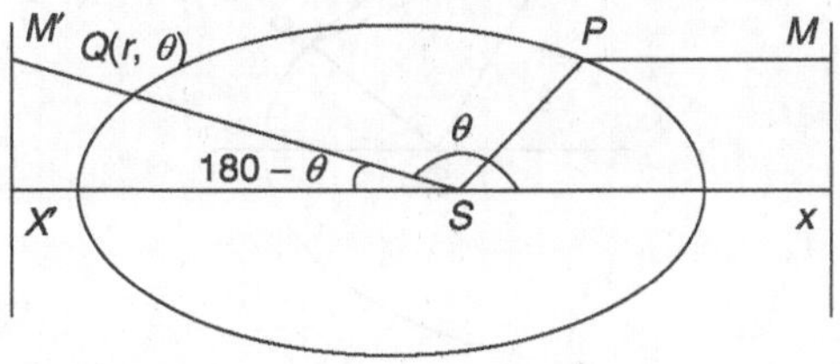

Then

$$\begin{aligned} SX' &= SQ\cos(180^\circ - \theta) \\ &= -SQ\cos\theta \\ &= -r\cos\theta \end{aligned}$$

But

$$\begin{aligned} SX' &= XX' - SX \\ &= \frac{2a}{e} - \frac{l}{e} = \frac{2a-l}{e} \end{aligned}$$

$$l = \frac{b^2}{a} = \frac{a^2(1-e^2)}{a} = a(1-e^2)$$

$$\begin{aligned} -r\cos\theta &= \frac{2l}{e(1-e^2)} - \frac{l}{e} \\ &= \frac{l}{e}\left[\frac{2}{1-e^2} - 1\right] \\ &= \frac{l}{e}\left[\frac{2-1+e^2}{1-e^2}\right] \\ &= \frac{l}{e}\left[\frac{1+e^2}{1-e^2}\right] \end{aligned}$$

$$\frac{l}{r} = \left(\frac{e^2-1}{e^2+1}\right)e\cos\theta$$

This is the required equation of the other directrix.

9.7.4 Equation of Chord Joining the Points whose Vectorial Angles are $\alpha - \beta$ and $\alpha + \beta$ on the Conic

Let the equation of the conic be $\frac{l}{r} = 1 + e\cos\theta$.

Let the equation of the chord PQ be $\frac{l}{r} = A\cos\theta + B\cos(\theta - \alpha)$.

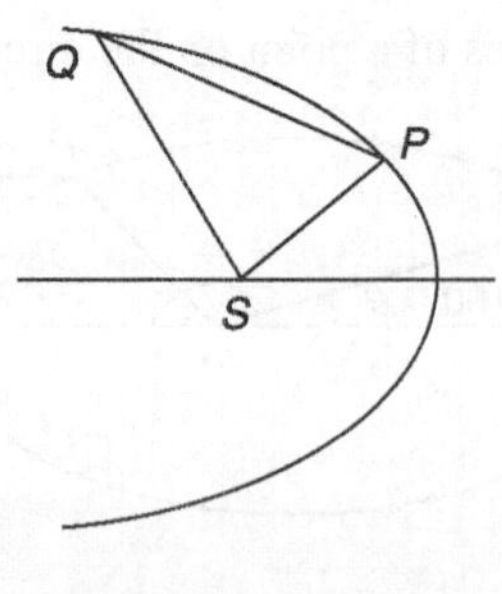

This chord passes through the point $(SP, \alpha-\beta)$ and $(SQ, \alpha+\beta)$.

$$\frac{l}{SP} = A\cos(\alpha-\beta) + B\cos\beta \tag{9.11}$$

$$\frac{l}{SQ} = A\cos(\alpha+\beta) + B\cos\beta \tag{9.12}$$

Also these two points lie on the conic $\frac{l}{r} = 1 + e\cos\theta$,

$$\frac{l}{SP} = 1 + e\cos(\alpha-\beta) \tag{9.13}$$

$$\frac{l}{SQ} = 1 + e\cos(\alpha+\beta) \tag{9.14}$$

From equations (9.11) and (9.13), we get

$$A\cos(\alpha-\beta) + B\cos\beta = 1 + e\cos(\alpha-\beta) \tag{9.15}$$

From equation (9.12) and (9.14), we get

$$A\cos(\alpha+\beta) + B\cos\beta = 1 + e\cos(\alpha+\beta) \tag{9.16}$$

Subtracting, we get

$$A[\cos(\alpha-\beta) - \cos(\alpha+\beta)] = e[\cos(\alpha-\beta) - \cos(\alpha+\beta)]$$

$$A = e$$

From equation (9.15), we get

$$e\cos(\alpha-\beta) + B\cos\beta = 1 + e\cos(\alpha-\beta)$$

$$B\cos\beta = 1 \quad \text{(i.e.)} \quad B = \sec\beta$$

The equation of the chord PQ is $\frac{l}{r} = e\cos\theta + \sec\beta\cos(\theta-\alpha)$.

9.7.5 Tangent at the Point whose Vectorial Angle is α on the Conic $\frac{l}{r} = 1 + e\cos\theta$

The equation of the chord joining the points with vectorial angles $\alpha-\beta$ and $\alpha+\beta$ is $\frac{l}{r} = e\cos\theta + \sec\beta\cos(\theta-\alpha)$.

This chord becomes the tangent at α if $\beta = 0$.

The equation of tangent at α is $\frac{l}{r} = 1 + e\cos\theta + \cos(\theta-\alpha)$.

9.7.6 Equation of Normal at the Point whose Vectorial Angle is α on the Conic

The equation of the conic is $\frac{l}{r} = 1 + e\cos\theta$

The equation of tangent at α on the conic $\frac{l}{r} = 1 + e\cos\theta$ is $\frac{l}{r} = e\cos\theta + \cos(\theta - \alpha)$.

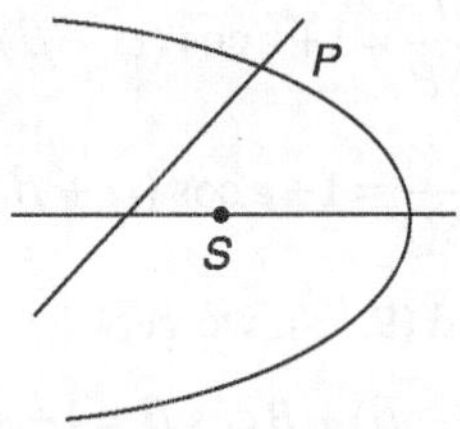

The equation of the line perpendicular to this tangent is $\frac{k}{r} = e\cos\left(\theta + \frac{\pi}{2}\right) + \cos\left(\theta + \frac{\pi}{2} - \alpha\right)$.

$$\text{(i.e.)} \quad \frac{k}{r} = -e\sin\theta - \sin(\theta - \alpha)$$

If this perpendicular line is normal at P, then it passes through the point (SP, α).

$$\frac{k}{SP} = -e\sin\alpha \text{ or } k = -SP \cdot e\sin\alpha \tag{9.17}$$

Since the point (SP, α) also lies on the conic $\frac{l}{r} = 1 + e\cos\theta$, we have

$$\frac{l}{SP} = 1 + e\cos\alpha$$

$$SP = \frac{l}{1 + e\cos\alpha}$$

From equation (9.17), we get $k = \frac{le\sin\alpha}{1 + e\cos\alpha}$.

Hence, the equation of the normal at α is

$$-\frac{1}{r}\frac{le\sin\alpha}{l+e\cos\alpha}=-e\sin\theta-\sin(\theta-\alpha)$$

$$\text{(i.e.)}\quad \frac{1}{r}\frac{le\sin\alpha}{l+e\cos\alpha}=e\sin\theta+\sin(\theta-\alpha)$$

9.7.7 Asymptotes of the Conic is $\frac{l}{r}=1+e\cos\theta\ (e>1)$

The equation of the conic is

$$\frac{l}{r}=1+e\cos\theta \tag{9.18}$$

The equation of the tangent at α is

$$\frac{l}{r}=e\cos\theta+\cos(\theta-\alpha) \tag{9.19}$$

This tangent becomes an asymptote if the point of contact is at infinity, that is, the polar coordinates of the point of contact are (∞,α). Since this point has to satisfy the equation of the conic (9.18) we have from equation (9.18),

$$0=1+e\cos\alpha \tag{9.20}$$

$$\text{or}\quad \cos\alpha=-\frac{1}{e}$$

The equation (9.19) can be written as $\frac{l}{r}=(e+\cos\alpha)\cos\theta+\sin\alpha\sin\theta$.

Substituting $\cos\alpha=-\frac{1}{e}$ and $\sin\alpha=\pm\sqrt{1-\frac{1}{e^2}}$, we get the equation of the asymptotes as $\frac{l}{r}=\left(e-\frac{1}{e}\right)\cos\theta\pm\sqrt{1-\frac{1}{e^2}}\sin\theta$.

$$\text{(i.e.)}\quad \frac{l}{r}=\frac{e^2-1}{e}\cos\theta\pm\frac{\sqrt{e^2-1}}{e}\sin\theta$$

$$\therefore\quad \frac{l}{r}=\frac{e^2-1}{e}\left[\cos\theta\pm\frac{1}{\sqrt{e^2-1}}\sin\theta\right]$$

9.7.8 Equation of Chord of Contact of Tangents from (r_1, θ_1) to the Conic

$$\frac{l}{r} = 1 + e\cos\theta$$

Let QR be the chord of contact of tangents from $P(r_1, \theta_1)$. Let vectorial angles of Q and R be $\alpha - \beta$ and $\alpha + \beta$. The equation of the chord QR is

$$\frac{l}{r} = e\cos\theta + \sec\alpha\cos(\theta - \beta) \tag{9.21}$$

The equations of tangents at Q and R are

$$\frac{l}{r} = e\cos\theta + \cos(\theta - \overline{\alpha - \beta}) \tag{9.22}$$

$$\frac{l}{r} = e\cos\theta + \cos(\theta - \overline{\alpha + \beta}) \tag{9.23}$$

These two tangents intersect at (r_1, θ_1).

$$\frac{l}{r_1} = e\cos\theta_1 + \cos(\theta_1 - \overline{\alpha - \beta}) \tag{9.24}$$

$$\frac{l}{r_1} = e\cos\theta_1 + \cos(\theta_1 - \overline{\alpha - \beta}) \tag{9.25}$$

From the above two equations, we get

$$\cos(\theta_1 - (\overline{\alpha - \beta}) = \cos(\theta_1 - \overline{\alpha + \beta})$$

$$\theta_1 - \overline{\alpha - \beta} = \pm(\theta_1 - \overline{\alpha + \beta})$$

$$\theta_1 - \overline{\alpha - \beta} = \theta_1 - \overline{\alpha + \beta} \Rightarrow \beta > 0 \text{ which is inadmissible}$$

$$\theta_1 - \overline{\alpha - \beta} = -\theta_1 + \alpha + \beta \Rightarrow \theta_1 = \alpha$$

Substituting this in equation (9.24), we get

$$\frac{l}{r_1} = e\cos\theta_1 + \cos\beta \tag{9.26}$$

Substituting equation (9.26) in (9.21), we get

$$\frac{l}{r} - e\cos\theta = \frac{1}{\left(\dfrac{l}{r_1} - e\cos\theta_1\right)} \cdot \cos(\theta - \theta_1)$$

$$\text{(i.e.)} \left(\frac{l}{r} - e\cos\theta\right)\left(\frac{l}{r_1} - e\cos\theta_1\right) = \cos(\theta - \theta_1)$$

This is the equation of the chord of contact.

9.7.9 Equation of the Polar of any Point (r_1, θ_1) with Respect to the Conic $\frac{l}{r} = 1 + e\cos\theta$

The polar of a point with respect to a conic is defined as the locus of the point of intersection of tangents at the extremities of a variable chord passing through the point $P(r_1, \theta_1)$.

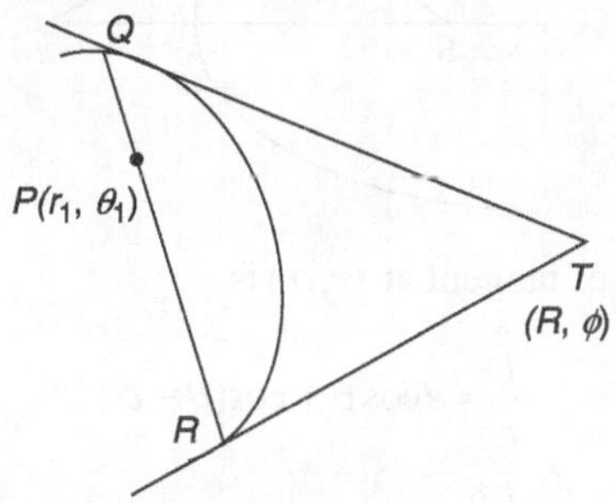

Let the tangents at Q and R intersect T. Since QR is the chord of contact of tangents from $T(R, \phi)$, its equation is

$$\left(\frac{l}{r} - e\cos\theta\right)\left(\frac{l}{R} - e\cos\phi\right) = \cos(\theta - \phi) \tag{9.27}$$

Since this passes through the point $P(r_1, \theta_1)$ we have

$$\left(\frac{l}{r_1} - e\cos\theta_1\right)\left(\frac{l}{R} - e\cos\phi\right) = \cos(\theta_1 - \phi) \tag{9.28}$$

Now the locus of the point $T(R, \phi)$ is polar of the (r_1, θ_1).

The polar of (r_1, θ_1) from equation (9.28) is

$$\left(\frac{l}{r} - e\cos\theta\right)\left(\frac{l}{r_1} - e\cos\theta_1\right) = \cos(\theta - \theta_1).$$

Example 9.7.1

Find the condition that the straight line $\frac{l}{r} = A\cos\theta + B\sin\theta$ may be a tangent to the conic $\frac{l}{r} = 1 + e\cos\theta$.

Solution

Let the line $\frac{l}{r} = A\cos\theta + B\sin\theta$ touches the conic at the point (r, α).

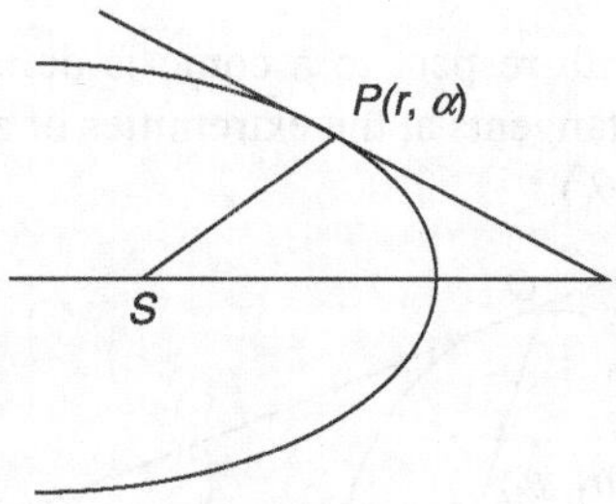

Then the equation of tangent at (r, α) is

$$\frac{l}{r} = e\cos\theta + \cos(\theta - \alpha) \tag{9.29}$$

$$\text{(i.e.)} \quad \frac{l}{r} = (e + \cos\alpha)\cos\theta + \sin\alpha\sin\theta \tag{9.30}$$

However, the equation of tangent is given as

$$\frac{l}{r} = A\cos\theta + B\sin\theta \tag{9.31}$$

Equations (9.30) and (9.31) represent the same line.
Identifying equations (9.30) and (9.31), we get

$$A = e + \cos\alpha \text{ (or) } A - e = \cos\alpha$$

$$B = \sin\alpha$$

Squaring and adding, we get $(A - e)^2 + B^2 = 1$

This is the required condition.

Example 9.7.2

Show that in a conic, semi latus rectum is the harmonic mean between the segments of a focal chord.

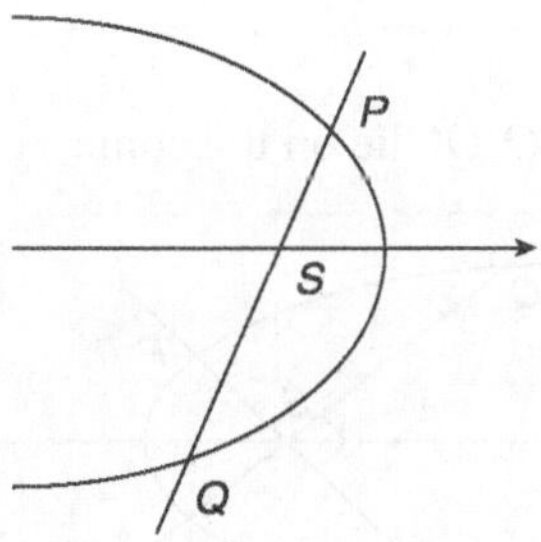

Solution

Let PQ be a focal chord of the conic $\frac{l}{r}=1+e\cos\theta$. Let P and Q have the polar coordinates (SP,α) and $(SQ,\alpha+\pi)$.

Since P and Q lie on the conic $\frac{l}{r}=1+e\cos\theta$.

We have
$$\frac{l}{SP}=1+e\cos\alpha \tag{9.32}$$
$$\frac{l}{SQ}=1-e\cos\alpha \tag{9.33}$$

Adding equations (9.32) and (9.33)
$$\frac{l}{SP}+\frac{l}{SQ}=2$$
$$\text{or}\quad \frac{1}{SP}+\frac{1}{SQ}=\frac{2}{l}$$

SP, l, SQ are in HP

(i.e.) l is the HM between SP and SQ.

Example 9.7.3

Show that in any conic the sum of the reciprocals of two perpendicular focal chords is a constant.

Solution

Let PP' and QQ' be perpendicular focal chords of the conic

$$\frac{l}{r} = 1 + e\cos\theta \tag{9.34}$$

Let P be the point (SP, α). The vectorial angles of Q, P', Q' are $\alpha + \frac{\pi}{2}, \alpha + \pi$ and $\alpha + \frac{3\pi}{2}$.

Since the points P, P', Q, Q' lie on the conic,

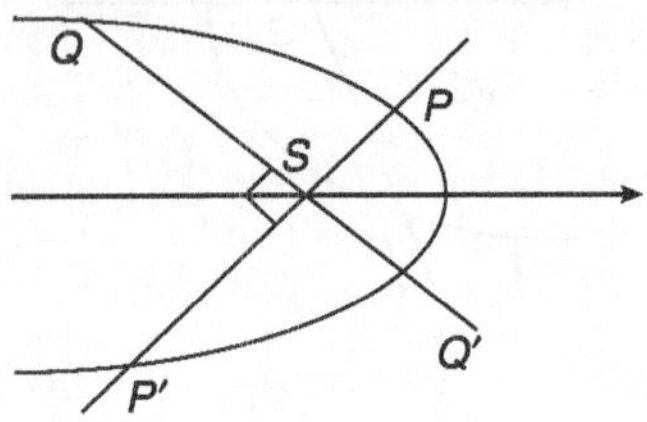

we have

$$\frac{l}{SP} = 1 + e\cos\alpha \qquad \therefore SP = \frac{1}{1 + e\cos\alpha}$$

$$\frac{l}{SP'} = 1 - e\cos\alpha \qquad \therefore SP' = \frac{l}{1 - e\cos\alpha}$$

$$\frac{l}{SQ} = 1 - e\sin\alpha \qquad \therefore SQ = \frac{l}{1 - e\sin\alpha}$$

$$\frac{l}{SQ'} = 1 + e\sin\alpha \qquad \therefore SQ' = \frac{l}{1 + e\sin\alpha}$$

$$PP' = SP + SP' = \frac{l}{1 + e\cos\alpha} + \frac{l}{1 - e\cos\alpha} = \frac{2l}{1 - e^2\cos^2\alpha}$$

$$QQ' = SQ' + SQ' = \frac{l}{1 - e\sin\alpha} + \frac{l}{1 + e\sin\alpha} = \frac{2l}{1 - e^2\sin^2\alpha}$$

$$\frac{1}{PP'} + \frac{1}{QQ'} = \frac{1 - e^2\cos^2\alpha}{2l} + \frac{1 - e^2\sin^2\alpha}{2l}$$

$$= \frac{2 - e^2}{2l} = \text{constant.}$$

Example 9.7.4

If a chord PQ of a conic whose eccentricity e and the semi latus rectum l subtends a right angle at the focus SP then prove that $\left(\frac{1}{SP}-\frac{1}{l}\right)^2+\left(\frac{1}{SQ}-\frac{1}{l}\right)^2=\frac{e^2}{l^2}$.

Solution

Let the equation of the conic be $\frac{l}{r}=1+e\cos\theta$. Let the vectorial angle of P be α. The vectorial angle of Q is $\alpha+\frac{\pi}{2}$.

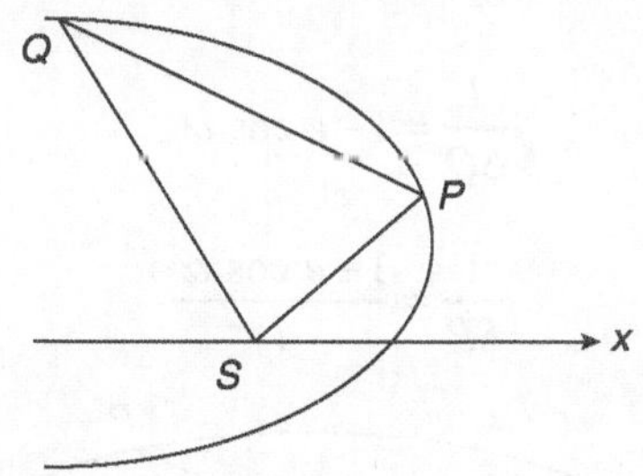

Since P and Q lie on the conic,

$$\frac{l}{SP}=1+e\cos\alpha \text{ and } \frac{l}{SQ}=1-e\sin\alpha$$

$$\frac{1}{SP}=\frac{1+e\cos\alpha}{l}$$

$$\frac{1}{SP}-\frac{1}{l}=\frac{e\cos\alpha}{l}$$

Similarly,

$$\frac{1}{SQ}-\frac{1}{l}=-\frac{e\sin\alpha}{l}$$

Squaring and adding, we get

$$\left(\frac{1}{SP}-\frac{1}{l}\right)^2+\left(\frac{1}{SQ}-\frac{1}{l}\right)^2=\frac{e^2}{l^2}.$$

Example 9.7.5

Let PSQ and $PS'R$ be two chords of an ellipse through the foci S and S'. Show that $\frac{SP}{SQ}+\frac{S'P}{S'R}$ is a constant.

Solution

Let the vectorial angle of P be α. Then the vectorial angle of Q is $\alpha+\pi$.

Since P and Q lie on the conic $\frac{l}{r}=1+e\cos\theta$

$$\frac{l}{SP}=1+e\cos\alpha \tag{9.35}$$

$$\frac{l}{SQ}=1-e\cos\alpha \tag{9.36}$$

$$\frac{1}{SP}=\frac{1+e\cos\alpha}{l}$$

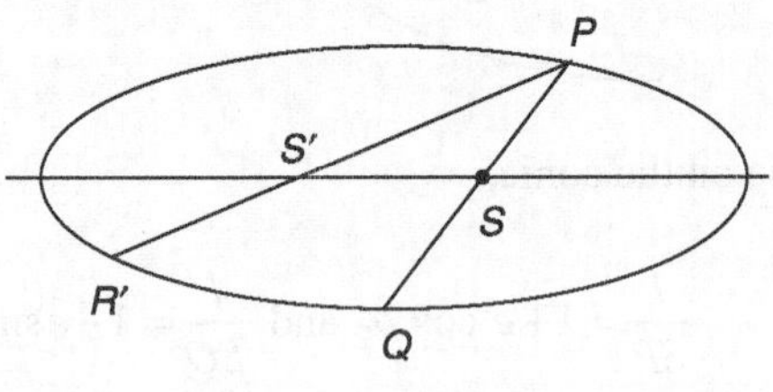

$$\frac{1}{SQ}=\frac{1-e\cos\alpha}{l}$$

$$\frac{1}{SP}+\frac{1}{SQ}=\frac{2}{l} \text{ or } \frac{l}{SP}+\frac{l}{SQ}=2 \tag{9.37}$$

Similarly, considering the other focal chord $PS'R$

$$\frac{l}{S'P}+\frac{l}{S'Q}=2 \tag{9.38}$$

Multiply equation (9.37) by $\frac{SP}{l}$, we get

$$\frac{SP}{SQ}=\frac{2SP}{l}-1 \tag{9.39}$$

Similarly from equation (9.38), we get

$$\frac{S'P}{S'R} = \frac{2S'P}{l} - 1 \tag{9.40}$$

Adding equations (9.39) and (9.40), we get $\frac{SP}{SQ} + \frac{S'P}{S'R} = \frac{2}{l}(SP + S'P) - 2$

$$= \frac{2}{l}(2a) - 2 = \frac{4a}{l} - 2 = \text{a constant}$$

Example 9.7.6

Prove that the perpendicular focal chords of a rectangular hyperbola are equal.

Solution

Let PSP' and QSQ' and be two perpendicular focal chords of a rectangular hyperbola. Then the vectorial angles of P and P' are α.

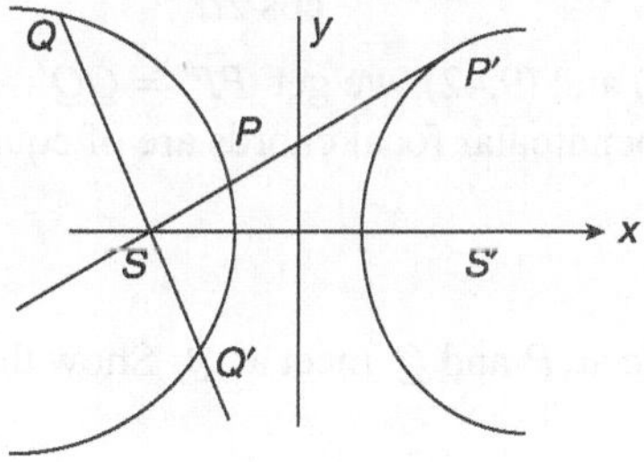

$$\frac{l}{SP} = 1 + e\cos\alpha$$

$$\frac{l}{SP'} = e\cos\alpha - 1$$

Since P' lies on the other branch of the hyperbola, the polar equation of the conic is $\frac{l}{r} = -1 + e\cos\alpha$.

$$SP = \frac{l}{1 + e\cos\alpha}$$

$$SP' = \frac{l}{e\cos\alpha - 1}$$

$$PP' = SP' - SP = \frac{l}{e\cos\alpha - 1} - \frac{1}{e\cos\alpha + 1}$$

$$= \frac{l[e\cos\alpha + 1 - e\cos\alpha + 1]}{e^2\cos^2\alpha - 1} = \frac{2l}{e^2\cos^2\alpha - 1}$$

$$= \frac{2l}{\cos 2\alpha} (\text{since in a RH } e = \sqrt{2}) \qquad (9.41)$$

Similarly,

$$SQ = \frac{l}{1 + e\sin\alpha}$$

$$SQ' = \frac{l}{1 - e\sin\alpha}$$

$$SQ + SQ' = \frac{2l}{1 - e^2\sin^2\alpha}$$

$$= \frac{2l}{\cos 2\alpha} \qquad (9.42)$$

From equations (9.41) and (9.42), we get $PP' = QQ'$.
That is, in a RH, perpendicular focal chords are of equal length.

Example 9.7.7

The tangents to a conic at P and Q meet at T. Show that if S is a focus then ST bisects $\angle PSQ$.

Solution

Let the equation of the conic be $\frac{1}{r} = 1 + e\cos\theta$. The equation of the tangent at P with vectorial angle α is $\frac{l}{r} = e\cos\theta + \cos(\theta - \alpha)$.

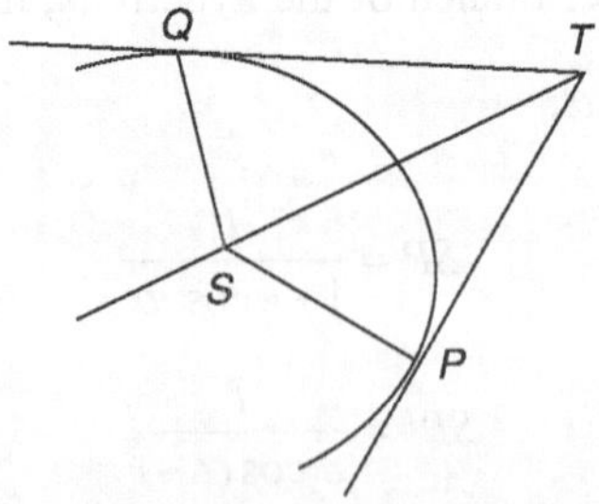

The equation of the tangent at Q with vectorial angle β is $\frac{l}{r} = e\cos\theta + \cos(\theta - \beta)$.
At the point of intersection of these two tangents,

$$\cos(\theta - \alpha) = \cos(\theta - \beta)$$

$$\therefore\ (\theta - \alpha) = \pm(\theta - \beta)$$

$$\text{(i.e.) } \alpha = \beta \text{ or } \theta = \frac{\alpha + \beta}{2}$$

$$\alpha = \beta \text{ is inadmissible } \therefore \theta = \frac{\alpha + \beta}{2}$$

(i.e.) ST bisects $\angle PSQ$.

Example 9.7.8

If the tangents at the extremities of a focal chord through the focus S of the conic $\frac{l}{r} = 1 + e\cos\theta$ meet the axis through S in T and T' show that $\frac{1}{ST} + \frac{1}{ST'} = \frac{2e}{l}$.

Solution

Let PSQ be a focal chord. Let the vectorial angles of P and Q be α and $\alpha + \pi$.
Then the equations of tangents at P and Q are

$$\frac{l}{r} = e\cos\theta + \cos(\theta - \alpha) \tag{9.43}$$

$$\frac{l}{r} = e\cos\theta + \cos(\theta - \overline{\alpha + \pi}) \tag{9.44}$$

$$\text{(i.e.) } \frac{l}{r} = e\cos\theta - \cos(\theta - \alpha) \tag{9.45}$$

When the tangents meet the axis, at those points $\theta = 0$.

$$\frac{l}{ST} = e + \cos\alpha$$

$$\frac{l}{ST'} = e - \cos\alpha$$

Adding $\frac{l}{ST'} + \frac{l}{ST'} = 2e$ or $\frac{l}{ST} + \frac{1}{ST'} = \frac{2e}{l}$

Example 9.7.9

If a chord of a conic $\frac{l}{r}=1+e\cos\theta$ subtends an angle 2α at the focus then show that the locus of the point where it meets the internal bisector of the angle is $\frac{l}{r}=\sec\alpha+e\cos\theta$.

Solution

Let PQ be a chord of the conic $\frac{l}{r}=1+e\cos\theta$ subtending an angle 2α at the focus. Let the internal bisector of PSQ meets PQ at T. Let the vectorial angles of P and Q be $\beta-\alpha$ and $\beta+\alpha$.
Let the polar coordinates of T be (r_1,β).

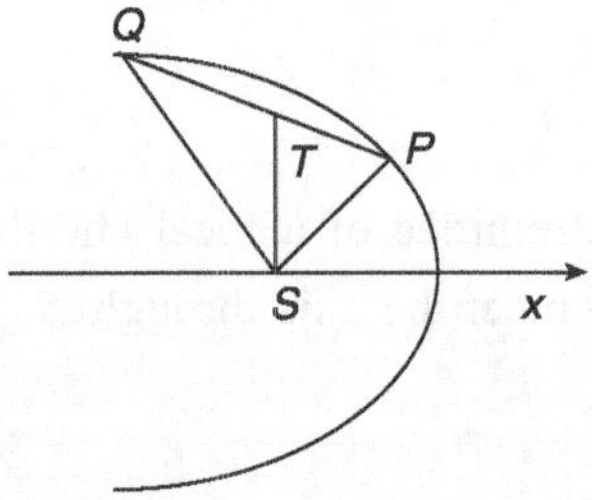

The equation of the chord PQ is

$$\frac{l}{r}=e\cos\theta+\sec\alpha\cos(\theta-\beta)$$

This passes through the point $T(r_1,\beta)$.

$$\therefore \frac{l}{r_1}=e\cos\beta+\sec\alpha$$

The locus of (r_1,β) is $\frac{l}{r}=\sec\alpha+e\cos\theta$.

Example 9.7.10

The tangents at two point P and Q of the conic meet in T and PQ subtends a constant angle 2α at the focus. Show that $\frac{1}{SP}+\frac{1}{SQ}-\frac{2\cos\alpha}{ST}$ is a constant.

Solution

Let the equation of the conic be $\frac{l}{r}=1+e\cos\theta$. Let the vectorial angles of P and Q be $\beta-\alpha$ and $\beta+\alpha$.

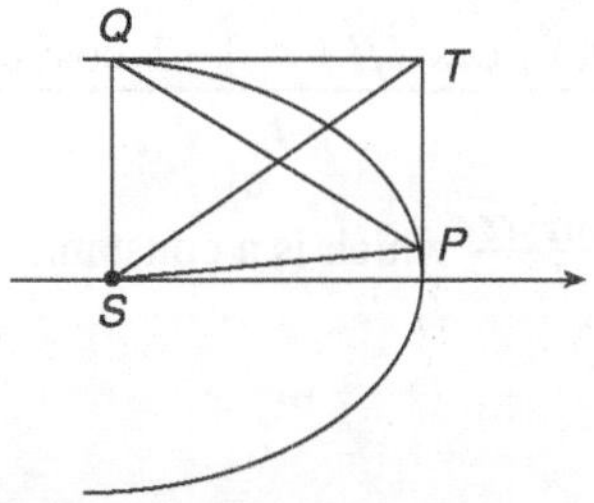

Since the points P and Q lie on the conic,

$$\frac{l}{SP} = 1 + e\cos(\beta - \alpha)$$

$$\therefore \frac{1}{SP} = \frac{1 + e\cos(\beta - \alpha)}{l} \tag{9.46}$$

$$\frac{l}{SQ} = 1 + e\cos(\beta + \alpha)$$

$$\therefore \frac{1}{SQ} = \frac{1 + e\cos(\beta + \alpha)}{l} \tag{9.47}$$

Also the equation of chord PQ is

$$\frac{l}{r} = e\cos\theta + \sec\alpha\cos(\theta - \beta)$$

$$\frac{l}{r} - e\cos\theta = \sec\alpha\cos(\theta - \beta) \tag{9.48}$$

PQ is also the polar of the point T and so its equation is

$$\left(\frac{l}{r} - e\cos\theta\right)\left(\frac{l}{ST} - e\cos\beta\right) = \cos(\theta - \beta) \tag{9.49}$$

Identifying equations (9.48) and (9.49), we get

$$\frac{1}{ST} = \frac{\cos\alpha + e\cos\beta}{l} \tag{9.50}$$

From equations (9.46) and (9.47) and (9.50), we get

$$\frac{1}{SP} + \frac{1}{SQ} - \frac{2\cos\alpha}{ST}$$

$$= \frac{1 + e\cos(\beta - \alpha)}{l} + \frac{1 + e\cos(\beta + \alpha)}{l} - \frac{2\cos\alpha(\cos\alpha + e\cos\beta)}{l}$$

$$= \frac{2+e[e\cos(\beta-\alpha)+\cos(\beta+\alpha)]-2\cos^2\alpha-2e\cos\alpha\cos\beta}{l}$$

$$= \frac{2(1-\cos^2\alpha)}{l} = \frac{2\sin^2\alpha}{l} \text{ which is a constant.}$$

Example 9.7.11

If a focal chord of an ellipse makes an angle α with the major axis then show that the angle between the tangents at its extremities is $\tan-1\left(\frac{2e\sin\alpha}{1-e^2}\right)$.

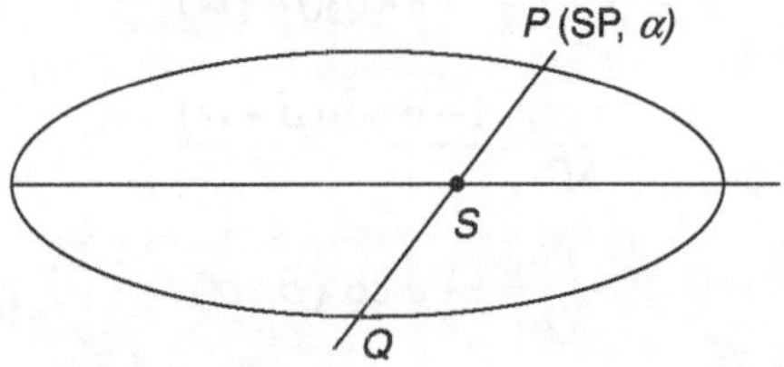

Solution

Let the equation of the conic be

$$\frac{l}{r} = 1+e\cos\theta \tag{9.51}$$

The equation of the tangent at P is

$$\frac{l}{r} = e\cos\theta+\cos(\theta-\alpha) \tag{9.52}$$

$$\text{(i.e.) } \frac{l}{r} = (e+\cos\alpha)\cos\theta+\sin\alpha\sin\theta \tag{9.53}$$

The equation of the tangent at Q is

$$\frac{l}{r} = e\cos\theta+\cos(\theta-\overline{\alpha+\pi}) \tag{9.54}$$

$$\frac{l}{r} = e\cos\theta-\cos(\theta-\alpha) \tag{9.55}$$

$$\frac{l}{r} = (e-\cos\alpha)\cos\theta-\sin\alpha\sin\theta \tag{9.56}$$

Transforming into cartesian coordinates by taking $x = r\cos\theta, y = \sin\theta$ Equations (9.53) and (9.54) becomes,

$$l = (e + \cos\alpha)x + y\sin\alpha$$
$$l = (e - \cos\alpha)x - y\sin\alpha$$

The slopes of the tangents are

$$m_1 = \frac{-(e+\cos\alpha)}{\sin\alpha}, m_2 = \frac{e-\cos\alpha}{\sin\alpha}$$

If θ is the angle between the tangents then

$$\tan\theta = \pm\left(\frac{m_1 - m_2}{1 + m_1 m_2}\right)$$

$$= \pm\frac{\frac{-(e+\cos\alpha)}{\sin\alpha} - \frac{e-\cos\alpha}{\sin\alpha}}{1 - \frac{e+\cos\alpha}{\sin\alpha}\cdot\frac{e-\cos\alpha}{\sin\alpha}}$$

$$= \pm\frac{(e+\cos\alpha+e-\cos\alpha)\sin\alpha}{\sin^2\alpha - (e^2 - \cos^2\alpha)}$$

$$= \pm\frac{2e\sin\alpha}{1-e^2}$$

The acute angle between the tangents is given by

$$\theta = \tan^{-1}\left(\frac{2e\sin\alpha}{1-e^2}\right)$$

Example 9.7.12

A focal chord SP of an ellipse is inclined at an angle α to the major axis. Prove that the perpendicular from the focus on the tangent at P makes with the axis an angle $\tan^{-1}\left(\frac{\sin\alpha}{e+\cos\alpha}\right)$.

Solution

Let the equation of the conic be

$$\frac{l}{r} = 1 + e\cos\theta \tag{9.57}$$

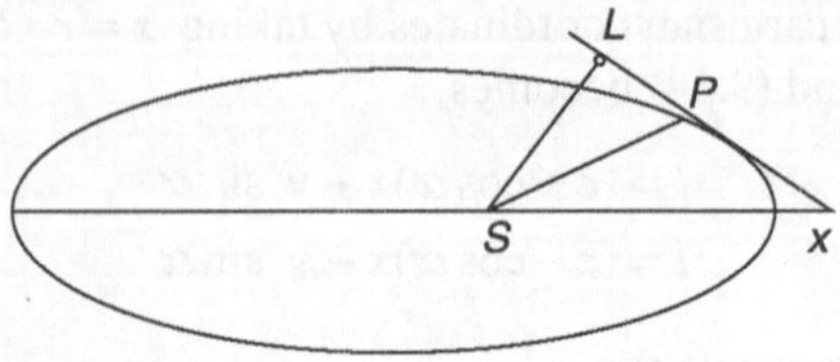

The equation of tangent at P is

$$\frac{l}{r}=1+e\cos\theta+\cos(\theta-\alpha) \tag{9.58}$$

The equation of the perpendicular line to the tangent at P is

$$\frac{k}{r}=e\cos\left(\theta+\frac{\pi}{2}\right)+\cos\left(\theta+\frac{\pi}{2}-\alpha\right)$$

$$\text{(i.e.)}\ \frac{k}{r}=-e\sin\theta-\sin(\theta-\alpha)$$

If the perpendicular passes through the focus then $k=0$

$$-e\sin\theta-\sin(\theta-\alpha)=0$$

$$\text{(i.e.)}\ e\sin\theta+\sin\theta\cos\alpha-\cos\theta\sin\alpha=0$$

$$\tan\theta=\frac{\sin\alpha}{e+\cos\alpha}$$

$$\text{or}\ \ \theta=\tan^{-1}\left(\frac{\sin\alpha}{e+\cos\alpha}\right)$$

Example 9.7.13 .

(i) If A circle passing through the focus of a conic whose latus rectum is $2l$ meets the conic in four points whose distances from the focus are r_1, r_2, r_3, r_4 then prove that $\frac{1}{r_1}+\frac{1}{r_2}+\frac{1}{r_3}+\frac{1}{r_4}=\frac{2}{l}$.

(ii) A circle of given radius passing through the focus S of a given conic intersects in A, B, C and D. Show that $SA\cdot SB\cdot SC\cdot SD$ is a constant.

Solution

Let the equations of the conic be

$$\frac{l}{r}=1+e\cos\theta \tag{9.59}$$

Let a be the radius of the circle and α be the angle the diameter makes with the initial line.

Then the equation of the circle is

$$r = 2a\cos(\theta - \alpha) \tag{9.60}$$

Eliminating θ between equations (9.59) and (9.60) we get an equation whose roots are the distances of the point of intersection from the focus.

Form equation (9.60), we get $r = 2\alpha(\cos\theta\cos\alpha + \sin\theta\sin\alpha)$.

From equation (9.59), we get $\cos\theta = \dfrac{l-r}{re}$.
Estimating θ, we get

$$r = 2a\left[\cos\alpha\frac{l-r}{re} + \sin\alpha\sqrt{1-\left(\frac{l-r}{re}\right)^2}\right]$$

$$\text{(i.e.)}\quad \frac{r}{2a} = \cos\alpha\frac{l-r}{re} + \sin\alpha\sqrt{1-\left(\frac{l-r}{re}\right)^2}$$

$$\left(\frac{r}{2a} - \frac{1-r}{re}\cos\alpha\right)^2 = \left[1-\left(\frac{l-r}{re}\right)^2\right]\sin^2\alpha$$

$$\frac{r^2}{4a^2} - \frac{l-r}{ae}\cos\alpha + \frac{(l-r)^2}{e^2r^2}\cos^2\alpha = \sin^2\alpha - \frac{(l-r)^2}{e^2r^2}\sin^2\alpha$$

$$\frac{r^2}{4a^2} - \frac{1-r}{ae}\cos\alpha + \frac{(l-r)^2}{e^2r^2} - \sin^2\alpha = 0$$

$$\frac{e^2r^4}{4a^2} + \frac{e}{a}r^3\cos\alpha + \left[-\frac{el}{a} + 1 - e^2\sin^2\alpha\right]r^2 - 2lr + l^2 = 0 \tag{9.61}$$

Dividing by r^4 and rewriting the equation in power of $\left(\frac{1}{r}\right)$, we get

$$\frac{l^2}{r^4} - \frac{2l}{r^3} + \left(1 - e^2\sin^2\alpha - \frac{el}{a}\right)\frac{1}{r^2} - \frac{e\cos\alpha}{ar} + \frac{e^2}{4a^2} = 0 \tag{9.62}$$

If r_1, r_2, r_3, r_4 are the distances of the points of intersection from the focus then $\frac{1}{r_1}, \frac{1}{r_2}, \frac{1}{r_3}, \frac{1}{r_4}$ are the roots of the above equation.

$$\frac{1}{r_1} + \frac{1}{r_2} + \frac{1}{r_3} + \frac{1}{r_4} = \frac{2l}{l^2} = \frac{2}{l}$$

Form equation (9.61), we get

$$r_1r_2r_3r_4 = \frac{4a^2l^2}{e^2} = a\text{ constant}.$$

(i.e.) $SA \cdot SB \cdot SC \cdot SD$ is a constant.

Example 9.7.14

If a chord of the conic $\frac{l}{r} = 1 + e\cos\theta$ subtends a constant angle 2β at the pole then show that the locus of the foot of the perpendicular from the pole to the chord $(e^2 - \sec^2\beta)r^2 - elr\cos\theta + l^2 = 0$.

Solution

Let the vectorial angles of P and Q be $\alpha - \beta$ and $\alpha + \beta$.

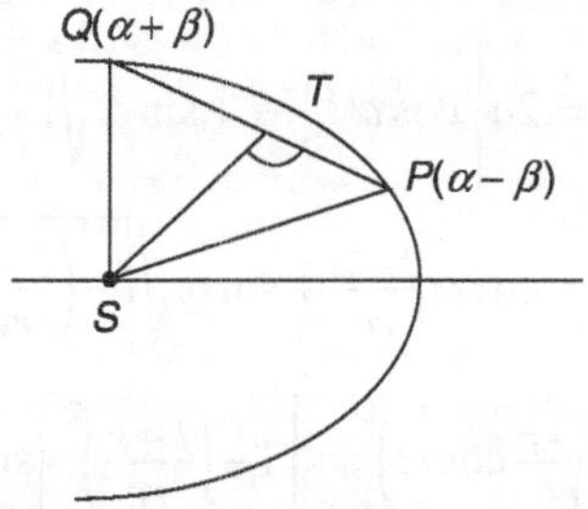

The equation of the chord PQ is

$$\frac{l}{r} = e\cos\theta + \sec\beta\cos(\theta - \alpha) \tag{9.63}$$

The equation of the line perpendicular to this chord is

$$\frac{k}{r} = -e\sin\theta - \sec\beta\sin(\theta - \alpha) \tag{9.64}$$

This line passes through the focus S and so $k = 0$.

$$0 = -e\sin\theta - \sec\beta\sin(\theta - \alpha) \tag{9.65}$$

From equation (9.63), we get

$$\frac{l}{r} - e\cos\theta = \sec\beta\cos(\theta - \alpha) \tag{9.66}$$

From equation (9.65), we get

$$-e\cos\theta = \sec\beta\sin(\theta - \alpha) \tag{9.67}$$

Squaring and adding (9.66) and (9.67), we get

$$\left(\frac{l}{r} - e\cos\theta\right)^2 + e^2\sin^2\theta = \sec^2\beta$$

$$l^2 - 2erl\cos\theta + e^2r^2\cos^2\theta + e^2r^2\sin^2\theta - r^2\sec^2\beta = 0$$

$$r^2(e^2 - \sec^2\beta) - 2elr\cos\theta + l^2 = 0$$

Example 9.7.15

A variable chord of conic subtends a constant angle 2β at the focus of the conic $\frac{l}{r}=1+e\cos\theta$. Show that the chord touches another conic having the same focus and directrix. Show also that the locus of poles of such chords of the conic is also a similar conic.

Solution

Let PQ be a chord of the conic $\frac{l}{r}=1+e\cos\theta$ subtending a constant angle 2α at the focus.

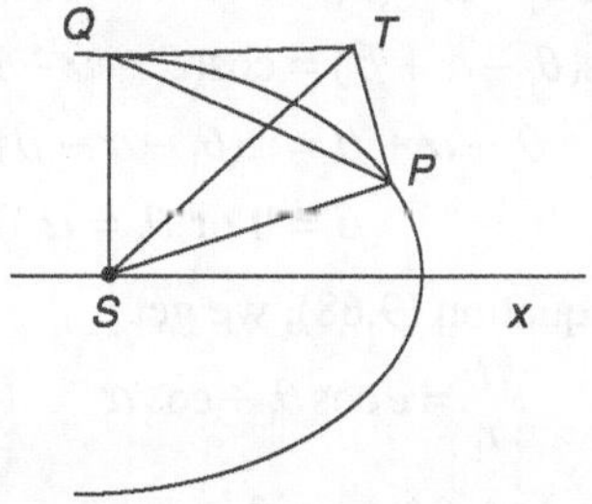

Let $T(r_1,\theta)$ be the point of intersection of tangents at P and Q. Then PQ is the polar of T and T is the pole of PQ. Let the vectorial angles of P and Q be $\alpha-\beta$ and $\alpha+\beta$.

Then the equation of chord PQ is

$$\frac{l}{r}=e\cos\theta+\sec\beta\cos(\theta-\alpha)$$

Dividing by sec β, $\frac{l}{r}\cos\beta=e(\sec\beta)\cos\theta+\cos(\theta-\alpha)$

$$\text{i.e. } \frac{L}{r}=E\cos\theta+\cos(\theta-\alpha)$$

where $L=e\cos\beta$ and $E=e\sec\theta$. This line is a tangent to the conic C'.

$$\frac{L}{r}=1+E\cos\theta$$

This equation has the same focus as $\frac{l}{r}=1+e\cos\theta$. Hence, the conic C' has the same focus and the same initial line as C. For the given conic $SX=\frac{l}{e}$.

For the conic C', $SX'=\frac{L}{e}=\frac{l\cos\beta}{e\cos\beta}=\frac{l}{e}$

$\therefore SX=SX'$. Hence X' coincides with X.

Hence, both the conics have the same focus and the same directrix. The equation of tangents at P and Q are $\frac{l}{r} = e\cos\theta + \cos(\theta - \alpha + \beta)$ and $\frac{l}{r} = e\cos\theta + \cos(\theta - \alpha - \beta)$.

These two tangents intersect at $T(r_1, \theta_1)$

$$\frac{l}{r_1} = e\cos\theta_1 + \cos(\theta_1 - \alpha - \beta) \tag{9.68}$$

$$\frac{l}{r_1} = e\cos\theta_1 + \cos(\theta_1 - \alpha + \beta) \tag{9.69}$$

From equations (9.68) and (9.69), we get

$$e\cos\theta_1 + \cos(\theta_1 - \alpha + \beta) = e\cos\theta_1 + \cos(\theta_1 - \alpha - \beta)$$

$$\cos(\theta_1 - \alpha + \beta) = \cos(\theta_1 - \alpha - \beta)$$

$$\theta_1 - \alpha + \beta = \pm(\theta_1 - \alpha - \beta)$$

$$\beta = 0 \text{ or } \theta_1 = \alpha \text{ But } \beta \neq 0$$

Substituting $\theta_1 = \alpha$ in equation (9.68), we get

$$\frac{l}{r_1} = e\cos\theta_1 + \cos\alpha$$

The locus of (r_1, θ_1) is $\frac{l}{r} = \cos\theta + e\cos\theta$.

$$\text{(i.e.)} \quad \frac{l\sec\alpha}{r} = 1 + e\sec\alpha\cos\theta$$

$$\frac{L'}{r} = 1 + E'\cos\theta \text{ where } L' = l\text{ secs }\beta, E' = e\sec\alpha$$

The locus of poles is the conic having the same focus and same directrix as the given conic.

Example 9.7.16

Show that the locus of the point of intersection of tangents at the extremities of a variable focal chord is the corresponding directrix.

Solution

Let the equation of the conic be $\frac{l}{r} = 1 + e\cos\theta$.

The equation of tangent at α is $\frac{l}{r} = 1 + e\cos\theta + \cos(\theta - \alpha)$.

The equation of tangent at $\alpha + \pi$ is $\frac{l}{r} = e\cos\theta + \cos(\theta - \overline{\alpha + \pi})$.

$$\therefore \frac{l}{r} = e\cos\theta - \cos(\theta - \alpha)$$

Let (r_1, θ_1) be the point of intersection of these two tangents.

Then, $\quad \dfrac{l}{r_1} = e\cos\theta_1 + \cos(\theta_1 - \alpha)$

$$\frac{l}{r_1} = e\cos\theta_1 - \cos(\theta_1 - \alpha)$$

Adding these two equations, we get

$$\frac{2l}{r_1} = 2e\cos\theta_1 \text{ or } \frac{l}{r_1} = e\cos\theta_1$$

Therefore, the locus is the corresponding directrix $\dfrac{l}{r} = e\cos\theta$.

Example 9.7.17

Show that the locus of the point of intersection of perpendicular tangents to a conic is a circle or a straight line.

Solution

Let the equation of the conic be

$$\frac{l}{r} = 1 + e\cos\theta \tag{9.70}$$

Let P and Q be the points on the conic whose vectorial angles are α and β. The equations of tangents at P and Q are

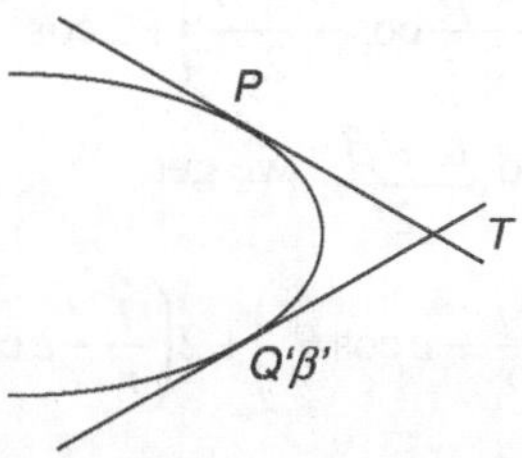

$$\frac{l}{r} = e\cos\theta + \cos(\theta - \alpha) \tag{9.71}$$

$$\frac{l}{r} = e\cos\theta + \cos(\theta - \beta) \tag{9.72}$$

Let (r_1, θ_1) be the point of intersection of tangents at P and Q. Then,

$$\frac{l}{r_1} = e\cos\theta_1 + \cos(\theta_1 - \alpha) \tag{9.73}$$

$$\frac{l}{r_1} = e\cos\theta_1 + \cos(\theta_1 - \beta) \tag{9.74}$$

From equations (9.73) and (9.74), we get

$$\cos\,(\theta_1 - \alpha) = (\theta_1 - \beta)$$
$$\theta_1 - \alpha = \pm(\theta_1 - \beta)$$
$$\alpha = \beta \text{ or } \theta_1 = \frac{\alpha+\beta}{2}$$

But $\alpha = \beta$ is not possible.

$$\therefore \theta_1 = \frac{\alpha+\beta}{2}$$

Form equation (9.73), we get $\cos\frac{\alpha-\beta}{2} = \frac{l - e\cos\theta}{r_1}$

Expanding equations (9.70) and (9.71), we get

$$\frac{l}{r} = (e+\cos\alpha)\cos\theta + \sin\alpha\sin\theta$$
$$\frac{l}{r} = (e+\cos\beta)\cos\theta + \sin\beta\sin\theta$$

Since these two lines are perpendicular, we have

$$(e+\cos\alpha)(e+\cos\beta) + \sin\alpha\sin\beta = 0$$
$$\text{(i.e.) } e^2 + e(\cos\alpha + \cos\beta) + \cos(\alpha-\beta) = 0$$
$$e^2 + e\left(2\cos\frac{\alpha+\beta}{2}\cos\frac{\alpha-\beta}{2}\right) + 2\cos^2\frac{\alpha-\beta}{2} - 1 = 0$$

Substituting for $\frac{\alpha+\beta}{2}$ and $\frac{\alpha-\beta}{2}$, we get

$$e^2 + 2e\cos\theta_1\left(\frac{l}{r_1} - e\cos\theta_1\right) + 2\left(\frac{l}{r_1} - e\cos\theta_1\right)^2 - 1 = 0$$

$$(e^2 - 1) - \frac{2el}{r_1}\cos\theta_1 + \frac{2l^2}{r_1^2} = 0$$
$$(1-e^2)r_1^2 + 2elr_1\cos\theta_1 - 2l^2 = 0$$

Therefore, the locus of (r_1, θ_1) is $(1-e^2)r^2 + 2elr\cos\theta - 2l^2 = 0$.

Example 9.7.18

Prove that points on the conic $\frac{l}{r} = 1 + e\cos\theta$ whose vectorial angles are α and β, respectively, will be the extremities of a diameter if $\frac{e+1}{e-1} = \tan\frac{\alpha}{2}\tan\frac{\beta}{2}$.

Solution

The equation of the conic is $\frac{l}{r}=1+e\cos\theta$.
Let α,β be the extremities of a diameter of the conic $\frac{l}{r}=1+e\cos\theta$. Then the tangents at α and β are parallel and hence their slopes are equal.
The equation of tangent at α is

$$\frac{l}{r}=e\cos\theta+\cos(\theta-\alpha)$$

$$\frac{l}{r}=(e+\cos\alpha)\cos\theta+\sin\alpha\sin\theta$$

$$l=(e+\cos\alpha)x+y\sin\alpha \text{ (in Cartesian form)}$$

The slope of this tangent is $-\left(\frac{e+\cos\alpha}{\sin\alpha}\right)$.
Since tangents at α and β are parallel,

$$-\left(\frac{e+\cos\alpha}{\sin\alpha}\right)=-\left(\frac{e+\cos\beta}{\sin\beta}\right)$$

$$(e+\cos\alpha)\sin\beta=(e+\cos\beta)\sin\alpha$$

$$e(\sin\beta-\sin\alpha)=\sin(\alpha-\beta)$$

$$e\left(2\cos\frac{\beta+\alpha}{2}\sin\frac{\beta-\alpha}{2}\right)=2\sin\frac{\alpha-\beta}{2}\cos\frac{\alpha-\beta}{2}$$

$$\frac{e}{1}=\frac{\cos\frac{\beta-\alpha}{2}}{-\cos\frac{\beta+\alpha}{2}}$$

$$\frac{e+1}{e-1}=\frac{\cos\frac{\beta-\alpha}{2}-\cos\frac{\beta+\alpha}{2}}{\cos\frac{\beta-\alpha}{2}+\cos\frac{\beta+\alpha}{2}}$$

$$=\frac{2\sin\frac{\alpha}{2}\sin\frac{\beta}{2}}{2\cos\frac{\alpha}{2}\cos\frac{\beta}{2}}$$

$$=\tan\frac{\alpha}{2}\tan\frac{\beta}{2}$$

Example 9.7.19

If a normal is drawn at one extremity of the latus rectum, prove that the distance from the focus of the other point in which it meets the conic is $\frac{(1+3e^2+e^4)l}{1+e^2-e^4}$.

Solution

The equation of the conic is

$$\frac{l}{r} = 1 + e\cos\theta \tag{9.75}$$

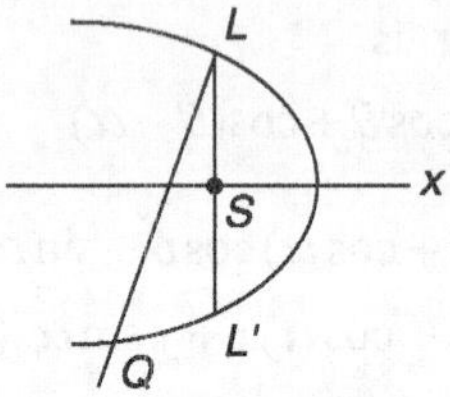

The equation of the normal at $L\left(l, \frac{\pi}{2}\right)$ is

$$\frac{le}{r} = e\sin\theta - \cos\theta \tag{9.76}$$

Solving equations (9.75) and (9.76), we get their point of intersection

$$e\sin\theta - \cos\theta = e(1 + e\cos\theta)$$

$$e\sin\theta = e + (1 + e^2)\cos\theta$$

$$\text{Squaring } e^2\sin^2\theta = [e + (1 + e^2)\cos\theta]^2$$

$$e^2(1 - \cos^2\theta) = e^2 + (1 + e^2)^2\cos^2\theta + 2e(1 + e^2)\cos\theta$$

$$(e^4 + 3e^2 + 1)\cos^2\theta + 2e(1 + e^2)\cos\theta = 0$$

$$\cos\theta = 0 \text{ or } \cos\theta = -\frac{2e(1 + e^2)}{e^4 + 3e^2 + 1}$$

If $\cos\theta = 0$ then $\theta = \frac{\pi}{2}$. This corresponds to the point L.

At the other end of the normal $\cos\theta = -\frac{2e(1 + e^2)}{e^4 + 3e^2 + 1}$

Substituting in equation (9.75) we get,

$$\frac{l}{r} = 1 - \frac{2e^2(1 + e^2)}{1 + 3e^2 + e^4}$$

$$r = \frac{l(1 + 3e^2 + e^4)}{1 + e^2 - e^4} = SQ$$

Example 9.7.20

If the tangents at the points P and Q on a conic intersect in T and the chord PQ meets the directrix at R then prove that the angle TSR is a right angle.

Solution

Let the vectorial angles of P and Q be α and β, respectively. Let the tangents at P and Q meet at $T(r_1,\theta_1)$.

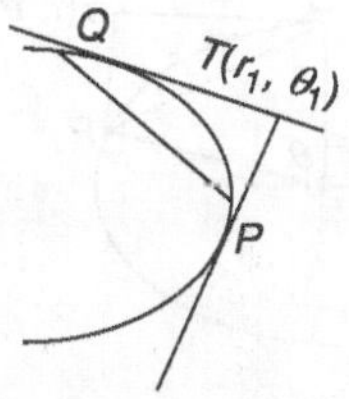

The equation of tangents at P and Q are $\frac{l}{r}=e\cos\theta+\cos(\theta-\alpha)$ and $\frac{l}{r}=e\cos\theta+\cos(\theta-\beta)$.

Since these tangents meet at (r_1,θ_1), we have $\frac{l}{r}=e\cos\theta_1+\cos(\theta_1-\alpha)$ and $\frac{l}{r}=e\cos\theta_1+\cos(\theta_1-\beta)$.

$$\therefore \cos(\theta_1-\alpha)=\cos(\theta_2-\beta)$$

$\theta_1-\alpha=\pm(\theta_2-\beta)$ which implies $\theta_1=\dfrac{\alpha+\beta}{2}$

Let θ be the vectorial angle of R. The equation of chord PQ is $\frac{l}{r}=\sec\frac{\beta-\alpha}{2}\cos\left(\theta-\frac{\alpha+\beta}{2}\right)+e\cos\theta$.

The equation of the directrix $\frac{l}{r}=e\cos\theta$.

Subtracting, we get $\sec\dfrac{\beta-\alpha}{2}\cos\left(\theta-\dfrac{\alpha+\beta}{2}\right)=0$

$$\cos\left(\theta-\frac{\alpha+\beta}{2}\right)=0$$

$$\theta-\frac{\alpha+\beta}{2}=\frac{\pi}{2}$$

$$\theta-\theta_1=\frac{\pi}{2}$$

$$\angle TSR=90^\circ$$

Example 9.7.21

A chord PQ of a conic subtends a constant 2γ at the focus S and tangents at P and Q meet in T. Prove that $\dfrac{1}{SP}+\dfrac{1}{SQ}-\dfrac{2\cos\gamma}{ST}=\dfrac{2\sin^2\gamma}{l}$.

Solution

Let the equation of the conic be $\frac{l}{r} = 1 + e\cos\theta$.

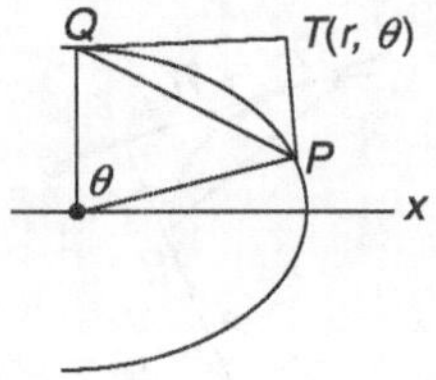

Let the vectorial angles of P and Q be α and β, respectively.

Since these points lie on the conic, $\frac{l}{SP} = 1 + e\cos\alpha$ and $\frac{l}{SQ} = 1 + \cos\beta$.

If the tangents at P and Q intersect at (r_1, θ_1) then $\theta_1 = \frac{\alpha+\beta}{2}$.

$$\frac{l}{r_1} = \cos\frac{\beta-\alpha}{2} + e\cos\frac{\alpha+\beta}{2}$$

Since PQ subtends an angle 2γ at S,

$$\alpha - \beta = 2\gamma$$

$$\frac{l}{ST} = \cos\gamma + e\cos\frac{\alpha+\beta}{2}$$

$$\frac{1}{SP} + \frac{1}{SQ} - \frac{2\cos\gamma}{ST}$$

$$= \frac{1}{l}\left[(1+e\cos\alpha) + (1+e\cos\beta) - 2\cos\gamma\left(\cos\gamma + e\cos\frac{\alpha+\beta}{2}\right)\right]$$

$$= \frac{1}{l}\left[2 + e\left((\cos\alpha + \cos\beta) - 2\cos^2\gamma - 2e\cos\gamma\cos\frac{\alpha+\beta}{2}\right)\right]$$

$$= \frac{1}{l}\left[2(1-\cos^2\gamma) + e\left(2\cos\frac{\alpha+\beta}{2}\cos\frac{\alpha-\beta}{2}\right) - e2\cos\gamma\sin\frac{\alpha+\beta}{2}\right]$$

$$= \frac{1}{l}\left[2\sin^2\gamma + e2\cos\frac{\alpha+\beta}{2}\cos\gamma - e.2\cos\gamma.\cos\frac{\alpha+\beta}{2}\right]$$

$$= \frac{2}{l}\sin^2\gamma$$

Exercises

1. If PSP' and QSQ' are two focal chords of a conic cutting each other at right angles then prove that $\frac{1}{SP.SP'}+\frac{1}{SQ.SQ'}=$ a constant.
2. If two conics have a common focus then show that two of their common chords pass through the point of intersection of their directrices.
3. Show that $\frac{1}{r}=1+e\cos\theta$ and $\frac{l}{r}=-1+e\cos\theta$ represent the same conic.
4. In a parabola with focus S, the tangents at any points P and Q on it meet at T. Prove that

   (i) $SP \cdot SQ = ST^2$

   (ii) The triangles SPT and SQT are similar.
5. If S be the focus, P and Q be two points on a conic such that the angle PSQ is constant, prove that the locus of the point of intersection of the tangents at P and Q is a conic section whose focus is S.
6. If the circle $r+2a\cos\theta=0$ cuts the conic $\frac{1}{r}=1+e\cos(\theta-\alpha)$ in four real points find the equation in r which determines the distances of these points from the pole. Also, show that if their algebraic sum equals $2a$ and the eccentricity of the conic is $2\cos\alpha$.
7. Prove that the two conics $\frac{l_1}{r}=1+e_1\cos\theta$ and $\frac{l_2}{r}=1+e\cos(\theta-\alpha)$ touch each other if $l_1^2(1-e_2^2)+l_2^2(1-e_1^2)=2l_1l_2(1-e_1e_2\cos\alpha)$.
8. P, Q and R are three points on the conic $\frac{l}{r}=1+e\cos\theta$. Tangents at Q meets SP and SR in M and N so that $SM=AN=l$ where S is the focus. Prove that the chord PQ touches the conic.
9. Prove that the portion of the tangent intercepted between the curve and directrix subtends a right angle at the focus.
10. Prove that the locus of the middle points of a system of focal chords of a conic section is a conic section which is a parabola, ellipse or hyperbola according as the original conic is a parabola ellipse or hyperbola.
11. Two equal ellipses of eccentricity e have one focus common and are placed with their axes at right angles. If PQ be a common tangent then prove that $\sin\frac{1}{2}\angle PSQ=\frac{e}{\sqrt{2}}$.
12. If the tangents at P and Q of a conic meet at a point T and S be the focus then prove that $ST^2=SP\cdot SQ$ if the conic is a parabola.
13. A conic is described having the same focus and eccentricity as the conic $\frac{l}{r}=1+e\cos\theta$ and the two conics touch at $\theta=\alpha$. Prove that the length of its latus rectum is $\frac{2l(1-e^2)}{e^2+2e\cos\alpha+1}$.

14. Prove that three normals can be drawn from a given point to a given parabola. If the normal at α,β,γ on the conic $\frac{l}{r}=1+\cos\theta$ meet at the point (ρ,ϕ) prove that $\phi=\frac{\alpha+\beta+\gamma}{2}$.

15. If the normals at three points of the parabola $r=a\cos ec^2\frac{\theta}{2}$ whose vectorial angles are α,β,γ meet in a point whose vectorial angle is ϕ then prove that $2\phi=\alpha+\beta+\gamma-\pi$.

16. If α,β,γ be the vectorial angles of three points on $\frac{2a}{r}=1+\cos\theta$, the normal at which are concurrent, prove that $\tan\frac{\alpha}{2}+\tan\frac{\beta}{2}+\tan\frac{\gamma}{2}=0$.

17. If the normal at P to a conic cuts the axis in G, prove that $SN=eSP$.

18. If SM and SN be perpendiculars from the focus S on the tangent and normal at any point on the conic $\frac{l}{r}=1+e\cos\theta$ and, ST the perpendicular on MN show that the locus to T is $r(e^2-1)=el\cos\theta$.

19. Show that if the normals at the points whose vectorial angles are $\theta_1,\theta_2,\theta_3$ and θ_4 on $\frac{l}{r}=1+e\cos\theta$ meet at the point (r',ϕ) then $\theta_1+\theta_2+\theta_3+\theta_4-2\phi=(2n+1)\pi$.

20. Prove that the chords of a rectangular hyperbola which subtend a right angle at a focus touch a fixed parabola.

21. If the tangent at any point of an ellipse make an angle α with its major axis and an angle β with the focal radius to the point of contact then show that $e\cos\alpha=\cos\beta$. **Ans.:** $A^2+B^2-2e(A\cos\gamma+B\sin\gamma)+e^2-1=0$

22. Prove that the exterior angle between any two tangents to a parabola is equal to half the difference of the vectorial angles of their points of contact.

23. Find the condition that the straight line $\frac{l}{r}=A\cos\theta+B\sin\theta$ may be a tangent to the conic $\frac{l}{r}=1-e\cos(\theta-\gamma)$.

24. Find the locus of poles of chords which subtend a constant angle at the focus.

25. Prove that if the chords of a conic subtend a constant angle at the focus, the tangents at the end of the chord will meet on a fixed conic and the chord will touch another fixed conic.

26. Find the locus of the point of intersection of the tangents to the conic $\frac{l}{r}=1+e\cos\theta$ at P and Q, where $\frac{1}{SP}+\frac{1}{SQ}=\frac{2}{k}$, k being a constant.

27. If the tangent and normal at any point P of a conic meet the transverse axis of T and G, respectively, and if S be the focus then prove that $\frac{1}{SG}-\frac{1}{ST}$ is a constant.

CHAPTER

10 Tracing of Curves

10.1 GENERAL EQUATION OF THE SECOND DEGREE AND TRACING OF A CONIC

In the earlier chapters, we studied standard forms of a conic namely a parabola, ellipse and hyperbola. In this chapter, we study the conditions for the general equation of the second degree to represent the different types of conic. In order to study these properties, we introduce the characteristics of change of origin and the coordinate axes, rotation of axes without changing the origin and reducing the second degree equation without xy-term.

10.2 SHIFT OF ORIGIN WITHOUT CHANGING THE DIRECTION OF AXES

Let Ox and Oy be two perpendicular lines on a plane. Let O' be a point in the xy-plane. Through O', draw $O'X$ and $O'Y$ parallel to Ox and Oy, respectively. Let the coordinates of O' be (h, k) with respect to the axes Ox and Oy. Draw $O'L$ perpendicular to Ox. Then $OL = h$ and $O'L = k$.

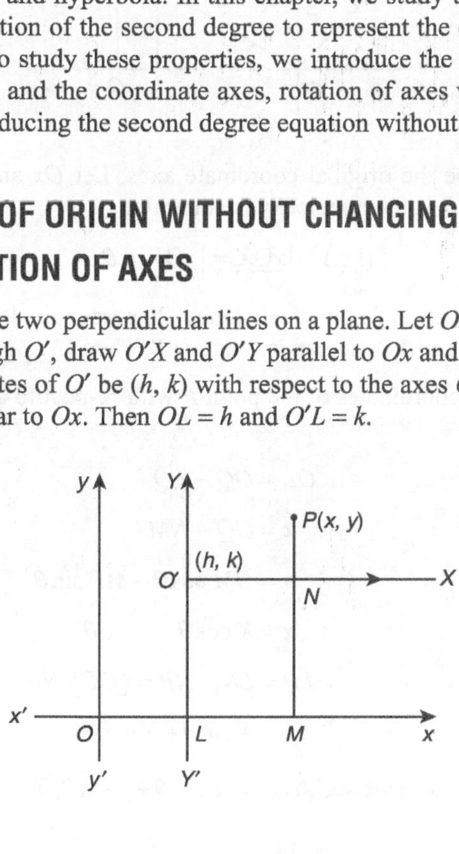

Let $P(x, y)$ be any point in the xy-plane. Draw PM perpendicular to Ox meeting axis at N. Then

$$x = OM = OL + LM = h + X = X + h$$

$$y = MP = MN + NP = LO' + NP = k + Y = Y + k$$

10.3 ROTATION OF AXES WITHOUT CHANGING THE ORIGIN

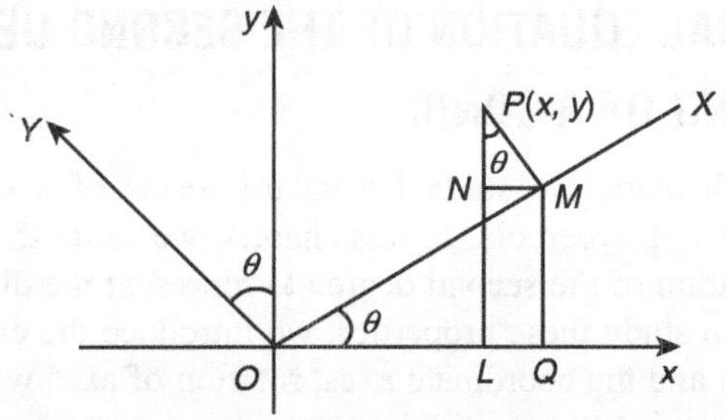

Let Ox and Oy be the original coordinate axes. Let Ox and Oy be rotated through an angle θ in the anticlockwise direction.

$$\text{(i.e.)} \quad \angle xOX = \angle yOY = \theta$$

Let $P(x, y)$ be a point in the xy-plane. Draw PL perpendicular to Ox, PM perpendicular to OX and MN perpendicular to LP. Then $\angle NPM = \theta$.

Let (X, Y) be the coordinates of the point P with respect to axes OX and OY. Then

$$OL = OQ - LQ$$

$$x = OQ - NM$$

$$= OM\cos\theta - MP\sin\theta$$

$$x = X\cos\theta - Y\sin\theta \qquad (10.1)$$

$$LP = LN + NP = QM + NP$$

$$y = X\sin\theta + Y\cos\theta \qquad (10.2)$$

From (10.1) and (10.2) we see that $X = x\cos\theta + y\sin\theta$, $Y = -x\sin\theta + y\cos\theta$.

10.4 REMOVAL OF XY-TERM

Here we want to transform the second degree equation $ax^2+2hxy+by^2+22x+2fy+c=0$ into a second degree curve without *XY*-term, where the axes are rotated through an angle θ without changing the origin we get

$$x = X\cos\theta - Y\sin\theta$$
$$y = X\sin\theta + Y\cos\theta$$

Then $$ax^2+2hxy+by^2+2gx+2fy+c$$

$$= a(X\cos\theta - Y\sin\theta)^2 + 2h(X\cos\theta - Y\sin\theta)(X\sin\theta + Y\cos\theta)$$
$$+ b(X\sin\theta + Y\cos\theta)^2 + 2g(X\cos\theta - Y\sin\theta) + 2f(X\sin\theta + Y\cos\theta) + c$$

$$= (a\cos^2\theta + 2hX\sin\theta\cos\theta + b\sin^2\theta)X^2 - 2[(a-b)\sin\theta\cos\theta$$
$$- h(\cos^2\theta - \sin^2\theta)]XY + (a\sin^2\theta - 2h\sin\theta\cos\theta + b\cos^2\theta)Y^2$$
$$+ (2g\cos\theta + 2f\sin\theta)X + (2f\cos\theta - 2g\sin\theta)Y + c$$

If *XY*-term has to be absent then

$$(a-b)\sin\theta\cos\theta - h(\cos^2\theta - \sin^2\theta) = 0$$

Then $\tan 2\theta = \dfrac{2h}{a-b}$ or $\theta = \dfrac{1}{2}\tan^{-1}\left(\dfrac{2h}{a-b}\right).$

Hence, by rotating the axes through an angle θ about O the general second degree expression will result into a second degree expression without *XY*-terms.

10.5 INVARIANTS

We will now prove that the expression $ax^2+2hxy+by^2$ will change to $Ax^2+2hXY+By^2$ if (i) $a+b=A+B$ and (ii) $ab-h^2=AB-H^2$.

Proof:

Let $P(x, y)$ be any point with respect to axes (ox, oy) and (X, Y) be its coordinates with respect to (OX, OY). Then $x = X\cos\theta - Y\sin\theta, y = X\sin\theta + Y\cos\theta$. Therefore, we get $x^2+y^2=X^2+Y^2$

Suppose $ax^2+2hxy+by^2 = AX^2+2HXY+BY^2$. Then $ax^2+2hxy+by^2+\lambda(x^2+y^2) = AX^2+2HXY+BY^2+\delta(X^2+Y^2)$.

If the LHS is of the form $(px+qy)^2$ then it will be changed into the form $[p(X\cos\theta - Y\sin\theta)+q(X\sin\theta+Y\cos\theta)]^2 = (p_1X+q_1Y)^2$.

$$\text{LHS} = (a+\lambda)^2 x^2 + 2hxy + (b+\lambda)^2 y^2$$

This will be a perfect square if $h^2 = (a+\lambda)(b+\lambda)$

$$\text{(i.e.)} \quad \lambda^2+\lambda(a+h)+ab-h^2 = 0 \tag{10.3}$$

RHS will be a perfect square if

$$\lambda^2+\lambda(A+B)+AB-H^2 = 0 \tag{10.4}$$

Comparing (10.3) and (10.4), we get $a+b = A+B$, $ab-h^2 = AB-H^2$.

10.6 CONDITIONS FOR THE GENERAL EQUATION OF THE SECOND DEGREE TO REPRESENT A CONIC

The general equation of the second degree

$$ax^2+2hxy+by^2+2gx+2fy+c = 0 \tag{10.5}$$

If the axes are rotated through an angle θ with the anticlockwise direction then $\tan 2\theta = \dfrac{2h}{a-b}$.

Then the equation transformed with the second degree in (X, Y) is

$$AX^2+BY^2+2GX+2FY = 0 \tag{10.6}$$

where $\quad a+b = A+B,\ ab-h^2 = AB.$

Now, we study several cases based on the values of A and B.

Case 1: If $ab-h^2 = 0$ then $A = 0$ or $B = 0$. Suppose $A = 0$ then the equation (10.6) takes the form

$$B\left(Y+\frac{F}{B}\right)^2 = -2GX-\frac{CF^2}{B} \tag{10.7}$$

If $G = 0$ then this equation will represent a pair of straight lines. If $G \neq 0$ then we have from (3).

$$\left(Y+\frac{F}{b}\right)^2=-\frac{2G}{B}\left(X-\frac{F^2}{2BG}+\frac{C}{2G}\right)$$

Shifting the origin to the point $\left(\frac{F^2}{2BG}-\frac{C}{2G},\frac{-F}{B}\right)$ the above equation can be written in the form $Y^{1^2}=-\frac{2G}{B}X^1$ which is a parabola.

Case 2: Suppose $ab-h^2\neq 0$ then neither $A=0$ nor $B=0$. Then equation (10.6) can be written as

$$A\left(X+\frac{G}{A}\right)^2+B\left(Y+\frac{F}{B}\right)^2=\frac{G^2}{A}+\frac{F^2}{B}-C$$

$$A\left(X+\frac{G}{A}\right)^2+B\left(Y+\frac{F}{B}\right)^2=K \qquad \text{where } K=\frac{G^2}{A}+\frac{F^2}{B}-C$$

Shifting the origin to the point $\left(-\frac{G}{A},-\frac{F}{B}\right)$ the above equation takes the form

$$AX'^2+BY'^2=K \tag{10.8}$$

If $B=0$ then equation (10.8) represents a form of straight lines real or imaginary. If $K\neq 0$ then equation (10.8) can be expressed in the form

$$\frac{X^{1^2}}{(K/A)}+\frac{Y^{1^2}}{(K/B)}=1 \tag{10.9}$$

which is an ellipse depending on A and B.

If A and B are of opposite signs, that is, $ab-h^2<0$ then the equation (10.9) will represent a hyperbola. If $B=-A$, that is, $a+b=0$ then equation (10.9) will represent a rectangular hyperbola. Hence we have the following condition for the nature of the second degree equation (10.3) to represent in different forms.

The conditions are as follows:

(1) It will represent a pair of straight line if $abc+2fgh-af^2-bg^2-ch^2=0$.
(2) It will represent a circle if $a=b$ and $h=0$.
(3) It will represent a parabola if $ab-h^2=0$.
(4) It will represent an ellipse if $ab-h^2>0$.
(5) It will represent a hyperbola if $ab-h^2<0$.
(6) It will represent a rectangular hyperbola if $a+b=0$.

10.7 CENTRE OF THE CONIC GIVEN BY THE GENERAL EQUATION OF THE SECOND DEGREE

The general equation of the second degree in x and y is

$$ax^2 + 2hxy + by^2 + 2gx + 2fy + c = 0 \tag{10.10}$$

Since this equation has x and y terms, the centre is not at the origin. Let us suppose the centre is at (x_1, y_1). Let us now shift the origin to (x_1, y_1) without changing the direction of axes. Then $X = x + x_1$, $Y = y + y_1$. Then equation (10.10) takes the form

$$a(X + x_1)^2 + 2h(X + x_1)(Y + y_1) + b(Y + y_1)^2 + 2g(X + x_1) + 2f(Y + y_1) + c = 0$$

$$\begin{aligned}&\text{(i.e.) } aX^2 + 2hXY + bY^2 + 2X(ax_1 + hy_1 + g) + 2Y(hx_1 + fy_1 + c)\\ &+ax_1^2 + 2hx_1y_1 + by_1^2 + 2gx_1 + 2fy_1 + c = 0\end{aligned} \tag{10.11}$$

Since the origin is shifted to the point (x_1, y_1) with respect to new axes, the coefficient of X and Y in (10.11) should be zero.

$$\therefore ax_1 + hy_1 + g = 0 \tag{10.12}$$

$$hx_1 + by_1 + f = 0 \tag{10.13}$$

Solving these two equations, we get

$$\frac{x_1}{hf - bg} = \frac{y_1}{gh - af} = \frac{1}{ab - h^2}$$

Then the coordinates of the centre are $\left(\frac{hf - bg}{ab - h^2}, \frac{gh - af}{ab - h^2}\right)$.

If $ab - h^2 = 0$, then the conic is a parabola.

10.8 EQUATION OF THE CONIC REFERRED TO THE CENTRE AS ORIGIN

From the result obtained in Section 10.7, the equation of the conic referred to centre as the origin is $ax^2 + 2hxy + by^2 + C_1 = 0$, where,

$$\begin{aligned}C_1 &= ax_1^2 + 2hx_1y_1 + by_1^2 + 2gx_1 + 2fy_1 + c\\ &= x_1(ax_1 + by_1 + g) + y_1(hx_1 + by_1 + f) + gx_1 + fy_1 + c\end{aligned}$$

$= gx_1 + fy_1 + c$ [Using (10.12) and (10.13) of Section 10.8]

$$= g\left(\frac{hf-bg}{ab-h^2}\right) + f\left(\frac{gh-af}{ab-h^2}\right) + c$$

$$= \frac{abc+2fgh-af^2-bg^2-ch^2}{ab-h^2} = \frac{\Delta}{ab-h^2}$$

Hence, the equation of the conic referred to centre as origin is $ax^2 + 2hxy + by^2 + C_1 = 0$ where $C_1 = \frac{\Delta}{ab-h^2}$.

Note 10.8.1: If $f = ax^2 + 2hxy + by^2 + 2gx + 2fy + c = 0$ then

$$\frac{\partial f}{\partial x} = 2(ax + hy + g)$$

$$\frac{\partial f}{\partial y} = 2(hx + by + f)$$

Therefore, the coordinates of the centre of the conic are given by solving the equations $\frac{\partial f}{\partial x} = 0$ and $\frac{\partial f}{\partial y} = 0$.

Example 10.8.1

Find the nature of the conic and find its centre. Also write down the equation of the conic referred to centre as origin:

(i) $2x^2 - 5xy - 3y^2 - x - 4y + 5 = 0$

(ii) $5x^2 - 6xy + 5y^2 + 22x - 26y + 29 = 0$

Solution

(i) Given: $2x^2 - 5xy - 3y^2 - x - 4y + 5 = 0$

Here, $a = 2, b = -3, h = \frac{-5}{2}$

$ab - h^2 = -6 - \frac{25}{4} < 0$. Therefore, the conic is a hyperbola.

The coordinates of the centre are given by $\frac{\partial f}{\partial x} = 0, \frac{\partial f}{\partial y} = 0$

$$\therefore 4x - 5y - 1 = 0$$

$$-5x - 6y - 4 = 0$$

Solving these two equations, we get $x = \frac{-2}{7}, y = \frac{-3}{7}$.

Therefore, the coordinates of the centre are $\left(\frac{-2}{7}, \frac{-3}{7}\right)$. The equation of the ellipse referred to centre as origin is

$$2x^2 - 5xy - 3y^2 + C_1 = 0 \quad \text{where } C_1 = gx_1 + fy_1 + c$$

$$= \frac{-1}{2}\left(\frac{-2}{7}\right) - \frac{4}{2}\left(\frac{-3}{7}\right) + 6$$

$$= \frac{1}{7} + \frac{6}{7} + 6 = 7$$

Therefore, the equation of the ellipse referred to the centre is $2x^2 - 5xy - 3y^2 + 7 = 0$.

(ii) $5x^2 - 6xy + 5y^2 + 22x - 26y + 29 = 0$ (10.14)

$a = 5, b = 5, h = 3, ab - h^2 = 25 - 9 = 16 > 0$

Therefore, the given equation represents an ellipse. The coordinates of the centre are given by $\frac{\partial f}{\partial x} = 0, \frac{\partial f}{\partial y} = 0$.

$$10x - 6y + 22 = 0$$

$$-6x + 10y - 26 = 0$$

Solving these equations we get the centre as (1, 2). The equation of the conic referred to centre as origin is $5x^2 - 6xy + 5y^2 + C_1 = 0$ where $C_1 = gx_1 + fy_1 + c$.

$$C_1 = 11 \times (-1) - 13(2) + 29 = -11 - 26 + 29 = -8$$

Therefore, the equation of the ellipse is $5x^2 - 6xy + 5y - 8 = 0$ or $5x^2 - 6xy + 5y = 8$.

10.9 LENGTH AND POSITION OF THE AXES OF THE CENTRAL CONIC WHOSE EQUATION IS $ax^2 + 2hxy + by^2 = 1$

Given:

$$ax^2 + 2hxy + by^2 = 0 \tag{10.15}$$

The equation of the circle concentric with this conic and radius r is

$$x^2 + y^2 = r^2 \tag{10.16}$$

Homogeneousing equation (10.15) with the help of (10.16) we get

$$ax^2 + 2hxy + by^2 = \frac{x^2 + y^2}{r^2} \tag{10.17}$$

$$\text{(i.e.)} \quad \left(a - \frac{1}{r^2}\right)x^2 + 2hxy + \left(b - \frac{1}{r^2}\right)y^2 = 0 \tag{10.18}$$

The two lines given by above homogeneous equation will be considered only if the radius of the circle is equal to length of semi-major axis or semi-minor axis. The condition for that is

$$\left(a - \frac{1}{r^2}\right)\left(b - \frac{1}{r^2}\right) = h^2 \tag{10.19}$$

This is a quadratic equation in r^2 and so it has two roots namely r_1^2 and r_2^2. For an ellipse the values r_1^2 and r_2^2 are both positive and the lengths of the semi-axes are $2r_1$ and $2r_2$. For a hyperbola one of the values is positive and the other is negative, that is, say r_1^2 is positive and r_2^2 is negative. Then the length of transverse axis is $2r_1$ and the length of conjugate axis is $2\sqrt{-r_2^2}$.

Using equation (10.18) in equation (10.17), we get

$$\left(a - \frac{1}{r^2}\right)x^2 + 2hxy + \left(\frac{h^2}{a - \frac{1}{r^2}}\right)y^2 = 0$$

$$\left(a - \frac{1}{r^2}\right)^2 x^2 + 2h\left(a - \frac{1}{r^2}\right)xy + h^2y^2 = 0$$

$$\text{(i.e.)} \quad \left[\left(a - \frac{1}{r^2}\right)x + hy\right]^2 = 0$$

Then the equations of axes are

$$\left(a - \frac{1}{r_1^2}\right)x + hy = 0$$

$$\left(a - \frac{1}{r_2^2}\right)x + hy = 0$$

The eccentricity of the conic can be determined from the length of the axes.

10.10 AXIS AND VERTEX OF THE PARABOLA WHOSE EQUATION IS $ax^2 + 2hxy + by^2 + 2gx + 2fy + c = 0$

This equation will represent a parabola if $ab - h^2 = 0$.

Given:

$$ax^2 + 2hxy + by^2 + 2gx + 2fy + c = 0. \tag{10.20}$$

Then equation (10.20) can be expressed in the form

$$(px + qy + \lambda)^2 + 2gx + 2fy + c = 0$$

(i.e.) $$(px + qy - \lambda)^2 + 2(\lambda p - g)x + 2(\lambda q - f)y + \lambda^2 + c = 0 \tag{10.21}$$

We choose λ such that

$$px + qy + \lambda = 0 \tag{10.22}$$

and

$$2(\lambda p - g)x + 2f(\lambda q - f)y + \lambda^2 - c = 0 \tag{10.23}$$

are perpendicular to each other.

$$\left(\frac{-p}{q}\right)\left(-\frac{\lambda p - g}{\lambda q - f}\right) = -1$$

or $$\lambda = \frac{pg + rf}{p^2 + q^2}$$

Now equation (10.21) can be written as

$$\left(\frac{px + qy + \lambda}{\sqrt{p^2 + q^2}}\right)^2 (p^2 + q^2) = 2\left(\frac{(\lambda p - q)x + (\lambda q - f)y + \frac{1}{2}(\lambda^2 - c)}{\sqrt{(\lambda p - g)^2 + (\lambda q - f)^2}}\right)$$

$$\times \sqrt{(\lambda p - g)^2 + (\lambda q - f)^2}$$

(i.e.) $$\left(\frac{px + qy + \lambda}{\sqrt{p^2 + q^2}}\right)^2 = \frac{2\sqrt{(\lambda p - q)^2 + (\lambda q - f)^2}}{p^2 + q^2}$$

$$\left(\frac{(\lambda p - q)x + (\lambda q - f)y + \frac{1}{2}(\lambda^2 - c)}{\sqrt{(\lambda p - g)^2 + (\lambda q - f)^2}}\right)$$

Since the above equation represents a parabola, the axis of the parabola is $px + qy + \lambda = 0$ and the tangent at the vertex is $(\lambda p - q)x + (\lambda q - f)y + \frac{1}{2}(\lambda^2 - c) = 0$ and the length of the latus rectum is $\frac{2\sqrt{(\lambda p - q)^2 + (\lambda q - f)^2}}{p^2 + q^2}$ where $\lambda = \frac{pg + qf}{p^2 + q^2}$.

Example 10.10.1

Trace the conic $36x^2 + 24xy + 29y^2 - 72x + 126y + 81 = 0$.

Solution

Given:

$$36x^2 + 24xy + 29y^2 - 72x + 126y + 81 = 0 \tag{10.24}$$

$$a = 36, b = 29, h = 12$$

$$ab - h^2 = 36 \times 29 - 144 > 0$$

Therefore, the given conic represents an ellipse. The coordinates of the centre are given by $\frac{\partial f}{\partial x} = 0$ and $\frac{\partial f}{\partial y} = 0$.

$$72x + 24y - 72 = 0 \quad \text{or} \quad 3x + y - 3 = 0 \tag{10.25}$$

$$24x + 58y + 126 = 0 \quad \text{or} \quad 12x + 29y + 63 = 0 \tag{10.26}$$

Solving equation (10.25) and (10.26) we get $x = 2, y = -3$.

Therefore, the centre of the ellipse is $(2, -3)$.

The equation of this ellipse in standard form is $36x^2 + 24xy + 29y^2 + C_1 = 0$ where $C_1 = gx_1 + fy_1 + c$.

$$C_1 = -36(2) + 63(-3) + 81 = -180.$$

$$\therefore 36x^2 + 24xy + 29y^2 = 180$$

$$\text{(i.e.)} \quad \frac{36}{180}x^2 + \frac{24}{180}xy + \frac{29}{180}y^2 = 1$$

The length of the axes are given by

$$\left(a-\frac{1}{r^2}\right)\left(b-\frac{1}{r^2}\right)=h^2$$

$$\left(36-\frac{1}{r^2}\right)\left(29-\frac{1}{r^2}\right)=144$$

Solving for r^2 we get $r^2 = 9$ or 4

$\therefore\ r_1 = 3 =$ Length of the semi-major axis

$r_2 = 2 =$ Length of the semi-minor axis

The equation of the major axis is $\left(a-\frac{1}{r^2}\right)x+hy=0.$

$$\text{(i.e.)}\quad \left(\frac{36}{180}-\frac{20}{180}\right)x+\frac{12}{180}y=0 \quad \text{or} \quad 4x+3y=0.$$

The equation of minor axis is $\left(\frac{36}{180}-\frac{45}{180}\right)x+\frac{12}{180}y=0$ or $3x-4y=0.$

Referring to the centre the equation of major and minor axes are

$$4(x-2)+3(y+3)=0$$

$$3(x-2)-4(y+3)=0$$

(i.e.) $4x + 3y + 1 = 0$ and $3x - 4y - 18 = 0$

The major axis $4x + 3y + 1 = 0$ meets the axes at $\left(0,-\frac{1}{3}\right)$ and $\left(-\frac{1}{4},0\right)$.

The minor axis meets the axes at the point $\left(-\frac{9}{2},0\right)$ and $(6,0)$.

The conic meets the axis at points are given by

$$36x^2 - 72x + 81 = 0$$

$$\text{(i.e.)}\quad 4x^2 - 8x + 9 = 0$$

$x=\dfrac{8\pm\sqrt{64-4\times4\times9}}{2\times4}$ which are imaginary.

Therefore, the conic does not meet the x-axis. Similarly, by substituting $y = 0$ in equation (10.24) we get $29y^2 + 126y + 81 = 0$.

$$y=\frac{-126\pm\sqrt{126^2-4\times29\times81}}{58}=(0,3.85),(0,0.49)$$

Therefore, the conic intersects y-axis in real points.

Example 10.10.2

Trace the conic $x^2 + 4xy + y^2 - 2x + 2y - 6 = 0$.

Solution

$$ab - h^2 = 1 \times 1 - 4 = -3 < 0.$$

Therefore, the conic is a hyperbola. The coordinates of the centre are given by $\frac{\partial f}{\partial x} = 0, \frac{\partial f}{\partial y} = 0$.

$$\text{(i.e.)} \quad 2x + 4y - 2 = 0 \quad \text{or} \quad x + 2y - 1 = 0$$

$4x + 2y + 2 = 0$ or $2x + y + 1 = 0$. Solving these two equations we get the centre.

Therefore, the coordinates of the centre are $(-1, 1)$.

The equation of the conic referred to the centre as origin without changing the directions of the axis is $x^2 + 4xy + y^2 + C_1 = 0$ where $C_1 = gx + fy + c$

$$= (-1) \times (-1) + 1 \times 1 - 6 = -4$$

$$x^2 + 4yx + y^2 = 4 \text{ (i.e.) } \frac{x^2}{4} + xy + \frac{y^2}{4} = 1$$

Therefore, the lengths of the axes are given by

$$\left(\frac{1}{4} - \frac{1}{r^2}\right)\left(\frac{1}{4} - \frac{1}{r^2}\right) = \left(\frac{1}{2}\right)^2$$

$$\text{or} \quad \frac{1}{4} - \frac{1}{r^2} = \pm\frac{1}{2} \quad \text{or} \quad r^2 = \frac{4}{3} \text{ or } -4$$

Hence, the semi-transverse axis is $r = \frac{2}{\sqrt{3}}$.

The length of semi-conjugate axis is $\sqrt{r_2^2} = 2$.

Semi-latus rectum $= \frac{r_2^2}{r_1} = 2\sqrt{3}$.

The equation of the transverse axis is

$$\left(\frac{1}{4}-\frac{1}{r_1^2}\right)(x+1)+\frac{1}{2}(y-1)=0$$

$$\text{(i.e.)}\quad \left(\frac{1}{4}-\frac{3}{4}\right)(x+1)+\frac{1}{2}(y-1)=0$$

$$(x+1)-(y-1)=0$$

$$\text{(i.e.)}\quad x-y+2=0$$

The equation of conjugate axes is $\left(\frac{1}{4}+\frac{1}{4}\right)(x+1)+\frac{1}{2}(y+1)=0.$

$$\text{(i.e.)}\quad x+y=0$$

The points where the hyperbola meets the x-axis are given by $x^2-2x-6=0.$

$$x=\frac{2\pm\sqrt{4+24}}{2}=-1.7 \text{ or } 3.7 \text{ nearly}$$

Eccentricity is given by $4=\frac{4}{3}(e^2-1) \qquad \therefore e=2$

So $CS=CS'=Ge=\frac{4}{\sqrt{3}}.$

When the curve meets the y-axis, $x=0$

$\therefore y^2+2y-6=0$

$\therefore y=1.7$ or -3.7 nearly

Hence, the curve passes through the points (–1.7, 0), (3.7, 0), (0, 1.7) and (0, –3.7). The curve is traced in following figure:

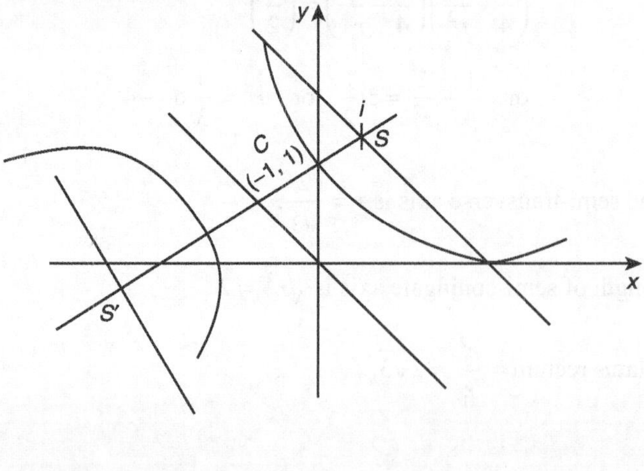

Example 10.10.3

Trace the conic $x^2 + 2xy + y^2 - 2x - 1 = 0$.

Solution

$$a = 1, b = 1, h = 1, g = -1, f = 0, c = -1$$

Here, $h^2 = ab$ and $abc + 2fgh - af^2 - bg^2 - ch^2 \neq 0$.

Therefore, the conic is a parabola.

The given equation can be written as $(x + y)^2 = 2x + 1$.

The equation can be written as

$$(x + y + 1)^2 = 2(\lambda + 1) + 2\lambda y + (\lambda^2 + 1) = 0 \qquad (10.27)$$

where λ is chosen such that $x + y + \lambda = 0$ and $2(\lambda + 1)x + 2\lambda y + (\lambda^2 + 1) = 0$ are perpendicular.

$$\therefore (-1)\left[\frac{-2(\lambda+1)}{2\lambda}\right] = -1 \text{ or } \lambda+1 = -\lambda \text{ or } \lambda = -\frac{1}{2}.$$

Now equation (10.27) can be written as $\left(x+y-\frac{1}{2}\right)^2 = \left(x-y+\frac{5}{4}\right)$

$$\text{(i.e.)} \quad \left(\frac{x+y-\frac{1}{2}}{\sqrt{2}}\right)^2 = \frac{1}{\sqrt{2}}\left(\frac{x-y+5}{\sqrt{2}}\right)$$

which is of the form $y^2 = 4ax$.

Therefore, lengths of latus rectum of the parabola is $\frac{1}{\sqrt{2}}$.

The axis of the parabola is $x + y - \frac{1}{2} = 0$

$$\text{or } 2x + 2y - 1 = 0$$

The equation of the tangent at the vertex is

$$x - y + \frac{5}{4} = 0 \text{ or } 4x - 4y + 5 = 0$$

Vertex of the parabola is $\left(-\frac{3}{8}, \frac{7}{8}\right)$.

When the parabola meets the x-axis, $y = 0$.

$$x^2 - 2x - 1 = 0 \quad \therefore x = 2.4, -0.4$$

Therefore, the points on the curve are $(-0.4, 0)$ and $(2.4, 0)$.

When the curve meets the y-axis, $x = 0$.

$$y^2 = 1 \text{ or } y = \pm 1$$

Hence $(0, -1)$ and $(0, 1)$ are points on the curve.
The graph of the curve is given as follows:

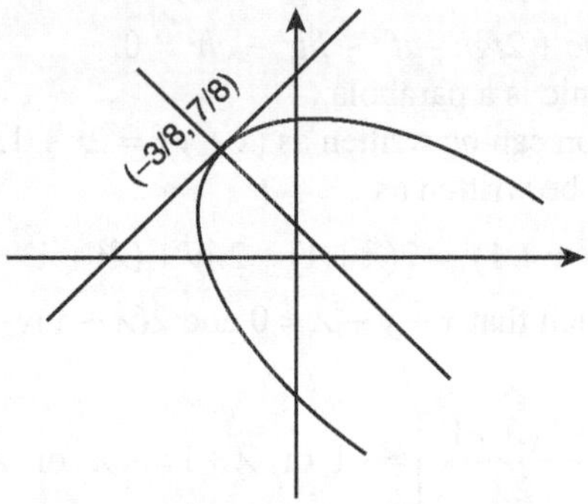

Exercises

Trace the following conics:

1. $9x^2 + 24xy + 16y^2 - 44x + 108y - 124 = 0$
2. $5x^2 - 6xy + 5y^2 + 22x - 26y + 29 = 0$
3. $32x^2 + 52xy - 7y^2 - 64x - 52y - 148 = 0$
4. $x^2 + 24xy + 16y^2 - 86x + 52y - 139 = 0$
5. $43x^2 + 48xy + 57y^2 + 10x + 180y + 25 = 0$
6. $x^2 - 4xy + 4y^2 - 6x - 8y + 1 = 0$
7. $x^2 + 2xy + y^2 - 4x - y + 4 = 0$
8. $5x^2 - 2xy + 5y^2 + 2x - 10y - 7 = 0$
9. $22x^2 - 12xy + 17y^2 - 112x + 92y + 178 = 0$.

CHAPTER

11 Three Dimension

11.1 RECTANGULAR COORDINATE AXES

To locate a point in a plane, two numbers are necessary. We know that any point in the xy plane can be represented as an ordered pair (a, b) of real numbers where a is called x-coordinate of the point and b is called the y-coordinate of the point. For this reason, a plane is called two dimensional.

To locate a point in space, three numbers are required. Any point in space is represented by an ordered triple (a, b, c) of real numbers. To represent a point in space we first choose a fixed point 'O' (called the origin) and then three directed lines through O which are perpendicular to each other (called coordinate axes) and labelled x-axis, y-axis as being horizontal and z-axis as vertical and we take the orientation of the axes. In order to do this, we first choose a fixed point O. In looking at the figure, you can think of y- and z-axes as lying in the plane of the paper and x-axis as coming out of the paper towards y-axis. The direction of z-axis is determined by the neighbourhood rule. If you curl the fingers of your right-hand around the z-axis in the direction of a 90° counter clockwise rotation from the positive x-axis to the positive y-axis then your thumb points in the positive direction of the z-axis.

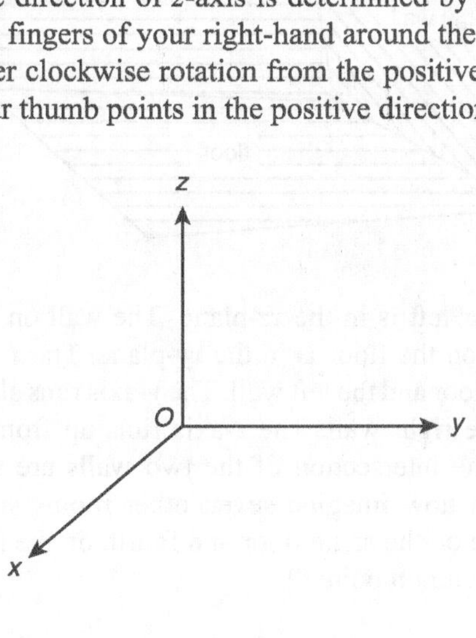

The three coordinate axes are determined by the three coordinate planes. The *xy*-plane is the plane that contains *x* and *y*-axes, the *yz*-plane contains *y* and *z*-axes and the *xz*-plane contains *x*- and *z*-axes.

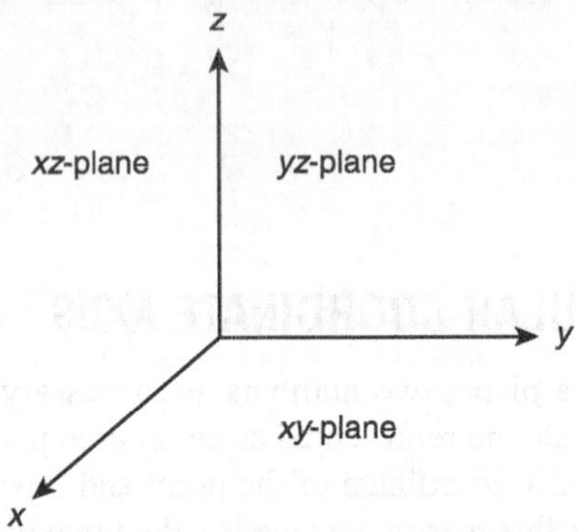

These three coordinate planes divide the space into eight parts called octants. The first octant is determined by the positive axes. Look at any bottom corner of a room and call the corner as origin.

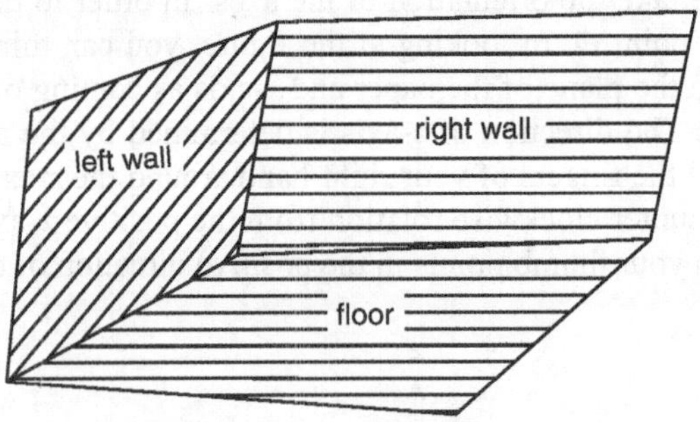

The wall on your left is in the *xz*-plane. The wall on your right is in the *yz*-plane. The wall on the floor is in the *xy*-plane. The *x*-axis runs along the intersection of the floor and the left wall. The *y*-axis runs along the intersection of the floor and the right wall. The *z*-axis runs up from the floor towards the ceiling along the intersection of the two walls are situated in the first octant and you can now imagine seven other rooms situated in the other seven octants (three on the same floor and fourth on the floor below), all are connected by the common point *O*.

If P is any point in space, and a be the directed distance in the first octant from the yz-plane to P. Let the directed distance from the xz-plane be b and let c be the distance from xy-plane to P.

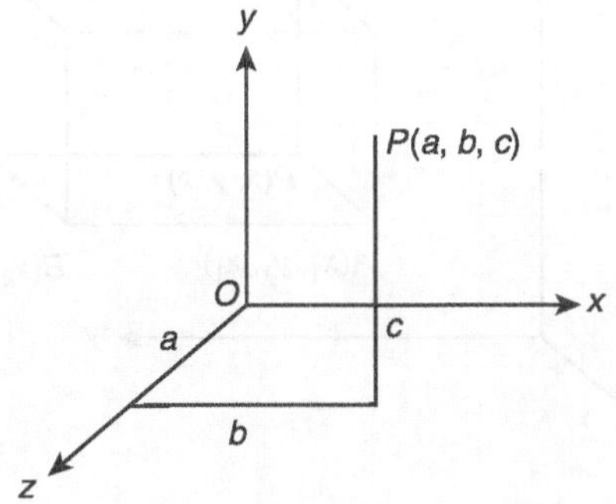

We represent the point P by the ordered triple (a, b, c) of real numbers and we call a, b and c the coordinates of P. a is the x-coordinate, b is the y-coordinate and c is the z-coordinate. Thus, to locate the point (a, b, c) we can start at the origin O and move a units along x-axis then b units parallel to y-axis and c units parallel to the z-axis as shown in the above figure.

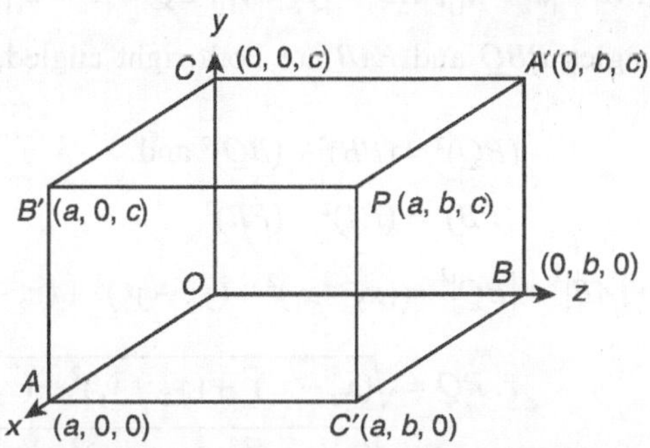

The point $P(a, b, 0)$ determines a rectangular box as in the above figure. If we drop a perpendicular from P to the xy-plane we get a point C' with coordinates $P(a, b, 0)$ called the projection of P on the xy-plane. Similarly, B' $(a, 0, c)$ and $A'(0, b, c)$ are the projections of P on xz-plane and yz-plane, respectively. The set of all ordered triples of real numbers is the Cartesian product $R \times R \times R = \{(x, y, z) | x, y, z \in R\}$, which is R^3. We have a one-to-one correspondence between the points P in space and ordered triples (a, b, c) in R^3. It is called a three-dimensional rectangular coordinate system. We notice that in terms of coordinates, the first octant can be described as the set $\{(x, y, z) | x \geq 0, y \geq 0, z \geq 0\}$.

11.2 FORMULA FOR DISTANCE BETWEEN TWO POINTS

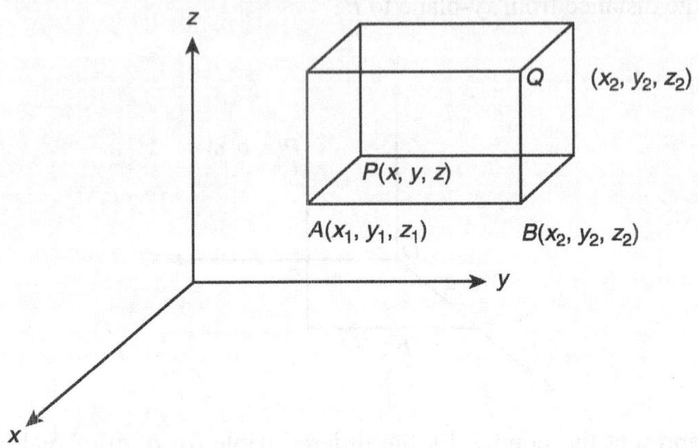

Consider a rectangular box, where P and Q are the opposite corners and the faces of the box are parallel to the coordinate planes. If $A(x_1, y_1, z_1)$ and $B(x_2, y_2, z_2)$ are the vertices of the box indicated in the above figure, then

$$|PA| = |x_2 - x_1|,\ |AB| = |y_2 - y_1|,\ BQ = |z_2 - z_1|$$

Since the triangles PBQ and PAB are both right angled, by Pythagoras theorem,

$$(PQ)^2 = (PB)^2 + (BQ)^2 \text{ and}$$

$$(PB)^2 = (PA)^2 + (PB)^2$$

$$\therefore |PQ|^2 = |PA|^2 + |AB|^2 + |BQ|^2 = (x_2 - x_1)^2 + (y_2 - y_1)^2 + (z_2 - z_1)^2$$

$$\therefore PQ = \sqrt{(x_2 - x_1)^2 + (y_2 - y_1)^2 + (z_2 - z_1)^2}$$

$$= \sqrt{(x_1 - x_2)^2 + (y_1 - y_2)^2 + (z_1 - z_2)^2}$$

Example 11.2.1

Find the distance between the points (2, 1, –5) and (4, –7, 6).

Solution

The distance between the points (2, 1, –5) and (4, –7, 6) is $PQ = \sqrt{(2-4)^2 + (1+7)^2 + (-5-6)^2} = \sqrt{4 + 64 + 121} = \sqrt{189}$ units

Aliter:

Let $P(x_1, y_1, z_1)$ and (x_2, y_2, z_2) be two points. Draw PA, QB perpendicular to xy-plane. Then the coordinates of A and B are $(x_1, y_1, 0)$ and $(x_2, y_2, 0)$.

(i.e.) (x_1, y_1) and (x_2, y_2) in the xy-plane.

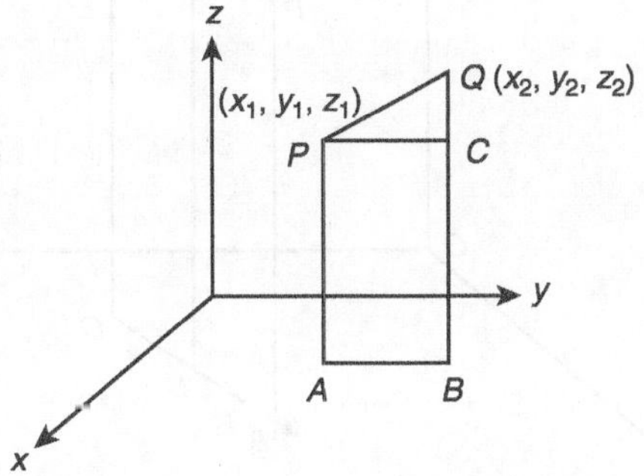

$\therefore AB^2 = (x_2 - x_1)^2 + (y_2 - y_1)^2$

Draw PC perpendicular to xy-plane. PA and PB being perpendicular to xy-plane, PA and QB are also perpendicular to AB.

$\therefore PABC$ is a rectangle and so $PC = AB$ and $PA = CB$.

$$\text{From triangle } PCQ,\ PQ^2 = PC^2 + CQ^2 = MN^2 + (AQ - AC)^2$$
$$= MN^2 + (CQ - PA)^2$$
$$= (x_2 - x_1)^2 + (y_2 - y_1)^2 + (z_2 - z_1)^2$$
$$\therefore PQ = \sqrt{(x_2 - x_1)^2 + (y_2 - y_1)^2 + (z_2 - z_1)^2}$$

Example 11.2.2

If O is the origin and P is the point (x, y, z) then $OP^2 = x^2 + y^2 + z^2$ or $OP = \sqrt{x^2 + y^2 + z^2}$.

11.2.1 Section Formula

The coordinates of a point that divides the line joining the points (x_1, y_1, z_1) and (x_2, y_2, z_2) are in the ratio $l{:}m$.

Let $R(x, y, z)$ divide the line joining the points $P(x_1, y_1, z_1)$ and $Q(x_2, y_2, z_2)$ in the ratio $l{:}m$.

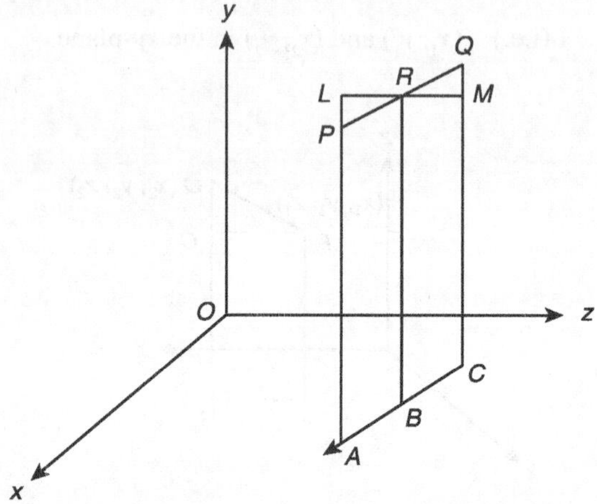

Draw PL, QC and RB perpendicular to the xy-plane. Then ACB is a straight line since projection of a straight line on a plane is a straight line. Through R draw a straight line LRM parallel to ACB to meet AP (produced) in A and CQ in M. Then triangles LPR and MRQ are similar.

$$\frac{PL}{MQ} = \frac{PR}{RQ} = \frac{l}{m} \tag{11.1}$$

However, $PL = LA - AP = BR - AP = z - z_1$

$$MQ = CQ - CM = CQ - BR = z_2 - z$$

Therefore, from (11.1), $\dfrac{z - z_1}{z_2 - z} = \dfrac{l}{m} \Rightarrow m(z - z_1) = l(z_2 - z)$

$$mz - mz_1 = lz_2 - lz \Rightarrow (l + m)z = lz_2 + mz_1$$

$$\Rightarrow z = \frac{lz_2 + m_1 z_1}{l + m}$$

Similarly, by drawing perpendiculars on xz and yz planes we can prove that $y = \dfrac{ly_2 + my_1}{l + m}, x = \dfrac{lx_2 + mx_1}{l + m}$.

Therefore, the coordinates of R are $\left(\dfrac{lx_2 + mx_1}{l + m}, \dfrac{ly_2 + my_1}{l + m}, \dfrac{lz_2 + mz_1}{l + m}\right)$.

Note 11.2.1.1: If R' divides PQ externally in the ratio $l{:}m$ then, $\frac{PR'}{R'Q} = \frac{l}{m}$ or $\frac{PQ'}{QR'} = \frac{l}{-m}$.

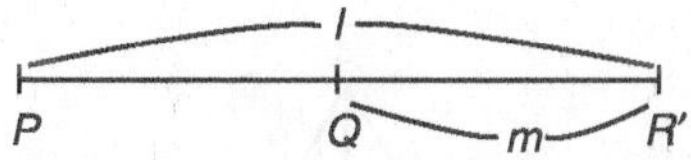

∴ Therefore, change m into $-m$ to get the coordinates of R', the external point of division. The coordinates of external point of division are $\left(\frac{lx_2 - mx_1}{l-m}, \frac{ly_2 - my_1}{l-m}, \frac{lz_2 - mz_1}{l-m}\right)$.

Note 11.2.1.2: To find the midpoint of PQ take $l{:}m = 1{:}1$.

Then the coordinates of midpoint are $\left(\frac{x_1 + x_2}{2}, \frac{y_1 + y_2}{2}, \frac{z_1 + z_2}{2}\right)$.

11.3 CENTROID OF TRIANGLE

Let ABC be a triangle with vertices $A(x_1, y_1, z_1)$, $B(x_2, y_2, z_2)$ and $C(x_3, y_3, z_3)$.

Then the midpoint of BC is $D\left(\frac{x_2 + x_3}{2}, \frac{y_2 + y_3}{2}, \frac{z_2 + z_1}{2}\right)$.

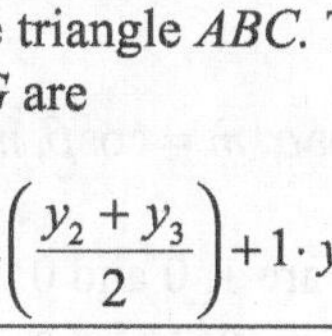

Let G be the centroid of the triangle ABC. Then G divides AD in the ratio 2:1. Then the Coordinates of G are

$$\left(\frac{2\left(\frac{x_2 + x_3}{2}\right) + 1\cdot x_1}{2+1}, \frac{2\left(\frac{y_2 + y_3}{2}\right) + 1\cdot y_1}{2+1}, \frac{2\left(\frac{z_2 + z_3}{2}\right) + 1\cdot z_1}{2+1}\right).$$

$$\text{(i.e.)} \quad \left(\frac{x_1 + x_2 + x_3}{3}, \frac{y_1 + y_2 + y_3}{3}, \frac{z_1 + z_2 + z_3}{3}\right)$$

Hence, the centroid of ΔABC is $\left(\frac{x_1 + x_2 + x_3}{3}, \frac{y_1 + y_2 + y_3}{3}, \frac{z_1 + z_2 + z_3}{3}\right)$.

11.4 CENTROID OF TETRAHEDRON

Let *OBCD* be a tetrahedron with vertices (x_i, y_i, z_i), $i = 1, 2, 3, 4$.

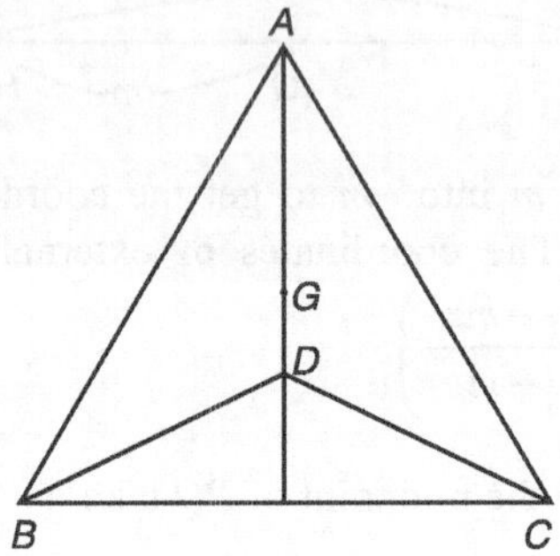

The centroid of the tetrahedron divides *AD* in the ratio 3:1. Therefore, the coordinates of *G* are

$$\left(\frac{x_1 + x_2 + x_3 + x_4}{4}, \frac{y_1 + y_2 + y_3 + y_4}{4}, \frac{z_1 + z_2 + z_3 + z_4}{4}\right)$$

11.5 DIRECTION COSINES

Direction cosines in three-dimensional coordinate geometry play a role similar to slope in two-dimensional coordinate geometry.

Definition 11.5.1: If a straight line makes angles α, β and γ with the positive directions of *x*-, *y*- and *z*-axes then $\cos\alpha$, $\cos\beta$ and $\cos\gamma$ are called the direction cosines of the line. The directional cosines are denoted by *l*, *m* and *n*.

$$\therefore\ l = \cos\alpha,\ m = \cos\beta,\ n = \cos\gamma.$$

The direction cosines of *x*-axis are 1, 0 and 0.

The direction cosines of *y*-axis are 0, 1 and 0.

The direction cosines of *z*-axis are 0, 0 and 1.

If *O* is the origin and *P*(*x*, *y*, *z*) be any point in space and *OP* = *r*, then the direction cosines of the line are *lr*, *mr*, *nr*.

Let *O* be the origin and *P*(*x*, *y*, *z*) is any point in space. Draw *PN* perpendicular to *XOY* plane. Draw *NL*, *NM* parallel to *y*- and *x*-axes.

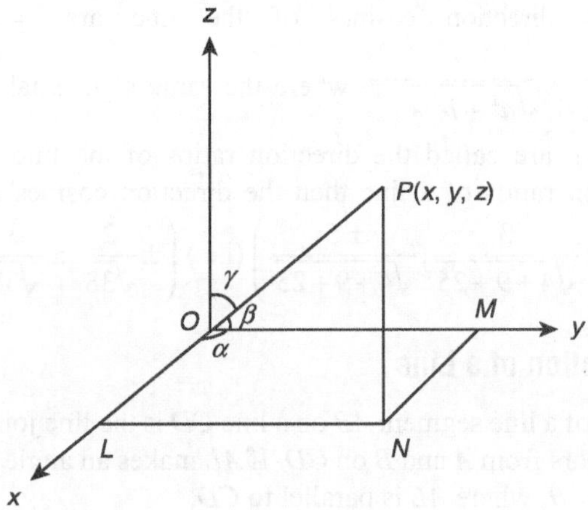

Then $OL = x_1$, $OM = y_1$, $PN = z_1$.

Also, $\frac{OM}{OP} = \cos\beta$(i.e.) $y = r\cos\beta$

Similarly, $x = r\cos\alpha$ and $z = r\cos\gamma$.

Then the coordinates of P are

$(r\cos\alpha, r\cos\beta, r\cos\gamma)$ (i.e.) (lr, mr, nr).

Note 11.5.2: The direction cosines of the line OP are $\frac{x}{r}, \frac{y}{r}, \frac{z}{r}$.

Note 11.5.3: If $OP = 1$ unit then the direction cosines of the line are (x, y, z). That is, the coordinates of the point P are the same as the direction cosines of the line OP.

Note 11.5.4: If $OP = 1$ unit and P is the point (x, y, z) then $OP^2 = x^2 + y^2 + z^2$ or $x^2 + y^2 + z^2 = 1$.

$$\therefore l^2 + m^2 + n^2 = 1$$

Therefore, direction cosines satisfy the property $\cos^2\alpha + \cos^2\beta + \cos^2\gamma = 1$.

11.5.1 Direction Ratios

Suppose a, b and c are three numbers proportional to l_1, m_1 and n_1 (the direction cosines of a line), then

$$\frac{l}{a} = \frac{m}{b} = \frac{n}{c} = \pm\frac{\sqrt{l^2 + m^2 + n^2}}{\sqrt{a^2 + b^2 + c^2}} = \pm\frac{1}{\sqrt{a^2 + b^2 + c^2}}$$

$$\therefore l = \pm\frac{a}{\sqrt{a^2 + b^2 + c^2}}, m = \pm\frac{b}{\sqrt{a^2 + b^2 + c^2}}, n = \pm\frac{c}{\sqrt{a^2 + b^2 + c^2}}$$

Therefore, the direction cosines of the line are $\pm\frac{a}{\sqrt{a^2+b^2+c^2}}$, $\pm\frac{b}{\sqrt{a^2+b^2+c^2}}, \pm\frac{c}{\sqrt{a^2+b^2+c^2}}$ where the same sign is taken throughout.

Here a, b and c are called the direction ratios of the line. If 2, 3 and 5 are the direction ratios of a line then the direction cosines of the line are $\left(\frac{\pm 2}{\sqrt{4+9+25}}, \pm\frac{3}{\sqrt{4+9+25}}, \frac{\pm 5}{\sqrt{4+9+25}}\right)$ (i.e.) $\left(\pm\frac{2}{\sqrt{38}}, \pm\frac{3}{\sqrt{38}}, \pm\frac{5}{\sqrt{38}}\right)$.

11.5.2 Projection of a Line

The projection of a line segment *AB* on a line *CD* is the line joining the feet of the perpendiculars from *A* and *B* on *CD*. If *AL* makes an angle θ with the line *CD* then $\underline{|LAB} = \theta$, where *AL* is parallel to *CD*.

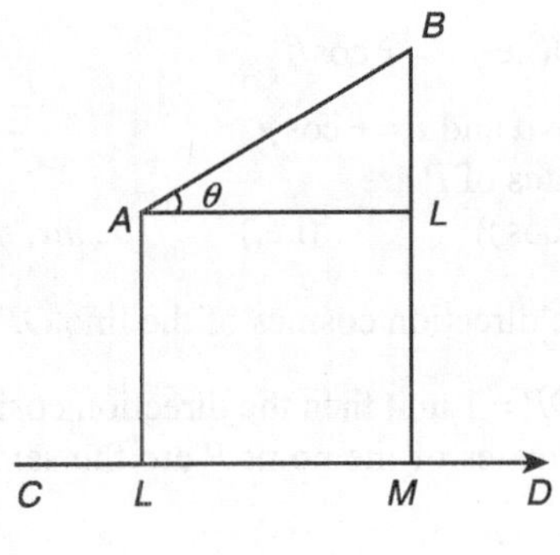

$$AL = AB\cos\theta$$
$$LM = AB\cos\theta$$

Therefore, the projection of *AB* on *CD* is $LM = AB\cos\theta$.

Note 11.5.2.1: The projection of broken lines *AB*, *BC* and *CD* on the line *CD* is *LM*, *MN* and *ND*.

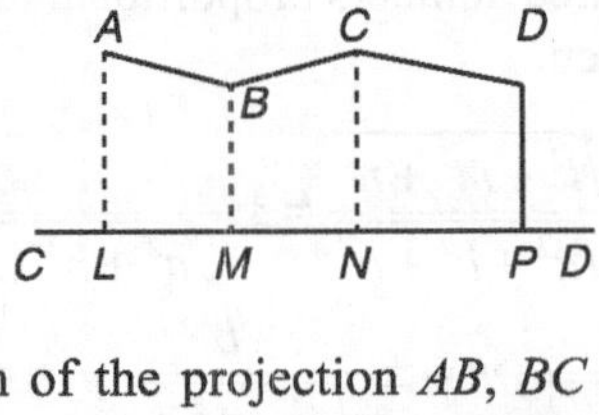

∴ Therefore, the sum of the projection *AB*, *BC* and *CD* is $LM + MN + ND = LP$.

11.5.3 Direction Cosines of the Line Joining Two Given Points

Let $P(x_1, y_1, z_1)$ and $Q(x_2, y_2, z_2)$ be the given points. We easily see that the projection of PQ on x-, y- and z-axes are $x_2 - x_1, y_2 - y_1$ and $z_2 - z_1$. However, the projections of PQ on x-, y- and z-axes are also $PQ\cos\alpha$, $PQ\cos\beta$ and $PQ\cos\gamma$.

$$\therefore\ PQ \cdot l = x_2 - x_1 \qquad\qquad \therefore\ l = \frac{x_2 - x_1}{PQ}$$

In addition, $m = \dfrac{y_2 - y_1}{PQ}, n = \dfrac{z_2 - z_1}{PQ}$. Since PQ is of constant length, the direction ratios of PQ are

$$(x_2 - x_1, y_2 - y_1, z_2 - z_1).$$

11.5.4 Angle between Two Given Lines

Let (l_1, m_1, n_1), (l_2, m_2, n_2) be the direction cosines of the two given lines namely AB and CD. Through O draw lines parallel to AB and CD and take points P and Q such that $OP = OQ = 1$ unit.

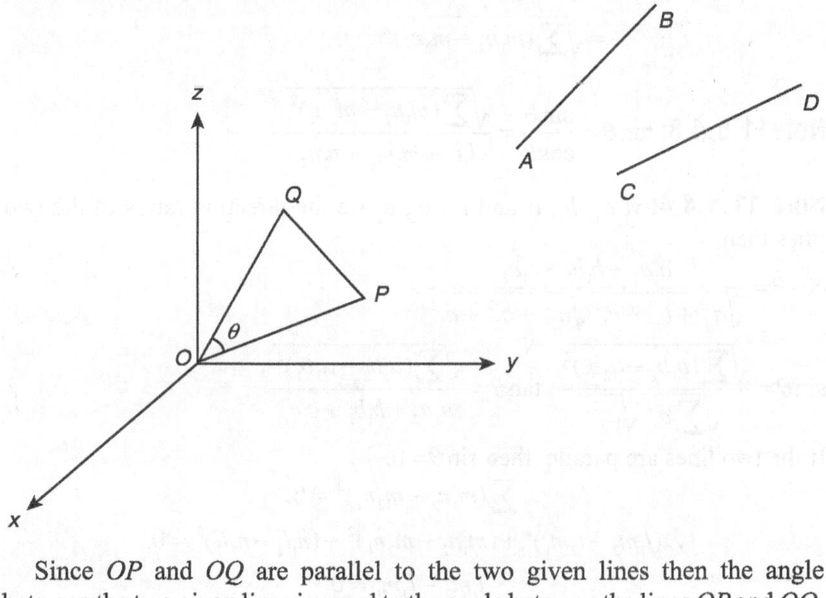

Since OP and OQ are parallel to the two given lines then the angle between the two given lines is equal to the angle between the lines OP and OQ. Since $OP = OQ = 1$ unit, the coordinates of P and Q are (l_1, m_1, n_1) and (l_2, m_2, n_2). Let $\angle POQ = \theta$.

Then $PQ^2 = OP^2 + OQ^2 - 2 \cdot OP \cdot OQ \cos\theta = 1 + 1 - 2 \cdot 1 \cdot 1 \cos\theta$

$$PQ^2 = 2 - 2\cos\theta \tag{11.2}$$

Also, $PQ^2 = (l_2 - l_1)^2 + (m_2 - m_1)^2 + (n_2 - n_1)^2$

$$= (l_1^2 + m_1^2 + n_1^2) + (l_2^2 + m_2^2 + n_2^2) - 2(l_1 l_2 + m_1 m_2 + n_1 n_2)$$

$$PQ^2 = 2 - 2(l_1 l_2 + m_1 m_2 + n_1 n_2) \tag{11.3}$$

From (11.2) and (11.3),

$$2 - 2\cos\theta = 2 - 2(l_1 l_2 + m_1 m_2 + n_1 n_2)$$

$$\cos\theta = l_1 l_2 + m_1 m_2 + n_1 n_2 \tag{11.4}$$

Note 11.5.4.1: If the two lines are perpendicular then $\theta = 90°$ and $\cos 90° = 0$.

$$\therefore \text{ from (11.3), } l_1 l_2 + m_1 m_2 + n_1 n_2 = 0$$

Note 11.5.4.2: $\sin\theta = \sqrt{1 - \cos^2\theta}$

$$= \sqrt{(l_1^2 + m_1^2 + n_1^2)(l_2^2 + m_2^2 + n_2^2) - (l_1 l_2 + m_1 m_2 + n_1 n_2)^2}$$

$$= \sqrt{(l_1 m_2 - l_2 m_1)^2 + (m_1 n_2 - m_2 n_1)^2 + (n_1 l_2 - n_2 l_1)^2}$$

$$= \sqrt{\sum (m_1 n_2 - m_2 n_1)^2}$$

Note11.5.4.3: $\tan\theta = \dfrac{\sin\theta}{\cos\theta} = \dfrac{\sqrt{\sum (m_1 n_2 - m_2 n_1)^2}}{l_1 l_2 + m_1 m_2 + n_1 n_2}$

Note 11.5.4.4: If a_1, b_1, c_1 and a_2, b_2, c_2 are the direction ratios of the two lines then

$$\cos\theta = \frac{a_1 a_2 + b_1 b_2 + c_1 c_2}{\sqrt{a_1^2 + b_1^2 + c_1^2}\sqrt{a_2^2 + b_2^2 + c_2^2}}.$$

$$\sin\theta = \frac{\sqrt{\sum (a_1 b_2 - a_2 b_1)^2}}{\sqrt{\sum a_1^2}\sqrt{a_2^2}}, \quad \tan\theta = \frac{\sqrt{\sum (a_1 b_2 - a_2 b_1)^2}}{a_1 a_2 + b_1 b_2 + c_1 c_2}$$

If the two lines are parallel then $\sin\theta = 0$.

$$\therefore \sum (m_1 n_2 - m_2 n_1)^2 = 0.$$

$$\therefore (l_1 m_2 - l_2 m_1)^2 + (m_1 n_2 - m_2 n_1)^2 + (n_1 l_2 - n_2 l_1)^2 = 0.$$

$$l_1 m_2 - l_2 m_1 = 0$$

$$m_1 n_2 - m_2 n_1 = 0$$

$$n_1 l_2 - n_2 l_1 = 0$$

$$\therefore \frac{l_1}{l_2} = \frac{m_1}{m_2}$$

$$\frac{m_1}{m_2} = \frac{n_1}{n_2}$$

$$\frac{n_1}{n_2} = \frac{l_1}{l_2}$$

$$\therefore \frac{l_1}{l_2} = \frac{m_1}{m_2} = \frac{n_1}{n_2}$$

Also, if a_1, b_1, c_1 and a_2, b_2, c_2 are the direction ratios of two parallel lines then $\frac{a_1}{a_2} = \frac{b_1}{b_2} = \frac{c_1}{c_2}$.

ILLUSTRATIVE EXAMPLES

Example 11.1

Show that the points (–2, 5, 8), (–6, 7, 4) and (–3, 4, 4) form a right-angled triangle.

Solution

The given points are $A(-2, 5, 8)$, $B(-6, 7, 4)$ and $C(-3, 4, 4)$.

$$AB^2 = (x_1 - x_2)^2 + (y_1 - y_2)^2 + (z_2 - z_1)^2 = (-2+6)^2 + (5-7)^2 + (8-4)^2$$
$$= 16 + 4 + 16 = 36$$

$$BC^2 = (-6+3)^2 + (7-4)^2 + (4-4)^2 + 0 = 9 + 9 = 18$$

$$AC^2 = (-2+3)^2 + (5-4)^2 + (8-4)^2 = 1 + 1 + 16 = 18$$

Since $BC^2 + AC^2 = AB^2$, the triangle is right angled. Since $BC = AC$, the triangle is also isosceles.

Example 11.2

Show that the points (3, 2, 5), (2, 1, 3), (–1, 2, 1) and (0, 3, 3) taken in order form a parallelogram.

Solution

Let the four points be $A(3, 2, 5)$, $B(2, 1, 3)$, $C(-1, 2, 1)$ and $D(0, 3, 3)$.

Then,

$$AB^2 = (3-2)^2 + (2-1)^2 + (5-3)^2 = 1+1+4 = 6.$$
$$BC^2 = (2+1)^2 + (1-2)^2 + (3-1)^2 = 9+1+4 = 14.$$
$$CD^2 = (-1-0)^2 + (2-3)^2 + (1-3)^2 = 1+1+4 = 6.$$
$$AD^2 = (3-0)^2 + (2-3)^2 + (5-3)^2 = 9+1+4 = 14$$

Since $AB = CD$ and $BC = AD$, the four points form a parallelogram.

Aliter:

The midpoint of AC is (1, 2, 3). The midpoint of BD is (1, 2, 3).

Therefore, in the figure $ABCD$, the diagonals bisect each other. Hence $ABCD$ is a parallelogram.

Example 11.3

Show that the points (−1, 2, 5), (1, 2, 3) and (3, 2, 1) are collinear.

Solution

The three given points are $A(-1, 2, 5)$, $B(1, 2, 3)$ and $C(3, 2, 1)$.

$$AB^2 = (-1-1)^2 + (2-2)^2 + (5-3)^2 = 4+0+4 = 8$$
$$BC^2 = (1-3)^2 + (2-2)^2 + (3-1)^2 = 4+0+4 = 8$$
$$AC^2 = (-1-3)^2 + (2-2)^2 + (5-1)^2 = 16+0+16 = 32$$
$$\therefore AB = 2\sqrt{2}, BC = 2\sqrt{2} \text{ and } AC = 4\sqrt{2}$$
$$\therefore AB + BC = AC$$

Hence, the three given points are collinear.

Example 11.4

Show that the points (3, 2, 2), (−1, 1, 3), (0, 5, 6) and (2, 1, 2) lie on a sphere whose centre is (1, 3, 4). Also find its radius.

Solution

Let the given points be $S(2, 1, 2)$, $P(3, 2, 2)$, $Q(-1, 1, 3)$, $R(0, 5, 6)$ and $C(1, 3, 4)$.

$$CP^2 = (1-3)^2 + (3-2)^2 + (4-2)^2 = 4+1+4 = 9$$
$$CQ^2 = (1+1)^2 + (3-1)^2 + (4-3)^2 = 4+4+1 = 9$$
$$CR^2 = (1-0)^2 + (3-5)^2 + (4-6)^2 = 1+4+4 = 9$$
$$CS^2 = (1-2)^2 + (3-1)^2 + (4-2)^2 = 1+4+4 = 9$$
$$\therefore CP = CQ = CP = CS = \sqrt{9} = 3.$$

Therefore, the points P, Q, R and S lie on a sphere whose centre is $C(1, 3, 4)$ and whose radius is 3 units.

Example 11.5

Find the ratio in which the straight line joining the points $(1, -3, 5)$ and $(7, 2, 3)$ is divided by the coordinate planes.

Solution

Let the line joining the points $P(1, -3, 5)$ and $Q(7, 2, 3)$ be divided by XY, YZ and ZX planes in the ratio l:1, m:1 and n:1, respectively. When the line PQ meets the XY planes, the Z-coordinates of the point of meet is 0.

$$\therefore \frac{7l+1}{l+1} = 0$$

$$\therefore 7l+1 = 0 \text{ or } l = \frac{-1}{7}.$$

(i.e.) The ratio in which PQ divides the plane YZ-plane is 1:7 externally.

Similarly, $\dfrac{2m-3}{7m+1} = 0.$

$$\therefore 2m-3 = 0 \text{ or } m = \frac{3}{2}$$

Since XZ-plane divides PQ in the ratio 3:2 internally.

Also $\dfrac{3n+5}{n+1} = 0$ or $n = \dfrac{-5}{3}$. Therefore, XY-plane divides PQ in the ratio 5:3 externally.

Example 11.6

P and Q are the points (3, 4, 12) and (1, 2, 2). Find the coordinates of the points in which the bisector of the angle POQ meets PQ.

Solution

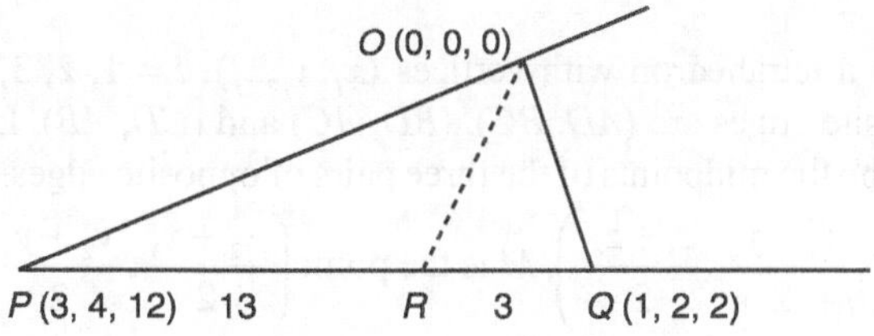

$$QP = \sqrt{9+16+144} = \sqrt{169} = 13$$

$$OQ = \sqrt{1+4+4} = \sqrt{9} = 3$$

We know that $\dfrac{PR}{RQ} = \dfrac{13}{3}$

R divides PQ internally in the ratio 13:3 and S divides PQ externally in the ratio 13:3.

Therefore, the coordinates of R are $\left(\dfrac{13\times1+3\times3}{16}, \dfrac{26+12}{16}, \dfrac{26+24}{16}\right)$.

(i.e.) $\left(\dfrac{11}{8}, \dfrac{19}{8}, \dfrac{25}{8}\right)$

S divides PQ externally in the ratio 13:3.

Therefore, the coordinates of S are $\left(\dfrac{13-9}{10}, \dfrac{26-12}{10}, \dfrac{26-24}{10}\right)$.

(i.e.) $\left(\dfrac{2}{5}, \dfrac{7}{5}, \dfrac{1}{5}\right)$

Example 11.7

Prove that the three lines which join the midpoints of opposite edges of a tetrahedron pass through the same point and are bisected at that point.

Solution

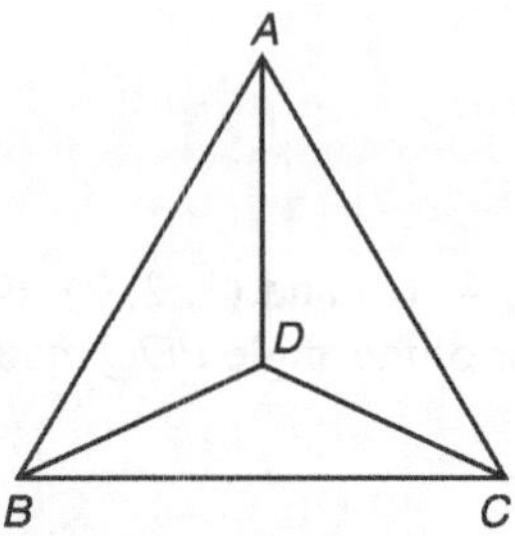

Let $ABCD$ be a tetrahedron with vertices (x_i, y_i, z_i), $i = 1, 2, 3, 4$. The three pairs of opposite edges are (AD, BC), (BD, AC) and (CD, AB). Let (L, N), (P, Q) and (R, S) be the midpoints of the three pairs of opposite edges. Then L is the point $\left(\dfrac{x_1+x_4}{2}, \dfrac{y_1+y_4}{2}, \dfrac{z_1+z_4}{2}\right)$, M is the point $\left(\dfrac{x_2+x_3}{2}, \dfrac{y_2+y_3}{2}, \dfrac{z_2+z_3}{2}\right)$.

The midpoint of LM is $\left(\frac{x_1+x_2+x_3+x_4}{4}, \frac{y_1+y_2+y_3+y_4}{4}, \frac{z_1+z_2+z_3+z_4}{4}\right)$.

By symmetry, this is also the midpoint of the lines PQ and RS.

Therefore, the lines LM, PQ and RS are concerned and are bisected at that point.

Example 11.8

A plane triangle of sides a, b and c is placed so that the midpoints of the sides are on the axes. Show that the lengths l, m and n intercepted on the axes are given by $8l^2 = b^2 + c^2 - a^2$, $8m^2 = c^2 + a^2 - b^2$ and $8n^2 = a^2 + b^2 - c^2$ and that the coordinates of the vertices of the triangle are $(-l, m, n)$, $(l, -m, n)$ and $(l, m, -n)$.

Solution

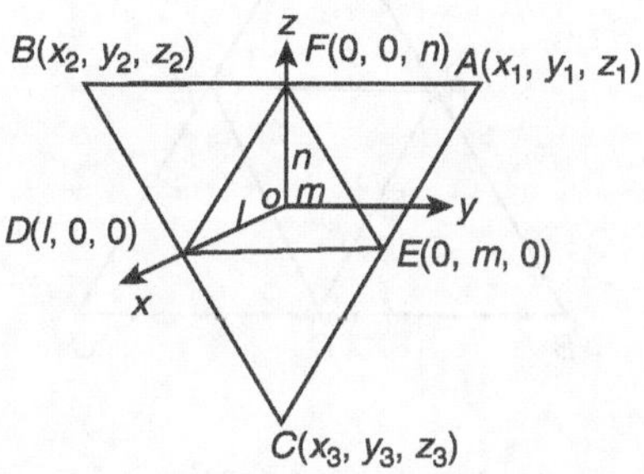

Let D, E and F be the midpoints of the sides BC, CA and AB, respectively. D, E and F are the points $(l, 0, 0)$, $(0, m, 0)$, $(0, 0, n)$, respectively. Let A, B and C be the points (x_1, y_1, z_1), (x_2, y_2, z_2) and (x_3, y_3, z_3), respectively.

Then,
$$\frac{x_2+x_3}{2} = l, \frac{y_2+y_3}{2} = 0, \frac{z_2+z_3}{2} = 0$$
$$\frac{x_3+x_1}{2} = 0, \frac{y_3+y_1}{2} = m, \frac{z_3+z_1}{2} = 0$$
$$\frac{x_1+x_2}{2} = 0, \frac{y_1+y_2}{2} = 0, \frac{z_1+z_2}{2} = n$$
$$x_1 + x_2 + x_3 = l,\ x_1 + x_3 = 0, x_1 + x_2 = 0$$
$$\therefore\ x_2 = l$$
$$x_3 = l$$
$$x_1 = -l$$

Similarly, $y_1 = m \quad z_1 = n$

$y_2 = -m \quad z_2 = n$

$y_3 = m \quad z_3 = -n$

Therefore, the vertices are $(-l, m, n)$, $(l, -m, n)$ and $(l, m, -n)$.

Also, $EF = \frac{1}{2}BC \Rightarrow \sqrt{l^2 + m^2} = \frac{1}{2}a$

$\Rightarrow a^2 = 4(m^2 + n^2)$

$b^2 = 4(n^2 + l^2)$

and $c^2 = 4(l^2 + m^2)$

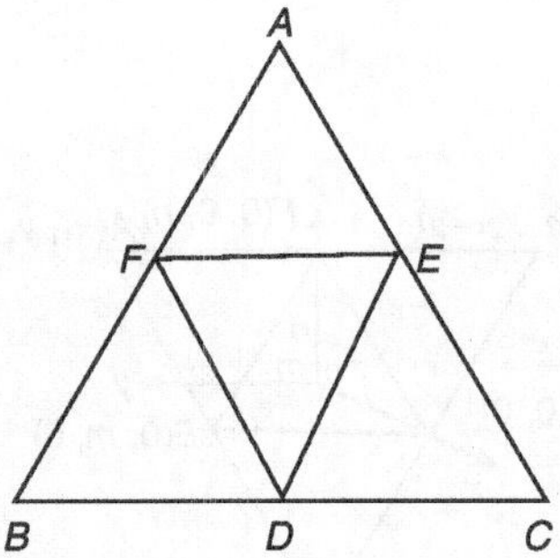

$$\therefore b^2 + c^2 - a^2 = 8l^2$$

$$c^2 + a^2 - b^2 = 8m^2$$

$$a^2 + b^2 - c^2 = 8n^2$$

Example 11.9

A directed line makes angles 60° and 60° with *x*- and *y*-axes, respectively. Find the angle it makes with *z*-axis.

Solution

If a line makes angles α, β and γ with *x*-, *y*- and *z*-axes, respectively then $\cos^2\alpha + \cos^2\beta + \cos^2\gamma = 1$.

Here, $\alpha = 60°, \beta = 60°$

$$\therefore \cos^2 60° + \cos^2 60° + \cos^2 \gamma = 1$$

$$\frac{1}{4}+\frac{1}{4}+\cos^2\gamma=1 \Rightarrow \cos^2\gamma=1-\frac{1}{2}=\frac{1}{2}$$

$$\therefore \cos\gamma=\pm\frac{1}{\sqrt{2}} \text{ or } \gamma=45^\circ \text{ or } 135^\circ.$$

Example 11.10

Find the acute angle between the lines whose direction ratios are 2, 1, −2 and 1, 1, 0.

Solution

The direction cosines of the two lines are

$$\frac{2}{3},\frac{1}{3},\frac{-2}{3} \text{ and } \frac{1}{\sqrt{2}},\frac{1}{\sqrt{2}},0.$$

If θ is the angle between the lines then

$$\cos\theta=l_1l_2+m_1m_2+n_1n_2=\frac{2}{3\sqrt{2}}+\frac{1}{3\sqrt{2}}=\frac{1}{\sqrt{2}}$$

$$\therefore \theta=45^\circ$$

Example 11.11

Find the angle between any two diagonals of a unit cube.

Solution

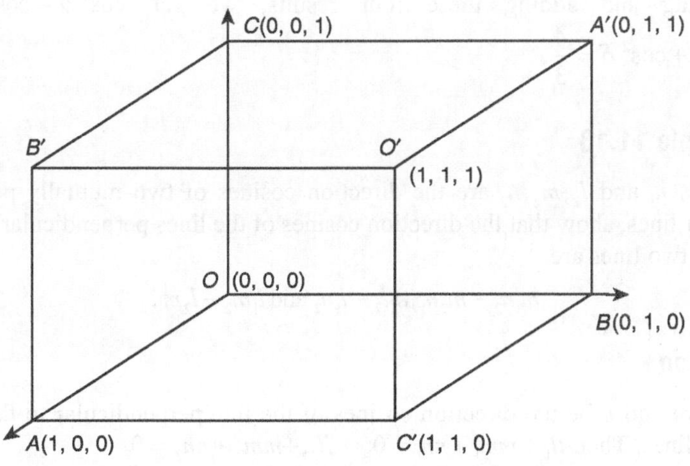

The four diagonals of the cube are OO', AA', BB' and CC'. Then the direction ratios of

OO' and AA' are (1, 1, 1) and (–1, 1, 1). The direction cosines of OO' and AA' are $\left(\frac{1}{\sqrt{3}}, \frac{1}{\sqrt{3}}, \frac{1}{\sqrt{3}}\right)$ and $\left(\frac{-1}{\sqrt{3}}, \frac{1}{\sqrt{3}}, \frac{1}{\sqrt{3}}\right)$. If θ is the angle between these two diagonals then $\cos\theta = \frac{-1}{3} + \frac{1}{3} + \frac{1}{3} = \frac{1}{3}$. $\therefore \theta = \cos^{-1}\left(\frac{1}{3}\right)$

Similarly, the angle between any two diagonals is $\cos^{-1}\left(\frac{1}{3}\right)$.

Example 11.12

If α, β, γ and δ are the angles made by a line with the four diagonals of a cube, prove that $\cos^2\alpha + \cos^2\beta + \cos^2\gamma + \cos^2\delta = \frac{4}{3}$.

Solution

The four diagonals are OO', AA', BB' and CC' (refer figure given in Example 11.11). Let l, m, n be the direction cosines of the line making angles α, β, γ and δ with the four diagonals.

$$\text{Then, } \cos\alpha = \frac{l}{\sqrt{3}} + \frac{m}{\sqrt{3}} + \frac{n}{\sqrt{3}}, \quad \cos\beta = -\frac{l}{\sqrt{3}} + \frac{m}{\sqrt{3}} + \frac{n}{\sqrt{3}}$$

$$\cos\gamma = \frac{l}{\sqrt{3}} - \frac{m}{\sqrt{3}} + \frac{n}{\sqrt{3}}, \quad \cos\delta = \frac{l}{\sqrt{3}} + \frac{m}{\sqrt{3}} - \frac{n}{\sqrt{3}}$$

Squaring and adding these four results, we get $\cos^2\alpha + \cos^2\beta + \cos^2\gamma + \cos^2\delta = \frac{4}{3}$.

Example 11.13

If l_1, m_1, n_1 and l_2, m_2, n_2 are the direction cosines of two mutually perpendicular lines, show that the direction cosines of the lines perpendicular to the above two lines are

$$m_1n_2 - m_2n_1, n_1l_2 - l_1n_2 \text{ and } l_1m_2 - l_2m_1.$$

Solution

Let l, m and n be the direction cosines of the line perpendicular to the two given lines. Then $ll_1 + mm_1 + nn_1 = 0$; $\quad ll_2, + mm_2 + nn_2 = 0$

$$\therefore \frac{l}{m_1n_2 - m_2n_1} = \frac{m}{n_1l_2 - n_2l_1} = \frac{n}{l_1m_2 - l_2m_1} = \frac{\sqrt{l^2 + m^2 + n^2}}{\sqrt{\sum (m_1n_2 - m_2n_1)^2}}$$

$$= \frac{1}{\sqrt{\sum (m_1n_2 - m_2n_1)^2}}$$

But $\sum \sqrt{(m_1n_2 - m_2n_1)^2} = \sin\theta$ and since the two lines are perpendicular $\theta = \frac{\pi}{2}$ and $\sin\frac{\pi}{2} = 1.$

Therefore, the direction cosines of the line perpendicular to the given two lines are $m_1n_2 - m_2n_1, n_1l_2 - n_2l_1, l_1m_2 - l_2m_1$.

Example 11.14

Show that three concurrent straight lines with direction cosines l_1, m_1, n_1; l_2, m_2, n_2 and l_3, m_3, n_3 are coplanar if

$$\begin{vmatrix} l_1 & m_1 & n_1 \\ l_2 & m_2 & n_2 \\ l_3 & m_3 & n_3 \end{vmatrix} = 0.$$

Solution

Let *l, m* and *n* be the direction cosines of the line which is perpendicular to the given three lines. If the lines are coplanar then the line with direction cosines *l, m* and *n* is normal to the given coplanar line.

$$ll_1 + mm_1 + nn_1 = 0$$
$$ll_2 + mm_2 + nn_2 = 0$$
$$ll_3 + mm_3 + nn_3 = 0$$

Eliminating *l, m* and *n* we get $\begin{vmatrix} l_1 & m_1 & n_1 \\ l_2 & m_2 & n_2 \\ l_3 & m_3 & n_3 \end{vmatrix} = 0.$

Example 11.15

Prove that the straight lines whose direction cosines are given by the equations $al + bm + cn = 0$ and $fmn + gnl + hlm = 0$ are perpendicular if $\frac{f}{a} + \frac{g}{h} + \frac{h}{c} = 0.$

Solution

The direction cosines of two lines are given by

$$al + bm + cn = 0 \tag{11.5}$$

$$fmn + gnl + hlm = 0 \tag{11.6}$$

From (11.5), $n = \dfrac{-(al+bm)}{c}$

Substituting in (11.6), we get

$$-(fm+gl)\left(\frac{al+bm}{c}\right) + hlm = 0$$

(i.e.) $$agl^2 + (af + hg - ch)lm + bfm^2 = 0$$

Dividing by m^2, we get

$$ag\left(\frac{l}{m}\right)^2 + (af + hg - ch)\frac{l}{m} + hf = 0 \tag{11.7}$$

If l_1, m_1, n_1 and l_2, m_2, n_2 are the direction cosines of the two given lines then $\dfrac{l_1}{m_1}$ and $\dfrac{l_2}{m_2}$ are the roots of the equation (11.7), then $\dfrac{l_1 l_2}{m_1 n_2} = \dfrac{hf}{ag}$.

Similarly, $\dfrac{m_1 m_2}{n_1 n_2} = \dfrac{cg}{bh}$ and $\dfrac{n_1 n_2}{l_1 l_2} = \dfrac{ah}{cf}$

But $l_1 l_2 + m_1 m_2 + n_1 n_2 = 0$

Dividing $n_1 n_2$, $\dfrac{l_1 l_2}{n_1 n_2} + \dfrac{m_1 m_2}{n_1 n_2} + 1 = 0$

$$\frac{cf}{ah} + \frac{cg}{bh} + 1 = 0 \Rightarrow \frac{f}{a} + \frac{g}{h} + \frac{h}{c} = 0$$

Example 11.16

If two pair of opposite edges of a tetrahedron are at right angles then show that the third pair is also at right angles.

Solution

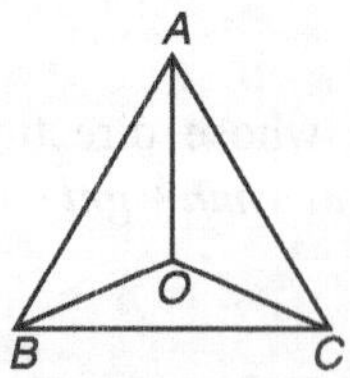

Let (*OA, BC*), (*OB, CA*) and (*OC, AB*) be three pair of opposite edges. Let *O* be the origin. Let the coordinates of *A, B* and *C* be (x_1, y_1, z_1), (x_2, y_2, z_2) and (x_3, y_3, z_3), respectively. Then the direction ratios of *OA* and *BC* are x_1, y_1, z_1 and $x_2 - x_3, y_2 - y_3, z_2 - z_3$.

Since *OA* is perpendicular to *BC,* we get

$$x_1(x_2 - x_3) + y_1(y_2 - y_3) + z_1(z_2 - z_3) = 0 \tag{11.8}$$

Since *OB* is perpendicular to *AC* we get

$$x_2(x_3 - x_1) + y_3(y_3 - y_1) + z_2(z_3 - z_1) = 0 \tag{11.9}$$

Adding (11.8) and (11.9), we get

$$x_3(x_1 - x_2) + y_3(y_2 - y_1) + z_3(z_1 - z_2) = 0 \tag{11.10}$$

This shows that *OC* is perpendicular to *AB.*

Example 11.17

If l_1, m_1, n_1; l_2, m_2, n_2 and l_3, m_3, n_3 be the direction cosines of the mutually perpendicular lines then show that the line whose direction ratios $l_1 + l_2 + l_3, m_1 + m_2 + m_3$ and $n_1 + n_2 + n_3$ make equal angles with them.

Solution

If l_1, m_1, n_1; l_2, m_2, n_2 and l_3, m_3, n_3 are the direction cosines of three mutually perpendicular lines $l_1^2 + m_1^2 + n_1^2 = 1$; $l_2^2 + m_2^2 + n_2^2 = 1$; $l_3^2 + m_3^2 + n_3^2 = 1$ also

$$l_1 l_2 + m_1 m_2 + n_1 n_2 = 0$$
$$l_2 l_3 + m_2 m_3 + n_2 n_3 = 0$$
$$l_3 l_1 + m_3 m_1 + n_3 n_1 = 0$$

Let θ be the angle between the lines with the direction cosines l_1, m_1, n_1 and direction ratios $l_1 + l_2 + l_3, m_1 + m_2 + m_3$ and $n_1 + n_2 + n_3$. Then,

$$\cos\theta = \frac{l_1(l_1 + l_2 + l_3) + m_1(m_1 + m_2 + m_3) + n(n_1 + n_2 + n_3)}{\sqrt{\sum (l_1 + l_2 + l_3)^2}}$$

$$= \frac{1}{\sqrt{\sum (l_1 + l_2 + l_3)^2}}$$

Similarly, the other two angles are equal to the same value of θ.
Therefore, the lines with the direction ratios $l_1 + l_2 + l_3, m_1 + m_2 + m_3, n_1 + n_2 + n_3$ are equally inclined to the line with direction cosines l_1, m_1, n_1; l_2, m_2, n_2 and l_3, m_3, n_3.

Example 11.18

Show that the straight lines whose direction cosines are given by $a^2l+b^2m+c^2n=0$, $mn+nl+lm=0$ will be parallel if $a\pm b\pm c=0$.

Solution

Given the direction cosines of two given lines satisfy the equations

$$a^2l+b^2m+c^2n=0 \tag{11.11}$$

$$mn+nl+lm=0 \tag{11.12}$$

From (11.11), $n=\dfrac{-(a^2l+b^2m)}{c^2}$.

Substituting this value of n in (11.12), we get

$$\frac{-(a^2l+b^2m)(m+l)}{c^2}+lm=0$$

(i.e.) $$a^2l^2+(a^2+b^2+c^2)lm+b^2m^2=0$$

Dividing by m^2, $$\frac{a^2l^2}{m^2}+(a^2+b^2-c^2)\frac{l}{m}+b^2=0 \tag{11.13}$$

If l_1, m_1, n_1 and l_2, m_2, n_2 are the direction cosines of the two given lines then $\dfrac{l_1}{m_1}, \dfrac{l_2}{m_2}$ are the roots of the equation (11.13).

Also, if the lines are parallel then $\dfrac{l_1}{m_1}=\dfrac{l_2}{m_2}$ then the roots of the equation (11.13) are equal. The condition for that is the discriminant is equal to zero.

$$\therefore\ (a^2+b^2-c^2)^2-4a^2b^2=0$$

$$(a^2+b^2-c^2)=\pm 2ab$$

$$(a\pm b)^2=c^2$$

$$a\pm b=\pm c$$

$$\therefore a\pm b\pm c=0$$

Example 11.19

The projections of a line on the axes are 3, 4, 12. Find the length and direction cosines of the line.

Solution

Let (l, m, n) be the direction cosines of the line and (x_1, y_1, z_1) and (x_2, y_2, z_2) be the extremities of the line. The direction cosines of x-, y and z-axes are (1, 0, 0), (0, 1, 0) and (0, 0, 1), respectively. The projection of the line on the axis is 3.

$$\therefore\ x_2 - x_1 = 3.$$

Similarly, $y_2 - y_1 = 4$, $z_2 - z_1 = 12$

$$\text{Then the length of the line } = \sqrt{(x_2 - x_1)^2 + (y_2 - y_1)^2 + (z_2 - z_1)^2}$$

$$= \sqrt{3^2 + 4^2 + 12^2} = 13$$

Then direction ratios of the line are 3, 4, 12.

Therefore, the direction cosines of the line are $\left(\frac{3}{13}, \frac{4}{13}, \frac{12}{13}\right)$.

Exercises

1. Show that the points (10, 7, 0), (6, 6, –1) and (6, 9, –4) form an isosceles right-angled triangle.
2. Show that the points (2, 3, 5), (7, 5, –1) and (4, –3, 2) form an isosceles triangle.
3. Show that the points (1, 2, 3), (2, 3, 1) and (3, 1, 2) form an equilateral triangle.
4. Show that the points (4, 0, 5), (2, 1, 3) and (1, 3, 2) are collinear.
5. Show that the points (1, –1, 1), (5, –5, 4), (5, 0, 8) and (1, 4, 5) form a rhombus.
6. Prove that the points (2, –1, 0), (0, –1, –1), (1, 1, –3) and (3, 1, –2) form the vertices of a rectangle.
7. Show that the points (1, 2, 3), (–1, 2, –1), (2, 3, 2) and (4, 7, 6) form a parallelogram.
8. Show that the points (–4, 3, 6), (–5, 2, 2), (–8, 5, 2), (–7, 6, 6) form a rhombus.
9. Show that the points (4, –1, 2), (0, –2, 3), (1, –5, –1) and (2, 0, 1) lie on a sphere whose centre is (2, –3, 1) and find its radius.
10. Find the ratio in which the line joining points (2, 4, 5) and (3, 5, –4) is divided by the xy-plane.

Ans.: (5, 4)

11. The line joining the points $A(-2, 6, 4)$ and $B(1, 3, 7)$ meets the YOZ-plane at C. Find the coordinates of C.

Ans.: (0, 4, 6)

12. Three vertices of a parallelogram $ABCD$ are $A(3, -4, 7)$, $B(-5, 3, -2)$ and $C(1, 2, -3)$. Find the coordinates of D.

Ans.: (9, –5, 6)

13. Show that the points (–5, 6, 8), (1, 8, 11), (4, 2, 9) and (–2, 0, 6) are the vertices of a square.

14. Show that the points $P(3, 2, -4)$, $Q(9, 8, -10)$ and $R(5, 4, -6)$ are collinear. Find the ratio in which R divides PQ.

Ans.: (1, 2)

15. Find the ratio in which the coordinate planes divide the line joining the points (–2, 4, 7) and (3, –5, 8).

Ans.: 7:9; 4:5; –7: –8

16. Prove that the line drawn from the vertices of a tetrahedron to the centroids of the opposite faces meet in a point which divides them in the ratio 3:1.

17. Find the coordinate of the circumcentre of the triangle formed by the points with vertices (1, 2, 1), (–2, 2, –1) and (1, 1, 0).

Ans.: $\left(\frac{-1}{2}, 2, 0\right)$

18. A and B are the points (2, 3, 5) and (7, 2, 4). Find the coordinates of the points which the bisectors of the angles AOB meet AB.

19. Find the length of the median through A of the triangle $A(2, -1, 4)$, $B(3, 7, -6)$ and $C(-5, 0, 2)$.

Ans.: 7

20. Prove that the locus of a point, the sum of whose distances from the points $(a, 0, 0)$ and $(-a, 0, 0)$ is a constant $2k$, is the curve $\frac{x^2}{k^2} + \frac{y^2 + z^2}{k^2 - a} = 1$.

21. What are the direction cosines of the line which is equally inclined to the axes?

Ans.: $\left(\pm\frac{1}{\sqrt{3}}, \pm\frac{1}{\sqrt{3}}, \pm\frac{1}{\sqrt{3}}\right)$

22. Find the angle between the lines whose direction ratios are (2, 3, 4) and (1, –2, 1).

Ans.: $\frac{\pi}{2}$

23. A variable line in two adjacent positions has direction cosines (l, m, n), $(l+\delta l, m+\delta m, n+\delta n)$. Prove that the small angle $\delta\theta$ between two positives is given by $\delta^2\theta = (\delta l)^2 + (\delta m)^2 + (\delta n)^2$.

24. Find the angle between the lines *AB* and *CD*, where *A*, *B*, *C* and *D* are the points (3, 4, 5), (4, 6, 3), (–1, 2, 4) and (1, 0, 5), respectively.

Ans.: $\cos^{-1}\left(\dfrac{4}{9}\right)$

25. Prove by direction cosines the points (1, –2, 3), (2, 3, –4) and (0, –7, 10) are collinear.

26. Find the angle between the lines whose direction ratios are (2, 1, –2) and (1, –1, 0).

Ans.: $\theta = \dfrac{\pi}{4}$

27. Show that the line joining the points (1, 2, 3) and (1, 5, 7) is parallel to the line joining the points (–4, 3, –6) and (2, 9, 2).

28. *P*, *Q*, *R* and *S* are the points (2, 3, –1), (3, 5, 3), (1, 2, 3) and (2, 5, 7). Show that *PQ* is perpendicular to *RS*.

29. Prove that the three lines with direction ratios (1, –1, 1), (1, –3, 0) and (1, 0, 3) lie in a plane.

30. Show that the lines whose direction cosines are given by $al + bm + cn = 0$ and $al^2 + bm^2 + cn^2 = 0$ are parallel if $\sqrt{af} \pm \sqrt{hg} \pm \sqrt{ch} = 0$.

31. Show that the lines whose direction cosines are given by the equations $al + vm + wn = 0$ and $al^2 + bm^2 + cn^2 = 0$ are parallel if $u^2(b+c) + v^2(c+a) + w^2(a+b) = 0$ and perpendicular if $\dfrac{u^2}{a} + \dfrac{v^2}{b} + \dfrac{w^2}{c} = 0$.

32. If the edges of a rectangular parallelepiped are *a*, *b* and *c*, show that the angle between the four diagonals are given by $\cos^{-1}\left(\dfrac{\pm a^2 \pm b^2 \pm c^2}{a^2 + b^2 + c^2}\right)$.

33. If in a tetrahedron the sum of the squares of opposite edges is equal, show that its pairs of opposite sides are at right angles.

34. Find the angle between the lines whose direction cosines are given by the equations:

(i) $l + m + n = 0$ and $l^2 + m^2 - n^2 = 0$

(ii) $l + m + n = 0$ and $2lm - 2nl - mn = 0$

Ans.: (i) $\dfrac{\pi}{3}$

(ii) $\dfrac{2\pi}{3}$

35. If (l_1, m_1, n_1) and (l_2, m_2, n_2) are the direction cosines of two lines inclined at an angle q, show that the actual direction cosines of the direction between the lines are $\left(\dfrac{l_1 + l_2}{2}\sec\dfrac{\theta}{2}, \dfrac{m_1 + m_2}{2}\sec\dfrac{\theta}{2}, \dfrac{n_1 + n_2}{2}\sec\dfrac{\theta}{2}\right)$.

36. AB, BC are the diagonals of adjacent faces of a rectangular box with centre at the origin O its edges being parallel to axes. If the angles AOB, BOC and COA are θ, ϕ and ω, respectively then prove that $\cos\theta + \cos\phi + \cos\omega = -1$.

37. If the projections of a line on the axes are 2, 3, 6 then find the length of the line.

Ans.: 7

38. The distance between the points P and Q and the lengths of the projections of PQ on the coordinate planes are d_1, d_2 and d_3, show that $2d^2 = d_1^2 + d_2^2 + d_3^2$.

39. Show that the three lines through the origin with direction ratios (1, –1, 7), (1, –1, 0) and (1, 0, 3) lie on a plane.

40. Show that the angle between the lines whose direction cosines are given by $l + m + n = 0$ and $fmn + gnl + hlm = 0$ is $\frac{\pi}{3}$ if $\frac{1}{f} + \frac{1}{g} + \frac{1}{h} = 0$.

CHAPTER

12 Plane

12.1 INTRODUCTION

In three-dimensional coordinate geometry, first we define a plane and from a plane we define a straight line. In this chapter, we define a plane and obtain its equation in different forms. We also derive formula to find the perpendicular distance from a given point to a plane. Also, we find the ratio in which a plane divides the line joining two given points.

Definition 12.1.1: A plane is defined to be a surface such that the line joining any two points wholly lies on the surface.

12.2 GENERAL EQUATION OF A PLANE

Every first degree equation in *x, y* and *z* represents a plane.

Consider the first degree equation in *x, y* and *z* as

$$ax + by + cz + d = 0 \tag{12.1}$$

where *a, b, c* and *d* are constants. Let $P(x_1, y_1, z_1)$ and $Q(x_2, y_2, z_2)$ be two points on the locus of equation (12.1). Then the coordinates of the points that divide line joining these two points in the ratio $\lambda{:}1$ are $\left(\dfrac{x_1 + \lambda x_2}{\lambda + 1}, \dfrac{y_1 + \lambda y_2}{\lambda + 1}, \dfrac{z_1 + \lambda z_2}{1 + \lambda}\right)$.

If this point lies on the locus of equation (12.1) then

$$\frac{a(x_1 + \lambda x_2)}{\lambda + 1} + \frac{b(y_1 + \lambda y_2)}{\lambda + 1} + \frac{c(z_1 + \lambda x_2)}{\lambda + 1} + d = 0\cdot$$

$$\text{(i.e.)} \quad a(x_1 + \lambda x_2) + b(y_1 + \lambda y_2) + c(z_1 + \lambda z_2) + d(\lambda + 1) = 0 \tag{12.2}$$

Since $P(x_1, y_1, z_1)$ and $Q(x_2, y_2, z_2)$ are two points on the locus of the equation (12.1) these two points have to satisfy the locus of the equation (12.1).

$$\text{(i.e.)} \quad ax_1 + by_1 + cz_1 + d = 0 \qquad (12.3)$$

$$ax_2 + by_2 + cz_2 + d = 0 \qquad (12.4)$$

Multiplying (12.4) by λ and adding with (12.3), we get $(ax_1 + by_1 + cz_1 + d) + \lambda(ax_2 + by_2 + cz_2 + d) = 0$ which is the equation (12.2).

Therefore, the point $\left(\frac{x_1 + \lambda x_2}{\lambda+1}, \frac{y_1 + \lambda y_2}{\lambda+1}, \frac{z_1 + \lambda x_2}{1+\lambda}\right)$ lies on the locus of equation (12.1).

Hence, this shows that if two points lies on the locus of equation (12.1) then every point on this line is also a point on the locus of equation (12.1). Hence, the equation (12.1) represents a plane and thus we have shown that every first degree equation in *x, y* and *z* represents a plane.

12.3 GENERAL EQUATION OF A PLANE PASSING THROUGH A GIVEN POINT

Let the equation of the plane passing through a given point (x_1, y_1, z_1) be

$$ax + by + cz + d = 0 \qquad (12.5)$$

since (x_1, y_1, z_1) lies on the plane (12.5).

$$ax_1 + by_1 + cz_1 + d = 0 \qquad (12.6)$$

Subtracting (12.6) from (12.5), we get $a(x - x_1) + b(y - y_1) + c(z - z_1) = 0$. This is the general equation of the plane passing through the given point (x_1, y_1, z_1).

12.4 EQUATION OF A PLANE IN INTERCEPT FORM

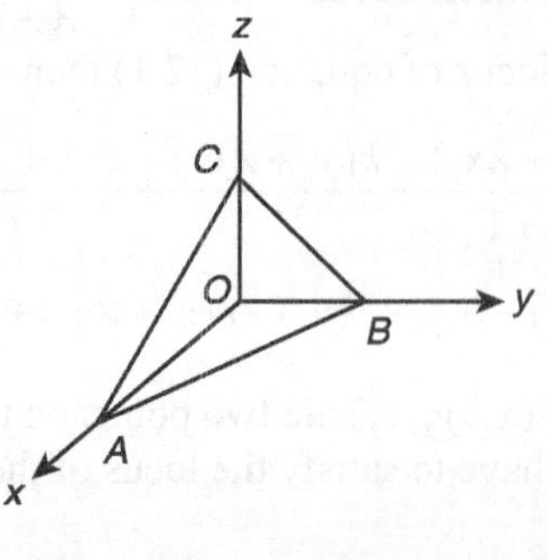

Let the equation of a plane be

$$Ax + By + Cz + D = 0 \tag{12.7}$$

Let this plane make intercepts *a*, *b* and *c* on the axes of coordinates. If this plane meets the *x*-, *y*- and *z*-axes at *A*, *B* and *C* then their coordinates are (*a*, 0, 0), (0, *b*, 0) and (0, 0, *c*), respectively. Since these points lie on the plane $Ax + By + Cz + D = 0$, the coordinates of the points have to satisfy the equation $Ax + By + Cz + D = 0$.

$$\therefore Aa + D = 0$$

$$Bb + D = 0$$

$$Cc + D = 0$$

$$\therefore A = \frac{-D}{a}, B = \frac{-D}{b}, C = \frac{-D}{c}$$

By replacing the values of *A*, *B* and *C*, we get $\frac{x}{a} + \frac{y}{b} + \frac{z}{c} = 1.$

This equation is called the intercept form of a plane.

12.5 EQUATION OF A PLANE IN NORMAL FORM

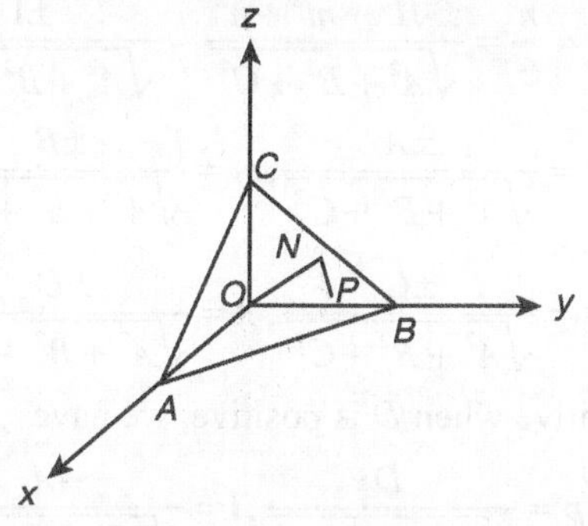

Let a plane meet the coordinate axes at *A*, *B* and *C*. Draw *ON* perpendicular to the plane *ABC* and let $ON = p$. Let the direction cosines of *ON* be ($\cos\alpha$, $\cos\beta$, $\cos\gamma$). Since $ON = p$, the coordinates of *N* are ($p\cos\alpha, p\cos\beta, p\cos\gamma$). Let $p(x_1, y_1, z_1)$ be any point in the plane *ABC*. If a line is perpendicular to a plane then it is perpendicular to every line to the plane. Therefore, *ON* is perpendicular to *OP*. Since the coordinates of *P* and *N* are (x_1, y_1, z_1) and ($p\cos\alpha, p\cos\beta, p\cos\gamma$) the direction ratios of *N* are $(x_1 - p\cos\alpha, y_1 - p\cos\beta, z_1 - p\cos\gamma)$ since *N* is perpendicular to *ON*.

$$\cos\alpha(x_1 - p\cos\alpha) + \cos\beta(y_1 - p\cos\beta) + \cos\gamma(z_1 - p\cos\gamma) = 0$$

(i.e.) $x_1 \cos\alpha + y_1 \cos\beta + z_1 + \cos\gamma = p(\cos^2\alpha + \cos^2\beta + \cos^2\gamma) = p$

since $\cos^2\alpha + \cos^2\beta + \cos^2\gamma = 1$

Therefore, the locus of (x_1, y_1, z_1) is $x\cos\alpha + y\cos\beta + z\cos\gamma = p$. This equation is called the normal form of a plane.

Note 12.5.1: Here, the coefficients of x, y and z are the direction cosines of normal to the plane and p is the perpendicular distance from the origin on the plane.

Note 12.5.2: Reduction of a plane to normal form: the equation of plane in general form is

$$Ax + By + Cz + D = 0 \tag{12.8}$$

Its equation in normal form is

$$lx + my + nz = p \tag{12.9}$$

Identifying (12.8) and (12.9), we get $\dfrac{A}{l} = \dfrac{B}{m} = \dfrac{C}{n} = \dfrac{D}{p}$

$$\frac{l}{A} = \frac{m}{B} = \frac{n}{C} = \frac{\pm\sqrt{l^2 + m^2 + n^2}}{\sqrt{A^2 + B^2 + C^2}} = \frac{\pm 1}{\sqrt{A^2 + B^2 + C^2}}.$$

$$l = \frac{\pm A}{\sqrt{A^2 + B^2 + C^2}}, m = \frac{\pm B}{\sqrt{A^2 + B^2 + C^2}},$$

$$\therefore\ n = \frac{\pm C}{\sqrt{A^2 + B^2 + C^2}}, p = \frac{\pm D}{\sqrt{A^2 + B^2 + C^2}}$$

Since p has to be positive when D is positive, we have

$$p = \frac{D}{\sqrt{A^2 + B^2 + C^2}}, l = \frac{-A}{\sqrt{A^2 + B^2 + C^2}},$$

$$m = \frac{-B}{\sqrt{A^2 + B^2 + C^2}}, n = \frac{-C}{\sqrt{A^2 + B^2 + C^2}}$$

When D is negative, $l = \dfrac{-A}{\sqrt{l^2 + m^2 + n^2}}, m = \dfrac{-B}{\sqrt{l^2 + m^2 + n^2}},$

$$n = \frac{-C}{\sqrt{l^2 + m^2 + n^2}}$$

12.6 ANGLE BETWEEN TWO PLANES

Let the equation of two planes be

$$a_1x + b_1y + c_1z + d = 0 \tag{12.10}$$

$$a_2x + b_2y + c_2z + d_2 = 0 \tag{12.11}$$

The direction ratios of the normals to the above planes are $\pm\frac{a_1}{\sqrt{a_1^2+b_1^2+c_1^2}}, \pm\frac{b_1}{\sqrt{a_1^2+b_1^2+c_1^2}}, \pm\frac{c_1}{\sqrt{a_1^2+b_1^2+c_1^2}}$ and $\pm\frac{a_2}{\sqrt{a_2^2+b_2^2+c_2^2}}, \pm\frac{b_2}{\sqrt{a_2^2+b_2^2+c_2^2}}, \pm\frac{c_2}{\sqrt{a_2^2+b_2^2+c_2^2}}$.

The angle between two planes is defined to be the angle between the normals to the two planes. Let θ be the angle between the planes.

Then, $\cos\theta = l_1l_2 + m_1m_2 + n_1n_2 = \pm\frac{(a_1a_2 + b_1b_2 + c_1c_2)}{\sqrt{a_1^2+b_1^2+c_1^2}\sqrt{a_2^2+b_2^2+c_2^2}}$

$$\therefore \theta = \cos^{-1}\left[\pm\frac{(a_1a_2 + b_1b_2 + c_1c_2)}{\sqrt{a_1^2+b_1^2+c_1^2}\sqrt{a_2^2+b_2^2+c_2^2}}\right]$$

Note 12.6.1: The positive sign of $\cos\theta$ gives the acute angle between the planes and negative sign gives the obtuse angle between the planes.

Note 12.6.2: If the planes are perpendicular then $\theta = 90°$.

$$\therefore a_1a_2 + b_1b_2 + c_1c_2 = 0$$

Note 12.6.3: If the planes are parallel then direction cosines of the normals are proportional.

$$\therefore \frac{a_1}{a_2} = \frac{b_1}{b_2} = \frac{c_1}{c_2}$$

Note 12.6.4: The equation of plane parallel to $ax + by + cz + d = 0$ can be expressed in the form $ax + by + cz + k = 0$.

12.7 PERPENDICULAR DISTANCE FROM A POINT ON A PLANE

Let the equation of the plane be

$$ax + by + cz + d = 0 \tag{12.12}$$

and $P(x_1, y_1, z_1)$ be the given point. We have to find the perpendicular distance from P to the plane. The normal form of the plane (12.12) is

$$lx + my + nz = p \tag{12.13}$$

Draw PM perpendicular from P to the plane (12.12). Draw the plane through P to the given plane (12.12). The equation of this plane is

$$lx + my + nz = p_1 \tag{12.14}$$

where p_1 is the perpendicular distance from the origin to the plane (12.13). This plane passes through (x_1, y_1, z_1).

$$\therefore lx_1 + my_1 + nz_1 = p_1$$

Draw BN perpendicular to the plane (12.3) meeting the plane (12.12) at M. Then $ON = p_1$ and $OM = p$.

$$MN = OM - ON = p - p_1 = p - (lx_1 + my_1 + nz_1).$$

Comparing equations (12.12) and (12.13),

$$\frac{l}{a} = \frac{m}{b} = \frac{n}{c} = \frac{-p}{d} = \frac{\pm\sqrt{l^2 + m^2 + n^2}}{\sqrt{a^2 + b^2 + c^2}}$$

$$\therefore l = \pm\frac{a}{\sqrt{a^2 + b^2 + c^2}}, m = \pm\frac{b}{\sqrt{a^2 + b^2 + c^2}},$$

$$n = \pm\frac{c}{\sqrt{a^2 + b^2 + c^2}}, p = \mp\frac{d}{\sqrt{a^2 + b^2 + c^2}}$$

The perpendicular distance $= p - (lx_1 + my_1 + nz_1)$

$$= \pm\frac{ax_1 + by_1 + cz_1 + d}{\sqrt{a^2 + b^2 + c^2}}$$

Therefore, the perpendicular distance from (x_1, y_1, z_1) is $= \left|\frac{ax_1 + by_1 + cz_1 + d}{\sqrt{a^2 + b^2 + c^2}}\right|$.

Aliter:

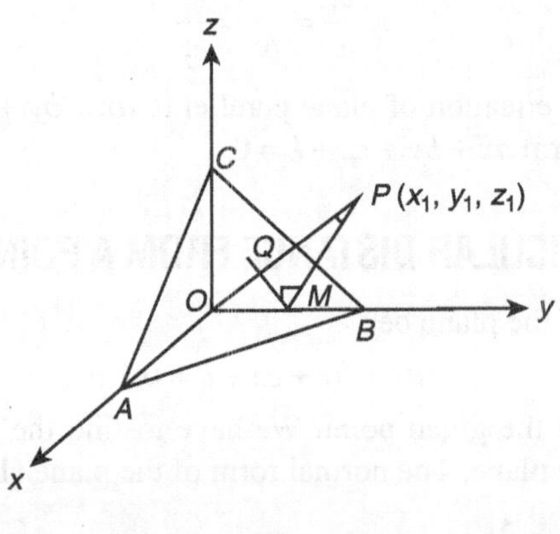

Let PM be the perpendicular from P on the plane $ax + by + cz + d = 0$. Let $P(x_1, y_1, z_1)$ and $M(x_2, y_2, z_2)$ be a point on the plane (12.12). Then QM and PM are perpendicular. Let θ be the $\underline{|MPQ}$. The direction ratios of PM and PN are (a, b, c) and $(x_1 - x_{2,} y_1 - y_2, z_1 - z_2)$.

$$\text{Then, } \cos\theta = \frac{a(x_1 - x_2) + b(y_1 - y_2) + c(z_1 - z_2)}{\sqrt{a^2 + b^2 + c^2}\sqrt{\sum (x_1 - x_2)^2}}$$

Also $PM = PQ\cos\theta$

$$= \frac{\pm a(x_1 - x_2) + b(y_1 - y_2) + c(z_1 - z_2)}{\sqrt{\sum a^2}\sqrt{\sum (x_1 - x_2)^2}}$$

$$= \pm \frac{a(x_1 - x_2) + b(y_1 - y_2) + c(z_1 - z_2)}{\sqrt{a^2 + b^2 + c^2}}$$

$$= \pm \frac{(ax_1 + by_1 + cz_1) - (ax_2 + by_2 + cz_2)}{\sqrt{a^2 + b^2 + c^2}} = \pm \frac{ax_1 + by_1 + cz_1 + d}{\sqrt{a^2 + b^2 + c^2}}$$

Note 12.7.1: The perpendicular distance from the origin to the plane $ax + by + cz + d = 0$ is $\dfrac{|d|}{\sqrt{a^2 + b^2 + c^2}}$.

12.8 PLANE PASSING THROUGH THREE GIVEN POINTS

Let $(x_1, y_1, z_1), (x_2, y_2, z_2)$ and (x_3, y_3, z_3) be three given points on a plane. Let the equation of the plane be

$$ax + by + cz + d = 0 \tag{12.15}$$

Any plane through (x_1, y_1, z_1) is

$$a(x_1 - x_2) + b(y_1 - y_2) + c(z_1 - z_2) = 0 \tag{12.16}$$

This plane also passes through (x_2, y_2, z_2) and (x_3, y_3, z_3).

$$a(x_2 - x_1) + b(y_2 - y_1) + c(z_2 - z_1) = 0 \tag{12.17}$$

$$a(x_3 - x_4) + b(y_3 - y_4) + c(z_3 - z_4) = 0 \tag{12.18}$$

Eliminating a, b and c from (12.6), (12.7) and (12.18), we get

$$\begin{vmatrix} x - x_1 & y - y_1 & z - z_1 \\ x_2 - x_1 & y_2 - y_1 & z_2 - z_1 \\ x_3 - x_1 & y_3 - y_1 & z_3 - z_1 \end{vmatrix} = 0.$$

This is the equation of the required plane.

Aliter:

Let the equation of the plane be

$$ax + by + cz + d = 0 \tag{12.19}$$

This plane passes through the points $(x_1, y_1, z_1), (x_2, y_2, z_2)$ and (x_3, y_3, z_3).

$$ax_1 + by_1 + cz_1 + d = 0 \tag{12.20}$$
$$ax_2 + by_2 + cz_2 + d = 0 \tag{12.21}$$
$$ax_3 + by_3 + cz_3 + d = 0 \tag{12.22}$$

Eliminating a, b, and c from (12.20), (12.21) and (12.22), we get

$$\begin{vmatrix} x & y & z & 1 \\ x_1 & y_1 & z_1 & 1 \\ x_2 & y_2 & z_2 & 1 \\ x_3 & y_3 & z_3 & 1 \end{vmatrix} = 0.$$

This is the equation of the required plane.

12.9 TO FIND THE RATIO IN WHICH THE PLANE JOINING THE POINTS (x_1, y_1, z_1) AND (x_2, y_2, z_2) IS DIVIDED BY THE PLANE $ax + by + cz + d = 0$.

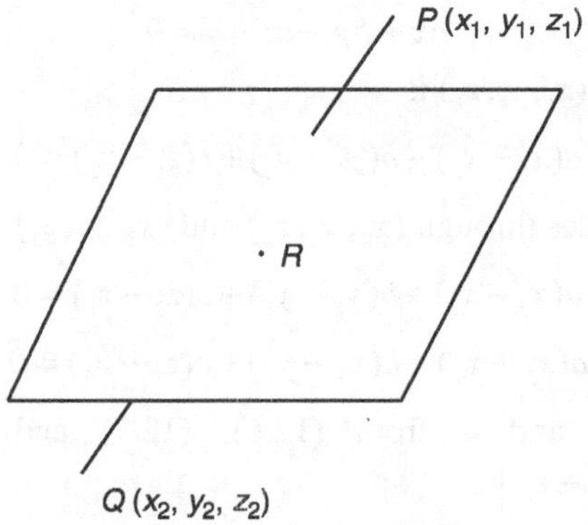

The equation of the plane is

$$ax + by + cz + d = 0 \tag{12.23}$$

Let the line joining the points $P(x_1, y_1, z_1)$ and $Q(x_2, y_2, z_2)$ meet the plane at R. Let R divided PQ in the ratio $\lambda:1$. Then the coordinates of R are $\left(\frac{x_1+\lambda x_2}{1+\lambda}, \frac{y_1+\lambda y_2}{1+\lambda}, \frac{z_1+\lambda z_2}{1+\lambda}\right)$. This point lies on the plane given by (1).

$$\therefore \frac{a(x_1+\lambda x_2)}{1+\lambda}, \frac{b(y_1+\lambda y_2)}{1+\lambda}, \frac{c(z_1+\lambda z_2)}{1+\lambda}+d=0$$

$$\text{(i.e.)} \quad a(x_1+\lambda x_2)+b(y_1+\lambda y_2)+c(z_1+\lambda z_2)+d(1+\lambda)=0$$

$$\text{(i.e.)} \quad ax_1+by_1+cz_1+d+\lambda(ax_2+by_2+cz_2+d)=0$$

$$\therefore \lambda = -\frac{ax_1+by_1+cz_1+d}{ax_2+by_2+cz_2+d}$$

Note 12.8.1: If $(ax_1+by_1+cz_1+d)$ and $(ax_2+by_2+cz_2+d)$ are of the same sign then λ is negative. Then the point R divides PQ externally and so the points P and Q lie on the same side of the plane.

Note 12.8.2: If $P(x_1, y_1, z_1)$ and the origin lie on the same side of the plane $ax+by+cz+d=0$ if $ax_1+by_1+cz_1+d$ and d of the same sign.

12.10 PLANE PASSING THROUGH THE INTERSECTION OF TWO GIVEN PLANES

Let the two given planes be

$$ax_1+by_1+cz_1+d_1=0 \qquad (12.24)$$

$$ax_2+by_2+cz_2+d_2=0 \qquad (12.25)$$

Then consider the equation $(ax_1+by_1+cz_1+d_1)+\lambda(ax_2+by_2+cz_2+d_2)=0$. This being the first degree equation in x, y and z, represents a plane. Let (x_1, y_1, z_1) be the point on the line of the intersection of planes given by equations (12.24) and (12.25). Then (x_1, y_1, z_1) lies on the two given planes.

$$\therefore a_1x_1+b_1y_1+c_1z_1+d_1=0 \qquad (12.26)$$

$$a_2x_1+b_2y_1+c_2z_1+d_2=0 \qquad (12.27)$$

Then, clearly $(a_1x_1+b_1y_1+c_1z_1+d_1)+\lambda(a_2x_1+b_2y_1+c_2z_1+d_2)=0$.

From this equation, we infer that the point (x_1, y_1, z_1) lies on the plane given by (12.26). Similarly, every point in the line of intersection of the planes (12.24) and (12.25) lie on the planes (12.24) and (12.25).

Hence, equation (12.26) is the plane passing through the intersection of the two given planes.

12.11 EQUATION OF THE PLANES WHICH BISECT THE ANGLE BETWEEN TWO GIVEN PLANES

Find the equation of the planes which bisect the angle between two given planes.

Let the two given planes be

$$a_1x + b_1y + c_1z + d_1 = 0 \qquad (12.28)$$

$$a_2x + b_2y + c_2z + d_2 = 0 \qquad (12.29)$$

Let $P(x_1, y_1, z_1)$ be a point on either of the bisectors of the angle between the two given planes. Then the perpendicular distance from P to the two given planes are equal in magnitude.

$$\text{(i.e.)} \quad \frac{a_1x_1 + b_1y_1 + c_1z_1 + d_1}{\sqrt{a_1^2 + b_1^2 + c_1^2}} = \pm \frac{a_2x_1 + b_2y_1 + c_2z_1 + d_2}{\sqrt{a_2^2 + b_2^2 + c_2^2}}$$

The locus of (x_1, y_1, z_1) is $\dfrac{a_1x + b_1y + c_1z + d_1}{\sqrt{a_1^2 + b_1^2 + c_1^2}} = \pm \dfrac{a_2x + b_2y + c_2z + d_2}{\sqrt{a_2^2 + b_2^2 + c_2^2}}$.

By taking the positive sign, we get the equation of one of the bisectors and by taking the negative sign, we get the equation to the other bisector.

Note 12.11.1: We can determine which of the two planes bisects the acute angle between the planes. For this, we have to find the angle θ between the bisector planes and one of the two given planes. If $\tan\theta < 1$ ($\theta < 45°$), then the bisector plane taken is the internal bisector and the other bisector plane is the external bisector. If $\tan\theta > 1$ then the bisector plane taken is the external bisector and the other bisector plane is the internal bisector.

Note 12.11.2: We can also determine the equation of the plane bisecting the angle between the planes that contain the origin. Suppose the equation of the two planes are $a_1x + b_1y + c_1z + d_1 = 0$ and $a_2x + b_2y + c_2z + d_2 = 0$ where d_1 and d_2 are positive. Let $P(x_1, y_1, z_1)$ be a point on the bisector between the angles of the planes containing the origin. Then d_1 and

$a_1x+b_1y+c_1z+d$ are of the sign. Since d_1 is positive, $a_1x + b_1y + c_1z$ is also positive. Similarly, $a_2x+b_2y+c_2z+d_2$ is also positive. Therefore, the equation of the plane bisecting the angle containing the origin is $\frac{a_1x+b_1y+c_1z+d_1}{\sqrt{a_1^2+b_1^2+c_1^2}} = +\frac{a_2x_1+b_2y_1+c_2z_1+d_2}{\sqrt{a_2^2+b_2^2+c_2^2}}$.

The equation of bisector plane not containing the origin is $\frac{a_1x+b_1y+c_1z+d_1}{\sqrt{a_1^2+b_1^2+c_1^2}} = -\frac{a_2x_1+b_2y+c_2z+d_2}{\sqrt{a_2^2+b_2^2+c_2^2}}$.

12.12 CONDITION FOR THE HOMOGENOUS EQUATION OF THE SECOND DEGREE TO REPRESENT A PAIR OF PLANES

The equation that represents a pair of planes be $ax^2+by^2+cz^2+2fyz+2gzx+2hxy=0$.

Let the two planes represented by the above homogenous equation of the second degree in x, y and z be $lx+my+nz=0$ and $l_1x+m_1y+n_1z=0$. Then

$$ax^2+by^2+cz^2+2fyz+2gzx+2hxy = (lx + my + nz)(l_1x+m_1y+n_1z).$$

Comparing the like terms on both sides, we get

$$ll_1 = a, mm_1 = b, nn_1 = c$$

$$\therefore lm_1 + l_1m = 2h$$

$$ln_1 + l_1n = 2g$$

$$mn_1 + m_1n = 2f$$

Now, consider the result $\begin{vmatrix} l & l_1 & 0 \\ m & m_1 & 0 \\ n & n_1 & 0 \end{vmatrix} \times \begin{vmatrix} l_1 & l & 0 \\ m_1 & m & 0 \\ n_1 & n & 0 \end{vmatrix} = 0$

$$\text{(i.e.)} \begin{vmatrix} ll_1+l_1l & lm_1+l_1m & ln_1+l_1n \\ lm_1+l_1m & mm_1+m_1m & mn_1+m_1n \\ ln_1+l_1n & mm_1+m_1m & nn_1+n_1n \end{vmatrix} = 0$$

$$\text{(i.e.)} \begin{vmatrix} 2a & 2h & 2g \\ 2h & 2b & 2f \\ 2g & 2f & 2c \end{vmatrix} = 0 \quad \text{(i.e.)} \begin{vmatrix} a & h & g \\ h & b & f \\ g & f & c \end{vmatrix} = 0$$

$$\text{(i.e.)}\quad a(bc - g^2) - h(ch - fg) + g(hf - hg) = 0$$
$$abc + 2fgh - af^2 - bg^2 - ch^2 = 0$$

This is the required condition.

Note 12.12.1: To find the angle between the two planes:

Let θ be the angle between the two planes. Then $\tan\theta = \dfrac{\sin\theta}{\cos\theta}$.

$$\tan\theta = \frac{\sqrt{1-\cos^2\theta}}{\cos\theta} = \pm\frac{\sqrt{(l^2+m^2+n^2)(l_1^2+m_1^2+n_1^2)-(ll_1+mm_1+nn_1)^2}}{ll_1+mm_1+nn_1}$$

$$= \pm\frac{\sqrt{(mn_1-m_1n)^2+(nl_1-n_1l)^2+(lm_1-l_1m)^2}}{ll_1+mm_1+nn_1}$$

$$\tan\theta = \pm\frac{\sqrt{\sum(mn_1-m_1n)^2}}{ll_1+mm_1+nn_1} = \pm\frac{\sqrt{\sum(mn_1+m_1n)^2-4mm_1nn_1}}{a+b+c}$$

$$= \pm\frac{\sqrt{\sum 4f^2-4bc}}{a+b+c} = \pm\frac{2\sqrt{f^2+g^2+h^2-bc-ca-ab}}{a+b+c}$$

$$\theta = \tan^{-1}\left(\frac{\pm 2\sqrt{f^2+g^2+h^2-bc-ca-ab}}{a+b+c}\right)$$

Note 12.12.2: If the planes are perpendicular then $\theta = 90°$ and the condition for that is $a + b + c = 0$.

ILLUSTRATIVE EXAMPLES

Example 12.1

The foot of the perpendicular from the origin to a plane is (13, –4, –3). Find the equation of the plane.

Solution

The line joining the points (0, 0, 0) and (13, –4, –3) is normal to the plane. Therefore, the direction ratios of the normal to the plane are (13, –4, –3).

The equation of the plane is $a(x-x_1)+b(y-y_1)+c(z-z_1)=0$

$$\text{(i.e.)}\quad 13(x-x_1)-4(y-y_1)-3(z-z_1)=0$$
$$13(x-13)-4(y+4)-3(z+3)=0$$
$$13x-4y-3z-169-16-9=0$$
$$13x-4y-3z-194=0$$

Example 12.2

A plane meets the coordinate axes at A, B and C such that the centroid of the triangle is the point (a, b, c). Find the equation of the plane.

Solution

Let the equation of the plane be $\frac{x}{\alpha}+\frac{y}{\beta}+\frac{z}{\gamma}=1$. Then the coordinates of A, B and C are $(\alpha, 0, 0)$, $(0, \beta, 0)$ and $(0, 0, \gamma)$. The centroid of the triangle ABC is $\left(\frac{\alpha}{3},\frac{\beta}{3},\frac{\gamma}{3}\right)$. But the centroid is given as (a, b, c).

$$a=\frac{\alpha}{3}, b=\frac{\beta}{3}, c=\frac{\gamma}{3} \text{ or } \alpha=3a, \beta=3b, \gamma=3c$$

Therefore, thc cquation of the plane is $\frac{x}{3a}+\frac{y}{3b}+\frac{z}{3c}=1$ or $\frac{x}{a}+\frac{y}{b}+\frac{z}{c}=3$.

Example 12.3

Find the equation of the plane passing through the points $(2, 2, 1)$, $(2, 3, 2)$ and $(-1, 3, 1)$.

Solution

The equation of the plane passing through the point $(2, 2, 1)$ is of the form $a(x-2)+b(y-2)+c(z-1)=0$.

This plane passes through the points $(2, 3, 2)$ and $(-1, 3, 1)$.

$$\therefore 0a+b+c=0 \text{ and } -3a+b+0c=0$$

Solving we get $\frac{a}{0-1}=\frac{b}{-3-0}=\frac{c}{0+3}$ or $\frac{a}{1}=\frac{b}{3}=\frac{c}{-3}$

Therefore, the equation of the plane is $1(x-2)+3(y-2)-3(z-1)=0$.

$$\therefore x+3y-3z-5=0$$

Example 12.4

Find the equation of the plane passing through the point $(2, -3, 4)$ and parallel to the plane $2x-5y-7z+15=0$.

Solution

The equation of the plane parallel to $2x-5y-7z+15=0$ is $2x-5y-7z+k=0$. This plane passes through the point $(2, -3, 4)$.

$$\therefore 4 + 15 - 28 + k = 0 \text{ or } k = 9$$

Hence, the equation of the required plane is $2x - 5y - 7z + 9 = 0$.

Example 12.5

Find the equation of the plane passing through the point (2, 2, 4) and perpendicular to the planes $2x - 2y - 4z - 3 = 0$ and $3x + y + 6z - 4 = 0$.

Solution

Any plane passing through (2, 2, 4) is $a(x - 2) + b(y - 2) + c(z - 4) = 0$.

This plane is perpendicular to the planes $2x - 2y - 4z - 3 = 0$ and $3x + y + 6z - 4 = 0$.

$$\text{(i.e.)} \quad 2a - 2b - 4c = 0$$
$$3a + b + 6c = 0$$

$$\therefore \frac{a}{-12+4} = \frac{b}{-12-12} = \frac{c}{2+6} \text{ or } \frac{a}{-8} = \frac{b}{-24} = \frac{c}{+8}$$

Therefore, the direction ratios of the normal to the required plane are 1, 3, –1.

Therefore, the equation of the plane is $(x - 2) + 0 + (z - 4) = 0$ (i.e.) $(x - 2) + 3(y - 2) - (z - 4) = 0$.

$$x + 3y - z - 4 = 0$$

Example 12.6

Determine the constants k so that the planes $x - 2y + kz = 0$ and $2x + 5y - z = 0$ are at right angles and in that case find the plane through the point (1, –1, –1) and perpendicular to both the given planes.

Solution

The planes $x - 2y + kz = 0$ and $2x + 5y - z = 0$ are perpendicular.

Therefore, $2 - 10 - k = 0 \therefore k = -8$.

Any plane passing through (1, –1, –1) is a $(x - 1) + b(y + 1) + c(x + 1) = 0$. This plane is perpendicular to the planes $x - 2y - 8z = 0$ and $2x + 5y - z = 0$.

$$\therefore a - 2b - 8c = 0 \text{ and } 2a + 5b - c = 0$$

$$\frac{a}{2+40} = \frac{b}{-16+1} = \frac{c}{5+4} \text{ or } \frac{a}{42} = \frac{b}{-15} = \frac{c}{9}$$

$$\Rightarrow \frac{a}{14} = \frac{b}{-5} = \frac{c}{3}$$

Therefore, the equation of the required plane is $14(x - 1) - 5(y + 1) + 3(z + 1) = 0$ or $14x - 5y + 3z - 16 = 0$.

Example 12.7

A variable plane is at a constant distance p from the origin and meets the axes in A, B and C. Show that the locus of the centroid of the tetrahedron $OABC$ is $x^{-2} + y^{-2} + z^{-2} = 16p^{-2}$.

Solution

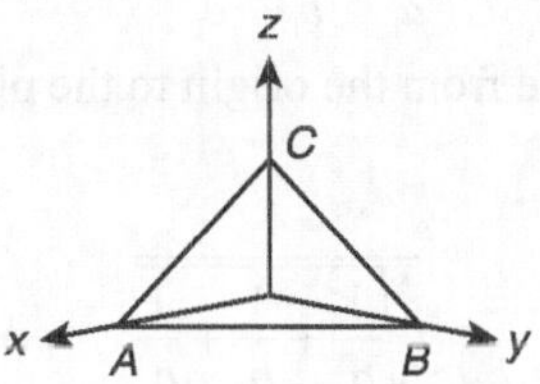

Let the equation of the plane ABC be $\frac{x}{a} + \frac{y}{b} + \frac{z}{c} = 1$. Then the coordinates of O, A, B and C are $(0, 0, 0)$, $(a, 0, 0)$, $(0, b, 0)$ and $(0, 0, c)$.

Let the centroid of the tetrahedron $OABC$ be (x_1, y_1, z_1). But the centroid of the tetrahedron is $\left(\frac{a}{4}, \frac{b}{4}, \frac{c}{4}\right)$

$$\therefore \frac{a}{4} = x_1, \frac{b}{4} = y_1, \frac{c}{4} = z_1$$

$$\therefore a = 4x_1, a = 4y_1, a = 4z_1$$

The perpendicular distance from O and the plane ABC is p.

$$\frac{1}{\sqrt{\frac{1}{a^2} + \frac{1}{b^2} + \frac{1}{c^2}}} = p \text{ or } \frac{1}{a^2} + \frac{1}{b^2} + \frac{1}{c^2} = \frac{1}{p^2}$$

$$\therefore \frac{1}{16x_1^2} + \frac{1}{16y_1^2} + \frac{1}{16z_1^2} = \frac{1}{p^2} \text{ or } x_1^{-2} + y_1^{-2} + z_1^{-2} = 16p^{-2}$$

The locus of (x_1, y_1, z_1) is $x^{-2} + y^{-2} + z^{-2} = 16p^{-2}$.

Example 12.8

Two systems of rectangular axes have the same origin. If a plane cuts them at distances (a, b, c) and (a_1, b_1, c_1) respectively, from the origin, prove that $\frac{1}{a^2} + \frac{1}{b^2} + \frac{1}{c^2} = \frac{1}{a_1^2} + \frac{1}{b_1^2} + \frac{1}{c_1^2}$.

Solution

Let (o, x, y, z) and (O, X, Y, Z) be the two system of coordinate axes. The equation of the plane with respective first system of coordinate axis is

$$\frac{x}{a}+\frac{y}{b}+\frac{z}{c}=1 \tag{12.30}$$

The equation of the same plane with respect to the second system of coordinate axes is

$$\frac{X}{a_1}+\frac{Y}{b_1}+\frac{Z}{c_1}=1 \tag{12.31}$$

The perpendicular distance from the origin to the plane given by the equation (12.30) is

$$\frac{1}{\sqrt{\frac{1}{a_1}+\frac{1}{b_1}+\frac{1}{c_1}}} \tag{12.32}$$

The perpendicular distance from the origin to the plane is given by the equation (12.31) is $\dfrac{1}{\sqrt{\frac{1}{a_1^2}+\frac{1}{b_1^2}+\frac{1}{c_1^2}}}$.

Since the equations (12.30) and (12.31) represent the same plane these two perpendicular distances are equal.

(i.e.)

$$\frac{1}{\sqrt{\frac{1}{a_1}+\frac{1}{b_1}+\frac{1}{c_1}}}=\frac{1}{\sqrt{\frac{1}{a_1^2}+\frac{1}{b_1^2}+\frac{1}{c_1^2}}}$$

or

$$\frac{1}{a^2}+\frac{1}{b^2}+\frac{1}{c^2}=\frac{1}{a_1^2}+\frac{1}{b_1^2}+\frac{1}{c_1^2}$$

Example 12.9

A variable plane passes through a fixed point (a, b, c) and meets the coordinate axes in A, B and C. Prove that the locus of the point of intersection of planes through A, B and C parallel to the coordinate planes is $\frac{a}{x}+\frac{b}{y}+\frac{c}{z}=1$.

Solution

Let the equation of the plane be $\frac{x}{\alpha}+\frac{y}{\beta}+\frac{z}{\gamma}=1$.

This plane passes through the point (a, b, c).

$$\therefore \frac{a}{\alpha}+\frac{b}{\beta}+\frac{c}{\gamma}=1 \tag{12.33}$$

Then the equation of the planes through *A, B* and *C* parallel to the coordinate planes are $x = \alpha$, $y = \beta$ and $z = \gamma$. Let (x_1, y_1, z_1) be the point of intersection of these planes. Then $x_1 = \alpha, y_1 = \beta$ and $z_1 = \gamma$

Therefore, from equation (12.33), we get $\frac{a}{x_1}+\frac{b}{y_1}+\frac{c}{z_1}=1$. The locus of (x_1, y_1, z_1) is $\frac{a}{x}+\frac{b}{y}+\frac{c}{z}=1$.

Example 12.10

A variable plane makes intercepts on the coordinate axes, the sum of whose squares is constant and is equal to k^2. Prove that the locus of the foot of the perpendicular from the origin to the plane is $(x^2+y^2+z^2)(x^{-2}+y^{-2}+z^{-2})=k^2$.

Solution

Let the equation of the plane be

$$\frac{x}{a}+\frac{y}{b}+\frac{z}{c}=1 \tag{12.34}$$

where *a, b* and *c* are the intercepts on the coordinate axes. Given that

$$a^2+b^2+c^2-k^2 \tag{12.35}$$

Let $P(x_1, y_1, z_1)$ be the foot of the perpendicular from *O* on this plane. The direction ratios of the normal *OP* are $\frac{1}{a},\frac{1}{b},\frac{1}{c}$. Therefore, the equation of the normal *OP* are $ax = by = cz$. Since (x_1, y_1, z_1) lies on the normal, $ax_1 = by_1 + cz_1 = t$ (say).

$$\therefore a=\frac{t}{x_1}, b=\frac{t}{y_1}, c=\frac{t}{z_1} \tag{12.36}$$

From (12.34) and (12.35), we get

$$t^2\left(\frac{1}{x_1^2}+\frac{1}{y_1^2}+\frac{1}{z_1^2}\right)=k^2 \tag{12.37}$$

The point (x_1, y_1, z_1) also lies on the plane.

$$\therefore \frac{x_1}{a}+\frac{y_1}{b}+\frac{z_1}{c}=1 \tag{12.38}$$

Eliminating *a, b, c* from (12.36) and (12.38)

$$\frac{1}{t}(x_1^2+y_1^2+z_1^2)=1 \tag{12.39}$$

Eliminating t from (12.38) and (12.39) $(x_1^2+y_1^2+z_1^2)(x_1^{-2}+y_1^{-2}+z_1^{-2})=k^2$

Therefore, the locus of (x_1, y_1, z_1) is $(x^2+y^2+z^2)(x^{-2}+y^{-2}+z^{-2})=k^2$.

Example 12.11

Find the equation of the plane which cuts the coordinate axes at A, B, and C such that the centroid of ΔABC is at the point (–1, –2, –4).

Solution

Let the equation of the plane be $\frac{x}{a}+\frac{y}{b}+\frac{z}{c}=1$. Then the coordinates of A, B and C are $(a, 0, 0)$, $(0, b, 0)$, $(0, 0, c)$. The centroid of ΔABC is $\left(\frac{a}{3},\frac{b}{3},\frac{c}{3}\right)$. But the centroid is given as (–1, –2, –4). $\therefore\ \frac{a}{3}=-1, \frac{b}{3}=-2, \frac{c}{3}=-4$

$$\therefore a=-3, b=-6, c=-12$$

Hence the equation of the plane ABC is $\frac{x}{-3}+\frac{y}{-6}+\frac{z}{-12}=1$.

$$\text{(i.e.)}\quad 4x+2y+z+12=0$$

Example 12.12

Find the equation of the plane passing through the point (–1, 3, 2) and perpendicular to the planes $x + 2y + 2z = 5$ and $3x + 3y + 2z = 8$.

Solution

The equation of the plane passing through the point (–1, 3, 2) is $A(x + 1) + B(y-3)+C(z-2)=0$. This plane is perpendicular to the planes $x+2y+2z=5$ and $3x + 3y + 2z = 8$. If two planes are perpendicular then their normals are perpendicular. The direction ratios of the normal to the required plane are A, B and C. The direction ratios of the normals to the given planes are 1, 2, 2 and 3, 3, 2.

$$A+2B+2C=0$$

$$3A+3B+2C=0$$

$$\frac{A}{4-6}=\frac{B}{6-2}=\frac{C}{3-6} \text{ or } \frac{A}{-2}=\frac{B}{4}=\frac{C}{-3}$$

Therefore, the direction ratios of the normal to the required plane are 2, –4, 3. The equation of the required plane is $2(x+1)-4(y-3)+3(z-2)=0$ (i.e.) $2x-4y+3z+8=0$.

Example 12.13

Find the equation of the plane passing through the points (9, 3, 6) and (2, 2, 1) and perpendicular to the plane $2x + 6y + 6z - 9 + 0$.

Solution

Any plane passing through the point (9, 3, 6) is

$$A(x-9) + B(y-3) + C(z-6) = 0 \tag{12.40}$$

The plane also passes through the point (2, 2, 1).

$$\therefore \quad \rightarrow -7A - B - 5C = 0 \text{ or } 7A + B + 5C = 0 \tag{12.41}$$

The plane (12.40) is perpendicular to the plane.

$$2x + 6y + 6z - 9 = 0 \tag{12.42}$$

$$\therefore \quad 2A + 6B + 6C = 0 \tag{12.43}$$

$$\text{(i.e.)} \quad 7A + B + 5C = 0$$

$$2A + 6B + 6C = 0$$

$$\frac{A}{6-30} = \frac{B}{10-42} = \frac{C}{42-2} \Rightarrow \frac{A}{-24} = \frac{B}{-32} = \frac{C}{40} \text{ or } \frac{A}{3} = \frac{B}{4} = \frac{C}{-5}$$

Therefore, the equation of the required plane is $3(x-9) + 4(y-3) - 5(z-6) = 0$.

$$\therefore \quad 3x + 4y - 5z = 9$$

Example 12.14

Show that the following points (0, –1, 0), (2, 1, –1), (1, 1, 1) and (3, 3, 0) are coplanar.

Solution

The equation of the plane passing through the point (0, –1, 0) is $Ax + B(y+1) + Cz = 0$.

This plane passes through the points (2, 1, –1) and (1, 1, 1).

$$\therefore \quad 2A + 2B - C = 0$$

$$A + 2B + C = 0$$

$$\therefore \quad \frac{A}{4} = \frac{B}{-3} = \frac{C}{2}$$

Therefore, the equation of the plane is $4x - 3(y + 1) + 2z = 0$.

$$\therefore \quad 4x - 3y + 2z - 3 = 0$$

Substituting $x = 3, y = 3, z = 0$ we get $12 - 9 - 3 = 0$ which is true.

Therefore, the plane passes through the points (3, 3, 0) and hence the four given points are coplanar.

Example 12.15

Find for what values of λ, the points $(0, -1, \lambda)$, (4, 5, 1), (3, 9, 4) and (−4, 4, 4) are coplanar.

Solution

The equation of the plane passing through the point (4, 5, 1) is $A(x - 4) + B(y - 5) + C(z - 1) = 0$.

This plane passes through the points (3, 9, 4) and (−4, 4, 4).

$$\therefore \quad -A + 4B + 3C = 0$$

$$-8A - B + 3C = 0$$

$$\frac{A}{15} = \frac{B}{-21} = \frac{C}{33} \text{ or } \frac{A}{5} = \frac{B}{-7} = \frac{C}{11}$$

Therefore, the equation of plane is $5(x - 4) - 7(y - 5) + 11(z - 1) = 0$.

$$\text{(i.e.)} \quad 5x - 7y + 11z + 4 = 0$$

If this plane passes through the point $(0, -1, \lambda)$ then $0 + 7 + 11\lambda + 4 = 0$.
$\therefore \lambda = -1$

Example 12.16

A variable plane moves in such a way that the sum of the reciprocals of the intercepts on the coordinate axes is a constant. Show that the plane passes through a fixed point.

Solution

Let the equation of the plane be

$$\frac{x}{a} + \frac{y}{b} + \frac{z}{c} = 1 \tag{12.44}$$

Given that the sum of the reciprocals of the intercepts is a constant.

$$\therefore \quad \frac{1}{a} + \frac{1}{b} + \frac{1}{c} = k$$

$$\text{(i.e.)} \quad \frac{1}{ak}+\frac{1}{bk}+\frac{1}{ck}=1 \tag{12.45}$$

(12.44) – (12.45) gives $\frac{1}{a}\left(x-\frac{1}{k}\right)+\frac{1}{b}\left(y-\frac{1}{k}\right)+\frac{1}{c}\left(z-\frac{1}{k}\right)=0$

This planes passes through the fixed point $\left(\frac{1}{k},\frac{1}{k},\frac{1}{k}\right)$.

Example 12.17

A point P moves on fixed plane $\frac{x}{a}+\frac{y}{b}+\frac{z}{c}=1$ and the plane through P perpendicular to OP meets the axes in A, B and C. If the planes through A, B and C are parallel to the coordinates planes meet in a point then show that the locus of Q is $\frac{1}{x^2}+\frac{1}{y^2}+\frac{1}{z^2}=\frac{1}{ax}+\frac{1}{by}+\frac{1}{cz}$.

Solution

The given plane is

$$\frac{x}{a}+\frac{y}{b}+\frac{z}{c}=1 \tag{12.46}$$

Let P be the point (α, β, γ). The plane passes through P.

$$\therefore \quad \frac{\alpha}{a}+\frac{\beta}{b}+\frac{\gamma}{c}=1 \tag{12.47}$$

The equation of the plane normal to OP is

$$\alpha(x-\alpha)+\beta(y-\beta)+\gamma(z-\gamma)=0$$

$$\text{(i.e.)}\ \alpha x+\beta y+\gamma z=\alpha^2+\beta^2+\gamma^2 \tag{12.48}$$

The intercepts made by this plane on the coordinate axes are $x=\frac{\alpha^2+\beta^2+\gamma^2}{\alpha}, y=\frac{\alpha^2+\beta^2+\gamma^2}{\beta}, z=\frac{\alpha^2+\beta^2+\gamma^2}{\gamma}$.

If these planes meet at (x_1, y_1, z_1) then

$$x_1=\frac{\alpha^2+\beta^2+\gamma^2}{\alpha}, y_1=\frac{\alpha^2+\beta^2+\gamma^2}{\beta}, z_1=\frac{\alpha^2+\beta^2+\gamma^2}{\gamma} \tag{12.49}$$

Now we have to eliminate α, β, γ using (12.47) and (12.49).

From (12.49),

$$\frac{1}{x_1^2}+\frac{1}{y_1^2}+\frac{1}{z_1^2}=\frac{1}{\alpha^2+\beta^2+\gamma^2} \tag{12.50}$$

From (12.47) and (12.49), $\Sigma\, \alpha^2\left(\frac{1}{ax_1}+\frac{1}{by_1}+\frac{1}{cz_1}\right)=1$ or

$$\frac{1}{ax_1}+\frac{1}{by_1}+\frac{1}{cz_1}=\frac{1}{\alpha^2+\beta^2+\gamma^2} \tag{12.51}$$

Therefore, the locus of (x_1, y_1, z_1) is $\frac{1}{x^2}+\frac{1}{y^2}+\frac{1}{z^2}=\frac{1}{ax}+\frac{1}{by}+\frac{1}{cz}$ from (12.50) and (12.51).

Example 12.18

If from the point $P(a, b, c)$ perpendiculars PL, PM be drawn to YZ- and ZX-planes, find the equation of the plane OLM.

Solution

P is the point (a, b, c). PL is drawn perpendicular to YZ-plane. Therefore, the coordinates of L are $(0, b, 0)$. PM is drawn perpendicular to ZX-plane. Therefore, the coordinates of M are $(0, 0, c)$. We have to find the equation of the plane OLM. The equation of the plane passing through $(0, 0, 0)$ is $Ax + By + Cz = 0$.This plane also passes through $(0, b, c)$, $(a, 0, c)$.

$$\therefore \quad O \cdot A + Bb + Cc = 0$$
$$Aa + O \cdot B + Cc = 0$$
$$\therefore \quad \frac{A}{bc}=\frac{B}{ca}=\frac{C}{-ab}$$

The equation of the plane OLM is $bcx + cay - abz = 0$.

Example 12.19

Show that $\left(\frac{-1}{2}, 2, 0\right)$ is the circumcentre of the triangle formed by the points $(1, 1, 0)$, $(1, 2, 1)$ and $(-2, 2, -1)$.

Solution

A, B and C are the points $(1, 1, 0)$, $(1, 2, 1)$ and $(-2, 2, -1)$ and P is the point $\left(\frac{-1}{2}, 2, 0\right)$. To prove that P is the circumcentre of the triangle ABC, we have to show that:

(i) the points P, A, B and C are coplanar and
(ii) $PA = PB = PC$.

The equation of the plane through the point (1, 1, 0) is $A(x-1)+B(y-1)+C(z-0)=0$.

This plane also passes through (1, 2, 1) and (–2, 2, –1).

$$0A+B+C=0$$

$$-3A+B-C=0$$

$$\frac{A}{-2}=\frac{B}{-3}=\frac{C}{3}$$

Therefore, the equation of the plane ABC is $-2(x-1)-3(y-1)+3z=0$.

$$\text{(i.e.)}\quad 2x+3y-3z-5=0$$

Substituting the coordinates of $P\left(\frac{-1}{2}, 2, 0\right)$ we get $-1+6-0-5=0$ which is true.

Therefore, the points P, A, B and C are coplanar.

Now

$$PA^2=\left(-\frac{1}{2}-1\right)^2+(2-1)^2+(0-0)^2=\frac{13}{4}$$

$$PB^2=\left(-\frac{1}{2}-1\right)^2+(2-2)^2+(0-1)^2=\frac{13}{4}$$

$$PC^2=\left(-\frac{1}{2}+2\right)^2+(2-2)^2+(0-1)^2=\frac{13}{4}$$

$$\therefore PA=PB=PC=\sqrt{\frac{13}{4}}.$$

Therefore, P is the circumcentre of the triangle ABC.

Example 12.20

Find the ratio in which the line joining the points (2, –1, 4) and (6, 2, 4) is divided by the plane $x+2y+3z+5=0$.

Solution

Let the plane $x+2y+3z+5=0$ divide the line joining the points P (2, –1, 4) and Q (6, 2, 4) in the ratio $\lambda : 1$.

Then the point of division is $\left(\frac{6\lambda+2}{\lambda+1}, \frac{2\lambda-1}{\lambda+1}, \frac{4\lambda+4}{\lambda+1}\right)$

This point lies on the plane $x + 2y + 3z + 5 = 0$.

$$\therefore \quad \frac{6\lambda+2}{\lambda+1}+2\left(\frac{2\lambda-1}{\lambda+1}\right)+3\left(\frac{4\lambda+4}{\lambda+1}\right)+5=0$$

$$\text{(i.e.)} \quad (6\lambda+2)+2(2\lambda-1)+3(4\lambda+4)+5(\lambda+1)=0$$

$$6\lambda+2+4\lambda-2+12\lambda+12+5\lambda+5=0$$

$$27\lambda+17=0 \text{ or } \lambda=\frac{-17}{27}$$

Therefore, the plane divides the line externally in the ratio 17:27.

Example 12.21

A plane triangle whose sides are of length a, b, and c is placed so that the middle points of the sides are on the axes. If α, β and γ are intercepts on the axes then show that the equation of the plane is $\frac{x}{\alpha}+\frac{y}{\beta}+\frac{z}{\gamma}=1$ where $\alpha=\frac{1}{2\sqrt{2}}\sqrt{b^2+c^2-a^2}$, $\beta=\frac{1}{2\sqrt{2}}\sqrt{c^2+a^2-b^2}$, and $\gamma=\frac{1}{2\sqrt{2}}\sqrt{a^2+b^2-c^2}$.

Solution

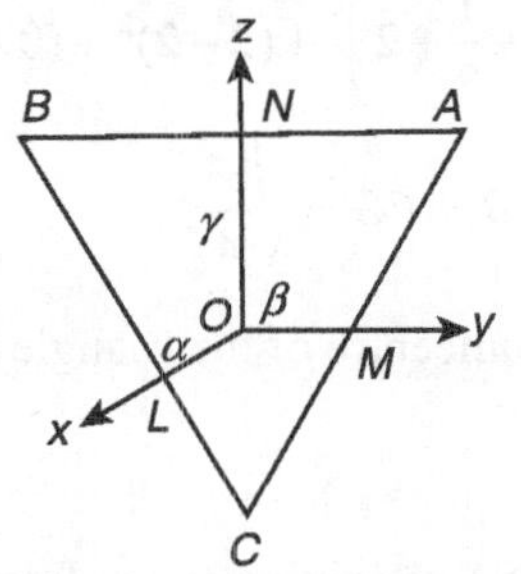

The equation of the plane is $\frac{x}{\alpha}+\frac{y}{\beta}+\frac{z}{\gamma}=1$.

Let the plane meet the axes at L, M, N respectively.
$L(\alpha, 0, 0)$, $M(0, \beta, 0)$, $N(0, 0, g)$

$$LM^2=\frac{1}{4}AB^2$$

$$\alpha^2 + \beta^2 = \frac{1}{4}c^2 \tag{12.52}$$

$$\beta^2 + \gamma^2 = \frac{1}{4}a^2 \tag{12.53}$$

$$\gamma^2 + \alpha^2 = \frac{1}{4}b^2 \tag{12.54}$$

(12.52) + (12.53) – (12.54) gives,

$$2\alpha^2 = \frac{1}{4}(b^2 + c^2 - a^2)$$

$$\alpha^2 = \frac{1}{8}(b^2 + c^2 - a^2)$$

$$\alpha = \frac{1}{2\sqrt{2}}\sqrt{b^2 + c^2 - a^2} \tag{12.55}$$

Similarly,
$$\beta = \frac{1}{2\sqrt{2}}\sqrt{c^2 + a^2 - b^2} \tag{12.56}$$

$$\gamma = \frac{1}{2\sqrt{2}}\sqrt{a^2 + b^2 - c^2} \tag{12.57}$$

Therefore, the equation of the plane is $\frac{x}{\alpha} + \frac{y}{\beta} + \frac{z}{\gamma} = 1$ where α, β, γ are given by (12.55), (12.56) and (12.57).

Let $(x_1, y_1, z_1), (x_2, y_2, z_2)$ and (x_3, y_3, z_3) be the vertices of the ΔABC. Then

$$x_2 + x_3 = 2\alpha$$

$$x_3 + x_1 = 0$$

$$x_1 + x_2 = 0$$

Adding we get $2(x_1 + x_2 + x_3) = 2\alpha$ or $x_1 + x_2 + x_3 = \alpha$

$$\therefore\ x_1 + 2\alpha = \alpha \quad \therefore x_1 = -\alpha$$

$$\therefore x_2 = \alpha \text{ and } x_3 = \alpha$$

Similarly,

$$y_1 = \beta, y_2 = -\beta, y_3 = \beta$$

$$z_1 = \gamma, z_2 = -\gamma, z_3 = \gamma$$

Therefore, the vertices of the triangle are $(-\alpha, \beta, \gamma)$ $(\alpha, -\beta, \gamma)$ and $(\alpha, \beta, -\gamma)$.

Example 12.22

Find the angle between the planes $2x - y + z = 6$, $x + y + 2z = 3$.

Solution

The direction ratios of the normal to the planes are 2, –1, 1 and 1, 1, 2. The direction cosines of the normal are $\frac{2}{\sqrt{6}}, \frac{-1}{\sqrt{6}}, \frac{1}{\sqrt{6}}, \frac{1}{\sqrt{6}}, \frac{1}{\sqrt{6}}, \frac{2}{\sqrt{6}}$. If θ is the angle between the planes, then

$$\cos\theta = ll_1 + mm_1 + nn_1 = \frac{2}{\sqrt{6}}.\frac{1}{\sqrt{6}} - \frac{1}{\sqrt{6}}.\frac{1}{\sqrt{6}} + \frac{1}{\sqrt{6}}.\frac{2}{\sqrt{6}} = \frac{3}{6} = \frac{1}{2}$$

$$\therefore \theta = \frac{\pi}{3}$$

Example 12.23

Prove that the plane $x + 2y + 2z = 0$, $2x + y - 2z = 0$ are at right angles.

Solution

The direction ratios of the normals to the planes are 1, 2, 2 and 2, 1, –2. If the lines are to be perpendicular then $a_1a_2 + b_1b_2 + c_1c_2 = 0$. Hence, $a_1a_2 + b_1b_2 + c_1c_2 = 2 + 2 - 4 = 0$.

Therefore, the normals are perpendicular and hence the planes are perpendicular.

Example 12.24

Find the equation of the plane containing the line of intersection of the planes $x + y + z - 6 = 0$, $2x + 3y + 4z + 5 = 0$ and passing through the point (1, 1, 1).

Solution

The equation of any plane containing the line is $x + y + z - 6 = \lambda(2x + 3y + 4z + 5) = 0$. If this line passes through the point (1, 1, 1) then, $1 + 1 + 1 - 6 + \lambda(2 + 3 + 4 + 5) = 0$.

$$14\lambda = 3 \text{ or } \lambda = \frac{3}{14}.$$

Therefore, the equation of the required plane is $x + y + z - 6 + \frac{3}{14}(2x + 3y + 4z + 5) = 0$.

$$\Rightarrow 14x + 14y + 14z - 84 + (6x + 9y + 12z + 15) = 0$$

$$20x + 23y + 26z - 69 = 0$$

Example 12.25

Find the equation of the plane which passes through the intersection of the planes $2x + 3y + 10z - 8 = 0$, $2x - 3y + 7z - 2 = 0$ and is perpendicular to the plane $3x - 2y + 4z - 5 = 0$.

Solution

The equation of any plane passing through the intersection of the planes $2x + 3y + 10z - 8 = 0$ and $2x - 3y + 7z - 2 = 0$ is $2x + 3y + 10z - 8 + \lambda(2x - 3y + 7z - 2) = 0$.

The direction ratios of the normal to this plane are $2 + 2\lambda$, $3 - 3\lambda$, $10 + 7\lambda$. The direction ratios of the plane $3x - 2y + 4z - 5 = 0$ are 3, –2, 4. Since these two planes are perpendicular, $3(2 + 2\lambda) - 2(3 - 3\lambda) + 4(10 + 7\lambda) = 0$.

$$6 + 6\lambda - 6 + 6\lambda + 40 + 28\lambda = 0 \text{ or } 40\lambda = -40 \text{ or } \lambda = -1$$

Therefore, the required plane is $2x + 3y + 10z - 8 - (2x - 3y + 7z - 2) = 0$.

$$\therefore\ 6y + 3z - 6 = 0 \text{ or } 2y + z - 2 = 0$$

Example 12.26

The plane $x - 2y + 3z = 0$ is rotated through a right angle about its line of intersection with the plane $2x + 3y - 4z + 2 = 0$. Find the equation of the plane in its new position.

Solution

The plane $x - 2y + 3z = 0$ is rotated about the line of intersection of the planes

$$x - 2y + 3z = 0 \qquad (12.58)$$

$$2x + 3y - 4z + 2 = 0 \qquad (12.59)$$

The new position of the plane (12.58) passes through the line of intersection of the two given planes. Therefore, its equation is

$$x - 2y + 3z + \lambda(2x + 3y - 4z + 2) = 0 \qquad (12.60)$$

The plane (12.60) is perpendicular to the plane (12.58).

$$1(1 + 2\lambda) - 2(-2 + 3\lambda) + 3(3 - 4\lambda) = 0 \text{ or } -16\lambda + 14 = 0 \text{ or } \lambda = \frac{7}{8}.$$

Therefore, the equation of the plane (12.58) in its new position is $x - 2y + 3z + \frac{7}{8}(2x + 3y - 4z + 2) = 0$.

$$\text{(i.e.)} \quad 22x + 5y - 42 + 14 = 0$$

Example 12.27

The line $lx + my = 0$ is rotated about its line of intersection with the plane $z = 0$ through an angle α. Prove that the equation of the plane is $lx + my \pm z\sqrt{l^2 + m^2}$ $\tan\alpha = 0$.

Solution

Any plane passing through the intersection of $lx + my = 0$ and $z = 0$ is

$$lx + my + \lambda z = 0 \qquad (12.61)$$

The plane $lx + my + \lambda z = 0$ is rotated through an angle α along the plane (12.61).

$$\therefore \cos\alpha = \frac{l.l + m.m + \lambda.0}{\sqrt{l^2 + m^2}\sqrt{l^2 + m^2 + \lambda^2}}$$

$$\cos\alpha = \frac{\sqrt{l^2 + m^2}}{\sqrt{l^2 + m^2 + \lambda^2}}$$

$$l^2 + m^2 + \lambda^2 = (l^2 + m^2)\sec^2\alpha \Rightarrow \lambda^2 = (l^2 + m^2)(\sec^2\alpha - 1)$$

$$= (l^2 + m^2)\tan^2\alpha$$

$$\lambda = \pm\sqrt{l^2 + m^2}\,\tan\alpha$$

Therefore, the equation of the plane in its new position is given by

$$lx + my \pm z\sqrt{l^2 + m^2}\,\tan\alpha = 0$$

Example 12.28

Find the equation of the plane passing through the line of intersection of the planes $2x - y + 5z - 3 = 0$ and $4x + 2y - z + 7 = 0$ and parallel to z-axis.

Solution

The equation of the plane passing through the line of intersection of the given planes is $2x - y + 5z - 3 + \lambda(4x + 2y - z + 7) = 0$. If the plane is parallel to

z-axis, its normal is perpendicular to z-axis. The directions of the normal to the plane are $2 + 4\lambda, -1 + 2\lambda, 5 - \lambda$. The direction ratios of the z-axis are 0, 0, 1.

$$\therefore \quad (2 + 4\lambda)0 + (-1 + 2\lambda)0 + (5 - \lambda)1 = 0.$$

$$\therefore \quad \lambda = 5$$

Hence, the equation of the required plane is $(2x - y + 5z - 3) + 5(4x + 2y - z + 7) = 0$.

(i.e.) $$22x + 9y + 32 = 0$$

Example 12.29

Find the distance of the points (2, 3, –5), (3, 4, 7) from the plane $x + 2y - 2z = 9$ and prove that these points lie on the opposite sides of the plane.

Solution

Let the line joining the points $P(2, 3, -5)$ and $Q(3, 4, 7)$ be divided by the plane in the ratio λ:1.

Then $$\lambda = -\frac{Ax_1 + By_1 + Cz_1 + D}{Ax_2 + By_2 + Cz_2 + D} = -\frac{2+6+10-9}{3+8-14-9} = \frac{-9}{-12} = \frac{3}{4}$$

= a positive quantity

Therefore, the points P and Q lie on the opposite side of the plane. The perpendicular distance from (2, 3, –5) to the plane $x + 2y - 2z - 9 = 0$ is

$$p_1 = \frac{2+6+10-9}{\sqrt{1+4+4}} = \frac{9}{3} = 3 \text{ units.}$$

The perpendicular distance from (3, 4, 7) to the plane is

$$p_2 = \frac{3+8-14-9}{\sqrt{1+4+4}} = \frac{-12}{3} = -4 \text{ units.}$$

$$\therefore \quad |p_2| = 4 \text{ units}$$

Note 12.29.1: Since p_1 and p_2 are of opposite signs the points are on the opposite sides of the plane.

Example 12.30

Prove that the points (2, 3, –5) and (3, 4, 7) lie on the opposite sides of the plane which meets the axes in A, B and C such that the centroid of the triangle A, B and C is the points (1, 2, 4).

Solution

Let the equation of the plane be $\frac{x}{a}+\frac{y}{b}+\frac{z}{c}=1$. Then the coordinates of *A, B* and *C* are $(a, 0, 0)$, $(0, b, 0)$ and $(0, 0, c)$. The centroid is $\left(\frac{a}{3},\frac{b}{3},\frac{c}{3}\right)$. But the centroid is given as (1, 2, 4).

$$\therefore \quad \frac{a}{3}=1 \text{ or } a=1$$

$$\frac{b}{3}=2 \text{ or } b=6$$

$$\frac{c}{3}=4 \text{ or } c=12.$$

Therefore, the equation of the plane *ABC* is $\frac{x}{3}+\frac{y}{6}+\frac{z}{12}=1.$

$$\text{(i.e.)} \quad 4x+2y+z=12$$

Let the line joining the points $P(2, 3, -5)$ and $Q(3, 4, 7)$ be divided by the plane in the ratio λ:1.

Then $\lambda=-\frac{8+6-5-12}{12+8+7-12}=\frac{-3}{15}=-\frac{1}{5}.$

Therefore, the points lie on the opposite sides of the plane.

Example 12.31

Find the distance between the parallel planes $2x-2y+z+3=0$, $4x-4y+2z+5=0$.

Solution

Let (x_1, y_1, z_1) be a point on the plane $2x-2y+z+3=0$

$$\therefore \quad 2x_1-2y_1+z_1+3=0 \tag{12.62}$$

Then the distance between the parallel planes is equal to the distance from (x_1, y_1, z_1) to the other plane.

$$=\pm\frac{4x_1-4y_1+2z_1+5}{\sqrt{4^2+4^2+2^2}}=\pm\frac{(-6+5)}{6}=\frac{1}{6} \quad \text{using (12.62)}$$

Note 12.31.1: The distance between the parallel planes $ax+by+cz+d=0$ and $ax+by+cz+d_1=0$ is $\frac{|d-d_1|}{\sqrt{a^2+b^2+c^2}}$.

On dividing the equation $4x - 4y + 2z + 5 = 0$ by 2, we get

$$2x - 2y + z + \frac{5}{2} = 0$$

$$\text{Distance between the planes} = \frac{3 - \frac{5}{2}}{\sqrt{9}} = \frac{\left(\frac{1}{2}\right)}{3} = \frac{1}{6}.$$

Example 12.32

A plane is drawn through the line of $x + y = 1, z = 0$ to make an angle $\sin^{-1}\left(\frac{1}{3}\right)$ with the plane $x + y + z = 0$. Prove that two such planes can be drawn. Find their equation. Show that the angle between the planes is $\cos^{-1}\left(\frac{7}{9}\right)$.

Solution

The equation of the plane through the line

$$x + y = 1, z = 0 \text{ is } x + y - 1 + \lambda z = 0 \qquad (12.63)$$

The direction ratios of this plane is 1, 1, λ. Also the direction ratios of the plane $x + y + z = 0$ are 1, 1, 1. If θ is the angle between these two planes then

$$\cos\theta = \frac{1+1+\lambda}{\sqrt{1^2+1^2+\lambda^2}\sqrt{1^2+1^2+1^2}} = \frac{\lambda+2}{\sqrt{3(\lambda^2+2)}}.$$

$$\cos^2\theta = \frac{(\lambda+2)^2}{\sqrt{3\lambda^2+6}} \Rightarrow 1 - \sin^2\theta = \frac{\lambda^2+4\lambda+4}{3\lambda^2+6}$$

$$1 - \frac{1}{9} = \frac{\lambda^2+4\lambda+4}{3\lambda^2+6} \Rightarrow \frac{8}{9} = \frac{\lambda^2+4\lambda+4}{3\lambda^2+6}$$

$$\text{or } 5\lambda^2 - 12\lambda + 4 = 0 \to (\lambda - 2)(5\lambda - 2) = 0 \to \lambda = 2 \text{ or } \frac{2}{5}.$$

From (12.63), the equations of the required planes are $x + y + 2z = 1$ and $5x + 5y + 2z - 5 = 0$. If θ is the angle between these two planes then

$$\cos\theta = \frac{5+5+4}{\sqrt{6}\sqrt{54}} = \frac{14}{18}.$$

$$\therefore \theta = \cos^{-1}\left(\frac{7}{9}\right)$$

Example 12.33

Find the bisectors of the angles between the planes $2x - y + 2z + 3 = 0$, $3x - 2y + 6z + 8 = 0$; also find out which plane bisects the acute angle.

Solution

The two given planes are

$$2x - y + 2z + 3 = 0 \tag{12.64}$$

$$3x - 2y + 6z + 8 = 0 \tag{12.65}$$

The equations of the bisectors are $\dfrac{2x-y+2z+3}{\sqrt{4+1+4}} = \pm\dfrac{3x-2y+6z+8}{\sqrt{9+4+36}}$

$$7(2x - y + 2z + 3) = \pm 3(3x - 2y + 6z + 8)$$

$$(14x - 7y + 14z + 21) = \pm(9x - 6y + 18z + 24)$$

$$\text{(i.e.)}\quad 5x - y - 4z - 3 = 0 \tag{12.66}$$

$$23x - 13y + 32z + 45 = 0 \tag{12.67}$$

Let θ be the angle between the planes (12.64) and (12.66) then

$$\cos\theta = \frac{10+1-8}{\sqrt{25+1+16}\sqrt{4+1+4}} = \frac{1}{\sqrt{42}}$$

$$\tan^2\theta = \sec^2\theta - 1 = 42 - 1 = 41$$

$$\therefore \tan\theta = \sqrt{41} > 1$$

Hence $\theta > 45°$. The plane $5x - y - 4z - 3 = 0$ bisects the obtuse angle between the planes (12.64) and (12.65).

Therefore, $23x - 13y + 32z + 45 = 0$ bisects the acute angle between the planes (12.64) and (12.65).

Example 12.34

Prove that the equation $2x^2 - 2y^2 + 4z^2 + 2yz + 6zx + 3xy = 0$ represents a pair of planes and angle between them is $\cos^{-1}\left(\frac{4}{9}\right)$.

Solution

Here $a = 2, b = -2, c = 4, f = 1, g = 3, h = \frac{3}{2}$.

Now, $abc + 2fgh - af^2 - bg^2 - ch^2 = 0$

$$\Rightarrow -16+9-2+18-9=0$$

Hence, the given equation represents a pair of planes. Let θ be the angle between the planes. Then $\tan\theta = \pm\dfrac{2\sqrt{f^2+g^2+h^2-bc-ca-ab}}{a+b+c}$

$$\tan\theta = \pm\frac{2\sqrt{1+9+\frac{9}{4}+8-8+4}}{4} = \frac{\sqrt{65}}{4}$$

$$\sec^2\theta = 1+\tan^2\theta = \frac{81}{16} \text{ and } \cos\theta = \frac{4}{9}$$

$$\therefore \theta = \cos^{-1}\left(\frac{4}{9}\right)$$

Exercises

Section A

1. If P is the point (2, 3, –1), find the equation of the plane passing through P and perpendicular to OP.

   Ans.: $2x+3y-z-14=0$

2. The foot of the perpendicular from the origin to a plane is (12, –4, –3). Find its equation.

   Ans.: $12x-4y-3z+69=0$

3. Find the intercepts made by the plane $4x-3y+2z-7=0$ on the coordinate axes.

   Ans.: $\frac{7}{4}, \frac{-7}{3}, \frac{7}{2}$

4. A plane meets the coordinate axes at A, B and C such that the centroid of the triangle is the point (a, b, c). Show that the equation of the plane is $\frac{x}{a}+\frac{y}{b}+\frac{z}{c}=3$.

5. Find the equation of the plane that passes through the point (2, –3, 1) and is perpendicular to the line joining the points (3, 4, –1) and (2, –1, 5).

   Ans.: $x+5y-6z+19=0$

6. O is the orgin and A is the point (a, b, c). Find the equation of the plane perpendicular to A.

   Ans.: $ax+by+cz-(a^2+b^2+c^2)=0$

7. Find the equation of the plane passing through the points:
   (i) (8, −2, 2), (2, 1, −4), (2, 4, −6)
   (ii) (2, 2, 1), (2, 3, 2), (−1, 3, 0)
   (iii) (2, 3, 4), (−3, 5, 1), (4, −1, 2)

   Ans.: (i) $2x - 2y - 2z = 14$, (ii) $2x + 3y - 3z - 7 = 0$, (iii) $x + y - z - 1 = 0$

8. Show that the points (0, −1, −1), (4, 5, 1),(3, 9, 4) and (−4, 4, 4) lie on a plane.

9. Show that the points (0, −1, 0), (2, 1, −1), (1, 1, 1) and (−3, 3, 0) are coplanar.

10. Find the equation of the plane through the three points (2, 3, 4), (−3, 5, 1) and (4, −1, 2). Also find the angles which the normal to the plane makes with the axes of reference.

    Ans.: $x + y - z - 1 = 0; \cos^{-1}\left(\frac{1}{\sqrt{3}}\right), \cos^{-1}\left(\frac{1}{\sqrt{3}}\right), \cos^{-1}\left(\frac{-1}{\sqrt{3}}\right)$

11. Find the equation of the plane which passes through the point (2, −3, 4) and is parallel to the plane $2x - 5y - 7z + 15 = 0$.

    Ans.: $2x - 5y - 7z + 9 = 0$

12. Find the equation of the plane through (1, 3, 2) and perpendicular to the planes $x + 2y + 2z - 5 = 0$ and $3x + 3y + 3z - 8 = 0$.

    Ans.: $2x - 4y + 3z + 8 = 0$

13. Find the equation of the plane which passes through the point (2, 2, 4) and perpendicular to the planes $2x - 2y - 4z + 3 = 0$ and $3x + y + 6z - 4 = 0$.

    Ans.: $x - 3y - z - 4 = 0$

14. Find the equation of the plane which passes through the points (9, 3, 6) and (2, 2, 1) and perpendicular to the plane $2x + 4y + 6z - 9 = 0$.

    Ans.: $3x + 4y - 5z - 9 = 0$

15. Find the equation of the straight line passing through the points (−1, 1, 1) and (1, −1, 1) and perpendicular to the plane $x + 2y + 2z - 5 = 0$.

    Ans.: $2x + 2y - 3z + 3 = 0$

16. Find the equation of the plane which passes through the points (2, 3, 1), (4, −5, 3) and are parallel to the coordinate axes.

    Ans.: $y + 4z - 7 = 0, x - z - 1 = 0, 4x + y - 11 = 0$

17. Find the equation of the plane which passes the point (1, 2, 3) and parallel to $3x + 4y - 5z = 0$.

    Ans.: $3x + 4y - 5z + 4 = 0$

18. Find the equation of the plane bisecting the line joining the points (2, 3, −1) and (−5, 6, 3) at right angles.

    Ans.: $x - y - z + 7 = 0$

19. A variable plane is at a constant distance p from the origin and meets the axes in A, B and C. Show that the locus of centroid of the tetrahedron $OABC$ is $x^{-2} + y^{-2} + z^{-2} = 16p^{-2}$.

20. $OABC$ is a tetrahedron in which OA, OB and OC are mutually perpendicular. Prove that the perpendicular from O to the base ABC meets it at its orthocentre.

21. Through the point $P(a, b, c)$ a plane is drawn at right angles to OP to meet the axes in A, B and C. Prove that the area of the triangle ABC is $\frac{p^5}{2abc}$ where p is the length of OP.

22. A plane contains the points $A(-4, 9, -9)$ and $B(5, -9, 6)$ and is perpendicular to the line which joins B and $C(4, -6, k)$. Obtain k and the equation of the plane.

Ans.: $k = \frac{51}{5}$, $5x - 15y - 21z = 34$

23. Find the distance between the parallel planes $2x + y + 2z - 8 = 0$ and $4x + 2y + 4z + 5 = 0$.

Ans.: $\frac{7}{2}$

24. Find the locus of the point, the sum of the squares of whose distances from the planes $x + y + z = 0$, $x = z = 0$, $x - 2y + z = 0$ is 9.

Ans.: $x^2 + y^2 + z^2 = 9$

25. Find the equation of the plane which is at a distance 1 unit from the origin and parallel to the plane $3x + 2y - z + 2 = 0$.

Ans.: $3x + 2y \quad z \pm \sqrt{14} = 0$

26. The plane $\frac{x}{a} + \frac{y}{b} + \frac{z}{c} = 1$ meets the coordinate axes in A, B and C, respectively.

Show that the area of the triangle ABC is $\frac{1}{2}[b^2c^2 + c^2a^2 + a^2b^2]^{\frac{1}{2}}$.

27. Show that the equations $by + cz + d = 0$, $cz + ax + d = 0$, $ax + by + d = 0$ represent planes parallel to OX, OY and OZ, respectively.

28. Show that the points $(2, 3, -5)$ and $(3, 4, 7)$ lie on the opposite sides of the plane meeting the axes in A, B and C such that the centroid of the triangle ABC is the point $(1, 2, 4)$.

29. Find the locus of the point such that the sum of the squares of its distances from the planes $x + y + z = 0$ and $x - 2y + z = 0$ is equal to its distance from the plane $x - z = 0$.

Ans.: $y^2 - 2xz = 0$

30. Find the locus of the point whose distance from the origin is 7 times its distance from the plane $2x + 3y - 6z = 2$.

Ans.: $3x^2 + 8y^2 + 53z^2 - 36yz - 24zx + 12xy - 8x - 12y + 24z + 14 = 0$

31. Prove that the equation of the plane passing through the points $(1, 1, 1)$, $(1, -1, 1)$ and $(-7, -3, -5)$ and is parallel to axis of y.

32. Determine the constant k so that the planes $x - 2y + kz = 0$ and $2x + 5y - z = 0$ are at right angles, and in that case find the plane through the point $(1, -1, -1)$ and perpendicular to both the given planes.

Ans.: $k = -8$, $14x - 5y + 3z - 16 = 0$

33. Prove that $3x - y - z + 11 = 0$ is the equation of the plane through $(-1, 6, 2)$ and perpendicular to the join of the points $(1, 2, 3)$ and $(-2, 3, 4)$.

34. A, B and C are points $(a, 0, 0)$, $(0, b, 0)$ and $(0, 0, c)$. Find the equation of the plane through BC which bisects OA. By symmetry write down the equations of the plane through CA bisecting OB and through AB bisecting OC. Show that these planes pass through $\left(\frac{a}{4}, \frac{b}{4}, \frac{c}{4}\right)$.

Section B

1. Find the equation of the plane through the intersection of the planes $x + 3y + 6 = 0$ and $3x - y - 4z = 0$ whose perpendicular distance from the origin is unity.

Ans.: $2x + y - 2z + 3 = 0$, $x - 2y - 2z - 3 = 0$

2. Find the equation of the plane through the intersection of the planes $x - 2y + 3z + 4 = 0$ and $2x - 3y + 4z - 7 = 0$ and the point $(1, -1, 1)$.

Ans.: $9x - 13y - 17z - 39 = 0$

3. Find the equation of the plane through the intersection of the planes $x + 2y + 3z + 4 = 0$ and $4x + 3y + 3z + 1 = 0$ and perpendicular to the plane $x + y + z + 9 = 0$ and show that it is perpendicular to xz-plane.

Ans.: $x - z = 2$

4. Find the equation of the plane through the point $(1, -2, 3)$ and the intersection of the planes $2x - y + 4z - 7 = 0$ and $x + 2y - 3z + 8 = 0$.

Ans.: $17x + 14y + 11z + 44 = 0$

5. Find the equation of the plane passing through the intersection of the planes $x + 2y + 3z + 4 = 0$ and $4x + 3y + 2z + 1 = 0$ and through the point $(1, 2, 3)$.

Ans.: $11x + 4y - 3z - 10 = 0$

6. Find the equation of the plane passing through the line of intersection of the planes $x - 2y - z + 3 = 0$ and $3x + 5y - 2z - 1 = 0$ which is perpendicular to the yz-plane.

Ans.: $11y + z - 10 = 0$

7. The plane $x + 4y - 5z + 2 = 0$ is rotated through a right angle about its line of intersection with the plane $3x + 2y + z + 1 = 0$. Find the equation of the plane in its new position.

Ans.: $20x + 10y + 12z + 5 = 0$

8. Are the planes given by the equations $3x + 4y + 5z + 10 = 0$ and $9x + 12y + 15z + 20 = 0$ parallel? If so find the distance between them.

Ans.: $\frac{\sqrt{2}}{3}$

9. Find the equation of the plane passing through the line of intersection of the planes $2x - y = 0$ and $3x - y = 0$ and are perpendicular to the plane $4x + 3y - 3z = 8$.

Ans.: $24x - 17y + 15z = 0$

10. Find the equation of the plane passing through the line of intersection of the planes $ax + by + cz + d = 0$ and $a_1x + b_1y + c_1z + d_1 = 0$ perpendicular to xy-plane.

Ans.: $(ac_1 - a_1c)x + (bc_1 - b_1c)y + (dc_1 - d_1c)z = 0$

11. Find the equation of the plane passing through the line of intersection of the planes $2x + 3y + 10z - 8 = 0$, $2x - 3y + 7z - 2 = 0$ and is perpendicular to the plane $3x - 2y + 4z - 5 = 0$.

Ans.: $2y + z - 2 = 0$

12. Obtain the equation of the planes bisecting the angles between the planes $x + 2y - 2z + 1 = 0$ and $12x - 4y + 3z + 5 = 0$. Also show that these two planes are at right angles.

Ans.: $23x - 38y + 35z + 2 = 0 \quad 49x + 14y - 17z + 28 = 0$

13. Find the equation of the plane that bisects the angle between the planes $3x - 6y - 2z + 5 = 0$ and $4x - 12y + 3z - 3 = 0$ which contain the origin. Does this plane bisect the acute angle?

Ans.: yes, $67x + 162y + 47z + 44 = 0$

14. Find the equation of the plane that bisects the acute angle between the planes $3x - 4y + 12z - 26 = 0$ and $x + 2y - 2z - 9 = 0$.

Ans.: $22x + 14y + 10z - 195 = 0$

15. Find the equation of the plane that bisects the obtuse angle between the planes $4x + 3y - 5z + 1 = 0$ and $12x + 5y - 13 = 0$.

Ans.: $8x - 14y - 13 = 0$

16. Show that the origin lies in the acute angle between the planes $x + 2y - 2z - 9 = 0$, $4x - 3y + 12z + 13 = 0$. Find the planes bisecting the angle between them and find the plane which bisects the acute angle.

Ans.: $x + 35y - 10z - 156 = 0$

17. Find the equation of the plane which bisects the acute angle between the planes $x + 2y + 2z - 3 = 0$ and $3x + 4y + 12z + 1 = 0$.

Ans.: $11x + 19y + 31z - 18 = 0$

18. **Prove that the equation $2x^2 - 6y^2 - 12z^2 + 18yz + 2zx + xy = 0$ represents a pair of planes and show that the angle between them is $\cos^{-1}\left(\frac{16}{21}\right)$.**

19. **Prove that the equation $\frac{3}{y-z} + \frac{4}{z-x} + \frac{5}{x-y} = 0$ represents a pair of planes.**

20. **If the equation $\phi(x, y, z) = ax^2 + by^2 + cz^2 + 2fyz + 2gzx + 2hxy = 0$ represents a pair of planes then prove that the product of the distances of the planes from (α, β, γ) is $\frac{\phi(\alpha,\beta,\gamma)}{\sqrt{\sum a^2 + 4\sum f^2 - 2\sum bc}}$.**

CHAPTER

13 Straight Line

13.1 INTRODUCTION

The intersection of two planes P_1 and P_2 is the locus of all the common points on both the planes P_1 and P_2. This locus is a straight line. Any given line can be uniquely determined by any of the two planes containing the line. Thus, a line can be regarded as the locus of the common points of two intersecting planes.

Let us consider the two planes

$$ax + by + cz + d = 0 \tag{13.1}$$

$$a_1x + b_1y + c_1z + d_1 = 0 \tag{13.2}$$

Any set of coordinates (x, y, z) which satisfy these two equations simultaneously will represent a point on the line of intersection of these two planes. Hence these two equations taken together will represent a straight line. It can be noted that the equation of x–axis are $y = 0, z = 0$. The equation of the y–axis is $x = 0, z = 0$ and the equation of the z–axis is $x = 0, y = 0$. The representation of the straight line by the equations $ax + by + cz + d = 0$ and $a_1x + b_1y + c_1z + d_1 = 0$ is called non–symmetrical form. Let us now derive the equations of a straight line in the symmetrical form.

13.2 EQUATION OF A STRAIGHT LINE IN SYMMETRICAL FORM

Let $A(x_1, y_1, z_1)$ be a point on the straight line and $P(x, y, z)$ be any point on the straight line. Let l, m, n be the direction cosines of the straight line. Let $OP = r$. The projections of AP on the coordinate axes are $x - x_1, y - y_1, z - z_1$. Also the projections of AP on the coordinate axes are given by lr, mr, nr. Then $x - x_1 = lr$, $y - y_1 = mr$ and $z - z_1 = nr$.

$$\therefore \frac{x-x_1}{l} = \frac{y-y_1}{m} = \frac{z-z_1}{n} = r$$

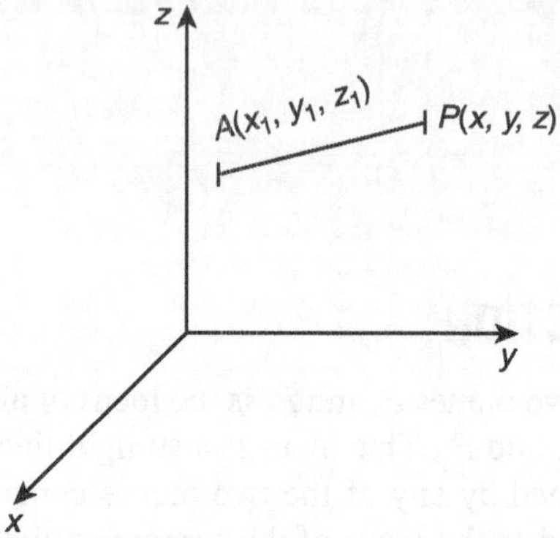

These equations are called the symmetrical form of the straight line.

Aliter:

We now derive the equations in symmetrical form from the vector equation of the straight line passing through a point and parallel to a vector.

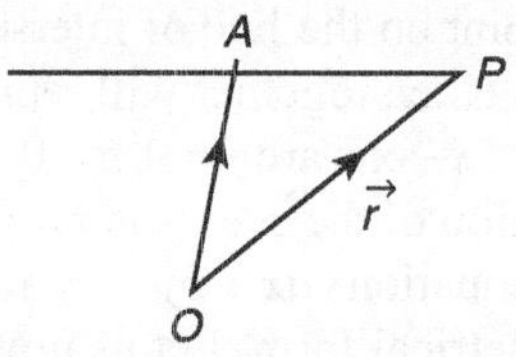

Let A be a given point on a straight line and P be any point on the straight line. Let $\vec{b}$ be a vector parallel to the line. Let O be the origin and $\overrightarrow{OA} = \vec{a}, \overrightarrow{OP} = \vec{r}$. Then

$$\overrightarrow{AP} = \overrightarrow{OP} - \overrightarrow{OA} = \vec{r} - \vec{a} \tag{13.3}$$

But

$$\overrightarrow{AP} = t\vec{b} \tag{13.4}$$

where t is a scalar.

From (13.3) and (13.4),

$$\vec{r} - \vec{a} = t\vec{b} \text{ or } \vec{r} = \vec{a} + t\vec{b} \tag{13.5}$$

This equation is true for all positions of P on the straight line and therefore this is the vector equation of the straight line. Let $\overrightarrow{OP} = x\vec{i} + y\vec{j} + z\vec{k}, \overrightarrow{OA} = x_1\vec{i} + y_1\vec{j} + z_1\vec{k}$ and $\vec{b} = l\vec{i} + m\vec{j} + n\vec{k}$.

Then from equation (13.4), we have

$$(x - x_1)\vec{i} + (y - y_1)\vec{j} + (z - z_1)\vec{k} = t\vec{b} = t(l\vec{i} + m\vec{j} + n\vec{k})$$

Equating the coefficients of $\vec{i}, \vec{j}$ and $\vec{k}$, we have

$$x - x_1 = tl, \quad y - y_1 = tm, \quad z - z_1 = tn$$

$$\therefore \frac{x - x_1}{l} = \frac{y - y_1}{m} = \frac{z - z_1}{n} = t$$

These are the cartesian equations of the straight line in symmetrical form.

Note 13.2.1: To express the equations of a straight line in symmetrical form we require (i) the coordinate of a point on the line and (ii) the direction cosines of the straight line.

Note 13.2.2: Any point on this line is $(x_1 + lr, y_1 + mr, z_1 + nr)$. Even if l, m and n are the direction ratios of the line, $(x_1 + lr, y_1 + mr, z_1 + nr)$ will represent a point on the line but r will not be distance between the points (x, y, z) and (x_1, y_1, z_1).

13.3 EQUATIONS OF A STRAIGHT LINE PASSING THROUGH THE TWO GIVEN POINTS

Let $P(x_1, y_1, z_1)$ and $Q(x_2, y_2, z_2)$ be two given points. The direction ratios of the line are $x_2 - x_1, y_2 - y_1, z_2 - z_1$.

Therefore, the equations of the straight line are $\dfrac{x - x_1}{x_2 - x_1} = \dfrac{y - y_1}{y_2 - y_1} = \dfrac{z - z_1}{z_2 - z_1}$.

Aliter:

Let O be the origin and P and Q be the points on the straight line and R be any point on the straight line.

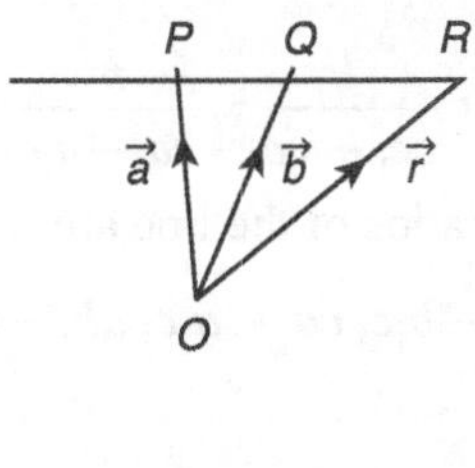

$$\overrightarrow{OP} = \vec{a}, \overrightarrow{OQ} = \vec{b}, \overrightarrow{OR} = \vec{r}$$

Then $\overrightarrow{PQ} = \overrightarrow{OQ} - \overrightarrow{OP} = \vec{b} - \vec{a}$. Also $\overrightarrow{PR} = \overrightarrow{OR} - \overrightarrow{OP} = \vec{r} - \vec{a}$.
But $\overrightarrow{PR} = t\overrightarrow{PQ}$ (i.e.) $\vec{r} - \vec{a} = t(\vec{b} - \vec{a})$

$$\vec{r} = \vec{a} + t(\vec{b} - \vec{a}) \tag{13.6}$$

This is the vector equation of the straight line.
Let $\vec{r} = x\vec{i} + y\vec{j} + z\vec{k}, \vec{a} = x_1\vec{i} + y_1\vec{i} + z_1\vec{k}, \vec{b} = x_2\vec{i} + y_2\vec{j} + z_2\vec{k}$.
Then from (13.6), we get

$$(x - x_1)\vec{i} + (y - y_1)\vec{j} + (z - z_1)\vec{k} = t[(x_2 - x_1)\vec{i} +$$

Equating the coefficients of $\vec{i}, \vec{j}$ and $\vec{k}$, we get

$$x - x_1 = t(x_2 - x_1); y - y_1 = t(y_2 - y_1); z - z_1 = k(z_2 - z_1)$$

$$\therefore \frac{x - x_1}{x_2 - x_1} = \frac{y - y_1}{y_2 - y_1} = \frac{z - z_1}{z_2 - z_1} = t$$

13.4 EQUATIONS OF A STRAIGHT LINE DETERMINED BY A PAIR OF PLANES IN SYMMETRICAL FORM

We have already seen that a straight line is determined by a pair of planes $ax + by + cz + d = 0$ and $a_1x + b_1y + c_1z + d_1 = 0$ we now express these equations in symmetrical form.

To find it we need to find the direction cosines of the line and the coordinates of a point on the line. Let l, m, n be the direction cosines of the line. This line is perpendicular to the normal to the two given planes since the line lies on the plane. The direction ratios of the two normals are a_1, b_1, c_1 and a_2, b_2, c_2. The direction cosines of the line are l, m, n. Since the normals are perpendicular to the line we have,

$$al + bm + cn = 0 \tag{13.7}$$

$$a_1l + b_1m + c_1n = 0 \tag{13.8}$$

Solving for l, m, n we get $\dfrac{l}{bc_1 - b_1c} = \dfrac{m}{ca_1 - a_1c} = \dfrac{n}{ab_1 - b_1a}$.

Therefore, the direction ratios of the line are

$$bc_1 - b_1c, ca_1 - a_1c, ab_1 - a_1b \tag{13.9}$$

To find a point on the line, let us find the point where the line meets the plane. $z = 0$ and $a_1x + b_1y + d = 0$ and $a_1x + b_1y + d_1 = 0$. Solving the last two equations, we get

$$\frac{x}{bd_1 - b_1d} = \frac{y}{a_1d - ad_1} = \frac{1}{ab_1 - a_1b}$$

Therefore, a point on the line is $\left(\frac{bd_1 - b_1d}{ab_1, a_1b}, \frac{a_1d - ad_1}{ab_1 - a_1b}, 0\right)$.

Then the equations of the straight lines are $\dfrac{x - \dfrac{bd_1 - b_1d}{ab_1 - ba_1}}{bc_1 - b_1c} = \dfrac{y - \dfrac{a_1d - ad_1}{ab_1 - a_1b}}{ca_1 - a_1c} = \dfrac{z - 0}{ab_1 - a_1b}$.

Note 13.4.1: We can also find the point where the line meets the *yz*–plane or *zx*–plane.

13.5 ANGLE BETWEEN A PLANE AND A LINE

Let the equation of the plane be $ax + by + cz + d = 0$.

Let the equation of the line be $\frac{x - x_1}{l} = \frac{y - y_1}{m} = \frac{z - z_1}{n}$.

Let θ be the angle between the plane and the line. The direction ratios of the normal to the plane are *a*, *b*, *c*. The direction ratios of the line are *l*, *m*, *n*. Since θ is the angle between the plane and the line, $\left(\frac{\pi}{2} - \theta\right)$ is the angle between the normal to the plane and the line.

$$\therefore \cos\left(\frac{\pi}{2} - \theta\right) = \frac{al + bm + cn}{\sqrt{a^2 + b^2 + c^2}\sqrt{l^2 + m^2 + n^2}}$$

$$\sin\theta = \frac{al + bm + cn}{\sqrt{a^2 + b^2 + c^2}\sqrt{l^2 + m^2 + n^2}} \text{ or } \theta = \sin^{-1}\left(\frac{al + bm + cn}{\sqrt{a^2 + b^2 + c^2}\sqrt{q^2 + m^2 + n^2}}\right)$$

Note 13.5.1: If the line is parallel to the plane, $\theta = 0$.

$$\therefore al + bm + cn = 0$$

13.6 CONDITION FOR A LINE TO BE PARALLEL TO A PLANE

Let the equation of the plane be

$$ax + by + cz + d = 0 \tag{13.10}$$

Let the equation of the line be

$$\frac{x - x_1}{l} = \frac{y - y_1}{m} = \frac{z - z_1}{n} \tag{13.11}$$

If the line is parallel to the plane then the normal to the plane is perpendicular to the line. The condition for this is

$$al + bm + cn = 0 \tag{13.12}$$

Since (x_1, y_1, z_1) is a point on the line and does not lie on the plane given by (13.10).

$$\therefore ax_1 + by_1 + cz_1 + d \neq 0$$

Hence the conditions for the line (13.11) to be parallel to the plane (13.10) are $al + bm + cn = 0$ and $ax_1 + by_1 + cz_1 + d \neq 0$.

13.7 CONDITIONS FOR A LINE TO LIE ON A PLANE

Let the equation of the line be

$$\frac{x - x_1}{l} = \frac{y - y_1}{m} = \frac{z - z_1}{n} \tag{13.13}$$

Let the equation of the plane be

$$ax + by + cz + d = 0 \tag{13.14}$$

Since the line lies on the plane,

$$al + bm + cn = 0 \tag{13.15}$$

Since the line lies on the plane every point on the line is also a point on the plane. (x_1, y_1, z_1) is a point on the line and therefore it should also lie on the plane given by (13.14). Hence, $ax_1 + by_1 + cz_1 + d = 0$. Therefore, the conditions for the line (13.13) to be parallel to the plane (13.14) are $al + bm + cn = 0$ and $ax_1 + by_1 + cz_1 + d = 0$.

13.8 TO FIND THE LENGTH OF THE PERPENDICULAR FROM A GIVEN POINT ON A LINE

Let the given point be $P(p, q, r)$ and the given line QR be

$$\frac{x-x_1}{l}=\frac{y-y_1}{m}=\frac{z-z_1}{n} \tag{13.16}$$

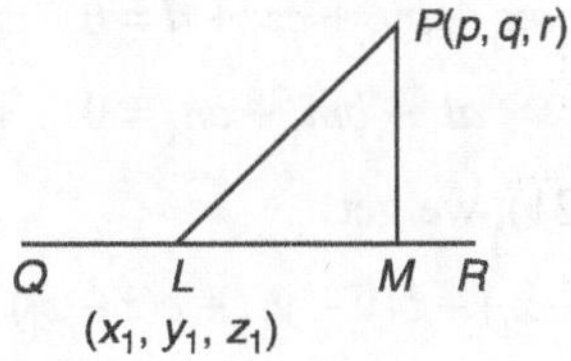

Then $L(x_1, y_1, z_1)$ is a point on the line. Draw PM perpendicular to the line.

$$PM^2 = PL^2 - LM^2 \tag{13.17}$$

$$PL^2 = (x_1 - p)^2 + (y_1 - q)^2 + (z_1 - r)^2$$

Also LM is the projection of PL on QR.

$$\therefore\ LM \text{ is } (x_1 - p)l + (y_1 - q)m + (z_1 - r)n \tag{13.18}$$

Then from (13.17),

$$PM^2 = (x_1 - p)^2 + (y_1 - q)^2 + (z_1 - r)^2 - [l(x_1 - p) + m(y_1 - q) + n(z_1 - r)]^2$$

$$PM^2 = [(y_1 - q)n + (z_1 - r)m]^2 + [(z_1 - r)l - (x_1 - r)n]^2 - [(x_1 - p)m\,(y_1 - q)l]^2 \tag{13.19}$$

(using Lagrange's identity)

13.9 COPLANAR LINES

Find the condition for the lines $\frac{x-x_1}{l_1}=\frac{y-y_1}{m_1}=\frac{z-z_1}{n_1}=r_1$ and $\frac{x-x_2}{l_s}=\frac{y-y_2}{m_2}=\frac{z-z_2}{n_2}=r_2$ to be coplanar and also find the equation of the plane containing these two lines.

Consider equations,

$$\frac{x-x_1}{l_1}=\frac{y-y_1}{m_1}=\frac{z-z_1}{n_1}=r_1 \tag{13.20}$$

$$\frac{x-x_2}{l_2}=\frac{y-y_2}{m_2}=\frac{z-z_2}{n_2}=r_2 \quad (13.21)$$

Let the equation of the plane be

$$ax+by+cz+d=0 \quad (13.22)$$

Since the planes contains lines (13.20), we have

$$ax_1+by_1+cz_1+d=0 \quad (13.23)$$

$$al_1+bm_1+cn_1=0 \quad (13.24)$$

From (13.20) and (13.21), we get

$$a(x-x_1)+b(y-y_1)+c(z-z_1)=0 \quad (13.25)$$

Since the plane also contains the line (13.21) the point (x_2, y_2, z_2) lies on the plane (13.22).

$$\therefore\ a(x_2-x_1)+b(y_2-y_1)+c(z_2-z_1)=0 \quad (13.26)$$

$$\text{Also } al_2+bm_2+cn_2=0 \quad (13.27)$$

Eliminating a, b, c from equation (13.24), (13.26) and (13.27), we get

$$\begin{vmatrix} x_2-x_1 & y_2-y_1 & z_2-z_1 \\ l_1 & m_1 & n_1 \\ l_2 & m_2 & n_2 \end{vmatrix}=0$$

This is the required condition for the lines (13.20) and (13.21) to be coplanar.

Eliminating a, b, c from the equation (13.23), (13.24) and (13.25), we get the equation of the plane containing the two given lines as

$$\begin{vmatrix} x-x_1 & y-y_1 & z-z_1 \\ l_1 & m_1 & n_1 \\ l_2 & m_2 & n_2 \end{vmatrix}=0.$$

Aliter:

If the planes are coplanar they may intersect. Any point on the line (13.20) is $x_1+l_1r_1, y_1+m_1r_1, z_1+n_1r_1$. Any point on the line (13.21) is $x_2+l_2r_2, y_2+m_2r_2, z_2+n_2r_2$.

If the two lines intersect then the two points are the same.

$$\therefore\ x_1+l_1r_1=x_2+l_2r_2$$

$$y_1+m_1r_1=y_2+m_2r_2$$

$$z_1 + n_1 r_1 = z_2 + n_2 r_2$$

$$\therefore\ (x_1 - x_2) + (lr_1 - l_2 r_2) = 0$$

$$(y_1 - y_2) + (m_1 r_1 - m_2 r_2) = 0$$

$$(z_1 - z_2) + (n_1 r_1 - n_2 r_2) = 0$$

Eliminating r_1 and r_2 from the above equations, we get

$$\begin{vmatrix} x_1 - x_2 & l_1 & l_2 \\ y_1 - y_2 & m_1 & m_2 \\ z_1 - z_2 & n_1 & n_2 \end{vmatrix} = 0 \quad \text{(i.e.)} \quad \begin{vmatrix} x_2 - x_1 & y_2 - y_1 & z_2 - z_1 \\ l_1 & m_1 & n_1 \\ l_2 & m_2 & n_2 \end{vmatrix} = 0$$

This is the required condition for coplanar lines.

13.10 SKEW LINES

Two non–intersecting and non–parallel lines are called skew lines. There also exists a shortest distance between the skew lines and the line of the shortest distance which is common perpendicular to both of these.

13.10.1 Length and Equations of the Line of the Shortest Distance

Let the equation of the skew lines be

$$\frac{x - x_1}{l_1} = \frac{y - y_1}{m_1} = \frac{z - z_1}{n_1} \qquad (13.28)$$

$$\frac{x - x_2}{l_2} = \frac{y - y_2}{m_2} = \frac{z - z_2}{n_2} \qquad (13.29)$$

Let PQ be the line of the shortest distance between lines (13.28) and (13.29). Let l, m, n be the direction cosines of the lines of the shortest distance PQ.

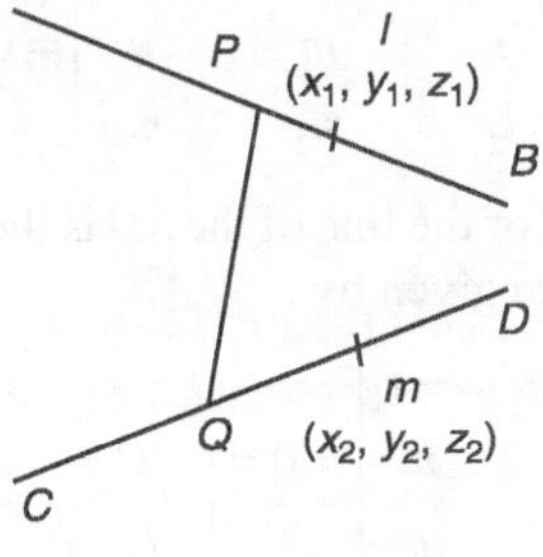

The condition for PQ to be perpendicular to AB and CD are

$$ll_1 + mm_1 + nn_1 = 0$$
$$ll_2 + mm_2 + nn_2 = 0$$

Solving these two, we get $\dfrac{l}{m_1n_2 - m_2n_1} = \dfrac{m}{n_1l_2 - n_2l_1} = \dfrac{n}{l_1m_2 - l_2m_1}$.

Therefore, the direction ratios of the line PQ are $m_1n_2 - m_2n_1, n_1l_2 - n_2l_1, l_1m_2 - l_2m_1$.

Therefore, the direction ratios of the line of the SD are $\dfrac{m_1n_2 - m_2n_1}{\sqrt{\sum(m_1n_2 - m_2m_1)^2}}, \dfrac{n_1l_2 - n_2l_1}{\sqrt{\sum(m_1n_2 - m_2m_1)^2}}, \dfrac{l_1m_2 - l_2m_1}{\sqrt{\sum(m_1n_2 - m_2m_1)^2}}$.

$l(x_1, y_1, z_1)$ and $m(x_2, y_2, z_2)$ are points on the lines (13.28) and (13.29).

Then the length of the $SD = PQ =$ Projection of LM on $PQ = (x_1 - x_2)l + (y_1 - y_2)m + (z_1 - z_2)n$, where l, m, n are the direction cosines of the line PQ.

$$= \frac{(x_1 - x_2)(m_1n_2 - m_2n_1) + (y_1 - y_2)(n_1l_2 - n_2l_1) + (z_1 - z_2)(l_1m_2 - l_2m_1)}{\sqrt{\sum(m_1n_2 - m_2n_1)^2}}$$

$$= \begin{vmatrix} x_1 - x_2 & y_1 - y_2 & z_1 - z_2 \\ l_1 & m_1 & n_1 \\ l_2 & m_1 & n_2 \end{vmatrix} \div \sqrt{\sum(m_1n_2 - m_2n_1)^2} \tag{13.30}$$

The equation of the plane containing the lines AB and PQ is

$$\begin{vmatrix} x - x_1 & y - y_1 & z - z_1 \\ l & m & n \\ l_1 & m_1 & n_1 \end{vmatrix} = 0 \tag{13.31}$$

The equation of the plane containing the lines CD and PQ is

$$\begin{vmatrix} x - x_2 & y - y_2 & z - z_2 \\ l & m & n \\ l_2 & m_2 & n_2 \end{vmatrix} = 0 \tag{13.32}$$

Therefore, the equation of the line of the SD is the intersection of these two planes and its equations are given by

$$\begin{vmatrix} x - x_1 & y - y_1 & z - z_1 \\ l & m & n \\ l_1 & m_1 & n_1 \end{vmatrix} = 0 = \begin{vmatrix} x - x_2 & y - y_2 & z - z_2 \\ l & m & n \\ l_2 & m_2 & n_2 \end{vmatrix}$$

Note 13.10.1: If the lines (13.28) and (13.29) are coplanar then the SD between the lines is zero. Hence the condition for the lines (13.28) and (13.29) to be coplanar, from (13.30) is $\begin{vmatrix} x-x_1 & y-y_1 & z-z_1 \\ l_1 & m_1 & n_1 \\ l_2 & m_2 & n_2 \end{vmatrix} = 0.$

Aliter:

Let the vector equations of the two lines be

$$\vec{r} = \vec{a}_1 + t\vec{b}_1 \tag{13.33}$$

$$\vec{r} = \vec{a}_2 + s\vec{b}_2 \tag{13.34}$$

where t and s are scalars.

$$\vec{a}_1 = x_1\vec{i} + y_1\vec{j} + z_1\vec{k},\ \vec{a}_2 = x_2\vec{i} + y_2\vec{j} + z_2\vec{k}$$

$$\vec{b}_1 = l_1\vec{i} + m_1\vec{j} + n_1\vec{k},\ \vec{b}_2 = l_2\vec{i} + m_2\vec{j} + n_2\vec{k}$$

If the lines (13.33) and (13.34) are coplanar then the plane is parallel to the vectors $\vec{b}_1$ and $\vec{b}_2$. Thereby $\vec{b}_1 \times \vec{b}_2$ is perpendicular to the plane containing $\vec{b}_1$ and $\vec{b}_2$. Also as $\vec{a}_1$ and $\vec{a}_2$ are the points on the plane, $\vec{a}_1 - \vec{a}_2$ is a line on the plane and is perpendicular to $\vec{b}_1 \times \vec{b}_2$. The condition for this is $(\vec{a}_1 - \vec{a}_2)\cdot(\vec{b}_1 \times \vec{b}_2) = 0$ or $[\vec{a}_1 - \vec{a}_2, \vec{b}_1, \vec{b}_2] = 0$.

The scalar form of the equation is $\begin{vmatrix} x_1-x_2 & y_1-y_2 & z_1-z_2 \\ l_1 & m_1 & n_1 \\ l_2 & m_2 & n_2 \end{vmatrix} = 0.$ The vector equation of the plane containing the two lines is $(\vec{r} - \vec{a})\cdot(\vec{b}_1 \times \vec{b}_2) = 0$ or $[\vec{r} - \vec{a}, \vec{b}_1, \vec{b}_2] = 0$.

But $\vec{r} - \vec{a} = (x - x_1)\vec{i} + (y - y_1)\vec{j} + (z - z_1)\vec{k}$.

Therefore, the scalar equation of the plane is $\begin{vmatrix} x-x_1 & y-y_1 & z-z_1 \\ l_1 & m_1 & n_1 \\ l_2 & m_2 & n_2 \end{vmatrix} = 0.$

Aliter:

Let the vector equation of the two lines be $\vec{r} = \vec{a}_1 + t\vec{b}_1$ and $\vec{r} = \vec{a}_2 + s\vec{b}_2$.

Let $\vec{a}_1 = x_1\vec{i} + y_1\vec{j} + z_1\vec{k}$, $\vec{a}_2 = x_2\vec{i} + y_2\vec{j} + z_2\vec{k}$ and $\vec{b}_1 = l_1\vec{i} + m_1\vec{j} + n_1\vec{k}$, $\vec{b}_2 = l_2\vec{i} + m_2\vec{j} + n_2\vec{k}$.

Let DQ be the SD between the lines AB at CD. Then $\overrightarrow{PQ}$ is perpendicular to both $\vec{b}_1$ and $\vec{b}_2$. Then $\overrightarrow{PQ}$ is parallel to $\vec{b}_1 \times \vec{b}_2$. Let $\vec{a}_1$ and $\vec{a}_2$ be the position vectors of points L and M on AB and CD, respectively.

$$SD = PQ = \text{projection of } LM \text{ on } PQ$$
$$= \text{the projection of } (\vec{a}_1 - \vec{a}_2) \text{ on } (\vec{b}_1 \times \vec{b}_2)$$
$$= \frac{\left[(\vec{a}_1 - \vec{a}_2)\cdot(\vec{b}_1 \times \vec{b}_2)\right]}{\vec{b}_1 \times \vec{b}_2}$$
$$= \frac{\left[(\vec{a}_1 - \vec{a}_2)\vec{b}_1, \vec{b}_2)\right]}{\vec{b}_1 \times \vec{b}_2}$$
$$= \begin{vmatrix} x_1 - x_2 & y_1 - y_2 & z_1 - z_2 \\ l_1 & m_1 & n_1 \\ l_2 & m_2 & n_2 \end{vmatrix} \div \left|\vec{b}_1 \times \vec{b}_2\right|$$

$$\vec{b}_1 \times \vec{b}_2 = \begin{vmatrix} \vec{i} & \vec{j} & \vec{k} \\ l_1 & m_1 & m_1 \\ l_2 & m_2 & n_2 \end{vmatrix} = \vec{i}(m_1n_2 - m_2n_1) + \vec{j}(l_1n_2 - l_2n_1) + \vec{k}(l_1m_2 - l_2m_1)$$

$$\left|\vec{b}_1 \times \vec{b}_2\right| = \sqrt{\Sigma(m_1n_2 - m_2n_1)^2}$$

$$\therefore\ SD = \begin{vmatrix} x_1 - x_2 & y_1 - y_2 & z_1 - z_2 \\ l_1 & m_1 & n_1 \\ l_2 & m_2 & n_2 \end{vmatrix} \div \sqrt{\Sigma(m_1n_2 - m_2n_1)^2}$$

13.10.2 Equation of the Line of SD

The equation of the line of the shortest distance is the equation of the line of intersection of the planes through the given lines and the SD. The equation of the plane containing the line $\vec{r} = \vec{a}_1 + t\vec{b}_1$ and the SDPQ is parallel to $\vec{b}_1 \times \vec{b}_2$ and therefore perpendicular to $\vec{b}_1 \times (\vec{b}_1 \times \vec{b}_2)$is

$$(\vec{r} - \vec{a}_1)\cdot\left[\vec{b}_1 \times (\vec{b}_1 \times \vec{b}_2)\right] = 0$$

$$\left[\vec{r}_1 - \vec{a}_1, b_1, \vec{b}_1 \times \vec{b}_2\right] = 0 \tag{13.35}$$

Similarly the equation of the plane containing the line $\vec{r} = \vec{a}_2 + s\vec{b}_2$ and PQ is

$$\left[\vec{r} - \vec{a}_2, \vec{b}_2 \times (\vec{b}_1 \times \vec{b}_2)\right] = 0$$

$$\left[\vec{r} - a_2, b_2, \vec{b}_1 \times \vec{b}_2\right] = 0 \tag{13.36}$$

The equation of the line of SD is the equation of the line of intersection of (13.35) and (13.36).

(i.e.) $$\left[\vec{r} - \vec{a}_1, \vec{b}_2, \vec{b}_1 \times \vec{b}_2\right] = 0$$

In scalar forms, the equations of the line are

$$\begin{vmatrix} x-x_1 & y-y_1 & z_1 \\ l_1 & m_1 & n_1 \\ m_1n_2-m_2n_1 & n_1l_2-n_2l_1 & l_1m_2-l_2m_1 \end{vmatrix} = 0$$

and

$$\begin{vmatrix} x-x_2 & y-y_2 & z-z_2 \\ l_2 & m_2 & n_2 \\ m_1n_2-m_2n_1 & n_1l_2-n_2l_1 & l_1m_2-l_2m_1 \end{vmatrix} = 0.$$

13.11 EQUATIONS OF TWO NON-INTERSECTING LINES

We will now show that the equations of any two skew lines can be part into the form $y = mx$, $z = c$ and $y = -mx$, $z = -c$.

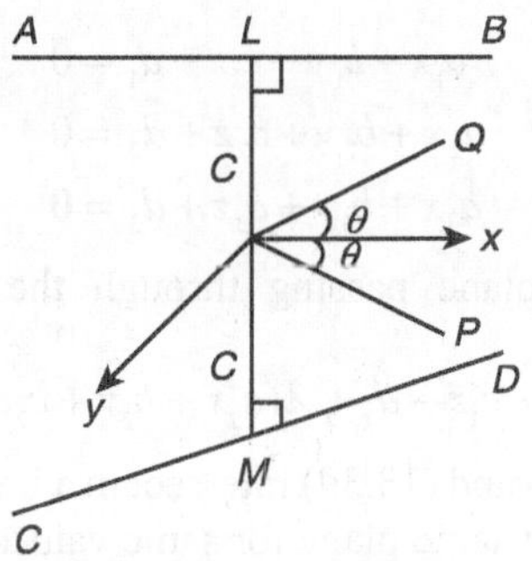

Let AB and CD be two skew lines.

Let LM be the common perpendicular to the skew lines. Let $LM = 2c$ and θ be its middle point. Choose O be the origin and draw lines OP and OQ parallel to AB and CD, respectively. Let the bisectors of $\underline{|POQ}$ be chosen as axes of x and y. Let OE be taken as z–axis. Let $\underline{|POQ} = 2\theta$ so that $\underline{|POX} = \underline{|XOQ} = \theta$.

Then the line OP makes angle θ, $\left(\dfrac{\pi}{2}-\theta\right)$ and $\dfrac{\pi}{2}$with x–, y–, z–axes. Its direction cosines are $\cos\alpha$, $\sin\alpha$, 0. The coordinates of L are $(0, 0, c)$. AB is a straight line passing through L and parallel to AB. The equations of the line OP are $\dfrac{x-0}{\cos\alpha} = \dfrac{y-0}{\sin\theta} = \dfrac{z-c}{0}$ or $y = x\tan\theta$, $z = c$

(i.e.) $y = mx$, $z = c$ where $m = \tan\theta$

The line OQ makes angles $-\theta, \frac{\pi}{2}-\theta$ and $\frac{\pi}{2}$with x-, y-, z-axes.

The direction cosines of the line OQ are $\cos\theta, -\sin\theta, \theta$.

The coordinates of M are $(0, 0, -c)$. CD is a straight line passing through F and parallel to CD. Its equations are $\frac{x-0}{\cos\theta}=\frac{y-0}{-\sin\theta}=\frac{z+c}{0}$.

(i.e.) $y=-mx, z=-c$ where $m=\tan\theta$

Note 13.11.1: Any point on the line AB is (r_1, mr_1, c) and on axis $(r_1, -mr_1, -c)$.

13.12 INTERSECTION OF THREE PLANES

Three planes may intersect in a line or a point. Let us find the conditions for three given planes to intersect (i) in a line and (ii) in a point.

Let the equations of three given planes be

$$a_1x+b_1y+c_1z+d_1=0 \tag{13.37}$$

$$a_2x+b_2y+c_2z+d_2=0 \tag{13.38}$$

$$a_3x+b_3y+c_3z+d_3=0 \tag{13.39}$$

The equation of any plane passing through the intersection of planes (13.37) and (13.38) is

$$a_1x+b_1y+c_1z+d_1+\lambda(a_2x+b_2y+c_2z+d_2)=0 \tag{13.40}$$

If planes (13.37), (13.38) and (13.39) intersect in a line then equations (13.39) and (13.40) represent the same plane for same values of λ.

Identifying equations (13.40) and (13.39), we get

$$\frac{a_1+\lambda a_2}{a_3}=\frac{b_1+\lambda b_2}{b_3}=\frac{c_1+\lambda c_2}{c_3}=\frac{d_1+\lambda d_2}{d_3}=\mu(say)$$

$$a_1+\lambda a_2-\mu a_3=0$$

$$b_1+\lambda b_2-\mu b_3=0$$

$$c_1+\lambda c_2-\mu c_3=0$$

$$d_1+\lambda d_2-\mu d_3=0$$

Eliminating λ and μ from the equation taken three at a time, we get

$$\begin{vmatrix} a_1 & a_2 & a_3 \\ b_1 & b_2 & b_3 \\ c_1 & c_2 & c_3 \end{vmatrix}=0, \begin{vmatrix} a_1 & a_2 & a_3 \\ b_1 & b_2 & b_3 \\ d_1 & d_2 & d_3 \end{vmatrix}=0$$

$$\begin{vmatrix} a_1 & a_2 & a_3 \\ c_1 & c_2 & c_3 \\ d_1 & d_2 & d_3 \end{vmatrix} = 0 \text{ and } \begin{vmatrix} b_1 & b_2 & b_3 \\ c_1 & c_2 & c_3 \\ d_1 & d_2 & d_3 \end{vmatrix} = 0$$

(i.e.) $$\Delta_1 = \begin{vmatrix} a_1 & b_1 & c_1 \\ a_2 & b_2 & c_2 \\ a_3 & b_3 & c_3 \end{vmatrix} = 0, \quad \Delta_3 = \begin{vmatrix} a_1 & b_1 & d_1 \\ a_2 & b_2 & d_2 \\ a_3 & b_3 & d_3 \end{vmatrix} = 0$$

$$\Delta_2 = \begin{vmatrix} a_1 & c_1 & d_1 \\ a_2 & c_2 & d_2 \\ a_3 & c_3 & d_3 \end{vmatrix} = 0 \text{ and } \Delta_4 = \begin{vmatrix} b_1 & c_1 & d_1 \\ b_2 & d_2 & c_2 \\ b_3 & c_3 & d_3 \end{vmatrix} = 0$$

Therefore, the conditions for the three planes to intersect in a line are $\Delta_1 = 0$, $\Delta_2 = 0$, $\Delta_3 = 0$ and $\Delta_4 = 0$.

Note 13.12.1: Of these four conditions only two are independent since if two planes have two points in common then they show the line joining these two points should also have in common. It can be proved if any two of these conditions are satisfied, then the other two will also satisfy.

Aliter:

The equations of the line of intersection of (13.37) and (13.38) are given by

$$\frac{x - \dfrac{a_1d_2 - b_2d_1}{a_1b_2 - a_2b_1}}{b_1c_2 - b_2c_1} = \frac{y - \dfrac{a_2d_1 - a_1d_2}{a_1b_2 - a_2b_1}}{a_1c_1 - a_2c_1} = \frac{z - 0}{a_1b_2 - a_2b_1} \tag{13.41}$$

If the planes (13.37), (13.38) and (13.39) intersect in a plane then the conditions are (i) the line (13.41) must be parallel to the plane (13.39) and (ii) the point $\left(\dfrac{b_1d_2 - b_2d_1}{a_1b_2 - a_2b_1}, \dfrac{a_2d_1 - a_1d_2}{a_1b_2 - a_2b_1}, 0\right)$ must lie on the plane given by (13.39). The conditions for (13.37) is $a_3(b_1c_2 - b_2c_1) + b_3(a_2c_1 - a_1c_2) + c_3(a_1b_2 - a_2b_1) = 0$.

$$\text{(i.e.)} \quad \begin{vmatrix} a_1 & b_1 & c_1 \\ a_2 & b_2 & c_2 \\ a_3 & b_3 & c_3 \end{vmatrix} = 0 \text{(i.e.) } \Delta_4 = 0$$

The condition (ii) is given by

$$a_3\left(\frac{b_1d_2 - b_2d_1}{a_1b_2 - a_2b_1}\right) + b_3\left(\frac{a_2d_1 - a_1d_2}{a_1b_2 - a_2b_1}\right) + d_3 = 0$$

(i.e.) $$a_3(b_1d_2 - b_2d_1) + b_3(a_2d_1 - a_1d_1 - a_1d_2) + d_3(a_1b_2 - a_2b_1) = 0$$

$$\text{(i.e.)} \begin{vmatrix} a_1 & b_1 & d_1 \\ a_2 & b_2 & d_2 \\ a_3 & b_3 & d_3 \end{vmatrix} = 0 \text{ (i.e.) } \Delta_3 = 0$$

Therefore, the conditions for planes to intersect in a line are $\Delta_3 = 0$ and $\Delta_4 = 0$.

(ii) **Condition for the plane to intersect at a point:**

Solving equations (13.37), (13.38) and (13.39), we get

$$\frac{x}{\Delta_1} = \frac{y}{\Delta_2} = \frac{z}{\Delta_3} = \frac{1}{\Delta_4} \qquad \therefore x = \frac{\Delta_1}{\Delta_4}, y = \frac{\Delta_2}{\Delta_4}, z = \frac{\Delta_3}{\Delta_4}$$

If the planes intersect at a point then $\Delta_4 \neq 0$. Hence the condition for a plane to intersect at a point is $\Delta_4 \neq 0$.

Aliter:

If the planes meet at a point then the line of intersection of any two planes is non–parallel to the third plane. Let l, m, n be the direction cosines of the intersection of planes (13.37) and (13.38). Then

$$a_1 l + b_1 m + c_1 n = 0$$
$$a_2 l + b_2 m + c_2 n = 0$$

Solving the two equations for l, m, n we get,

$$\frac{l}{b_1c_2 - b_2c_1} = \frac{m}{a_2c_1 - a_1c_2} = \frac{n}{a_1b_2 - a_2b_1}$$

Therefore, the direction ratios of the lines are $b_1c_2 - b_2c_1, a_2c_1 - a_1c_2, a_1b_2 - a_2b_1$. Also the line of intersection will not be parallel to the third plane.

$$\therefore\ a_3(b_1c_2 - b_2c_1) + b_3(a_2c_1 - a_1c_2) + c_3(a_1b_2 - a_2b_1) \neq 0$$

$$\text{(i.e.)} \begin{vmatrix} a_1 & b_1 & c_1 \\ a_2 & b_2 & c_2 \\ a_3 & b_3 & c_3 \end{vmatrix} \neq 0 \text{ or } \Delta_4 \neq 0$$

This is the required condition.

13.13 CONDITIONS FOR THREE GIVEN PLANES TO FORM A TRIANGULAR PRISM

The line of intersection of planes (13.37) and (13.38) is given by

$$\frac{x - \dfrac{b_1d_2 - b_2d_1}{a_1b_2 - a_2b_1}}{b_1c_2 - b_2c_1} = \frac{y - \dfrac{a_1d_2 - a_2d_1}{a_1b_2 - a_2b_1}}{a_2c_1 - a_1c_2} = \frac{z - 0}{a_1b_2 - a_2b_1}$$

The three planes form a triangular prism if the line is parallel to the third plane. The conditions for this are the line is normal to the plane (13.39) and the point $\left(\frac{b_1d_2 - b_2d_1}{a_1b_2 - a_2b_1}, \frac{a_2d_1 - a_1d_2}{a_1b_2 - a_2b_1}, 0\right)$ does not lie on the plane (13.39).

$$\therefore\ a_3(b_1c_2 - b_2c_1) + b_3(c_1a_2 - c_2a_1) + c_3(a_1b_2 - a_2b_1) = 0$$

and

$$a_3\left(\frac{b_1d_2 - b_2d_1}{a_1b_2 - a_2b_1}\right) + b_3\left(\frac{a_2d_1 - a_1d_2}{a_1b_2 - a_2b_1}\right) + d_3 \neq 0$$

$$\begin{vmatrix} a_1 & b_1 & c_1 \\ a_2 & b_2 & c_2 \\ a_3 & b_3 & c_3 \end{vmatrix} = 0 \text{ and } \begin{vmatrix} a_1 & b_1 & d_1 \\ a_2 & b_2 & d_2 \\ a_3 & b_3 & d_3 \end{vmatrix} \neq 0$$

(i.e.) $\Delta_4 = 0$ and $\Delta_3 \neq 0$. These are the required conditions.

ILLUSTRATIVE EXAMPLES

Example 13.1

Find the equation of the line joining the points (2, 3, 5) and (–1, 2, –4).

Solution

The direction ratios of the line are 2 + 1, 3 – 2, 5 + 4 (i.e.) 3, 1, 9. Therefore, the equations of the line are $\frac{x-2}{3} = \frac{y-3}{1} = \frac{z-5}{9}$.

Example 13.2

Find the equation of the line passing through the point (3, 2, –6) and perpendicular to the plane $3x - y - 2z + 2 = 0$.

Solution

The direction ratios of the line are the direction ratios of the normal to the plane. Therefore, the direction ratios of the line are 3, –1, –2. Given that (4, 2, –6) is a point on the plane.

Therefore, the equations of the line are $\frac{x-3}{3} = \frac{y-2}{-1} = \frac{z+6}{-2}$.

Example 13.3

Find the equations of the line passing through the point (1, 2, 3) and perpendicular to the planes $x - 2y - z + 5 = 0$ and $x + y + 3z + 6 = 0$.

Solution

Let l, m, n be the direction ratios of the line of intersection of the planes $x - 2y - z + 5 = 0$ and $x + y + 3z + 6 = 0$.

Then $\quad l - 2m - n = 0$

and $\quad l + m + 3n = 0$

$$\therefore \frac{l}{-6+1} = \frac{m}{-1-3} = \frac{n}{1+2}$$

$$\text{(i.e.)} \quad \frac{l}{-5} = \frac{m}{-4} = \frac{n}{3}.$$

Since the line also passes through the point (–1, 2, 3), its equations is $\frac{x-1}{5} = \frac{y-2}{4} = \frac{z-3}{-3}$.

Example 13.4

Express the symmetrical form of the equations of the line $x + 2y + z - 3 = 0$, $6x + 8y + 3z - 13 = 0$.

Solution

To express the equations of a line in symmetrical form we have to find (i) the direction ratios of the line and (ii) a point on the line.

Let l, m, n be the direction ratios of line. Then $l + 2m + n = 0$ and $6l + 8m + 3n = 0$.

$$\therefore \frac{l}{6-8} = \frac{m}{6-3} = \frac{n}{8-12}$$

$$\text{(i.e.)} \quad \frac{l}{-2} = \frac{m}{3} = \frac{n}{-4} \text{ or } \frac{l}{2} = \frac{m}{-3} = \frac{n}{4}$$

Let us find the point where the line meets the xy–plane (i.e.) $z = 0$.

$$\therefore x + 2y - 3 = 0$$

$$6x + 8y - 13 = 0$$

$$\frac{x}{-26+24} = \frac{y}{-18+13} = \frac{1}{8-12} \quad \text{or} \quad \frac{x}{2} = \frac{y}{5} = \frac{1}{4}$$

$$\therefore x = \frac{1}{2}, y = \frac{5}{4}, z = 0$$

Therefore, the equations of the line are $\dfrac{x-\frac{1}{2}}{2} = \dfrac{y-\frac{5}{4}}{-3} = \dfrac{z-0}{4}$.

Example 13.5

Find the perpendicular distance from the point (1, 3, −1) to the line $\dfrac{x-13}{5} = \dfrac{y+8}{-8} = \dfrac{z-31}{1}$.

Solution

The equations of the line are

$$\frac{x-13}{5} = \frac{y+8}{-8} = \frac{z-31}{1} = r$$

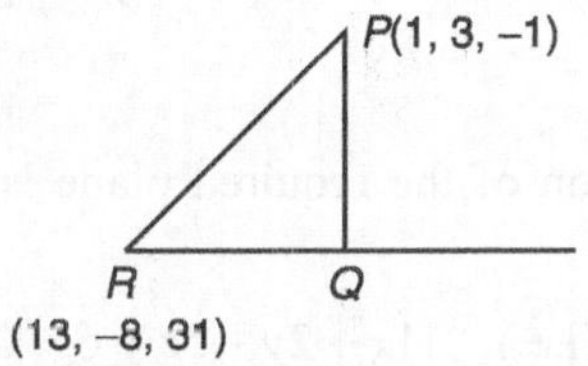

Any point on this line are $(5r+13, -8r-8, r+31)$. Draw PQ perpendicular to the plane. The direction ratios of the line are $(5r+12, -8r-11, r+32)$.

Since the line PQ is perpendicular to QR, we have

$$5(5r+12) - 8(-8r-11) + 1(r+32) = 0$$

$$25r + 60 + 64r + 88 + r + 32 = 0$$

$$\text{or } 90r + 180 = 0 \quad \Rightarrow r = -2$$

Q is the point (3, 8, 29) and P is (1, 3, −1)

$$\therefore PQ^2 = (3-1)^2 + (8-3)^2 + (29+1)^2 = 4 + 25 + 900 = 929$$

$$\therefore PQ^2 = 929 \text{ units}$$

Example 13.6

Find the equation of plane passing through the line $\dfrac{x-1}{2} = \dfrac{y+1}{-1} = \dfrac{z-3}{4}$ and parallel to the line $\dfrac{x}{1} = \dfrac{y}{2} = \dfrac{z}{3}$.

Solution

Any plane containing the line

$$\frac{x-1}{2}=\frac{y+1}{-1}=\frac{z-3}{4} \tag{13.42}$$

is

$$A(x-1)+B(y+1)+C(z-3)=0 \tag{13.43}$$

where

$$2A-B+4C=0 \tag{13.44}$$

Also the line is parallel to the plane

$$\frac{x}{1}=\frac{y}{2}=\frac{z}{3} \tag{13.45}$$

$$\therefore A+2B+3C=0 \tag{13.46}$$

Solving for A, B and C from (13.44) and (13.46), we get $\frac{A}{-3-8}=\frac{B}{4-6}=\frac{C}{4+1}$ or $\frac{A}{11}=\frac{B}{2}=\frac{C}{-5}$

Therefore, the equation of the required plane is $11(x-1)+2(y+1)-5(z-3)=0$.

$$\text{(i.e.)} \quad 11x+2y-5z+6=0$$

Example 13.7

Find the image of the point (2, 3, 5) on the plane $2x+y-z+2=0$.

Solution

Let Q be the image of the point $P(2, 3, 5)$ on the plane $2x+y-z+2=0$. The equation of the line PQ is $\frac{x-2}{2}=\frac{y-3}{1}=\frac{z-5}{-1}=r$.

Any point on this line is $(2r+2, r+3, -r+5)$. When the line meets the plane, this point lies on the plane $2x+y-z+2=0$.

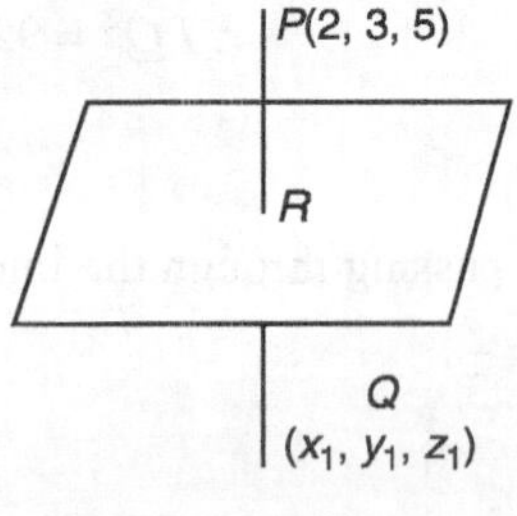

$$\therefore 2(2r+2)+r+3-(-r+5)+2=0$$

Or $r+4+3+r+3+r-5+2=0$ or $6r+4=0$ or $r=\dfrac{-2}{3}$

Therefore, the coordinates of R are $\left(\dfrac{-4}{3}+2, \dfrac{-2}{3}+3, \dfrac{2}{3}+5\right)$.

$$\text{(i.e.)} \quad \left(\frac{2}{3}, \frac{7}{3}, \frac{17}{3}\right).$$

If Q is the point (x_1, y_1, z_1) then

$$\frac{2+x_1}{2}=\frac{2}{3}, \frac{3+y_1}{2}=\frac{7}{3}, \frac{5+z_1}{2}=\frac{17}{3}$$

$$\therefore x_1=\frac{-2}{3}, y_1=\frac{5}{3}, z_1=\frac{19}{3}.$$

$$\therefore Q \text{ is } \left(\frac{-2}{3}, \frac{5}{3}, \frac{19}{3}\right)$$

Example 13.8

Find the image of the line $\dfrac{x-1}{3}=\dfrac{y-3}{5}=\dfrac{z-4}{2}=r$ in the plane $2x-y+z+3=0$.

Solution

The equations of the line are

$$\frac{x-1}{3}=\frac{y-3}{5}=\frac{z-4}{2}=r \qquad (13.47)$$

Any point on this line is $(3r+1, 5r+3, 2r+4)$.

As this point lies on the plane,

$$2x-y+z+3=0 \qquad (13.48)$$

we have $2(3r+1)-(5r+3)+(2r+4)+3=0$.

(i.e.) $3r+6=0$ or $r=-2$.

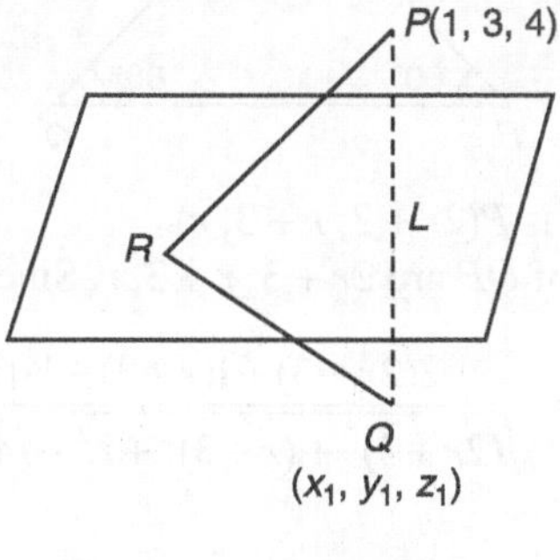

Hence the coordinates of R are (–5, –7, 0). The equations of the line PL perpendicular to the plane are

$$\frac{x-1}{2}=\frac{y-3}{-1}=\frac{z-4}{1}=r_1$$

Any point on this line is $(2r_1+1, -r_1+3, r_1+4)$. If this point lies on the plane (13.48), we get $2(2r_1+1)-(-r_1+3)+(r_1+4)+3=0$.

$$\text{(i.e.)} \quad 6r_1+6=0 \text{ or } r_1=-1.$$

Therefore, the coordinates of L where this line meets the plane (13.47) are (–1, 4, 3). If $Q\ (x_1, y_1, z_1)$ is the image of P in the plane

$$\frac{1+x_1}{2}=-1, \frac{3+y_1}{2}=4, \frac{z_1+4}{2}=3$$

or
$$x_1=-3, y_1=5, z_1=2.$$
$$\therefore Q \text{ is } (-3, 5, 2).$$

Hence the equations of the reflection line RQ are $\dfrac{x+5}{1}=\dfrac{y+7}{6}=\dfrac{z}{1}$.

Example 13.9

Find the equation of the straight lines through the origin each of which intersects the straight line $\dfrac{x-3}{2}=\dfrac{y-3}{1}=\dfrac{z}{1}$ and are inclined at an angle of 60° to it.

Solution

The equations of the line PQ are

$$\frac{x-3}{2}=\frac{y-3}{1}=\frac{z}{1}=r$$

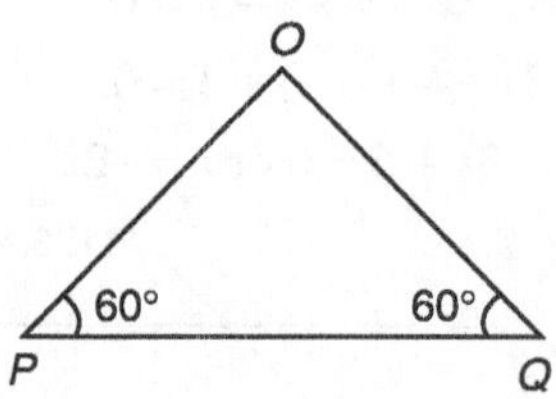

The point P on this line is $P(2r+3, r+3, r)$.

The direction ratios of OP are $2r+3, r+3, r$. Since $\angle POQ=\dfrac{\pi}{3}$,

$$\cos\frac{\pi}{3}=\frac{2(2r+3)+1(r+3)+1\cdot(r)}{\sqrt{(2r+3)^2+(r+3)^2+r^2}\cdot\sqrt{4+1+1}}$$

$$\text{(i.e.)} \quad \frac{6r+9}{\sqrt{(6r^2+18r+18)}\sqrt{6}} = \frac{1}{2}$$

$$\text{(i.e.)} \quad \frac{9(2r+3)^2}{36(r^2+3r+3)} = \frac{1}{2} \Rightarrow \frac{4r^2+12r+9}{4(r^2+3r+3)} = \frac{1}{4}$$

$$\text{(i.e.)} \quad 4r^2 + 12r + 9 = r^2 + 3r + 3$$

$$\text{(i.e.)} \quad 3r^2 + 9r + 6 = 0$$

or $r^2 + 3r + 2 = 0$ or $r = -1, -2$. Therefore, the coordinates of P and Q are $(1, -2, -1)$ and $(-1, 1, -2)$.

Hence the equations of the lines OP and OQ are $\frac{x}{1} = \frac{y}{2} = \frac{z}{-1}$ and $\frac{x}{1} = \frac{y}{1} = \frac{z}{2}$.

Example 13.10

Find the coordinates of the point where the line given by $x + 3y - z = 6, y - z = 4$ cuts the plane $2x + 2y + z = 6$.

Solution

Let l, m, n be the direction cosines of the line $x + 3y - z = 6, y - z = 4$. Then

$$l + 3m - n = 0$$

$$m - n = 0.$$

$\therefore \frac{l}{-2} = \frac{m}{1} = \frac{n}{1}$ Therefore, the direction ratios of the line are 2, −1, −1. When the line meets the xy–plane whose equation is $z = 0$, we have $x + 3y = 6, y = 4$. Therefore, the point where the line meets xy–plane is $(-6, 4, 0)$.

Therefore, the equations of the line are $\frac{x+6}{2} = \frac{y-4}{-1} = \frac{z}{-1} = r$.

Any point on this line is $(2r - 6, -r + 4, -r)$. This point lies on the plane $2x + 2y + z = 0$.

$$2(2r - 6) + 2(-r + 4) - r = 0$$

$$\therefore r - 4 = 0 \text{ or } r = 4$$

Hence the required point is $(2, 0, -4)$.

Example 13.11

Find the distance of the point (1, –2, 3) from the plane $x - y + z = 5$ measured parallel to the line $\frac{x}{2} = \frac{y}{3} = \frac{z}{-6}$.

Solution

The equations of the line through (1, –2, 3) and parallel to the line $\frac{x}{2} = \frac{y}{3} = \frac{z}{-6}$ are $\frac{x-1}{2} = \frac{y+2}{3} = \frac{z-3}{-6} = r$. Any point on this line is $(2r + 1, 3r - 2, -6r + 3)$. If this point lies on the plane $x - y + z = 5$ then $(2r + 1) - (3r - 2) + (-6r + 3) = 5$.

(i.e.) $-7r + 1 = 0$ or $r = \frac{1}{7}$.

Therefore, the point P is $\left(\frac{9}{7}, \frac{-11}{7}, \frac{15}{7}\right)$.

Therefore, the distance between the points A (1, –2, 3) and $\left(\frac{9}{7}, \frac{-11}{7}, \frac{15}{7}\right)$ is

$$AP^2 = \left(1 - \frac{9}{7}\right)^2 + \left(-2 + \frac{11}{7}\right)^2 + \left(3 - \frac{15}{7}\right)^2 = \frac{4}{49} + \frac{9}{49} + \frac{36}{49} = 1.$$

Example 13.12

Prove that the equation of the line through the points (a, b, c) and (a', b', c') passes through the origin if $aa' + bb' + cc' = pp'$ where p and p' are the distances of the points from the origin.

Solution

The equations of the line through (a, b, c) and (a', b', c') are

$$\frac{x-a}{a'-a} = \frac{y-b}{b'-b} = \frac{z-c}{c'-c}$$

If this passes through the origin then

$$\frac{-a}{a'-a} = \frac{-b}{b'-b} = \frac{-c'}{c}$$

$$\text{(i.e.)} \quad \frac{a'-a}{a} = \frac{b'-b}{b} = \frac{c'-c}{c} \quad \text{or} \quad \frac{a'}{a} = \frac{b'}{b} = \frac{c'}{c}$$

$$\left.\begin{aligned} \therefore\ ab' - a'b &= 0 \\ bc' - b'c &= 0 \\ ca' - a'a &= 0 \end{aligned}\right\} \quad (13.49)$$

Let p and p' be the distances of the points (a, b, c) and (a', b', c') from the origin.

$$\therefore p^2 = a^2 + b^2 + c^2$$
$$p'^2 = a'^2 + b'^2 + c'^2$$
$$p^2p'^2 = (a^2 + b^2 + c^2)(a'^2 + b'^2 + c'^2)$$

By Lagrange's identity,

$$(a^2 + b^2 + c^2)(a'^2 + b'^2 + c'^2) - (aa' + bb' + cc')^2$$

$$= (ab' - a'b)^2 + (bc' - b'c)^2 + (ca' - c'a)^2 = 0 \text{ using (3).}$$

$$p^2p'^2 - (aa' + bb' + cc')^2 = 0 \quad \text{or} \quad p^2p'^2 = (aa' + bb' + cc')^2$$

$$\therefore pp' = aa' + bb' + cc'$$

Example 13.13

If from the point $P(x, y, z)$, PM is drawn perpendicular to the line $\frac{x}{l} = \frac{y}{m} = \frac{z}{n}$ and is produced to Q such that $PM = MQ$ then show that $\frac{x + x_1}{l} = \frac{y + y_1}{m} = \frac{z + z_1}{n} = \frac{2(lx + my + nz)}{l^2 + m^2 + n^2}$.

Solution

The equation of the line OA is $\frac{x}{l} = \frac{y}{m} = \frac{z}{n} = r$.

Any point on this line is (lr, mr, nr).

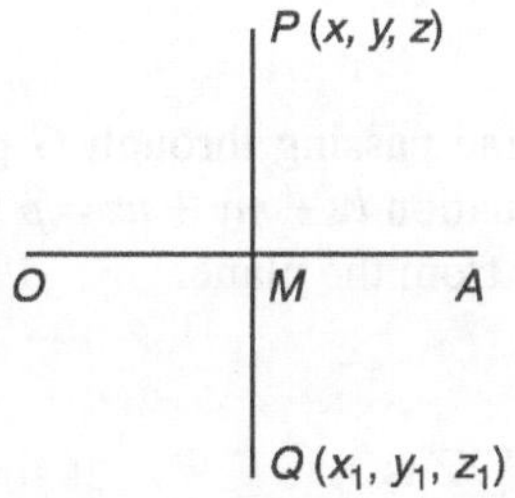

If M is this point then

$$lr = \frac{x + x_1}{2}, mr = \frac{y + y_1}{2}, nr = \frac{z + z_1}{2} \text{ or } \frac{x + x_1}{l} = \frac{y + y_1}{m} = \frac{z + z_1}{n} = 2r \quad (13.50)$$

The direction ratios of the line MP are $x - lr, y - mr, z - nr$.

Since MP is perpendicular to OA,

$$l(x - lr) + m(y - mr) + n(z - nr) = 0 \quad \text{or}$$
$$lx + my + nz = r(l^2 + m^2 + n^2)$$
$$r = \frac{lx + my + nz}{l^2 + m^2 + n^2}$$

From (13.50), we get $\dfrac{x + x_1}{l} = \dfrac{y + y_1}{m} = \dfrac{z + z_1}{z} = \dfrac{2(lx + my + nz)}{l^2 + m^2 + n^2}$.

Example 13.14

Reduce the equations of the lines $x = ay + b, z = cy + d$ to symmetrical form and hence find the condition that the line be perpendicular to the line whose equations are $x = a'y + b', z = c'y + d'$.

Solution

The line

$$x = ay + b, z = cy + d \tag{13.51}$$

can be expressed in the symmetrical form as

$$\frac{x - b}{a} = \frac{y}{1} = \frac{z - d}{c} \tag{13.52}$$

The line

$$x = a'y + b', z = cy' + d' \tag{13.53}$$

in symmetrical form is

$$\frac{x - b'}{a'} = \frac{y}{1} = \frac{z - d'}{c'} \tag{13.54}$$

If the lines (13.51) and (13.52) are perpendicular then $aa' + bb' + cc' = 0$.

This is the required condition.

Example 13.15

Find the equation of the line passing through G perpendicular to the plane XYZ represented by the equation $lx + my + nz = p$ where $l^2 + m^2 + n^2 = 1$ and calculate the distance of G from the plane.

Solution

The equation of the plane XYZ is

$$lx + my + nz = p \tag{13.55}$$

where $l^2 + m^2 + n^2 = 1$. When this plane meets the x–axis, $y = 0$ and $z = 0$.

$$\therefore x = \frac{p}{l}$$

Hence X is the point $\left(\frac{p}{l}, 0, 0\right)$. Similarly, Y is $\left(0, \frac{p}{m}, 0\right)$ and Z is $\left(0, 0, \frac{p}{n}\right)$.

The centroid of ΔXYZ is $\left(\frac{p}{3l}, \frac{p}{3m}, \frac{p}{3n}\right)$.

The equation of the line through G perpendicular to the plane XYZ is
$$\frac{x-\frac{p}{3l}}{l}=\frac{y-\frac{p}{3m}}{m}=\frac{z-\frac{p}{3n}}{n}=r.$$

When this line meets the YOZ plane, $x = 0$. (13.56)

Then $-\frac{p}{3l^2}=r$

Here, r is the distance of G from the plane (13.55) since p and l^2 are positive, $r = GA$

or
$$GA=\frac{p}{3l^2}.$$

Similarly,
$$GB=\frac{p}{3m^2},\ GC=\frac{p}{3n^2}$$

$$\therefore \frac{1}{GA}+\frac{1}{GB}+\frac{1}{GC}=\frac{3}{p}(l^2+m^2+n^2)=\frac{3}{p}.$$

Example 13.16

Find the perpendicular distance of angular points of a cube from a diagonal which does not pass through the angular point.

Solution

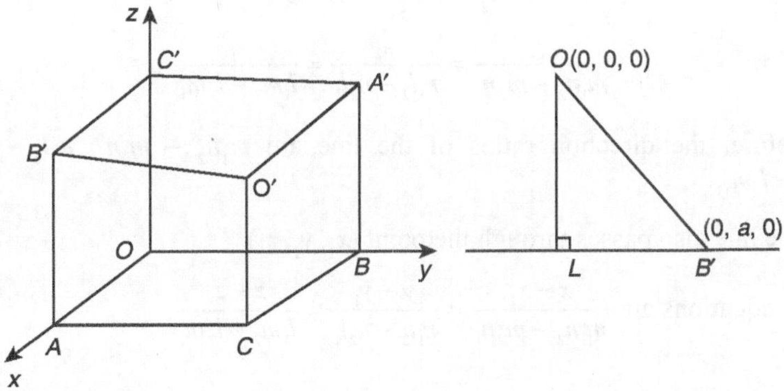

Let a be the side of the cube. BB' is a diagonal of the cube not passing through O. The direction ratios of BB' are $a, -a, a$. (i.e.) $1, -1, 1$. The direction cosines of BB' are $\frac{1}{\sqrt{3}}, \frac{-1}{\sqrt{3}}, \frac{1}{\sqrt{3}}$. The projections of OB' on BB'

$$= LB' = (x_1 - x_2)l + (y_1 - y_2)m + (z_1 - z_2)n$$

$$= (a-0)\frac{1}{\sqrt{3}} + (0-a)\left(-\frac{1}{\sqrt{3}}\right) + (0-a)\frac{1}{\sqrt{3}} = \frac{a}{\sqrt{3}}$$

$$OB = \sqrt{0 + a^2 + 0} = a \quad OL'^2 = OB'^2 - LB'^2 = a^2 - \frac{a^2}{3}$$

$$OL'^2 = \frac{2a^2}{3}$$

$$\therefore OL = \sqrt{\frac{2}{3}}a = \frac{\sqrt{6}}{3}a$$

Example 13.17

Prove that the equations of the line through the point (α, β, γ) and at right angles to the lines $\frac{x}{l_1} = \frac{y}{m_1} = \frac{z}{n_1}, \frac{x}{l_2} = \frac{y}{m_2} = \frac{z}{n_2}$ are $\frac{x - x_1}{m_1n_2 - m_2n_1} = \frac{y - y_1}{n_1l_2 - n_2l_1} = \frac{z - z_1}{l_1m_2 - l_2m_1}$.

Solution

Let l, m, n be the direction cosines of the line perpendicular to the two given lines. Then we have

$$ll_1 + mm_1 + nn_1 = 0$$

$$ll_2 + mm_2 + nn_2 = 0$$

$$\frac{l}{m_1n_2 - m_2n_1} = \frac{m}{n_1l_2 - n_2l_1} = \frac{n}{l_1m_2 - l_2m_1}$$

Therefore, the direction ratios of the line are $m_1n_2 - m_2n_1$, $n_1l_2 - n_2l_1$, $l_1m_2 - l_2m_1$.

The line also passes through the point (x_1, y_1, z_1).

Its equations are $\frac{x - x_1}{m_1n_2 - m_2n_1} = \frac{y - y_1}{n_1l_2 - n_2l_1} = \frac{z - z_1}{l_1m_2 - l_2m_1}$.

Exercises 1

1. Show that the line $\frac{x-1}{2} = \frac{y-1}{-1} = \frac{z}{1}$ is parallel to the plane $2x + 3y - z + 4 = 0$.

2. Find the equation of the plane through the line $\frac{x+1}{-3} = \frac{y-3}{2} = \frac{z+2}{1}$ and the point (0, 7, –7). Show further the plane contains the line $\frac{x}{1} = \frac{y-7}{-3} = \frac{z+7}{2}$.

Ans.: $x + y + z = 0$

3. Find the equation of the plane which passes through the line $3x + 5y + 7z - 5 = 0 = x + y + z - 3$ and parallel to the line $4x + y + z = 0 = 2x - 3y - 5z$.

Ans.: $2x + 4y + y + 6z = 2$

4. Find the equations of the line through the point (1, 0, 7) which intersect each of the lines $x = y = z, x - 1 = y - 5 = \frac{z+1}{2}$.

Ans.: $7x - 6y - z = 0, 9x - 7y - z - 2 = 0$

5. Find the equation of the plane which passes through the point (5, 1, 2) and is perpendicular to the line $\frac{x-2}{1} = \frac{y-4}{2} = \frac{z-5}{2}$. Find also the coordinates of the point in which this line cuts the plane.

Ans.: $x - 2y - 2z - 1 = 0$; (1, 2, 3)

6. Find the equation of the plane through (1, 1, 2) and (2, 10, –1) and perpendicular to the straight line $\frac{x-4}{3} = \frac{y-2}{1} = \frac{z-6}{7}$.

Ans.: $3x - y - 7z + 2 = 0$

7. Find the projection of the line $3x - y + 2z = 1, x + 2y - z = 2$ on the plane $3x + 2y + z = 0$.

Ans.: $3x + 2y + z = 0, 3x - 8y + 7z + 4 = 0$

8. Find the projection of the line $x = 3 - 6t, y = 2t, z = 3 + 2t$ in the plane $3x + 4y - 5z - 26 = 0$.

Ans.: $\frac{x-9}{18} = \frac{y+2}{-1} = \frac{z-1}{-10}$

9. Find the equation of the plane which contains the line and is perpendicular to the plane $x + 2y + z = 12$.

Ans.: $9x - 2y - 5z + 4 = 0$

10. Find the equation of the plane which passes through the z–axis and is perpendicular to the line $\frac{x-1}{\cos\alpha} = \frac{y}{\sin\alpha} = \frac{z-3}{0}$.

Ans.: $x\cos\alpha + y\sin\alpha = 0$

11. Find the equations of two planes through the origin which are parallel to the line $\frac{x-1}{2} = \frac{y+3}{-1} = \frac{z+1}{-2}$ and distant $\frac{5}{3}$ from it. Show also that the two planes are perpendicular.

Ans.: $x + 2y - 2z = 0, 2x + 2y + z = 0$

12. Find the equations to the line of the greatest slope through the point (1, 2, –1) in the plane $x - 2y + 3z = 0$ assuming that the axes are so placed that the plane $2x + 3y - 4z = 0$ is horizontal.

Ans.: $\frac{x-7}{2z} = \frac{y-2}{5} = \frac{z+1}{-4}$

13. Assuming the line $\frac{x}{4}=\frac{y}{-3}=\frac{z}{7}$ as vertical, find the equation of the line of the greatest slope in the plane $2x+y-5z=12$ and passing through the point $(2, 3, -1)$.

Ans.: $\frac{x-2}{3}=\frac{y-3}{-1}=\frac{z+1}{1}$

14. With the given axes rectangular the line $\frac{x}{2}=\frac{y}{-3}=\frac{z}{1}$ is vertical. Find the direction cosines of the line of the greatest slope in the plane $3x-2y+z=0$ and the angle of this line makes with the horizontal plane.

Ans.: $\frac{11}{\sqrt{378}},\frac{16}{\sqrt{378}},\frac{-1}{\sqrt{378}};\sin^{-1}\left(\frac{-9}{1453}\right)$

15. Show that the lines $\frac{x}{\alpha}=\frac{y}{\beta}=\frac{z}{\gamma};\frac{x}{a\alpha}=\frac{y}{b\beta}=\frac{z}{c\gamma};\frac{x}{l}=\frac{y}{m}=\frac{z}{n}$ will be coplanar if $\frac{l(b-c)}{\alpha}+\frac{m(c-a)}{\beta}+\frac{n(a-b)}{\gamma}=0$.

16. Show that the equation of the plane through the line $\frac{x}{l}=\frac{y}{m}=\frac{z}{n}$ and which is perpendicular to the plane containing the lines $\frac{x}{l}=\frac{y}{n}=\frac{z}{l}$ and $\frac{x}{n}=\frac{y}{l}=\frac{z}{m}$ is $\Sigma(m-n)x=0$.

17. Show that the line $\frac{\alpha x}{l}=\frac{\beta y}{m}=\frac{\gamma z}{n};\frac{x}{l}=\frac{y}{m}=\frac{z}{n}$ and $\frac{x}{\alpha l}=\frac{y}{\beta m}=\frac{z}{\gamma n}$ will lie in a plane if $\alpha=\beta$ or $\beta=\gamma$ or $\gamma=\alpha$.

18. Find the equation of the plane passing through the line $\frac{x-1}{2}=\frac{y+1}{-1}=\frac{z-3}{4}$ and perpendicular to the plane $x+2y+z=12$.

Ans.: $9x-2y+5z+4=0$

19. Find the equations of the line through $(3, 4, 0)$ and perpendicular to the plane $2x+4y+7z=8$.

Ans.: $\frac{x-3}{2}=\frac{y-4}{4}=\frac{z}{7}$

20. Find the equation of the plane passing through the line $\frac{x-1}{2}=\frac{y+1}{-1}=\frac{z-3}{4}$ are parallel to the line $\frac{x}{1}=\frac{y}{2}=\frac{z}{3}$.

Ans.: $4y-3z+1=0, 2x-7z+1=0, 3x-2y+1=0$.

21. Show that the equation of the planes through the line which bisect the angle between the lines $\frac{x}{l}=\frac{y}{m}=\frac{z}{n};\frac{x}{l'}=\frac{y}{m'}=\frac{z}{n'}$ (where l, m, n and l', m', n' are direction cosines) and perpendicular to the plane containing them are $(l+l')x+(m+m')y+(n+n')z=0$.

22. Find the equation of the plane through the line $\frac{x-1}{2}=\frac{y-2}{3}=\frac{z-3}{4}$ and parallel to the coordinate planes.

Ans.: $x\cos\theta+y\sin\theta=0$

23. Prove that the plane through the point (α, β, γ) and the line $x = py + q = zx + r$ is given by $\begin{vmatrix} x & py+q & rz+s \\ \alpha & p\beta+q & r\gamma+s \\ 1 & 1 & 1 \end{vmatrix} = 0$.

24. The line L is given by $\frac{x-1}{2} = \frac{y}{-1} = \frac{z+2}{2}$. Find the direction cosines of the projections of L on the plane $2x + y - 3z = 4$ and the equation of the plane through L parallel to the line $2x + 5y + 3z = 4$, $x - y - 5z = 6$.

Ans.: $\frac{2}{\sqrt{6}}, \frac{-1}{\sqrt{6}}, \frac{1}{\sqrt{6}}; 3x + 4y - 2z = 7$

Exercises 2

1. Find the equation of the line joining the points
   (i) (2, 3, 5) and (−1, 2, −4)
   (ii) (1, −1, 3) and (3, 3, 1)

Ans.: (i) $\frac{x-2}{3} = \frac{y-3}{1} = \frac{z}{9}$; (ii) $\frac{x-1}{2} = \frac{y+1}{4} = \frac{z-3}{-2}$

2. Find the equations of the line passing through the point (3, 2, −8) and is perpendicular to the plane $3x - y - 2z + 2 = 0$.

Ans.: $\frac{x-3}{3} = \frac{y-2}{-1} = \frac{z+8}{-2}$

3. Find the equations of the line passing through the point (3, 1, −6) and parallel to each of the planes $x + y + 2z - 4 = 0$ and $2x - 3y + z + 5 = 0$.

Ans.: $\frac{x-3}{3} = \frac{y-1}{3} = \frac{z+6}{-5}$

4. Find the equations of the line through the point (1, 2, 3) and parallel to the line of intersection of the planes $x - 2y - z + 5 = 0$, $x + y + 3z - 6 = 0$.

Ans.: $\frac{x-2}{5} = \frac{y-3}{4} = \frac{z-1}{-3}$

5. Find the point at which the line $\frac{x-1}{2} = \frac{y-2}{3} = \frac{z+3}{4}$ meets the plane $2x + 4y - z + 1 = 0$.

Ans.: $\left(\frac{10}{3}, \frac{-3}{2}, \frac{5}{3}\right)$

6. Find the coordinates of the point at which the line $\frac{x}{1} = \frac{y-1}{2} = \frac{z+2}{3}$ meets the plane $2x + 3y + z = 0$.

Ans.: $\left(\frac{-1}{11}, \frac{9}{11}, \frac{25}{11}\right)$

7. Prove that the equations of the normal to the plane $ax + by + cz + d = 0$ through the point (α, β, γ) are $\dfrac{x-\alpha}{a} = \dfrac{y-\beta}{b} = \dfrac{z-\gamma}{c}$.

8. Express in symmetrical form the following lines:
   (i) $x + 2y + z = 3,\ 6x + 8y + 3z = 13$
   (ii) $x - 2y + 3z - 4 = 0,\ 2x - 3y + 4z - 5 = 0$
   (iii) $x + 3y - z - 15 = 0,\ 5x - 2y + 4z + 8 = 0$

Ans.: (i) $\dfrac{x-2}{2} = \dfrac{y+1}{-3} = \dfrac{z-3}{4}$

(ii) $\dfrac{x+2}{1} = \dfrac{y+3}{2} = \dfrac{z}{1}$

(iii) $\dfrac{x-2}{2} = \dfrac{y-1}{-1} = \dfrac{z}{-3}$

9. Prove that the lines $3x + 2y + z - 5 = 0,\ x + y - 2z - 3 = 0$ and $8x - 4y - 4z = 0,\ 7x + 10y - 8z = 0$ are at right angles.

10. Prove that the lines $x - 4y + 2z = 0,\ 4x - y - 3z = 0$ and $x + 3y - 5z + 9 = 0,\ 7x - 5y - z + 7 = 0$ are parallel.

11. Find the point at which the perpendicular from the origin on the line joining the points (–9, 4, 5) and (11, 0, –1) meets it.

Ans.: (1, 2, 2).

12. Prove that the lines $2x + 3y - 4z = 0,\ 3x - 4y + 7 = 0$ and $5x - y - 3z + 12 = 0,\ x - 7y + 5z - 6 = 0$ are parallel.

13. Find the perpendicular from the point (1, 3, 9) to the line $\dfrac{x-13}{5} = \dfrac{y+8}{-8} = \dfrac{z-31}{1}$.

Ans.: 21

14. Find the distance of the point (–1, –5, –10) from the point of intersection of the line $\dfrac{x-2}{3} = \dfrac{y+1}{4} = \dfrac{z-2}{12}$ and the plane $x - y + z = 5$.

Ans.: 13

15. Find the length of the perpendicular from the point (5, 4, –1) to the line $\dfrac{x-1}{2} = \dfrac{y}{9} = \dfrac{z}{5}$.

Ans.: $\sqrt{\dfrac{2109}{110}}$

16. Find the foot of the perpendicular from the point (–1, 11, 5) to the line $\dfrac{x-2}{3} = \dfrac{y+3}{2} = \dfrac{z}{2}$.

Ans.: $\left(\dfrac{17}{191}, \dfrac{7}{17}, \dfrac{58}{17}\right)$

17. Obtain the coordinates of the foot of the perpendicular from the origin on the line joining the points (–9, 4, 5) and (11, 0, –1).

18. Find the image of the point (4, 5, –2) in the plane $x - y + 3z - 4 = 0$.

Ans.: (6, 3, 4)

19. Find the image of the point (1, 3, 4) in the plane $2x - y + z + 3 = 0$.

Ans.: (1, 0, 7)

20. Find the image of the point (2, 3, 5) in the plane $2x + y - z + 2 = 0$.

Ans.: $\left(\frac{-2}{3}, \frac{5}{3}, \frac{19}{3}\right)$

21. Find the image of the point (p, q, r) in the plane $2x + y + z = 6$ and hence find the image of the line $\frac{x-1}{2} = \frac{y-2}{1} = \frac{z-3}{-4}$.

Ans.: $\left(\frac{12-p-2q-2r}{3}, \frac{6-2p+2q-r}{3}, \frac{6-2p-q+2r}{3}, \frac{3x-1}{4} = \frac{3y-5}{2} = \frac{3z-8}{-13}\right)$

22. Find the coordinates of the foot of the perpendicular from (1, 0, 2) to the line $\frac{x+1}{3} = \frac{y-2}{-3} = \frac{z+1}{-1}$. Also find the length of the perpendicular.

Ans.: $\left(\frac{8}{19}, \frac{11}{19}, \frac{-28}{19}\right); \frac{11}{19}\sqrt{38}$

23. Find the equation in symmetrical form of the projection of the line $\frac{x-1}{2} = \frac{y+1}{-1} = \frac{z-3}{4}$ on the plane $x + 2y + z = 12$.

Ans.: $\frac{5x-4}{4} = \frac{5y-28}{-7} = \frac{z}{2}$

24. Prove that the point which the line $\frac{x-1}{3} = \frac{y}{-2} = \frac{z-3}{1}$ meets the plane $2x + 35y - 39z + 12 = 0$ is equidistant from the planes $12x - 15y + 16z = 28$ and $6x + 6y - 7z = 8$.

25. Find the equation of the projection of the straight line $\frac{x-1}{1} = \frac{y+1}{2} = \frac{z}{3}$ on the plane $x + y + 2z = 5$ in symmetrical form.

Ans.: $\frac{x}{1} = \frac{3y-5}{-3} = \frac{3z-5}{0}$

26. Prove that two lines in which the planes $3x - 7y - 5z = 1$ and $5x - 13y + 3z + 2 = 0$ cut the plane $8x - 11y + 2z = 0$ include a right angle.

27. Reduce to symmetrical form the line given by the equations $x + y + z + 1 = 0$, $4x + y - 2z + 2 = 0$. Hence find the equation of the plane through (1, 1, 1) and perpendicular to the given line.

Ans.: $\frac{3x+1}{3} = \frac{3y+2}{-6} = \frac{z}{1}; x - 2y + z = 0$

28. Show that the line $x + 2y - z - 3 = 0, x + 3y - z - 4 = 0$ is parallel to the xz–plane and find the coordinates of the point where it meets yz–plane.

Ans.: (0, 1, –1)

29. Find the angle between the lines $x - 2y + z = 0, x + y - z - 3 = 0$, and $x + 2y + z - 5 = 0, 8x + 12y + 5z = 0$.

Ans.: $\cos^{-1}\left(\frac{8}{\sqrt{406}}\right)$

30. Find the equation of the plane passing through the line $\frac{x-1}{2}=\frac{y+1}{-1}=\frac{z-3}{4}$ and parallel to the line $\frac{x}{1}=\frac{y}{2}=\frac{z}{3}$.

Ans.: $11x+2y-5z+6=0$

31. The plane $\frac{x}{a}+\frac{y}{b}+\frac{z}{c}=1$ meets the axes in A, B and C. Find the coordinates of the orthocentre of the ΔABC.

Ans.: $\left(\frac{a^{-1}}{\Sigma a^{-2}}, \frac{b^{-1}}{\Sigma a^{-2}}, \frac{c^{-1}}{\Sigma a^{-2}}\right)$

32. The equation to a line AB are $\frac{x}{1}=\frac{y}{-2}=\frac{z}{3}$. Through a point $P(1, 2, 3)$, PN is drawn perpendicular to AB and PQ is drawn parallel to the plane $2x+3y+4z=0$ to meet AB in Q. Find the equations of PN and PQ and the coordinates of N and Q.

Ans.: $\frac{x-1}{1}=\frac{y-2}{5}=\frac{z-3}{3}; \frac{x-1}{3}=\frac{y-2}{-14}=\frac{z-3}{9}; \left(\frac{5}{2}, -5, \frac{15}{2}\right)$

ILLUSTRATIVE EXAMPLES (COPLANAR LINES AND SHORTEST DISTANCE)

Example 13.18

Prove that the lines $\frac{x+1}{-3}=\frac{y+10}{8}=\frac{z-1}{2}$ and $\frac{x+3}{-4}=\frac{y+1}{7}=\frac{z-4}{1}$ are coplanar and find the equation of the plane continuing these two lines.

Solution

(−1, −10, 1) is a point on the first line and −3, 8, 2 are the direction ratios of the first line. (−3, −1, 4) is a point on the second line and −4, 7, 1 are the direction ratios of the second line. If the lines are coplanar then

$$\begin{vmatrix} x_2-x_1 & y_2-y_1 & z_2-z_1 \\ l_1 & m_1 & n_1 \\ l_2 & m_2 & n_2 \end{vmatrix}=0.$$

Now $$\begin{vmatrix} 2 & -9 & -3 \\ -3 & 8 & 2 \\ -4 & 7 & 1 \end{vmatrix}=2(8-14)+9(-3+8)-3(-21+32)$$

$$=-12+45-33=0$$

Therefore, the two lines are coplanar. The equation of the plane containing the lines is $\begin{vmatrix} x+1 & y+10 & z-1 \\ -3 & 8 & 2 \\ -4 & 7 & 1 \end{vmatrix} = 0.$

$$\text{(i.e.)} \quad (x+1)(8-14) - (y+10)(-3+8) + (z-1)(-21+32) = 0$$

$$\text{(i.e.)} \quad -6x - 6 - 5y - 50 + 11z - 11 = 0$$

$$6x + 5y - 11z + 67 = 0$$

Example 13.19

Show that the lines $\frac{x+1}{-3} = \frac{y-3}{2} = \frac{z+2}{1}$ and $\frac{x}{1} = \frac{y-7}{-3} = \frac{z+7}{2}$ intersect. Find the point of intersection and the equation of the plane containing these two lines.

Solution

The two given lines are

$$\frac{x+1}{-3} = \frac{y-3}{2} = \frac{z+2}{1} = r \tag{13.57}$$

$$\frac{x}{1} = \frac{y-7}{-3} = \frac{z+7}{2} = r_1 \tag{13.58}$$

Any point on the first line is $(-3r - 1, 2r + 3, r - 2)$. Any point on the second line is $(r_1, -3r_1 + 7, 2r_1 - 7)$. If the two lines intersect then the two points are one and the same.

$$-3r - 1 = r_1 \qquad 3r + r_1 = -1 \tag{13.59}$$

$$2r + 3 = -3r_1 + 7 \quad \text{or} \quad 2r + 3r_1 = 4 \tag{13.60}$$

$$r - 2 = 2r_1 - 7 \qquad r - 2r_1 = -5 \tag{13.61}$$

$$2r + 3r_1 = 4$$

$$2r - 4r_1 = -10$$

$$7r_1 = 14$$

$$r_1 = 2, r = -1$$

Solving (13.60) and (13.61), we get $r = -1$ and $r_1 = 2$. These values satisfy equation (13.59). The point of intersection is (2, 1, −3).

The equation of the plane containing the two lines is

$$\begin{vmatrix} x+1 & y-3 & z+2 \\ -3 & +2 & 1 \\ 1 & -3 & 2 \end{vmatrix} = 0 \text{ (i.e.) } (x+1)(4+3)-(y-3)(-6-1)+(z+2)(9-2)=0$$

$$\text{(i.e.)} \quad 7x+7+7y-21+7z+14=0$$

$$\text{(i.e.)} \quad 7x+7y+7z=0 \Rightarrow x+y+z=0$$

Example 13.20

Show that the lines $\frac{x+1}{1}=\frac{y+1}{2}=\frac{z+1}{3}$ and $x + 2y + 3z - 8 = 0$, $2x + 3y + 4z - 11 = 0$ are coplanar. Find the equation of the plane containing these two lines.

Solution

The two lines are

$$\frac{x+1}{1}=\frac{y+1}{2}=\frac{z+1}{3}=r \tag{13.62}$$

$$x + 2y + 3z - 8 = 0$$

$$2x + 3y + 4z - 11 = 0 \tag{13.63}$$

Any plane containing the second line is

$$x + 2y + 3z - 8 + \lambda(2x + 3y + 4z - 11) = 0 \tag{13.64}$$

If the line given by (13.62) lies on this plane then the point (−1, −1, −1) also lies on the plane.

$$-1 -2 -3 -8 + \lambda(-2 - 3 - 4 -11) = 0$$

$$\Rightarrow -14 + \lambda(-20) = 0$$

$$\therefore \lambda = \frac{-7}{10}$$

The equation of the plane (13.64) is

$$x+2y+3z-8-\frac{7}{10}(2x+3y+4z-11)=0$$

$$10x + 20y + 30z - 80 - 14x - 21y - 28z + 77 = 0$$

or

$$4x + y - 2z + 3 = 0 \tag{13.65}$$

Also the normal to this plane should be perpendicular to the line (13.62). The direction ratios of the normal to the plane are 4, 1, −2. The direction ratios of the line (13.62) are 1, 2, 3. Also $ll_1 + mm_1 + nn_1 = 4 + 2 - 6 = 0$ which is true. Hence, the plane containing the two given lines is $4x + y - 2z + 3 = 0$. Any point on the first line is $(r - 1, 2r - 1, 3r - 1)$. If the two given lines intersect at this point then it should lie on the second line and hence on the plane $x + 2y + 3z - 8 = 0$.

$$\therefore r - 1 + 2(2r - 1) + 3(3r - 1) - 8 = 0 \quad \text{(i.e.)} \quad 14r - 14 = 0 \text{ or } r = 1$$

Therefore, the point of intersection of the two given lines is (0, 1, 2).

Example 13.21

Show that the lines $x + 2y + 3z - 4 = 0$, $2x + 3y + 4z - 5 = 0$ and $2x + 3y + 3z - 5 = 0$, $3x - 2y + 4z - 6 = 0$ are coplanar and find thc cquation of thc plane containing the two lines.

Solution

Let us express the first line in symmetrical form. Let l, m, n be the direction cosines of the first line. Then this line is perpendicular to the normals of the planes $x + 2y + 3z - 4 = 0$ and $2x + 3y + 4z - 5 = 0$.

$$\therefore l + 2m + 3n = 0$$

$$2l + 3m + 4n = 0$$

Solving, we get $\dfrac{l}{8-9} = \dfrac{m}{6-4} = \dfrac{n}{3-4}$ or $\dfrac{l}{-1} = \dfrac{m}{2} = \dfrac{n}{-1}$

Therefore, the direction ratios of the first line are 1, −2, 1.

To find a point on the first line let us find where this line meets the XOY plane (i.e.) $z = 0$.

$$x + 2y = 4$$

$$2x + 3y = 5$$

Solving these two equations we get the point as (−2, 3, 0).
Therefore, the equations of the first line are

$$\frac{x+2}{1} = \frac{y-3}{-2} = \frac{z}{1} \tag{13.66}$$

Any plane containing the second line is

$$2x - 3y + 3z - 5 + \lambda(3x - 2y + 4z - 6) = 0 \tag{13.67}$$

If the plane contains the second line then the point (–2, 3, 0) should lie on the plane (13.67).

$$\therefore -4-9-5+\lambda(-6-6-6)=0 \text{ or } -18-18\lambda=0$$

or $$\lambda=-1$$

Hence the equations of the plane (13.67) becomes

$$x+y-1=0 \tag{13.68}$$

Also it should satisfy the condition. That the normal to the plane should be perpendicular to the line (13.66).

The direction ratios of the normal to the plane (13.68) are 1, 1, 1.

The direction ratios of the line are 1, –2, 1. Also 1 – 2 + 1 = 0 which is satisfied. Hence the equation of the required plane is $x+y+z-1=0$.

Example 13.22

Prove that the lines $x = ay + b = cz + d$ and $x = \alpha y + \beta = \gamma z + \delta$ are coplanar if $(r-c)(\alpha\beta - bd) - (\alpha - a)(\alpha\delta - \delta\gamma) = 0$.

Solution

First let us express the given lines in symmetrical form. The two given lines are $\frac{x}{ac}=\frac{y+\frac{b}{a}}{c}=\frac{z+\frac{d}{c}}{a}$ and $\frac{x}{\alpha\gamma}=\frac{y+\frac{\beta}{\alpha}}{\gamma}=\frac{z+\frac{\delta}{\gamma}}{\alpha}$.

Then two lines are coplanar if $\begin{vmatrix} 0 & \frac{b}{a}-\frac{\beta}{\alpha} & \frac{d}{c}-\frac{\delta}{\gamma} \\ ac & c & a \\ \alpha\gamma & \gamma & \alpha \end{vmatrix}=0.$

$$\text{(i.e.)}\quad -\left(\frac{b}{a}-\frac{\beta}{\alpha}\right)(ac\alpha-a\alpha\gamma)+\left(\frac{d}{c}-\frac{\delta}{\gamma}\right)(ac\gamma-\alpha c\gamma)=0$$

$$\text{(i.e.)}\quad -\left(\frac{b\alpha-a\beta}{a\alpha}\right)a\alpha(c-\gamma)+\left(\frac{d\gamma-e\delta}{c\gamma}\right)c\gamma(a-\alpha)=0$$

$$\text{(i.e.)}\quad (\gamma-c)(a\beta-b\alpha)-(a\delta-d\gamma)(\alpha-a)=0$$

Example 13.23

Prove that the lines $a_1x+b_1y+c_1z+d_1=0=a_2x+b_2y+c_2z+d_2$ and $a_3x+b_3y+c_3z+d_3=0=a_4x+b_4y+c_4z+d_4$ are coplanar if $\begin{vmatrix} a_1 & b_1 & c_1 & d_1 \\ a_2 & b_2 & c_2 & d_2 \\ a_3 & b_3 & c_3 & d_3 \\ a_4 & b_4 & c_4 & d_4 \end{vmatrix}=0.$

Solution

Let the two lines intersect at (x_1, y_1, z_1). Then (x_1, y_1, z_1) should lie on the planes containing these lines.

$$\begin{aligned} a_1x_1+b_1y_1+c_1z_1+d_1&=0 \\ a_2x_1+b_2y_1+c_2z_1+d_2&=0 \\ a_3x_1+b_3y_1+c_3z_1+d_3&=0 \\ a_4x_1+b_4y_1+c_4z_1+d_4&=0 \end{aligned}$$

Eliminating (x_1, y_1, z_1) from the above equations we get

$$\begin{vmatrix} a_1 & b_1 & c_1 & d_1 \\ a_2 & b_2 & c_2 & d_2 \\ a_3 & b_3 & c_3 & d_3 \\ a_4 & b_4 & c_4 & d_4 \end{vmatrix}=0.$$

This is the required condition.

Example 13.24

Find the shortest distance and the equation to the line of shortest distance between the two lines $\frac{x+7}{3}=\frac{y+4}{4}=\frac{z+3}{-2}$ and $\frac{x-21}{6}=\frac{y+5}{-4}=\frac{z-2}{-1}$.

Solution

The two given lines are $\frac{x+7}{3}=\frac{y+4}{4}=\frac{z+3}{-2}=r$ and $\frac{x-21}{6}=\frac{y+5}{-4}=\frac{z-2}{-1}=r_1$.

The coordinates of any point P on the first line are $(3r-7, 4r-4, -2r-3)$.

The coordinates of any point Q on the second line are $(6r_1+21, -4r_1-5, -r_1+2)$.

The direction ratios of the line PQ are $3r - 6r_1 - 28$, $4r + 4r_1 + 1$, $-2r + r_1 - 5$.

If PQ is the line of the shortest distance then the two lines are perpendicular. The direction ratios of the two lines are 3, 4, −2 and 6, −4, −1. Then $3(3r - 6r_1 - 28) + 4(4r + 4r_1 + 1) - 2(-2r + r_1 - 5) = 0$ and $6(3r - 6r_1 - 2r) - 4(4r + 4r_1 + 1) - 1(-2r + r_1 - 8) = 0$

$$\text{(i.e.)} \quad 29r - 4r_1 - 90 = 0$$

$$4r - 53r_1 - 167 = 0$$

Solving for r and r_1, we get

$$\frac{r}{-3042} = \frac{r_1}{4563} = \frac{1}{-1521} \Rightarrow \frac{r}{-2} = \frac{r_1}{3} = \frac{1}{-1}$$

$$\therefore r = 2 \text{ and } r_1 = -3.$$

The coordinates of P and Q are given by $P(-1, 4, -7)$ and $Q(3, 7, 5)$.

$$\therefore PQ^2 = (3+1)^2 + (7-4)^2 + (5+7)^2 = 16 + 9 + 144 = 169.$$

$$\therefore PQ = 13 \text{ units}$$

The equations of the line of the shortest distance are $\frac{x+1}{3+1} = \frac{y+4}{7-4} = \frac{z+7}{5+7}$

(i.e.) $\frac{x+1}{4} = \frac{y+4}{3} = \frac{z+7}{12}$.

Example 13.25

Show that the shortest distance between z–axis and the line of intersection of the plane $2x + 3y + z - 1 = 0$ with $3x + 2y + z - 2 = 0$ is $\frac{7\sqrt{5}}{25}$ units.

Solution

The equations of the plane containing the given line is

$$2x + 3y + 4z - 1 + \lambda(3x + 2y + z - 2) = 0 \qquad (13.69)$$

The direction ratios of the normal to this plane are $2 + 3\lambda$, $3 + 2\lambda$, $4 + \lambda$.

The direction ratios of the z-axis are 0, 0, 1. If z-axis is parallel to the line then $0(2 + 3\lambda)$, $0(3 + 2\lambda) + 1(4 + \lambda) = 0$.

$$\therefore \lambda = -4$$

Therefore, the equation of the plane (13.69) is $2x + 3y + 4z - 1 - 4(3x + 2y + z - 2) = 0$

$$\text{(i.e.)} \quad -10x - 5y + 7 = 0 \text{ or } 10x + 5y - 7 = 0 \qquad (13.70)$$

Now the SD = the perpendicular distance from any point on the z–axis to the plane

= perpendicular distance from (0, 0, 0) to the plane

$$= \frac{1-71}{\sqrt{100+25}} = \frac{7}{5\sqrt{5}} = \frac{7\sqrt{5}}{25} \text{ units}$$

Example 13.26

Find the points on the lines $\frac{x-6}{3} = \frac{y-7}{-1} = z-4$ and $\frac{x}{-3} = \frac{y+9}{2} = \frac{z-2}{4}$ which are nearest to each other. Hence find the shortest distance between the lines and also its equation.

Solution

The given lines are

$$\frac{x-6}{3} = \frac{y-7}{-1} = \frac{z-4}{1} = r \qquad (13.71)$$

$$\frac{x}{-3} = \frac{y+9}{2} = \frac{z-2}{4} = r_1 \qquad (13.72)$$

Any point on the line (13.71) is $P(3r+6, -r+7, r+4)$. Any point on the line (13.72) is $Q(-3r_1, 2r_1-9, 4r_1+2)$. The direction ratios of PQ are $(3r + 3r_1 + 6, -r - 2r_1 + 16, r - 4r_1 + 2)$.

Since PQ is perpendicular to the two given lines.

$$3(3r + 3r_1 + 6) - 1(-r - 2r_1 + 16) + 1(r - 4r_1 + 2) = 0$$

$$-3(3r + 3r_1 + 6) + 2(-r - 2r_1 + 16) + 4(r - 4r_1 + 2) = 0$$

$$11r + 7r_1 + 4 = 0 \quad \times 7 \Rightarrow \quad 77r + 49r_1 = -28$$

$$\text{(i.e.)} \quad 7r + 29r_1 - 22 = 0 \quad \times 11 \Rightarrow \quad 77r + 319r_1 = 242$$

$$\Rightarrow r_1 = 1, r = -1.$$

Therefore, the points P and Q are (3, 8, 3) and (–3, –7, 6). The SD is the distance PQ.

$$\therefore PQ^2 = (3+3)^2 + (8+7)^2 + (3-6)^2 = 36 + 225 + 9 = 270.$$

$$PQ = \sqrt{270} = 3\sqrt{30} \text{ units}$$

The direction ratios of PQ are 6, 15, –3 (i.e.) 2, 5, –1. P is (3, 8, 3).

Therefore, the equations of the line of SD are $\frac{x-3}{2} = \frac{y-8}{5} = \frac{z-3}{-1}$.

Example 13.27

Find the shortest distance between the lines $\frac{x-8}{3}=\frac{y+9}{-16}=\frac{z-10}{7}$ and $\frac{x-15}{3}=\frac{y-29}{8}=\frac{z-15}{-5}$. Find also the equation of the line of the shortest distance.

Solution

Let l, m, n be the direction ratios of the line of the SD. Since it is perpendicular to both the lines

$$3l-16m+7n=0$$
$$3l+8m-5n=0$$

Solving for l, m, n, we get

$$\frac{l}{80-56}=\frac{m}{21+15}=\frac{n}{24+48}$$

$$\text{(i.e.)}\quad \frac{l}{24}=\frac{m}{36}=\frac{n}{72}$$

$$\text{(i.e.)}\quad \frac{l}{2}=\frac{m}{3}=\frac{n}{6}$$

The direction ratios of the line of SD are 2, 3, 6. The direction cosines of the line of SD are $\frac{2}{7},\frac{3}{7},\frac{6}{7}$.

The length of the line of the SD $=|(x_1-x_2)l+(y_1-y_2)m+(z_1-z_2)n|$ where (x_1, y_1, z_1) and (x_2, y_2, z_2) are the direction cosines of the line of SD.

$$\therefore (x_1, y_1, z_1) \text{ is } (8, -9, 10) \text{ and } (x_2, y_2, z_2) \text{ is } (15, 29, 15)$$

$$l, m, n \text{ are } \frac{2}{7},\frac{3}{7},\frac{6}{7}.$$

$$\therefore SD = \left|(8-15)\frac{2}{7}+(-9-29)\frac{3}{7}+(10-15)\frac{6}{7}\right| = \frac{158}{7} \text{ units.}$$

The equation of the plane containing the first line and the line of SD is

$$\begin{vmatrix} x-8 & y+9 & z-10 \\ 3 & -16 & 7 \\ 2 & 3 & 6 \end{vmatrix} = 0$$

$$\text{(i.e.)}\quad 117x+4y-41z-490=0$$

The equation of the plane containing the second line and the line of *SD* is

$$\begin{vmatrix} x-18 & y-29 & z-15 \\ 3 & 8 & -5 \\ 2 & 3 & 6 \end{vmatrix} = 0.$$

(i.e.) $63x-28y+7z-238=0$

Therefore, the equations of the line of *SD* are $117x+4y+71z-490=0$, $63x-28y+7z-238=0$.

Example 13.28

If $2d$ is the shortest distance between the lines $x=0, \frac{y}{b}+\frac{z}{c}=1$ and $y=0, \frac{x}{a}-\frac{z}{c}=1$ then prove that $\frac{1}{d^2}=\frac{1}{a^2}+\frac{1}{b^2}+\frac{1}{c^2}$.

Solution

The two given lines are

$$x=0, \frac{y}{b}+\frac{z}{c}=1 \tag{13.73}$$

$$y=0, \frac{x}{a}-\frac{z}{c}=1 \tag{13.74}$$

The equation of any plane containing the first line is

$$\frac{y}{b}+\frac{z}{c}-1+\lambda x=0 \tag{13.75}$$

The equation of the second line in symmetrical form is

$$\frac{x}{a}=\frac{y}{0}=\frac{z+c}{c} \tag{13.76}$$

The plane given by equation (13.75) is parallel to the line (13.76). If

$$\lambda a+\frac{1}{b}\cdot 0+c\cdot\frac{1}{c}=0 \Rightarrow \lambda=\frac{-1}{a}$$

Hence from (13.75), the equation of the plane containing line (13.73) and parallel to the plane (13.74) is $\frac{x}{a}-\frac{y}{b}-\frac{z}{c}+1=0$.

Then the *SD* between the given lines = the perpendicular distance from any point on the line (13.74) to the plane (13.75).

(0, 0, –c) is a point on the line (13.76).

$$\therefore SD=2d=\frac{0+0-(-1)+1}{\sqrt{\frac{1}{a^2}+\frac{1}{b^2}+\frac{1}{c^2}}} \Rightarrow \frac{1}{d}=\sqrt{\frac{1}{a^2}+\frac{1}{b^2}+\frac{1}{c^2}}$$

or $$\frac{1}{d^2} = \frac{1}{a^2} + \frac{1}{b^2} + \frac{1}{c^2}$$

Example 13.29

Show that the shortest distance between any two opposite edges of the tetrahedron formed by the planes $y + z = 0$, $z + x = 0$, $x + y = 0$ and $x + y + z = a$ is $\frac{2a}{\sqrt{6}}$ and the three lines of the shortest distance intersect at the point $x = y = z = a$.

Solution

The equations of the edge determined by the planes $y + z = 0$, $z + x = 0$ is

$$\frac{x}{1} = \frac{y}{1} = \frac{z}{-1} \tag{13.77}$$

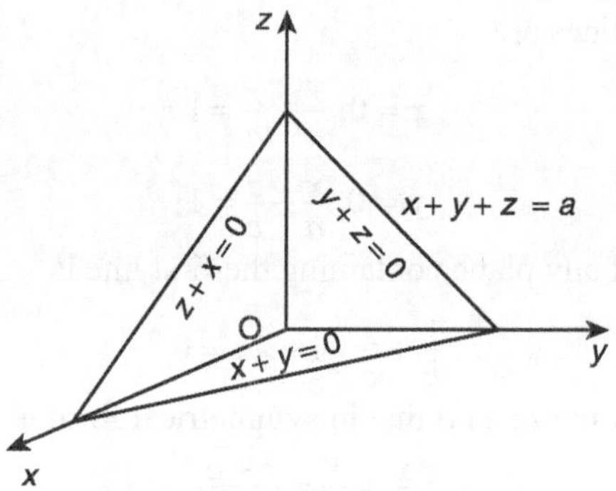

The equation of the opposite edges are $x + y = 0$, $x + y + z = a$

$$\text{(i.e.)} \quad \frac{x}{1} = \frac{y}{-1} = \frac{z-a}{0} \tag{13.78}$$

Let l, m, n be the direction cosines of the line of the SD between two lines. Then $l + m - n = 0$, $l - m + 0.\,n = 0$.

Solving for l, m, n we get $\frac{l}{1} = \frac{m}{1} = \frac{n}{2}$.

Therefore, the direction cosines of the line of the SD are $\frac{1}{\sqrt{6}}, \frac{1}{\sqrt{6}}, \frac{2}{\sqrt{6}}$.

$$\therefore SD = |(x_1 - x_2)l + (y_1 - y_2)m + (z_1 - z_2)n|$$

$$= \left|(0-0)\frac{1}{\sqrt{6}} + (0-0)\frac{1}{\sqrt{6}} + (a-0)\frac{2}{\sqrt{6}}\right| = \frac{2a}{\sqrt{6}}$$

The equation of the plane containing the edge given by (13.77) and the line of the SD is $\begin{vmatrix} x & y & z \\ 1 & 1 & -1 \\ 1 & 1 & 2 \end{vmatrix} = 0.$

$$\text{(i.e.)} \quad x + y - z + a = 0$$

Therefore, the equations of the SD are given by

$$x - y = 0, x + y - z + a = 0 \tag{13.79}$$

This line passes through the point (a, a, a). Similarly, by symmetry we note that the other two lines of SD also pass through the point (a, a, a).

Example 13.30

A square $ABCD$ of diagonal $2a$ is folded along the diagonal AC, so that the planes DAC, BAC are at right angles. Find the shortest distance between DC and AB.

Solution

Let a be the side of the square. Let us take OB, OC, OD as the axes of coordinates. The coordinates of B, C, D and A are $(a, 0, 0)$, $(0, a, 0)$, $(0, 0, a)$, $(0, 0, -a)$.

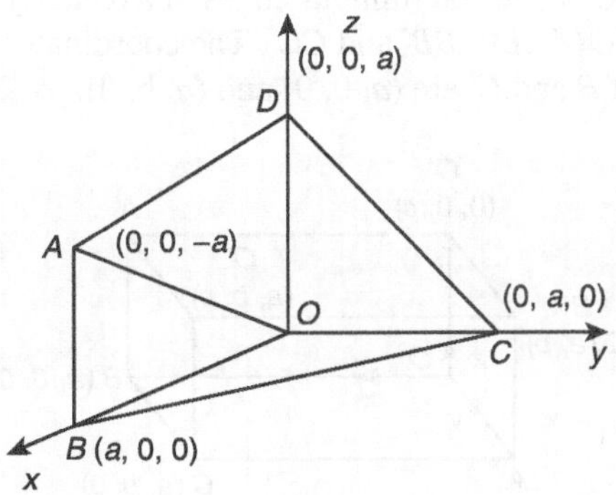

The equations of AB are

$$\frac{x-a}{1} = \frac{y}{1} = \frac{z}{0} \tag{13.80}$$

The equations of CD are

$$\frac{x}{0} = \frac{y-a}{1} = \frac{z}{-1} \tag{13.81}$$

The equations of the plane passing through the straight line (13.80) and parallel to (13.81) is $\begin{vmatrix} x-a & y & z \\ 1 & 1 & 0 \\ 0 & 1 & -1 \end{vmatrix} = 0.$

$$-(x-a) + y + z = 0 \text{ or } x - y - z - a = 0 \tag{13.82}$$

Therefore, the required shortest distance = perpendicular from the point $(0, a, 0)$ to the plane (13.82).

$$= \left|\frac{-2a}{\sqrt{3}}\right| = \frac{2a}{\sqrt{3}} \text{ units}$$

Example 13.31

Prove that the shortest distance between the diagonal of rectangular parallelepiped and the edge not meeting it is $\frac{bc}{\sqrt{b^2+c^2}}, \frac{ca}{\sqrt{a^2+b^2}}, \frac{ab}{a^2+b^2}$ where a, b, c are the edges of the parallelepiped.

Solution

Let OA, OB and OC be the coterminous edges of a rectangular parallelepiped. The diagonals are OO', AA', BB' and CC'. The coordinates of O' are (a, b, c). The coordinates of B and C' are $(a, 0, 0)$ and $(a, b, 0)$.

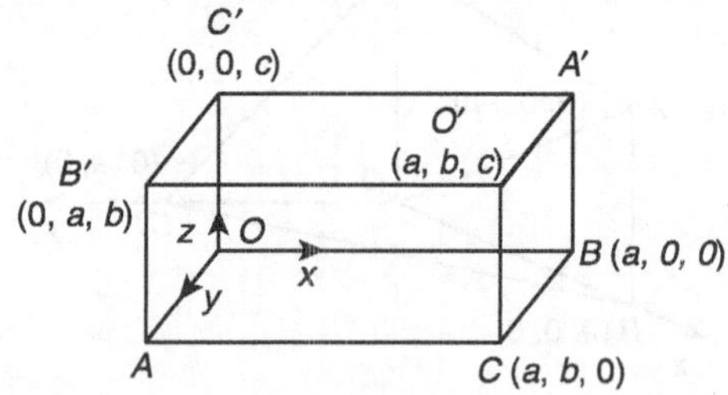

The equations of OO' are $\frac{x}{a} = \frac{y}{b} = \frac{z}{c}$.

The equations of BC are $\frac{(x-a)}{0} = \frac{y}{b} = \frac{z}{0}$.

Let l, m, n be the direction cosines of the line of the SD. Then

$$la + mb + nc = 0$$

$$0 + mb + 0 = 0 \quad \Rightarrow m = 0 \text{ and } \frac{l}{n} = \frac{-c}{a}$$

Hence, l, m, n are $-c, o, a$.

The direction cosines of the line of the SD are $\left(\frac{-c}{\sqrt{c^2+a^2}}, 0, \frac{a}{\sqrt{a^2+c^2}}\right)$

$$\text{Length of the } SD = \left[(0-a)\cdot\left(\frac{(-c)}{\sqrt{c^2+a^2}}\right) + (0-b)(0) + (0-0)\cdot\frac{a}{\sqrt{a^2+c^2}}\right]$$

$$= \frac{ac}{\sqrt{c^2+a^2}}$$

Similarly we can prove that the other two SD are $\frac{bc}{\sqrt{b^2+c^2}}$ and $\frac{ab}{\sqrt{a^2+b^2}}$.

Exercises 3

1. Prove that the lines $\frac{x+1}{-3} = \frac{y-3}{2} = \frac{z+2}{1}$ and $\frac{x}{1} = \frac{y-7}{-3} = \frac{z+7}{2}$ are coplanar and find the equation of the plane containing the line.

   Ans.: $x - y + z = 0$

2. Prove that the lines $\frac{x+1}{3} = \frac{y+3}{5} = \frac{z+5}{7}$ and $\frac{x-2}{2} = \frac{y-4}{3} = \frac{z=6}{5}$ intersect. Find the point of intersection and the plane containing the line.

   Ans.: $\left(\frac{1}{2}, \frac{-1}{2}, \frac{-3}{2}\right), x - 2y + z = 0$

3. Show that the lines $\frac{x-4}{1} = \frac{y+3}{-4} = \frac{z+1}{2}$ and $\frac{x-1}{7} = \frac{y+1}{-3} = \frac{z+10}{3}$ intersect and find the equation of the plane containing the lines.

   Ans.: $(5, -7, 6)$

4. Prove that the lines $\frac{x-4}{5} = \frac{y-3}{-2} = \frac{z-2}{-6}$ and $\frac{x-3}{4} = \frac{y-2}{-3} = \frac{z-1}{-7}$ are coplanar. Find also the point of intersection and the equation of the plane through them.

   Ans.: $(-1, 5, 8), 4x - 11y + 7z + 3 = 0$

5. Show that the lines $\frac{x+3}{2} = \frac{y+5}{3} = \frac{z-7}{-3}$ and $\frac{x+1}{4} = \frac{y+1}{5} = \frac{z+1}{-1}$ are coplanar. Find the equation of the plane containing the line.

   Ans.: $x - 2y + z = 0$

6. Show that the lines $\frac{x-5}{4}=\frac{y-7}{4}=\frac{z+3}{-5}$ and $\frac{x-8}{1}=\frac{y-4}{1}=\frac{z-5}{3}$ are coplanar. Find the point of intersection and the equation of the plane containing them.

Ans.: $(1, 3, 2)$, $17x - 47y - 24z + 172 = 0$

7. Show that the lines $\frac{x-1}{2}=\frac{y-2}{3}=\frac{z-3}{4}$ and $\frac{x-2}{3}=\frac{y-3}{4}=\frac{z-4}{5}$ are coplanar and find the equation of the plane containing them.

Ans.: $x - 2y + z = 0$

8. Show that the lines $\frac{x+1}{2}=\frac{y+2}{3}=\frac{z-4}{-3}$ and $\frac{x-1}{4}=\frac{y-2}{5}=\frac{z+4}{-1}$ are coplanar and find the equation of the plane containing them.

Ans.: $6x - 5y - z = 0$

9. Show that the lines $\frac{x-1}{1}=\frac{y-1}{2}=\frac{z+2}{3}$ and $\frac{x+2}{2}=\frac{y-5}{-1}=\frac{z+3}{3}$ intersect and find the equation of the plane containing these lines.

Ans.: $(2, 3, 1)$, $\frac{x-2}{7}=\frac{y-3}{4}=\frac{z-1}{-5}$

10. Show that the lines $\frac{x+5}{3}=\frac{y+4}{1}=\frac{z-7}{-1}$ and $3x + 2y + z - 2 = 0$, $x - 3y + 2z - 13 = 0$ intersect. Find also the equation of the plane containing them.

Ans.: $(-1, 2, 3)$, $6x - 5y - z = 0$.

11. Show that the lines $x - 3y + 2z + 4 = 0$, $2x + y + 4z + 1 = 0$ and $3x + 2y + 5z - 1 = 0$, $2y + z = 0$ are coplanar. Find their point of intersection and the equation of the plane containing these lines.

Ans.: $(3, 1, -2)$, $3x + 4y + 6z - 1 = 0$.

12. Show that the lines $x + y + z - 3 = 0$, $2x + 3y + 4z - 5 = 0$ and $4x - y + 5z - 7 = 0$, $2x - 5y - z - 3 = 0$ are coplanar. Find the equation of the plane containing these lines.

Ans.: $x + 2y + 3z - 2 = 0$.

13. Show that the lines $7x - 4y + 7z + 16 = 0$, $4x + 3y - 2z + 3 = 0$ and $x - 3y + 4z + 6 = 0$, $x - y + z + 1 = 0$ are coplanar.

14. Show that the lines $7x - 2y - 2z + 3 = 0$, $9x - 6y + 3 = 0$ and $5x - 4y + z = 0$, $6y - 5z = 0$ are coplanar. Find the equation of the plane in which they lie.

Ans.: $x - 2y + z = 0$

15. Show that the lines $\frac{x-1}{2}=\frac{y-1}{-1}=\frac{z+6}{-3}$ and $x+2y+z+2=0$, $4x+5y+3z+6=0$ are coplanar. Find the point of intersection of these two lines.

Ans.: $\left(\frac{5}{3},\frac{5}{3},-7\right)$

16. Show that the lines $\frac{x+1}{2}=\frac{y+1}{3}=\frac{z+1}{4}$ and $x+2y+3z-14=0$, $3x+4y+5z-26=0$ are coplanar. Find their point of intersection and the equation of the plane containing them.

Ans.: $(1, 2, 3)$, $11x+2y-7z+6=0$.

17. Show that the lines $3x-y-z+2=0$, $x-2y+3z-6=0$ and $3x-4y+3z-4=0$, $2x-2y+z-1=0$ are coplanar. Find their point of intersection and the equation of the plane containing these lines.

Ans.: $(1, 2, 3)$, $x-z+2=0$

18. Show that the lines $2x-y-z-3=0$, $x-3y+2z-4=0$ and $x-y+z-2=0$, $4x+y-6z-3=0$ are coplanar and find the equation of the plane containing these two lines.

Ans.: $(1, -1, 0)$, $x-z-1=0$

19. Show that the lines $x+2y+3z-4=0$; $2x+3y+4z-5=0$ and $2x-3y+3z-5=0$, $3x-2y+4z-6=0$ are coplanar. Find the equation of the plane containing these two lines.

Ans.: $x+y+z-1=0$

20. Show that the equation of the plane through the line $\frac{x}{l}=\frac{y}{m}=\frac{z}{n}$ and which is perpendicular to the plane containing the lines $\frac{x}{m}=\frac{y}{n}=\frac{z}{l}$ and $\frac{x}{n}=\frac{y}{l}=\frac{z}{m}$ is $(m-n)\,x+(n-l)\,y+(l-m)\,z=0$.

21. Prove that the lines $\frac{x-x_1}{l}=\frac{y-y_1}{m}=\frac{z-z_1}{n}$ and $ax+by+cz+d=0$, $a_1x+b_1y+c_1z+d_1=0$, are coplanar if $\frac{ax_1+by_1+cz_1+d}{al+bm+cz}=\frac{a_1x_1+b_1y_1+c_1z_1+d_1}{a_1l+b_1m+c_1z}$.

22. Show that the lines $\frac{x}{\alpha}=\frac{y}{\beta}=\frac{z}{\gamma}$, $\frac{x}{a\alpha}=\frac{y}{b\beta}=\frac{z}{c\gamma}$ and $\frac{x}{l}=\frac{y}{m}=\frac{z}{n}$ are coplanar if $\frac{l}{\alpha}(b-c)+\frac{m}{\beta}(c-a)+\frac{n}{\gamma}(a-b)=0$.

23. *A*, *A′*; *B*, *B′* and *C*, *C′* are points on the axes, show that the lines of intersection of the planes (*A′BC*, *AB′C′*), (*B′CA*, *BC′A′*) and (*C′AB*, *CA′B′*) are coplanar.

24. Find the shortest distance between the lines $\frac{x-0}{2}=\frac{y}{-3}=\frac{z}{1}$ and $\frac{x-2}{3}=\frac{y-1}{-5}=\frac{z+2}{2}$ and also the equations of the line of the *SD*.

Ans.: $\frac{1}{\sqrt{3}}$, $4x+y-5z=0, 9x+y-8z-31=0$

25. Find the shortest distance between the lines $\frac{x-3}{-3}=\frac{y-8}{1}=\frac{z-3}{-1}$ and $\frac{x+3}{5}=\frac{y+7}{-2}=\frac{z-6}{-4}$ and find the equation of the line of the shortest distance.

Ans.: $3\sqrt{3}$, $4x-5y-17z+79=0, 22x-5y+19z-83=0$

26. Find the shortest distance between the lines $\frac{x-3}{1}=\frac{y-5}{-2}=\frac{z-7}{1}$ and $\frac{x+1}{7}=\frac{y+1}{-6}=\frac{z+1}{1}$. Find also the equation of the line of *SD* and the points where the line of *SD* intersect the two given lines.

Ans.: $2\sqrt{29}$, (3, 5, 7), (–1, –1, –1) $\frac{x-3}{2}=\frac{y-5}{3}=\frac{z-7}{4}$

27. Show that the shortest distance between *z*–axes and the line of intersection of the plane $2x+3y+4z-1=0$ with $3x+2y+z-2=0$ is $\frac{7\sqrt{5}}{25}$.

28. Show that the shortest distance between the lines $\frac{x-1}{2}=\frac{y-2}{3}=\frac{z-3}{4}$ and $\frac{x-2}{3}=\frac{y-4}{4}=\frac{z-5}{5}$ is $\frac{1}{\sqrt{6}}$ and its equation are $11x+2y-7z+6=0, 7x+y-3z+7=0$.

29. Find the length of the shortest distance between the lines $\frac{x-2}{2}=\frac{y+1}{3}=\frac{z}{4}$ and $2x+3y-6z-6=0, 3x-2y-z+5=0$.

Ans.: $\frac{97\sqrt{3}}{78}$

30. Find the shortest distance between *z*–axis and the line $ax+by+cz+d=0, a'x+b'y+c'z+d'=0$.

Ans.: $\frac{dc'-d'c}{\sqrt{(ac'-a'c)^2+(bc'-b'c)^2}}$

31. Find the shortest distance between an edge of a cube and a diagonal which does not meet it.

Ans.: $\frac{a}{\sqrt{3}}$

32. A line with direction cosines proportional to 1, 7, –5 is drawn to intersect the lines $\frac{x-5}{3}=\frac{y-7}{-1}=\frac{z+2}{1}$ and $\frac{x+3}{-3}=\frac{y-3}{2}=\frac{z-6}{4}$. Find the coordinates of the point of intersection and the length intercepted on it.

Ans.: (2, 8, –3), (0, 1, 2), $\sqrt{78}$

33. A line with direction cosines proportional to 2, 7, –5 is drawn to intersect the lines $\frac{x-5}{3}=\frac{y-7}{1}=\frac{z+2}{1}, \frac{x+3}{-3}=\frac{y-3}{2}=\frac{z}{4}$. Find the coordinates of the points of intersection and the length intercepted on it.

Ans.: (2, 8, –3), (0, 1, 2); $\sqrt{78}$

34. The two lines $\frac{x-\alpha}{l}=\frac{y-\beta}{m}=\frac{z-\gamma}{n}$ and $\frac{x-\alpha'}{l'}=\frac{y-\beta'}{m'}=\frac{z-\gamma'}{n'}$ are cut by a third line whose direction cosines are λ, μ, γ. Show that the length intercepted on the third line is given by $\begin{vmatrix} \alpha-\alpha' & \beta-\beta' & \gamma-'\gamma \\ l & m & n \\ l' & m' & n' \end{vmatrix} \div$

$\begin{vmatrix} l & m & n \\ l' & m' & n' \\ \delta & \mu & v \end{vmatrix}$ and show that the length of the shortest distance is

$$\begin{vmatrix} \alpha-\alpha' & \beta-\beta' & \gamma-\gamma' \\ l & m & n \\ l' & m' & n' \end{vmatrix} \div \sqrt{\sum (mn'-m'n)^2}.$$

35. The lengths of two opposite edges of a tetrahedron are *a*, *b*, *c*; the shortest distance is equal to *d* and the angle between them is θ. Prove that the volume of the tetrahedron is $\frac{1}{6}\, abd \sin\theta$.

36. Show that the equation of the plane containing the line $x=0, \frac{y}{b}+\frac{z}{c}=1$ and parallel to the line $y=0, \frac{x}{a}-\frac{z}{c}=1$ is $\frac{x}{a}-\frac{y}{b}-\frac{z}{c}+1=0$. If *d* is the shortest distance between the lines then show that $\frac{4}{d^2}=\frac{1}{a^2}-\frac{1}{b^2}+\frac{1}{c^2}$.

37. Show that the shortest distance between the lines $y=az+b, z=\alpha x+\beta$ and $y=a'z+b, z=\alpha' x+\beta' y$ is $\frac{(\alpha-\alpha')(b-b')-(\alpha'\beta-\alpha\beta')(a-a')}{[\alpha^2-\alpha'^2(a-a')^2+(\alpha-\alpha^2)+(a\alpha-a'\alpha')^2]^{1/2}}$.

38. Find the shortest distance between the lines $x=2z+3, y=3z+4$ and $x=4z+5$, $y=5z+6$. What conclusion do you draw from your answer?

Ans.: Zero; Coplanar lines

CHAPTER

14 Sphere

14.1 DEFINITION OF SPHERE

The locus of a moving point in space such that its distance from a fixed point is constant is called a sphere. The fixed point is called the centre of the sphere. The constant distance is called the radius.

14.2 THE EQUATION OF A SPHERE WITH CENTRE AT (*a*, *b*, *c*) AND RADIUS *r*

Let $P(x, y, z)$ be any point on the sphere. Let $C(a, b, c)$ be the centre.

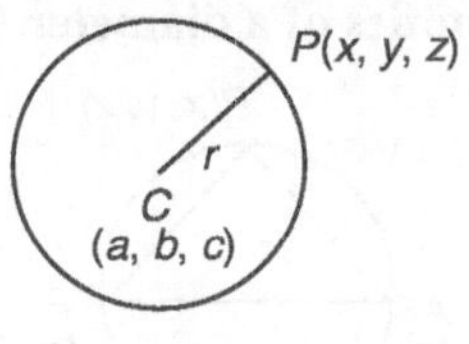

Then,

$$CP = r$$

$$CP^2 = r^2$$

$$(x - a)^2 + (y - b)^2 + (z - c)^2 = r^2$$

This is the equation of the required sphere.

Show that the equation $x^2 + y^2 + z^2 + 2ux + 2vy + 2wz + d = 0$ always represents a sphere. Find its centre and radius.

$$x^2 + y^2 + z^2 + 2ux + 2vy + 2wz + d = 0$$

$$x^2 + y^2 + z^2 + 2ux + 2vy + 2wz = -d$$

Add $u^2 + v^2 + w^2$ to both sides.

$$x^2 + y^2 + z^2 + 2ux + 2vy + 2wz + u^2 + v^2 + w^2 = u^2 + v^2 + w^2 - d$$

$$(x^2 + 2ux + u^2) + (y^2 + 2vy + v^2) + (z^2 + 2wz + w^2) = u^2 + v^2 + w^2 - d$$

$$\text{(i.e.)} \quad (x + u)^2 + (y + v)^2 + (z + w)^2 = \left(\sqrt{u^2 + v^2 + w^2 - d}\right)^2$$

This equation shows that this is the locus of a point (x, y, z) moving from the fixed point $(-u, -v, -w)$ keeping a constant distance $\sqrt{u^2 + v^2 + w^2 - d}$ from it.

Therefore, the locus is a sphere whose centre is $(-u, -v, -w)$ and whose radius is $\sqrt{u^2 + v^2 + w^2 - d}$.

Note 14.2.1: A general equation of second degree in x, y, z will represent a sphere if (i) coefficients of x^2, y^2, z^2 are the same and (ii) the coefficients of xy, yz, zx are zero.

14.3 EQUATION OF THE SPHERE ON THE LINE JOINING THE POINTS (x_1, y_1, z_1) AND (x_2, y_2, z_2) AS DIAMETER

Find the equation of the sphere on the line joining the points (x_1, y_1, z_1) and (x_2, y_2, z_2) as the extremities of a diameter.

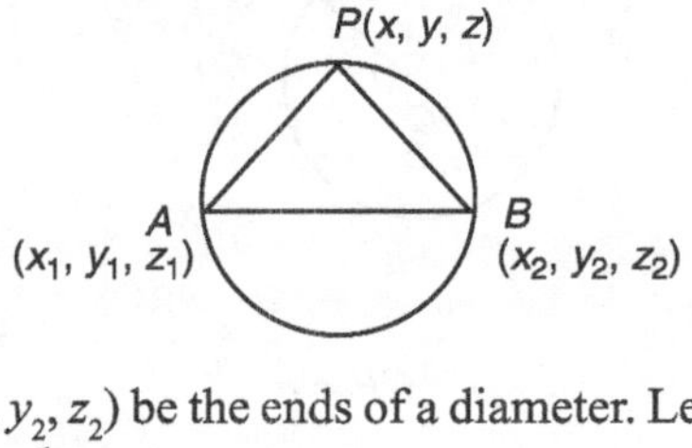

$A(x_1, y_1, z_1)$ and $B(x_2, y_2, z_2)$ be the ends of a diameter. Let (x, y, z) be any point on the surface of the sphere.

Then $\angle APB = 90°$

Therefore, AP is perpendicular to BP.

The direction ratios of AP are $x - x_1, y - y_1, z - z_1$.

The direction ratios of BP are $x - x_2, y - y_2, z - z_2$.

Since AP is perpendicular to BP,

$$(x - x_1)(x - x_2) + (y - y_1)(y - y_2) + (z - z_1)(z - z_2) = 0$$

This is the equation of the required sphere.

14.4 LENGTH OF THE TANGENT FROM $P(x_1, y_1, z_1)$ TO THE SPHERE $x^2 + y^2 + z^2 + 2ux + 2vy + 2wz + d = 0$

Find the length of the tangent from $P(x_1, y_1, z_1)$ to the sphere $x^2 + y^2 + z^2 + 2ux + 2vy + 2wz + d = 0$.

The centre of the sphere is $(-u, -v, -w)$.

The radius of the sphere is $\sqrt{u^2 + v^2 + w^2 - d}$.

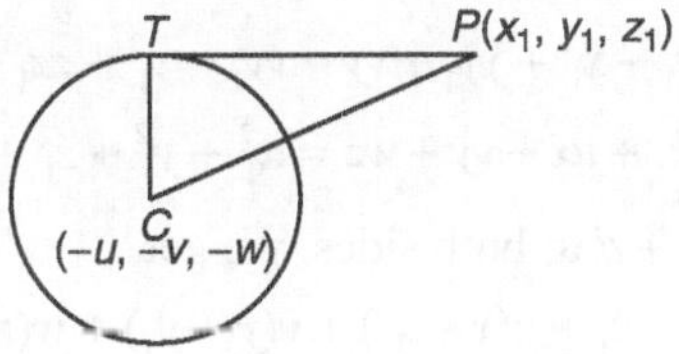

$$PC^2 = (x_1 + u)^2 + (y_1 + v)^2 + (z_1 + w)^2$$

$$\angle CTP = 90°$$

$$PT^2 = PC^2 - CT^2$$

$$= (x_1 + u)^2 + (y_1 + v)^2 + (z_1 + w)^2 - (u^2 + v^2 + w^2 - d)$$

$$= x_1^2 + y_1^2 + z_1^2 + 2ux_1 + 2vy_1 + 2wz_1 + d$$

$$\therefore PT = \sqrt{x_1^2 + y_1^2 + z_1^2 + 2ux_1 + 2vy_1 + 2wz_1 + d}$$

Note 14.4.1: If $PT^2 > 0$, the point P lies outside the sphere.
If $PT^2 = 0$, then the point P lies on the sphere.
If $PT^2 < 0$, then the point P lies inside the sphere.

14.5 EQUATION OF THE TANGENT PLANE AT (x_1, y_1, z_1) TO THE SPHERE $x^2 + y^2 + z^2 + 2ux + 2vy + 2wz + d = 0$

Find the equation of the tangent plane at (x_1, y_1, z_1) to the sphere $x^2 + y^2 + z^2 + 2ux + 2vy + 2wz + d = 0$.

The centre of the sphere is $(-u, -v, -w)$.

$P(x_1, y_1, z_1)$ is a point on the sphere and the required plane is a tangent plane to the sphere at P.

Therefore, the direction ratios of CP are $x_1 + u, y_1 + v, z_1 + w$.

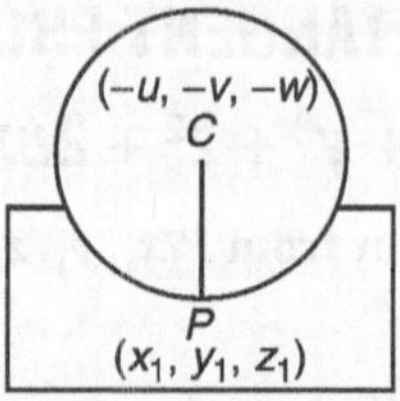

Therefore, the equation of the tangent plane at (x_1, y_1, z_1) is $(x_1 + u)(x - x_1) + (y_1 + v)(y - y_1) + (z_1 + w)(z - z_1) = 0$.

(i.e.) $xx_1 + ux - ux_1 - x_1^2 + yy_1 + vy - vy_1 - y_1^2 + zz_1 + wz - wz_1 - z_1^2 = 0$

(i.e.) $xx_1 + yy_1 + zz_1 + ux + vy + wz = x_1^2 + y_1^2 + z_1^2 + ux_1 + vy_1 + wz_1$

Adding $ux_1 + vy_1 + wz_1 + d$ to both sides, we get

$$xx_1 + yy_1 + zz_1 + u(x + x_1) + v(y + y_1) + w(z + z_1) + d$$

$$= x_1^2 + y_1^2 + z_1^2 + 2ux_1 + 2vy_1 + 2wz_1 + d$$

$= 0$ since (x_1, y_1, z_1) lies on the sphere.

Therefore, the equation of the tangent plane at (x_1, y_1, z_1) is $xx_1 + yy_1 + zz_1 + u(x + x_1) + v(y + y_1) + w(z + z_1) + d = 0$.

14.6 SECTION OF A SPHERE BY A PLANE

Let C be the centre of the sphere and P be any point on the section of the sphere by the plane.

Draw CN perpendicular to the plane. Then N is the foot of the perpendicular from P on the plane section.

Join CP. Since CN is perpendicular to NP, CNP is a right angled triangle.

$$CP^2 = CN^2 + NP^2 \text{ or}$$

$$NP^2 = CP^2 - CN^2$$

$$NP = \sqrt{CP^2 - CN^2} = \text{a constant}$$

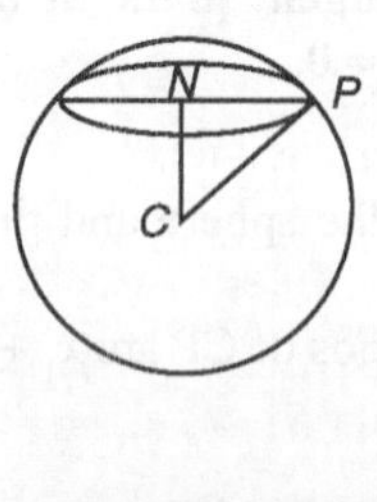

Since CP and CN are constants, NP = constant shows that the locus of P is a circle with centre at N and radius equal to NP.

Note 14.6.1: If the radius of the circle is less than the radius of the sphere then the circle is called a small circle. In other words, a circle of the sphere not passing through the centre of the sphere is called a small circle.

Note 14.6.2: If the radius of the circle is equal to the radius of the sphere then the circle is called a great circle of the sphere. In other words, a circle of the sphere passing through the centre of the sphere is called a great circle.

14.7 EQUATION OF A CIRCLE

The section of a sphere by a plane is a circle. Suppose the equation of the sphere is

$$x^2 + y^2 + z^2 + 2ux + 2vy + 2wz + d = 0 \tag{14.1}$$

and the plane section is

$$ax + by + cz + k = 0 \tag{14.2}$$

Then any point on the circle lie on the sphere (14.1) as well as the plane section (14.2). Hence, the equations of the circle of the sphere are given by $x^2 + y^2 + z^2 + 2ux + 2vy + 2wz + d = 0$ and $ax + by + cz + k = 0$.

14.8 INTERSECTION OF TWO SPHERES

The curve of intersection of two spheres is a circle.

Let the two spheres be

$$S = x^2 + y^2 + z^2 + 2ux + 2vy + 2wz + d = 0$$

$$S_1 = x^2 + y^2 + z^2 + 2u_1x + 2v_1y + 2w_1z + d = 0$$

$$S - S_1 = 2(u - u_1)x + 2(v - v_1)y + 2(w - w_1)z + d - d_1 = 0 \tag{14.3}$$

Equation (14.3) is a linear equation in x, y, z and therefore represents a plane and this plane passes through the point of intersection of the given two spheres.

In addition, we know that section of the sphere by a plane is a circle. Hence the curve of intersection of the spheres is given by $S_1 - S_2 = 0$.

14.9 EQUATION OF A SPHERE PASSING THROUGH A GIVEN CIRCLE

Let the given circle be $S = x^2 + y^2 + z^2 + 2ux + 2vy + 2wz + d = 0$ and

$$P = ax + by + cz + k = 0 \tag{14.4}$$

Consider the equation $S + \lambda P = 0$.

(i.e.) $x^2 + y^2 + z^2 + 2ux + 2vy + 2wz + d + \lambda(ax + by + cz + k) = 0$ (14.5)

This equation represents a sphere. Suppose (x_1, y_1, z_1) is a point on the given circle. Then

$$x_1^2 + y_1^2 + z_1^2 + 2ux_1 + 2vy_1 + 2wz_1 + d = 0 \tag{14.6}$$

$$ax_1 + by_1 + cz_1 + k = 0 \tag{14.7}$$

Equations (14.6) and (14.7) show that the point (x_1, y_1, z_1) lies on the sphere given by equation (14.5).

Since (x_1, y_1, z_1) is an arbitrary point on the circle, it follows that every point on the circle is a point on the sphere given by (14.5).

Hence equation (14.5) represents the equation of a sphere passing through the circle (14.4).

14.10 CONDITION FOR ORTHOGONALITY OF TWO SPHERES

Let the two given spheres be

$$x^2 + y^2 + z^2 + 2ux + 2vy + 2wz + d = 0 \tag{14.8}$$

$$x^2 + y^2 + z^2 + 2u_1x + 2v_1y + 2w_1z + d_1 = 0 \tag{14.9}$$

The centres of the spheres are $A(-u, -v, -w)$ and $B(-u_1, -v_1, -w_1)$. The radius $r = \sqrt{u^2 + v^2 + w^2 - d}$.

$$r_1 = \sqrt{u_1^2 + v_1^2 + w_1^2 - d_1}$$

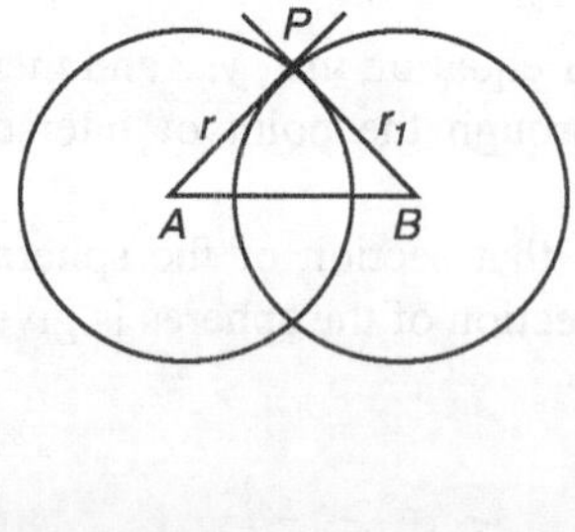

Two spheres are said to be orthogonal, if the tangent planes at this point of intersection are at right angles.

(i.e.) The radii drawn through the point of intersection are at right angles.

$$\therefore AB^2 = AP^2 + BP^2$$

(i.e.) $(-u+u_1)^2+(-v+v_1)^2+(-w+w_1)^2=(u^2+v^2+w^2-d)+(u_1^2+v_1^2+w_1^2-d_1)$

(i.e.) $2uu_1 + 2vv_1 + 2ww_1 = d + d_1$

This is the required condition.

14.11 RADICAL PLANE

The locus of a point whose powers with respect to two spheres are equal is called the radical plane of the two spheres.

14.11.1 Obtain the Equations to the Radical Plane of Two Given Spheres

Let the two given spheres be

$$x^2 + y^2 + z^2 + 2ux + 2vy + 2wz + d = 0 \quad (14.10)$$

$$x^2 + y^2 + z^2 + 2u_1x + 2v_1y + 2w_1z + d_1 = 0 \quad (14.11)$$

Let (x_1, y_1, z_1) be a point such that the power of this point with respect to spheres (14.10) and (14.11) be equal. Then

$$x_1^2 + y_1^2 + z_1^2 + 2ux_1 + 2vy_1 + 2wz_1 + d$$

$$= x_1^2 + y_1^2 + z_1^2 + 2u_1x_1 + 2v_1y_1 + 2w_1z_1 + d_1$$

(i.e.) $2(u - u_1)x_1 + 2(v - v_1)y_1 + 2(w - w_1)z_1 + (d - d_1) = 0$

The locus of (x_1, y_1, z_1) is

$$2(u - u_1)x + 2(v - v_1)y + 2(w - w_1)z + (d - d_1) = 0 \quad (14.12)$$

This is a linear equation in x, y and z and hence this equation represents a plane.

Hence equation (14.12) is the equation to the radical plane of the two given spheres.

Note 14.11.1.1: When two spheres intersect, the plane of their intersection is the radical plane.

Note 14.11.1.2: When the two spheres touch, the common tangent plane through the point of contact is the radical plane.

14.11.2 Properties of Radical Plane

1. **The radical plane of two spheres is perpendicular to the line joining their centres.**

Proof:

Let the equations of the two spheres be

$$x^2 + y^2 + z^2 + 2u_1x + 2v_1y + 2w_1z + d_1 = 0 \quad (14.13)$$

$$x^2 + y^2 + z^2 + 2u_2x + 2v_2y + 2w_2z + d_2 = 0 \quad (14.14)$$

The centres of the two spheres are

$$C_1(-u_1, -v_1, -w_1) \text{ and } C_2(-u_2, -v_2, -w_2).$$

The direction ratios of the line of centres are $u_1 - u_2, v_1 - v_2, w_1 - w_2$.

The radical plane of spheres (14.13) and (14.14) is $2(u_1 - u_2)x + 2(v_1 - v_2)y + 2(w_1 - w_2)z + (d_1 - d_2) = 0$.

The direction ratios of the normal to the plane are $u_1 - u_2, v_1 - v_2, w_1 - w_2$.

Therefore, the line of centre is parallel to the normal to the radical plane.

Hence, the radical plane of two spheres is perpendicular to the line joining the centres.

2. **The radical planes of three spheres taken in pairs pass through a line.**

Proof:

Let $S_1 = 0, S_2 = 0, S_3 = 0$ be the equations of the three given spheres in each of which the coefficients of x^2, y^2 and z^2 are unity.

Then the equations of the radical planes taken in pairs are $S_1 - S_2 = 0$, $S_2 - S_3 = 0$, $S_3 - S_1 = 0$.

These equation show that the radical planes of the three spheres pass through the line $S_1 = S_2 = S_3$.

Hence the result is proved.

Note 14.11.2.1: The line of concurrence of the three radical planes is called radical line of the three spheres.

3. **The radical planes of four spheres taken in pairs meet in a point.**

Proof:

Let $S_1 = 0$, $S_2 = 0$, $S_3 = 0$ and $S_4 = 0$ be the equations of the four given spheres, in each of which the coefficients of x^2, y^2, z^2 are unity.

Then the equations of the radical planes taken two by two are

$$S_1 - S_2 = 0,\ S_1 - S_3 = 0,\ S_1 - S_4 = 0$$
$$S_2 - S_3 = 0,\ S_2 - S_4 = 0 \text{ and } S_3 - S_4 = 0$$

These equations show that the radical planes of the four spheres meet in at a point given by $S_1 = S_2 = S_3 = S_4$.

Note 14.11.2.2: The point of concurrence of the radical planes of four spheres is called the radical centre of the four spheres.

14.12 COAXAL SYSTEM

Definition 14.12.1: A system of spheres is said to be coaxal if every pair of spheres of the system has the same radical plane.

14.12.1 General Equation to a System of Coaxal Spheres

Let $S = x^2 + y^2 + z^2 + 2ux + 2vy + 2wz + d = 0$ and $S_1 = x^2 + y^2 + z^2 + 2u'x + 2v'y + 2w'z + d' = 0$ be the equation of any two spheres.

Now consider the equation

$$S + \lambda S' = 0 \tag{14.15}$$

where λ is a constant. Clearly this equation represents a sphere.

Consider two different spheres of this system for two different values of λ.

$$\text{(i.e.)}\quad S + \lambda_1 S' = 0 \tag{14.16}$$

and

$$S + \lambda_2 S' = 0 \tag{14.17}$$

The coefficients of x^2, y^2, z^2 terms in (14.15) are $1 + \lambda$.

$$\therefore \frac{S + \lambda_1 S'}{1 + \lambda_1} = 0 \tag{14.18}$$

$$\frac{S + \lambda_2 S'}{1 + \lambda_2} = 0 \tag{14.19}$$

represent two spheres of the system with unit coefficients for x^2, y^2, z^2 terms. Therefore, the equation of the radical plane of (14.18) and (14.19) is $\dfrac{S + \lambda_1 S'}{1 + \lambda_1} - \dfrac{S + \lambda_2 S'}{1 + \lambda_2} = 0$.

$$(1 + \lambda_2)(S + \lambda_1 S') - (1 + \lambda_1)(S + \lambda_2 S') = 0$$

$$\text{(i.e.)}\quad (\lambda_2 - \lambda_1)(S - S') = 0$$

Since $\lambda_2 \neq \lambda_1$, $S - S' = 0$ which is the equation to the radical plane of spheres (14.16) and (14.17).

Since this equation is independent of λ, every pair of the system of spheres (14.15) has the same radical plane. Hence equation (14.15) represents the general equation to the coaxal system of the spheres.

14.12.2 Equation to Coaxal System is the Simplest Form

In a coaxal system of spheres, the line of centres is normal to the common radical plane.

Therefore, let us choose the x-axis as the line of centres and the common radical plane as the yz-plane, that is, $(x = 0)$.

Let the equation to a sphere of the coaxal system be $x^2 + y^2 + z^2 + 2ux + 2vy + 2wz + d = 0$.

Since the line of centres is the x-axis, is y and z coordinates are zero $v = 0$, $w = 0$.

Then the equation of the above sphere reduces to the form $x^2 + y^2 + z^2 + 2ux + d = 0$.

Let us now consider two sphere of this system say, $x^2 + y^2 + z^2 + 2ux + d = 0$ and $x^2 + y^2 + z^2 + 2u_1x + d_1 = 0$.

The radical plane of these two spheres is

$$2(u - u_1)x + d - d_1 = 0 \qquad (14.20)$$

But the equation of the radical plane is $x = 0$.

Therefore, from (14.20), $d - d_1 = 0$ or $d_1 = d$

Hence the equation to any sphere of the coaxal system is of the form $x^2 + y^2 + z^2 + 2\lambda x + d = 0$ where λ is a variable and d is a constant.

14.12.3 Limiting Points

Limiting points are defined to be the centres of point spheres of the coaxal system.

Let the equation to a coaxal system be

$$x^2 + y^2 + z^2 + 2\lambda x + d = 0 \qquad (14.21)$$

Centre is $(-\lambda, 0, 0)$ and radius is $\sqrt{\lambda^2 - d}$.

For point sphere radius is zero.

$$\therefore \sqrt{\lambda^2 - d} = 0$$

$$\therefore \lambda = \pm\sqrt{d}$$

Therefore, the limiting points of the system of spheres given by (14.21) are $(\sqrt{d},0,0)$ and $(-\sqrt{d},0,0)$.

Note 14.12.3.1: Limiting points are real or imaginary according as d is positive or negative.

14.12.4 Intersection of Spheres of a Coaxal System

Let the equation to a coaxal system of sphere be $x^2+y^2+z^2+2\lambda x+d=0$. Now consider two spheres of the system say

Now consider two spheres of the system say

$$S_1 = x^2+y^2+z^2+2\lambda_1 x+d=0 \qquad (14.22)$$

$$S_2 = x^2+y^2+z^2+2\lambda_2 x+d=0 \qquad (14.23)$$

The intersection of these two spheres is $S_1 - S_2 = 0$.

$$\text{(i.e.)} \quad 2(\lambda_1-\lambda_2)x=0$$

(i.e.) $x=0$ since $\lambda_1 \neq \lambda_2$substituting $x=0$ in (14.22) or (14.23)

we get,

$$y^2+z^2=-d$$

$$y^2+z^2=\left(\sqrt{-d}\right)^2$$

Therefore, this equation is a circle in the *yz*-plane and also it is independent of λ. Hence every sphere of the system meets the radical plane with same circle.

Note 14.12.4.1: This circle is called the common circle of the coaxal system.

Note 14.12.4.2: If $d < 0$, the common circle is real and the system of spheres are said to be intersecting type. If $d = 0$, the common circle is a point circle and in this case any two spheres of the system touch each other.

If $d > 0$, the common circle is imaginary and the spheres are said to be of non-intersecting type.

ILLUSTRATIVE EXAMPLES

Example 14.1

Find the equation of the sphere with centre at (2, –3, –4) and radius 5 units.

Solution

The equation of the sphere whose centre is (a, b, c) and radius r is $(x - a)^2 + (y - b)^2 + (z - c)^2 = r^2$. Therefore, the equation of the sphere whose centre is $(2, -3, -4)$ and radius 5 is $(x-2)^2+(y+3)^2+(z-4)^2=5^2$.

$$\text{(i.e.)}\quad x^2+y^2+z^2-4x+6y-8z+4+9+16-25=0$$

$$\text{(i.e.)}\quad x^2+y^2+z^2-4x+6y-8z+4=0$$

Example 14.2

Find the coordinate of the centre and radius of the sphere $16x^2+16y^2+16z^2-16x-8y-16z-35=0$.

Solution

The equation of the sphere is $16x^2+16y^2+16z^2-16x-8y-16z-35=0$.

Dividing by 16, $x^2+y^2+z^2-x-\frac{1}{2}y-z-\frac{35}{16}=0$

Centre of the sphere is $\left(\frac{1}{2},\frac{1}{4},\frac{1}{2}\right)$.

Radius is $r=\sqrt{\frac{1}{4}+\frac{1}{16}+\frac{1}{4}+\frac{35}{16}}$.

$$=\sqrt{\frac{4+1+4+35}{16}}=\sqrt{\frac{44}{16}}$$

$$=\sqrt{\frac{11}{4}}=\frac{\sqrt{11}}{2}\text{ units}$$

Example 14.3

Find the equation of the sphere with the centre at (1, 1, 2) and touching the plane $2x - 2y + z = 5$.

Solution

The radius of the sphere is equal to the perpendicular distance from the centre (1, 1, 2) on the plane $2x - 2y + z - 5 = 0$.

$$r=\frac{|2-2+2-5|}{\sqrt{4+4+1}}=\frac{3}{3}=1\text{ unit}$$

The equation of the sphere with centre at (1, 1, 2) and radius 1 unit is $(x-1)^2+(y-1)^2+(z-2)^2=1$.

$$\text{(i.e.)}\quad x^2+y^2+z^2-2x-2y-4z+5=0$$

Example 14.4

Find the equation of the sphere passing through the points (1, 0, 0), (0, 1, 0), (0, 0, 1) and (0, 0, 0).

Solution

Let the equation of the sphere be $x^2+y^2+z^2+2ux+2vy+2wz+d=0$.

This passes through the points (1, 0, 0), (0, 1, 0), (0, 0, 1) and (0, 0, 0).

$$\therefore\ 2u+d+1=0$$
$$2v+d+1=0$$
$$2w+d+1=0$$
$$d=0$$
$$\therefore\ 2u=-1,\ 2v=-1 \text{ and } 2w=-7,\ d=0$$

The equation of the sphere is $x^2+y^2+z^2-x-y-z=0$.

Example 14.5

Find the equation of the sphere which passes through the points (1, 0, 0), (0, 1, 0) and (0, 0, 1) and has its centre on the plane $x+y+z=6$.

Solution

Let the equation of the sphere be $x^2+y^2+z^2+2ux+2vy+2wz+d=0$.
This sphere passes through the points (1, 0, 0), (0, 1, 0) and (0, 0, 1).

$$\therefore\ 2u+d=0$$
$$2v+d=0$$
$$2w+d=0$$

The centre of the sphere is $(-u, -v, -w)$.
This lies on the plane $x+y+z-6=0$.

$$\therefore -u-v-w=6$$
$$\therefore\ \frac{d}{2}+\frac{d}{2}+\frac{d}{2}=6 \text{ or } \frac{3d}{2}=6 \text{ or } d=4$$
$$\therefore\ u=v=w=-2$$

The equation of the sphere is $x^2+y^2+z^2-4x-4y-4z+4=0$.

Example 14.6

Find the equation of the sphere touching the plane $2x + 2y - z = 1$ and concentric with the sphere $2x^2 + 2y^2 + 2z^2 + x + 2y - z = 0$.

Solution

$2x^2 + 2y^2 + 2z^2 + x + 2y - z = 0$ or $x^2 + y^2 + z^2 + \frac{1}{2}x + y - \frac{1}{2}z = 0$

Centre is $\left(\frac{-1}{4}, \frac{-1}{2}, \frac{1}{4}\right)$.

The sphere touches the plane $2x + 2y - z - 1 = 0$.

$$\therefore r = \frac{\left|\frac{-2}{4} - \frac{2}{2} - \frac{1}{4} - 1\right|}{\sqrt{4+4+1}} = \frac{11}{12} \text{ units.}$$

The equation of the sphere is $\left(x + \frac{1}{4}\right)^2 + \left(y + \frac{1}{2}\right)^2 + \left(z - \frac{1}{4}\right)^2 = \left(\frac{11}{12}\right)^2$

Example 14.7

Find the equation of the sphere which passes through the points (2, 7, –4) and (4, 5, –1) has its centre on the line joining the these two points as diameter.

Solution

$$\text{Centre of the sphere} = \left(\frac{2+4}{2}, \frac{7+5}{2}, \frac{-4-1}{2}\right)$$

$$= \left(3, 6, \frac{-5}{2}\right)$$

$$\text{Radius} = \frac{1}{2}\sqrt{(2-4)^2 + (7-5)^2 + (-4+1)^2} = \frac{1}{2}\sqrt{4+4+9} = \frac{\sqrt{17}}{2}.$$

Therefore, the equation of the sphere is $(x-3)^2 + (y-6)^2 + \left(z + \frac{5}{2}\right)^2 = \frac{17}{4}$.

$$x^2 + y^2 + z^2 - 6x - 12y + 5z + 9 + 36 + \frac{25}{4} - \frac{17}{4} = 0$$

$$x^2 + y^2 + z^2 - 6x - 12y + 5z + 47 = 0$$

Aliter:

The two given points are the extremities of a diameter of the sphere.

Therefore, the equation of the sphere is

$$(x-2)(x-4)+(y-7)(y-5)+(z+4)(z+1)=0$$
$$\text{(i.e.)}\quad x^2+y^2+z^2-6x-12y+5z+47=0$$

Example 14.8

The plane $\frac{x}{a}+\frac{y}{b}+\frac{z}{c}=1$ cuts the coordinate axes in A, B and C. Find the equation of the sphere passing through A, B, C and O. Find also its centre and radius.

Solution

The plane $\frac{x}{a}+\frac{y}{b}+\frac{z}{c}=1$ cuts the coordinates of A, B and C. The coordinates of A, B and C are (a, 0, 0), (0, b, 0) and (0, 0, c).

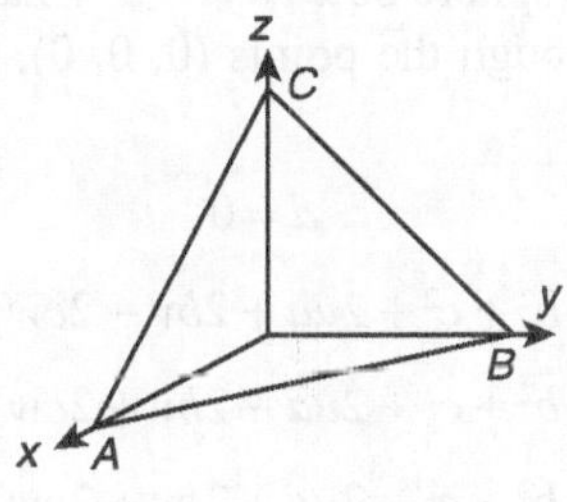

Let the equation of the sphere passing through A, B and C be $x^2+y^2+z^2+2ux+2vy+2wz+d=0$.

Since this passes through origin O, $d=0$. Since this passes through A, B and C.

$$a^2+2ua=0,\ b^2+2vb=0,\ c^2+2wc=0.$$
$$\therefore\quad 2u=-a,\ 2v=-b \text{ and } 2w=-c.$$

Hence the equation of the sphere is $x^2+y^2+z^2-2ax-2by-2cz=0$.

Centre of the sphere is (a, b, c) and radius of the sphere $=\sqrt{a^2+b^2+c^2}$.

Example 14.9

Find the equation of the sphere circumscribing the tetrahedron whose faces are $\frac{y}{b}+\frac{z}{c}=0$, $\frac{z}{c}+\frac{x}{a}=0$, $\frac{x}{a}+\frac{y}{b}=0$ and $\frac{x}{a}+\frac{y}{b}+\frac{z}{c}=1$.

Solution

The faces of the tetrahedron are

$$\frac{y}{b}+\frac{z}{c}=0 \qquad (14.24)$$

$$\frac{z}{c}+\frac{x}{a}=0 \qquad (14.25)$$

$$\frac{x}{a}+\frac{y}{b}=0 \qquad (14.26)$$

$$\frac{x}{a}+\frac{y}{b}+\frac{z}{c}-1=0 \qquad (14.27)$$

Now easily seen that the vertices of the tetrahedron are (0, 0, 0), $(a, b, -c)$, $(a, -b, c)$ and $(-a, b, c)$.

Let the equation of the sphere be $x^2 + y^2 + z^2 + 2ux + 2vy + 2wz + d = 0$.

This sphere passes through the points (0, 0, 0), $(a, b,-c,)$. $(a, -b, c)$ and $(-a, b,c)$.

$$\therefore d=0$$

$$a^2 + b^2 + c^2 + 2ua + 2bv - 2cw = 0 \qquad (14.28)$$

$$a^2 + b^2 + c^2 + 2ua - 2bv + 2cw = 0 \qquad (14.29)$$

$$a^2 + b^2 + c^2 - 2ua + 2bv + 2cw = 0 \qquad (14.30)$$

Adding (14.28) and (14.29), $2(a^2 + b^2 + c^2) + 4ua = 0$

$$\therefore 2u = \frac{-(a^2+b^2+c^2)}{a}$$

Similarly, $2v = \dfrac{-(a^2+b^2+c^2)}{b}$, $2w = \dfrac{-(a^2+b^2+c^2)}{c}$

Therefore, the equation of the sphere is $x^2 + y^2 + z^2 - (a^2 + b^2 + c^2)\left(\dfrac{x}{a}+\dfrac{y}{b}+\dfrac{z}{c}\right) = 0$.

Example 14.10

A sphere is inscribed in a tetrahedron whose faces are $x = 0, y = 0, z = 0$ and $2x + 6y + 3z = 14$. Find the equation of the sphere. Also find its centre and radius.

Solution

Let the equation of the sphere be $x^2 + y^2 + z^2 + 2ux + 2vy + 2wz + d = 0$.

Since the sphere touches the plane $x = 0$, the perpendicular distance from the centre $(-u, -v, -w)$ on this plane is equal to the radius.

$$\therefore \quad -u = r, -v = r, -w = r.$$

Also the sphere touches the plane $2x + 6y + 3z - 14 = 0$.

$$\therefore \quad \left|\frac{-2u - 6v - 3w - 14}{\sqrt{4+36+9}}\right| = r$$

$$\text{(i.e.)} \quad \frac{|11r - 14|}{7} = r$$

$$\therefore \quad 7r = \pm(11r - 14)$$

$$7r = 11r - 14 \quad \text{or} \quad 7r = -11r + 14$$

$$\therefore \quad 4r = 14 \quad \text{or} \quad r = \frac{14}{18} = \frac{7}{9}$$

$$r = \frac{7}{9}; r = \frac{7}{2}$$

When $r = \frac{7}{9}$, the equation of the sphere is $\left(x - \frac{7}{9}\right)^2 + \left(y - \frac{7}{9}\right)^2 + \left(z - \frac{7}{9}\right)^2 = \left(\frac{7}{9}\right)^2$.

$$\text{(i.e.)} \quad x^2 + y^2 + z^2 - \frac{14x + 14y + 14z}{9} + 2 \times \frac{49}{81} = 0$$

$$81(x^2 + y^2 + z^2) - 126\,(x + y + z) + 98 = 0$$

For this sphere, centre is $\left(\frac{7}{9}, \frac{7}{9}, \frac{7}{9}\right)$ and radius $= \frac{7}{9}$.

When $r = \frac{7}{2}$,

$$\left(x - \frac{7}{2}\right)^2 + \left(y - \frac{7}{2}\right)^2 + \left(z - \frac{7}{2}\right)^2 = \left(\frac{7}{2}\right)^2$$

$$x^2 + y^2 + z^2 - 7(x + y + z) + \frac{49}{2} = 0$$

$$\text{(i.e.)} \quad 2(x^2 + y^2 + z^2) - 14(x + y + z) + 49 = 0$$

Example 14.11

Find the equation of the sphere passing through the points (1, 0, –1), (2, 1, 0), (1, 1, –1) and (1, 1, 1).

Solution

Let the equation of the sphere be $x^2 + y^2 + z^2 + 2ux + 2vy + 2wz + d = 0$.
This sphere passes through (1, 0, –1), (2, 1, 0), (1, 1, –1) and (1, 1, 1).

$$\therefore \quad 1 + 0 + 1 + 2u + 0 - 2w + d = 0$$

$$4 + 1 + 0 + 4u + 2v + 0 + d = 0$$

$$1 + 1 + 1 + 2u + 2v - 2w + d = 0$$

$$1 + 1 + 1 + 2u + 2v + 2w + d = 0$$

$$2u - 2w + d = -2 \quad (14.31)$$

$$4u + 2v + d = -5 \quad (14.32)$$

$$2u + 2v - 2w + d = -3 \quad (14.33)$$

$$2u + 2v + 2w + d = -3 \quad (14.34)$$

(14.33) – (14.34) gives $-4w = 0 \quad \therefore \quad w = 0$
(14.32) – (14.33) gives $2u + 2w = -2$

$$\therefore 2u = -2 \Rightarrow u = -1$$

(14.32) – (14.31) gives $2u + 2v + 2w = -3 \Rightarrow -2 + 2v + 0 = -3$

$$2v = -1 \Rightarrow v = \frac{-1}{2}$$

From (14.31), $-2 + d = -2 \Rightarrow d = 0$.
Therefore, the required equation of the circle is $x^2 + y^2 + z^2 - 2x - y = 0$.

Example 14.12

Find the equation of the sphere which touches the coordinate axes, whose centre lies in the positive octant and has a radius 4.

Solution

Let the equation of the sphere be $x^2 + y^2 + z^2 + 2xu + 2vy + 2wz + d = 0$.

The equation of the x- axis is $\frac{x}{1} = \frac{y}{0} = \frac{z}{0} = t$.

Any point on this line is $(t, 0, 0)$.

The point lies on the given sphere $t^2 + 2ut + d = 0$.

Since the sphere touches the x-axis the two roots of this equation are equal.

$$\therefore\ 4u^2 - 4d = 0 \quad \text{or} \quad u^2 = d.$$

Similarly, $v^2 = d$ and $w^2 = d$

The radius of the sphere is $\sqrt{u^2 + v^2 + w^2 - d} = \sqrt{3d - d} = \sqrt{2d}$.

$$\sqrt{2d} = 4 \text{ or } d = 8 \quad u^2 = v^2 = w^2 = 8.$$

Since the centre lies on the x-axis, $-u = -v = -w = 2\sqrt{2}$.

Therefore, the required equation is $x^2 + y^2 + z^2 - 4\sqrt{2}(x + y + z) + 8 = 0$.

Example 14.13

Find the radius and the equation of the sphere touching the planc $2x + 2y - z = 0$ and concentric with the sphere $2x^2 + 2y^2 + 2z^2 + x + 2y - z = 0$.

Solution

Since the required sphere is concentric with the sphere $2x^2 + 2y^2 + 2z^2 + x + 2y - z = 0$ its centre is the same as that of the given sphere $x^2 + y^2 + z^2 + \frac{1}{2}x + y - \frac{1}{2}z - 0$.

Centre is $\left(\frac{-1}{4}, \frac{-1}{2}, \frac{1}{4}\right)$. The radius of the required sphere is equal to the perpendicular distance from this point to the plane $2x + 2y - z = 0$.

$$\text{(i.e.)} \quad r = \left| \frac{2\left(\frac{-1}{4}\right) + 2\left(\frac{-1}{2}\right) - \frac{1}{4}}{\sqrt{4+4+1}} \right| = \frac{7}{12}$$

The equation of the required sphere is $\left(x + \frac{1}{4}\right)^2 + \left(y + \frac{1}{2}\right)^2 + \left(z - \frac{1}{4}\right)^2 = \left(\frac{7}{12}\right)^2$.

Example 14.14

Find the equation of the sphere which passes through the points (1, 0, 0), (0, 2, 0), (0, 0, 3) and has its radius as small as possible.

Solution

Let the equation of the sphere be $x^2 + y^2 + z^2 + 2ux + 2vy + 2wz + k = 0$. This sphere passes through the points (1, 0, 0), (0, 2, 0) and (0, 0, 3).

$$\therefore \quad 2u + k + 1 = 0 \quad \therefore u = -\left(\frac{k+1}{2}\right)$$

$$4v + k + 4 = 0 \quad \therefore v = -\left(\frac{k+4}{4}\right)$$

$$6w + k + 9 = 0 \quad \therefore w = -\left(\frac{k+9}{6}\right)$$

The radius of the sphere is given by $r^2 = u^2 + v^2 + w^2 - k$.

$$r^2 = \frac{(k+1)^2}{4} + \frac{(k+4)^2}{16} + \frac{(k+9)^2}{36} - k = \frac{36(k+1)^2 + 9(k+4)^2 + 4(k+9)^2 - 144k}{144}$$

$$= \frac{1}{144}[49k^2 + 72k + 504] = \frac{49}{144}\left[k^2 + \frac{72}{49}k + \frac{504}{49}\right]$$

$$= \frac{49}{144}\left[\left(k + \frac{36}{49}\right)^2 + \frac{504}{49} - \left(\frac{36}{49}\right)^2\right]$$

r is minimum if $k = \dfrac{-36}{49}$

$$\therefore u = -\left[\frac{\frac{-36}{49}+1}{2}\right] = \frac{13}{98}; v = -\left[\frac{\frac{-36}{49}+4}{4}\right] = \frac{40}{49}; w = -\left[\frac{\frac{-36}{49}+9}{6}\right] = \frac{135}{98}$$

The required equation of the sphere is $x^2 + y^2 + z^2 - \frac{13}{49}x - \frac{80}{49}y - \frac{135}{49}z - \frac{36}{49} = 0$.

$$\text{(i.e.)} \quad 49(x^2 + y^2 + z^2) - 13x - 80y - 135z - 36 = 0$$

Example 14.15

Find the equation of the sphere tangential to the plane $x - 2y - 2z = 7$ at $(3, -1, -1)$ and passing through the point $(1, 1, -3)$.

Solution

The equation of normal at A is $\dfrac{x-3}{1}=\dfrac{y+1}{-2}=\dfrac{z+1}{-2}=r$.

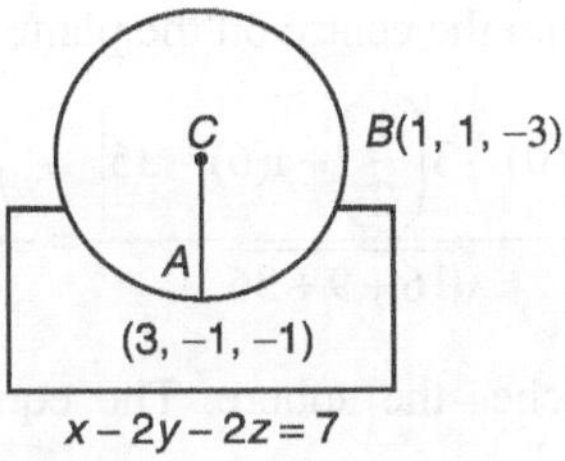

Any point in this line is $(r+3, -2r-1, -2r-1)$. If this point is the centre of the sphere then $CA = CB$.

$$r^2+4r^2+4r^2=(r+2)^2+(-2r-2)^2+(-2r+2)^2$$

$$\therefore 4r=-12 \quad \text{or} \quad r=-3.$$

Therefore, centre of the sphere is (0.5, 5).

Radius $=\sqrt{9r^2}=\sqrt{81}=9$

Therefore, the equation of the sphere is $(x-0)^2+(y-5)^2+(z-5)^2=81$.

$$\text{(i.e.)} \quad x^2+y^2+z^2-10y-10z-31=0$$

Example 14.16

Show that the plane $4x-3y+6z-35=0$ is a tangent plane to the sphere $x^2+y^2+z^2-y-2z-14=0$ and find the point of contact.

Solution

If the plane is a tangent plane to the sphere then the radius is equal to the perpendicular distance from the centre on the plane.

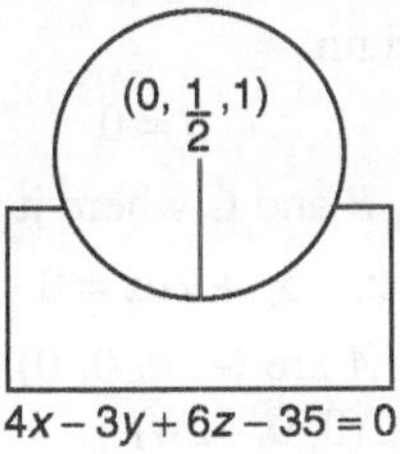

The centre of the sphere $x^2+y^2+z^2-y-2z-14=0$ is $\left(0,\frac{1}{2},1\right)$.

$$r=\sqrt{0+\frac{1}{4}+1+14}=\sqrt{\frac{61}{4}}$$

Perpendicular distance from the centre on the plane is

$$\frac{\left|4(0)-3\left(\frac{1}{2}\right)+1(6)-35\right|}{\sqrt{16+9+36}}=\frac{\sqrt{61}}{2}.$$

Therefore, the plane touches the sphere. The equations of the normal to the tangent plane are $\frac{x}{4}=\frac{y-\frac{1}{2}}{-3}=\frac{z-1}{6}=t.$

Any point on this line is $\left(4t,-3t+\frac{1}{2},6t+1\right)$.

If this point lies on the plane $4x-3y+6z-35=0$ then,

$$4(4t)-3\left(-3t+\frac{1}{2}\right)+6(6t+1)-35=0.$$

$$16t+9t+36t-\frac{3}{2}+6-35=0 \quad \text{(i.e.)}\ 61t-\frac{61}{2}=0,\quad t=\frac{1}{2}$$

Therefore, the point of contact is $(2,-1,4)$.

Example 14.17

A sphere of constant radius r passes through the origin O and cuts the axes in A, B and C. Find the locus of the foot of the perpendicular from O to the plane ABC.

Solution

Let the equation of the sphere be $x^2+y^2+z^2+2ux+2vy+2wz+d=0$.
This passes through the origin.

$$\therefore\quad d=0$$

The sphere cuts the axes at A, B and C where it meets the x-axis.

$$y=0,z=0\quad\therefore\quad x^2+2ux=0\quad\therefore\quad x=-2u$$

Therefore, the coordinates of A are $(-2u,0,0)$. Similarly the coordinates of B and C are $B(0,-2v,0)$ and $C(0,0,-2w)$.

Therefore, the equations of the sphere is $x^2 + y^2 + z^2 - 2ux - 2vy - 2wz = 0$.
Radius $= r$

$$\therefore \quad u^2 + v^2 + w^2 = r^2 \tag{14.35}$$

The equations of the plane ABC is

$$\frac{x}{-2u} + \frac{y}{-2v} + \frac{z}{-2w} = 1 \tag{14.36}$$

The direction ratios of the normal to this plane are $\left(\frac{-1}{2u}, \frac{-1}{2v}, \frac{-1}{2w}\right)$.
The equations of the normal are

$$-2ux = -2vy = -2wz = t \tag{14.37}$$

Let (x, y, z) be the foot of the perpendicular from O on the plane.
Then (x_1, y_1, z_1) lies on (14.37).

$$\therefore ux_1 = vy_1 = wz_1 = \frac{-t}{2} \text{ or } u = \frac{-t}{2x_1}, v = \frac{-t}{2y_1}, w = \frac{-t}{2z_1}$$

Substituting in (14.35), $\frac{t^2}{4}(x_1^{-2} + y_1^{-2} + z_1^{-2}) = r^2$

$$t^2(x_1^{-2} + y_1^{-2} + z_1^{-2}) = 4r^2 \tag{14.38}$$

The point (x_1, y_1, z_1) also lies on the plane (14.36)

$$\therefore \frac{x_1}{-2u} + \frac{y_1}{-2v} + \frac{z_1}{-2w} = 1 \quad \Rightarrow \frac{x_1^2}{t} + \frac{y_1^2}{t} + \frac{z_1^2}{t} = 1$$

or

$$\frac{1}{t}(x_1^2 + y_1^2 + z_1^2) = 1 \quad \text{or} \quad \frac{1}{t^2}(x_1^2 + y_1^2 + z_1^2)^2 = 1 \tag{14.39}$$

Multiplying (14.38) and (14.39), we get

$$(x_1^2 + y_1^2 + z_1^2)^2(x_1^{-2} + y_1^{-2} + z_1^{-2}) = 4r^2$$

The locus of (x_1, y_1, z_1) is $(x^2 + y^2 + z^2)^2\,(x^{-2} + y^{-2} + z^{-2}) = 4r^2$.

Example 14.18

A sphere of constant radius $2k$ passes through the origin and meets the axes in A, B and C. Show that the locus of the centroid of the tetrahedron $OABC$ is $x^2 + y^2 + z^2 = k^2$.

Solution

Let the equation of the sphere be $x^2 + y^2 + z^2 + 2ux + 2vy + 2wz + d = 0$.
This passes through the origin.

$$\therefore d = 0$$

When this sphere meets the x-axis, $y = 0, z = 0$.

$$\therefore \quad x^2 + 2ux = 0.$$

Since $x \neq 0, x = -2u$.
Therefore, the coordinates of A are $(-2u, 0, 0)$.
Similarly the coordinates of B and C are $(0, -2v, 0)$ and $(0, 0, -2w)$.
Let (x_1, y_1, z_1) be the centroid of the tetrahedron $OABC$.

$$x_1 = \frac{-2u}{4}, \quad u = -2x_1$$

$$y_1 = \frac{-2v}{4}, \quad v = -2y_1 \qquad (14.40)$$

$$z_1 = \frac{-2w}{4}, \quad w = -2z_1$$

The radius of the sphere is r.

$$\therefore \quad u^2 + v^2 + w^2 = 4r^2$$

Using (14.40), $x_1^2 + y_1^2 + z_1^2 = r^2$
The locus of (x_1, y_1, z_1) is the sphere $x^2 + y^2 + z^2 = r^2$.

Example 14.19

A sphere of constant radius r passes through the origin and meets the axes in A, B and C. Prove that the centroid of the triangle ABC lies on the sphere $9(x^2 + y^2 + z^2) = 4r^2$.

Solution

Let the equation of the sphere be $x^2 + y^2 + z^2 + 2ux + 2vy + 2wz + d = 0$.
This line passing through the origin.

$$\therefore \quad d = 0.$$

When the circle meets the x-axis, $y = 0, z = 0$

$$\therefore \quad x^2 + 2ux = 0$$

as $$x \neq 0, x = -2u$$

$\therefore$ A *is* the point (–2u, 0, 0).

Similarly B and C are the points (0, $-2v$, 0) and (0, 0, $-2w$). Also given the radius is r.

$$\therefore u^2 + v^2 + w^2 = r^2 \tag{14.41}$$

Let (x_1, y_1, z_1) be the centroid of the triangle ABC.

But the centroid is $\left(\frac{-2u}{3}, \frac{-2v}{3}, \frac{-2w}{3}\right)$.

$$\therefore \quad \frac{-2u}{3} = x_1 \quad \therefore \quad u = \frac{-3x_1}{2}$$

$$\therefore \quad \frac{-2v}{3} = y_1 \quad \therefore \quad v = \frac{-3y_1}{2}$$

$$\therefore \quad \frac{-2w}{3} = z_1 \quad \therefore \quad w = \frac{-3z_1}{2}.$$

$\therefore$ from (14.41): $\frac{9}{4}(x_1^2 + y_1^2 + z_1^2) = r^2$.

The locus of (x_1, y_1, z_1) is $9(x^2 + y^2 + z^2) = 4r^2$.

Example 14.20

A plane passes through the fixed point (a, b, c) and meets the axes in A, B, C. Prove that the locus of the centre of the sphere is $\frac{a}{x} + \frac{b}{y} + \frac{c}{z} = 2$.

Solution

Let the equation of the sphere be $x^2 + y^2 + z^2 + 2ux + 2vy + 2wz + d = 0$. This passes through the origin.

$$\therefore \quad d = 0$$

When this sphere meets the x-axis, $y = 0$ and $z = 0$.

$$\therefore \quad x^2 + 2ux = 0. \text{ As } x \neq 0, x = -2u.$$

Therefore, the coordinates of A are $(-2u, 0,0)$.

Similarly the coordinates of B and C are $(0, -2v, 0)$ and $(0, 0, -2w)$.

The equation of the plane ABC is $\frac{x}{-2u} + \frac{y}{-2v} + \frac{z}{-2w} = 1$.

This plane passes through the point (a, b, c).

$$\therefore \quad \frac{a}{-2u}+\frac{b}{-2v}+\frac{c}{-2w}=1 \quad \text{(i.e.)} \quad \frac{a}{-u}+\frac{b}{-v}+\frac{c}{-w}=2$$

The locus of the centre $(-u, -v, -w)$ is $\frac{a}{x}+\frac{b}{y}+\frac{c}{z}=2$.

Example 14.21

Find the centre and radius of the circle $x^2 + y^2 + z^2 - 8x + 4y + 8z - 45 = 0$, $x - 2y + 2z = 3$.

Solution

The centre of the sphere $x^2 + y^2 + z^2 - 8x + 4y + 8z - 45 = 0$ is $(4, -2, -4)$.

$$r = \sqrt{16+4+16+45} = \sqrt{81} = 9.$$

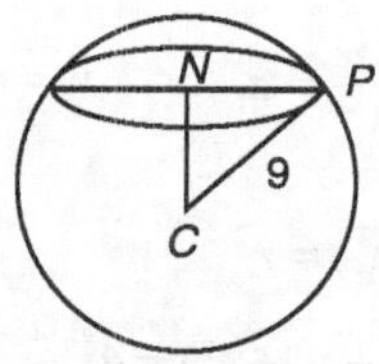

CN is the perpendicular from the centre of the sphere on the plane $x - 2y + 2z = 3$.

$$= \frac{|4+4-8-3|}{\sqrt{1+4+4}} = 1$$

$$\therefore \quad NP = \sqrt{CP^2 - CN^2} = \sqrt{81-1} = \sqrt{80}.$$

Therefore, the radius of the circle is $\sqrt{80}$ units. The equation of the line CN is $\frac{x-4}{1}=\frac{y+2}{-2}=\frac{z+4}{2}=t$.

Any point on this line is $t + 4, -2t - 2, 2t - 4$. This point is the centre of the circle then this lies on the plane $x - 2y + 2z - 3 = 0$ then $t + 4 - 2(2t - 2) + 2(2t - 4) - 3 = 0$.

$$9t - 3 = 0$$

$$\therefore \quad t = 1/3.$$

Therefore, the centre of the circle is

$$\left(4+\frac{1}{3}, -2-\frac{2}{3}, -4+\frac{2}{3}\right) \quad \text{(i.e.)} \quad \left(\frac{13}{3}, \frac{-8}{3}, \frac{-10}{3}\right).$$

Example 14.22

Show that the centres of all sections of the sphere $x^2 + y^2 + z^2 = r^2$ by planes through the point (α, β, γ) lie on the sphere $x(x-\alpha)+y(y-\beta)+z(z-\gamma)=0$.

Solution

Let (x_1, y_1, z_1) be a centre of a section of the sphere $x^2 + y^2 + z^2 = r^2$ by a plane through (α, β, γ). Then the equation of the plane is $x_1(x - x_1) + y_1(y - y_1) + z_1(z - z_1) = 0$. This plane passes through the point (α, β, γ).

$$x_1(\alpha - x_1) + y_1(\beta - y_1) + z_1(\gamma - z_1) = 0$$

Therefore, the locus of (x_1, y_1, z_1) is $x(\alpha - x) + y(\beta - y) + z(\gamma - z) = 0$.
(i.e.) $x(x-\alpha)+y(\beta - y)+z(\gamma - z)=0$ which is a sphere.

Example 14.23

Find the equation of the sphere having the circle $x^2 + y^2 + z^2 = 5, x - 2y + 2z = 5$ for a great circle. Find its centre and radius.

Solution

Any sphere containing the given circle is $x^2 + y^2 + z^2 - 5 + 2\lambda(x - 2y + 2z - 5) = 0$.

The centre of this sphere is $(-\lambda, 2\lambda, -2\lambda)$. Since the given circle is a great circle, the centre of the sphere should lie on the plane section $x - 2y + 2z = 5$.

$$-\lambda - 4\lambda - 4\lambda = 5 \text{ or } \lambda = \frac{-5}{9}$$

Therefore, the equation of the sphere is $x^2 + y^2 + z^2 - 5 - \frac{10}{9}(x - 2y + 2z - 5) = 0$.

$$9(x^2 + y^2 - z^2 - 5) - 10(x - 2y + 2z - 5) = 0.$$

Centre of the sphere is $\left(\frac{5}{9}, \frac{-10}{9}, \frac{10}{9}\right)$.

$$r = \sqrt{\frac{25+100+100-45}{81}} = \sqrt{\frac{180}{81}} = \frac{2}{3}\sqrt{5}$$

Example 14.24

Find the equations of the spheres which passes through the circle $x^2 + y^2 + z^2 = 5, x + 2y + 3z = 3$ and touch the plane $4x + 3y = 15$.

Solution

Any sphere containing the given circle is $x^2 + y^2 + z^2 - 5 + \lambda(x + 2y + 3z - 3) = 0$.

Centre is $\left(\frac{-\lambda}{2}, -\lambda, \frac{-3\lambda}{2}\right)$.

$$r = \sqrt{\frac{\lambda^2}{4} + \lambda^2 + \frac{9\lambda^2}{4} + 5 + 5\lambda} = \sqrt{\frac{14\lambda^2}{4} + 5\lambda + 5}$$

If the sphere touches the plane $4x + 3y = 15$ then the radius of the sphere is equal to the perpendicular distance from the centre on the plane.

$$\therefore \sqrt{\frac{7\lambda^2}{2} + 5\lambda + 3} = \pm\frac{-2\lambda - 3\lambda - 15}{\sqrt{4^2 + 3^2}}$$

$$= \pm\frac{-5(\lambda + 3)}{5}$$

$$\therefore \frac{7\lambda^2}{2} + 3\lambda + 3 = (\lambda + 3)^2$$

$$7\lambda^2 + 6\lambda + 10 = 2(\lambda^2 + 6\lambda + 9)$$

$$5\lambda^2 - 6\lambda - 8 = 0$$

$$5\lambda^2 - 10\lambda + 4\lambda - 8 = 0$$

$$5\lambda(\lambda - 2) + 4(\lambda - 2) = 0$$

$$(\lambda - 2)(5\lambda + 4) = 0$$

$$\lambda = 2, \frac{-4}{5}$$

There are two spheres touching the given plane whose equations are $x^2 + y^2 + z^2 - 5 + 2(x + 2y + 3z - 3) = 0$ and $x^2 + y^2 + z^2 - 5 - \frac{4}{5}(x + 2y + 3z - 3) = 0$

(i.e.) $x^2 + y^2 + z^2 + 2x + 4y + 6z - 11 = 0$

$5(x^2 + y^2 + z^2) - 4x - 8y - 12z - 13 = 0$

Example 14.25

Prove that the circles $x^2 + y^2 + z^2 - 2x + 3y + 4z - 5 = 0$, $5y + 6z + 1 = 0$ and $x^2 + y^2 + z^2 - 3x - 4y + 5z - 6 = 0$, $x + 2y - 7z = 0$ lie on the same sphere and find its equation.

Solution

The equation of the sphere through the first circle is

$$x^2 + y^2 + z^2 - 2x + 3y + 4z - 5 + \lambda(5y + 6z + 1) = 0 \qquad (14.42)$$

The equation of the sphere through the second circle is

$$x^2 + y^2 + z^2 - 3x - 4y + 5z - 6 + \mu(x + 2y - 7z) = 0 \qquad (14.43)$$

The given circles will lie on the same sphere if equation (14.42) and (14.43) are identical.

Therefore, comparing equations (14.42) and (14.43) we get,

$$-2 = -3 + \mu \qquad (14.44)$$

$$3 + 5\lambda = -4 + 2\mu \qquad (14.45)$$

$$4 + 6\lambda = 5 - 7\mu \qquad (14.46)$$

$$-5 + \lambda = -6 \qquad (14.47)$$

From (14.47), $\lambda = -1$

From (14.46), $\mu = 1$

These two values λ and μ satisfy (14.42), the equations (14.45) and (14.46).

Hence, the two given circles lie on the same sphere. The equation of the sphere is $x^2 + y^2 + z^2 - 3x - 4y + 5z - 6 + x + 2y - 7z = 0$.

$$\text{(i.e.)} \quad x^2 + y^2 + z^2 - 2x - 2y - 2z - 6 = 0$$

Example 14.26

The plane $\frac{x}{a} + \frac{y}{b} + \frac{z}{c} = 1$ meets the circle O, A, B and C. Find the equations of the circumcircle of the triangle ABC and also find its centre.

Solution

The equation of the plane ABC is $\frac{x}{a} + \frac{y}{b} + \frac{z}{c} = 1$.

Therefore, the coordinates of A, B and C are $(a, 0, 0)$, $(0, b, 0)$ and $(0, 0, c)$ respectively.

Also we know that the equation of the sphere $OABC$ is $x^2 + y^2 + z^2 - ax - by - cz = 0$.

Therefore, the equation of the circumcircle of the triangle ABC are $x^2 + y^2 + z^2 - ax - by - cz = 0$ and $\frac{x}{a} + \frac{y}{b} + \frac{z}{c} = 1$.

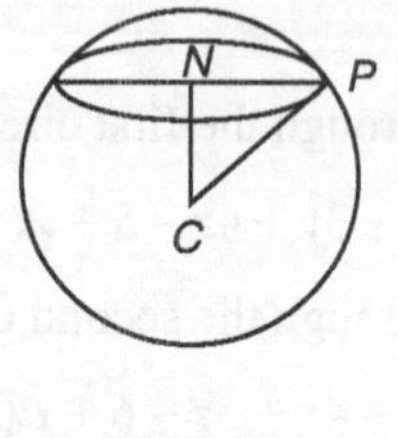

The centre of the sphere $OABC$ is $\left(\frac{a}{2},\frac{b}{2},\frac{c}{2}\right)$.

The equation of the normal CN is $\frac{x-\frac{a}{2}}{\frac{1}{a}}=\frac{y-\frac{b}{2}}{\frac{1}{b}}=\frac{z-\frac{c}{2}}{\frac{1}{c}}=t.$

Any point on this line is $\left(\frac{a}{2}+\frac{t}{a},\frac{b}{2}+\frac{t}{b},\frac{c}{2}+\frac{t}{c}\right)$.

Thus, point lies on the plane $\frac{x}{a}+\frac{y}{b}+\frac{z}{c}=1.$

$$\frac{1}{a}\left(\frac{a}{2}+\frac{t}{a}\right)+\frac{1}{b}\left(\frac{b}{2}+\frac{t}{b}\right)+\frac{1}{c}\left(\frac{c}{2}+\frac{t}{c}\right)=1$$

$$\therefore t\left(\frac{1}{a^2}+\frac{1}{b^2}+\frac{1}{c^2}\right)=-\frac{1}{2}$$

$$\therefore t=-\frac{1}{2(a^{-2}+b^{-2}+c^{-2})}$$

Hence the centre of the circle is $\left(\frac{a}{2}-\frac{\frac{1}{a}}{2\Sigma a^{-2}},\frac{b}{2}-\frac{\frac{1}{b}}{2\Sigma a^{-2}},\frac{c}{2}-\frac{\frac{1}{c}}{2\Sigma a^{-2}}\right)$.

(i.e.) $\left(\frac{a(b^{-2}+c^{-2})}{2\Sigma a^{-2}},\frac{b(c^{-2}+a^{-2})}{2\Sigma a^{-2}},\frac{c(a^{-2}+b^{-2})}{2\Sigma a^{-2}}\right)$

Example 14.27

Obtain the equations to the sphere through the common circle of the sphere $x^2+y^2+z^2+2x+2y=0$ and the plane $x+y+z+4=0$ which intersects the plane $x+y=0$ in circle of radius 3 units.

Solution

The equation of the sphere containing the given circle is $x^2 + y^2 + z^2 + 2x + 2y + \lambda(x + y + z + 4) = 0$.

Centre of this sphere is $\left[-\left(1+\frac{\lambda}{2}\right), -\left(1+\frac{\lambda}{2}\right), -\frac{\lambda}{2}\right]$.

$$\text{Radius} = \sqrt{\left(1+\frac{\lambda}{2}\right)^2 + \left(1+\frac{\lambda}{2}\right)^2 + \left(\frac{\lambda}{2}\right)^2 - 4\lambda}$$

$$= \frac{1}{2}\sqrt{3\lambda^2 - 8\lambda + 8}$$

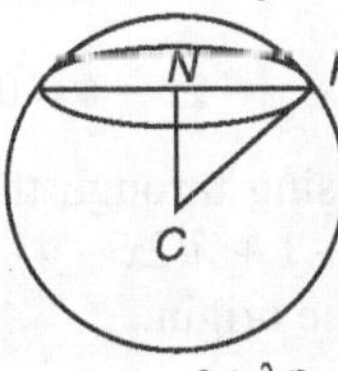

CN = Perpendicular from the centre C on the plane $x + y = 0$.

$$= \left|\frac{-\left(1+\frac{\lambda}{2}\right)-\left(1+\frac{\lambda}{2}\right)}{\sqrt{2}}\right| = \frac{2+\lambda}{\sqrt{2}}$$

$$CP^2 = CN^2 + NP^2$$

$$\frac{1}{4}(3\lambda^2 - 8\lambda + 8) = \frac{(2+\lambda)^2}{2} + 9$$

$$3\lambda^2 - 8\lambda + 8 = 8 + 8\lambda + 2\lambda^2 + 36$$

$$\text{(i.e.)} \quad \lambda^2 - 16\lambda - 36 = 0$$

$$(\lambda + 2)(\lambda - 18) = 0$$

$$\therefore \quad \lambda = -2 \text{ or } 18$$

Therefore,the equations of the required spheres are $x^2 + y^2 + z^2 + 2x + 2y - 2(x + y + z + 4) = 0$ and $x^2 + y^2 + z^2 + 2x + 2y + 18(x + y + z + 4) = 0$.

$$\text{(i.e.)}\quad x^2 + y^2 + z^2 - 2z - 8 = 0$$

$$x^2 + y^2 + z^2 + 20x + 20y + 18z + 72 = 0$$

Example 14.28

Find the equation of the sphere which touches the sphere $x^2 + y^2 + z^2 + 2x - 6y + 1 = 0$ at the point (1, 2, –2) and passes through the origin.

Solution

$$x^2 + y^2 + z^2 + 2x - 6y + 1 = 0 \tag{14.48}$$

The equation of the tangent plane at (1, 2, –2) is $x + 2y - 2z + (x + 1) - 3(y + 2) + 1 = 0$.

$$2x - y - 2z - 4 = 0 \tag{14.49}$$

The equation of the sphere passing through the intersection of (14.48) and (14.49) is $x^2 + y^2 + z^2 + 2x - 6y + 1 + \lambda(2x - y - 2z - 4) = 0$.

This sphere passes through the origin.

$$\therefore\quad 1 - 4\lambda = 0 \text{ or } \lambda = \frac{1}{4}$$

Therefore, the equation of the required sphere is $4(x^2 + y^2 + z^2 + 2x - 6y + 1) + (2x - y - 2z - 4) = 0$.

$$\text{(i.e.)}\quad 4(x^2 + y^2 + z^2) + 10x - 25y - 2z = 0$$

Example 14.29

Show that the condition for the sphere $x^2 + y^2 + z^2 + 2ux + 2vy + 2wz + d = 0$ to cut the sphere $x^2 + y^2 + z^2 + 2u_1x + 2v_1y + 2w_1z + d_1 = 0$ in a great circle is $2uu_1 + 2vv_1 + 2ww_1 - (d + d_1) = 2r_1^2$ where r_1 is the radius of the latter sphere.

Solution

$$S: x^2 + y^2 + z^2 + 2ux + 2vy + 2wz + d = 0$$

$$S_1: x^2 + y^2 + z^2 + 2u_1x + 2v_1y + 2w_1z + d_1 = 0$$

The intersection of these two sphere is $S - S_1 = 0$.

$$\text{(i.e.)}\quad 2(u - u_1)x + 2(v - v_1)y + 2(w - w_1)z + d - d_1 = 0.$$

The centre of the sphere $S_1 = 0$ is $(-u_1, -v_1, -w_1)$.

Since $S_1 = 0$ cuts $S_2 = 0$ in a great circle, the centre of the sphere lies on the plane of intersection $S_1 - S_2 = 0$.

$$\therefore \quad 2(u - u_1)(-u_1) + 2(v - v_1)(-v_1) + 2(w - w_1)(-w_1) + d - d_1 = 0$$

$$\text{(i.e.)} \quad -2uu_1 - 2vv_1 - 2ww_1 + 2(u_1^2 + v_1^2 + w_1^2) + d - d_1 = 0$$

$$2uu_1 + 2vv_1 + 2ww_1 - (d + d_1) = 2(u_1^2 + v_1^2 + w_1^2 - d_1)$$

$$\text{(i.e.)} \quad 2uu_1 + 2vv_1 + 2ww_1 - (d + d_1) = 2r_1^2$$

Example 14.30

A tangent plane to the sphere $x^2 + y^2 + z^2 = r^2$ makes intercepts a, b and c on the coordinate axes. Prove that $a^{-2} + b^{-2} + c^{-2} = r^{-2}$.

Solution

Let $P(x_1, y_1, z_1)$ be a point on the sphere $x^2 + y^2 + z^2 = r^2$.

$$\therefore x_1^2 + y_1^2 + z_1^2 = r^2 \tag{14.50}$$

The equation of the tangent plane at P is $xx_1 + yy_1 + zz_1 = r^2$.

$$\text{(i.e.)} \quad \frac{xx_1}{r^2} + \frac{yy_1}{r^2} + \frac{zz_1}{r^2} = 1$$

Therefore, the intercepts made by the plane on the coordinate axes are $a = \frac{r^2}{x_1}$, $b = \frac{r^2}{y_1}$, $c = \frac{r^2}{z_1}$.

$$\frac{1}{a^2} + \frac{1}{b^2} + \frac{1}{c^2} = \frac{x_1^2 + y_1^2 + z_1^2}{r^4} = \frac{r^2}{r^4} \text{ from (14.50)}$$

$$= \frac{1}{r^2}$$

$$\therefore \quad a^{-2} + b^{-2} + c^{-2} = r^{-2}$$

Example 14.31

Two spheres of radii r_1 and r_2 intersect orthogonally. Prove that the radius of the common circle is $\frac{r_1 r_2}{\sqrt{r_1^2 + r_1^2}}$.

Solution

Let the equation of the common circle be

$$x^2 + y^2 = r^2, z = 0 \tag{14.51}$$

Then the equation of the sphere through the given circle is $x^2 + y^2 + z^2 - r^2 + \lambda z = 0$ where λ is arbitrary.
Let the equation of the two spheres through the given circle be $x^2 + y^2 + z^2 - r^2 + \lambda_1 z = 0$ and $x^2 + y^2 + z^2 - r^2 + \lambda_2 z = 0$
If r_1 and r_2 are the radii of the above two spheres then

$$r_1 = \sqrt{\frac{\lambda_1^2}{4} + r^2} = \frac{1}{2}\sqrt{\lambda_1^2 + 4r^2}$$

$$r_2 = \sqrt{\frac{\lambda_1^2}{4} + r^2} = \frac{1}{2}\sqrt{\lambda_2^2 + 4r^2} \tag{14.52}$$

Since the two spheres cut orthogonally.

$$2uu_1 + 2vv_1 + 2ww_1 = d + d_1$$

$$\therefore \quad 2\frac{\lambda_1}{2}\frac{\lambda_2}{2} = -r^2 - r^2$$

$$\lambda_1\lambda_2 = -4r^2 \tag{14.53}$$

Eliminating λ_1 and λ_2 from (14.52) and (14.53), we get

$$\sqrt{4r_1^2 - 4r^2}\sqrt{4r_2^2 - 4r^2} = -4r^2$$

$$16(r_1^2 - r^2)(r_2^2 - r^2) = 16r^4$$

$$r_1^2 r_2^2 - r^2(r_1^2 + r_2^2) + r^4 = r^4$$

$$\therefore r = \frac{r_1 r_2}{\sqrt{r_1^2 + r_2^2}}$$

Example 14.32

Find the equation of the sphere which touches the plane $3x + 2y - z + 2 = 0$ at the point $(1, -2, 1)$ and cuts orthogonally the sphere $x^2 + y^2 + z^2 - 4x + 6y + 4 = 0$.

Solution

Let the equation of the required sphere be

$$x^2 + y^2 + z^2 + 2ux + 2vy + 2wz + d = 0 \tag{14.54}$$

This sphere touches the plane $3x + 2y - z + 2 = 0$ at $(1, -2, 1)$.
The equation of the tangent plane at $(1, -2, 1)$ is $xx_1 + yy_1 + zz_1 + u(x + x_1) + v(y + y_1) + w(z + z_1) + d = 0$.

$$\text{(i.e.)} \quad x(u + 1) + y(v - 2) + z(w + 1) + u - 2v + w + d = 0 \quad (14.55)$$

But the tangent plane is given as

$$3x + 2y - z + 2 = 0 \quad (14.56)$$

Identifying equations (14.55) and (14.56) we get,

$$\frac{u+1}{3} = \frac{v-2}{2} = \frac{w+1}{-1} = \frac{u-2v+w+d}{2} = k(\text{say})$$

$$\therefore \quad u = 3k - 1 \quad (14.57)$$

$$v = 2k + 2 \quad (14.58)$$

$$w = -k - 1 \quad (14.59)$$

$$u - 2v + w + d = 2k$$

$$(3k - 1) - 2(2k + 2) + (-k - 1) + d = 2k$$

$$d = 4k + 6 \quad (14.60)$$

The sphere (14.54) cuts orthogonally the sphere

$$x^2 + y^2 + z^2 - 4x + 6y + 4 = 0 \quad (14.61)$$

$$\therefore \quad 2u(-2) + 2v(3) + d + 4 = 1$$

$$-4u + 6v + d + 4 = 1$$

$$-4(3k - 1) + 6(2k + 2) + 4k + 6 + 4 = 1$$

$$-12k + 4 + 12k + 12 = 4k + 10$$

$$16 = 4k + 10$$

$$6 = 4k \text{ or } k = \frac{3}{7}$$

$$\therefore u = 3k - 1 = \frac{7}{2}$$

$$v = 2k + 2 = 5$$

$$w = -k - 1 = -\frac{5}{2}$$

$$d = 4k + 6 = 12$$

Therefore, the equation of the sphere is $x^2 + y^2 + z^2 + 7x + 10y - 5z + 12 = 0$.

Example 14.33

Find the equations of the radical planes of the spheres $x^2 + y^2 + z^2 + 2x + 2y + 2z + 2 = 0$, $x^2 + y^2 + z^2 + 4y = 0$ and $x^2 + y^2 + z^2 + 3x - 2y + 8z + 6 = 0$. Also find the radical line and the radical centre.

Solution

Consider the equations,

$$x^2 + y^2 + z^2 + 2x + 2y + 2z + 2 = 0 \qquad (14.62)$$
$$x^2 + y^2 + z^2 + 4y = 0 \qquad (14.63)$$
$$x^2 + y^2 + z^2 + 3x - 2y + 8z + 6 = 0 \qquad (14.64)$$

The radical plane of the spheres (14.62) and (14.63) is $S_1 - S_2 = 0$.

$$\text{(i.e.)} \quad 2x - 2y + 2z + 2 = 0$$
$$x - y + z + 1 = 0$$

The radical plane of the spheres (14.63) and (14.64) is $S_2 - S_3 = 0$.

$$\text{(i.e.)} \quad 3x - 6y - 8z + 6 = 0$$

The radical plane of the sphere (14.62) and S_3 is $S_1 - S_3 = 0$.

$$\text{(i.e.)} \quad x - 4y + 6z + 4 = 0$$

The equation of the radical line of the spheres are given by

$$S_1 - S_2 = 0, S_2 - S_3 = 0$$
$$\text{or} \quad x - y + z + 1 = 0, 3x - 6y + 8z + 6 = 0$$

Also the radical line is given by

$$3x - 6y + 8z + 6 = 0, 2x - 3y + 7z + 4 = 0.$$

The radical centre is the point of intersection of the above two lines. So we have to solve the equations

$$x - y + z = -1$$
$$3x - 6y + 8z = 6$$
$$2x - 3y + 7z = -4$$

Solving these equations we get

$$x = \frac{-1}{6}, y = \frac{1}{2}, z = \frac{-3}{10}$$

Therefore, the radical centre is $\left(\frac{-1}{6}, \frac{1}{2}, \frac{-3}{10}\right)$.

Example 14.34

Find the equation of the sphere through the origin and coaxal with the spheres $x^2 + y^2 + z^2 = 1$ and $x^2 + y^2 + z^2 + x + 2y + 3z - 5 = 0$.

Solution

The radical plane of the two given spheres is $S - S_1 = 0$.

$$\text{(i.e.)} \quad x + 2y + 3z - 4 = 0$$

The equation of any sphere coaxal with given spheres is $S + \lambda P = 0$.

$$\text{(i.e.)} \quad x^2 + y^2 + z^2 - 1 + \lambda(x + 2y + 3z - 4) = 0$$

This sphere passes through the origin.

$$\therefore \quad -1 - 4\lambda = 0$$

$$\therefore \quad \lambda = -\frac{1}{4}$$

Therefore, the equation of the required sphere is $x^2 + y^2 + z^2 - 1 - \frac{1}{4}(x + 2y + 3z - 4) = 0$.

$$\text{(i.e.)} \quad 4(x^2 + y^2 + z^2) - 4 - x - 2y - 3z + 4 = 0$$

$$\therefore 4(x^2 + y^2 + z^2) - x - 2y - 3z = 0$$

Example 14.35

Find the limiting points of the coaxal system of spheres determined by $x^2 + y^2 + z^2 + 4x - 2y + 2z + 6 = 0$ and $x^2 + y^2 + z^2 + 2x - 4y - 2z + 6 = 0$.

Solution

The radical plane of the two given spheres is $2x + 2y + 4z = 0$.

The equation to any sphere of the coaxal system is $x^2 + y^2 + z^2 + 4x - 2y + 2z + 6 + \lambda(x + y + 2z) = 0$.

The centre is $\left(-2-\frac{\lambda}{2}, 1-\frac{\lambda}{2}, -1-\lambda\right)$.

Radius is $\sqrt{\left(2+\frac{\lambda}{2}\right)^2+\left(1-\frac{\lambda}{2}\right)^2+(1+\lambda)^2-6}.$

For limiting point of the coaxal system radius = 0.

$$\frac{(4+\lambda)^2}{4}+\frac{(2-\lambda)^2}{4}+(1+\lambda)^2-6=0$$

$$(4+\lambda)^2+(2-\lambda)^2+4(1+\lambda)^2-24=0$$

$$6\lambda^2+12\lambda=0$$

$$6\lambda(\lambda+2)=0\ \lambda=0,\ \lambda=-2$$

Therefore, the limiting points are the centres of point spheres of the coaxal system.

Therefore, the limiting points are (–2, 1, –1) and (–1, 2, 1).

Example 14.36

The point (–1, 2, 1) is a limiting point of a coaxal system of spheres of which $x^2+y^2+z^2+3x-2y+6=0$ is a member. Find the equation of the radical plane of this system and the coordinates of other limiting point.

Solution

The point gives belonging to the coaxal system corresponding to the limiting point (–1, 2, 1) is $(x+1)^2+(y-2)^2+(z-1)^2=0$.

$$\text{(i.e.)}\quad x^2+y^2+z^2+2x-4y-2z+6=0$$

Two members of the system of the system are $x^2+y^2+3x-3y+6=0$ and $x^2+y^2+z^2+2x-4y-2z+6=0$.

The radical plane of the coaxal system is $x+y+2z=0$. Any member of the system is $x^2+y^2+2x-4y-2z+6+\lambda(x+y+2z)=0$.

Centre is $\left(-1-\frac{\lambda}{2},2-\frac{\lambda}{2},1-\lambda\right)$

$$r=\sqrt{\frac{(2+\lambda)^2}{4}+\frac{(4-\lambda)^2}{4}+(1-\lambda)^2-6}$$

For limiting points radius = 0

$$(2 + \lambda)^2 + (4 - \lambda)^2 + 4(1 - \lambda)^2 - 24 = 0$$

$$6\lambda^2 - 12\lambda = 0 \therefore \lambda = 0, \lambda = 2$$

When $\lambda = 0$ centre is $(-1, 2, 1)$ which is the given limiting point.
When $\lambda = 2$, the centre is $(-2, 1, -1)$ which is other limiting point.

Example 14.37

Show that the spheres $x^2 + y^2 + z^2 = 25$ and $x^2 + y^2 + z^2 - 24x - 40y - 18z + 225 = 0$ touch externally. Find their point of contact.

Solution

$$x^2 + y^2 + z^2 = 25 \tag{14.65}$$

$$x^2 + y^2 + z^2 - 24x - 40y - 18z + 225 = 0 \tag{14.66}$$

The centre and radius of sphere (14.65) are

$$C_1(0, 0, 0), r_1 = 5$$

Centre and radius of the sphere (14.66) are

$$C_2(12, 20, 9) \text{ and } r = \sqrt{144 + 400 + 81 - 225} = 20$$

The distance between the centres

$$C_1C_2 = \sqrt{144 + 400 + 81} = \sqrt{625} = 25$$

$$\therefore r_1 + r_2 = C_1C_2$$

Hence the two given spheres touch externally.

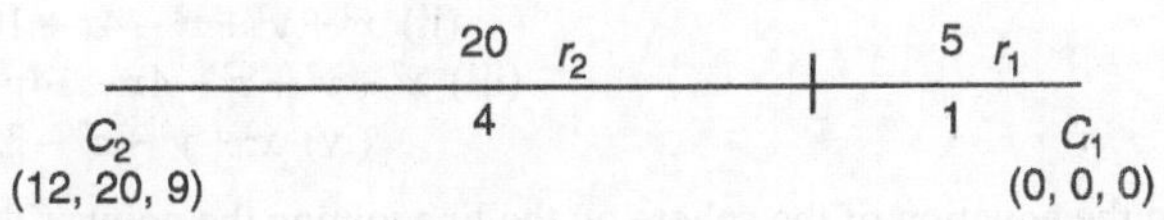

Therefore, the point of contact divides the lines of centres in the ratio 4 : 1
Therefore, the coordinates of the point of contact is

$$\left(\frac{4\times0+1\times12}{5}, \frac{4\times0+1\times20}{5}, \frac{4\times0+1\times9}{5}\right)$$

$$\text{(i.e.)} \left(\frac{12}{5}, 4, \frac{9}{5}\right)$$

Exercises 1

1. Find the equation of the sphere with
   (i) centre at (1, –2, 3) and radius 5 units.
   (ii) centre at $\left(-\frac{1}{3}, \frac{2}{3}, \frac{1}{3}\right)$ and radius 1 unit.
   (iii) centre at (1, 2, 3) and radius 4 units.

   Ans.: (i) $x^2 + y^2 + z^2 - 2x + 4y - 6z - 11 = 0$
   (ii) $3(x^2 + y^2 + z^2) + 2x - 4y - 2z - 1 = 0$
   (iii) $x^2 + y^2 + z^2 - 2x - 4y - 6z - 2 = 0$

2. Find the coordinates of the centre and radius of the following spheres:
   (i) $x^2 + y^2 + z^2 + 2x - 4y - 6z + 15 = 0$
   (ii) $2x^2 + 2y^2 + 2z^2 - 2x - 4y - 6z - 1 = 0$
   (iii) $ax^2 + ay^2 + az^2 + 2ux + 2vy + 2wz + d = 0$

   Ans.: (i) (–1, 2, 3), $r = 3$
   (ii) $\left(\frac{1}{2}, -1, \frac{3}{2}\right)$, $r = 2$
   (iii) $\left(\frac{-u}{a}, \frac{-v}{a}, \frac{-w}{a}\right)$, $\sqrt{\frac{u^2 + v^2 + w^2 - ad}{a^2}}$

3. Find the equation of the sphere whose centre is at the point (1, 2, 3) and which passes through the point (3, 1, 2).

   Ans.: $x^2 + y^2 + z^2 - 2x - 4y - 6z + 8 = 0$

4. Find the equation of the sphere passing through points:
   (i) (0, 0, 0), (0, 1, –1), (–1, 2, 0) and (1, 2, 3)
   (ii) (2, 0, 1), (1 – 5, –1), (0, –2, 3) and (4, –1, 3)
   (iii) (0, –1, 2), (0, –2, 3), (1, 5, –1) and (2, 0, 1)
   (iv) (–1, 1, 1), (1, –1, 1), (1, 1, –1), (0, 0, 0)

   Ans.: (i) $(x^2 + y^2 + z^2) - 13x - 25y - 11z = 0$
   (ii) $x^2 + y^2 + z^2 - 4x + 10y - 2z + 5 = 0$
   (iii) $x^2 + y^2 + z^2 - 4x - 14y - 22z + 25 = 0$
   (iv) $x^2 + y^2 + z^2 - 3x - 3y - 3z = 0$

5. Find the equation of the sphere on the line joining the points (2, –3, –1) and (1, –2, –1) at the ends of a diameter.

   Ans.: $x^2 + y^2 + z^2 - 3x + 5y + 7 = 0$

6. Find the radius of the sphere touching the plane $2x + 2y - z - 1 = 0$ and concentric with the sphere $2x^2 + 2y^2 + 2z^2 + x + 2y - z = 0$.

   Ans.: $\frac{7}{12}$ units

7. Find the equation of the sphere passing through the points (0, 2, 3), (1, 1, –1), (–5, 4, 2) and having its centre on the plane $3x + 4y + 2z - 6 = 0$.

   Ans.: $9(x^2 + y^2 + z^2) + 28x + 7y - 20z - 96 = 0$

8. Prove that a sphere can be made to pass through the midpoints of the edges of a tetrahedron whose faces are $x = 0, y = 0, z = 0$ and $\frac{x}{a}+\frac{y}{b}+\frac{z}{c} = 2$. Find its equation.

Ans.: $x^2 + y^2 + z^2 - ax - by - cz = 0$

9. Find the condition that the plane $lx + my + nz = p$ may touch the sphere $x^2 + y^2 + z^2 + 2ux + 2vy + 2wz + d = 0$.

Ans.: $(ul + vm + wn + p)^2 = (l^2 + m^2 + n^2)\ (u^2 + v^2 + w^2 \quad d)$

10. Prove that the sphere circumscribing the tetrahedron whose faces are $y + z = 0$, $z + x = 0, x + y = 0$ and $x + y + z = 1$ is $x^2 + y^2 + z^2 - 3(x + y + z) = 0$.

11. A point moves such that, the sum of the squares of its distances from the six faces of a cube is a constant. Prove that its locus is the sphere $x^2 + y^2 + z^2 = 3(k^2 - a^2)$.

12. Prove that the spheres $x^2 + y^2 + z^2 = 100$ and $x^2 + y^2 + z^2 - 12x + 4y - 6z + 40 = 0$ touch internally and find the point of contact.

Ans.: $\left(\frac{60}{2}, \frac{-20}{7}, \frac{30}{7}\right)$

13. Prove that the spheres $x^2 + y^2 + z^2 = 25$ and $x^2 + y^2 + z^2 - 24x - 40y - 18z + 225 = 0$ touch externally. Find the point of contact.

Ans.: $\left(\frac{9}{5}, \frac{12}{5}, 4\right)$

14. Find the condition that the plane $lx + my + nz = p$ to be a tangent to the sphere $x^2 + y^2 + z^2 = r.^2$

Ans.: $r^2\ (l^2 + m^2 + n^2) = p^2$

15. Find the equation of the sphere which touches the coordinate planes and whose centre lies in the first octant.

Ans.: $x^2 + y^2 + z^2 - 2vx - 2vy - 2vz + 2v^2 = 0$

16. Find the equation of the sphere with centre at $(1, -1, 2)$ and touching the plane $2x - 2y + z = 3$.

Ans.: $x^2 + y^2 + z^2 - 2x + 2y + z + 5 = 0$

17. Find the equation of the sphere which has the points $(2, 7, -4)$ and $(4, 5, -1)$ as the extremities of a diameter.

Ans.: $x^2 + y^2 + z^2 - 6x - 12y + 5z + 47 = 0$

18. Find the equation of the sphere which touches the three coordinate planes and the plane $2x + y + 2z = 6$ and being in the first octant.

Ans.: $x^2 + y^2 + z^2 - 6x - 6y - 6z + 18 = 0$

19. A point P moves from two points $A(1, 3, 4)$ and $B(1, -2, -1)$ such that 3.PA = 2.PB. Show that the locus of P is the sphere $x^2 + y^2 + z^2 - 2x - 4y - 16z + 42 = 0$. Show also that this sphere divides A and B internally and externally in the ration 2:3.

20. A plane passes through a fixed point (a, b, c). Show that the locus of the foot of the perpendicular to it from the origin is the sphere $x^2 + y^2 + z^2 - ax - by - cz = 0$.

21. A variable sphere passes through the origin O and meets the coordinate axes in A, B and C so that the volume of the tetrahedron $OABC$ is a constant. Find the locus of the centre of the sphere.

Ans.: $xyz = k^2$

22. Find the equation of the sphere on the line joining the points:

(i) $(4, -1, 2)$ and $(2, 3, 6)$ as the extremities of a diameter

(ii) $(2, -3, 4)$ and $(-5, 6, -7)$ as the extremities of a diameter

Ans.: $x^2 + y^2 + z^2 - 6x - 2y - 8z + 17 = 0$
$x^2 + y^2 + z^2 + 3x - 3y + 3z - 56 = 0$

23. A plane passes through a fixed point (a, b, c) and cuts the axes in A, B and C. Show that the locus of centre of the sphere ABC is $\frac{a}{x} + \frac{b}{y} + \frac{c}{z} = 2$.

24. A sphere of constant radius $2k$ passes through the origin and meets the axes in A, B and C. Prove that the locus of the centroid of ΔABC is $9(x^2 + y^2 + z^2) = a^2$.

25. The tangent plane at any point of the sphere $x^2 + y^2 + z^2 = a^2$ meets the coordinate axes at A, B and C. Find the locus of the point of intersection of the planes drawn parallel to the coordinate planes through A, B and C.

Ans.: $x^{-2} + y^{-2} + z^{-2} = a^{-2}$

26. OA, OB and OC are three mutually perpendicular lines through the origin and their direction cosines are l_1, m, n; l_2, m_2, n_2 and l_3, m_3, n_3. If $OA = a$, $OB = b$, $OC = c$ then prove that the equation of the sphere $OABC$ is $x^2 + y^2 + z^2 - x(al_1 + bl_2 + cl_3) - y(am_1 + bm_2 + cm_3) - z(an_1 + bn_2 + cn_3) = 0$.

Exercises 2

1. Find the centre and radius of a section of the sphere $x^2 + y^2 + z^2 = 1$ by the plane $lx + my + nz = 1$.

Ans.: $\left(\frac{l}{\sum l^2}, \frac{m}{\sum l^2}, \frac{n}{\sum l^2}\right), \sqrt{\frac{l^2 + m^2 + n^2 - 1}{l^2 + m^2 + n^2}}$.

2. Find the equation of the sphere through the circle $x^2 + y^2 + z^2 = 5$, $x + 2y + 3z = 3$ and the point $(1, 2, 3)$.

Ans.: $5(x^2 + y^2 + z^2) - 4x - 8y - 12z - 13 = 0$.

3. Prove that the plane $x + 2y - z = 4$ cuts the sphere $x^2 + y^2 + z^2 - x + z - 2 = 0$ in a circle of radius unity and find the equation of the sphere which has this circle for one of its great circles.

Ans.: $x^2 + y^2 + z^2 - 2x - 2y + 2z - 2 = 0$

4. Find the centre and radius of the circle in which the sphere $x^2 + y^2 + z^2 = 25$ is cut by the plane $2x + y + 2z = 9$.

Ans.: (2, 1, 2); 4

5. Show that the intersection of the sphere $x^2 + y^2 + z^2 - 2x - 4y - 6z - 2 = 0$ and the plane $x - 2y + 2z - 20 = 0$ is a circle of radius $\sqrt{7}$ with its centre at (2, 4, 5).

6. Find the centre and radius of the circle $x^2 + y^2 + z^2 - 2x - 4z + 1 = 0, x + 2y + 2z = 11$.

Ans.: $\left(\frac{5}{9}, \frac{19}{9}, \frac{28}{9}\right), \frac{\sqrt{11}}{3}$

7. Prove that the radius of the circle $x^2 + y^2 + z^2 + x + y + z = 4, x + y + z = 0$ is 2.

8. Find the centre and radius of the circle $x^2 + y^2 + z^2 + 2x - 2y - 4z - 19 = 0, x + 2y + 2z + 7 = 0$.

Ans.: $\left(\frac{-7}{3}, \frac{-5}{3}, \frac{-2}{3}\right), 5$

9. Find the radius of the circle $3x^2 + 3y^2 + 3z^2 + x - 5y - 2 = 0, x + y = 2$.

10. Show the two circles $2(x^2 + y^2 + z^2) + 8x + 13y + 17z - 17 = 0, 2x + y - 3z + 1 = 0$ and $x^2 + y^2 + z^2 + 3x - 4y + 3z = 0, x - y + 2z - 4 = 0$ lie on the same sphere and find its equation.

Ans.: $x^2 + y^2 + z^2 + 5x - 6y + 7z - 8 = 0$

11. Prove that the two circles $x^2 + y^2 + z^2 - 2x + 3y + 4z - 5 = 0, 5y + 6z + 1 = 0$ and $x^2 + y^2 + z^2 - 3x - 4y + 5z - 6 = 0, x + 2y - 7 = 0$ lie on the same sphere and find its equation.

Ans.: $x^2 + y^2 + z^2 - 2x - 2y - 2z - 6 = 0$

12. Find the area of the section of the sphere $x^2 + y^2 + z^2 + 12x - 2y - 6z + 30 = 0$ by the plane $x - y + 2z + 5 = 0$.

Ans.: $\frac{49}{8}\pi$

13. Find the equation of the sphere which has its centre on the plane $5x + y - 4z + 3 = 0$ and passing through the circle $x^2 + y^2 + z^2 - 3x + 4y - 2z + 8 = 0, 4x - 5y + 3z - 3 = 0$.

Ans.: $x^2 + y^2 + z^2 + 9x - 11y + 7z - 1 = 0$

14. Find the equation of the sphere having the circle $x^2 + y^2 + z^2 + 10x - 4z - 8 = 0, x + y + z - 3 = 0$ as a great circle.

Ans.: $x^2 + y^2 + z^2 + 6x - 4y - 3z + 4 = 0$

15. A variable plane is parallel to a given plane $\frac{x}{a}+\frac{y}{b}+\frac{z}{c}=0$ and meets the axes at A, B and C. Prove that the circle ABC lies on the surface

$$yz\left(\frac{b}{c}+\frac{c}{b}\right)+zx\left(\frac{c}{a}+\frac{a}{c}\right)+xy\left(\frac{a}{b}+\frac{b}{a}\right)=0$$

16. Find the equation of the spheres which pass through the circle $x^2+y^2+z^2-4x-y+6z+12=0$, $2x+3y-7z=10$ and touch the plane $x-2y-2z=1$.

Ans.: $x^2+y^2+z^2-2x+2y-4z+2=0$
$x^2+y^2+z^2-6x-4y+10z+22=0$

17. Find the equation of the sphere which pass through the circle $x^2+y^2+z^2=5$, $x+2y+3z=5$ and touch the plane $z=0$.

Ans.: $x^2+y^2+z^2-2x+y+5z+5=0$
$5(x^2+y^2+z^2)-(2x-4y+5z+1=0)$

18. Find the centre and radius of the circle $x^2+y^2+z^2-2x+4y+2z-6=0$, $x+2y+2z-4=0$.

Ans.: $(2, 0, 1)$, $\sqrt{3}$

19. Find the equation of the sphere which passes through the point $(3, 1, 2)$ and meets XOY plane in a circle of radius 3 units with the centre at the point $(1, -2, 0)$.

Ans.: $x^2+y^2+z^2-2x+4y-4z-4=0$

20. Find the centre and radius of the circle $x^2+y^2+z^2+12x-12y-16z+111=0$, $2x+2y+z=17$.

Ans.: $(-4, 8, 9)$, $r=4$

21. Find the centre and radius of the circle $x^2+y^2+z^2+2x+2y-4z-19=0$, $x+2y+2z+7=0$.

Ans.: $\left(\frac{-7}{3},\frac{-10}{3},\frac{-4}{3}\right),3$

22. Find the centre and radius of the circle $x^2+y^2+z^2=9$, $x+y+z=1$.

Ans.: $\left(\frac{1}{3},\frac{1}{3},\frac{1}{3}\right)$, $\sqrt{\frac{26}{3}}$

23. Find the equation of the sphere through the circle $x^2+y^2+z^2=9$, $2x+3y+4z=5$ and the point $(1, 2, 1)$.

Ans.: $3(x^2+y^2+z^2)-2x-2y-4z-22=0$

24. Find the equation of the sphere containing the circle $x^2+y^2+z^2-2x=9$, $z=0$ and the point $(4, 5, 6)$.

Ans.: $x^2+y^2+z^2-2x-10z-9=0$

25. Find the equation of the sphere passing through the circle $x^2+y^2=a^2$, $z=0$ and the point (α, β, λ).

Ans.: $r(x^2+y^2+z^2-a^2)-z(\alpha^2+\beta^2+\gamma^2-a^2)=0$

26. Find the equation of the sphere through the circle $x^2+y^2+z^2+2x+3y+6=0$, $x-2y+4z-9=0$ and the centre of the sphere.

Ans.: $x^2+y^2+z^2-2x+4y-6z+5=0$

27. Find the equations of the sphere through the circle $x^2+y^2+z^2=1$, $2x+4y+5z=6$ and touching the plane $z=0$.

Ans.: $x^2+y^2+z^2-2x-4y-5z+5=0$
$5(x^2+y^2+z^2)-2x-4y-5z+1=0$

28. Find the equation of the sphere having the circle $x^2+y^2+z^2+10y-4z-8=0$, $x+y+23=0$ as a great circle.

Ans.: $x^2+y^2+z^2-4x+6y-8z+4=0$

29. Show that the two circles $x^2+y^2+z^2-y+2z=0$, $x-y+z-2=0$ and $x^2+y^2+z^2+x-3y+z-5=0$, $2x-y+4z-1=0$ lie on the same sphere and find its equation.

Ans.: $x^2+y^2+z^2+3x-4y+5z-6=0$

30. Prove that the circles $x^2+y^2+z^2-2x+3y+4z-5=0$, $5y+6z+1=0$ and $x^2+y^2+z^2-3x+4y+5z-6=0$, $x+2y-7z=0$ lie on the same sphere. Find its equation.

Ans.: $x^2+y^2+z^2-2x-2y-2z-6=0$

31. Find the conditions that the circles $x^2+y^2+z^2+2ux+2vy+2wz+d=0$, $lx+my+nz=p$ and $(x^2+y^2+z^2)\ 2u'x+2v'y+2w'z+d'=0$, $l'x+m'y+n'z=p'$ to lie on the same circle.

Ans.: $\Delta \begin{vmatrix} 2(u-u') & 2(v-v') & 2(w-w') & d-d' \\ l & m & n & -p \\ l' & m' & n' & -p' \end{vmatrix} = 0$

32. Find the centre and radius of the circle formed by the intersection of the sphere $x^2+y^2+z^2=2225$ and the plane $2x-2y+z=27$.

Ans.: $(6, -6, 3)$, 12

33. Find the centre and radius of the circle $x^2+y^2+z^2=25$, $x+2y+2z=9$.

Ans.: $(1, 2, 2)$, 4

34. Find the equation of the circle which lies on the sphere $x^2+y^2+z^2=25$ and has the centre at $(1, 2, 3)$.

Ans.: $x^2+y^2+z^2=25$, $x+2y+3z=14$

35. A plane passes through a point (α, β, γ) and intersects the sphere $x^2+y^2+z^2=a^2$. Show that the locus of the centre of the circle of intersection is the sphere $x(x-\alpha)+y(y-\beta)+z(z-\gamma)=0$.

36. Find the equation of the sphere through the circle $x^2+y^2+z^2-4=0$ and the point $(2, 1, 1)$.

Ans.: $x^2+y^2+z^2-2x+y-2z-1=0$

37. Find the equation of the sphere through the circle $x^2 + y^2 + z^2 = 9$, $2x + 3y + 4z = 5$ and through the origin.

Ans.: $5(x^2 + y^2 + z^2) - 18x - 27y - 36z = 0$

38. Show that the two circles $2(x^2 + y^2 + z^2) + 8x - 13y + 17z - 17 = 0$, $2x + y - 3z + 1 = 0$ and $x^2 + y^2 + z^2 + 3x - 4y + 3z = 0$, $x - y + 2z - 4 = 0$ lie on the same sphere and find its equation.

Ans.: $x^2 + y^2 + z^2 + 5x - 6y - 7z - 8 = 0$

39. Find the equation of the sphere which has its centre on the plane $5x + y - 4z + 3 = 0$ and passing through the circle $x^2 + y^2 + z^2 - 3x + 4y - 2z + 8 = 0$, $4x - 5y + 3z - 3 = 0$.

Ans.: $x^2 + y^2 + z^2 + 9x - 11y + 7z - 1 = 0$

40. Find the equation of the sphere which has the circle $S = x^2 + y^2 + z^2 + 2x + 4y + 6z - 11 = 0$, $2x + y + 2z + 1 = 0$ as great circle.

Ans.: $x^2 + y^2 + z^2 - 2x + 2y + 2z - 13 = 0$.

41. Find the equation of the sphere whose radius is 1 and which passes through the circle of intersection of the spheres $x^2 + y^2 + z^2 + 2x + 2y + 2z - 6 = 0$ and $x^2 + y^2 + z^2 + 3x + 3y - z - 1 = 0$.

Ans.: $3x^2 + 3y^2 + 3z^2 + 16x + 16y + 4z + 32 = 0$

42. If r is the radius of the circle $x^2 + y^2 + z^2 + 2ux + 2vy + 2wz + d = 0$, $lx + my + nz = 0$ then prove that $(r^2 + d^2)(l^2 + m^2 + n^2) = (mw - nv)^2 + (nv - lw)^2 + (lv - mu)^2$.

43. Find the equation of the sphere through the circle $x^2 + y^2 = 4$, $z = 0$ meeting the plane $x + 2y + 2z = 0$ in a circle of radius 3.

Ans.: $x^2 + y^2 + z^2 - 6z - 4 = 0$

44. Find the equation of the sphere through the circle $x^2 + y^2 + z^2 = 1$, $2x + 3y + 4z = 5$ and which intersect the sphere $x^2 + y^2 + z^2 + 3(x - y + z) - 56 = 0$ orthogonally, $x^2 + y^2 + z^2 - 12x - 18y - 24z + 29 = 0$.

45. The plane ABC whose equation is $\frac{x}{a} + \frac{y}{b} + \frac{z}{c} = 1$ meets the axes in A, B and C. Find the equation to determine the circumcircle of the triangle ABC and obtain the coordinates of its centre.

Ans.: $x^2 + y^2 + z^2 - ax - by - cz = 0$, $\frac{x}{a} + \frac{y}{b} + \frac{z}{c} = 1$

$$\left(\frac{a(b^{-2} + c^{-2})}{2\sum a^{-2}}, \frac{b(c^{-2} + a^{-2})}{2\sum a^{-2}}, \frac{c(c^{-2} + b^{-2})}{2\sum a^{-2}}\right)$$

46. Find the equation of the circle circumscribing the triangle formed by the three points $(a, 0, 0)$, $(0, b, 0)$, $(0, 0, c)$. Obtain the coordinates of the centre of the circle.

Ans.: $x^2 + y^2 + z^2 - x - 2y - 3z = 0$, $6x - 3y - 2z - 6 = 0$

$$\text{Centre} = \left(\frac{-13}{98}, \frac{40}{49}, \frac{135}{98}\right)$$

47. Find the equation of the sphere through the circle $x^2 + y^2 + z^2 + 2x + 3y + 6 = 0$, $x - 2y + 4z - 9 = 0$ and the centre of the sphere $x^2 + y^2 + z^2 - 2x + 4y - 6z + 5 = 0$.

Ans.: $x^2 + y^2 + z^2 + 7y - 8z + 24 = 0$

48. Find the equation of the sphere having its centre on the plane $4x - 5y - z - 3 = 0$ and passing through the circle $x^2 + y^2 + z^2 - 2x - 3y + 4z + 8 = 0$, $x^2 + y^2 + z^2 + 4x + 5y - 6z + 2 = 0$.

Ans.: $x^2 + y^2 + z^2 + 7x + 9y - 11z - 1 = 0$

49. A circle with centre (2, 3, 0) and radius unity is drawn on the plane $z = 0$. Find the equation of the sphere which passes through the circle and the point (1, 1, 1).

Ans.: $x^2 + y^2 + z^2 - 4x - 6y - 6z + 12 = 0$

50. Find the equation of the sphere which passes through the circle $x^2 + y^2 = 4$, $z = 0$ and is cut by the plane $x + 2y + 2z = 0$ in a circle of radius 3.

Ans.: $x^2 + y^2 + z^2 + 6z - 4 = 0$
$x^2 + y^2 + z^2 - 6z - 4 = 0$

51. Prove that the plane $x + 2y - z - 4 = 0$ cuts the sphere $x^2 + y^2 + z^2 - x + z - 2 = 0$ in a circle of radius unity and also find the equation of the sphere which has this circle as great circle.

Ans.: $x^2 + y^2 + z^2 - 2x - 2y + 2z + 2 = 0$

52. Find the equation of the sphere having the circle $x^2 + y^2 + z^2 + 10y - 4z - 3 = 0$, $x + y + z - 3 = 0$ as a great circle.

Ans.: $x^2 + y^2 + z^2 - 4x + 6y - 8z + 4 = 0$

53. P is a variable point on a given line and A, B and C are projections on the axes. Show that the sphere $OABC$ passes through a fixed circle.

Exercises 3

1. Find the equations of the spheres which pass through the circle $x^2 + y^2 + z^2 = 5$, $x + 2y + 3z = 5$ and touch the plane $z = 0$.

Ans.: $x^2 + y^2 + z^2 - 2x + y + 5z + 5 = 0$
$5(x^2 + y^2 + z^2) - 2x - 4y + 5z + 1 = 0$

2. Find the equations of the spheres which pass through the circle $x^2 + y^2 + z^2 - 4x - y + 3z + 12 = 0$, $2x + 3y - 8z = 10$ and touch the plane $x - 2y - 2z = 1$.

Ans.: $x^2 + y^2 + z^2 - 2x + 2y - 4z + 2 = 0$
$x^2 + y^2 + z^2 - 6x - 4y + 10z + 22 = 0$

3. Show that the tangent plane at (1, 2, 3) to the sphere $x^2 + y^2 + z^2 + x + y + z - 20 = 0$ is $3x + 5y + 7z - 34 = 0$.

4. Find the equation of the sphere which touches the sphere $x^2 + y^2 + z^2 - x + 3y + 2z - 3 = 0$ at $(1, 1, -1)$ and passes through the origin.

Ans.: $2x^2 + 2y^2 + 2z^2 - 3x + y + 4z = 0$

5. Find the equation of the sphere which touches the sphere $x^2 + y^2 + z^2 + 2x - 6y + 1 = 0$ at $(1, 2, -2)$ and passes through the origin.

Ans.: $4(x^2 + y^2 + z^2) + 10x - 28y - 2z = 0$

6. Show that the plane $2x - 2y + z + 16 = 0$ touches the sphere $x^2 + y^2 + z^2 + 2x + 4y + 2z - 3 = 0$ and find the point of contact.

Ans.: $(-3, 4, -2)$

7. Find the equation of the tangent plane at the origin to the sphere $x^2 + y^2 + z^2 - 8x - 6y + 4z = 0$.

Ans.: $4x - 3y + 2z = 0$

8. Find the equation of the tangent planes to the sphere $x^2 + y^2 + z^2 + 2x - 4y + 6z - 7 = 0$ which passes through the line $6x - 3y - 23 = 0$, $3z + 2 = 0$.

Ans.: $8x - 4y + z - 34 = 0$, $4x - 2y - z - 16 = 0$

9. Show that the plane $2x - 2y + z - 12 = 0$ touches the sphere $x^2 + y^2 + z^2 - 2x - 4y + 2z - 3 = 0$.

10. Show that the point $P(1, -3, 1)$ lies on the sphere $x^2 + y^2 + z^2 + 2x + 2y - 7 = 0$ and obtain the equation of the tangent plane at P.

Ans.: $2x - 2y + z = 9$

11. If the point $(5, 1, 4)$ is one extremities of a diameter of the sphere $x^2 + y^2 + z^2 - 2x - 2y - 2z - 22 = 0$ and find the coordinates of the other extremity. Find the equation to the tangent planes at the two extremities and show that they are parallel.

Ans.: $(-3, 1, -2)$; $4x + 3y - 22 = 0$, $4x + 3y + 8 = 0$

12. Find the value of k for which the plane $x + y + z = k$ touches the sphere $x^2 + y^2 + z^2 - 2x - 4y - 6z + 11 = 0$. Find the point of contact in each case.

Ans.: $k = 3$ or 9; $(0, 1, 2)$, $(2, 3, 4)$

13. Find the equation to the tangent planes to the sphere $x^2 + y^2 + z^2 - 2x - 4y - 6z - 2 = 0$ which are parallel to the plane $x + 2y + 2z - 20 = 0$.

Ans.: $x + 2y + 2z - 23 = 0$
$x + 2y + 2z - 1 = 0$

14. A sphere touches the plane $x - 2y - 2z - 7 = 0$ at the point $(3, -1, -1)$ and passes through the point $(1, 1, -3)$. Find the equation.

Ans.: $x^2 + y^2 + z^2 - 10y - 10z - 31 = 0$

15. Show that the line $\frac{x-7}{2} = \frac{y-4}{7} = \frac{z+3}{10}$ touches the sphere $x^2 + y^2 + z^2 - 6x + 2y - 4z + 5 = 0$. Find the point of contact.

Ans.: $(5, -3, 3)$

16. Tangent planes at any point of the sphere $x^2 + y^2 + z^2 = r^2$ meets the coordinate axes at A, B and C. Show that the locus of the point of intersection of the planes drawn parallel to the coordinate planes through the points A, B and C is the surface $x^{-2} + y^{-2} + z^{-2} = r^{-2}$.

17. Find the condition that the line $\frac{x-a}{l} = \frac{y-b}{m} = \frac{z-c}{n}$ where l, m and n are the direction cosines of a line, should touch the sphere $x^2 + y^2 + z^2 + 2ux + 2vy + 2wz + d = 0$. Show that there are two spheres through the points (0, 0, 0), (2a, 0, 0), (0, 2b, 0) and (0, 0, 2c) which touch the above line and that the distance between their centres is $\frac{2}{n^2}[c^2 - (a^2 + b^2 + c^2)n^2]^{\frac{1}{2}}$.

18. Find the equation of the sphere which touches the sphere $x^2 + y^2 + z^2 + 3y - x + 2z - 3 = 0$ at (1, 1, –1) and passes through the origin.

Ans.: $2x^2 + 2y^2 + 2z^2 - 3x + y + 4z = 0$

19. Find the equation of the tangent line in symmetrical form to the circle $x^2 + y^2 + z^2 + 5x - 7y + 2z - 8 = 0$, $3x - 2y + 4z + 3 = 0$.

Ans.: $\frac{x+3}{32} = \frac{y-5}{34} = \frac{z-7}{-7}$

20. Show that the plane $2x - 2y - z + 12 = 0$ touches the sphere $x^2 + y^2 + z^2 - 2x - 4y + 2z - 3 = 0$ and find the point of contact.

Ans.: (–1, 4, –2)

21. Find the equation of the sphere which touches the sphere $x^2 + y^2 + z^2 + 2x - 6y + 1 = 0$ at the point (1, 2, –2) and passes through the origin.

Ans.: $4(x^2 + y^2 + z^2) + 10x - 25y - 22 = 0$

22. Find the equations of the spheres which pass through the circle $x^2 + y^2 + z^2 = 1$, $2x + 4y + 5z - 6 = 0$ and touch the plane $z = 0$.

Ans.: $5(x^2 + y^2 + z^2) - 2x - 4y - 5z + 6 = 0$
$5(x^2 + y^2 + z^2) - 2x - 4y - 5z + 1 = 0$

23. Find the equations of the sphere passing through the circle $x^2 + y^2 + z^2 - 5 = 0$, $2x + 3y + z - 3 = 0$ and touching the plane $3x + 4z - 15 = 0$.

Ans.: $x^2 + y^2 + z^2 + 4x + 6y - 2z - 11 = 0$
$5(x^2 + y^2 + z^2) - 8x - 12y - 4z - 37 = 0$

24. Find the point of intersection of the line and the sphere $x^2 + y^2 + z^2 - 4x + 6y - 2z + 5 = 0$.

Ans.: (4, –1, 2), (0, –2, 3)

25. Prove that the sum of the squares of the intercepts made by a given line on any three mutually perpendicular lines through a fixed point is constant.

Exercises 4

1. Prove that the spheres $x^2 + y^2 + z^2 + 6y + 2z + 8 = 0$ and $x^2 + y^2 + z^2 + 6x + 8y + 4z + 2 = 0$ intersect orthogonally.

2. Find the equation of the sphere which passes through the circle $x^2 + y^2 + z^2 - 2x + 3y - 4z + 6 = 0$, $3x - 4y + 5z - 15 = 0$ and which cuts orthogonally the sphere $x^2 + y^2 + z^2 + 2x + 4y - 6z + 11 = 0$.

   Ans.: $5(x^2 + y^2 + z^2) - 13x + 19y - 25z + 45 = 0$

3. Find the equation of the sphere that passes through the circle $x^2 + y^2 + z^2 - 2x + 3y - 4z + 6 = 0$, $3x - 4y + 5z - 15 = 0$ and which cuts the sphere $x^2 + y^2 + z^2 + 2x + 4y + 6z + 11 = 0$ orthogonally.

   Ans.: $x^2 + y^2 + z^2 + x - y + z - 9 = 0$

4. Prove that every sphere through the circle $x^2 + y^2 - 2ax + r^2 = 0$, $z = 0$ cuts orthogonally every sphere through the circle $x^2 + z^2 = r^2$, $y = 0$.

5. Find the equation of the sphere which touches the plane $3x + 2y - z + 7 = 0$ at the point $(1, -2, 1)$ and cuts the sphere $x^2 + y^2 + z^2 - 4x + 6y + 4 = 0$ orthogonally.

   Ans.: $3(x^2 + y^2 + z^2) + 6x + 20y - 10z + 36 = 0$

6. Find the equation of the sphere that passes through the points (a, b, c) and $(-2, 1, -4)$ and cuts orthogonally the two spheres $x^2 + y^2 + z^2 + x - 3y + 2 = 0$ and $(x^2 + y^2 + z^2) + x + 3y + 4 = 0$.

   Ans.: $x^2 + y^2 + z^2 + 2x - 2y + 4z - 3 = 0$

7. Find the equation of the sphere which touches the plane $3x + 2y - z + 2 = 0$ at the point $P(1, -2, 1)$ and also cuts orthogonally.

   Ans.: $x^2 + y^2 + z^2 + 7x + 10y - 5z + 12 = 0$

8. If d is the distance between the centres of the two spheres of radii r_1 and r_2 then prove that the angle between them is $\cos^{-1}\left(\dfrac{r_1^2 + r_2^2 - d^2}{2r_1r_2}\right)$.

9. Find the condition that the sphere $a(x^2 + y^2 + z^2) + 2lx + z - y + 2nz + p = 0$ and $b(x^2 + y^2 + z^2)k^2$ may cut orthogonally.

   Ans.: $ak^2 = bp^2$

10. Find the equation of the radical planes of the spheres $x^2 + y^2 + z^2 + 2x + 2y + 2z - 2 = 0$, $x^2 + y^2 + z^2 + 4y = 0$, $x^2 + y^2 + z^2 + 3x - 2y + 8z - 6 = 0$.

    Ans.: $x - y + z - 1 = 0$, $3x - 6y + 8z - 6 = 0$

    $x - 4y + 6z + 4 = 0$

11. Find the equation of the radical line of the spheres $(x - 2)^2 + y^2 + z^2 = 1$, $x^2 + (y - 3)^2 + z^2 = 6$ and $(x + 2)^2 + (y + 1)^2 + (z - 2)^2 = 6$.

    Ans.: $\dfrac{x}{3} = \dfrac{y}{2} = \dfrac{z}{7}$

12. Find the equation of the radical line of the spheres $x^2 + y^2 + z^2 + 2x + 2y + 2z + 2 = 0$, $x^2 + y^2 + z^2 + 4y = 0$, $x^2 + y^2 + z^2 + 3x - 2y + 8z + 6 = 0$.

Ans.: $x - y + z + 1 = 0,\ 3x - 6y + 8z + 6 = 0$

13. Find the radical plane of the spheres $x^2 + y^2 + z^2 - 8x + 4y + 4z + 12 = 0$, $x^2 + y^2 + z^2 - 6x + 3y + 3z + 9 = 0$.

Ans.: $2x - y - z - 3 = 0$

14. Find the spheres coaxal with the spheres $x^2 + y^2 + z^2 + 2x + y + 3z - 8 = 0$ and $x^2 + y^2 + z^2 - 5 = 0$ and touching the plane $3x + 4y = 15$.

Ans.: $5(x^2 + y^2 + z^2) - 8x - 4y - 12z - 13 = 0$

15. Find the limiting points of the coaxal system defined by the spheres $x^2 + y^2 + z^2 + 3x - 3y + 6 = 0$, $x^2 + y^2 + z^2 - 6y - 6z + 6 = 0$.

Ans.: $(-1, 2, 1), (-2, 1, -1)$

16. Find the limiting points of the coaxal system determined by the two spheres whose equations are $x^2 + y^2 + z^2 - 8x + 2y - 2z + 32 = 0$, $x^2 + y^2 + z^2 - 7x + z + 23 = 0$.

Ans.: $(3, 1, -2), (5, -3, 4)$

17. Find the equations of the spheres whose limiting points are $(-1, 2, 1)$ and $(-2, 1, -1)$ and which touches the plane $2x + 3y + 6z + 7 = 0$.

18. Find the equation of the sphere which touches the plane $3x + 2y - z + 2 = 0$ at the point $(1, -2, 1)$ and also cut orthogonally the sphere $x^2 + y^2 + z^2 - 4x + 6y + 4 = 0$.

Ans.: $x^2 + y^2 + z^2 + 7x + 10y - 5z + 12 = 0$

19. Find the limiting points of the coaxal system two of whose members are $x^2 + y^2 + z^2 - 3x - 3y + 6 = 0$, $x^2 + y^2 + z^2 - 4y - 6z + 6 = 0$.

Ans.: $(2, -3, 4)$ and $(-2, 3, -4)$

20. The point $(-1, 2, 1)$ is a limiting point of a coaxal system of spheres of which the sphere $x^2 + y^2 + z^2 + 3x - 2y + 6 = 0$ is a member. Find the coordinates of the other limiting point.

Ans.: $(-2, 1, -1)$

CHAPTER

15 Cone

15.1 DEFINITION OF CONE

A cone is a surface generated by a straight line which passes through a fixed point and intersects a fixed curve or touches a given curve.

The fixed point is called the *vertex* of the cone and the fixed curve is called a *guiding curve* of the cone. The straight line is called a generator.

15.2 EQUATION OF A CONE WITH A GIVEN VERTEX AND A GIVEN GUIDING CURVE

Let (α, β, γ) be the given vertex and $ax^2 + 2hxy + by^2 + 2gx + 2fy + c = 0$, $z = 0$ be the guiding curve.

The equations of any line passing through the point (α, β, γ) are

$$\frac{x-\alpha}{l} = \frac{y-\beta}{m} = \frac{z-\gamma}{n} \tag{15.1}$$

When this line meets the plane at $z = 0$ we get,

$$\frac{x-\alpha}{l} = \frac{y-\beta}{m} = \frac{-\gamma}{n}$$

$$\therefore x = \alpha - \frac{l\gamma}{n},\ y = \beta - \frac{m\gamma}{n},\ z = 0$$

This point lies on the given curve $ax^2 + 2hxy + by^2 + 2gx + 2fy + c = 0, z = 0$.

$$\therefore a\left(\alpha - \frac{l\gamma}{n}\right)^2 + 2h\left(\alpha - \frac{l\gamma}{n}\right)\left(\beta - \frac{m\gamma}{n}\right) + b\left(\beta - \frac{m\gamma}{n}\right)^2 +$$

$$2g\left(\alpha - \frac{l\gamma}{n}\right) + 2f\left(\beta - \frac{m\gamma}{n}\right) + c = 0 \tag{15.2}$$

Eliminating l, m, n from (15.1) and (15.2) we get the equation of the cone.

From (15.1) $\quad \dfrac{l}{n} = \dfrac{x-\alpha}{z-\gamma}$ and $\dfrac{m}{n} = \dfrac{y-\beta}{z-\gamma}$

$$a\left(\alpha - \frac{x-\alpha}{z-\gamma}\gamma\right)^2 + 2h\left(\alpha - \frac{x-\alpha}{z-\gamma}\gamma\right)\left(\beta - \frac{y-\beta}{z-\gamma}\gamma\right) + \beta\left(\beta - \frac{y-\beta}{z-\gamma}\gamma\right)^2$$
$$+2g\left(\alpha - \frac{x-\alpha}{z-\gamma}\gamma\right) + 2f\left(\beta - \frac{y-\beta}{z-\gamma}\gamma\right) + c = 0$$

Multiplying throughout by $(z-\gamma)^2$, we get

$a(\alpha z - \gamma x)^2 + 2h(\alpha z - \gamma x)(\beta z - \gamma y) + b(\beta z - \gamma y)^2 + 2g(\alpha z - \gamma x)(z-\gamma) + zf(\beta z - \gamma y)(z-\gamma) + c(z-\gamma)^2 = 0.$

This is the required equation of the cone.

Example 15.2.1

Find the equation of the cone with its vertex at (1, 1, 1) and which passes through the curve $x^2 + y^2 = 4$, $z = 2$.

Solution

Let V be the vertex of the cone and P be any point on the surface of the cone. Let the equations of the generator VP be

$$\frac{x-1}{l} = \frac{y-1}{m} = \frac{z-1}{n} \tag{15.3}$$

This line intersects the plane $z = 2$.

$$\therefore \quad \frac{x-1}{l} = \frac{y-1}{m} = \frac{1}{n}$$

$$\therefore \quad x = 1 + \frac{l}{n},\ y = 1 + \frac{m}{n}$$

This point lies on the curve $x^2 + y^2 = 4$.

$$\therefore \quad \left(1 + \frac{l}{n}\right)^2 + \left(1 + \frac{m}{n}\right)^2 = 4$$

or

$$(l+n)^2 + (m+n)^2 = 4n^2 \tag{15.4}$$

Eliminating l, m, n from (15.3) and (15.4) we get

$$\left(1+\frac{x-1}{z-1}\right)^2+\left(1+\frac{y-1}{z-1}\right)^2=4$$

or

$$(z-1+x-1)^2+(z-1+y-1)^2=4(z-1)^2$$

$$(x+z-2)^2+(y+z-2)^2=4(z-1)^2$$

This is required equation of the cone.

Example 15.2.2

Find the equation of the cone whose vertex is (a, b, c) and whose base is the curve $\alpha x^2+\beta y^2=1, z=0$.

Solution

The vertex of the cone is $V(a, b, c)$.

The guiding curve is

$$\alpha x^2+\beta y^2=1, z=0 \tag{15.5}$$

Let l, m, n be the direction cosines of the generator VP.

Then the equations of VP are

$$\frac{x-a}{l}=\frac{y-b}{m}=\frac{z-c}{n} \tag{15.6}$$

When this line meets $z=0$ we have

$$\frac{x-a}{l}=\frac{y-b}{m}=\frac{-c}{n}$$

$$\therefore x=a-\frac{cl}{n},\ y=b-\frac{cm}{n}$$

The point lies on the curve $\alpha x^2+\beta y^2=1$.

$$\therefore \alpha\left(a-\frac{cl}{n}\right)^2+\beta\left(b-\frac{cm}{n}\right)^2=1 \tag{15.7}$$

We have to eliminate l, m, n from (15.6) and (15.7)

$$\alpha\left(a-c\frac{x-a}{z-c}\right)^2+\beta\left(b-c\frac{y-b}{z-c}\right)^2=1$$

or $\alpha[a(z-c)-c(x-a)]^2+\beta[b(z-c)-c(y-b)]^2=(z-c)^2$

(i.e.) $\alpha c^2x^2+\beta c^2y^2+(\alpha a^2+\beta b^2-1)^2z^2-2bc\beta yz-2ac\alpha xz+2zc-c^2=0$

This is the equation of the required cone.

15.3 EQUATION OF A CONE WITH ITS VERTEX AT THE ORIGIN

To show that the equation of a cone with its vertex at the origin is homogeneous, let

$$f(x, y, z)=0 \tag{15.8}$$

be the equation of a cone with its vertex at the origin.

Let $P(\alpha, \beta, \gamma)$ be any point on the surface. Then, OP is a generator of the cone. Since (α, β, γ) lies on the cone

$$f(x, y, z)=0 \tag{15.9}$$

The equations of OP are

$$\frac{x}{\alpha}=\frac{y}{\beta}=\frac{z}{\gamma}=t \tag{15.10}$$

Any point on this line is $(t\alpha, t\beta, t\gamma)$.
The point lies on the cone $f(x, y, z)=0$.

$$\therefore \quad f(t\alpha, t\beta, t\gamma)=0 \tag{15.11}$$

From equations (15.8) and (15.11), we observe that the equation $f(x, y, z)=0$ is homogeneous.

Conversely, every homogeneous equation in (x, y, z) represents a cone with its vertex at the origin.

Let $f(x, y, z)=0$ be a homogeneous equation in x, y, z.
Since $f(x, y, z)=0$ is a homogeneous equation, $f(x, y, z)=0$ for any real number. In particular $f(0, 0, 0)=0$.

Therefore, the origin lies on the locus of the equation (15.8). As $f(tx, ty, tz)=0$, any point on the line through the origin lies on the equation (15.8). In other words, the locus of (15.8) is a surface generated by the line through the origin. Hence equation represents a cone with its vertex at the origin.

Note 15.3.1: If $f(x, y, z)$ can be expressed as the product of n linear factors then $f(x, y, z)=0$ represents n planes through the origin.

Note 15.3.2: If $f(x, y, z) = 0$ is a homogeneous equation of second degree in x, y and z then $f(x, y, z) = 0$ is a quadric cone. If it can be factored into two linear factors then it represents a pair of planes through the origin; then we regard equation $f(x, y, z) = 0$ as degenerate cone, the vertex being any point on the line of intersection of the two planes.

Generators: The line $\frac{x}{l} = \frac{y}{m} = \frac{z}{n}$ is a generator of the cone $f(x, y, z) = 0$ with its vertex at the origin if and only if $f(l, m, n) = 0$.

Let $\frac{x}{l} = \frac{y}{m} = \frac{z}{n} = r$ be a generator of the cone $f(x, y, z) = 0$ then the point (lr, mr, nr) lies on the cone. Taking $r = 1$, the point (l, m, n) lies on the cone $f(x, y, z) = 0$.

$$\therefore \quad f(l, m, n) = 0$$

Converse: Let $f(x, y, z) = 0$ be the equation of the cone with its vertex at the origin and $\frac{x}{l} = \frac{y}{m} = \frac{z}{n} = r$ be a line through the origin such that $f(l, m, n) = 0$.

Since the vertex is at the origin, $f(x, y, z) = 0$ is a homogeneous equation in x, y and z.

Now we will prove that $\frac{x}{l} = \frac{y}{m} = \frac{z}{n}$ is a generator of the cone.

Any point on the generator is (lr, mr, nr).

Since $f(x, y, z) = 0$ and $f(l, m, n) = 0$, it follows that $f(lr, mr, nr) = 0$.

$$\therefore \frac{x}{l} = \frac{y}{m} = \frac{z}{n} \text{ is a generator of the cone } f(x, y, z) = 0.$$

Example 15.3.1

Find the equation of the cone with its vertex at the origin and which passes through the curve $ax^2 + by^2 + cz^2 - 1 = 0 = \alpha x^2 + \beta y^2 - 2z$.

Solution

Let the equation of the generator be

$$\frac{x}{l} = \frac{y}{m} = \frac{z}{n} = r \tag{15.12}$$

Any point on this line is (lr, mr, nr). This point lies on the curve

$$ax^2 + by^2 + cz^2 = 1$$

$$\alpha x^2 + \beta y^2 - 2z = 0$$

$$r^2(al^2 + bm^2 + cn^2) = 1 \tag{15.13}$$

$$r(\alpha rl^2 + \beta rm^2 - 2n) = 0 \tag{15.14}$$

From (15.14), $$r = \frac{2n}{\alpha l^2 + \beta m^2}$$

Substituting this in (15.13) we get

$$\frac{4n^2}{(\alpha l^2 + \beta m^2)^2}(al^2 + bm^2 + cn^2) = 1$$

$$\text{(i.e.)} \quad 4n^2(al^2 + bm^2 + cn^2) = (\alpha l^2 + \beta m^2)^2$$

As l, m, n are proportional to x, y, z the equation of the cone is $4z^2(ax^2 + by^2 + cz^2) = (\alpha x^2 + \beta y^2)^2$.

Example 15.3.2

Find the equation of the cone whose vertex is at the origin and the guiding curve is $\frac{x^2}{4} + \frac{y^2}{9} + z^2 = 1,\ x + y + z = 1.$

Solution

Since the vertex of the cone is the origin its equation must be a homogeneous equation of second degree.

The equations of the guiding curve are

$$\frac{x^2}{4} + \frac{y^2}{9} + z^2 = 1 \tag{15.15}$$

$$x + y + z = 1 \tag{15.16}$$

Homogenizing the equation (15.15) with the help of (15.16) we get the equation of the required cone. Hence the equation of the cone is $\frac{x^2}{4} + \frac{y^2}{9} + z^2 = (x + y + z)^2.$

$$\text{(i.e.)} \quad 27x^2 + 32y^2 + 7z^2\ (xy + yz + zx) = 0$$

Example 15.3.3

The plane $\frac{x}{a}+\frac{y}{b}+\frac{z}{c}=1$ meets the coordinate axes at A, B and C. Prove that the equation of the cone generated by lines drawn from O to meet the circle ABC is $yz\left(\frac{b}{c}+\frac{c}{b}\right)+zx\left(\frac{c}{a}+\frac{a}{c}\right)+xy\left(\frac{a}{b}+\frac{b}{a}\right)=0.$

Solution

The points A, B, C are $(a, 0, 0)$, $(0, b, 0)$ and $(0, 0, c)$, respectively.

The equation of the sphere $OABC$ is $x^2+y^2+z^2-ax-by-cz=0$.

The equations of the circle ABC are

$$x^2+y^2+z^2-ax-by-cz=0 \tag{15.17}$$

$$\frac{x}{a}+\frac{y}{b}+\frac{z}{c}=1 \tag{15.18}$$

Homogenizing equation (15.17) with the help of (15.18) we get the equation of the required cone.

$$\text{(i.c.)}\quad x^2+y^2+z^2-(ax+by+cz)\left(\frac{x}{a}+\frac{y}{b}+\frac{z}{c}\right)=0$$

$$\text{(i.e.)}\quad \left(\frac{b}{c}+\frac{c}{b}\right)yz+\left(\frac{c}{a}+\frac{a}{c}\right)zx+\left(\frac{a}{b}+\frac{b}{a}\right)xy=0$$

15.4 CONDITION FOR THE GENERAL EQUATION OF THE SECOND DEGREE TO REPRESENT A CONE

Let the general equation of the second degree be

$$ax^2+by^2+cz^2+2fyz+2gzx+2hxy+2ux+2vy+2wz+d=0 \tag{15.19}$$

Let (x_1, y_1, z_1) be the vertex of the cone. Shift the origin to the point (x_1, y_1, z_1). Then

$$x=X+x_1$$

$$y=Y+y_1$$

$$z=Z+z_1$$

Then the equation (15.19) becomes

$a(X+x_1)^2+b(Y+y_1)^2+c(Z+z_1)^2+2f(Y+y_1)(Z+z_1)+2g(Z+z_1)(X+x_1)+2h(X+x_1)(Y+y_1)+2u(X+x_1)+2v(Y+y_1)+2w(Z+z_1)+d=0$

$$\text{(i.e.)}\quad aX^2+bY^2+cZ^2+2fYZ+2gZX+2hXY+2(ax_1+hy_1+gz_1+u)X$$
$$+2(hx_1+by_1+fz_1+v)Y+2(gx_1+fy_1+cz_1+w)Z+2ux_1+2vy_1$$
$$+2wz_1+(ax_1^2+by_1^2+cz_1^2+2fy_1z_1+2gz_1x_1+2hx_1y_1+d)=0$$

Since this equation has to be a homogeneous equation in X, Y and Z.
Coefficient of $X=0$
Coefficient of $Y=0$
Coefficient of $Z=0$ and constant term $=0$.

$$\therefore\quad ax_1+hy_1+gz_1+u=0 \tag{15.20}$$

$$hx_1+by_1+fz_1+v=0 \tag{15.21}$$

$$gx_1+fy_1+cz_1+w=0 \tag{15.22}$$

$$ax_1^2+by_1^2+cz_1^2+2fy_1z_1+2gz_1x_1+2hx_1y_1+2ux_1+2vy_1+2wz_1+d=0 \tag{15.23}$$

(i.e.) $x_1(ax_1+hy_1+gz_1+u)+y_1(hx_1+by_1+fz_1+v)+z_1(gx_1+fy_1+cz_1+w)+ux_1+vy_1+wz_1+d=0$

$$=0+0+0+ux_1+vy_1+wz_1+d=0$$

$$\therefore\quad ux_1+vy_1+wz_1+d=0 \tag{15.24}$$

Eliminating x, y, z from (15.21), (15.22), (15.23) and (15.24), we get,

$$\begin{vmatrix} a & h & g & u \\ h & b & f & v \\ g & f & c & w \\ u & v & w & d \end{vmatrix}=0$$

This is the required condition.

Note 15.4.1: If the given equation of the second degree is $f(x, y, z)=0$ then make it homogeneous by introducing the variable t where $t=1$.

Then

$$\frac{\partial F}{\partial x} = 0, \frac{\partial F}{\partial y} = 0, \frac{\partial F}{\partial z} = 0, \frac{\partial F}{\partial t} = 0 \text{ where } t = 1$$

Solving any three of these four equations, we get the vertex of the cone. Test whether these values of x, y, z satisfy the fourth equation.

Example 15.4.1

Find the equation of the cone of the second degree which passes through the axes.

Solution

The cone passes through the axes. Therefore, the vertex of the cone is the origin.

The equations of the cone is a homogeneous equation of second degree in x, y and z.

$$\text{(i.e.)} \quad ax^2 + by^2 + cz^2 + 2fyz + 2gzx + 2hxy = 0 \tag{15.25}$$

Given that x-axis is a generator.

Then $y = 0, z = 0$ must satisfy the equation (15.25)

$$\therefore \quad a = 0$$

Since y-axis is a generator $b = 0$.
Since z-axis is a generator $c = 0$.
Hence the equation of the cone is $fyz + gzx + hxy = 0$.

Example 15.4.2

Show that the lines through the point (α, β, γ) whose direction cosines satisfy the relation $al^2 + bm^2 + cn^2 = 0$, generates the cone $a(x - \alpha)^2 + b(y - \beta)^2 + c(z - \gamma)^2 = 0$.

Solution

The equations of any line through (α, β, γ) are

$$\frac{x}{l} = \frac{y}{m} = \frac{z}{n} \tag{15.26}$$

where

$$al^2 + bm^2 + cn^2 = 0 \tag{15.27}$$

Eliminating l, m, n from (15.26) and (15.27)

we get,

$$a(x-\alpha)^2 + b(y-\beta)^2 + c(z-\gamma)^2 = 0$$

Example 15.4.3

Find the equation to the quadric cone which passes through the three co-ordinate axes and the three mutually perpendicular lines $\frac{x}{1} = \frac{y}{-2} = \frac{z}{3}$, $\frac{x}{1} = \frac{y}{-1} = \frac{z}{-1}$ and $\frac{x}{5} = \frac{y}{4} = \frac{z}{1}$.

Solution

We have seen that the equation of the cone passing through the axes is

$$fyz + gzx + hxy = 0 \tag{15.28}$$

This cone passes through line $\frac{x}{1} = \frac{y}{-2} = \frac{z}{3}$.

$$\therefore \quad f(-6) + g(3) + h(-2) = 0$$

or

$$6f - 3g + 2h = 0 \tag{15.29}$$

Since the cone also passes through the line $\frac{x}{1} = \frac{y}{-1} = \frac{z}{-1}$ we have

$$f - g - h = 0 \tag{15.30}$$

From (15.29) and (15.30) we get

$$\frac{f}{5} = \frac{g}{8} = \frac{h}{-3} \tag{15.31}$$

From (15.28) and (15.31) we get $5yz + 8zx - 3xy = 0$.

Since the cone passes through the line $\frac{x}{5} = \frac{y}{4} = \frac{z}{1}$,

we get $4f + 5g + 20h = 0$ and $4(5) + 5(8) + 20(-3) = 0$ is also true.

Therefore, the equation of the required cone is $5yz + 8zx - 3xy = 0$.

Example 15.4.4

Prove that the equation $2x^2 + 2y^2 + 7z^2 - 10yz - 10zx + 2x + 2y + 26z - 17 = 0$ represents a cone whose vertex is (2, 2, 1).

Solution

Let $F(x, y, z, t) = 2x^2 + 2y^2 + 7z^2 - 10yz - 10zx - 2xt + 2yt + 26zt - 17t^2 = 0$

$$\frac{\partial F}{\partial x} = 4x - 10z + 2t$$

$$\frac{\partial F}{\partial y} = 4y - 10z + 2t$$

$$\frac{\partial F}{\partial z} = 14z - 10y - 10x + 26t$$

$$\frac{\partial F}{\partial t} = 2x + 2y + 26z - 34t$$

at $t = 1$, $\frac{\partial F}{\partial x} = 0, \frac{\partial F}{\partial y} = 0, \frac{\partial F}{\partial z} = 0$ and $\frac{\partial F}{\partial t} = 0$

give the equations

$$2x - 5z + 1 = 0$$

$$2y - 5z + 1 = 0$$

$$5x + 8y - 7z - 13 = 0$$

$$x + y + 12z - 17 = 0$$

Solving the first three equations we get $x = 2, y = 2, z = 1$.
These values also satisfy the fourth equation.
Therefore, the given equation represents a cone with its vertex at the point (2, 2, 1).

Exercises 1

1. Find the equation of the cone whose vertex is at the origin and which passes through the curve of intersection of the plane $lx + my + nz = p$ and the surface, $ax^2 + by^2 + cz^2 = 1$.

   Ans.: $p^2(ax^2 + by^2 + cz^2) = (lx + my + nz)^2$

2. Find the equation of the cone whose vertex is (α, β, γ) and whose guiding curve is the parabola $y^2 = 4ax$, $z = 0$.

Ans.: $(ry - \beta z)^2 = 4a(z - \gamma)(\alpha z - rx)$

3. Prove that the line $\frac{x}{l} = \frac{y}{m} = \frac{z}{n}$ where $2l^2 + 3m^2 - 5n^2 = 0$ is a generator of the cone $2x^2 + 3y^2 - 5z^2 = 0$.

4. Find the equation of the cone whose vertex is at the point (1, 1, 0) and whose guiding curve is $x^2 + z^2 = 4$, $y = 0$.

Ans.: $x^2 - 3y^2 + z^2 - 2xy + 8y - 4 = 0$

5. Find the equation of the cone whose vertex is the point (0, 0, 1) and whose guiding curve is the ellipse $\frac{x^2}{15} + \frac{y^2}{9} = 1$, $z = 3$. Also obtain section of the cone by the plane $y = 0$ and identify its type.

Ans.: $36x^2 + 100y^2 - 225z^2 + 450z - 225 = 0$

$x = \pm\frac{5}{2}(z-1)$; pair of straight lines

6. Find the equations of the cones with vertex at the origin and passing through the curves of intersection given by the equations:

(i) $ax^2 + by^2 = 22$, $lx + my + nz = 1$

(ii) $\frac{x^2}{a^2} + \frac{y^2}{b^2} + \frac{z^2}{c^2} = 1, \frac{x^2}{\alpha^2} + \frac{y^2}{\beta^2} = 2z$

(iii) $x^2 + y^2 + z^2 + x - 2y + 3z - 4 = 0$
$x + y + z = 2$

(iv) $x^2 + y^2 + z^2 + x - 2y + 3z = 4$

(v) $x^2 + y^2 + z^2 + 2x - 3y + 4 = 5$

Ans.: (i) $p(ax^2 + by^2) = 2z(lx + my + nz)$

(ii) $4z^2\left(\frac{x^2}{a^2} + \frac{y^2}{b^2} + \frac{z^2}{c^2}\right) = \left(\frac{x^2}{\alpha^2} + \frac{y^2}{\beta^2}\right)^2$

(iii) $2x^2 + 3y^2 + 4z^2 - xy - 3yz - 2zx = 0$

(iv) $8x^2 + y^2 - 5xy - 3yz - 4zx = 0$

7. The plane $x + y + z = 1$ meets the coordinate axes in A, B and C. Prove that the equation to the cone generated by the lines through O, to meet the circle ABC is $yz + zx + xy = 0$.

8. A variable plane is parallel to the plane $\frac{x}{a}+\frac{y}{b}+\frac{z}{c}=0$ and meets the axes in A, B and C. Prove that, the circle ABC lies on the cone $yz\left(\frac{b}{c}+\frac{c}{a}\right)+zx\left(\frac{c}{a}+\frac{a}{c}\right)+xy\left(\frac{a}{b}+\frac{b}{a}\right)=0.$

9. **i) Find the equation of the quadric cone which passes through the three coordinate axes and three mutually perpendicular lines** $\frac{x}{2}=\frac{y}{1}=\frac{z}{-1}, \frac{x}{1}=\frac{y}{3}=\frac{z}{5}, \frac{x}{8}=\frac{y}{-11}=\frac{z}{5}.$

Ans.: $16yz-33zx-25xy=0$

ii) Prove that the equation of the cone whose vertex is (0, 0, 0) and the base curve $z=k, f(x,y)=0$ is $f\left(\frac{xk}{z},\frac{yk}{z}\right)=0$, where $f(x,y)=ax^2+2hxy+by^2+2gx+2fy+c=0$.

10. Find the equation to the cone whose vertex is the origin and the base circle $x=a, y^2+z^2=b^2$ and show that the section of the cone by a plane parallel to the xy-plane is hyperbola.

Ans.: $a^2(y^2+z^2)=b^2x^2$

11. Planes through OX and OY include an angle α. Show that the line of intersection lies on the cone $z^2(x^2+y^2+z^2)=x^2y^2\tan^2\alpha$.

12. Prove that a cone of second degree can be found to pass through two sets of rectangular axes through the same origin.

13. Prove that the equation $x^2-2y^2+3z^2+5yz-6zx-4xy+8x-19y-2z-20=0$ represents a cone with its vertex at $(1, -2, 3)$.

14. Prove that the equation $2y^2-8yz-4zx-8xy+6x-4y-2z+5=0$ represents a cone whose vertex is $\left(\frac{-7}{6},\frac{1}{3},\frac{5}{6}\right)$.

15. Prove that the equation $ax^2+by^2+cz^2+2ux+2vy+2wz+d=0$ represents a cone if $\frac{u^2}{a}+\frac{v^2}{b}+\frac{w^2}{c}=d$.

15.5 RIGHT CIRCULAR CONE

A right circular cone is a surface generated by a straight line which passes through a fixed point, and makes a constant angle with a fixed straight line through the fixed point. The fixed point is called the *vertex* of the cone and the constant angle is called the *semi-vertical angle* and fixed straight line is called the *axis of the cone.*

The section of right circular cone by any plane perpendicular to its axis is a circle.

15.5.1 Equation of a Right Circular Cone with Vertex $V(\alpha, \beta, \gamma)$, Axis VL with Direction Ratios l, m, n and Semi-vertical Angle θ

Let $P(x, y, z)$ be any point on the surface of the cone.

Then direction ratios of VP are $x - \alpha, y - \beta, z - \gamma$.

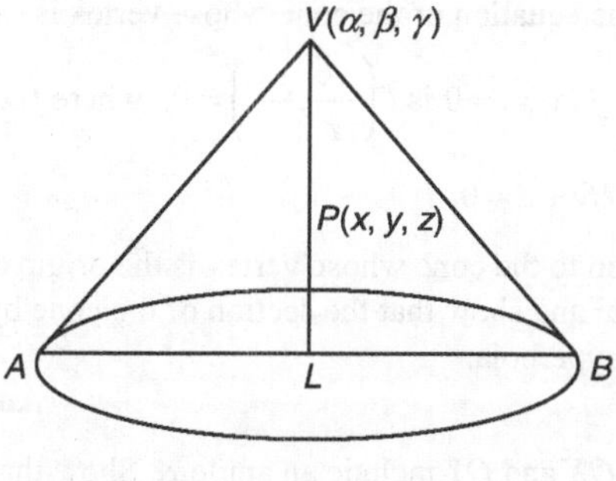

The direction ratios of the perpendicular VL are l, m, n.

Now, $$\cos\theta = \frac{(x-\alpha)l + (y-\beta)m + (z-\gamma)n}{\sqrt{(x-\alpha)^2 + (y-\beta)^2 + (z-\gamma)^2}\sqrt{l^2 + m^2 + n^2}}$$

(i.e.) $$[(x - \alpha)l + (y - \beta)m + (z - \gamma)n]^2 = [(x - \alpha)^2 + (y - \beta)^2 + (z - \gamma)^2](l^2 + m^2 + n^2)\cos^2\theta \quad (15.32)$$

This is the required equation of the cone.

Note 15.5.1.1:

(i) If the vertex is at the origin then the equation of the cone becomes $(lx + my + nz)^2 = [(x^2 + y^2 + z^2)(l^2 + m^2 + n^2)]\cos^2\theta$.

(ii) If l, m, n are the direction cosines of the line then

$$(lx + my + nz)^2 = (x^2 + y^2 + z^2)\cos^2\theta \quad (15.33)$$

(iii) If axis of cone is the z-axis then the equation (15.33) becomes

$$z^2 = (x^2 + y^2 + z^2)\cos^2\theta$$

$$z^2 \sec^2\theta = x^2 + y^2 + z^2$$

or $$x^2 + y^2 = z^2(\sec^2\theta - 1)$$

$$x^2 + y^2 = z^2 \tan\theta$$

15.5.2 Enveloping Cone

It has been seen in the two-dimensional analytical geometry that two tangents can be drawn from a given point to a conic. In analogy with that an infinite number of tangent lines can be drawn from a given point to a conicoid, in particular to a sphere. All such tangent lines generate a cone with the given point as vertex. Such a cone is called an enveloping cone.

Definition 15.5.2.1: The locus of tangent lines drawn from a given point to a given surface is called an enveloping cone of the surface. The given point is called the vertex of the cone.

Equation of enveloping cone: Let us find the equation of the enveloping cone of the sphere $x^2 + y^2 + z^2 = a^2$ with the vertex at (x, y, z).

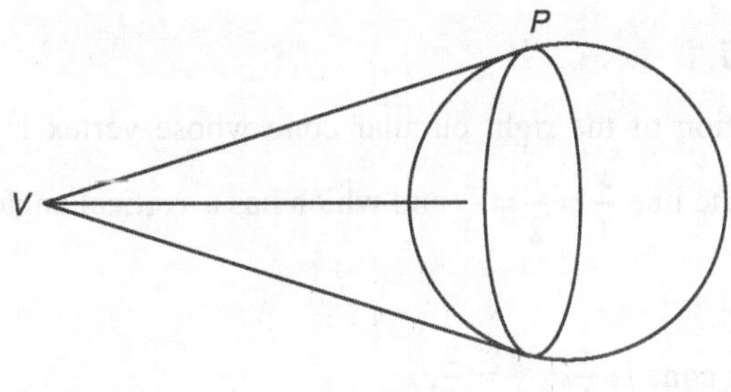

Let $P(x, y, z)$ be any point on the tangent drawn from $V(x_1, y_1, z_1)$ to the given sphere. Let Q be the point that divides PQ in the ratio $1:\lambda$. Then the coordinates of Q are $\left(\dfrac{\lambda x + x_1}{1+\lambda}, \dfrac{\lambda y + y_1}{1+\lambda}, \dfrac{\lambda z + z_1}{1+\lambda}\right)$.

If this point lies on the sphere then,

$$\left(\frac{\lambda x + x_1}{1+\lambda}\right)^2 + \left(\frac{\lambda y + y_1}{1+\lambda}\right)^2 + \left(\frac{\lambda z + z_1}{1+\lambda}\right)^2 = a^2$$

$$\text{(i.e.)} \quad (\lambda x + x_1)^2 + (\lambda y + y_1)^2 + (\lambda z + z_1)^2 = a^2(1+\lambda)^2 \text{ or}$$

$$(x^2 + y^2 + z^2 - a^2)\lambda^2 + 2(xx_1 + yy_1 + zz_1 - a^2)\lambda + (x_1^2 + y_1^2 + z_1^2 - a^2) = 0 \tag{15.34}$$

This is a quadratic equation in λ. There are two values of λ indicating that there are two points on VP which divides PQ in the ratio $1:\lambda$ and lie on the sphere. If PQ is a tangent to the sphere then these two points coincide and the point is the point of contact.

Therefore, the two values of λ of equation (15.34) must be equal and hence the discriminant must be zero.

$\therefore (xx_1 + yy_1 + zz_1 - a^2)^2 = (x^2 + y^2 + z^2 - a^2)(x_1^2 + y_1^2 + z_1^2 - a^2)$. This is the equation of the enveloping cone.

Note 15.5.2.2: The equation of the enveloping cone can be expressed in the form

$$T^2 = SS_1 \text{ where}$$
$$T = xx_1 + yy_1 + zz_1 - a^2$$
$$S = x^2 + y^2 + z^2 - a^2$$
$$S_1 = x_1^2 + y_1^2 + z_1^2 - a^2$$

Example 15.5.1

Find the equation of the right circular cone whose vertex is at the origin, whose axis is the line $\frac{x}{1} = \frac{y}{2} = \frac{z}{3}$ and which has a vertical angle of 60°.

Solution

The axis of the cone is $\frac{x}{1} = \frac{y}{2} = \frac{z}{3}$.

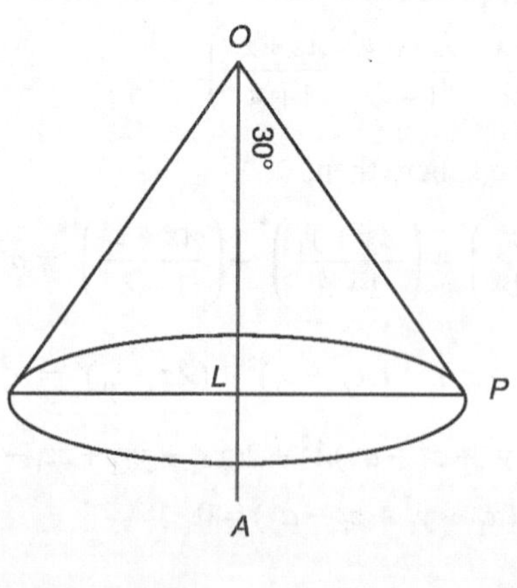

Therefore, the direction ratios of the axis of the cone are 1, 2, 3.

The direction cosines are $\frac{1}{\sqrt{14}}, \frac{2}{\sqrt{14}}, \frac{3}{\sqrt{14}}$.

Let $P(x, y, z)$ be any point on the surface of the cone.
Let PL be perpendicular to OA.

$$\angle POL = 30°$$

$$\frac{OL}{OP} = \cos 30° \text{ or } 2OL = \sqrt{3}OP$$

Also, $$OP^2 = x^2 + y^2 + z^2$$

$$OL = \text{Projection of } OP \text{ on } OA$$

$$= \frac{x}{\sqrt{14}} + y \times \frac{2}{\sqrt{14}} + z \times \frac{3}{\sqrt{14}} = \frac{x+2y+3z}{\sqrt{14}}$$

$$\therefore \frac{2(x+2y+3z)}{\sqrt{14}} = \sqrt{3}\sqrt{x^2+y^2+z^2}$$

$$4(x+2y+3z)^2 = 42(x^2+y^2+z^2)$$

or $$19x^2 + 13y^2 + 3z^2 - 8xy - 24yz - 12zx = 0$$

Example 15.5.2

Find the equation of the right circular cone with its vertex at the origin, axis along the z-axis and semi-vertical angle α.

Solution

The direction cosines of the axis of the cone are 0, 0, 1.

Let $P(x, y, z)$ be any point on the cone.

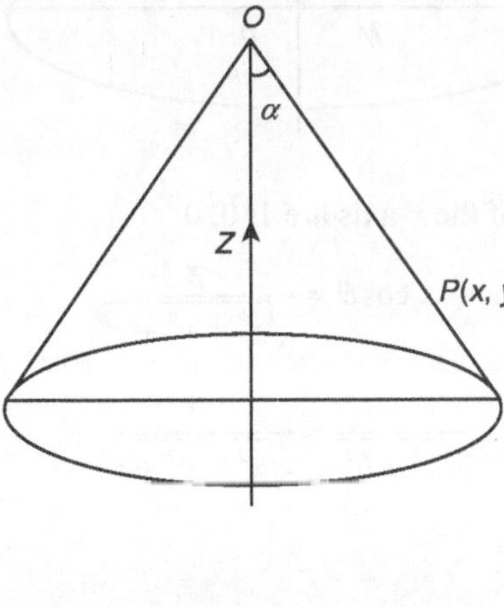

Then, $\cos\alpha = \dfrac{z}{\sqrt{x^2+y^2+z^2}}$ or $(x^2+y^2+z^2)\cos^2\alpha = z^2$ or $x^2+y^2 = z^2\tan^2\alpha$

This is the required equation of the cone.

Example 15.5.3

Find the semi-vertical angle and the equation of the right circular cone having its vertex at origin and passing through the circle $y^2 + z^2 = b^2, x = a$.

Solution

The guiding circle of the right circular cone is $y^2 + z^2 = b^2, x = a$.

Therefore, the axis of the cone is along x-axis.

If θ is the semi-vertical angle, then $\cos\theta = \dfrac{a}{\sqrt{a^2+b^2}}$.

Let $P(x, y, z)$ be any point on the surface of the cone.

The direction ratios of OP are x, y, z.

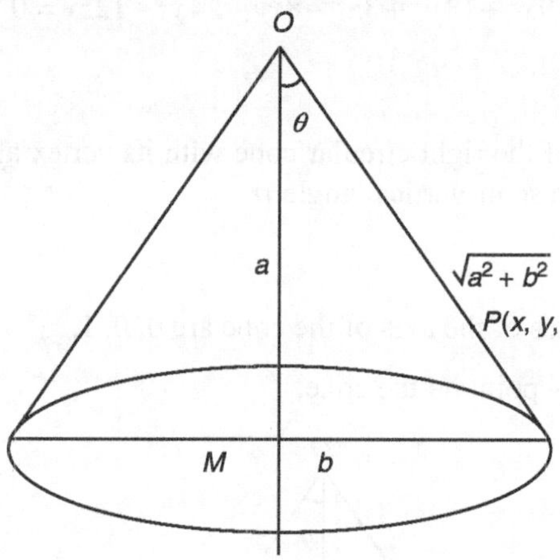

The direction cosines of the x-axis are 1, 0, 0.

$$\therefore \cos\theta = \frac{x}{\sqrt{x^2+y^2+z^2}}$$

or
$$\frac{a}{\sqrt{a^2+b^2}} = \frac{x}{\sqrt{x^2+y^2+z^2}}$$

$$a^2(x^2 + y^2 + z^2) = x^2(a^2 + b^2)$$

$$\text{(i.e.)} \quad b^2x^2 - a^2(y^2 + z^2) = 0$$

which is the required equation of the cone.

Example 15.5.4

A right circular cone has its vertex at (2, −3, 5). Its axis passes through $A(3, -2, 6)$ and its semi-vertical angle is 30°. Find its equation.

Solution

The axis is the line joining the points (2, −3, 5) and (3, −2, 6).

Therefore, its equations are $\dfrac{x-2}{1} = \dfrac{y+3}{1} = \dfrac{z-5}{1}$.

The direction ratios of the axes are 1, 1, 1.

The direction cosines are $\dfrac{1}{\sqrt{3}}, \dfrac{1}{\sqrt{3}}, \dfrac{1}{\sqrt{3}}$.

Let $P(x, y, z)$ be any point on the cone.

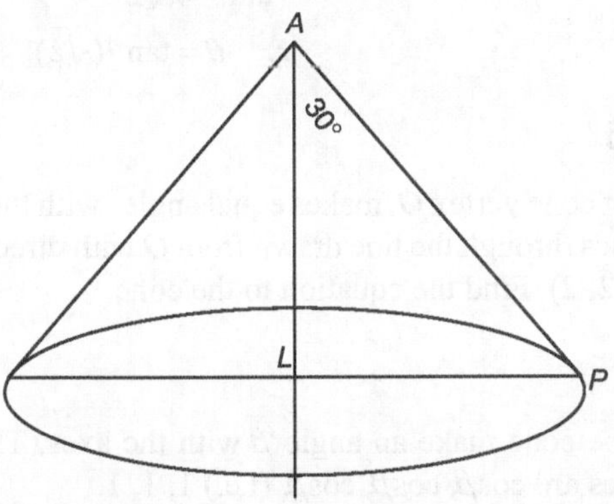

$$\angle PAL = 30°$$

$$\frac{AL}{AP} = \cos 30° = \frac{\sqrt{3}}{2}$$

$$AL = \frac{x-2}{\sqrt{3}} + \frac{y+3}{\sqrt{3}} + \frac{z-5}{\sqrt{3}}$$

$$AP = \sqrt{(x-2)^2 + (y+3)^2 + (z-5)^2}$$

$$\therefore \frac{1}{\sqrt{3}} \frac{(x+y+z-4)}{\sqrt{(x-2)^2+(y+3)^2+(z-5)^2}} = \frac{\sqrt{3}}{2}$$

Squaring, cross multiplying and simplifying we get, $5x^2 + 5y^2 + 5z^2 - 8xy - 8yz - 8zx - 4x + 86y - 58z + 278 = 0$.

Example 15.5.5

A right circular cone has three mutually perpendicular generators. Prove that the semi-vertical angle of the cone is $\tan^{-1}\sqrt{2}$.

Solution

The equation of the right circular cone with vertex at the origin, semi-vertical angle α and axis along z-axis is given by $x^2 + y^2 = z^2 \tan^2\alpha$.

This cone will have three mutually perpendicular generators if coefficient of x^2 + coefficient of y^2 + coefficient of $z^2 = 0$.

$$\therefore \quad 1 + 1 - \tan^2\theta = 0$$

or

$$\tan\theta = \sqrt{2}$$

$$\therefore \quad \theta = \tan^{-1}(\sqrt{2})$$

Example 15.5.6

The axis of a right cone vertex O, makes equal angles with the coordinate axes and the cone passes through the line drawn from O with direction cosines proportional to $(1, -2, 2)$. Find the equation to the cone.

Solution

Let the axis of the cone make an angle β with the axes. Then the direction cosines of the axes are $\cos\beta, \cos\beta, \cos\beta$ (i.e.) 1, 1, 1.

Let α be the semi-vertical angle of the axis of the cone.

The direction ratios of one of the generators are 1, –2, 2.

$$\therefore \cos\alpha \frac{1-2+2}{\sqrt{1+1+1}\sqrt{1+4+4}} = \frac{1}{3\sqrt{3}}$$

Let $P(x, y, z)$ be any point on the cone.

Then the direction ratios of OP are x, y, z.

The direction ratios of the axis are 1, 1, 1.

$$\therefore \cos\alpha = \frac{x+y+z}{\sqrt{3}\sqrt{x^2+y^2+z^2}}$$

$$\frac{1}{3\sqrt{3}} = \frac{x+y+z}{\sqrt{3}\sqrt{x^2+y^2+z^2}}$$

Squaring and cross multiplying we get,

$$9(x+y+z)^2 = x^2+y^2+z^2$$

or $\quad 4x^2 + 4y^2 + 4z^2 + 9xy + 9yz + 9zx = 0$

Example 15.5.7

Prove that the line drawn from the origin so as to touch the sphere $x^2 + y^2 + z^2 + 2ux + 2vy + 2wz + d = 0$ lie on the cone $d(x^2 + y^2 + z^2) = (ux + vy + wz)^2$.

Solution

The lines drawn from the origin to touch the sphere generates the enveloping cone.

The equation of the enveloping cone of the given sphere is $T^2 = SS_1$.

$$\text{(i.e.)} \quad (ux + vy + wz + d)^2 - (x^2 + y^2 + z^2 + 2ux + 2vy + 2wz + d)d$$

$$\text{(i.e.)} \quad d(x^2 + y^2 + z^2) = (ux + vy + wz)^2$$

Example 15.5.8

Show that the plane $z = 0$ cuts the enveloping cone of the sphere $x^2 + y^2 + z^2 = 11$ which has its vertex at (2, 4, 1) in a rectangular hyperbola.

Solution

The equation of the enveloping cone with its vertex at (2, 4, 1) is $T^2 = SS_1$.

$$S = x^2 + y^2 + z^2 - 11$$

$$S_1{}^2 = 4 + 16 + 1 - 11 = 10$$

$$T = 2x + 4y + z - 11$$

$$\therefore (2x + 4y + z - 11)^2 = 10(x^2 + y^2 + z^2 - 11)$$

The section of this cone by the plane $z = 0$ is $(2x + 4y - 11)^2 = 10(x^2 + y^2 - 11)$.
Coefficient of x^2 + coefficient of $y^2 = 6 - 6 = 0$
Hence, the plane $z = 0$ cuts the enveloping cone in a rectangular hyperbola.

Exercises 2

1. Find the equation of the right circular cone whose vertex is the line $\frac{x}{1}=\frac{y}{2}=\frac{z}{3}$ and which has a vertical angle of 60°.

   Ans.: $19x^2+13y^2+3z^2-8xy-24yz-12zx=0$

2. If (*x*, *y*, *z*) is any point on the cone whose vertex is (1, 0, 2) and semi-vertical angle is 30° and the equation to the axis is $\frac{x-1}{1}=\frac{y}{2}=\frac{z-2}{-2}$, show that the equation of the cone is $27[(x-1)^2+y^2+(z-2)^2]=4(x+2y-2z+3)^2$.

3. Find the equation to the right circular cone of semi-vertical angle 30°, whose vertex is (1, 2, 3) and whose axis is parallel to the line $x=y=z$.

   Ans.: $5(x^2+y^2+z^2)-8(yz+zx+xy)+30x+12y-6z-18=0$

4. Find the equation to the right circular cone whose vertex is (3, 2, 1), semi-vertical angle is 30° and axis is the line $\frac{x-3}{4}=\frac{y-2}{1}=\frac{z-1}{3}$.

   Ans.: $7x^2+37y^2+21z^2-16xy-12yz-48zx+38x-88y+126z-32=0$

5. Find the equation of the right circular cone with vertex at (1, –2, –1), semi-vertical angle 60° and axis $\frac{x-1}{3}=\frac{y+1}{-4}=\frac{z+1}{5}$.

   Ans.: $5[(5x+4y+14)^2+(3z-5x+8)^2+(4x+3y+2)^2]$
   $=75[(x-1)^2+(y+2)^2+(z+1)^2]$

6. Find the equation of the right circular cone which passes through the three lines drawn from the origin with direction ratios (1, 2, 2) (2, 1, –2) (2, –2, 1).

   Ans.: $8x^2-5y^2-4z^2+yz+5zx+5xy=0$

7. Lines are drawn through the origin with direction cosines proportional to (1, 2, 2), (2, 3, 6), (3, 4,12). Find the equation of the right circular cone through them. Also find the semi-vertical angle of the cone.

   Ans.: $xy-yz+zx=0;\cos^{-1}\left(\frac{1}{\sqrt{3}}\right)$

8. Find the equation of the cone generated when the straight line $2y+3z=6, x=0$ revolves about the *z*-axis.

   Ans.: $4x^2+4y^2-9z^2+36z-36=0$

9. Find the equation to the right circular cone which has the three coordinate axes as generators.

   Ans.: $xy+yz+zx=0$

10. Find the equation of the right circular cone with its vertex at the point (0, 0, 0), its axis along the y-axis and semi-vertical angle θ.

Ans.: $x^2 + z^2 = y^2 \tan^2\theta$

11. If α is the semi-vertical angle of the right circular cone which passes through the lines ox, oy, $x = y = z$, show that $\cos\alpha = \left(9 - 4\sqrt{3}\right)^{-\frac{1}{2}}$.

12. **Prove that $x^2 + y^2 + z^2 - 2x + 4y + 6z + 6 = 0$ represents a right circular cone whose vertex is the point (1, 2, –3), whose axis is parallel to oy and whose semi-vertical angle is 45°.**

13. **Prove that the semi-vertical angle of a right circular cone which has three mutually perpendicular tangent planes is $\cot^{-1}\sqrt{2}$.**

14. Find the enveloping cone of the sphere $x^2 + y^2 + z^2 - 2x - 2y = 2$ with its vertex at (1, 1, 1).

Ans.: $3x^2 - y^2 + 4zx - 10x + 2y - 4z + 6 = 0$

15. Find the enveloping cone of the sphere $x^2 + y^2 + z^2 = 11$ which has its vertex at (2, 4, 1) and show that the plane $z = 0$ cuts the enveloping cone in a rectangular hyperbola.

15.6 TANGENT PLANE

Tangent plane from the point (x_1, y_1, z_1) to the cone $ax^2 + by^2 + cz^2 + 2fyz + 2gzx + 2hxy = 0$.

The equations of any line through the point (x_1, y_1, z_1) are

$$\frac{x - x_1}{l} = \frac{y - y_1}{m} = \frac{z - z_1}{n} = r \qquad (15.35)$$

Any point on the line is $(x_1 + lr, y_1 + mr, z1 + nr)$.

If this point lies on the given cone then

$a(x_1 + lr)^2 + b(y_1 + mr)^2 + c(z_1 + nr)^2 + 2f(y_1 + mr)(z_1 + nr) + 2g(z_1 + nr)(x_1 + lr) + 2h(x_1 + lr)(y_1 + mr) = 0$

$$\text{(i.e.)}\quad (al^2 + bm^2 + cn^2 + 2fmn + 2gnl + 2hlm)r^2 + $$
$$2[(ax_1 + hy_1 + gz_1)l + (hx_1 + by_1 + fz_1)m + (gx_1 + fy_1 + cz_1)n]r + $$
$$(ax_1^2 + by_1^2 + cz_1^2 + 2fy_1z_1 + 2gz_1x_1 + 2hx_1y_1) = 0 \qquad (15.36)$$

This equation is a quadratic in r. Since (x_1, y_1, z_1) is a point on the cone

$$(ax_1^2 + by_1^2 + cz_1^2 + 2fy_1z_1 + 2gz_1x_1 + 2hx_1y_1) = 0 \qquad (15.37)$$

Therefore, one root of the equation (15.36) is zero. If line (15.35) is a tangent to the curve, then both the roots are equal and hence the other root must be zero.

The condition for this is the coefficient of $r = 0$.

$$\text{(i.e.)} \quad (ax_1 + hy_1 + gz_1)l + (hx_1 + by_1 + fz_1)m + (gx_1 + fy_1 + cz_1)n = 0 \qquad (15.38)$$

Hence this is the condition for the line (15.35) to be a tangent to the curve at the point (x_1, y_1, z_1).

Since equation (15.38) can be satisfied for infinitely many values of l, m, n there are infinitely many tangent lines at any point of the cone.

The locus of all such tangent lines is obtained by eliminating l, m, n from (15.35) and (15.38).

$$\therefore (ax_1 + hy_1 + gz_1)(x - x_1) + (hx_1 + by_1 + fz_1)(y - y_1)$$
$$+(gx_1 + fy_1 + cz_1)(z - z_1) = 0$$
$$\text{(i.e.)} (ax_1 + hy_1 + gz_1)x + (hx_1 + by_1 + fz_1)y + (gx_1 + fy_1 + cz_1)z$$
$$= ax_1^2 + by_1^2 + cz_1^2 + 2fy_1z_1 + 2gz_1x_1 + 2hx_1y_1 = 0 \text{ by } (15.37)$$

Therefore, the equation of the tangent plane at (x_1, y_1, z_1) is $(ax_1 + hy_1 + gz_1)x + (hx_1 + by_1 + fz_1)y + (gx_1 + fy_1 + cz_1)z = 0$.

Note 15.6.1: (0, 0, 0) satisfies the above equation and hence the tangent plane at any point of a cone passes through the vertex.

The tangent plane at any point of a cone touches the cone along the generator through *P*.

Proof:

Let the equation of the cone be $ax^2 + by^2 + cz^2 + 2fyz + 2gzx + 2hxy = 0$.
Let $P(x_1, y_1, z_1)$ be any point on the cone.

The equation of the tangent plane at P is $(ax_1 + hy_1 + gz_1)x + (hx_1 + by_1 + fz_1)y + (gx_1 + fy_1 + cz_1)z = 0$.

The equations of the generator through P are $\frac{x}{x_1} = \frac{y}{y_1} = \frac{z}{z_1} = r$.

Any point on this line is (rx_1, ry_1, rz_1). The equation of the tangent plane at (rx_1, ry_1, rz_1) is $(arx_1 + hry_1 + grz_1)x + (hrx_1 + bry_1 + frz_1)y + (grx_1 + fry_1 + crz_1)\, z = 0$.

Dividing by r,

$(ax_1 + hy_1 + gz_1)x + (hx_1 + by_1 + fz_1)y + (gx_1 + fy_1 + cz_1)z = 0$ which is also the equation of the tangent plane at (x_1, y_1, z_1). Therefore, the tangent plane at P touches the cone along the generator through P.

15.6.1 Condition for the Tangency of a Plane and a Cone

Let the equation of the cone be

$$ax^2 + by^2 + cz^2 + 2fyz + 2gzx + 2hxy = 0 \tag{15.39}$$

Let the equation of the plane be

$$lx + my + nz = 0 \tag{15.40}$$

Let the plane (15.40) touch the cone at (x_1, y_1, z_1).

The equation of the tangent plane at (x_1, y_1, z_1) is

$$(ax_1 + hy_1 + gz_1)x + (hx_1 + by_1 + fz_1)y + (gx_1 + fy_1 + cz_1)z = 0 \tag{15.41}$$

If the plane (15.40) touches the cone (15.39) then equations (15.40) and (15.41) are identical.

Therefore identifying (15.40) and (15.41) we get,

$$\frac{ax_1 + hy_1 + gz_1}{l} = \frac{hx_1 + by_1 + fz_1}{m} = \frac{gx_1 + fy_1 + cz_1}{n} = r_1 \tag{15.42}$$

$$\therefore\ ax_1 + hy_1 + gz_1 - r_1 l = 0 \tag{15.43}$$

$$hx_1 + by_1 + fz_1 - r_1 m = 0 \tag{15.44}$$

$$gx_1 + fy_1 + cz_1 - r_1 n = 0 \tag{15.45}$$

Also as (x_1, y_1, z_1) lies on the plane

$$lx_1 + my_1 + nz_1 = 0 \tag{15.46}$$

Eliminating (x_1, y_1, z_1), r_1 from (15.43), (15.44), (15.45) and (15.46) we get,

$$\begin{vmatrix} a & h & g & l \\ h & b & f & m \\ g & f & c & n \\ l & m & n & o \end{vmatrix} = 0$$

Simplifying this we get,

$$Al^2 + Bm^2 + Cn^2 + 2Fmn + 2Gnl + 2Hlm = 0 \tag{15.47}$$

where A, B, C, F, G, H are the cofactors of a, b, c, f, g, h in the determinant

$$\begin{vmatrix} a & h & g \\ h & b & f \\ g & f & c \end{vmatrix}$$

Hence (15.47) is the required condition for the plane (15.40) to touch the cone.

15.7 RECIPROCAL CONE

15.7.1 Equation of the Reciprocal Cone

Let us now find the equation of the cone reciprocal to the cone

$$ax^2 + by^2 + cz^2 + 2fyz + 2gzx + 2hxy = 0. \qquad (15.48)$$

Let a tangent plane to the cone be

$$lx + my + nz = 0 \qquad (15.49)$$

Then we have the condition

$$Al^2 + Bm^2 + Cn^2 + 2Fmn + 2Gnl + 2Hlm = 0 \qquad (15.50)$$

where A, B, C, F, G, H are the cofactors of a, b, c, f, g, h in the determinant

$$\begin{vmatrix} a & h & g \\ h & b & f \\ g & f & c \end{vmatrix}$$

The equation of the line through the vertex (0, 0, 0) of the cone (15.48) and normal to the tangent plane (15.49) are

$$\frac{x}{l} = \frac{y}{m} = \frac{z}{n} = t \qquad (15.51)$$

The locus of (4) which is got by eliminating l, m, n from (15.50) and (15.51) is

$$Ax^2 + By^2 + Cz^2 + 2Fyz + 2Gzx + 2Hxy = 0 \qquad (15.52)$$

This is the equation of the reciprocal cone.

Note 15.7.1.1: If we find the reciprocal cone of (15.52) we get the equation of cone as $ax^2 + by^2 + cz^2 + 2fyz + 2gzx + 2hxy = 0$.

Definition 15.7.1.2: Two cones are said to be reciprocal cones of each other if each one is the locus of the normal through the vertex to the tangent planes of the other.

15.7.2 Angle between Two Generating Lines in Which a Plane Cuts a Cone

Let the equation of the cone and the plane be

$$ax^2 + by^2 + cz^2 + 2fyz + 2gzx + 2hxy = 0. \tag{15.53}$$

and

$$lx + my + nz = 0 \tag{15.54}$$

Let one of the lines of the section be $\frac{x}{l} = \frac{y}{m} = \frac{z}{n}$.

Since this line lies on the plane we have,

$$al^2 + bm^2 + cn^2 + 2fmn + 2ngl + 2hlm = 0 \tag{15.55}$$

$$ul + vm + wn = 0 \tag{15.56}$$

from (15.56) $\quad l = \frac{-(vm + wn)}{u} (u \neq v)$

Substituting this in (15.55) we get,

$$\frac{a(vm + wn)^2}{u^2} + bm^2 + cn^2 - 2gn\frac{(vm + wn)}{u} + 2fmn - 2hm\frac{vm + wn}{u} = 0$$

$$\text{(i.e.)} \quad (av^2 + bu^2 - 2huv)m^2 + 2(avw - guv - fu^2 - huw)mn$$
$$+ (aw^2 + cu^2 - 2guw)n^2 = 0 \tag{15.57}$$

This is a quadratic equation in $\frac{m}{n}$ (assume $n \neq 0$).

There are two values for $\frac{m}{n}$ say $\frac{m_1}{n_1}$ and $\frac{m_2}{n_2}$.

Hence the given plane intersects the cone in two lines namely,

$$\frac{x}{l_1} = \frac{y}{m_1} = \frac{z}{n_1} \text{ and } \frac{x}{l_2} = \frac{y}{m_2} = \frac{z}{n_2}$$

where $\quad l_1 = -\frac{vm_1 + wn_1}{u} \quad l_2 = -\frac{vm_2 + wn_2}{u}$

Also since $\frac{m_1}{n_1}$ and $\frac{m_2}{n_2}$ are the roots of the equation (15.57)

$$\frac{m_1}{n_1}+\frac{m_2}{n_2}=\frac{m_1n_2+m_2n_1}{n_1n_2}$$

$$=\frac{-2(avw-guv+fu^2-huw)}{av^2+bu^2-2huv}$$

and

$$\frac{m_1m_2}{n_1n_2}=\frac{aw^2+cu^2-2guv}{av^2+bu^2-2huv}$$

$$l_1l_2=\frac{v^2m_1m_2+w^2n_1n_2+vw(m_1n_2+m_2n_1)}{u^2}$$

$$\therefore \frac{l_1l_2}{n_1n_2}=\frac{1}{u^2}\left[\frac{v^2m_1m_2}{n_1n_2}+w^2+vw\left(\frac{m_1}{n_1}+\frac{m_2}{n_2}\right)\right]$$

$$=\frac{1}{u^2}\left[\begin{array}{l}\frac{v^2(aw^2+cu^2-2guw)}{av^2+bu^2-2huv}+w^2-\\ 2vw\frac{avw-guv+fu^2-huw}{av^2+bu^2-2huv}\end{array}\right]$$

$$=\frac{cv^2+bw^2-2fwv}{av^2+bu^2-2huv}$$

$$\therefore \frac{l_1l_2}{bw^2+cv^2-2fvw}=\frac{m_1m_2}{cu^2+aw^2-2gwu}$$

$$=\frac{n_1n_2}{av^2+bu^2-2huv}$$

$$=\frac{m_1n_2+m_2n_1}{-2(avw-guv+fu^2-huw)}$$

$$=\frac{\sqrt{(m_1n_2+m_2n_1)^2-4m_1m_2n_1n_2}}{\pm 2\sqrt{\left[\begin{array}{l}(avw-guv+fu^2-huv)^2-\\ 4(av^2+bu^2-2huv)(aw^2+cu^2-2guw)\end{array}\right]}}$$

$$=\frac{m_1n_2-m_2n_1}{\pm 2u[-(Au^2+Bv^2+Cw^2+2Fvw+2Gwu+2Huv)]^{\frac{1}{2}}}$$

$$=\frac{m_1n_2-m_2n_1}{\pm 2uP}$$

where $P^2 = -(Au^2 + Bv^2 + Cw^2 + 2Fvw + 2Gwu + 2Huv)$ and A, B, C, F, G, H are the cofactors of a, b, c, f, g, h in the determinant $\begin{vmatrix} a & h & g \\ h & b & f \\ g & f & c \end{vmatrix}$.

It follows by symmetry $\dfrac{l_1 l_2}{bw^2 + cv^2 - 2fvw}$

$$= \frac{m_1 m_2}{cu^2 + aw^2 - 2gwu}$$

$$= \frac{n_1 n_2}{av^2 + bu^2 - 2huv}$$

$$= \frac{m_1 n_2 - m_2 n_1}{\pm 2up} = \frac{n_1 l_2 - n_2 l_1}{\pm 2vp} = \frac{l_1 m_2 - l_2 m_1}{\pm 2wp}$$

If θ is the acute angle between the lines then

$$\frac{\cos\theta}{l_1 l_2 + m_1 m_2 + n_1 n_2} = \frac{\sin\theta}{\sqrt{(l_1 m_2 - l_2 m_1)^2 + (m_1 n_2 - m_2 n_1)^2 (n_1 l_2 - n_2 l_1)^2}}$$

(i.e.) $$\frac{\cos\theta}{(a+b+c)(u^2+v^2+w^2) - f(u,v,w)} = \frac{\sin\theta}{\pm 2P\sqrt{u^2+v^2+w^2}}$$

$$\therefore \quad \tan\theta = \frac{\pm 2P\sqrt{u^2+v^2+w^2}}{(a+b+c)(u^2+v^2+w^2) - f(u,v,w)}$$

Note 15.7.2.1: If the two lines are perpendicular then $(a + b + c)(u^2 + v^2 + w^2) - f(u, v, w) = 0$.

(i.e.) $f(u, v, w) = (a + b + c)(u^2 + v^2 + w^2)$

Note 15.7.2.2: If the lines of intersection are coincident then $\theta = 0$.

$$\therefore 2P\sqrt{u^2+v^2+w^2} = 0$$

$$\therefore \; P = 0 \text{ since } u^2 + v^2 + w^2 \neq 0$$

or $$P^2 = 0$$

$$Au^2 + Bv^2 + Cw^2 + 2Fvw + 2Gwu + 2Huv = 0$$

$$\begin{vmatrix} a & h & g & u \\ h & b & f & v \\ g & f & c & w \\ u & v & w & o \end{vmatrix} = 0$$

This is the condition for the plane $ux + vy + wz = 0$ to be a tangent plane to the cone.

15.7.3 Condition for Mutually Perpendicular Generators of the Cone

$$f(x, y, z) = ax^2 + by^2 + cz^2 + 2fyz + 2gzx + 2hxy = 0$$

We have seen that the condition for the plane $ux + vy + wz = 0$ cut the cone in two perpendicular generators is that

$$(a + b + c)(u^2 + v^2 + w^2) - f(u, v, w) = 0 \tag{15.58}$$

If there is a third generator which is perpendicular to the above two lines of intersection then it must be a normal to the plane $ux + vy + wz = 0$.

Therefore, its equations are

$$\frac{x}{u} = \frac{y}{v} = \frac{z}{w} \tag{15.59}$$

Since (15.59) is a generator of the cone $f(x, y, z) = 0$, we get

$$f(u, v, u) = 0 \tag{15.60}$$

(15.58) and (15.60) holds if and only if $a + b + c = 0$.

Therefore, the condition for the cone to have three mutually perpendicular generators is $a + b + c = 0$.

Example 15.7.1

Find the angle between the lines of section of the plane $3x + y + 5z = 0$ and the cone $6yz - 2zx + 5xy = 0$.

Solution

Let the equations of the line of section be

$$\frac{x}{l} = \frac{y}{m} = \frac{z}{n} = r \tag{15.61}$$

As this line lies on the plane and also on the cone we get

$$3l + m + 5n = 0 \tag{15.62}$$

$$6mn - 2nl + 5lm = 0 \tag{15.63}$$

From (15.62) $\quad m = -(3l + 5n)$

Substituting this in (15.63) we get,

$$-6n(3l + 5n) - 2nl + 5l(3l + 5n) = 0$$

$$\text{(i.e.)} \quad 15l^2 + 45ln + 30n^2 = 0$$

$$\text{(i.e.)} \quad l^2 + 3ln + 2n^2 = 0$$

$$(l + n)(l + 2n) = 0$$

$$l + n = 0 \quad \text{or} \quad l + 2n = 0$$

Solving $l + n = 0$ and $3l + m + 5n = 0$ we get,

$$\frac{l}{1} = \frac{m}{2} = \frac{n}{-1}$$

Solving $l + 2n = 0$ and $3l + m + 5n = 0$ we get,

$$\frac{l}{-2} = \frac{m}{1} = \frac{n}{1}$$

Therefore, the direction ratios of the two lines are 1, 2, −1 and 2,−1, −1.

If θ is the angle between the lines $\cos\theta = \dfrac{2-2+1}{\sqrt{6}\sqrt{6}} = \dfrac{1}{6}$.

Therefore, the acute angle between the lines is $\theta = \cos^{-1}\left(\dfrac{1}{6}\right)$.

Example 15.7.2

Prove that the angle between the lines given by $x + y + z = 0$, $ayz + bzx + cxy = 0$ is $\dfrac{\pi}{2}$ if $a + b + c = 0$.

Solution

The plane

$$x + y + z = 0 \tag{15.64}$$

will cut the cone

$$ayz + bzx + cxy = 0 \tag{15.65}$$

in two lines through the vertex (0, 0, 0).

The equations of the lines of the section are

$$\frac{x}{l}=\frac{y}{m}=\frac{z}{n} \tag{15.66}$$

where l, m, n are the direction ratios of the lines.

Since the line given by (15.66) lies on the plane and the cone

$$l+m+n=0 \tag{15.67}$$

$$amn+bnl+clm=0 \tag{15.68}$$

Substituting $n=-(l+m)$ in (15.67) we get

$$-am(l+m)-b(l+m)l+clm=0$$

$$\text{(i.e.)}\quad bl^2+(a+b-c)lm+am^2=0$$

$$\text{or}\quad b\left(\frac{l^2}{m^2}\right)+(a+b-c)\frac{1}{m}+a=0$$

If l_1, m_1, n_1 and l_2, m_2, n_2 are the direction ratios of the two lines we get,

$$\frac{l_1l_2}{m_1m_2}=\frac{a}{b}$$

$$\therefore \frac{l_1l_2}{a}=\frac{m_1m_2}{b}$$

Similarly we can show that

$$\frac{m_1m_2}{b}=\frac{n_1n_2}{c}$$

$$\therefore \frac{l_1l_2}{a}=\frac{m_1m_2}{b}=\frac{n_1n_2}{c}=t \quad \text{say}$$

$$l_1l_2=ta$$

$$m_1m_2=tb$$

$$n_1n_2=tc$$

If θ is the angle between the lines

$$\cos\theta=l_1l_2+m_1m_2+n_1n_2=t(a+b+c)=0\ (\because\ a+b+c=0)$$

$$\therefore\quad \theta=\frac{\pi}{2}$$

Example 15.7.3

Prove that the cones $ax^2 + by^2 + cz^2 = 0$ and $\frac{x^2}{a} + \frac{y^2}{b} + \frac{z^2}{c} = 0$ are reciprocal.

Solution

The equation of the reciprocal cone $ax^2 + by^2 + cz^2 = 0$ is

$$Ax^2 + By^2 + Cz^2 + 2Fyz + 2Gzx + 2Hxy = 0$$

where A, B, C are the cofactors of a, b, c in $\begin{vmatrix} a & h & g \\ h & b & f \\ g & f & c \end{vmatrix}$.

$$\text{(i.e.)} \quad A = bc - f^2 \quad F = gh - af = 0$$
$$B = ca - g^2 \quad G = hf - bg = 0$$
$$C = ab - h^2 \quad H = fg - ch = 0$$

The equation of the reciprocal cone is $bcx^2 + cay^2 + abz^2 = 0$ or $\frac{x^2}{a} + \frac{y^2}{b} + \frac{z^2}{c} = 0$.

Similarly, we can show that the reciprocal cone of $\frac{x^2}{a} + \frac{y^2}{b} + \frac{z^2}{c} = 0$ is $ax^2 + by^2 + cz^2 = 0$.

Therefore, the two given cones are reciprocal to each other.

Example 15.7.4

Show that the equation $\sqrt{fx} \pm \sqrt{gy} \pm \sqrt{hz} = 0$ represents a cone which touches the coordinate planes.

Solution

$$\sqrt{fx} \pm \sqrt{gy} \pm \sqrt{hz} = 0$$

$$\left(\sqrt{fx} \pm \sqrt{gy}\right)^2 = \left(-\sqrt{hz}\right)^2$$

$$fx + gy - hz = \pm 2\sqrt{fgxy}$$

Squaring $(fx + gy - hz)^2 = 4fgxy$

$$\text{(i.e.)} \quad f^2x^2 + g^2y^2 + h^2z^2 + 2fgxy - 2hgyz - 2hfzx = 0$$

This being a homogeneous equation of second degree in x, y, z, it represents a cone. When this cone meets the plane $x = 0$ we get, $(gy - hz)^2 = 0$.

Hence the above cone is cut by the plane $x = 0$ in coincident lines and hence $x = 0$ touches the cone.

Similarly, $y = 0$, $z = 0$ also touch the cone.

Exercises 3

1. Find the angle between the lines of the section of the planes and cones:
   (i) $x + 3y - 2z = 0, x^2 + 9y^2 - 4z^2 = 0$
   (ii) $6x - 10y - 7z = 0, 108x^2 - 20y^2 - 7z^2 = 0$

   Ans.: (i) $\cos^{-1}\frac{3}{\sqrt{65}}$, (ii) $\cos^{-1}\frac{16}{21}$

2. Show that the angle between the lines in which the plane $x + y + z = 0$ cuts the cone $ayz + bzx + cxy = 0$ is $\frac{\pi}{3}$ if $\frac{1}{a}+\frac{1}{b}+\frac{1}{c}=0$.

3. Prove that the equation $a^2x^2 + b^2y^2 + c^2z^2 - 2bcyz - 2cazx - 2abxy = 0$ represents a cone which touches the coordinate plane.

4. If $\frac{x}{1}=\frac{y}{2}=\frac{z}{3}$ represents one of the generators of the three mutually perpendicular generators of the cone $5yz - 8zx - 3xy = 0$ then find the other two.

   Ans.: $\frac{x}{1}=\frac{y}{1}=\frac{z}{-1}; \frac{x}{5}=\frac{y}{-4}=\frac{z}{1}$

5. If $x=\frac{y}{2}=z$ represents one of the three mutually perpendicular generators of the cone $11yz + 6zx - 14xy = 0$ then find the other two.

   Ans.: $\frac{x}{2}=\frac{y}{-3}=\frac{z}{4}; \frac{x}{-11}=\frac{y}{2}=\frac{z}{7}$

CHAPTER

16 Cylinder

16.1 DEFINITION

The surface generated by a variable line which remains parallel to a fixed line and intersects a given curve (or touches a given surface) is called a cylinder.

The variable line is called the *generator*, the fixed straight line is called the *axis of the cylinder* and the given curve is called the *guiding curve of the cylinder*.

16.2 EQUATION OF A CYLINDER WITH A GIVEN GENERATOR AND A GIVEN GUIDING CURVE

Let us find the equation of the cylinder whose generators are parallel to the line

$$\frac{x}{l} = \frac{y}{m} = \frac{z}{n} \tag{16.1}$$

and whose guiding curve is the conic

$$ax^2 + 2hxy + by^2 + 2gx + 2fy + c = 0; \quad z = 0 \tag{16.2}$$

Let (α, β, γ) be any point on the cylinder. Then the equations of a generator are

$$\frac{x-\alpha}{l} = \frac{y-\beta}{m} = \frac{z-\gamma}{n} \tag{16.3}$$

Let us find the point where this line meets the plane $z = 0$. When $z = 0$, $\frac{x-\alpha}{l} = \frac{y-\beta}{m} = \frac{-\gamma}{n}$.

$$x = \alpha - \frac{l\gamma}{n},\ y = \beta - \frac{m\gamma}{n},\ z = 0$$

This point is $\left(\alpha-\frac{l\gamma}{n},\beta-\frac{m\gamma}{n},0\right)$.

When the generator meets the conic, this point lies on the conic.

$$\therefore a\left(\alpha-\frac{l\gamma}{n}\right)^2+2h\left(\alpha-\frac{l\gamma}{n}\right)\left(\beta-\frac{m\gamma}{n}\right)+b\left(\beta-\frac{m\gamma}{n}\right)^2$$
$$+2g\left(\alpha-\frac{l\gamma}{n}\right)+2f\left(\beta-\frac{m\gamma}{n}\right)+c=0$$

i.e. $a\,(n\alpha-l\gamma)^2+2h\,(n\alpha-l\gamma)\,(n\beta-m\gamma)+b\,(n\beta-m\gamma)^2+2gn\,(n\alpha-l\gamma)+2fn(n\beta-m\gamma)+cn^2=0$

The locus of the point (α,β,γ) is

$$a(nx-lz)^2+2h(nx-lz)(ny-mz)+b(ny-mz)^2+2gn(nx-lz)$$
$$+2fn(ny-mz)+cn^2=0$$

This is the required equation of the cylinder.

Note 1: If the generators are parallel to z-axis $l=0,\ m=0,\ n=l$ then the equation of the cylinder becomes,

$$f(x,y)=ax^2+2hxy+by^2+2gx+2fy+c=0$$

Note 2: The equation $f(x,y)=0$ in space represents a cylinder whose generators are parallel to z-axis.

16.3 ENVELOPING CYLINDER

The locus of the tangent lines drawn to a sphere and parallel to a given line

Let the given sphere be

$$x^2+y^2+z^2=a^2 \tag{16.4}$$

Let the given line be

$$\frac{x}{l}=\frac{y}{m}=\frac{z}{n} \tag{16.5}$$

Let (α,β,γ) be any point on the locus.

Then any line through (α,β,γ) parallel to line (16.5) is

$$\frac{x-\alpha}{l}=\frac{y-\beta}{m}=\frac{z-\gamma}{n}=r \tag{16.6}$$

Any point on this line is $(\alpha+lr,\ \beta+mr,\ \gamma+nr)$

If the point lies on the sphere (16.4), then

$$(\alpha + lr)^2 + (\beta + mr)^2 + (\gamma + nr)^2 = a^2$$

$$\text{(i.e.) } r^2(l^2 + m^2 + n^2) + 2r(\alpha l + \beta m + \gamma n) + \alpha^2 + \beta^2 + \gamma^2 - a^2 = 0 \quad (16.7)$$

This is a quadratic equation in r giving the two values for r corresponding to two points common to the sphere and the line. If the line is a tangent then the two values of r must be equal and hence the discriminant must be zero.

$$\therefore (\alpha l + \beta m + \gamma n)^2 = (l^2 + m^2 + n^2)(\alpha^2 + \beta^2 + \gamma^2 - a^2)$$

The locus (α, β, γ) is

$$(lx + my + nz)^2 = (l^2 + m^2 + n^2)(x^2 + y^2 + z^2 - a^2)$$

which is a cylinder. This cylinder is called the enveloping cylinder of the sphere.

Enveloping cylinder as a limiting form of an enveloping cone

Let $\frac{x}{l} = \frac{y}{m} = \frac{z}{n} = r$ be the axis of the enveloping cylinder. Any point on this line is (lr, mr, nr).

Let this point be the vertex of the enveloping cone. Then the equation of the enveloping cone is $T^2 = SS_1$.

$$\text{(i.e.) } (lrx + mry + nrz - a^2)^2 = (x^2 + y^2 + z^2 - a^2)(l^2r^2 + mr^2 + n^2r^2 - a^2)$$

$$\text{(i.e.) } \left(lx + my + nz - \frac{a^2}{r}\right) = (x^2 + y^2 + z^2 - a^2)\left(l + m + n - \frac{a^2}{r^2}\right)$$

When $r \to \infty$, the above equation becomes,

$$(lx + my + nz)^2 = (x^2 + y^2 + z^2 - a^2)(l^2 + m^2 + n^2)$$

which is the equation to the enveloping cylinder.

16.4 RIGHT CIRCULAR CYLINDER

A right circular cylinder is a surface generated by a straight line which remains parallel to a fixed straight line at a constant distance from it. The fixed straight line is called the axis of the cylinder and the constant distance is called the radius of the cylinder.

The equation of a right circular cylinder whose axis is the straight line $\frac{x-\alpha}{l}=\frac{y-\beta}{m}=\frac{z-\gamma}{n}$ and whose radius is a.

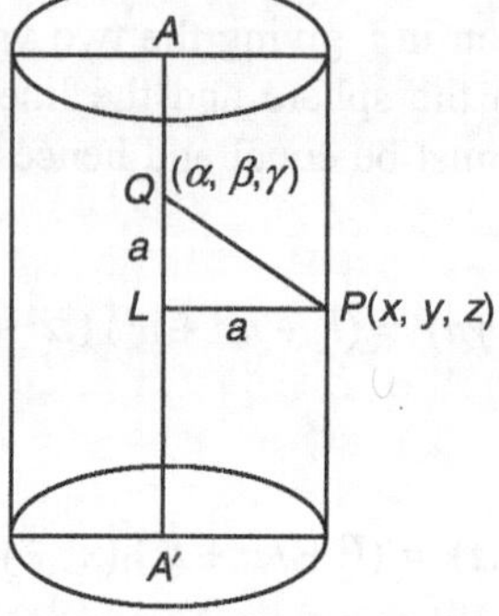

Let $P(x, y, z)$ be any point on the cylinder. Let AA' be the axis of the cylinder.

Draw PL perpendicular to the axis and $PL = a$.

Let $Q(\alpha, \beta, \gamma)$ be a point on the axis of the cylinder.

$$PQ^2 = (x-\alpha)^2 + (y-\beta)^2 + (z-\gamma)^2$$

$$QL = \text{Projection of } PQ \text{ on this axis of the cylinder}$$

$$= \frac{(x-\alpha)l+(y-\beta)m+(z-\gamma)n}{\sqrt{l^2+m^2+n^2}}$$

Also, $QL^2 = PQ^2 - PL^2$

$PQ^2 - QL^2 = a^2$

$$(x-\alpha)^2+(y-\beta)^2+(z-\gamma)^2-\left[\frac{(x-\alpha)l+(y-\beta)m+(z-\gamma)n}{\sqrt{l^2+m^2+n^2}}\right]^2 = a^2$$

This is the required equation of the right cylinder.

ILLUSTRATIVE EXAMPLES

Example 16.1

Find the equation of the cylinder whose generators are parallel to the line $\frac{x}{-1}=\frac{y}{2}=\frac{z}{3}$ and whose guiding curve is $x^2 + y^2 = 9$, $z = 1$.

Solution

Let $P(x, y, z)$ be a point on the cylinder.

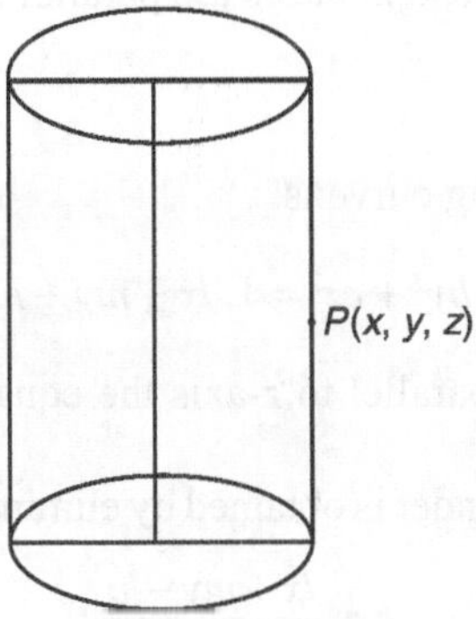

The equations of the generator through P and parallel to the line $\frac{x}{-1}=\frac{y}{2}=\frac{z}{3}$ are

$$\frac{x-x_1}{-1}=\frac{y-y_1}{2}=\frac{z-z_1}{3} \tag{16.8}$$

The guiding curve is

$$x^2+y^2=9, z=1 \tag{16.9}$$

When the generator through P meets the guiding curve,

$$\frac{x-x_1}{-1}=\frac{y-y_1}{2}=\frac{z-z_1}{3}$$

$$\therefore x=x_1-\frac{1-z_1}{3}=\frac{3x_1+z_1-1}{3}, y=y_1+\frac{2(1-z_1)}{3}=\frac{3y_1-2z_1+2}{3}$$

Since this point lies on the curve (16.9),

$$(3x_1+z_1-1)^2+(3y_1-2z_1+2)^2=81$$

The locus of (x_1, y_1, z_1) is $(3x+z-1)^2+(3y-2z+2)^2=81$

$$\text{(i.e.) } 9x^2+9y^2+5z^2+6xz-12yz-6x+12y-10z-76=0$$

This is the equation of the required cylinder.

Example 16.2

Find the equation of the cylinder which intersects the curve $ax^2 + by^2 + cz^2 = 1$, $lx + my + nz = p$ and whose generators are parallel to z-axis.

Solution

The equation of the guiding curve is

$$ax^2 + by^2 + cz^2 = 1,\ lx + my + nz = p \tag{16.10}$$

Since the generators are parallel to z-axis the equation of the cylinder is of the form $f(x, y) = 0$.

The equation of the cylinder is obtained by eliminating z in equation (16.10)

$$z = \frac{lx + my - p}{-n}$$

Substituting this in $ax^2 + by^2 + cz^2 = 1$, we get,

$$ax^2 + by^2 + c\left(\frac{lx + my - p}{-n}\right)^2 = 1$$

$$\therefore\ (an^2 + cl^2)\,x^2 + (bn^2 + cm^2)\,y^2 + 2clmxy - 2cpmy + cp^2 - n^2 = 0$$

This is the equation of the required cylinder.

Example 16.3

Find the equation of the cylinder whose generators are parallel to the line $\frac{x}{1} = \frac{y}{-2} = \frac{z}{3}$ and whose guiding curve is the ellipse $x^2 + 2y^2 = 1$, $z = 3$.

Solution

The equation to the guiding curve is

$$x^2 + 2y^2 = 1,\ z = 3 \tag{16.11}$$

Let (x_1, y_1, z_1) be a point on the cylinder. Then the equations of the generator through $P(x_1, y_1, z_1)$ are $\frac{x - x_1}{1} = \frac{y - y_1}{-2} = \frac{z - z_1}{3}$

When this line meets the plane $z = 3$, we have,

$$\frac{x - x_1}{1} = \frac{y - y_1}{-2} = \frac{3 - z_1}{3}$$

$$\therefore x = x_1 + \frac{3-z_1}{3} = \frac{3x_1 - z_1 + 3}{3}, y = y_1 - \frac{2(3-z_1)}{3} = \frac{3y_1 + 2z_1 - 6}{3}$$

This point lies on the curve $x^2 + 2y^2 = 1$.

$$\therefore \left(\frac{3x_1 - z_1 + 3}{3}\right)^2 + 2\left(\frac{3y_1 + 2z_1 - 6}{3}\right)^2 = 1$$

$$(3x_1 - z_1 + 3)^2 + 2(3y_1 + 2z_1 - 6)^2 = 9$$

$$9x_1^2 + 18y_1^2 + 9z_1^2 + 24y_1z_1 - 6z_1x_1 + 18x_1 - 72y_1 - 54z_1 + 72 = 0$$

The locus of (x_1, y_1, z_1) is

$$9x^2 + 18y^2 + 9z^2 + 24yz - 6zx + 18x - 72y - 54z + 72 = 0$$
$$\text{(i.e.) } 3x^2 + 6y^2 + 3z^2 + 8yz - 2zx + 6x - 24y - 18z + 24 = 0$$

Example 16.4

Find the equation of the surface generated by the straight line $y = mx$, $z = nx$ and intersecting the ellipse $\frac{x^2}{a^2} + \frac{y^2}{b^2} = 1, z = 0$.

Solution

The given line $y = mx$, $z = nx$ can be expressed in symmetrical form as

$$\frac{x}{1} = \frac{y}{m} = \frac{z}{n} \qquad (16.12)$$

Let $P(x_1, y_1, z_1)$ be any point on the cylinder.

Then the equations of the generator through P are

$$\frac{x - x_1}{1} = \frac{y - y_1}{m} = \frac{z - z_1}{n}$$

This meets the curve $\frac{x^2}{a^2} + \frac{y^2}{b^2} = 1$, $z = 0$

When $z = 0$, $\frac{x - x_1}{1} = \frac{y - y_1}{m} = \frac{-z_1}{n}$

$$\therefore x = x_1 - \frac{z_1}{n} = \frac{nx_1 - z_1}{n}, y = y_1 - \frac{mz_1}{n} = \frac{ny_1 - mz_1}{n}$$

This point lies on the curve $\frac{x^2}{a^2} + \frac{y^2}{b^2} = 1$.

$$\frac{1}{a^2}\left(\frac{nx_1 - z_1}{n}\right)^2 + \frac{1}{b^2}\left(\frac{ny_1 - mz_1}{n}\right)^2 = 1$$

$$b^2(nx_1 - z_1)^2 + a^2(ny_1 - mz_1)^2 = n^2a^2b^2$$

The locus of (x_1, y_1, z_1) is

$$b^2(nx - z)^2 + a^2(ny - mz)^2 = n^2a^2b^2$$

which is the equation of the required cylinder.

Example 16.5

Find the equation of the circular cylinder whose generating lines have the direction cosines l, m, n and which passes through the circumference of the fixed circle $x^2 + y^2 = a^2$ on the xoz plane.

Solution

Let $P(x_1, y_1, z_1)$ be any point on the cylinder. Then the equations of the generator through P are $\frac{x - x_1}{l} = \frac{y - y_1}{m} = \frac{z - z_1}{n}$.

This meets the plane $y = 0$.

$$\frac{x - x_1}{l} = \frac{-y_1}{m} = \frac{z - z_1}{n}$$

$$\therefore x = x_1 - \frac{ly_1}{m} = \frac{mx_1 - ly_1}{m}, z = z_1 - \frac{ny_1}{m} = \frac{mz_1 - ny_1}{m}$$

This point lies on the curve

$$x^2 + y^2 = a^2$$

$$\left(\frac{mx_1 - ly_1}{m}\right)^2 + \left(\frac{mz_1 - ny_1}{m}\right)^2 = a^2$$

$$(mx_1 - ly_1)^2 + (mz_1 - ny_1)^2 = m^2a^2$$

The locus of (x_1, y_1, z_1) is

$$(mx - ly)^2 + (mz - ny)^2 = m^2a^2$$

which is the required equation of the cylinder.

Example 16.6

Find the equations of the right circular cylinder of radius 3 with equations of axis as $\frac{x-1}{2} = \frac{y-3}{2} = \frac{z-5}{-1}$.

Solution

The equations of the axis are

$$\frac{x-1}{2} = \frac{y-3}{2} = \frac{z-5}{-1}$$

(1, 3, 5) is a point on the axis.
2, 2, –1 are the direction ratios of the axis.

$\therefore$ direction cosines are $\frac{2}{3}, \frac{2}{3}, \frac{-1}{3}$

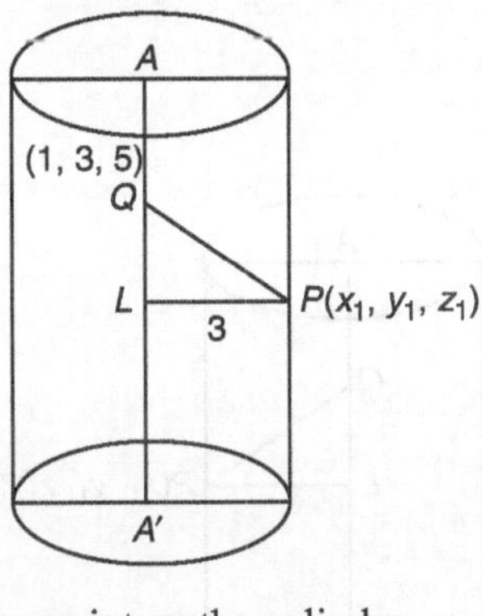

Let $P(x_1, y_1, z_1)$ be any point on the cylinder.

QL = Projection of PQ on the axis

$$= (x - x_1)l + (y - y_1)m + (z - z_1)n$$

$$= (x_1 - 1)\frac{2}{3} + (y_1 - 3)\frac{2}{3} - (z_1 - 5)\frac{1}{3}$$

$$= \frac{2x_1 + 2y_1 - z_1 - 3}{3}$$

Also, $PQ^2 = QL^2 + LP^2$

(i.e.) $(x_1-1)^2+(y_1-3)^2+(z_1-5)^2 = \left(\dfrac{2x_1+2y_1-z_1-3}{3}\right)^2+9$

The locus of (x_1, y_1, z_1) is

$$9(x^2-2x+1+y^2-6y+9+z^2-10z+25)$$
$$=4x^2+4y^2+z^2+9+8xy-4xz-12x-4yz-12y+6z+81$$

(i.e.) $5x^2+5y^2+8z^2-8xy+4xz+4yz-6x-42y-96z+225=0$

This is the equation of the required cylinder.

Example 16.7

Find the equation of the right circular cylinder whose axis is $x = 2y = -z$ and radius 4.

Solution

The equations of the axis of the cylinder are

$$\frac{x}{1}=\frac{y}{\frac{1}{2}}=\frac{z}{-1}$$

(i.e.) $\dfrac{x}{2}=\dfrac{y}{1}=\dfrac{z}{-2}$

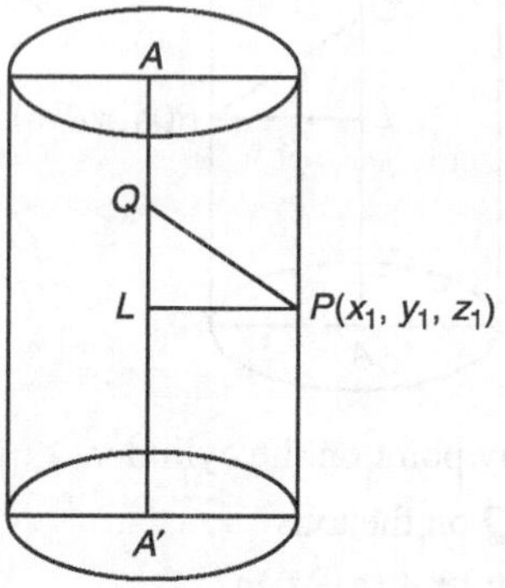

Let $P(x_1, y_1, z_1)$ be any point on the cylinder.
The equations of the generator through P are

$$\frac{x-x_1}{2}=\frac{y-y_1}{1}=\frac{z-z_1}{-2}$$

The direction cosines of the axis are $\frac{2}{3}, \frac{1}{3}, \frac{-2}{3}$.

QL = Projection of PQ on this axis.

$$= (x_1 - 0)\frac{2}{3} + (y_1 - 0)\frac{1}{3} + (z_1 - 0)\frac{-2}{3}$$

$$= \frac{2x_1 + y_1 - 2z_1}{3}$$

Also, $PQ^2 = QL^2 + LP^2$

$$x_1^2 + y_1^2 + z_1^2 = \left(\frac{2x_1 + 2y_1 - z_1 - 3}{3}\right)^2 + 16$$

The locus of (x_1, y_1, z_1) is

$$9(x^2 + y^2 + z^2) = (2x_1 + 2y_1 - z_1 - 3)^2 + 144$$

(i.e.) $$5x^2 + 5y^2 + 8z^2 - 4xy + 4yz + 8zx - 144 = 0$$

Example 16.8

Find the equation of the cylinder whose generators have direction cosines l, m, n and which passes through the circle $x^2 + z^2 = a^2, y = 0$.

Solution

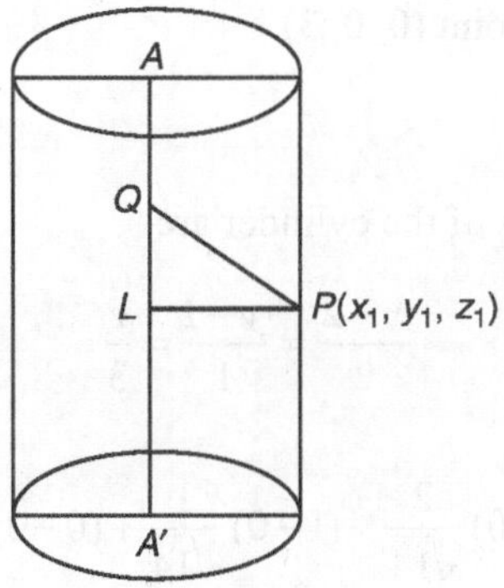

Let $P(x_1, y_1, z_1)$ be any point on the cylinder.

The equation of the generators through P are

$$\frac{x - x_1}{l} = \frac{y - y_1}{m} = \frac{z - z_1}{n}$$

This line meets the curve $y = 0, x^2 + z^2 = a^2$

When $y = 0$, $\dfrac{x - x_1}{l} = \dfrac{-y_1}{m} = \dfrac{z - z_1}{n}$

$$\therefore\ x = x_1 - \frac{ly_1}{m} = \frac{mx_1 - ly_1}{m},$$

$$z = z_1 - \frac{ny_1}{m} = \frac{mz_1 - ny_1}{m}$$

This point lies on $x^2 + z^2 = a^2$ or $\left(\dfrac{mx_1 - ly_1}{m}\right)^2 + \left(\dfrac{mz_1 - ny_1}{m}\right)^2 = a^2.$

The locus of (x_1, y_1, z_1) is

$$(mx - ly)^2 + (mz - ny)^2 = m^2a^2$$

This is the equation of the required cylinder.

Example 16.9

Find the equation of the right circular cylinder whose axis is $\dfrac{x-2}{2} = \dfrac{y-1}{1} = \dfrac{z}{3}$ and passes through the point (0, 0, 3).

Solution

The equations of the axis of the cylinder are

$$\frac{x-2}{2} = \frac{y-1}{1} = \frac{z}{3}$$

$$r^2 = 4 + 1 + 9 - \left[(2-0)\frac{2}{\sqrt{14}} + (1-0)\frac{1}{\sqrt{14}} + (0-3)\frac{3}{\sqrt{14}}\right]^2$$

$$= 14 - \left[\frac{4+1-9}{\sqrt{14}}\right]^2$$

$$= \frac{196-16}{14} = \frac{180}{14} = \frac{90}{7}$$

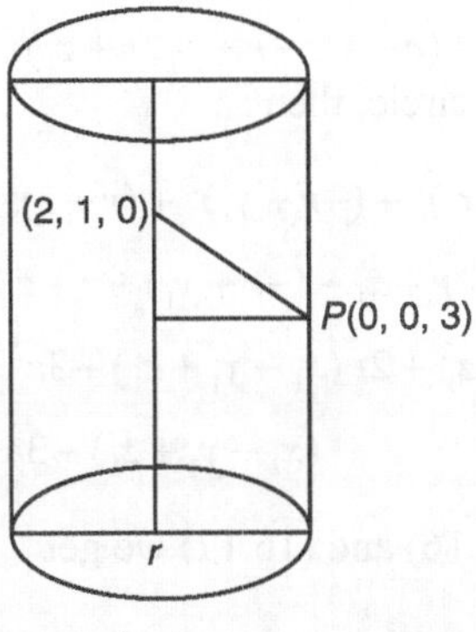

Let $P(x_1, y_1, z_1)$ be any point on the cylinder, then

$$\frac{90}{7} = (x_1 - 2)^2 + (y_1 - 1)^2 + (z_1 - 0)^2$$

$$-\left[(x_1 - 2)\frac{2}{\sqrt{14}} + (y_1 - 1)\frac{1}{\sqrt{14}} + (z_1 - 3)\frac{3}{\sqrt{14}}\right]^2$$

$$\frac{90}{7} = (x_1 - 2)^2 + (y_1 - 1)^2 + z_1^2 - \frac{(2x_1 + y_1 - 3z_1 - 14)^2}{14}$$

$$\frac{90}{7} \times 14 = 14[(x_1 - 2)^2 + (y_1 - 1)^2 + z_1^2] - (2x_1 + y_1 - 3z_1 - 14)^2$$

The locus of (x_1, y_1, z_1) is

$$10x^2 + 13y^2 + 5z^2 - 6yz - 12zx - 4xy - 36x - 18y + 30z - 135 = 0$$

This is the equation of the required cylinder.

Example 16.10

Find the equation to the right circular cylinder which passes through circle $x^2 + y^2 + z^2 = 9, x - y + z = 3$.

Solution

For the right circular cylinder, the guiding curve is the circle $x^2 + y^2 + z^2 = 9$, $x - y + z = 3$.

Therefore, the direction ratios of the axis of the cylinder are 1, −1, 1.

Let $P(x_1, y_1, z_1)$ be any point on the cylinder.

Then the equations of the generator through P are

$$\frac{x - x_1}{l} = \frac{y - y_1}{m} = \frac{z - z_1}{n} = r \tag{16.13}$$

Any point on this line is $(r-x_1, -r+y_1, r+z_1)$.
If this point lies on the circle, then

$$(r-x_1)^2+(-r+y_1)^2+(r+3)^2=9 \tag{16.14}$$

and

$$r+x_1-(-r+y_1)+r+z_1=3 \tag{16.15}$$

(i.e.)

$$x_1^2+y_1^2+z_1^2+2r(x_1-y_1+z_1)+3r^2=9 \tag{16.16}$$

$$(x_1-y_1+z_1)+3r=3 \tag{16.17}$$

Eliminating r from (16.16) and (16.17) we get

$$(x_1^2+y_1^2+z_1^2)+\frac{2}{3}(3-x_1+y_1-z_1)(x_1-y_1+z_1)+3.\frac{1}{9}(3-x_1+y_1-z_1)^2=9$$

Simplifying, the locus of (x_1, y_1, z_1) is

$$x^2+y^2+z^2+xy+yz+zx-9=0$$

which is the equation of the required cylinder.

Example 16.11

Find the equation to the right circular cylinder of radius a whose axis passes through the origin and makes equal angles with the coordinate axes.

Solution

Let l, m, n be the direction cosines of the axis of the cylinder.

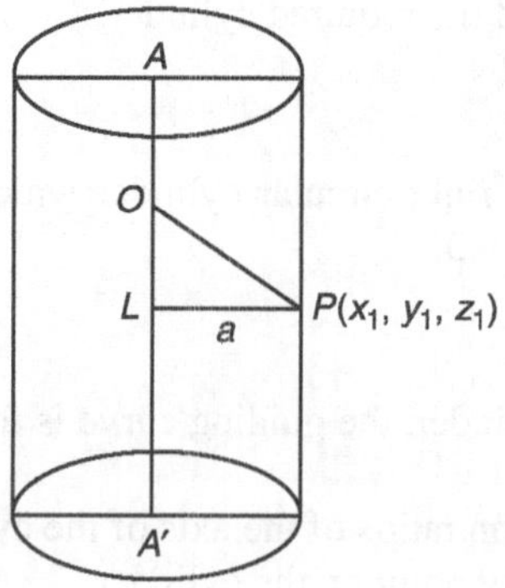

Given: $l=m=n$

$$\therefore\ l=m=n=\frac{\sqrt{l^2+m^2+n^2}}{\sqrt{3}}=\frac{1}{\sqrt{3}}$$

The axis passes through the origin.
Let $P\,(x_1, y_1, z_1)$ be any point on the cylinder.
OL = Projection of OP on the axes

$$= (x_1 - 0)\frac{1}{\sqrt{3}} + (y_1 - 0)\frac{1}{\sqrt{3}} + (z_1 - 0)\frac{-2}{\sqrt{3}}$$

$$= \frac{x_1 + y_1 + z_1}{\sqrt{3}}$$

$$OP^2 = OL^2 + LP^2$$

$$\left(\sqrt{x_1^2 + y_1^2 + z_1^2}\right)^2 = \left(\frac{x_1 + y_1 + z_1}{\sqrt{3}}\right)^2 + a^2$$

$$\therefore (x_1^2 + y_1^2 + z_1^2)^2 - 3(x_1^2 + y_1^2 + z_1^2 - 3a^2) - 0$$

The locus of (x_1, y_1, z_1) is

$$2x^2 + 2y^2 + 2z^2 - 2xy + 2yz - 2zx - 3a^2 = 0$$

This is the equation of the required cylinder.

Example 16.12

Find the equation to the right circular cylinder described on the circle through the points $A(1, 0, 0)$, $B(0, 1, 0)$ and $C(0, 0, 1)$ as the guiding curve $x^2 + y^2 + z^2 - yz - zx - xy = 1$.

Solution

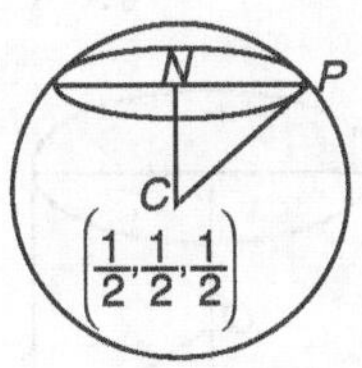

The equation of the sphere $OABC$ is

$$x^2 + y^2 + z^2 - x - y - z = 0 \tag{16.18}$$

The equation of the plane ABC is

$$x + y + z = 1 \tag{16.19}$$

Therefore, the equation of the circle ABC is

$$x^2+y^2+z^2-x-y-z=0,\ x+y+z=1 \tag{16.20}$$

The centre of the sphere is $\left(\frac{1}{2},\frac{1}{2},\frac{1}{2}\right)$.

$$r=\sqrt{\frac{1}{4}+\frac{1}{4}+\frac{1}{4}}=\frac{\sqrt{3}}{2}$$

CN = Perpendicular from the centre on the plane $x+y+z=1$

$$=\frac{\frac{1}{2}+\frac{1}{2}+\frac{1}{2}-1}{\sqrt{3}}=\frac{1}{2\sqrt{3}}$$

$$NP^2=CP^2-CN^2$$

$$=\frac{3}{4}-\frac{1}{12}=\frac{2}{3}$$

$$\therefore NP=\sqrt{\frac{2}{3}}$$

The equations of the line CN are

$$\frac{x-\frac{1}{2}}{1}=\frac{y-\frac{1}{2}}{1}=\frac{z-\frac{1}{2}}{1}$$

which is the axis of the cylinder.

The direction ratios of the axis are 1, 1, 1.

The direction cosines of the axis are $\frac{1}{\sqrt{3}},\frac{1}{\sqrt{3}},\frac{1}{\sqrt{3}}$.

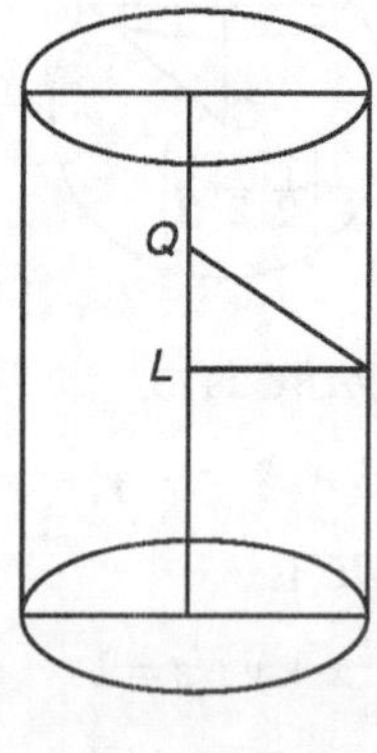

P is $\left(\frac{1}{2}, \frac{1}{2}, \frac{1}{2}\right)$.

Q is (x_1, y_1, z_1).

$$PL = \sqrt{\frac{2}{3}}$$

$$PQ = \left(x - \frac{1}{2}\right)^2 + \left(y - \frac{1}{2}\right)^2 + \left(z - \frac{1}{2}\right)^2$$

$$= x_1^2 + y_1^2 + z_1^2 - x_1 - y_1 - z_1 + \frac{3}{4}$$

QL = Projection of PQ on the axis

$$= \left(x - \frac{1}{2}\right)^2 \frac{1}{\sqrt{3}} + \left(y - \frac{1}{2}\right)^2 \frac{1}{\sqrt{3}} + \left(z - \frac{1}{2}\right)^2 \frac{1}{\sqrt{3}}$$

$$= \frac{x_1, y_1, z_1 - \frac{1}{3}}{\sqrt{3}}$$

$$PQ^2 = QL^2 + LP^2$$

$$x_1^2 + y_1^2 + z_1^2 - x_1 - y_1 - z_1 + \frac{3}{4} = \left(\frac{x_1 + y_1 + z_1 - \frac{3}{2}}{\sqrt{3}}\right)^2 + \frac{2}{3}$$

The locus of (x_1, y_1, z_1) is

$$3\left(x^2 + y^2 + z^2 - x - y - z + \frac{3}{4}\right) = \left(x + y + z - \frac{3}{2}\right)^2 + 2$$

$$\text{(i.e.) } x^2 + y^2 + z^2 - yz - zx - xy = 1$$

which is the required equation of the cylinder.

Example 16.13

Find the equation of the enveloping cylinder of the sphere $x^2 + y^2 + z^2 - 2x + 4y = 1$ having its generators parallel to the line $x = y = z$.

Solution

Let $P\ (x_1, y_1, z_1)$ be any point on a tangent which is parallel to the line

$$x = y = z \qquad (16.21)$$

Therefore, the equation of the tangent lines are

$$\frac{x-x_1}{1}=\frac{y-y_1}{1}=\frac{z-z_1}{1} \tag{16.22}$$

Any point on this line is (x_1+r, y_1+r, z_1+r).
This point lies in this sphere

$$x^2+y^2+z^2-2x+4y-1=0 \tag{16.23}$$

$$\therefore\ (x_1+r)^2+(y_1+r)^2+(z_1+r)^2-2(x_1+r)+4(y_1+r)-1=0$$

$$\text{(i.e.) } 3r^2+2r(x_1+y_1+z_1-1)+(x_1^2+y_1^2+z_1^2-2x_1+4y_1-1)=0$$

If equation (16.22) touches the sphere of equation (16.23), then the two values of r of this equation are equal.

$\therefore$ Discriminant = 0

$$\therefore 4(x_1+y_1+z_1-1)^2-12(x_1^2+y_1^2+z_1^2-2x_1+4y_1-1)=0$$

$$\text{(i.e.) } 3(x_1^2+y_1^2+z_1^2-2x_1+4y_1-1)-(x_1+y_1+z_1-1)^2=0$$

On simplifying we get the locus of (x_1, y_1, z_1) as

$$x^2+y^2+z^2-yz-zx-2x+7y+z-2=0$$

which is the required equation.

Exercises

1. Find the equation of the cylinder, whose guiding curve is $x^2+z^2-4x-2z+4=0$, $y=0$ and whose axis contains the point (0, 3, 0). Find also the area of the section of the cylinder by a plane parallel to the xz plane.
2. Find the equation of the cylinder, whose generators are parallel to the line $\frac{x}{1}=\frac{y}{-2}=\frac{z}{3}$ and passing through the curve $x^2+2y^2=1$, $z=0$.
3. Prove that the equation of the cylinder with generators parallel to z-axis and passing through the curve $ax^2+by^2=2cz$, $lx+my+nz=p$ is $n(ax^2+by^2)+2c(lx+my)-2pc=0$.
4. Find the equation of the cylinder, whose generators are parallel to the line $\frac{x}{1}=\frac{y}{2}=\frac{z}{3}$ and passes through the curve $x^2+y^2=16$, $z=0$.

5. Find the equation to the cylinder with generators parallel to z-axis which passes through the curve of intersection of the surface represented by $x^2 + y^2 + 2z^2 = 12$ and $lx + y + z = 1$.

6. Find the equation of the cylinder, whose generators intersect the conic $ax^2 + 2hxy + by^2 = 1, z = 0$ and are parallel to the line with direction cosines l, m, n.

7. A cylinder cuts the plane $z = 0$ with curve $x^2 + \frac{y^2}{4} = \frac{1}{4}$ and has its axis parallel to $3x = -6y = 2z$. Find its equation.

8. A straight line is always parallel to the yz plane and intersects the curves $x^2 + y^2 = a^2, z = 0$ and $x^2 = az, y = 0$. Prove that it generates the surface $x^4y^2 = (x^2 - az)^2 (a^2 - x^2)$.

9. Find the equation of a right circular cylinder of radius 2 whose axis passes through (1, 2, 3) and has direction cosines proportional to 2, 1, 2.

10. Find the equation of the right circular cylinder of radius 2 whose axis passes through (1, 2, 3) and has direction cosines proportional 2, –3, 6.

11. Find the equation of the right circular cylinder of radius 1 with axis as $\frac{x-1}{2} = \frac{y}{3} = \frac{z-3}{1}$.

12. Find the equation of the right circular cylinder whose generators are parallel to $\frac{x}{2} = \frac{y}{5} = \frac{z}{3}$ and which passes through the curve $3x^2 + 4y^2 = 12, z = 0$.

13. Find the equation of the right circular cylinder of radius 4 whose axis is $x = 2y = -z$.

14. Find the equation of the right circular cylinder whose guiding curve is the circle through the point $(a, 0, 0)$, $(0, b, 0)$, $(0, 0, c)$.

15. Find the equation of the enveloping cylinder of the sphere $x^2 + y^2 + z^2 - 2x + 4y = 1$ having its generators parallel to the line $x = y = z$.

16. Find the enveloping cylinder of the sphere $x^2 + y^2 + z^2 - 2y - 4z = 11$ having its generators parallel to the line $x = -2y = 2z$.

17. Find the equation of the right cylinder which envelopes a sphere of centre (a, b, c) and radius r and its generators parallel to the direction l, m, n.

Answers

1. $10x^2 + 5y^2 + 13z^2 + 12xy + 4xz + 6yz - 36x - 30y - 18z + 36 = 0$
2. $3x^2 + 6y^2 + 3z^2 - 2xz + 8yz - 3 = 0$

4. $9x^2 + 9y^2 + 5z^2 - 6xz - 12yz - 144 = 0$
5. $3x^2 + 3y^2 + 4xy - 4x - 4y - 10 = 0$
7. $36x^2 + 9y^2 + 17z^2 + 6yz - 48xz - 9 = 0$
9. $5x^2 + 8y^2 + 5z^2 - 4yz - 8zx - 4xy + 22x - 16y - 14z - 10 = 0$
10. $45x^2 + 40y^2 + 13z^2 + 36yz - 24zx + 12xy - 42x - 280y - 126z + 294 = 0$
11. $10x^2 + 5y^2 + 13z^2 - 12xy - 6yz - 4zx - 8x + 30y - 74z + 59 = 0$
12. $27x^2 + 36y^2 + 112z^2 - 36xz - 120yz - 180 = 0$
13. $5x^2 + 8y^2 + 5z^2 - 4xy + 4yz + 8zx - 144 = 0$
14. $\left(\frac{1}{a^2}+\frac{1}{b^2}+\frac{1}{c^2}\right)(x^2+y^2+z^2-ax-by-cz)$

$$=\left(\frac{1}{a}+\frac{1}{b}+\frac{1}{c}-1\right)\left(\frac{1}{a}+\frac{1}{b}+\frac{1}{c}-2\right)$$

15. $x^2 + y^2 + z^2 - xy - yz - zx - 4x + 5y - z - 2 = 0$
16. $5x^2 + 8y^2 + 8z^2 + 4xy + 2yz - 4xz + 4x - 18y - 36z = 99$
17. $(l^2 + m^2 + n^2)[(x-a)^2 + (y-b)^2 + (z-c)^2 - r^2]$

$$= [l(x-a) + m(y-b) + n(z-c)]^2$$

Index

A

asymptotes, 8-28–8-33
  angle between, 8-30–8-33
  equations of hyperbola, 8-28–8-29
auxillary circle, 7-5
axis of cylinder, 16-1
axis of parabola, 6-3

C

cartesian coordinates, 9-3
centre of ellipse, 7-3
centroid of tetrahedron, 11-8
centroid of triangle, 1-4, 11-7
chord in terms of midpoint, 6-14–6-15
chord joining two points, 6-16–6-17
chord of circle, 4-9–4-10
chord of contact of tangents, 4-5–4-6
circle, 4-1–4-36, 9-9–9-16
  coaxal system, 5-3–5-4
  definition of, 4-1
  equation of a tangent at, 4-3
  equation of, 14-5
  examples related to, 4-11–4-36
  length of a tangent to a point on, 4-2–4-3
  limiting points, 5-4–5-6
  orthogonal circle, 5-2–5-3
  parametric form of, 4-11
  polar equation of, 9-9–9-11
  radical axis of, 5-1–5-2
  system of, 5-1–5-36
coaxal system, 5-3–5-4, 14-9–14-11
  definition of, 5-3–5-4
  limiting points of, 5-4
combined equation of pair of tangents, 7-52–7-54
common chord of two circle, 5-2
concurrent straight lines, 2-10–2-11
concyclic points, 7-49–7-50, 8-45–8-46
condition for tangency, 6-8–6-9, 7-11–7-12
cone, 15-1–15-34
  condition for a general equation of second degree to represent, 15-7–15-11
  equation of a cone with a given vertex, 15-1–15-4
  equation of a cone with a vertex at origin, 15-4–15-7
conic, 6-1–6-2, 7-1
  general equation of second degree and tracing of, 10-1
  general equation of, 6-1–6-2
  polar equation of, 9-17–9-50
conjugate axis, 8-3
conjugate diameters, 7-54–7-74, 8-35–8-41
  definition of, 8-35
  equi-conjugate diameters, 7-84–7-85
  examples related to, 7-85–7-100, 8-38–8-41
  property of, 8-35–8-38
  semi-conjugate diameter, 7-81–7-82
conjugate hyperbola, 8-4–8-24
  examples related to, 8-5–8-24
conjugate lines, 4-8–4-9, 6-13–6-14, 7-18
  condition for, 7-18

conormal points, 7-48–7-49, 8-45
coordinate geometry, 1-1–1-27
  examples related to, 1-7–1-27
coplanar lines, 13-7–13-9
cylinder, 16-1–16-18
  definition of, 16-1
  equation with a given generator, 16-1–16-2

D

direction cosines, 11-8–11-13
  definition of, 11-8
direction ratios, 11-9–11-10
director circle of ellipse, 7-12–7-13
directrix, 6-3
  equation of, 6-3
distance form, 2-6–2-7

E

eccentric angle, 7-45
  equation of chord joining the points at, 7-45–7-46
ellipse, 7-1–7-100
  standard equation of, 7-1–7-4
enveloping cone, 15-15–15-16
enveloping cylinder, 16-2–16-3
equation of chord of contact, 6-12
equation of chord, 7-51–7-52
  in terms of its middle point, 7-51–7-52
equation of normal, 6-11–6-12, 7-14–7-16
equation of polar, 7-17
equation of tangent, 6-10–6-11, 7-13–7-14
equation to chord of contact, 7-16
equi-conjugate diameters, 7-84–7-85
external point of division, 1-4

F

focal distance, 7-4
focus directrix property, 6-5–6-7, 7-1
  examples related to, 6-5–6-7, 7-6–7-10

G

guiding curve of cylinder, 16-1

H

homogenous equation of second degree, 3-2
hyperbola, 8-1–8-60
  definition of, 8-1
  equation in parametric form, 8-3
  focus directrix property of, 8-2
  important property of, 8-3
  latus rectum of, 8-3
  parameter of, 8-5
  rectangular hyperbola, 8-3
  standard equation of, 8-1–8-3

I

intercept form, 2-4–2-5
intersection of parabola and circle, 6-19
invariants, 10-3–10-4

L

latus rectum, 603
  definition of, 6-3
  equation of, 6-3
  length of, 6-3
limiting points, 5-4–5-6
  definition of, 5-4
  examples related to, 5-16–5-26
  theorems related to, 5-4–5-6
locus of midpoint, 7-80
locus, 1-6–1-7

M

major axis of ellipse, 7-3
minor axis of ellipse, 7-3

N

non-intersecting lines, 13-13–13-14
normal form, 2-5–2-6

number of normals from a point, 6-18–6-19
number of tangents, 6-9

O

orthogonal circle, 5-2–5-3
  definition of, 5-2

P

pair of tangents, 4-10–4-11, 6-13–6-14
parabola, 6-1–6-61
  axes and vertex of, 10-10–10-16
  forms of, 6-3–6-4
  general equation of, 6-2–6-3
  important points related to, 6-3
parallel chords, 7-80–7-81
  definition of, 7-81
parameter, 6-16
  examples related to, 6-41–6-61
parametric form, 2-6–2-7
parametric representation, 6-16
perpendicular tangents, 6-9–6-10
plane, 12-1–12-33
  angle between two planes, 12-5
  definition of, 12-1
  equation of a plane in intercept form, 12-2–12-3
  equation of a plane in normal form, 12-3–12-4
  equation of planes which bisect the angle between two given planes, 12-10–12-12
  examples related to, 12-12–12-33
  general equation of a plane passing through a given point, 12-2
  general equation of, 12-1–12-2
  perpendicular distance from a point on a plane, 12-5–12-7
  plane passing through intersection of two given planes, 12-9–12-10
  plane passing through three given points, 12-7–12-8
point of intersection of normals, 6-18
point of intersection of tangents, 6-17–6-18
polar, 4-7–4-8
polar coordinates, 9-1–9-50
  definition of, 9-1–9-3
polar equation, 9-4–9-9
polar of a point, 6-13
  definition of, 6-13
pole, 4-7–4-8
  definition of, 4-7
position of a point, 7-4–7-5
projection of a line, 11-10

Q

quadrilateral, 1-6–1-7

R

radical axis, 5-1–5-2
  definition of, 5-1
  examples related to, 5-6–5-16
radical plane, 14-7–14-9
reciprocal cone, 15-26–15-34
rectangular coordinate axis, 11-1–11-3
rectangular hyperbola, 8-3, 8-41–8-60
  definition of, 8-3
  properties of, 8-44
  results concerning, 8-44–8-45
right circular cone, 15-13–15-21
right circular cylinder, 16-3–16-4

S

second directrix, 7-3
second focus, 7-3
section formulas, 1-3–1-7, 11-5–11-7
semi-conjugate diameter, 7-81–7-82
skew lines, 13-9–13-13
slope-intercept form, 2-3–2-4

slope-point form, 2-5
sphere, 14-1–14-39
  definition of, 14-1
  equation of, 14-1–14-2
  intersection of two, 14-5
  passing through a circle, 14-6
standard equation, 7-1
straight line, 2-1–2-45, 13-1–13-47
  angle between two, 2-11–2-12
  concurrent straight lines, 2-10–2-11
  condition of a general equation of a second degree to represent a pair of, 3-5–3-9
  determination of general equation of, 2-1–2-2
  equation in symmetrical form, 13-1–13-3
  equation passing through two given points, 13-3–13-4
  equations of bisectors of angle between, 2-12
  equations of two non-intersecting lines, 13-13–13-14
  examples related to, 2-12–2-45, 13-17–13-47
  intercept form, 2-4–2-5
  intersection of three planes, 13-14–13-16
  intersection of two, 2-8–2-10
  normal form, 2-5–2-6
  pair of, 3-1–3-35
  parametric form and distance form, 2-6–2-7
  perpendicular distance on, 2-7–2-8
  slope of, 2-3
  slope-intercept form of, 2-3–2-4
  slope-point form, 2-5
  two points form, 2-5

T

tangent plane, 15-23–15-26
three dimension, 11-1–11-25
tracing of curves, 10-1–10-15
transverse axis, 8-3
triangle, 1-5–16
triangular prism, 13-16–13-17
two circles, 5-1–5-2
  radical axis of, 5-2
two points form, 2-5

V

vertex of parabola, 6-3

X

XY-term, 10-3